Marine Mammals of the World

A Comprehensive Guide to Their Identification Second Edition

Thomas A. Jefferson
Marc A. Webber
Robert L. Pitman
Illustrations by **Uko Gorter**

AMSTERDAM • BOSTON • HEIDELBERG • LONDON • NEW YORK • OXFORD • PARIS
SAN DIEGO • SAN FRANCISCO • SINGAPORE • SYDNEY • TOKYO

Academic Press is an imprint of Elsevier

Academic Press is an imprint of Elsevier
125 London Wall, London EC2Y 5AS, UK
525 B Street, Suite 1800, San Diego, CA 92101-4495, USA
225 Wyman Street, Waltham, MA 02451, USA
The Boulevard, Langford Lane, Kidlington, Oxford OX5 1GB, UK

Second edition

Notices

Knowledge and best practice in this field are constantly changing. As new research and experience broaden our understanding,changes in research methods, professional practices, or medical treatment may become necessary. Practitioners and researchers must always rely on their own experience and knowledge in evaluating and using any information, methods, compounds, or experiments described herein. In using such information or methods they should be mindful of their own safety and the safety of others, including parties for whom they have a professional responsibility. To the fullest extent of the law, neither the Publisher nor the authors, contributors, or editors, assume any liability for any injury and/or damage to persons or property as a matter of products liability, negligence or otherwise, or from any use or operation of any methods, products, instructions, or ideas contained in the material herein.

Library of Congress Cataloging-in-Publication Data
A catalog record for this book is available from the Library of Congress

British Library Cataloguing-in-Publication Data
A catalogue record for this book is available from the British Library

ISBN: 978-0-12-409542-7

For information on all Academic Press publications
visit our website at http://store.elsevier.com/

Publisher: Janice Audet
Acquisition Editor: Kristi Gomez
Editorial Project Manager: Pat Gonzalez
Production Project Manager: Julia Haynes
Designer: Nina Lisowski

Printed and bound in the United States of America

Last digit is the print number: 9 8 7

Table of Contents

Preface and Acknowledgments...vii

1. Introduction...1

2. Basic Marine Mammal Biology...7

3. Taxonomic Groupings Above the Species Level...17

4. Cetaceans...24

North Atlantic right whale—*Eubalaena glacialis*...30
North Pacific right whale—*Eubalaena japonica*...34
Southern right whale—*Eubalaena australis*...37
Bowhead whale—*Balaena mysticetus*...42
Pygmy right whale—*Caperea marginata*...46
Blue whale—*Balaenoptera musculus*...49
Fin whale—*Balaenoptera physalus*...54
Sei whale—*Balaenoptera borealis*...59
Bryde's whale—*Balaenoptera edeni*...63
Omura's whale—*Balaenoptera omurai*...67
Common minke whale—*Balaenoptera acutorostrata*...70
Antarctic minke whale—*Balaenoptera bonaerensis*...75
Humpback whale—*Megaptera novaeangliae*...79
Gray whale—*Eschrichtius robustus*...84
Sperm whale—*Physeter macrocephalus*...88
Pygmy sperm whale—*Kogia breviceps*...95
Dwarf sperm whale—*Kogia sima*...99
Baird's beaked whale—*Berardius bairdii*...102
Arnoux's beaked whale—*Berardius arnuxii*...106
Cuvier's beaked whale—*Ziphius cavirostris*...109
Northern bottlenose whale—*Hyperoodon ampullatus*...114
Southern bottlenose whale—*Hyperoodon planifrons*...119
Shepherd's beaked whale—*Tasmacetus shepherdi*...122
Blainville's beaked whale—*Mesoplodon densirostris*...125

Note on Beaked Whales of the Genus *Mesoplodon*...130

Gray's beaked whale—*Mesoplodon grayi*...132
Ginkgo-toothed beaked whale—*Mesoplodon ginkgodens*...136
Deraniyagala's beaked whale—*Mesoplodon hotaula*...138
Hector's beaked whale—*Mesoplodon hectori*...140
Perrin's beaked whale—*Mesoplodon perrini*...142
Hubbs' beaked whale—*Mesoplodon carlhubbsi*...144
Pygmy beaked whale—*Mesoplodon peruvianus*...146
Sowerby's beaked whale—*Mesoplodon bidens*...149
Gervais' beaked whale—*Mesoplodon europaeus*...152
True's beaked whale—*Mesoplodon mirus*...155
Strap-toothed beaked whale—*Mesoplodon layardii*...158
Andrew's beaked whale—*Mesoplodon bowdoini*...162
Stejneger's beaked whale—*Mesoplodon stejnegeri*...164
Spade-toothed beaked whale—*Mesoplodon traversii*...167
Longman's beaked whale—*Indopacetus pacificus*...169
Narwhal—*Monodon monoceros*...172
Beluga whale—*Delphinapterus leucas*...176
Australian snubfin dolphin—*Orcaella heinsohni*...180
Irrawaddy dolphin—*Orcaella brevirostris*...183
Killer whale—*Orcinus orca*...186
Short-finned pilot whale—*Globicephala macrorhynchus*...193
Long-finned pilot whale—*Globicephala melas*...197
False killer whale—*Pseudorca crassidens*...200
Pygmy killer whale—*Feresa attenuata*...204
Melon-headed whale—*Peponocephala electra*...207
Risso's dolphin—*Grampus griseus*...210
Tucuxi—*Sotalia fluviatilis*...214
Guiana dolphin—*Sotalia guianensis*...217
Rough-toothed dolphin—*Steno bredanensis*...220
Atlantic humpback dolphin—*Sousa teuszii*...224
Indo-Pacific humpback dolphin—*Sousa chinensis*...227
Indian Ocean humpback dolphin—*Sousa plumbea*...230
Australian humpback dolphin—*Sousa sahulensis*...234
Common bottlenose dolphin—*Tursiops truncatus*...237
Indo-Pacific bottlenose dolphin—*Tursiops aduncus*...243
Pantropical spotted dolphin—*Stenella attenuata*...246
Atlantic spotted dolphin—*Stenella frontalis*...251
Spinner dolphin—*Stenella longirostris*...255
Clymene dolphin—*Stenella clymene*...261
Striped dolphin—*Stenella coeruleoalba*...264
Short-beaked common dolphin—*Delphinus delphis*...268
Long-beaked common dolphin—*Delphinus capensis*...273
Fraser's dolphin—*Lagenodelphis hosei*...278
White-beaked dolphin—*Lagenorhynchus albirostris*...282
Atlantic white-sided dolphin—*Lagenorhynchus acutus*...285
Pacific white-sided dolphin—*Lagenorhynchus obliquidens*...289
Dusky dolphin—*Lagenorhynchus obscurus*...293
Hourglass dolphin—*Lagenorhynchus cruciger*...297

Peale's dolphin—*Lagenorhynchus australis*...300
Northern right whale dolphin—
Lissodelphis borealis...303
Southern right whale dolphin—*Lissodelphis peronii*...306
Commerson's dolphin—
Cephalorhynchus commersonii...309
Heaviside's dolphin—*Cephalorhynchus heavisidii*...313
Hector's dolphin—*Cephalorhynchus hectori*...316
Chilean dolphin—*Cephalorhynchus eutropia*...319
Dall's porpoise—*Phocoenoides dalli*...322
Harbor porpoise—*Phocoena phocoena*...326
Spectacled porpoise—*Phocoena dioptrica*...330
Burmeister's porpoise—*Phocoena spinipinnis*...333
Vaquita—*Phocoena sinus*...336
Indo-Pacific finless porpoise—
Neophocaena phocaenoides...339
Narrow-ridged finless porpoise—
Neophocaena asiaeorientalis...342
South Asian river dolphin—*Platanista gangetica*...346
Amazon River dolphin—*Inia geoffrensis*...350
Franciscana—*Pontoporia blainvillei*...355

5. Pinnipeds...358

Northern fur seal—*Callorhinus ursinus*...360
Note on Hybrid Southern Hemisphere Fur Seals
Genus *Arctocephalus*...366
Antarctic fur seal—*Arctocephalus gazella*...368
Juan Fernandez and Guadalupe fur seals—
Arctocephalus philippii...374
Galapagos fur seal—*Arctocephalus galapagoensis*...381
South American fur seal—*Arctocephalus australis*...385
New Zealand fur seal—*Arctocephalus forsteri*...390
Subantarctic fur seal—*Arctocephalus tropicalis*...395
Cape and Australian fur seals—
Arctocephalus pusillus...399
Steller sea lion—*Eumetopias jubatus*...406
California sea lion—*Zalophus californianus*...413
Galapagos sea lion—*Zalophus wollebaeki*...421
South American sea lion—*Otaria byronia*...426
Australian sea lion—*Neophoca cinerea*...431
New Zealand sea lion—*Phocarctos hookeri*...435
Walrus—*Odobenus rosmarus*...440
Hawaiian monk seal—*Neomonachus schauinslandi*...446
Mediterranean monk seal—*Monachus monachus*...451
Ross seal—*Ommatophoca rossii*...455
Crabeater seal—*Lobodon carcinophaga*...458
Leopard seal—*Hydrurga leptonyx*...463
Weddell seal—*Leptonychotes weddellii*...468
Southern elephant seal—*Mirounga leonina*...472
Northern elephant seal—*Mirounga angustirostris*...478
Bearded seal—*Erignathus barbatus*...484
Hooded seal—*Cystophora cristata*...489
Gray seal—*Halichoerus grypus*...492
Ribbon seal—*Histriophoca fasciata*...496
Harp seal—*Pagophilus groenlandicus*...500
Harbor seal—*Phoca vitulina*...504
Spotted seal—*Phoca largha*...509
Ringed seal—*Pusa hispida*...513
Baikal seal—*Pusa sibirica*...518
Caspian seal—*Pusa caspica*...521

6. Sirenians...523

West Indian manatee—*Trichechus manatus*...524
African manatee—*Trichechus senegalensis*...527
Amazonian manatee—*Trichechus inunguis*...529
Dugong—*Dugong dugon*...532

7. Otters & Polar Bear...535

Sea otter—*Enhydra lutris*...536
Marine otter—*Lontra felina*...539
Polar bear—*Ursus maritimus*...542

8. Extinct Species...547

Baiji—*Lipotes vexillifer*...548
Japanese sea lion—*Zalophus japonicus*...551
Caribbean monk seal—*Neomonachus tropicalis*...553
Steller's sea cow—*Hydrodamalis gigas*...555

9. Identification Keys...557

A. *Key to Identification of Cetaceans of the World, Based on External Appearance*...557
B. *Key to Identification of Cetaceans of the World, Based on Skull Morphology*...567
C. *Key to Identification of Baleen Whales of the World, Based on Baleen Plates*...587
D. *Key to Identification of Pinnipeds of the World, Based on External Appearance*...588
E. *Key to Identification of Pinnipeds of the World, Based on Skull Morphology*...592
F. *Key to Identification of Sirenians of the World, Based on External Appearance and Distribution*...599
G. *Key to Identification of Sirenians of the World, Based on Skull Morphology*...599

10. Sources for More Information...601

11. Index—Common Names...603

12. Index—Scientific Names...605

Photo Credit Supplementary Information...608

Preface and Acknowledgments

A book such as this cannot be produced without the assistance of a large number of people with diverse knowledge and skills. In addition to all of the people and organizations that we thanked for assistance with preparation of the first edition of this book (all of whom contributed to our getting to this stage), many others have provided invaluable help with preparation and developing materials for this second edition. First and foremost we must thank our editors, Kristi Gomez and Pat Gonzalez, for their editorial assistance and in seeing this book through from the concept stage all the way to publication. We quite literally could not have done it without them.

Many people submitted photos for consideration for this book. For submission of photographs, we thank: G. Abel, J. M. Acebes, J. Acevedo, C. Ackerman, M. Alden, V. Alexeev, S. Allen, M. Arshod, B. Atkins, N. Ayliffe, D. Baetscher, R. W. Baird, E. Baker, R. Baldwin, M. Baran, J. Bastida, R. Bastida, J. Batten, S. Bauman-Pickering, A. Bautista, I. Beasley, V. Beaver, B. Becker, K. Bennett, S. Blanc, C. and G. Blanchard, L. Blight, Blue Dolphin Tours, P. Bordino, M. Borobia, L. Bouveret, P. Boveng, L. V. Brattstrom, J. Brack, G. Braulik, T. Brereton, R. Brewer, F. Broms, A. Brower, A. Brown, M. Brown, V. Burkanov, P. Caceres, D. Cagnazzi, M. Cameron, G. Cardenas, M. Carwardine, P. Carzon, A. Cecchetti, M. A. Cedenilla, S. Cerchio, G. Chan, K. Charlton-Robb, K. Chater, T. Cheeseman, B. Cheney, A. M. Chit, C. Christman, D. Clarke, T. Collins, T. Collopy, R. Constanine, A. M. Cosentino, M. Cotter, J. Cotton, D. Coughran, M. Cremer, C. F. Cross, I. Cumming, W. Davis, S. Dawson, M. de Boer, G. de Rango, M. Deakos, R. Digiovanni, L. Dinraths, L. Dolar, D. Donnelly, A. Douglas, P. Duley, C. Dunn, J. Durban, S. Ebbert, L. Edwards, D. Ellifrit, F. R. Elorriaga-Verplancken, S. Elwen, P. Ensor, I. Eriksson, P. Faiferman, H. Fearnbach, M. Fishbach, R. Floerke, H. Foley, K. Frampton, A. Friedlander, G. Friedrichsen, S. Fuhrmann, N. Funasahi, N. Funasaka, S. Gabdurakhmanov, C. Gallagher, C. Garrigue, G. Gemmell S. Gero, J. Gibbens, P. Gill, B. Gisborne, A. Glass, K. Goetz, H. Goldstein, C. Gomez, J. Gould, M. Greenfeilder, B. Griffith, N. Groves, D. Gualtieri, Y. Guibalt, J. Haelters, M. Hagbloom, P. Hamilton, S. Hansen, C. Harrington, J. Hart, M. Haya, E. Heard, J. Heikkiliä, A. Henry, E. Higuera, D. Holley, M. Honda, C. Howell, S. N. G. Howell, R. Hucke-Gaete, G. Huglin, S. K. Hung, T. Hunt, B. Hurley, J. Jansen, M. Jarman, M. Jenner, A. Jongeneel, M. Jüssi, C. Kahn, H. Kashevarof, C. Kahn, O. M. Kaya, B. Keener, L. Keith, M. Kelly, R. Kiefner, R. Kirkwood, A. Knowleton, W. Koski, H. Krakowski, D. Kreb, A. Kugler, M. Kunnasranta, K. Laidre, P. Laporta, S. Last, M. Lauf, H. Lawson, L. Lilly, R. Leeney, P. Livingstone, J. London, F. Lyman, M. Lynch, M. Lynn, K. Mahoney Robinson, M. Malleson, R. Mansur, D. Mantle, K. Matsuoka, G. McCullough, R. McDonell, B. McTavish, W. Meijilink, T. Melling, M. Meunier, M. Meyer, Mikurashima Tourist Association, M. Milligan, C. Millon, H. Minakuchi, G. Minton, S. Moore, H. Moors-Murphy, R. Moraga, L. Morse, H. Moshiri, D. Moussa, L. Mouysset, L. Murray, P. Naessig, J. Nahmens, R. Nanaykkara, A. Naranjit, N. Nathan, New South Wales Office of Environment, M. Niemeyer, I. Nimz, M. S. Nolan, K. O'Brien, G. Ocio, K. Oda, C. Oedekoven, J. Ohara, P. Olson, D. Orbach, C. D. Ortega-Ortiz, W. Osborne, D. Palacios-Alfaro, G. Parra, D. Perrine, P. Pichon, S. Plön, J. Poklen, F. Pongiglione, L. Ponnampalam, M. Pool, B. Pratt, T. Pusser, B. Raften, A. Ramalho, W. Rayment, A. Reckendorf, M. Reeves, M. Renner, F. Ritter, K. Rittmaster, K. Robinson, K. M. Robinson, M. Romano, B. Rone, A. Roos, M. Rosso, Rubiyat, G. Ryan, E. Sabater, S. Saino, G. P. Sanino, M. Santos, K. Sasamori, J. Scarff, A. Schaffer, M. Scheidat, A. Schulman-Janiger, M. Schuyl, K. Sekiguchi, L. Shaw, S. Sheehan, O. Shipley, A. S. Shirazi, V. Shore, O. Shpak, M. Sidwell, B. Siegel, D. Sinclair, B. D. Smith, J. Smith, M. Smultea, B. Solovyev, Southwest Fisheries Science Center/NOAA, T. Stenton, D. Still, K. Stockin, M. Sullivan, D. Sutaria, A. Suzuki, K. M. Sweeting, J-P. Sylvestre, I. Szczepaniak, T. Takahashi, N. Tanaka, N. Tapson, C. Tarver, A. Taylor, B. L. Taylor, J. Taylor, The Marine Mammal Center, G. Thomson, L. Thompson, F. Todd, J. Todderdell, J. Towers, N. Tregenza, F. Trujillo,

S. Tsang, A. U, J. Urban, R. Uyeyama, J. van Franeker, M. van der Linde, J. Varley, M. G. Velasco, I. Visser, W. Visser, P. Wade, D. Walsh, D. Wang, J. Y. Wang, T. Web, D. Weber, S. Webb, D. Webster, C. Weir, D. Weller, K. Whitaker, M. White, R. White, White Whale Program/Russian Academy of Sciences, H. Whitehead, D. Withrow, T. Whitty, B. Würsig, T. K. Yamada, M. Yoneda, U. Yoshikazu, M. Yoshioka, U. Yu, K. Yui, J. Zaera, and M. Zani.

Although we could not select photos from all of these people, we greatly appreciate each of them (and any others we have inadvertently forgotten) for the submission of their wonderful photos for our consideration. It was a great pleasure looking through all of this amazing photographic material!

We were fortunate to have the help of several good folks in proof reading the early text: C. Cunningham, V. James, and M. Moore. For expert review of chapters and individual species accounts for this edition, our appreciation goes out to E. Aliaga-Rossel, R. W. Baird, I. Beasley, J. Calambokidis, P. J. Clapham, B. Keener, L. Keith, T. R. Kieckhefer, M. Mangel, H. Marsh, S. L. Mesnick, P. A. Olson, W. F. Perrin, B. Powell, J. Y. Wang, L. Wedekin, D. Weller, and B. Würsig. Special thanks to Bill Keener, who read through the proofs of the entire book, and went above and beyond the call of duty to help us weed out errors, inaccuracies, and inconsistencies.

But, in the end, this product is ours. Despite a very extensive effort to make this guide as accurate and complete as possible, problems will inevitably have crept in, and we must take sole responsibility for any errors, misidentifications, oversights, or inconsistencies that remain. We apologize in advance for this, and assure the reader that any such problems that slipped through were not intentional. Since we expect to be producing future editions of this guide (if there is a demand after several years of "rest"), we again sincerely invite readers to contact us to alert us of any such mistakes or inaccuracies, so that they can be rectified in the future. Our email addresses are listed below.

Thomas A. Jefferson
Lakeside, California
sclymene@aol.com

Marc A. Webber
Homer, Alaska
marcwebber@sbcglobal.net

Robert L. Pitman
La Jolla, California
robert.pitman@noaa.gov

1. Introduction

The Need for This Guide

Interest in wildlife in general, and marine mammals in particular, has increased significantly in the past few decades, both in the general public and in the scientific and management communities. More people than ever are including wildlife watching in their activities, and this includes educational and adventure expeditions to see wild marine mammals up close. At the same time, there is increasing awareness of the integral importance of marine mammals to healthy aquatic ecosystems, and of the growing threats that a variety of human activities pose to these animals and their environments. Research and education programs are seeking to better understand and more clearly communicate the nature of these threats and appropriate steps to reduce or eliminate their impacts.

Good identification guides are integral to all these activities. Although there are many guides to limited geographical areas and some subsets of the world's marine mammal fauna, there are few comprehensive guides that cover all the world's whales, dolphins, porpoises, seals, sea lions, walruses, manatees, dugongs, marine and sea otters, and polar bears. Additionally, few of the existing guides provide special aids to identifying live animals, in-hand specimens, and skulls. This, the second edition of this identification guide, compiled after several years of work by the authors and illustrators, is intended as a significant step toward filling that need.

We have attempted to make this volume as complete, comprehensive, and up-to-date as possible. However, we are aware that it is limited by the differences in the amount and quality of information available on the various groups, as well as by the inadequacies of our approach towards representing what is available. Therefore, we prefer to think of this as somewhat of a middle point, to be improved by input from those who use it in the field and lab. In this second edition, we strived to improve on the first edition by correcting errors and inconsistencies, updating information, improving the format, and perhaps most importantly, to overhaul the photographs with the new crop of spectacular photos that are becoming increasingly available every day. Future editions (assuming that there will be future editions, which is mainly determined by how well this one sells) will be modified to correct errors and deficiencies revealed by extensive use. In the meantime, we hope this book helps both amateurs and professionals with the sometimes-difficult task of positively identifying species of marine mammals they see alive or encounter dead.

Most biologists use the term "marine mammal" to include members of five different mammalian groups: cetaceans (whales, dolphins, and porpoises), pinnipeds (seals, sea lions, and walruses), sirenians (manatees, dugongs, and sea cows), marine and sea otters, and the polar bear. These diverse groups are currently thought to represent five or six different recolonizations of the water by land-dwelling ancestors. The term marine mammal, therefore, implies no systematic or taxonomic relationship. In fact, the cetaceans are more closely related to camels and hippos than they are to other marine mammals, the pinnipeds have more in common with bears and weasels, and the sirenians are more closely allied to elephants and hyraxes. These differences not withstanding, however, all marine mammals have one thing in common—they derive all (or most) of their food from marine (or sometimes fresh) water.

All marine mammals have undergone major adaptations, which permit them to live in the water. The cetaceans and sirenians spend their entire lives in the water, while other marine mammals come ashore for various reasons, at particular times in their life cycle (most commonly to reproduce, molt, or rest). Major structural modifications to the bodies of cetaceans, sirenians, and pinnipeds involve the loss of hind limbs (e.g., cetaceans and sirenians), the adaptation of limbs for propulsion through water (e.g., pinnipeds), and the general streamlining of the body for hydrodynamic efficiency (all three groups). Structural modifications to the marine and sea otters and the polar bear by a marine existence are less apparent in body form; these animals in most ways still closely resemble their terrestrial counterparts. Like its predecessors (Jefferson et al. 1993, 2008), since this is an identification guide, we include mainly information useful for identifying marine mammal species. For good introductions to the biology of mammals in general, see Gould and McKay (1990) and Macdonald (1984). More detail specifically on the biology of marine mammals can be found for cetaceans in Leatherwood and Reeves (1983), Evans (1987), Harrison and Bryden (1988), and Martin (1990); for pinnipeds in King (1983), Bonner (1990), Riedman (1990a), and Reeves et al. (1992); for sirenians in Reynolds and Odell (1991) and Reeves et al. (1992); for marine and sea otters in Riedman (1990b) and Reeves et al. (1992); and for polar bears in Stirling (2011) and Reeves et al. (1992). The best sources for basic information on the biology and phylogeny of marine mammals are Reynolds and Rommel (1999), Twiss and Reeves (1999), Berta et al. (2006), Hoelzel (2002), and Perrin et al. (2002, 2009).

Marine Mammal Identification and How to Use This Guide

Most available marine mammal identification guides do not provide the most appropriate information for accurate identification, have limited geographic or taxonomic scope, or are badly out-of-date. Two very good recent ones are Reeves et al. (2002) and Shirihai (2006). Marine mammals can be difficult to identify at sea. Even under ideal conditions, an observer often gets little more than a brief view of a splash, blow, dorsal fin, or back, and this is often at a great distance. Rough weather, glare, fog, or other poor sighting conditions only compound the problem. The effects of lighting, in particular, must be kept in mind. Many diagnostic characters may only be visible under good lighting or at close range. One must always acknowledge the limitations of the particular set of conditions that they are exposed to when making a marine mammal identification.

Many species appear similar to another, especially in the brief glimpses typical at sea. Animals of some poorly-known groups (most notably, beaked whales and Southern Hemisphere fur seals) are especially difficult to identify to species, even with a good look at a live animal or an "in hand" specimen (and even to most marine mammal specialists). For all these reasons, even experts often must log a sighting as "unidentified." In all cases, this designation, accompanied by a detailed description is preferable to recording an incorrect identification. This point cannot be overemphasized!

In addition to the diagnostic variation among species, marine mammals exhibit other types of variation in morphology and coloration. These are important to keep in mind when making identifications, as such variation can mask diagnostic species characters and cause confusion and even misidentifications. The most common types of such variation in external appearance are discussed briefly below.

Intraspecific geographic variation Marine mammal species generally occur in populations that are (more-or-less) reproductively isolated from each other. If these populations have been separated for a long enough period of time, they may have evolved noticeable differences in their external morphology. Virtually every marine mammal species (with the possible exception of those that only occur as a single population, like the vaquita and the recently-extinct baiji), shows some geographic variation. Much of this variation is subtle and not very noticeable, and therefore will not significantly affect field identification, but sometimes distinct geographic forms may have evolved. These may differ quite strongly in overall size, body shape, coloration, etc. Some such variants have been formally described as subspecies (which, in many cases, are incipient species), and have been given trinomials (subspecific names), but most have not been formally recognized. This book attempts to provide descriptions and illustrations/photos of geographic forms that are well-described and may be recognizable in the field, regardless of whether or not they have been described as subspecies.

Sexual dimorphism Most marine mammal species show some sexual dimorphism, with one sex being somewhat larger than the other. In addition, many toothed whales and pinnipeds have males and females showing distinct differences in body shape and coloration. These differences usually remain insignificant until near the age of sexual maturity, but can become quite pronounced in adults. One must keep this dimorphism in mind when making identifications, especially in cases where only a single individual is involved. In the species accounts, we make every attempt to describe significant sexual dimorphism as it relates to marine mammal species identification.

Developmental variation Obviously, young marine mammals do not look exactly like adults. Clearly, they are smaller than adults, but they may also have very different body proportions and color patterns. The head and appendages of most newborn marine mammals are typically proportionately larger than they are in adults. It is not uncommon for cetacean calves and pinniped pups to show very different pigmentation than adults. For instance, most dolphins have a muted version of the adult color pattern when first born. Size and other external differences of young animals are described in this book, when adequately documented.

Seasonal variation Seasonal variation in external appearance is not nearly as important in marine mammals as it is in, for instance, birds, and for the most part is not an issue in species identification. However, there are some species in which seasonal differences are important. An example of this is the northern elephant seal, which has a seasonal molt (occurring at different times for different

Adults and young, besides being different in size, often have differences in their color patterns and even body shapes. Risso's dolphins. Southern California. PHOTO: M. SMULTEA.

Anomalous color patterns (sometimes one-off and other times regularly occurring) exist for most cetacean species. One must be very careful in identifying such unusual individuals. Common bottlenose dolphin. Southern California. PHOTO: A. SCHULMAN-JANIGER

Sexual dimorphism is common in many pinnipeds and cetaceans. Northern elephant seals show an extreme form, which involves both size and body shape differences. The large individual in the background is an adult male, and the smaller one in the foreground is a fully-grown adult female. PHOTO: T. A. JEFFERSON

The bodies of many cetacean species have round to oval-shaped white scars, which are caused by the bites of cookie-cutter sharks. It is thus important to distinguish these scars from natural coloration markings. Dwarf minke whale. PHOTO: W. OSBORNE

age classes). During the molting period, seals generally look ragged and have a quite different appearance than the more typical pelage of the rest of the year. We attempt to identify seasonal variants in this guide.

Anomalous color morphs Some marine mammal species are characterized by the existence of uncommonly-occurring color morphs. Some occur infrequently, with the same basic pattern showing up from time to time. This is the case for some species of dolphins (e.g., short-beaked common dolphins, Pacific white-sided dolphins, northern and southern right whale dolphins), porpoises (e.g., Dall's porpoise), and seals (e.g., subantarctic fur seals), for instance. In addition, melanistic (all-black) individuals, albinos, and other anomalously-white color morphs occasionally occur, and are a possibility for any marine mammal species. In particular, albinos (or other types of anomalous white patterns) are being seen and reported much more frequently these days. Where one or more uncommon color morphs are well-known, we attempt to describe and illustrate them in this book.

Hybrids and intergrades Sometimes animals of different species mate with each other and produce offspring. Hybrids are the result of such interspecific matings. It has even been recently determined that one species of dolphin (the Clymene dolphin) actually evolved through hybridization of two other species (spinner and striped dolphins). One must always be concerned about the possibility of a hybrid when making marine mammal identifications. Hybrids between baleen whale species, between narwhals and belugas, between dolphin species, between porpoise species, and between pinniped species have been documented. In fact, in the Southern Hemisphere fur seals (genus *Arctocephalus*), species are very similar in appearance and behavior and there is much overlap in ranges. As a result, hybridization is very common, and can be a major confounding factor in making species identifi-

Several species of the genus *Arctocephalus* are known to hybridize. This adult male is an example of a subantarctic X Antarctic fur seal that underscores the challenges of identifying species in closely related groups. PHOTO: P. J. N. DE BRUYN

Albinism and leucism has been reported for many species of marine mammals and probably occurs in low frequencies. This is an example of a leucistic harbor porpoise seen in San Francisco Bay. PHOTO: B. KEENER

cations. Within the delphinids, the common bottlenose dolphin (*Tursiops truncatus*) appears to be very non-particular about its mating partners, and hybrids between this species and a wide variety of others (e.g., long-beaked common dolphins, Risso's dolphins, false killer whales, pilot whales, etc.) have been documented in captivity and in nature. Although not yet documented, hybrids among the many, very similar species of the beaked whale genus *Mesoplodon* should also be considered a possibility.

In addition, intergrades may appear (these are the equivalent of hybrids, but result from a cross between subspecies, not full species). Intergrades are known to be very common in spinner dolphins, for instance, where a "geographic form" of the species in the eastern tropical Pacific (the whitebelly spinner) is now known to be an intergrade "swarm." One other type of situation is being discovered more and more often lately. That is introgression of DNA from one species into individuals of another species. This may result from one-off mating instances of different species, and again, usually involve bottlenose dolphins. Depending on the extent of the introgression, the appearance of the animals (the "phenotype") may not show any evidence of the "alien" DNA, but these cases serve as another reminder of the complexity and variability of what we sometimes simplistically think of as discrete units of biological diversity—species. We generally do not describe hybrids and intergrades in this guide, except in those few cases where they appear very commonly (such as whitebelly spinner dolphins, and Dall's X harbor porpoise hybrids). However, one must always be on the lookout for them, and aware that they do exist.

Individual variation Beyond all of the types of variation described above, there is individual variation in every species. No two individuals are exactly alike—this variation is actually the raw material of natural selection. There will naturally be some range in all of the species' diagnostic characters. Some species are more variable than others, and certain features (such as total length and tooth counts) tend to show great amounts of individual variation in most species. This should always be kept in mind when making identifications. We attempt to document and illustrate some of the individual variation present within any species in this book, but it should be recognized that we can only present a small fraction of what naturally occurs.

Scarring, injuries and deformities One must always remember that, in the course of evading predators, finding and securing food, interacting with conspecifics, and avoiding impacts of human activities, marine mammals become scarred and injured. Further, they may develop deformities, either as a result of congenital defects or disease. These may cause an animal to appear quite different than the classic "textbook specimen" illustrating all the diagnostic characters. Most such defects will have little or no impact on the ability to identify marine mammals to species, but some may cause problems. For the most part, we are unable to describe and illustrate the effects of injuries and deformities in this book, but we caution the reader to keep these factors in mind.

Finally, although it is not really a type of variation, the effects of lighting must also be considered. This is especially true when making identifications of living animals in the field, and when examining photographs. Some subtle color pattern components may only be visible in the best lighting, and glare on an animal's body can sometimes look like light-colored patches. This is especially a problem when making an identification from still photographs, where brief glints of light can be "frozen" in time and mislead observers.

Notes on the Format of the Species Accounts

The species accounts are presented in taxonomic order, with closely-related species grouped together, even though, in some cases the main species that may cause confusion are not especially closely related. The following describes how the species accounts are set-up:

Recently-used synonyms This is a list of synonyms of the scientific name that have been widely used in the past 50 years or so. The list is not intended to be comprehensive. Older, and more obscure, names are not listed here.

Common names For each species, the standard common names in English, Spanish, and French are given here. Note that other common names are used in these languages in some areas, and that there are a wide variety of common names used in other languages as well. However, we have made no attempt to compile a list of all the common names that the species goes by.

Taxonomic information After a brief list of the higher-level taxonomic groupings that the species belongs to, this section contains a very brief summary of any recent taxonomic controversies and taxonomic revisions, and mentions subspecies, where widely recognized.

Species characteristics This section includes the main characteristics of the species used in identification of whole specimens, including body shape, color pattern, size, and such things as tooth or baleen counts. Tooth counts of toothed whales given in this guide are for each tooth row (the upper and lower jaws each have two tooth rows), unless otherwise indicated. For some of the better-known species, there is also a listing of age/sex classes that animals may be divided into, along with descriptions of how to recognize them. However, the latter is only attempted for those situations where such age/sex classes have been well-described.

Recognizable geographic forms There is significant geographic variation in virtually every species of marine mammal. However, beyond this for some species, distinct geographic forms have been described and possess external characteristics that allow them to be identified in the field. When this is the case, we present a short description of such geographic forms in this section. If "none" is listed, this does not mean that geographic forms do not exist, but rather that we do not feel that they have been adequately described or that they are not possible to reliably recognize in the field.

Can be confused with Those species that are most commonly confused with the species of interest are listed here, along with tips on how they can be separated. We tend not to repeat all the diagnostic characters of the other species here, but simply list the types of features that should be paid attention to. The reader must generally consult the species account for those species to get details.

This leopard seal makes a meal of a gentoo penguin. Although these seals are famous for preying on penguins, they have a diverse diet that includes krill, fish, and penguins, as well as other species of seals. PHOTO: M. NOLAN

Distribution This section includes a short description of the species' range, to be used along with the distribution map provided for each species. We must emphasize that, although we have put considerable effort into providing the most useful range maps possible, the distribution maps should be considered approximate. The range limits shown, especially in offshore areas, are sometimes little more than educated guesses, based on incomplete data from that region, considered in light of available information on the species' distribution and habitat preferences elsewhere. In some cases, the limits of range indicated are probably more a reflection of search effort than of real distribution limits. Therefore, an absence of shading in a certain area does not necessarily mean that the species is not found there (this is especially true for many of the beaked whales, which are still known mostly from strandings—these tell us little about the species' true distribution and habitat preferences).

Ecology and behavior The basic ecology of the species is very briefly summarized here. Although it is often of less use in identification than morphological characteristics, behavior can nonetheless sometimes be used to help in identification. Group sizes, in particular, are often useful. However, it should be emphasized that the behavior of highly complex social mammals is highly variable, and thus coloration and morphology should always be used as the primary features for identification.

Feeding and prey This section contains a brief list of the types of prey items that the species feeds upon.

Threats and status The history of human exploitation of the species is briefly reviewed, and current conservation issues identified. Available population estimates are of variable accuracy, and should thus be taken cautiously. Techniques for estimating sizes of mammal populations at sea are still evolving and are far from standardized, and available tools have been used unevenly, often with violations of underlying assumptions. For these reasons, the density of shading on the distribution maps is intended to show only known or postulated range, and not population density.

IUCN status The legal status of each species is also given in the accounts. "Endangered Species Lists" are maintained by both the federal agencies of many countries (e.g., U.S. List of Threatened and Endangered Species) and the International Union for the Conservation of Nature and Natural Resources, now the World Conservation Union (IUCN Red List), among other international agencies. In this guide, we present only IUCN designations. "Endangered" status is assigned to those species considered to be in immediate danger of extinction. Species at risk of soon becoming endangered are generally listed as "Vulnerable" (or some similar designation). Because of incomplete information, political considerations, and the time lag in completing requirements for listing, these status designations do not always accurately reflect the true status of a species (for instance, some species listed as Endangered are at no immediate risk; others not listed may be on the verge of extinction). Nevertheless, they are helpful as warning flags that plans to exploit a given species must proceed only with great caution, and will give some idea of the degree of concern for the species' future.

References To save space, we have deleted the references section for each species, which appeared in the first edition. General references are given at the end of the book.

Notes on the Identification Keys

Marine mammals specimens "in hand" can best be identified by using the keys to external features, provided as appendices at the end of the book. With such specimens, it may be possible to view the entire body and to measure relative proportions of features. Various features of coloration and morphology are often useful in such considerations. We have used geographical information as little as possible to separate the species. This will help to avoid biasing observers toward making an identification based on what they think is "supposed" to be there.

Marine mammal skulls can be keyed-out using the keys provided in the appendices. We have assumed that no geographical information is available, so the key can be used to identify an untagged skull of unknown origin in a museum, for instance. It is clear from our own work and discussions with colleagues that is not yet possible to prepare a completely reliable and effective skull key for the non-specialist. Published keys and related literature are marred with errors and inconsistencies. Skulls of many species are sufficiently similar that it will be necessary to examine a full series of each to define reliable diagnostic features. Until that exercise is completed for each species, the skull keys must be considered to be works in progress.

It can sometimes be very difficult, or impossible, to identify marine mammals to species, whether based on at-sea sightings, "in hand" whole specimens, or an unlabeled skull. Great variability in behavior, coloration, body morphology, and bone structure can occur. Sometimes it may only be possible to label an animal or group as "unidentified long-beaked dolphin," "unidentified beaked whale," or "unidentified fur seal," for instance. If this guide helps lead to a specific identification in some cases and to narrow down the choices in others, then it will have served an important function. We are happy to share our experience with others to help them in this endeavor.

Request For Feedback from Users

Feedback from colleagues and readers has been critical in helping us to find and correct errors, and make important improvements to this second edition. Assuming that there will be future editions, we want to sincerely invite all of the users of this book to provide us with feedback on the accuracy of the information contained herein and leads on better photos. Mistakes and inaccuracies are inevitable when reviewing so much information, and if users of the book contact the authors with suggestions for changes in future editions, we promise to give those serious consideration.

2. Basic Marine Mammal Biology

In this chapter, we will introduce those readers who are unfamiliar with marine mammals to the subjects of this guide. We will not attempt to give a detailed summary of the biology of marine mammals, as that is not the purpose of this book. Besides, it has already been done much better than we could do, elsewhere (e.g., see Berta et al. 2006; Hoelzel 2002; Parsons 2013; Perrin et al. 2002, 2009; Reeves and Stewart 2003). Instead, we will simply provide a brief summary of the basic biology of the group of animals that we call marine mammals, primarily intended for use by those readers who are new to these animals.

What is a Marine Mammal?

It is important to recognize that marine mammals are not a natural biological grouping. Many people do not realize this, but the term "marine mammal" is somewhat of a "catch-all" phrase used for those groups of mammals that have returned to life in the "sea." The most important criterion is that they must get all or most of their food from the aquatic environment. It is not essential that they actually live in the sea. In fact, many species of marine mammals never encounter marine waters, living instead in various land-locked lakes and rivers. However, all of them are thought to have come from marine ancestors, and have structural and behavioral adaptations for aquatic life.

Marine mammals are not necessarily completely dependent on an aquatic existence. For instance, pinnipeds do not generally mate or give birth in the water, and polar bears may spend great amounts of time moving on land long distances away from the nearest marine waters. But, these mammals, along with the cetaceans and sirenians, do obtain most or all of their sustenance from the water, and this makes them marine mammals. One or two species of otters (the sea otter and marine otter) and the polar bear are also usually included as marine mammals.

In reality, there is no hard-and-fast rule of what is a marine mammal. Some people consider other mammals also to be in this group. Even some bats have been suggested to be marine mammals in the past, but by and large, this has not been accepted (they, after all, do not have the structural adaptations that we use to define other marine mammals), and the scheme introduced below is, by far, the most common in use. It originated with the list of "marine mammals" produced when the U.S. Congress passed the Marine Mammal Protection Act of 1972, and has been widely followed ever since.

Types of Marine Mammals

There are several different types of marine mammals. The two most commonly seen and best-known groups of marine mammals are the cetaceans (whales, dolphins, and porpoises) and the pinnipeds (seals, sea lions, and walruses). Most people are very familiar with these animals from seeing them in zoos and aquariums, or on television and in movies. They are both well-adapted to living in the oceans, although the pinnipeds must return to land for some life history stages (e.g., mating, breeding, and molting). Cetaceans are fully-adapted to live their entire lives in the water and never return to land for any significant period of time. The body plans of cetaceans and pinnipeds are radically modified from those of more familiar terrestrial mammals.

The sirenians are much less-often encountered by people, because there are only a few species (four today) and they occur only in certain parts of the world, mainly in the tropical zones. They are also well-adapted to a wholly-aquatic life, although they are largely creatures of the continental margins (and many even inhabit lakes and rivers). Sirenians are unique in being the only vegetarians among the marine mammals. They also have radically modified morphology. Finally, there are several species of fissipeds (the group of carnivores that have separate digits) that qualify as marine mammals, even though the other members of their taxonomic families are not considered among the marine mammals. These include one bear, the polar bear, and two (or three) otters, the sea and marine otters. In general, it can be said that these animals are much less completely adapted to living in the water than are the cetaceans, pinnipeds, and sirenians. They are only slightly diverged from their closest terrestrial relatives. However, that does not mean that they are not at home at sea. Quite to the contrary, especially the sea otter and polar bear are well adapted to living in the harsh conditions of the sea.

Some people have taken the definition to extremes, and considered additional species to be marine mammals, based largely on where they feed. This includes some species of bats, but such a broad definition of "marine mammal" is not generally recognized, and we restrict ourselves in this book to the species that have traditionally been considered marine mammals.

Basic Marine Mammal Biology

Evolutionary History The major groups of marine mammals have separate evolutionary origins, from different groups of terrestrial mammals. The cetaceans arose >50 MYA, and they are now universally thought to be monophyletic (all arising from the same ancestor), but there has been other controversy about their origins. The terrestrial mammal ancestors of cetaceans were previously thought to be a group of primitive, wolf-like, hoofed animals called mesonychid condrylarths. However, recent fossil and molecular evidence suggests that cetaceans are most closely related to the artiodactyls (a group of even-toed, hoofed mammals that includes ruminants such as cattle, camels, and hippos). In particular, hippos have been shown with molecular data (but not fossils) to be the closest modern mammal relative of the whales, and this has thrown doubt on the mesonychid hypothesis. However, some people believe that the mesonychids may actually have been primitive artiodactyls, which would bring all of the facts back into alignment.

There were three major phases of cetacean radiation. The first occurred about 45–53 MYA (Eocene) in the shallow, warm, tropical waters of the ancient Tethys Sea. It involved the initial radiation of the most primitive cetaceans, the archaeocetes (now extinct). It included the appearance of *Ambulocetus*, a 4 m walking protocetacean, which has been seen as a "missing link" in cetacean evolution. The second major phase resulted in the initial radiation of the odontocetes (toothed whales) and mysticetes (baleen whales), about 25–35 MYA (Oligocene). These two modern suborders included, at the time, a large array of unusual species that would later become extinct. The development of important modern adaptations, such as echolocation in the toothed whales, and filter-feeding in the baleen whales, occurred during this period. The final radiation, in the Miocene about 12–15 MYA, involved the appearance of modern cetaceans, especially the delphinoids and balaenopterids.

Pinniped evolution has also been plagued with controversy. Traditionally, a diphyletic origin of the pinnipeds was proposed, with walruses and eared seals evolving from ursid (bear-like) ancestors, and true seals originating from mustelid (otter-like) ancestors. However, current thinking favors a monophyletic origin of the pinnipeds from an aquatic carnivore (most probably an ursid) ancestor in the North Pacific about 30–35 MYA. The fossil record goes back to at least 25–27 MYA. There is still some controversy, most of it involving whether the walrus is most closely related to the phocids or the otariids.

There are five major lineages of pinnipeds, the three extant ones (walruses, eared seals, and true seals), as well as two extinct ones (the Desmatophocidae and Enaliarctos). These latter two groups did not survive to modern times, and joined the multitude of other "evolutionary experiments" that did not work out (resulting in the evolutionary dead-end of extinction).

The sirenians have a long evolutionary history, with a fossil record extending back >50 MYA. They evolved from proboscideans (represented by modern elephants and hyraxes), and have no connection with either cetaceans or pinnipeds. The earliest sirenians were pig-like, quadrupedal, amphibious mammals. The manatees originated in South America, and the dugongids began in the North Pacific, attaining a wide diversity in the fossil record there. Sirenians attained their maximum diversity in the Oligocene and Miocene but, due largely to the cooling of ocean waters, have since been reduced to the five recent species.

Compared to these ancient groups of marine mammals, the otters and bears are relative newcomers to the oceans. Sea otter evolutionary history is still somewhat unclear, but it is thought that they evolved only a few million years ago, and are closely-related to other otters. The polar bear originated from brown and grizzly bears in Siberia <1 MYA, and is still so closely-related to them that hybridization in zoos is common. The oldest known fossils are less than 100,000 years old.

Zoogeography, Distribution, and Migrations Marine mammals are not randomly distributed in the world's oceans. It has long been known, for example, that certain species are found exclusively or primarily in waters of a particular depth, temperature range, or oceanographic regime, and not in areas lacking one or all of these characteristics. For most species, however, little is known of the particular factors that cause them to be found in one area and not in another that appears, qualitatively at least, the same.

One major factor affecting productivity, and thus indirectly influencing the distribution of marine mammals, is the pattern of major ocean currents. These currents are driven largely by prevailing winds and are modified in their effects by the "Coriolis Force." Simply stated, the rotation of the earth causes major surface currents to move clockwise in the Northern Hemisphere and counterclockwise in the Southern Hemisphere. This has different implications for animals on east and west sides of ocean basins. In the Northern Hemisphere, warm tropical waters move further north along the east coasts of continental land masses, and warm-water species are often found unexpectedly far north. In the Southern Hemisphere, by contrast, cold polar waters move northward along the west coasts of continents, allowing cold-water marine mammals to range closer to the equator.

The interplay of these surface currents and subsurface movements of major water masses moves nutrients around by upwelling (the vertical turning over of bottom and surface waters) and indrift (the bringing in of nutrients by horizontal currents). As these nutrients and sunlight are the basic ingredients of productivity, areas of light mixing often are more productive than still areas of little or no mixing. Thus, the presence of marine mammals, and other high-order predators and consumers in an area is related primarily to prey and only secondarily to the water conditions supporting that productivity. Wherever oceanic conditions are such, it is likely that some species of marine mammal will be present to exploit that richness.

Each species of marine mammal has its particular habitat preference, which may comprise deep or shallow waters, tropical or polar regions, marine or estuarine regions, or any variation in between. Most species occur only in a distinct part of the world, although some are more cosmopolitan (found throughout the globe). The only species that can truly be said to be cosmopolitan, however, is probably the killer whale. It has been recorded in all oceans and seas, in many bays and channels, and even some rivers, from the equator to the pack ice in both hemispheres. It has habitat preferences, to be sure, but there is almost nowhere in the world's oceans and seas where a killer whale sighting would be considered out of the question.

Although seasonal movements are known for most species of marine mammals, not all undergo what could be called true migrations. The baleen whales; sperm whales, and some other large toothed whales; many pinnipeds, and a few other species of marine mammals do have distinct, predictable seasonal movements of large population segments (migrations). The ultimate and proximate reasons for these vary widely, but they are usually related to allowing maximum exploitation of food resources and meeting the needs of mating and breeding. Recently, the importance of predation pressure (especially from killer whales) has also become more appreciated.

A biologist examines a flipper tag on Browning Peninsula, near Casey Station, Wilkes Land, Antarctica. While some marine mammals are not very large compared to more familiar terrestrial mammals, others are enormous. This male southern elephant seal represents the largest species of pinniped. PHOTO: J. A. FRANEKER, IMARES

The majority of cetaceans have most of their cervical vertebrae fused and therefore cannot move their necks much at all. Irrawaddy and Australian snubfin dolphins (the latter pictured here) are exceptions, with free cervical vertebrae. Western Australia. PHOTO: S. ALLEN

Anatomy and Physiology Although they are all clearly mammals, and thus share the basic mammalian characteristics with others in this class, marine mammals have undergone various evolutionary changes to their body shape and function in order to adapt to their lives in the oceans and freshwaters of the world. All species of marine mammals are streamlined to varying degrees. The cetaceans and some of the pinnipeds show very strong streamlining and are correspondingly fast swimmers. Manatees (along with sea otters and possibly some of the pinnipeds) have taken their hydrodynamic changes to a much lesser extent, but even these species have reduced appendages, and smooth bodies for decreasing drag and turbulence. Most species can move fairly fast underwater, at least for short bursts.

All cetaceans and sirenians have lost their hind limbs (although bony vestiges buried in the abdominal muscles may still be present), and the hind limbs have been radically modified into flippers in pinnipeds. Polar bears and sea otters have only slightly modified hind limbs and can still walk reasonably well on land. The forelimbs have been modified into flippers in the three major groups of marine mammals, and additionally a dorsal fin has evolved for stability (and thermoregulation) in the cetaceans. Further, the cetaceans and sirenians have lost their mammalian fur, and only retain a small scattering of hairs on the body surface. Cetaceans, sirenians, and some pinnipeds also have internal testes.

The head shape of some marine mammals, primarily cetaceans and sirenians, has also been drastically modified from their terrestrial ancestors. The nostrils have moved to the top of the head to allow ease of breathing in water (not requiring the head to be lifted fully out of the water), and the jaws and teeth have been modified to assist in feeding, and in cetaceans, echolocation and batch feeding. The substitution of baleen plates made of keratin for teeth in modern mysticetes is arguably the most radical adaptation of all.

There are many internal adaptations to a marine existence that are difficult to see, unless a specimen is dissected. These include the development of a subcutaneous layer of fat called blubber (which serves functions in energy storage, insulation, and streamlining); varied modifications of the respiratory system to assist in extended breathholding and large pressure changes; modifications to the skull for breathing at sea, housing the baleen plates in mysticetes, and the production and reception of echolocation sounds in odontocetes.

Only bats can rival the cetaceans and sirenians for their evolutionary modifications from the traditional mammalian body plan. It seems that life in a medium different from living on land has required a large number of changes, and this explains why for many centuries, cetaceans were thought to be fish, rather than mammals. It wasn't until naturalists began to look past "skin deep" that they discovered the unmistakable similarities with other mammals.

Life History and Reproduction Some types of marine mammals are very long-lived, and are among the mammal species reaching the greatest ages. This includes the baleen whales, the larger toothed whales, some dolphins, a few pinnipeds, and the sirenians. It is thought that some large toothed whales, like the killer whale, may routinely live to be as old as humans (perhaps nearing 80–90 years in some cases). Though not without controversy, some bowhead whales have been reported to live to nearly 200 years! On the other hand, some species, such the porpoises, appear to have short life spans, rarely reaching the age of 20 years.

The reproductive patterns of marine mammals are not radically different from those of their terrestrial relatives in many ways. However, there are obviously some major adaptations that have been necessary to allow marine mammals to reproduce in the water. Cetaceans and sirenians are the only groups of marine mammals that undertake all of their reproductive activities in the water. Pinnipeds and the marine otters and polar bear all mate and give birth on land (with a few minor exceptions).

All marine mammal young are precocial, meaning that they are well-developed when they are born. They must be, to live in or close to the water very early in life. In fact, cetacean and sirenian newborns are capable of swimming as soon as they exit the womb. Seals and other marine mammals generally require a bit of learning before they are ready to tackle the marine environment.

The gestation period of most marine mammal species is around one year, and in pinnipeds, there is a period of delayed implantation of the embryo, which allows for annual reproduction when the natural gestation period may only be 8–9 months. The length of lactation is highly variable, lasting from only a few days in some seals, to many years in some toothed whales.

While seasonal breeding is the rule in all temperate and high-latitude species, some marine mammals that live in the tropics year-round have a protracted breeding period, with at least some births scattered throughout the year. Only a single newborn is born to most marine mammal species under normal circumstances, and twinning is considered rare.

Some of the marine species are unusual among mammals in having a post-reproductive period (or period of reproductive senescence) in females. Generally, this occurs in some of the longer-lived toothed cetaceans (such as killer and pilot whales). Other species may reproduce until they die.

Feeding Ecology Different groups of marine mammals have different feeding ecologies. Mysticetes (baleen whales), which are the largest species, feed on shoaling fishes and small invertebrates, some of the smallest. Baleen whales are batch feeders, taking in large amounts of prey and filtering them from the waters with the fringes on the inside of their baleen plates. Most balaenopterids are "gulpers," which means they lunge through a prey patch, taking in huge amounts of sea water with the aid of their expandable throats. Then they close the mouth and use muscular actions to force the water out between the baleen plates, leaving the tiny prey trapped on the inside, which they then swallow.

Right and bowhead whales are much less active in their feeding. They are "skimmers," which swim slowly through patches of prey and filter out the food items (mostly copepods) as the water flows through the baleen plates. The gray whale employs yet a different feeding strategy. It is a "sucker." Gray whales use suction inside their mouth to pull in a batch of amphipods or other invertebrates, as they roll on their sides (usually near the bottom).

The toothed whales, dolphins and porpoises take in individual prey items, generally one at a time. They feed mostly on fishes and squids, which are located and captured with the aid of echolocation, or sonar. Unlike most (but not all) species of baleen whales, toothed whales may use cooperative feeding techniques, sometimes involving dozens to thousands of individuals to corral and herd prey. The killer whale is the only cetacean species that regularly feeds on other marine mammals, and almost all marine mammal groups may become killer whale prey, at one time or another. They will even hunt whales much larger than themselves, while working in cooperative groups,

Most cetaceans have a dorsal fin, though a few species have none, with a smooth back. The narwhal is one of only a few species that has a low dorsal ridge, instead of a fin. PHOTO: C. WEIR

Some marine mammals use sperm competition as a reproductive strategy. The enormous penis of this leaping male harbor porpoise suggests that sperm competition is in use by this species. San Francisco Bay. PHOTO: B. KEENER

Most mysticetes feed individually, but some species, such as the humpback whale, use cooperative feeding techniques to herd their prey (in this case, Pacific herring). Southeast Alaska. PHOTO: T. A. JEFFERSON

Bowriding behavior is common among many dolphin and small whale species. This rough-toothed dolphin is heading to the bow wave of a research vessel in calm waters of the northern Gulf of Mexico. PHOTO: T. A. JEFFERSON

The Irrawaddy dolphin is one of the only dolphins that regularly spits water, and they even use this to assist in feeding. PHOTO: D. SUTARIA

reminiscent of those of wolves and other pack-hunting carnivores.

The pinnipeds generally feed on fishes and squids, although some take primarily invertebrates such as krill. Seals and sea lions take in prey items individually, and almost always feed solitarily. Individuals generally do not cooperate in feeding aggregations. There are few special adaptations for effective feeding in pinnipeds, with the possible exception of incredible deep, repeated diving abilities of some species. For instance, elephant seals can dive for 20 minutes and surface for three minutes, 24 hours a day, day after day, for weeks on end!

The sirenians are all herbivores, and therefore do not need to be particularly fast or maneuverable to catch prey. Manatees feed mostly on water hyacinths and other aquatic plants that may be submergent or emergent. Dugongs feed mostly on seagrasses, and leave feeding trails in seagrass beds where they have been active. The extinct Steller's sea cow ate mostly kelp.

The sea otter feeds mostly on invertebrates, such as crabs, clams, mussels, and sea urchins. It can only dive to relatively shallow depths to obtain such items, which are then brought to the surface and eaten as the animal lies on its back at the surface. Sea otters often use rocks as tools to break apart the hard shells of their prey. Polar bears feed mostly on seals, although they do also take beluga whales and even large fish. A common technique is for a polar bear to wait by a breathing hole and then snatch a seal from the hole as it comes up to breath.

Predation/Parasites/Disease Nearly all marine mammal species suffer from predation, with the killer whale being the only possible exception. Sharks are the major predators for many types of marine mammals, although killer whales (and even false killer whales) may be important predators for others. Predation by killer whales has been postulated to have been a major factor affecting the evolution of the migratory patterns of some of the baleen whales. Recently, killer whale predation has been suggested to be a primary pressure controlling the populations (and even causing depletion of) sea otters and Steller sea lions in the North Pacific. This is controversial, however, and clearly more work needs to be done to confirm or deny these hypotheses.

Disease affects all marine mammal species, although certainly some more than others. A large number of afflictions can affect marine mammals, both in captivity and in the wild. In recent years, morbilliviruses and related distemper viruses have been identified as the causes of some mass die-offs of marine mammals (primarily of cetaceans and pinnipeds). As a result, these diseases have received a great deal of attention from marine mammal biologists and veterinarians.

Although parasites may be present in marine mammals that are considered healthy and functioning normally in their societies, parasites may also cause disease that can result in death (either directly, or indirectly, by causing the animals to strand or be otherwise vulnerable to accidents). A large number of parasite species (mostly internal, but some external) have been identified from marine mammals. Parasites may be present in many organ systems in the body, but are generally found in the respiratory, digestive, circulatory, and reproductive systems.

Behavior and Social Organization The majority of cetacean species live in social groups, called schools, herds, or pods. A few species are relatively solitary, but even these species gather together for breeding or in feeding aggregations, at times. Pods of large whales generally number less than a dozen or so, but oceanic dolphin schools may number several thousand! The stability of social groups ranges from the very stable, long-term pods of killer whales to very ephemeral and short-lasting associations of many smaller dolphins and porpoises.

Cetaceans have keen senses, with the exception of smell (olfaction), which is probably nearly non-existent in this group. The toothed whales and dolphins, in particular, are acoustic creatures *par excellence*. In addition to a keen sense of hearing, and the ability to tell much about their environment through passive listening, odontocetes also have a highly-sophisticated sense of echolocation, or sonar. It is thought that acoustics is the dominant sense in odontocetes, but these animals also appear to have very good vision (both in air and under water).

The supposed high intelligence of dolphins is legendary, but in fact they are probably on a par with many other species of social predatory mammals. It is true that dolphins are very clever and can learn easily, but there is no solid evidence to believe that their intellectual capabilities are above those of other higher mammals. However, there are those in the marine mammal field who dispute this, and who see unparalleled intellectual powers among the dolphins and whales.

Pinnipeds tend to be much more solitary, at least at sea. On haul-outs and rookeries on shore, however, seals and sea lions often gather into huge groups, which have a very specific structure. Most pinnipeds are highly polygynous, with a single male controlling mating access to groups of females (harems). It is due to the highly polygynous nature of pinniped societies that most species have very strong sexual dimorphism (with males growing much larger than females).

Seals and sea lions have keen senses, including good hearing, vision, and smell. No species of pinniped is known to have a well-developed echolocation system, as the toothed whales do, but they can nonetheless tell much about their surroundings with sound. The sense of smell is so well-developed that many pinniped colonies have to be approached from downwind, to avoid stampeding the entire group of animals into the water.

The manatees and dugong generally live in relatively small groups, and are in fact often seen alone. They do gather into feeding and breeding groups, but these are generally short-lasting. The only stable social bond is likely between mother and calf.

Sirenians are generally relatively slow-moving, passive creatures (at least compared to the very active dolphins and seals). As large-bodied vegetarians, they have little need to be able to move very quickly, and have evolved bodies that are more geared towards negative buoyancy than speed (at least in the manatees). Their senses are less well-studied than those of dolphins and sea lions, but vision and hearing at least, are probably not as keen as in those other groups.

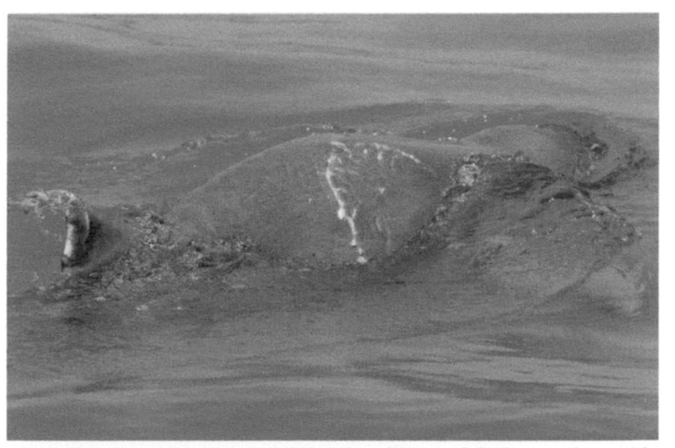

All marine mammals may exhibit deformities or injuries, such as this shark bite on an Australian snubfin dolphin. Care must be taken in the identification of these animals. PHOTO: A. BROWN

Strandings When marine mammals wash-up on shore unintentionally, whether alive or dead, this is called stranding (or sometimes, beaching). Most marine mammal strandings are of dead animals, but live specimens can also strand, and will usually die unless humans intervene and either push them back to sea or move them to a rehabilitation facility.

Single strandings are most common, but mass strandings of two or more individuals (not including a mother and calf/pup) also occur. In fact, sometimes entire pods of whales strand (usually alive), and witnessing such an event is truly one of the most spectacular (if somewhat sad) sights in nature!

Fishing nets, in particular various forms of gillnets and driftnets, kill hundreds of thousands of marine mammals every year—in this case a Dall's porpoise. Fishing net entanglement is now widely recognized as the single largest threat to the continued existence of many marine mammal populations and even some species. PHOTO: T. A. JEFFERSON

The causes of marine mammal strandings are not always known, but sometimes they are obvious. Single strandings usually involve an animal that is sick or injured and often too weak to swim against the currents and other forces that inevitably bring it to toward shore. In some cases, they die before they reach the beach, but in others they may still be alive. Such live-stranded individuals are unlikely to survive, even with good veterinary care.

Mass strandings usually involve multiple members of a social grouping (sometimes the entire school or pod), and almost invariably consist mostly of living individuals. Very often, one or only a few individuals in the school are sick or hurt, and most of the group is perfectly healthy. Therefore, the chances of "saving" or rehabilitating mass-stranded animals are much better. The cause of the mass beaching is generally related to the strong social bonds of the species that frequently mass strand (most often medium to large odontocetes, like the sperm, pilot, and false killer whales). More and more often, mass strandings and mortalities are being linked to the use of military sonars, which seem to disrupt the behavioral and/or physiological mechanisms for safely dealing with deep diving.

Strandings are extremely important phenomena for several groups of marine mammals, especially cetaceans. For many species of whales and dolphins, much of what we know of their biology may come from information gleaned from stranded specimens. For some of the poorly-known beaked whales, stranding records may be all that we have!

Discarded and lost fishing line, rope and net fragments are a hazardous type of marine pollution that injures and kills pinnipeds of many species such as this adult female Juan Fernandez fur seal. PHOTO: M. GOEBEL/NMFS

Table 1. Recent Marine Mammals of the World—132 species

Cetacea (whales, dolphins, and porpoises)

Mysticeti (baleen whales)

Family Balaenidae (right and bowhead whales)
North Atlantic right whale—*Eubalaena glacialis*
North Pacific right whale—*Eubalaena japonica*
Southern right whale—*Eubalaena australis*
Bowhead whale—*Balaena mysticetus*

Family Neobalaenidae (pygmy right whale)
Pygmy right whale—*Caperea marginata*

Family Balaenopteridae (rorquals)
Blue whale—*Balaenoptera musculus*
Fin whale—*Balaenoptera physalus*
Sei whale—*Balaenoptera borealis*
Bryde's whale —*Balaenoptera edeni*
Omura's whale—*Balaenoptera omura*i
Common minke whale—*Balaenoptera acutorostrata*
Antarctic minke whale—*Balaenoptera bonaerensis*
Humpback whale—*Megaptera novaeangliae*

Family Eschrichtiidae (gray whale)
Gray whale—*Eschrichtius robustus*

Odontoceti (toothed whales)

Family Physeteridae (sperm whale)
Sperm whale—*Physeter macrocephalus*

Family Kogiidae (pygmy and dwarf sperm whales)
Pygmy sperm whale—*Kogia breviceps*
Dwarf sperm whale—*Kogia sima*

Family Monodontidae (narwhal and beluga)
Narwhal—*Monodon monoceros*
Beluga whale—*Delphinapterus leucas*

Family Ziphiidae (beaked whales)
Baird's beaked whale—*Berardius bairdii*
Arnoux's beaked whale—*Berardius arnuxii*
Cuvier's beaked whale—*Ziphius cavirostris*
Northern bottlenose whale—*Hyperoodon ampullatus*
Southern bottlenose whale—*Hyperoodon planifrons*
Shepherd's beaked whale—*Tasmacetus shepherdi*
Blainville's beaked whale—*Mesoplodon densirostris*
Gray's beaked whale—*Mesoplodon grayi*
Ginkgo-toothed beaked whale—*Mesoplodon ginkgodens*
Deraniyagala's beaked whale—*Mesoplodon hotaula*
Hector's beaked whale—*Mesoplodon hectori*
Perrin's beaked whale—*Mesoplodon perrini*
Hubbs' beaked whale—*Mesoplodon carlhubbsi*
Pygmy beaked whale—*Mesoplodon peruvianus*
Sowerby's beaked whale—*Mesoplodon bidens*
Gervais' beaked whale—*Mesoplodon europaeus*
True's beaked whale—*Mesoplodon mirus*
Strap-toothed beaked whale—*Mesoplodon layardii*
Andrews' beaked whale—*Mesoplodon bowdoini*
Stejneger's beaked whale—*Mesoplodon stejnegeri*
Spade-toothed beaked whale—*Mesoplodon traversii*
Longman's beaked whale—*Indopacetus pacificus*

Family Delphinidae (marine dolphins)
Irrawaddy dolphin—*Orcaella brevirostris*
Australian snubfin dolphin—*Orcaella heinsohni*
Killer whale—*Orcinus orca*
Short-finned pilot whale—*Globicephala macrorhynchus*
Long-finned pilot whale—*Globicephala melas*
False killer whale—*Pseudorca crassidens*
Pygmy killer whale—*Feresa attenuata*
Melon-headed whale—*Peponocephala electra*
Tucuxi—*Sotalia fluviatilis*
Guiana Dolphin—*Sotalia guianensis*
Atlantic humpback dolphin—*Sousa teuszii*
Indian Ocean humpback dolphin—*Sousa plumbea*
Indo-Pacific humpback dolphin—*Sousa chinensis*
Australian humpback dolphin—*Sousa sahulensis*
Rough-toothed dolphin—*Steno bredanensis*
Pacific white-sided dolphin—*Lagenorhynchus obliquidens*
Dusky dolphin—*Lagenorhynchus obscurus*
White-beaked dolphin—*Lagenorhynchus albirostris*
Atlantic white-sided dolphin—*Lagenorhynchus acutus*
Hourglass dolphin—*Lagenorhynchus cruciger*
Peale's dolphin—*Lagenorhynchus australis*
Risso's dolphin—*Grampus griseus*
Common bottlenose dolphin—*Tursiops truncatus*
Indo-Pacific bottlenose dolphin—*Tursiops aduncus*
Pantropical spotted dolphin—*Stenella attenuata*
Atlantic spotted dolphin—*Stenella frontalis*
Spinner dolphin—*Stenella longirostris*
Clymene dolphin—*Stenella clymene*
Striped dolphin—*Stenella coeruleoalba*
Short-beaked common dolphin—*Delphinus delphis*
Long-beaked common dolphin—*Delphinus capensis*
Fraser's dolphin—*Lagenodelphis hosei*
Northern right whale dolphin—*Lissodelphis borealis*
Southern right whale dolphin—*Lissodelphis peronii*
Commerson's dolphin—*Cephalorhynchus commersonii*
Heaviside's dolphin—*Cephalorhynchus heavisidii*
Hector's dolphin—*Cephalorhynchus hectori*
Chilean dolphin—*Cephalorhynchus eutropia*

Family Phocoenidae (porpoises)
Dall's porpoise—*Phocoenoides dalli*
Harbor porpoise—*Phocoena phocoena*
Spectacled porpoise—*Phocoena dioptrica*
Burmeister's porpoise—*Phocoena spinipinnis*
Vaquita—*Phocoena sinus*
Indo-Pacific finless porpoise—*Neophocaena phocaenoides*
Narrow-ridged finless porpoise—*Neophocaena asiaeorientalis*

Family Platanistidae (South Asian river dolphin)
South Asian river dolphin—*Platanista gangetica*

Family Iniidae (boto)
Boto or Amazon River dolphin—*Inia geoffrensis*

Family Lipotidae (baiji)
Baiji—*Lipotes vexillifer* (extinct)

Family Pontoporiidae (franciscana)
Franciscana—*Pontoporia blainvillei*

Sirenia (sea cows)

Family Trichechidae (manatees)
West Indian manatee—*Trichechus manatus*
African manatee—*Trichechus senegalensis*
Amazonian manatee—*Trichechus inunguis*

Family Dugongidae (dugongs)
Dugong—*Dugong dugon*
Steller's sea cow—*Hydrodamalis gigas* (extinct)

Carnivora (carnivores)

Family Mustelidae (otters)
Sea otter—*Enhydra lutris*
Marine otter—*Lontra felina*

Family Ursidae (bears)
Polar bear—*Ursus maritimus*

Pinnipedia (sea lions, walrus, seals)

Family Otariidae (fur seals and sea lions)
Northern fur seal—*Callorhinus ursinus*
Antarctic fur seal—*Arctocephalus gazella*
Juan Fernandez and Guadalupe fur seals—*Arctocephalus philippii*
Galapagos fur seal—*Arctocephalus galapagoensis*
South American fur seal—*Arctocephalus australis*
New Zealand fur seal—*Arctocephalus forsteri*
Subantarctic fur seal—*Arctocephalus tropicalis*
Cape and Australian fur seals—*Arctocephalus pusillus*
Steller sea lion—*Eumetopias jubatus*
California sea lion—*Zalophus californianus*
Galapagos sea lion—*Zalophus wollebaeki*
Japanese sea lion—*Zalophus japonicus* (extinct)
South American sea lion—*Otaria byronia*
Australian sea lion—*Neophoca cinerea*
New Zealand sea lion—*Phocarctos hookeri*

Family Odobenidae (walrus)
Walrus—*Odobenus rosmarus*

Family Phocidae (true seals)
Hawaiian monk seal—*Neomonachus schauinslandi*
Caribbean monk seal—*Neomonachus tropicalis* (extinct)
Mediterranean monk seal—*Monachus monachus*
Ross seal—*Ommatophoca rossii*
Crabeater seal—*Lobodon carcinophaga*
Leopard seal—*Hydrurga leptonyx*
Weddell seal—*Leptonychotes weddellii*
Southern elephant seal—*Mirounga leonina*
Northern elephant seal—*Mirounga angustirostris*
Bearded seal—*Erignathus barbatus*
Hooded seal—*Cystophora cristata*
Gray seal—*Halichoerus grypus*
Ribbon seal—*Histriophoca fasciata*
Harp seal—*Pagophilus groenlandicus*
Harbor seal—*Phoca vitulina*
Spotted seal—*Phoca largha*
Ringed seal—*Pusa hispida*
Baikal seal—*Pusa sibirica*
Caspian seal—*Pusa caspica*

Exploitation and Conservation Marine mammals have long been highly-prized targets of humans looking for a good source of food, furs, oil, and later a whole host of other products. Because they are large, they were attractive subjects for human exploitation, but their relatively inaccessible habitats made them hard to hunt until the last few hundred years. Although there is evidence that prehistoric humans may have at least taken advantage of the fortuitous stranding of a fresh whale or seal on their shores, most marine mammal species were relatively safe from large-scale human exploitation until recent centuries.

The first known large-scale hunting of large whales was by the Basques, starting in the first millennium AD. They mainly targeted the North Atlantic right whale, and were so effective and persistent that they decimated that species. Norse and Icelandic whalers also hunted in the North Atlantic, and the Japanese began their culture of whale hunting in the 1600s. In the 1700s, the "Yankee whaling" era, focused largely on sperm whales, began and the United States became a major player in the commercial whaling game. Finally, in the late 1800s, the development of steam-powered vessels and the exploding harpoon heralded the modern era of commercial whaling. Fast-swimming species, such as the rorquals (blue, fin, sei, Bryde's and minke whales) were now within the reach of commercial whalers. It didn't take long for them to decimate species after species, starting with the largest and working their way down, to commercial extinction.

In recent decades, the direct killing of whales and dolphins has become much less important, and the indirect deaths of especially dolphins and porpoises have increased dramatically. There is now no doubt that more cetaceans die incidentally in fishing nets each year than from any other threat, including whale and dolphin hunting. In the last few decades, we have also seen the development of other major threats to these animals in the form of such things as habitat degradation and loss, environmental contamination, noise pollution and damage, and even live captures for captive display and research. The baiji was the first cetacean species known to have been wiped-out by humans (and we hope it will be the last), but several other species are now on the very verge of that same fate (e.g., the vaquita in Mexico; the North Pacific and North Atlantic right whales).

A relatively new threat facing cetaceans comes in the form of human-made noise that can be potentially disturbing or even damaging to the animals. Shipping and boat noise is certainly an issue in some cases, but the major concerns nowadays have to do with seismic survey noise (generally created with airguns in the search for petrochemicals), and intense sounds created by military sonars (generally used to detect submarines). These sounds have been shown to cause serious problems for some species, and many mass strandings of whales and dolphins have been linked to low- and mid-frequency sonar.

By the end of the 19th century, most species of fur seals were nearly wiped-out by sealing. Most species have made a remarkable comeback. Juvenile and pup Antarctic fur seals are seen here in the shadow of an old whaling station. South Gerogia Island.
PHOTO: M. NOLAN

Seals and sea lions also have a long history of human exploitation. Because they were so easy to kill when hauled out on land, some local pinniped populations may have been extirpated by sealers in early times. Species that occur in more remote areas (for instance, the Antarctic seals), on the other hand, were not really exploited until the late 1700s or early 1800s. While most species survived the exploitation and even recovered to pre-exploitation numbers and beyond (witness the northern elephant seal, for instance), some were wiped out. The West Indian monk seal was so heavily hunted that the last remaining seals probably died-out in the middle part of the 20th century. The Japanese sea lion is also thought to be extinct, although this has not been confirmed through a range-wide survey (and the eternal optimists in us would like to believe that there is still some hope that a remnant of this species may survive in Japanese or Korean waters).

Like cetaceans, pinnipeds have seen the development of other, more insidious, threats in recent years. Fishing gear entanglements are certainly a major one, and very recently the depletion of prey and other ecosystem effects of large-scale fishing operations have been blamed for the crashes of some pinniped populations (e.g., the Steller sea lion in Alaska waters).

Sirenians seem to be hunted wherever they occur, an unfortunate result of the apparently excellent taste of their flesh and their relative ease of capture. All three species of manatees and the dugong have been brought to levels that threaten extinction in the next few decades. This is largely due to hunting for food, but other threats also apply. Especially in Florida, collisions with high-speed vessels and entrapment in human-made structures also take their toll.

The fifth species of recent sirenian, the giant Steller's sea cow of the cold waters of the North Pacific and Bering Sea, was wiped out by sealers and fur traders in the late 1700s, only a few decades after its discovery. This is a truly sad tale of human greed and stupidity, and it should serve as a reminder of the frailty of nature.

Sea otters were also hunted to near extinction throughout much of their range in the North Pacific by fur traders in the 1700–1800s. In fact, they were thought to be extinct along the west coast of the continental US until a small remnant was discovered in the late 1930s. Luckily, this remnant has increased and expanded its range, and still survives today. The polar bear was never driven to near extinction, but it has been heavily exploited by native peoples and westerners. It has a relatively healthy population, and appears to be in no immediate danger, but the effects of global warming are casting some doubts on its ability to survive for more than a few decades into the future.

The public perception that all large whales are endangered is wrong. The truth is that most large whales are recovering from past exploitation (the North Atlantic and North Pacific right whales are the major exceptions), and the most serious conservation problems now lie with several of the smaller species. All of the sirenians, a few seals (e.g., the Mediterranean and Hawaiian monk seals), and several of the dolphins and porpoises (e.g., the vaquita, Indus susu, Maui's dolphin, and Atlantic humpback dolphin) are probably in the worst shape. It is our sincere hope that this book will help people to appreciate the diversity of the world's marine mammals, and inspire them to work towards their effective preservation.

3. Taxonomic Groupings Above the Species Level

In this chapter, we provide a brief overview of the higher-level (above species) classification and taxonomic groupings of marine mammals. It should be noted that marine mammal taxonomy is somewhat controversial and is constantly in a state of flux. We are actually in a splitting phase now, so there are more species in this edition than in the first edition. So, rather than trying to present the "flavor of the month" and having that perhaps be out-of-date when the user actually reads it, we have instead favored a more classic and conservative system of classification.

For example, a new phylogenetic classification published in 2005, and based on multiple lines of evidence, suggested some changes in ranking of various taxa. It considered the cetaceans (whales, dolphins, and porpoises) to be part of a more inclusive Order Cetartiodactyla (which would include hippopotamuses, among other artidactyls). In such a classification, the suborders Mysticeti and Odontoceti might be demoted to infraorders (taxa below the level of suborder). The Society for Marine Mammalogy's Committee on Taxonomy has adopted this basic plan, with some modifications. In their recent list (Committee on Taxonomy 2014), cetaceans are listed in the Order Cetartiodactyla, but Cetacea is listed as an unranked taxon (as are Odontoceti, Mysticeti, and Pinnipedia). It is mentioned in a footnote that, "many marine mammalogists and paleontologists favor retention of Order Cetacea in the interest of taxonomic stability." We tend to be in the latter category. Although it is confusing having so many unranked (but nonetheless valid) taxa, we do note that the current information from molecular studies supports a system that unites the cetaceans and other cetartiodactyls into a single taxonomic order. Clearly, there is much more work to do in marine mammal taxonomy!

Whether this arrangement will be accepted and supported by future work remains to be seen, although it currently appears to be well supported by genetic and morphological data and is gaining favor. However, for now, in the species accounts we will present the major higher-level groupings without their rankings, to avoid controversy and confusion.

Cetacea—*Whales, dolphins, and porpoises*

The 90 living species (one extinct) currently grouped in the Cetacea (either an order or taxon below order—see above about the status of Cetacea) are divided into two groups—Odontoceti (toothed whales) and Mysticeti (baleen whales). All representatives of a third group, Archaeoceti (ancient whales), are extinct. It is generally agreed that cetaceans are the most derived of all mammals (with the possible exception of bats). Evolved from terrestrial ancestors, they have largely adapted to living in the water, and have no need to come ashore, even for resting or reproduction. All cetaceans share a similar general body plan: a streamlined (albeit some more-so than others) spindle-shaped torso; flattened paddle-like foreflippers; telescoped skull; nasal openings on top (rather than on the front) of the head; dense (pachyostotic) ear bones, a well-developed blubber layer; internal reproductive organs; newly-derived boneless structures in the form of tail flukes and a dorsal fin or ridge (secondarily lost in some species); and the loss of such aquatic hindrances as hind limbs (present, if at all, only as vestiges), external ear flaps, and fur (although all have hair at some time during their early development and some retain a few rostral hairs for life).

Marine mammal taxonomy is not fully settled by any means. Revisions take place regularly, and new species are still being discovered. These beaked whales, previously thought to be Baird's, are now in the process of being described as a new species. North Pacific. PHOTO: D. NAGASHIMA

Bubble-netting is a technique used by humpback whales to corral their prey—various species of schooling fish or krill. PHOTO: W. DAVIS

Family Balaenidae—Right and bowhead whales The right and bowhead whales (four species in two genera) are large and chunky, with heads that comprise up to one-third of their body length. They lack a dorsal fin or any trace of a dorsal ridge. Overall, they tend to be far less streamlined than other baleen whales. Right and bowhead whales have developed a relatively passive skim-feeding technique, and tend to be slower than other whales. The baleen plates are the longest and have some of the finest fringes of the four mysticete families—both are adaptations to taking very small prey. Viewed in profile, the mouthline is extremely arched and the skull profile is highly convex; all seven cervical (neck) vertebrae are fused together.

Family Neobalaenidae—Pygmy right whale The single species in this family (though it has recently been proposed as the last remnant of the otherwise-extinct family Cetotheridiae), the pygmy right whale of the temperate Southern Hemisphere, is poorly known. Although it is in some ways intermediate between the Balaenopteridae and Balaenidae, the pygmy right whale may not be closely related to the Balaenidae. Much smaller than the right and bowhead whales (<7 m), it is slender, with a moderately-arched mouthline. The head represents only about one-quarter of the total length, and there is a short falcate dorsal fin set behind mid-back. There is also a pair of shallow throat depressions, which have been hypothesized to be incipient throat grooves. The skull is also somewhat intermediate in form; the rostrum is moderately arched (reminiscent of balaenids), but is much wider at its base (reminiscent of balaenopterids).

Although they may somewhat resemble fish externally, the cetaceans' internal anatomy effectively betrays their terrestrial mammalian ancestry. For instance, their flippers contain reduced counterparts of all or most of the hand and arm bones characteristic of other mammals; and pelvic rudiments (and occasionally hind limb remnants) are present. The internal anatomy of cetaceans is surprisingly like that of more familiar land mammals, with such interesting exceptions as the presence of a four-chambered stomach and cartilaginous reinforcements of the airways all the way down to the alveoli.

Mysticeti—Baleen whales Mysticeti has traditionally been considered a suborder, but its ranking is currently undetermined. There are four families of baleen whales. Mysticetes are universally large (with females growing larger than males); the smallest is the pygmy right whale (<7 m long), and the largest is the blue whale (the largest animal ever to live, up to 33 m or more in length and 145,000 kg in weight). The baleen whales have a double blowhole, a symmetrical skull, lack of a bony mandibular symphysis, and a sternum consisting of a single bone. In the mouth, the upper jaw is hung with baleen (stiff plates of keratin with fringes on the inside), instead of teeth. Baleen whales are batch feeders, taking in great quantities of water in a single gulp, and then using the fringes on their baleen plates to filter small schooling fish or invertebrates from the water. Nearly all mysticetes make long-range seasonal migrations. They generally occur in smaller groups than most odontocetes, and tend to have a simpler social organization. There are 14 species in six genera.

Family Balaenopteridae—Rorquals This family contains eight (or possibly nine) species in two genera of the largest animals ever to live; all balaenopterids have adult body lengths of over 7 m, and some are much larger. The rorquals are streamlined animals (the humpback whale somewhat less so than the others), with a series of long pleats extending from the snout tip to as far back as the navel on the ventral surface. Balaenopterids are generally fast and active lunge feeders; their morphology allows them to open their jaws very widely (>90° in some cases) and distend their throats to take in huge mouthfuls of food and water during feeding. The baleen plates are of moderate length and fringe fineness. Density and fringe diameter vary among species and, along with plate number and width/length ratio, are diagnostic characters. Rorquals have dorsal fins (varying in size and shape) set behind the midpoint of the back. The upper jaw has a relatively flat profile, a feature reflecting the structure of the skull. Within a given feature, differences among balaenopterids are often subtle variations on a theme, rather than class distinctions. Therefore, information on many features may be needed to distinguish among them and reliance on a single character for identification is strongly discouraged.

Family Eschrichtiidae—Gray whale The single species of gray whale was once present in both the North Atlantic and North Pacific oceans, but was exterminated in the Atlantic in the last few hundred years. This monospecific family is in some ways intermediate between the Balaenidae and the Balaenopteridae. The gray whale is stocky and has an arched upper jaw, but neither of these characters is as pronounced as in the right whales. Gray whales are slow-moving coastal animals that suck prey from the bottom sediments. Gray whales have the shortest and coarsest baleen of all species, a feature that probably reflects both the size of their prey and their tendency to take in gravel, sand, and other debris during feeding. There are 2–5 short throat creases, a dorsal hump followed by a series of knobs or knuckles along the dorsal surface of the tail stock, and only four digits in the flipper. The bodies of adults are usually covered with patches of barnacles and whale lice.

Odontoceti—Toothed whales Odontoceti has traditionally been considered a suborder, but its ranking is currently undetermined. With the exception of the sperm whale (males of which can reach lengths of at least 18 m), odontocetes are small to medium-sized cetaceans. Sexual size dimorphism is the rule. Toothed whales are characterized by the presence of homodont teeth throughout life[1] (although teeth are buried in the gum or jawbone in some species, worn, broken, or lost in others, and take peculiar shapes in still others), a single blowhole (U-shaped and with the open side facing forward in all except three species), an asymmetrical skull (generally with a concave profile), nasal bones elevated above the rostrum, presence of antorbital notches, a thin-walled "pan bone" at the posterior end of the mandible, a sternum with three or more parts, a complex system of nasal sacs, and a fatty organ in the forehead area called the melon. All are hypothesized to be capable of echolocation (i.e., producing specialized sounds, and receiving and processing the echoes from these sounds to navigate, find food, avoid predators, etc.), although this ability has been experimentally verified for only a handful of species held successfully in captivity. Odontocetes capture individual prey, which consists largely of various species of fishes and squids. There are 77 species in 33 genera.

Family Physeteridae—Sperm whale The single species of sperm whale is the largest toothed cetacean and has the highest degree of sexual size dimorphism. There is a low dorsal hump, followed by a series of crenulations. It has a large head with a squarish profile, narrow underslung lower jaw, and functional teeth only in the lower jaw (these fit into sockets in the upper jaw). The blowhole is located at the left front of the head. Internally, the head is highly modified, and is divided into sections called the "junk" and the spermaceti organ, or "case." The spermaceti organ is a large oil-filled reservoir, the function(s) of which are controversial. There is a dish shape to the facial area of the skull, extreme cranial asymmetry, and a long rostrum. Sperm whales are known to be capable of very long, deep dives to over 3,200 meters.

Family Kogiidae—Pygmy and dwarf sperm whales This family contains two species in a single genus. The pygmy and dwarf sperm whales are much smaller and share only a superficial resemblance to the "great" sperm whale. They have blunt squarish heads, with narrow, underslung lower jaws (like their larger counterparts), but the head is relatively much smaller than in the sperm whale, and the blowhole is not located at the front of the head as it is in the sperm whale. The skull structure is curious: it shares a basin- like facial area and great asymmetry with the sperm whale, but is much shorter. The dorsal fin in both species is relatively larger than that of the sperm whale, but is similarly followed by a series of small bumps or knobs. Both species are characterized by a vertical pigmentation mark on the side of the neck, similar in appearance to a shark's gill slit, a feature totally unique among cetaceans. The biology of these animals is very poorly known,

Family Monodontidae—Narwhal and beluga This is a family of small whales (<6 m in length), with stocky bodies, blunt bulbous heads, broad rounded flippers, and no dorsal fins. There are two species in two genera. Both species are inhabitants of high arctic areas of the Northern Hemisphere, often living among the ice. The skull is unique in that, in profile, it is very flat, with little or no rise in the area of the nares. Unlike the situation in most cetaceans, the cervical vertebrae are generally not fused, allowing monodontids a great range of neck flexibility. The two species are both restricted to high-latitude waters of the Arctic, often living among the ice.

Family Ziphiidae—Beaked whales The beaked whales (a large family, with 22 species in six genera) are medium-sized cetaceans (4–13 m long), which as a rule, have reverse sexual dimorphism (females larger than males).

[1] Tooth counts of toothed whales given in this guide are for each tooth row (the upper and lower jaws each have two tooth rows), unless otherwise indicated.

A variety of feeding techniques are used by marine mammals, and some species of delphinids are known to regularly chase fish and marine mammal prey onto sloping beaches—risking stranding and death in order to get a meal. Australian humpback dolphin. PHOTO: F. WARDLE, COURTESY D. CAGNAZZI

In general, beaked whales have a pronounced beak, relatively small dorsal fin set far back on the body, small flippers that fit into depressions on the sides, two short throat grooves, flukes lacking a median notch, and no more than 1–2 pairs of functional teeth (tusks) that erupt from the lower jaw of males only (major exceptions of the latter are *Berardius,* in which both adult males and females have two pairs of exposed teeth, and *Tasmacetus,* in which both sexes have long rows of slender functional teeth). Beaked whales are poorly known as a rule; however, most are thought to be deep-diving squid eaters. They generally travel in small groups.

Family Delphinidae—Marine dolphins The family Delphinidae has been called a "taxonomic trash basket," because many small to medium-sized odontocetes of various forms have been lumped together in this group for centuries. It is the largest marine mammal family, with 38 species in 17 genera. Needless to say, then, the so-called delphinids are extremely diverse in form. They range in size from the 1–1.8 m dolphins of the genera *Sotalia* and *Cephalorhynchus* to the killer whale, in which males can reach lengths of at least 9.8 m. Most delphinids, however, share the following characteristics: a marine habitat, a noticeable beak, conical teeth, and a large falcate dorsal fin set near the middle of the back. There are exceptions to every one of these rules, except the presence of basically-conical teeth, however. Generally, delphinids have a complex social organization, and they form the largest groups of any marine mammal (sometimes in the thousands).

Family Phocoenidae—Porpoises Porpoises (seven species in three genera) are small cetaceans (all less than 2.5 m) that some taxonomists in the past have classified with the delphinids. They tend to be coastal in distribution (but Dall's and spectacled porpoises are exceptions), rather stocky in form, with either a short, indistinct beak or no beak at all. Most have a short triangular dorsal fin; all have spade-shaped teeth, and bony protuberances on the skull in front of the bony nares. In some species, females are larger than males. They exhibit paedomorphosis (especially in the skull), a condition in which adults retain juvenile characteristics. Phocoenids appear to live in smaller groups and have a simpler social structure than do most delphinids.

Family Platanistidae—South Asian river dolphin This family includes the susu and bhulan of the Ganges and Indus rivers, which were previously classified as separate species, but are now recognized as subspecies of *Platanista gangetica.* Animals in this family are nearly blind, and apparently rely largely on echolocation to navigate and find food. The body is fairly small (to about 2.6 m) and is soft and "mushy." There is a long forceps-like beak, with front teeth that can extend outside the closed mouth. The blowhole is a longitudinal slit. The susu and bhulan have no true dorsal fin, only a low dorsal ridge. The most characteristic feature of the skull is a pair of enlarged maxillary crests that overhang the rostrum. Distribution is restricted to several large river systems of South Asia. Some evidence suggests that this may be the most closely-related odontocete group to the mysticetes.

Family Iniidae—Boto The single species in this family (two more have recently been described, but currently not recognized), the boto and bufeo of the Amazon and Orinoco drainages in South America, are unique in several ways. They are large river dolphins, each with a moderately long, thick beak dotted with sparse hairs. The dorsal ridge is very low and usually indistinct. Many adults are nearly totally white/pink in color. The rear teeth are flattened and the zygomatic arches of the skull are incomplete. These strange animals often swim in the flooded forest – they are truly forest dolphins.

Family Lipotidae—Baiji This family contains a single recently-extinct species, the baiji of the Yangtze River in China. It was a mid-sized river dolphin. The dorsal fin was reduced to a low structure, with a short base. The rostrum was upturned, and there was a distinct constriction at its base. The baiji was the most endangered cetacean in the world until 2006, when it was declared to be probably extinct. The lack of documented reports since then leaves little doubt that is has indeed gone extinct.

Family Pontoporiidae—Franciscana This family contains only one species, a coastal marine species of the east coast of South America, known as the franciscana. This is the smallest of the platanistoids (true river dolphins), rarely reaching 1.8 m in length. Females are larger than males in this species. Franciscanas have extremely long beaks (proportionately the longest of any cetacean) and rather low, triangular dorsal fins.

Unlike most other marine mammals, manatees have teats that are located in the axillary region—in most others they are more posterior. Thus, the manatee calf needs to nurse from a site near the breast, and this is considered to have led to the mermaid myth. Crystal Springs, Florida. PHOTO: T. PUSSER

Order Sirenia—Manatees and Dugongs

There are four living species of sirenians: three manatees and the dugong. A fifth species, Steller's sea cow of the North Pacific and Bering Sea, was exterminated by over-hunting in the 1700s. Sirenians, like cetaceans, are totally aquatic. They are the only marine mammalian herbivores. As a consequence, even the most marine species, the dugong, tends to be less oceanic than members of other marine mammal groups. In fact, most manatees spend much or all of their lives in fresh or brackish water. All four living species are restricted to tropical/subtropical habitats. Steller's sea cow was unique. It inhabited cold temperate to subarctic waters of the North Pacific. Sirenians have the following morphological characteristics in common: robust body; tough, thick skin with little hair; two nostrils on top or at the front of a thick muzzle; no ear pinnae; no hind limbs; mammary nipples located near the axillae; forelimbs modified into flippers; a horizontally flattened tail; and dense (pachyostotic) bones.

Family Trichechidae—Manatees The three species of manatees (all in a single genus) are found in tropical/subtropical areas of the Atlantic Ocean (and appended freshwater systems) and are very sensitive to cold. They are characterized by a horizontally-flattened, rounded tail (as opposed to the whale-like flukes of dugongs). With only six cervical vertebrae, manatees are among the few groups of mammals that diverge from the normal mammalian number of seven. They are also unusual in that their teeth are replaced throughout life with new ones from the rear of the mouth (like a 'conveyor belt'). The external auditory meatus (external ear opening) is very broad and shallow.

Family Dugongidae—Dugong There is only one living species in the family Dugongidae. The other recent member, Steller's sea cow *(Hydrodamalis gigas)*, was exterminated by overhunting in 1768. The dugong is a tropical/subtropical inhabitant of the Indo-Pacific, but Steller's sea cow was an inhabitant of cold temperate to subarctic waters. In members of this family, the flattened tail is expanded into flukes, similar to those of cetaceans. Other characteristics include a rostrum that is deflected downwards, the presence of erupted tusks in males (dugong only; Steller's sea cow had no functional teeth), a more streamlined body than those of manatees, and the absence of nails on the flippers.

Order Carnivora—*Carnivorous mammals* (including pinnipeds, marine otters, and polar bear)

By far, most carnivores are terrestrial mammals. Besides pinnipeds, the Order Carnivora contains seven families of largely meat-eating mammals, including cats, dogs, bears, raccoons, weasels, otters, civets, and hyenas. Of these, only two families contain marine mammal representatives, the Mustelidae (otters and weasels) and the Ursidae (bears). Below we discuss only the marine species within this Order.

Family Mustelidae—Otters The mustelids are the otters, weasels, and their kin (minks, polecats, martens, wolverines, skunks, and badgers). Although four other species of freshwater otters may obtain some of their food from the sea, only two of the approximately 67 species in this family are truly marine, the sea otter and the marine otter. Thus, we restrict our treatment to these two species usually considered among marine mammals. Otters are often classified in their own subfamily, the Lutrinae (containing about 13 species). Marine and sea otters are largely restricted to the Pacific Ocean (two marine species in two genera). The extinct sea mink (*Neovison macrodon*) is sometimes considered a marine mammal as well, but we do not cover it in this book.

Polar bears are considered marine mammals because they derive almost their entire sustenence from marine animals. PHOTO: S. EBBERT

Family Ursidae—Bears There are seven species of bears in the world; six are wholly terrestrial and only one qualifies as a marine mammal. Bears are very familiar animals to many people; in particular, the grizzly/brown and black bears of the Northern Hemisphere are often exhibited in zoos and are well known (grizzly/brown bears are most closely related to the polar bear). The single marine species, the polar bear, ranks as the least aquatic and least derived of all marine mammals. It was the most recently-evolved marine mammal, only diverging from brown bears 1.0–1.5 MYA. Polar bears may spend long periods of time on shore. They are restricted in distribution to arctic regions of the Northern Hemisphere.

Pinnipedia—*Seals, sea lions, and walruses*

Pinnipedia was traditionally considered a suborder, but its current ranking is undetermined. There are 35 species of pinnipeds, all of which are assigned to three families of the mammalian order Carnivora: the Otariidae, Phocidae, and Odobenidae. Although technically the taxonomic suborder Pinnipedia is not recognized by many marine mammal biologists anymore, we will make use of Pinnipedia in this book, for clarity of presentation. The otariids are the 15 species of sea lions and fur seals, sometimes referred to as the eared seals. The phocids are the 19 species of true seals, sometimes referred to as the "earless" seals (they have ears, just no external pinnae). Odobenids are reduced to just a single living species, the walrus. There has been controversy as to whether the pinnipeds are monophyletic (i.e., evolved from a single ancestor) or biphyletic (from two separate ancestors). However, recent evidence indicates that they are indeed monophyletic, along with an extinct group called the desmatophocids.

Pinnipeds are highly specialized aquatic carnivores that live in a diversity of marine habitats, and some freshwater ones as well. One unifying feature of the group is that all must return to a solid substrate, such as land or ice, to bear their pups. Almost without exception, females give birth to a single offspring per reproductive effort. All species are amphibious, though the otariids are the most agile and mobile of the pinnipeds on land. In general, phocids are more capable divers and breath-holders, although there is overlap in the capabilities of some otariids and phocids. In many ways, the walrus is intermediate between the phocids and otariids.

All pinnipeds have fur (but also use blubber for thermoregulation), two sets of limbs (called foreflippers and hindflippers), long whiskers (vibrissae), nasal openings at the tip of the snout, and strongly reduced or lost ear flaps. Pinnipeds molt every year, some gradually over several weeks or months, others dramatically in a short time. In the species accounts, pinniped coloration is described in somewhat more detail than for cetaceans, because for identification, there is often more of an emphasis on the subtle shading often visible on hauled-out pinnipeds.

Family Otariidae—Eared seals and sea lions All 15 species (one extinct, in seven genera) of sea lions and fur seals have a polygynous mating system and pronounced sexual dimorphism. Characteristics of this family are: small external ear flaps (pinnae), smooth vibrissae, light skin, a double layer of fur with short underfur and longer guard hairs, hairless hindflippers, four teats in females, scrotal testes, and skulls with shelf-like supraorbital processes and sagittal crests (the latter enlarged in adult males only). Eared seals swim with their large foreflippers and can rotate their hindflippers forward to walk on all fours on land. Southern Hemisphere fur seals rest in a characteristic posture, with head down and flippers swaying gently.

Family Odobenidae—Walrus While there were multiple species in the past, today only a single walrus species persists. Walruses are enormous animals that combine features of both otariids (e.g., moderately long foreflippers that can lift the forward part of the body off the ground) and phocids (e.g., lack of ear pinnae). The neck is long and the hindflippers can rotate under the body and permit walking, although walruses are so bulky they cannot walk as easily as most otariids can. The tail is sheathed in skin and not readily visible or free, as it is in other pinnipeds. The tusks are a unique feature, and are important in feeding and assisting with hauling out. The walrus skull is very dense (pachyostotic) and has antorbital processes that are composed of both frontal and maxilla bones. Walruses have numerous short, smooth vibrissae on their thick fleshy mystacial ("moustache") pads. The testes of walruses are internal, not scrotal, and females have four retractable mammary teats. The skin is dark in younger animals and lightens with age. Walruses swim with phocid-like side-to-side strokes of the hindflippers, with assistance from the foreflippers. They only occur in high latitudes of the Northern Hemisphere.

These Galapagos sea lions show the basic body shape that is typical of all eared seals (otariids). Their ability to bring their hindflippers up under their body and walk on land reminds one of their terrestrial ancestry. PHOTO: M. NOLAN

Family Phocidae—True seals This is the largest family of the pinnipeds, with 19 species in 14 genera (one species is extinct). The true, or earless, seals include the largest of the pinnipeds, the elephant seals. Species within the group have variable degrees of sexual dimorphism (in some species, females are the larger sex). Phocids are characterized by the absence of external ear pinnae, a short muzzle, beaded vibrissae, dark skin, short fur, generally two teats in females, internal testes, furred hindflippers, pachyostotic (dense) ear bones, inflated tympanoperiotic bones, and the absence of supraorbital processes or an enlarged sagittal crest on the skull. Propulsion in water is provided by figure-eight movements of the hindflippers, and movement on land is achieved by inch-worming or "galumphing," without much help from the relatively small foreflippers. They lack the ability to draw the hindflippers under the body to lift themselves off the ground. As a rule, true seals are more aquatic than eared seals, spending proportionately less time on land or ice.

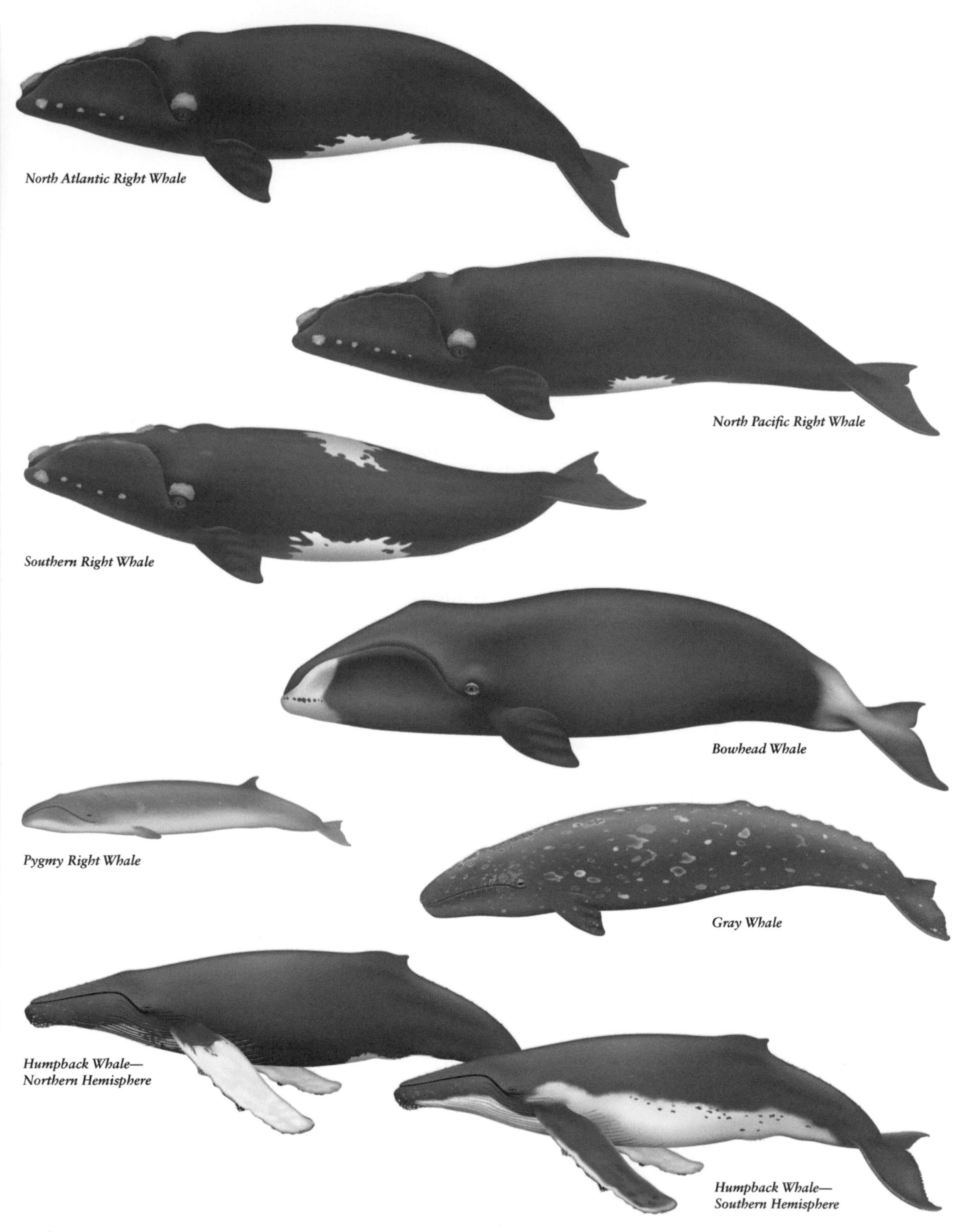
North Atlantic Right Whale
North Pacific Right Whale
Southern Right Whale
Bowhead Whale
Pygmy Right Whale
Gray Whale
Humpback Whale—
Northern Hemisphere
Humpback Whale—
Southern Hemisphere

4. Cetaceans

Blue Whale
Pygmy Blue Whale
Fin Whale
Sei Whale
Bryde's Whale
Common Minke Whale
Omura's Whale
Dwarf Minke Whale
Antarctic Minke Whale

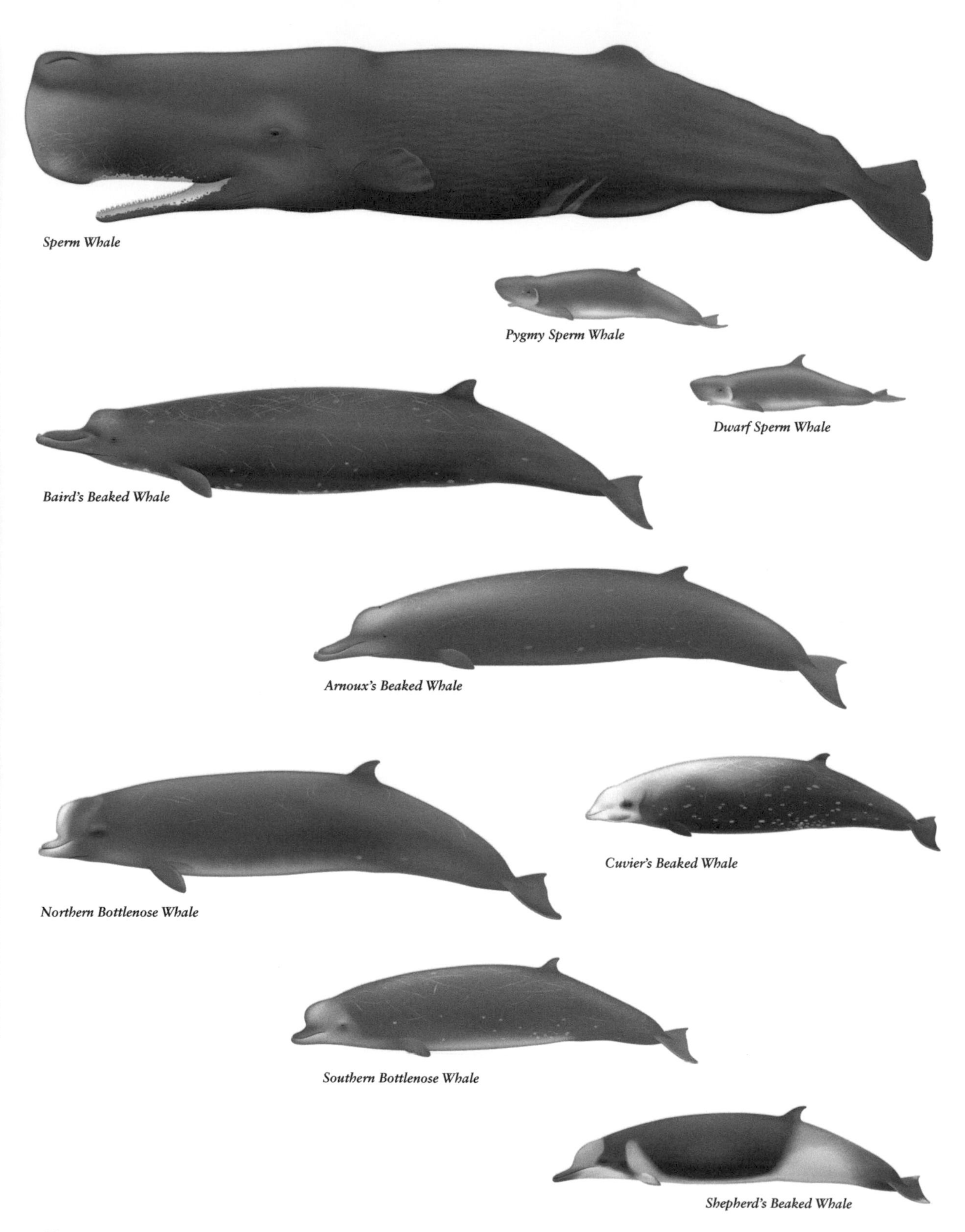
Sperm Whale
Pygmy Sperm Whale
Dwarf Sperm Whale
Baird's Beaked Whale
Arnoux's Beaked Whale
Northern Bottlenose Whale
Cuvier's Beaked Whale
Southern Bottlenose Whale
Shepherd's Beaked Whale

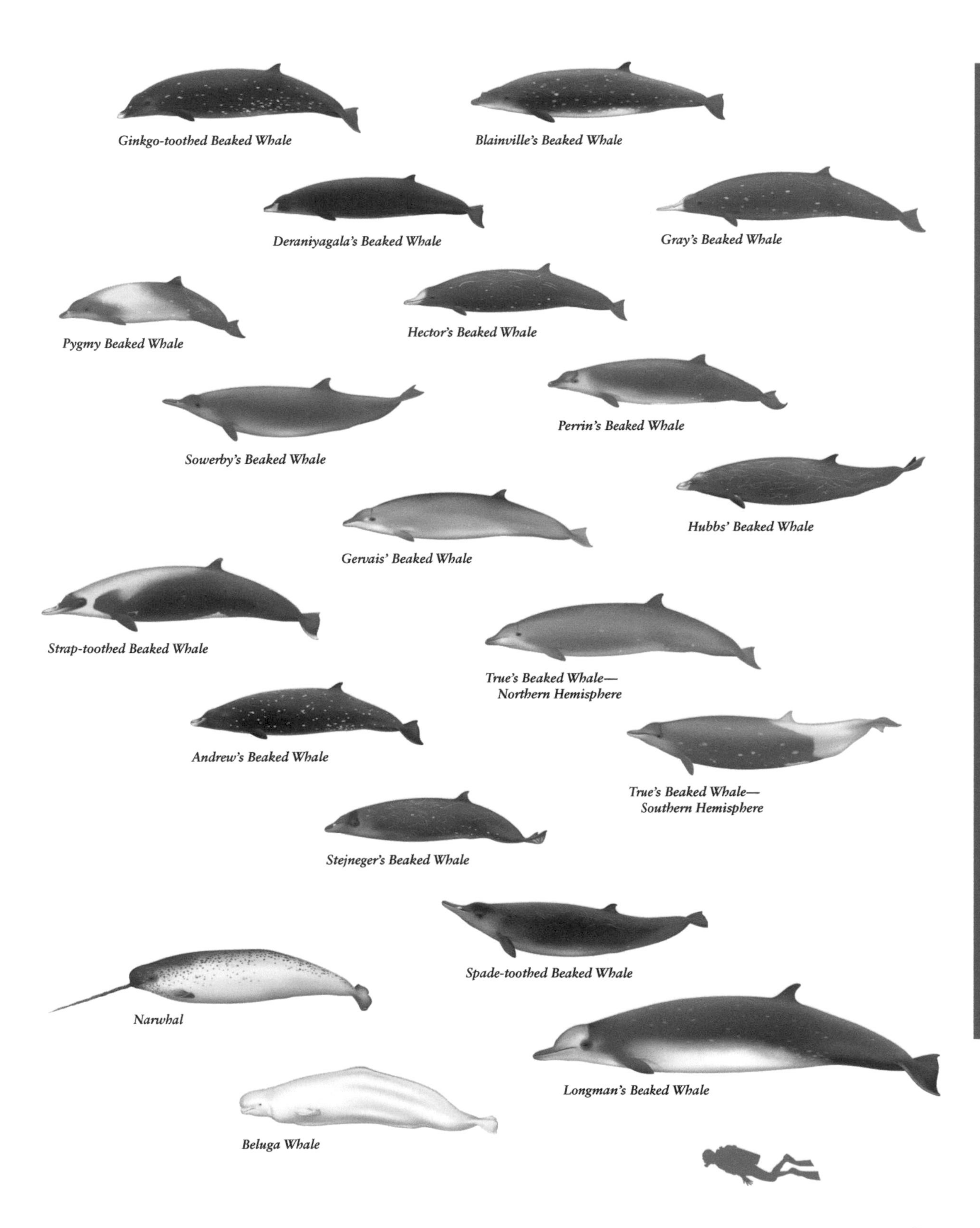
Ginkgo-toothed Beaked Whale
Blainville's Beaked Whale
Deraniyagala's Beaked Whale
Gray's Beaked Whale
Pygmy Beaked Whale
Hector's Beaked Whale
Sowerby's Beaked Whale
Perrin's Beaked Whale
Hubbs' Beaked Whale
Gervais' Beaked Whale
Strap-toothed Beaked Whale
True's Beaked Whale—Northern Hemisphere
Andrew's Beaked Whale
True's Beaked Whale—Southern Hemisphere
Stejneger's Beaked Whale
Spade-toothed Beaked Whale
Narwhal
Longman's Beaked Whale
Beluga Whale

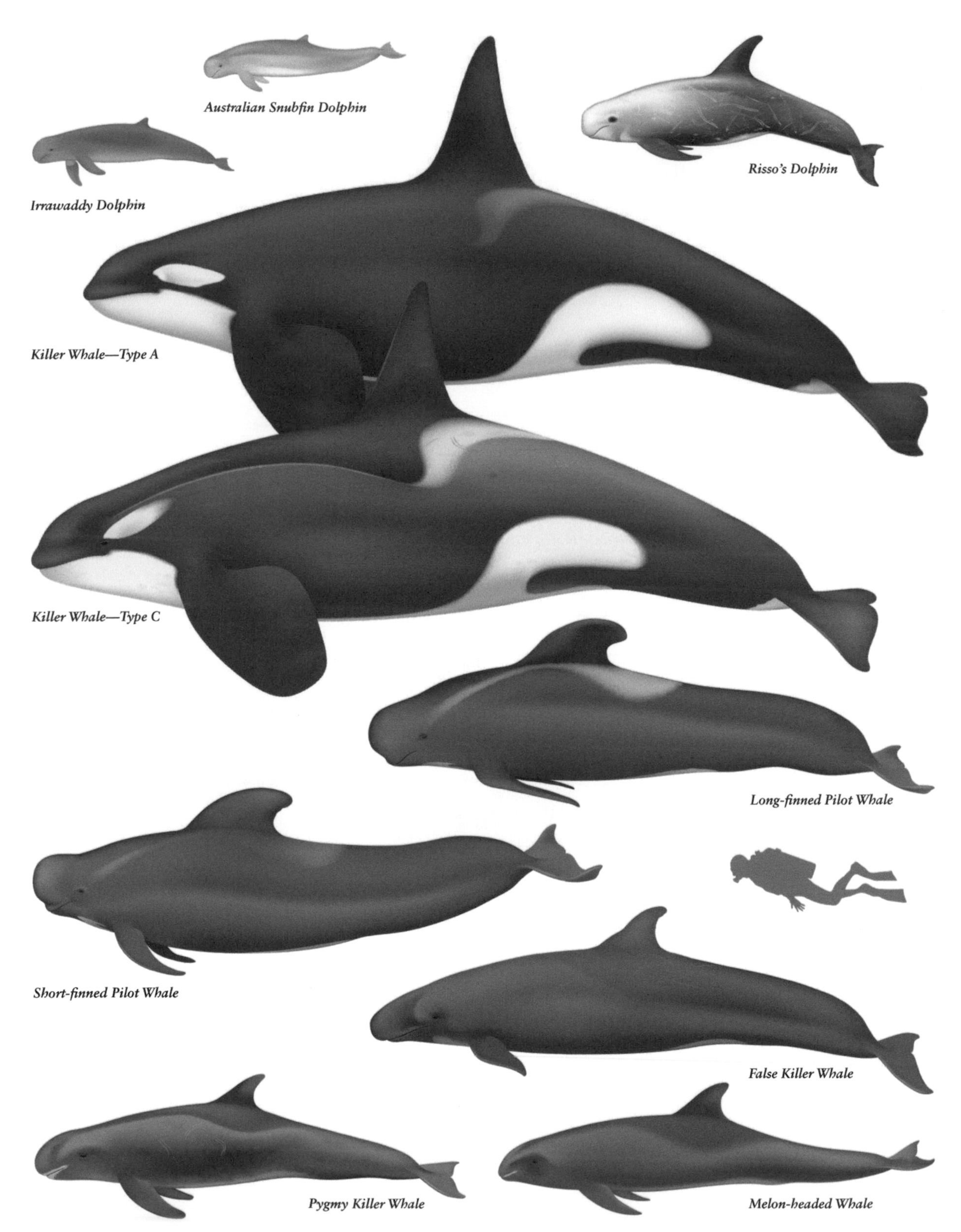

Australian Snubfin Dolphin
Irrawaddy Dolphin
Risso's Dolphin
Killer Whale—Type A
Killer Whale—Type C
Long-finned Pilot Whale
Short-finned Pilot Whale
False Killer Whale
Pygmy Killer Whale
Melon-headed Whale

Tucuxi
Guiana Dolphin
Rough-toothed Dolphin
Atlantic Humpback Dolphin
Indo-Pacific Humpback Dolphin
Indian Ocean Humpback Dolphin
Australian Humpback Dolphin
Pantropical Spotted Dolphin
Common Bottlenose Dolphin
Indo-Pacific Bottlenose Dolphin
Atlantic Spotted Dolphin
Spinner Dolphin
Clymene Dolphin
Striped Dolphin
Short-beaked Common Dolphin
Long-beaked Common Dolphin
Fraser's Dolphin
White-beaked Dolphin
Atlantic White-sided Dolphin
Pacific White-sided Dolphin
Dusky Dolphin
Hourglass Dolphin
Peale's Dolphin
Northern Right Whale Dolphin
Southern Right Whale Dolphin
Commerson's Dolphin
Heaviside's Dolphin
Hector's Dolphin
Chilean Dolphin
Dall's Porpoise
Harbor Porpoise
Spectacled Porpoise
Burmeister's Porpoise
Vaquita
Indo-Pacific Finless Porpoise
Narrow-ridged Finless Porpoise
South Asian River Dolphin
Franciscana
Amazon River Dolphin

North Atlantic Right Whale—*Eubalaena glacialis*

(Müller, 1776)

Recently-used synonyms None.
Common names En.–North Atlantic right whale; Sp.–*ballena franca*; Fr.–*baleine de Biscaye, baleine noire*.
Taxonomic information Cetartiodactyla, Cetacea, Mysticeti, Balaenidae. A recent study demonstrated that right whales in the three major ocean basins where they occur (North Pacific, North Atlantic, and Southern Ocean) represent separate lineages. As such, three species of right whales are now recognized: *E. japonica*, *E. glacialis*, and *E. australis*, respectively.

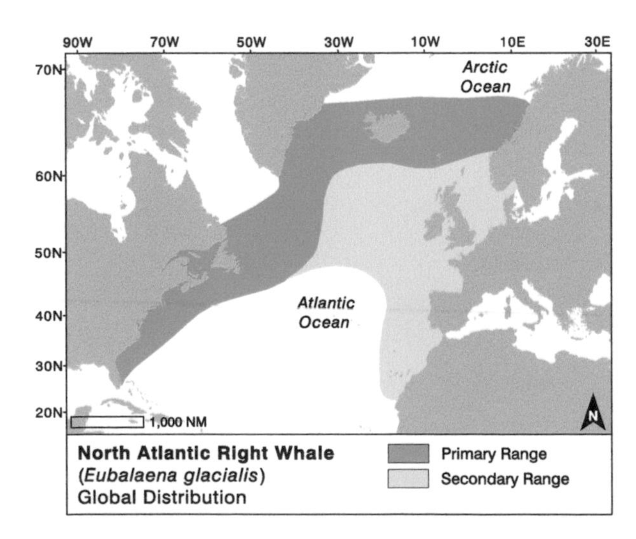

Species characteristics The right whales are among the stockiest of all whales. The axillary girth may be more than half the total length. The North Atlantic right whale has a massive head that can be one-fourth to one-third of its body length. The jawline is arched and the upper jaw is very narrow in dorsal view. It is narrowest near the middle and it widens slightly toward the tip. The upper edges of the lips are often scalloped. The eyes are located just above the corners of the mouth, and the blowholes are located at the highest point of the rear part of the head; they point slightly to the sides. The flippers are broad and are more paddle-shaped than the pointed flippers of most other cetaceans. They may be up to 1.7 m long. There is no dorsal fin or dorsal ridge on the broad back. The back is smooth and wide. The flukes are very wide (up to 35% or more of total length) and gracefully tapered, with a smooth trailing edge and a deep notch.

Most North Atlantic right whales are predominantly black, but there may be large white splotches of varying extent on the underside of the belly and the chin (about 20–30% of whales have them) and sometimes elsewhere on the body (not as extensive as in the southern right whale, however). Some individuals may appear mottled, which is caused by uneven sloughing of patches of skin. Callosities are scattered about the head; these are areas of roughened skin to which whale lice attach. The largest of these callosities, on the top of the rostrum, is called the bonnet. The upper jaw callosities tend to be more continuous than in the southern species. There are generally

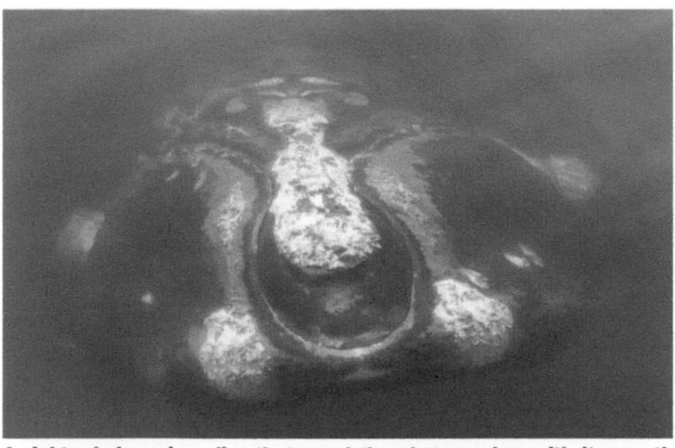

A right whale swims directly toward the photographer with its mouth closed; this perspective clearly shows the distribution of the callosity patches on its head and high-arching lips. New England. PHOTO: Y. GUILBAULT

A North Atlantic right whale skim-feeding at the surface and probably taking copepods; the lips on either side are spread open and the pink tongue is visible between the two rows of baleen plates. New England. PHOTO: M. MOORE

Right whales sometimes occur in "surface-active groups" for socializing and mating. PHOTO: M. BROWN

There is often a lot of physical contact in these "surface-active" aggregations. PHOTO: M. BROWN

The elegantly-sculpted flukes of right whales are "cleaner" than those of humpbacks; they are all black with a smooth trailing edge. New England. PHOTO: M. BROWN

A breaching North Atlantic right whale falling on its back and showing its rotund body, massive head, high-arching lip, and broad flippers. New England. PHOTO: M. HAGBLOOM

few or no callosities on the upper margin of the lower lips (however, these are fairly common in southern right whales). The skin of the callosity areas is more smooth on young animals, and it becomes more roughened with age. The patterns of the callosities are individually distinctive, and these are used, together with prominent scarring, to identify individual whales.

The widely-separated blowholes produce a bushy V-shaped blow up to 5 m high. Inside the mouth are 200–270 (average 250) long, thin baleen plates, which may reach nearly 3 m in length. They are brownish-gray to black in color. The fringes of these plates are very fine, reflecting the small prey taken by this species. Adult North Atlantic right whales range in length to 17 m, but may

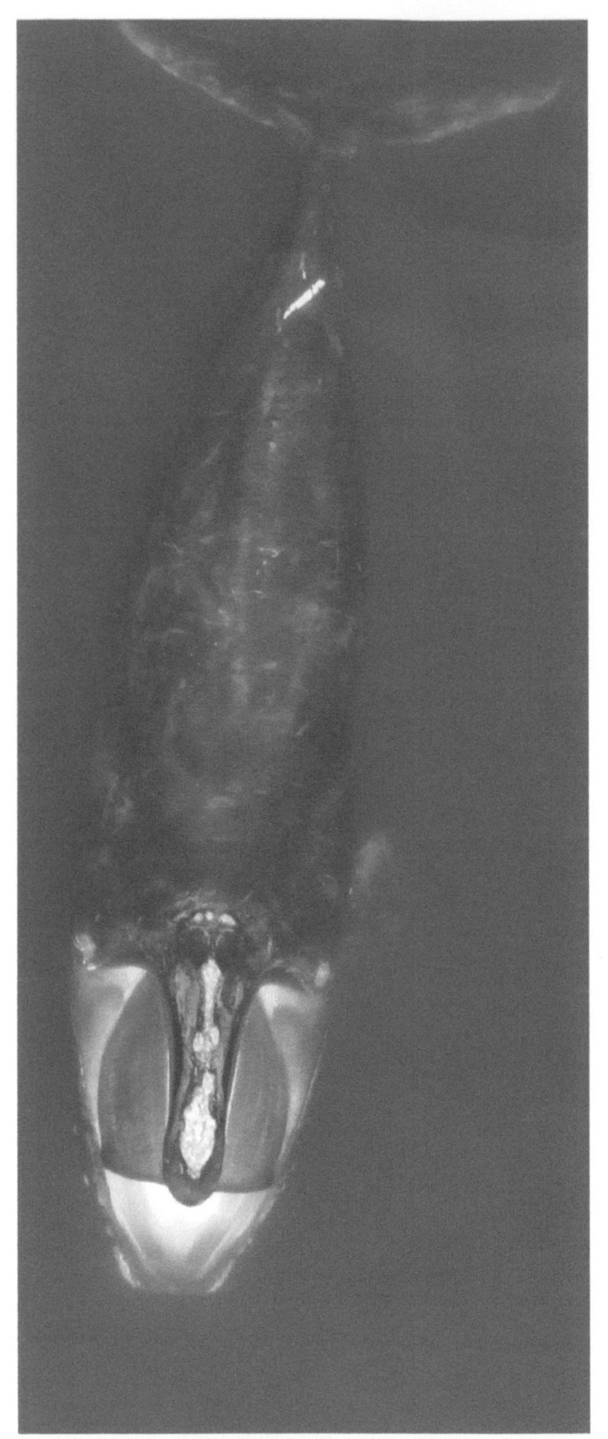

A right whale skim-feeding just below the surface; the mouth is open and a rack of baleen plates is visible on either side of the narrow rostrum. The pale area in front of the rostrum is the tongue. New England. PHOTO: C. CRISTMAN

A mother and calf right whale share a moment; in addition to the absence of a dorsal fin, the white callosity patches and distinctly arching lips are easily discernable from the air. New England. PHOTO: K. MAHONEY

occasionally exceed 18 m. Females reach slightly larger sizes than males. Adults may reach weights of 90,000 kg. Newborns are 4.0–4.5 m long.

Recognizable geographic forms None.

Can be confused with Although it is nearly impossible to distinguish the three right whale species from observations at sea, their distributions are widely separated, and they can be attributed to species based on their location. The North Atlantic right whale can be easily distinguished from other large whale species that occur in the Atlantic Ocean (primarily rorquals and sperm whales) by its robust body, large head, arched mouthline, long baleen plates, and V-shaped blow. North Atlantic right and bowhead whales once overlapped in distribution, and may again if populations recover. The presence/absence of callosities and coloration differences should make it easy to distinguish them.

Distribution North Atlantic right whales from two populations primarily inhabit temperate and subpolar waters of the North Atlantic Ocean. Historically, the two populations were presumably largely isolated from each other, and the eastern stock is now thought to be functionally extinct. The western North Atlantic stock breeds off the southeastern US (Florida and Georgia) and feeds in the Gulf of Maine and off eastern Canada, as far north as Nova Scotia. The eastern Atlantic stock occurred in northern European waters, from Spain to Norway and Iceland, and probably moved south to waters off West Africa to breed. There are a few extralimital records from the Mediterranean. However, records in most parts of the eastern North Atlantic in recent years have been extremely rare. The population identity of right whales seen on rare occasions off Iceland and Greenland is unclear.

Ecology and behavior Right whales in the North Atlantic are mostly seen in groups of fewer than about a dozen (most often singles or pairs). Larger groups may form in areas of aggregation for feeding and breeding. Right whales are generally larger and more slow moving than rorquals. They can be aerially active, sometimes fluke-slapping and breaching repeatedly, and they generally raise their flukes before a deep dive. Right whale individuals can be identified by the patterns of callosities and scars (the latter mainly on the head and back). The mating system appears to involve sperm competition (males competing to inseminate females, not so much by physical aggression as by delivering large loads of sperm, thereby displacing that of other males). Male right whales have the largest testes in the animal kingdom. Groups of males are often seen surrounding a female, and mating with multiple males in these groups appears to be common. Interestingly, mating appears to occur throughout the year, but calving is seasonal. After a pregnancy lasting about a year, young are born in winter (mostly from December to March) on subtropical or mid-latitude breeding grounds, nowadays primarily off the coasts of Florida and Georgia. The inter-calving interval averages about three years in this species, longer than for most other species of whales.

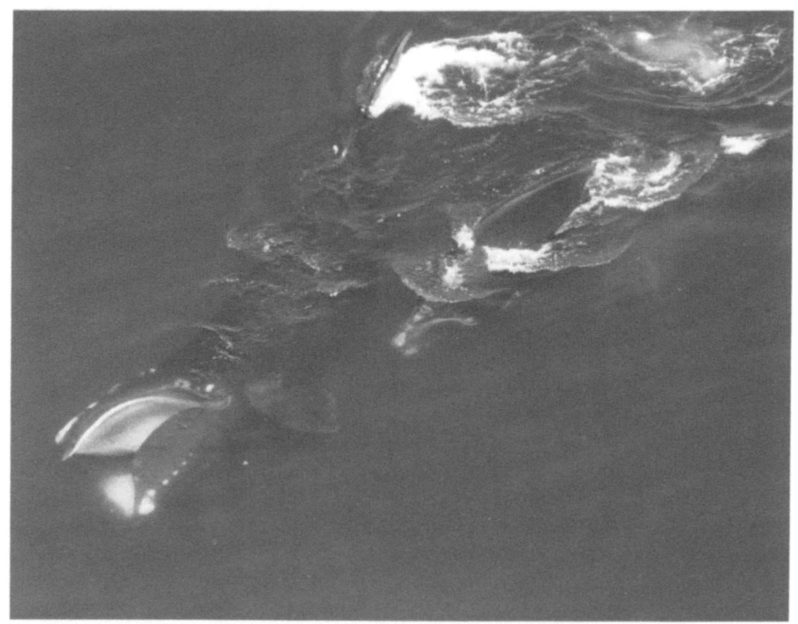

A feeding right whale mother, with her large calf, rolls on her right side with her cavernous mouth agape. New England. PHOTO: P. DULEY, NOAA FISHERIES, NEFSC

The functions of the callosities, which only right whales have, are unknown. It has been suggested that they are used by males in sexual selection for access to females. Some evidence appears to support this hypothesis, although females have callosities too. Other functions cannot be ruled out. North Atlantic right whales may live to be at least 70 years of age.

Feeding and prey Right whales feed on calanoid copepods and other small invertebrates (e.g., smaller copepods, krill, pteropods, and larval barnacles), generally by slowly skimming through patches of concentrated prey at or below the surface. The most common prey species throughout most of the North Atlantic range is the copepod *Calanus finmarchicus*.

Threats and status This was the first species of whale to be commercially hunted, starting as early as the 11th century with a Basque fishery in the Bay of Biscay. The right whale was slow, yielded large amounts of oil and baleen, and floated when dead, thus making it the "right" whale to kill. Heavy hunting left it severely depleted by the late 1600s, but some direct exploitation continued into the 20th century. Right whales are now fully protected, but current known threats include vessel collisions and entanglement in fishing gear. More than 60% of living North Atlantic right whales bear scars from fishing gear; about 6% have non-lethal injuries from vessel collisions, and these incidents have regularly killed other right whales. In addition, habitat destruction, pollution, and disturbance by vessel traffic may also be factors. This is one of the world's most endangered large whales, with only 450 whales thought to remain. Although the population was apparently growing in the 1980s and early 1990s, it began to drop in the late 1990s. It is now thought to be stable or increasing slightly. Past declines were almost certainly due to high anthropogenic mortality, and the species clearly remains in very real danger of extinction.

IUCN status Endangered.

North Pacific Right Whale—*Eubalaena japonica*

(Lacépéde, 1818)

Recently-used synonyms None.
Common names En.–North Pacific right whale; Sp.–*ballena franca del Pacífico*; Fr.–*baleine franche du Pacifique.*
Taxonomic information Cetartiodactyla, Cetacea, Mysticeti, Balaenidae. This species was only recently split off from *E. glacialis*, due to indications of its genetic distinctness. There are few known morphological differences between the two Northern Hemisphere right whale species.
Species characteristics Right whales in the North Atlantic and North Pacific are not known to differ significantly in their external morphology or coloration. North Pacific right whales are extremely stocky animals; the maximum girth is about 75% of the total length. They have massive heads that can be up to nearly one-third of their body length. The eyes are set just above the corners of the mouth. The jawline is strongly arched and the upper jaw is very narrow in dorsal view, although it widens somewhat toward the tip. The upper margin of the lower jaw is generally somewhat serrated. The flippers are broad and tend to be more squarish or paddle-shaped than the pointed flippers of most other cetaceans, and they may be up to 1.7 m long. There is no hint of a dorsal fin or dorsal ridge on the broad back. The flukes are very wide (up to 35% or more of total length) and gently tapered, with a smooth trailing edge and a deep median notch.

Most North Pacific right whales are predominantly black, but there may be large white splotches of varying extent on the underside and sometimes elsewhere on the body (not as extensive as in the southern right whale, however). Callosities are scattered about the head; these are areas of roughened skin to which yellowish-white whale lice attach. The largest of these callosities, on the tip of the rostrum, is called the bonnet. There are generally few or no callosities on the upper margin of the lower lips (however these are fairly common in southern right whales). Inside the mouth are 200–270 long, thin baleen plates, which may reach nearly 3 m in length. They are brownish-gray to black in color. The fringes of these plates are very fine, reflecting the small prey taken by this species.

This appears to be the largest of the three right whale species. Adults range in length to over 19 m, with females larger than males. Based on data from the North Atlantic species, newborns are thought to be 4.0–4.5 m long and adults may weigh up to 90,000 kg. The widely-separated blowholes typically produce a bushy V-shaped blow up to 5 m high.
Recognizable geographic forms None
Can be confused with In the northern extremes of their range, especially in the Bering and Okhotsk seas, North Pacific right whales may be readily confused with bowhead whales from a distance. A closer look will provide the means to distinguish them. Bowhead whales lack callosities and right whales generally have white patches only on the belly. North Pacific right whales may also be confused with humpback or gray whales, but only from a great distance.

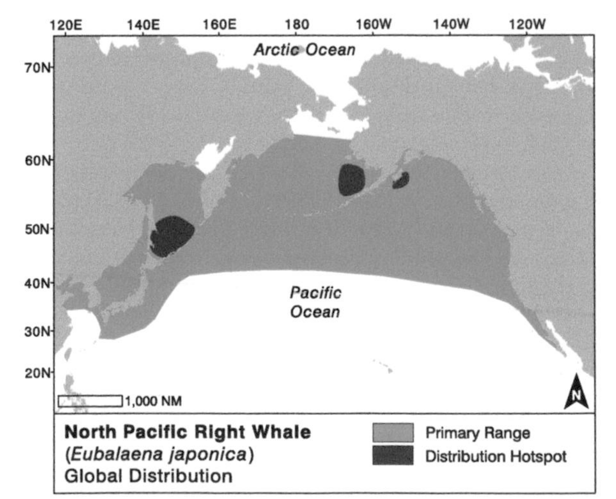

North Pacific right whales, like all members of their genus, have a broad back with no hint of a dorsal fin or ridge. Kamchatka. PHOTO: S. BLANC

Same animal as on the left swimming away and the showing strongly-arched lower lip. Kamchatka, Russia. PHOTO: S. BLANC

A rare image of a right whale apparently lunge-feeding—notice the water draining from the left corner of the mouth. Northwestern Pacific. PHOTO: V. BURKANOV

The callosity pattern on the head of each right whale is unique and allows for the identification of every clearly-photographed individual, as in this North Pacific right whale photographed off Japan. PHOTO: K. ODA

A North Pacific right whale swims toward the photographer showing the callosity patterning on its arching rostrum. Kamchatka, Russia. PHOTO: S. BLANC

Unlike gray and humpback whales, which sometimes show V-shaped blows, right whales always have V-blows, which are a bit more widely spaced at the base. Bering Sea. PHOTO: J. DURBAN

The V-shape of the blow of right whales can sometimes be seen even from the side; note the arched lip is discernable even in this distant photo. Kamchatka, Russia. PHOTO: S. BLANC

The flukes of right whales are smoother and straighter (less scalloped) along the trailing edge than those of the somewhat-similar humpback whale—both species regularly raise their flukes on diving. Northwest Pacific. PHOTO: V. BURKANOV

Distribution Right whales are now extremely rare in the North Pacific and little is understood of their current distribution there. Most of what is known of their historical range is based on whaling records. They previously had an extensive distribution in offshore waters (some >2,000 m water depth). There was evidence of a general northward migration in spring and a southward shift in autumn. However, records anywhere in winter are sparse and the breeding grounds are not known. The only areas where North Pacific right whales are detected often today are the Bering and Okhotsk seas; they are regularly seen over the southeastern Bering Sea shelf in April to September. Sightings and whaling catches have occurred in the Okhotsk and Bering seas, Gulf of Alaska, and around the Kuril and Aleutian islands. There have also been recent reports further south off Japan (this may be part of the migration route), the western US, and Baja California, Mexico. A sighting has been reported from Hawaiian waters, which is certainly extralimital. The absence of evidence for coastal winter breeding grounds suggests that (unlike their congeners) North Pacific right whales may breed in offshore areas.

Ecology and behavior Not much is known of the ecology and behavior of the North Pacific right whale, since these animals have been rarely observed alive in the last few decades. Much of what is known comes from whaling records. These whales generally occur most often as singles or pairs. Larger groups or aggregations may form on feeding or breeding grounds. They perform breaches and other aerial behaviors and generally raise their flukes before a long dive.

As in the North Atlantic species, the mating system appears to involve sperm competition. Calving areas have not yet been identified in the North Pacific; the absence of many records from coastal areas in winter may suggest that they breed offshore. Feeding areas are most often in shallow coastal regions of the temperate to subpolar zones. In general, this species appears to have a more oceanic distribution than does the North Atlantic right whale, which may have to do with the different availability of prey resources between the two ocean basins; however, it may also simply reflect lack of recent sighting and whaling effort in the offshore North Atlantic. There is good evidence to suggest the existence of two separate stocks of North Pacific right whales in the eastern and western Pacific. Based on information from the North Atlantic, these whales may live to be over 70 years old. The sighting of cow/calf pairs in the Bering Sea in recent years provides some renewed hope for the recovery of this species.

Feeding and prey Right whales feed on calanoid copepods and other small invertebrates (smaller copepods, krill, pteropods, and larval barnacles), generally by slowly skimming through patches of prey, either near the surface or at depth.

Threats and status Hunting by the Japanese began as early as the late 1500s, and by Europeans and Americans in 1835. North Pacific right whales were heavily exploited in the 1800s, and any recovery was impeded by large illegal Soviet catches (see below). They have been protected for over 65 years, but illegal hunting continued into the 1970s. Soviet whalers illegally killed at least 660 right whales in the eastern North Pacific mostly in the 1960s; this may have been the bulk of the remaining population. With the exception of a small area in Alaskan waters, sightings are now rare, and major threats may be largely the same as for *E. glacialis* (primarily fishing gear entanglements and vessel strikes). Loss of sea ice resulting in increased trans-arctic ship traffic is a possible future threat. There were once probably over 11,000 North Pacific right whales, possibly as many as 30,000. It is thought that there are almost certainly no more than 400–500 alive today, and these are mostly found in the western Pacific. The eastern population was devastated by the Soviet catches and was recently estimated at only about 30–50 whales; thus it is critically endangered. Population trends are not known with any certainty, although there is no evidence for a significant recovery of numbers in the last few decades and until recently, sightings of young calves in the eastern North Pacific have been non-existent. Photos from any sightings of this rare species (even opportunistic ones by non-scientists!) can provide critical information.

IUCN status Endangered (overall species), Critically Endangered (Northeast Pacific population).

Southern Right Whale—*Eubalaena australis*

(Desmoulins, 1822)

Recently-used synonyms None.
Common names En.–southern right whale; Sp.–*ballena franca austral*; Fr.–*baleine australe*.
Taxonomic information Cetartiodactyla, Cetacea, Mysticeti, Balaenidae.
Species characteristics Southern right whales are enormous animals. These stocky whales have extremely large heads, which can be one-fourth to one-third of the body length. The mouthline is strongly bowed and the rostrum is arched and very narrow when viewed from above. As is true for all right and bowhead whales, there is no trace of a dorsal fin or ridge in the southern right whale. The back is smooth and broad. The large (up to 1.7 m long) flippers are fan- or paddle-shaped. The flukes are very wide (up to 35% or more of total length) and gracefully tapered, with a smooth trailing edge and a deep notch. All right whales have callosities on their heads, the largest of which is called the bonnet. These upper jaw callosities tend to be more discontinuous in this species than in their Northern Hemisphere counterparts. In addition, callosities often appear on the upper margin of the lower lips in southern right whales. The callosity patterns are individually distinctive and are used by researchers to identify individuals.

Southern right whales are black over most of the body, but most have white patches of variable shape and size on the belly and sometimes on the chin. They sometimes appear to have light mottling (generally caused by uneven sloughing of skin) and about 3–6% of animals have white or light gray blazes on the back (the latter are uncommon in the northern species). Color variants have been noted; these include blue-black, light brown, and partially white individuals (the latter are generally heavily mottled). In addition to those on the callosities, whale lice are common in creases and folds on the bodies of southern right whales.

Southern right whale adults reach up to 17 m in length; as in the other species in the genus, females grow

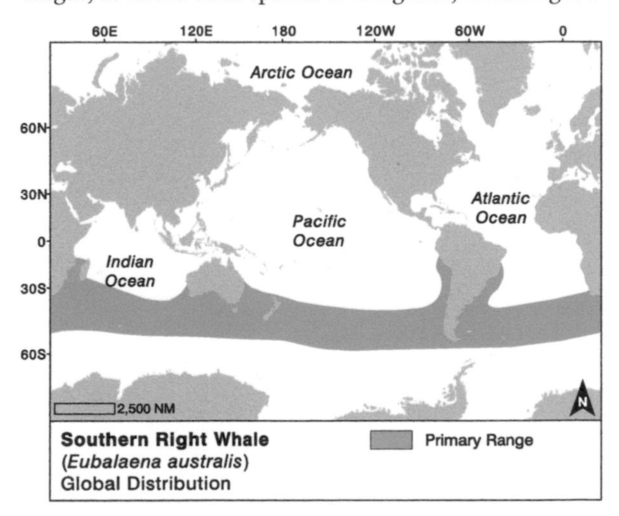

Southern Right Whale
(Eubalaena australis)
Global Distribution

The hairs on the chin of this southern right whale are clear evidence that whales evolved from fur-bearing terrestrial counterparts. Notice the position of the "bonnet" on the tip of the rostrum, as the whale hangs in the water. Argentina. PHOTO: M. NOLAN

A right whale is just beginning its V-shaped blow as it faces the photographer. Argentina. PHOTO: M. NOLAN

The broad, finless back of a southern right whale at rest. South Africa. PHOTO: C. WEIR

During the breeding season, right whales head into shallow, nearshore waters for calving—probably for protection from killer whales and large sharks. Argentina. PHOTO: M. NOLAN

Some right whales show varying degrees of silvery-white coloration versus the all-black coloration of most individuals. South Africa. PHOTO: L. EDWARDS

A pair of adult southern right whales swimming in tandem—the bubble train of the near animal may possibly have some communication function. The eye is just below the large, posterior callosity patch. PHOTO: S. DAWSON

larger than males. They tend to be slightly smaller than those in the Northern Hemisphere. These animals can reach weights of at least 80,000 kg. Newborn animals are 4–5 m long. The 200–270 baleen plates per side are narrow and long, up to 3 m in length. The plates tend to be dark gray to black (although some can be nearly white) and have fine gray to black fringes. The blow of the southern right whale is relatively short and V-shaped, making this species identifiable at a distance, especially if seen from ahead or behind.

Recognizable geographic forms None.

Can be confused with The southern right whale is the only whale in its range with a smooth, finless back and callosities; this should make misidentifications unlikely. From a distance the bushy, somewhat V-shaped, blows of humpback whales can be mistaken for those of right whales. At close range, however, the two species are unmistakable.

Distribution Southern right whales have a circumpolar distribution in the Southern Hemisphere, from approximately 20°S to 55°S, although they have been observed as far south as 65°S. They are migratory, moving south to feed in summer months. For winter, they migrate north and much of the distribution is concentrated near coastlines. Major mating and breeding areas are nearshore off southern Australia (including Tasmania), southern New Zealand/Campbell Island, southern South America, and South Africa/Mozambique/Madagascar, as well as around several subantarctic islands. Multiple stocks are considered to exist. A few southern right whales have been sighted in Antarctic waters in summer.

Ecology and behavior Southern right whales have been well studied on their winter breeding grounds, especially at Peninsula Valdes, Argentina, and in Australia and South Africa. The Argentine study has been ongoing since the early 1970s and is one of the longest-running behavioral studies of any cetacean population. Researchers have used callosity patterns to identify individuals on these grounds, and have learned much about the right

A portrait of a southern right whale eyeing the photographer; it is thin, perhaps from too much time fasting on the breeding grounds at Peninsula Valdes, Argentina. PHOTO: H. MINAKUCHI

A southern right whale falls back into the water on its back; the uniquely shaped flippers of right whales are mitten-like. South Africa. PHOTO: J. SCARFF

Although right whales are almost entirely shiny black, they do have variable amounts of white on the belly. PHOTO: J.-P. SYLVESTRE

whale's behavior, communication, and reproduction. Southern right whales often seem slow and lumbering, but at times can be surprisingly quick and active. They often breach, and slap their flippers and flukes on the surface. Southern right whales often raise their flukes on a dive. They have even been seen apparently "sailing" by catching the wind with their raised flukes. Typical feeding dives last 10–20 min.

Most of the breeding in Argentina takes place in August and September, but mating has been observed in most months of the year. Male right whales have huge testes (up to 2 m and 500 kg each) and long penises, two characteristics predicted in species in which males compete for females primarily through sperm competition, rather than by direct aggression. In fact, observations of mating right whales often appear to support this hypothesis, with one female often engaging in sequential mating with multiple males, which do not appear to show much overt aggression.

Feeding and prey Surface and subsurface skim-feeding is the rule in this species. They move forward through patches of prey with the mouth open, allowing water to move through and catching the prey items on the finely-fringed inner surfaces of the baleen plates.

Southern right whales prey mostly on copepods and krill, apparently sometimes feeding near the bottom.

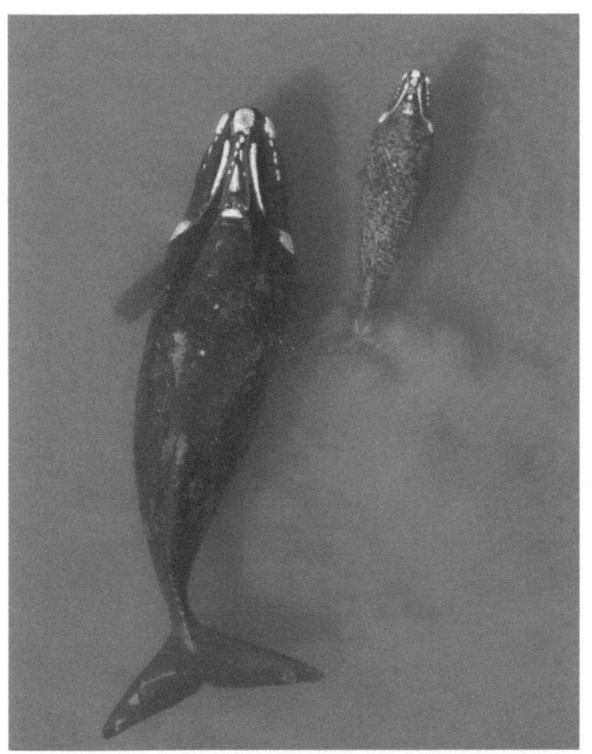

The visible shadows of these whales and the mud plume of the calf indicate just how shallow the water is where the mother right whale keeps her calf while it is young and vulnerable. Peninsula Valdes, Argentina. PHOTO: C. JOHNSON

Threats and status After North Atlantic right whale stocks began to be reduced, European and American whaling activity shifted to the Southern Hemisphere. Heavy exploitation there left stocks badly depleted. Unauthorized catches by the Soviets of more than 3,000 right whales in the 20th century added to the depletion. Fisheries interactions, as well as potential vessel disturbance and collisions, continue to threaten southern right whale populations today. In Argentina, attacks by gulls cause damage to the whales and are also considered to be a threat. Although there are thought to be only about 8,000 animals remaining, the southern right whale is not seriously endangered as the two northern species are. Some populations appear to be recovering, such as those that breed off Argentina (although the gull attacks may change this), South Africa, and western Australia. All three of these appear to be increasing at 7–8% per year, and clearly this means the prospects of long-term survival are much better for this species than for either of the Northern Hemisphere species.
IUCN status Least Concern.

The distinctive V-shape of the right whale blow; notice the lips reaching almost to the top of the head. South Africa. PHOTO: C. WEIR

In good light, the finger bones are clearly evident in the "hand" of the right whale, which has been repurposed for a life at sea. Argentina. PHOTO: M. NOLAN

For unknown reasons, southern right whales sometimes hang vertically, upside down in the water for extended periods of time with their flukes suspended in the air. Argentina. PHOTO: M. NOLAN

Bowhead Whale—*Balaena mysticetus*

Linnaeus, 1758

Recently-used synonyms None.
Common names En.–bowhead whale, Greenland right whale; Sp.–*ballena de cabeza arqueada*; Fr.–*baleine du Groenland*.
Taxonomic information Cetartiodactyla, Cetacea, Mysticeti, Balaenidae.
Species characteristics Bowhead whales superficially resemble their close relatives, the right whales. They are extremely rotund overall, but usually have a distinct "neck" region. The head is large (up to two-fifths of the body length); the upper jaw is arched (thus the name, bowhead) and narrow when viewed from above. There is a prominent muscular bulge in the blowhole area (sometimes called the "crown" or "stack" by whalers), with a distinct depression behind. The mouthline is strongly bowed, and the eye is placed just above the corner of the mouth. There is no dorsal fin or ridge, and the back is very broad and smooth. The flippers are large and fan-shaped, and have blunt tips. The flukes are wide (2–6 m across) and tapered toward the tips, with smooth contours. Bowheads have extremely thick skin (up to 2.5 cm) and blubber (up to 28 cm) layers.

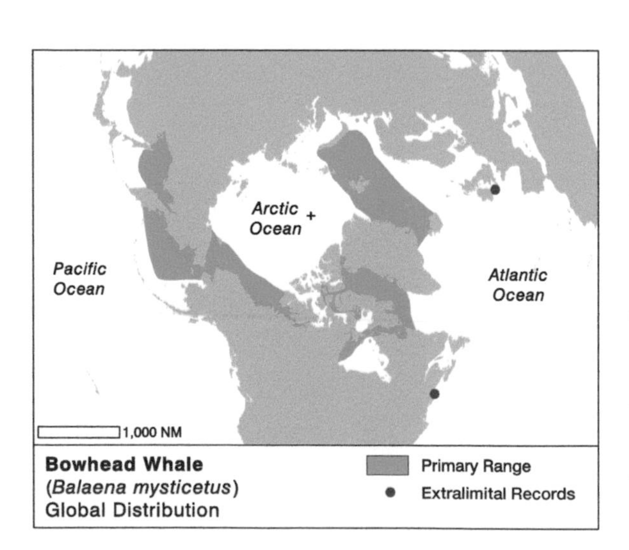

Bowhead Whale
(*Balaena mysticetus*)
Global Distribution

Predominantly black in color (although some lighter-colored animals are occasionally seen), bowheads have varying amounts of white beneath, usually showing dorsally as a white patch at the front of the lower jaw. This patch often has several dark gray to black spots, each indicating the position of a chin hair. There is also often a light gray to white band around the tail stock, just in front of the flukes, and sometimes other white or light gray areas on the body (such as on the belly and flukes). The white on the tail expands with age, and very large, old bowheads may have an almost completely white tail.

Bowheads have the longest baleen plates of any whale. The 230–360 plates in each side of the jaw can reach lengths of 5.2 m; they have long, fine fringes. The plates are dark gray to brownish-black, generally with slightly lighter fringes. As is true for the closely-related right whales, the blow is V-shaped and bushy. Females

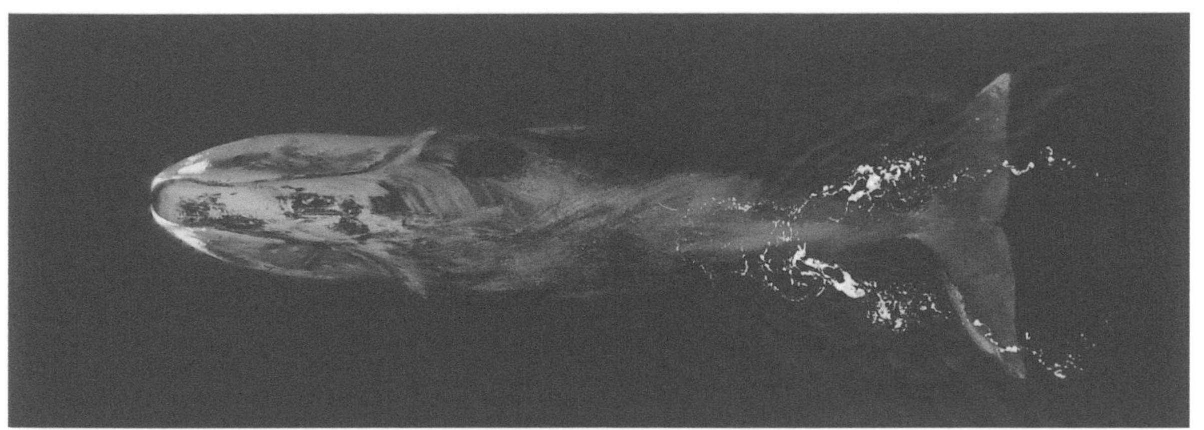

The massive head and white chin of the bowhead whale are still clearly visible on this mud-covered individual that has probably been feeding on the bottom. Alaska. PHOTO: V. BEAVER

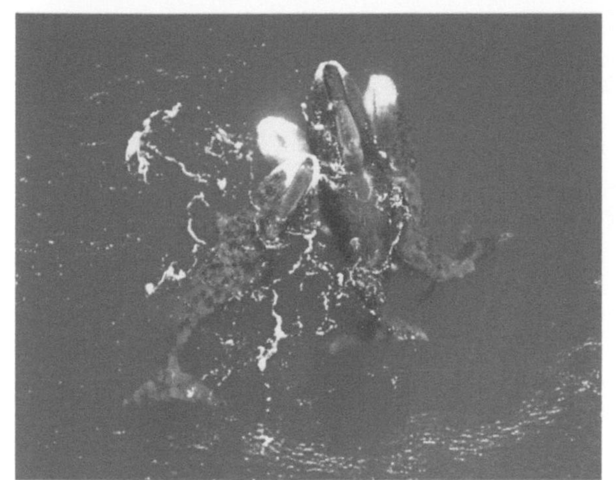

A mating group of bowheads, with at least three males vying for access to a presumably receptive female in the center. PHOTO: A. BROWER

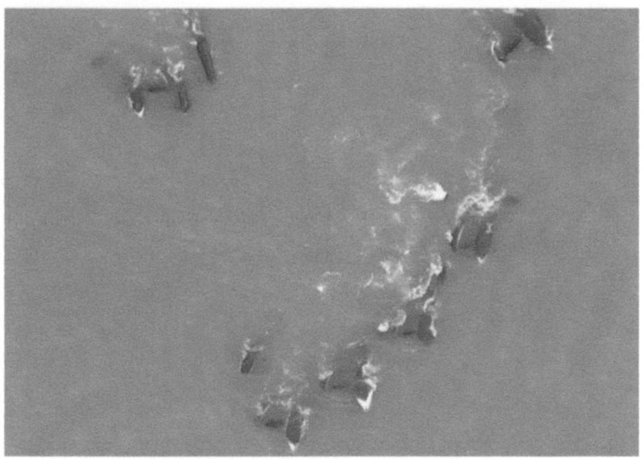

In this photo, there are nine bowheads; six are rolled over on their sides and skim-feeding at the surface with their mouths wide open; four have formed a phalanx, like grain harvesters. Alaska. PHOTO: K. GOETZ AND L. V. BRATTSTROM

grow somewhat larger than males. Male bowhead whales range to 18 m in length, females to 20 m. Weights of large individuals have been estimated at about 90,000 kg. Calves are about 4.0–4.5 m long at birth.

Recognizable geographic forms None.

Can be confused with Bowheads should be very easy to identify, as the only other large whales with a smooth back are the right whales, which are rare in the bowhead's northern haunts. The right whales' callosities and absence of white on the chin and peduncle will allow them to be distinguished from bowheads. Gray whales use some of the same summer range as bowheads, but the gray whale's dorsal hump and knuckles, weakly-arched lower jaw, and more slender body shape and lighter coloration should make them distinguishable in most circumstances.

Distribution Bowheads are the most boreal of all great whale species, and are found only in arctic and subarctic regions generally between about 55° and 85°N. Once there might have been a single panarctic population. Currently, there are five stocks recognized in the North Atlantic Ocean, and the Bering, Beaufort, Chukchi, and Okhotsk seas. These animals live much of their lives in and near sea ice, migrating to the high arctic in summer, but retreating southward in autumn with the advancing ice edge.

Ecology and behavior Bowhead whales are usually seen in groups of three or fewer, but larger aggregations form during migrations and on the feeding grounds. Although often slow-moving, bowheads do breach, fluke slap, and engage in other aerial behavior, sometimes for extended periods. They lift their flukes before a steep dive, but not as often as some other great whales. Complex low frequency calls (some arranged into songs) are common, at least during migration. Bowheads are closely associated with ice for most of the year. They are adapted to living among the arctic ice, and they can break through ice as thick at 60 cm. Their

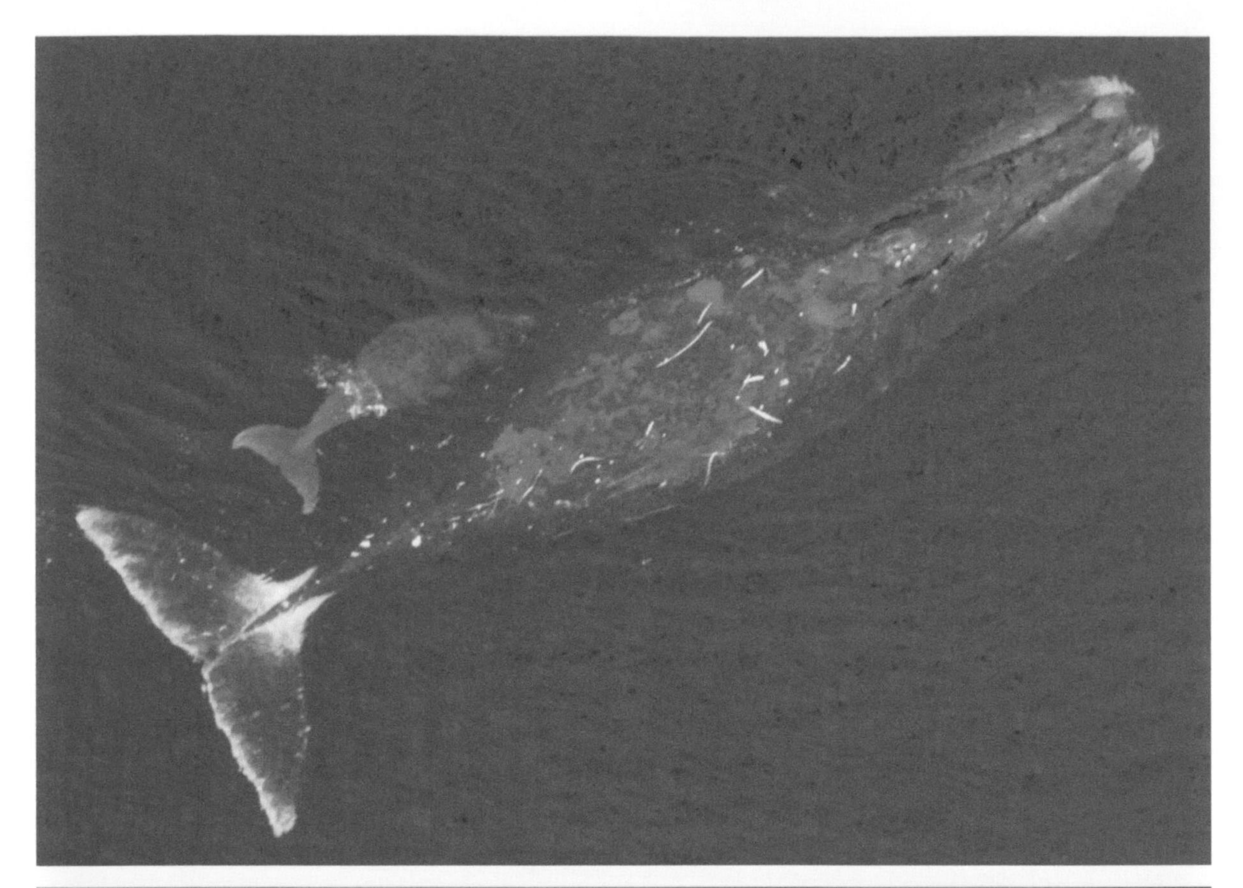

The extent of the white on the chin and at the base of the flukes is variable among bowheads; although the long, narrow, arching rostrum is similar to that of the right whale, bowheads lack any callosities. PHOTOS: V. BEAVER

A ventral view of a breaching bowhead, showing extensive white on the chin. Alaska. PHOTO: V. BEAVER

The flukes of the bowhead whale are usually all black with a smooth trailing edge. Sea of Okhotsk, Russia. PHOTO: R. L. PITMAN

The high-arching lip and narrow, bowed rostrum free of any callosities readily distinguish the bowhead from other whales in its range. Baffin Island, Canada. PHOTO: M. NOLAN

diving abilities, while not as well studied as for many other species of large whales, are thought to be exceptional—they can possibly stay submerged for over an hour.

Bowhead whales probably mate in late winter or early spring. Calves are born mainly the following spring (about 13–14 months later), as whales migrate toward their summering grounds. The breeding system is thought to be similar to that of the right whales, with males using a form of sperm competition. Mating groups often are composed of the female and multiple male suitors, similar to the situation in right whales. Bowheads may be among the longest-lived of all mammals. There is some evidence from aspartic acid physiology and harpoon head recovery records to suggest that some individuals may live for more than a century (possibly even over 200 years)! Their major natural predator is the killer whale. Between 4–8% of Alaskan bowheads killed by Eskimos bore scars from killer whale attacks.

Feeding and prey Small to medium-sized (mostly 3–30 mm long) crustaceans, especially krill and copepods, form the bulk of the bowhead's diet. They also feed on mysids and gammarid amphipods; their diet includes at least 60 species. Bowheads skim-feed at the surface and in the water column. It has recently been suggested that they also feed near the bottom, but probably do not directly ingest benthic sediments, as gray whales routinely do. During surface skim-feeding, coordinated group patterns have been observed, including whales feeding in echelon (V-shaped) formation.

Threats and status All bowhead whale stocks were depleted by commercial hunting, which began in the 1500s, but was most intense in the 1800s and early 1900s. Heavy hunting has left four of the five recognized bowhead stocks highly endangered (less than a few hundred animals each). However, the Bering-Chukchi-Beaufort stock (also known as the Western Arctic stock) is thought to number over 10,000 animals, and is apparently increasing at about 3% per year. The stock that occurs off West Greenland may also be increasing. Limited whaling by native peoples in Alaskan and Russian waters is allowed under a quota system set and overseen by the International Whaling Commission and the US Government. This, along with small (but not insignificant) kills by Russian and Canadian native peoples may be a factor in the recovery abilities of the affected stocks. The major non-direct threats to these animals are disturbance from oil and gas exploration and extraction activities in the arctic and subarctic regions, along with entanglement in fishing gear and pollution. Climate change (i.e., warming of polar regions) may also be a factor in their long-term recovery.

IUCN status Least Concern (overall species), Critically Endangered (Svalbard/Barents Sea stock), Endangered (Okhotsk Sea stock), Least Concern (Bering/Chukchi/Beaufort stock).

Pygmy Right Whale—*Caperea marginata*

(Gray, 1846)

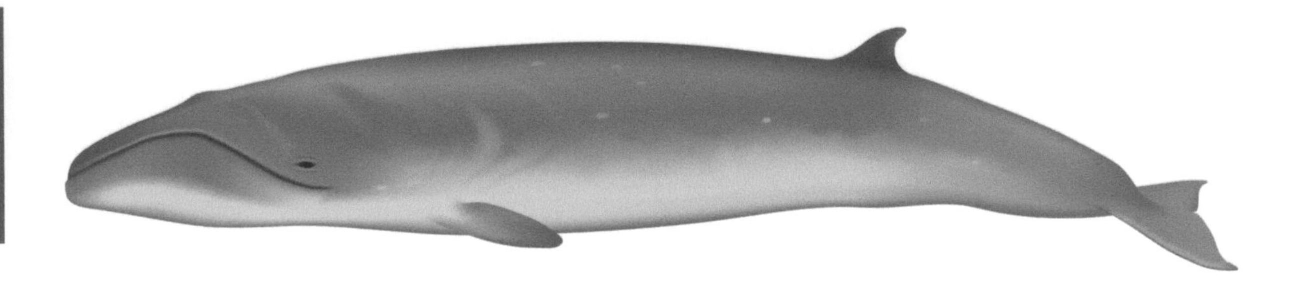

Recently-used synonyms *Neobalaena marginata*.

Common names En.–pygmy right whale; Sp.–*ballena franca pigmea*; Fr.–*baleine pygmée*.

Taxonomic information Cetartiodactyla, Cetacea, Mysticeti, Neobalaenidae. The pygmy right whale is considered by some to be the last remaining species of the otherwise-extinct family Cetotheridae.

Species characteristics Although its name might imply that it is a "right whale," the pygmy right whale is actually now placed in a separate family and is not commonly regarded as one of the right whales. The falcate dorsal fin is set about 70–75% of the way back from the snout tip. This species is rather slender, resembling more the streamlined rorquals than the chunky right and bowhead whales, and the head is not large (less than one-quarter of the body length). The pygmy right whale is like the right whales in that it has an arched jaw line; also the upper jaw curves downward toward the tip, although not as much as in right and bowhead whales. There is a single mid-dorsal ridge extending down the top of the head from the blowholes to the rostrum tip. The flippers are small and slender with rounded tips. The flukes are broad, with a deep median notch. There are two shallow throat creases, reminiscent of those in gray whales.

The color of the body is dark gray above, shading to white below. Large adults may have a brownish tinge. There are typically white to light gray streaks and bands sweeping up from the belly onto the back. Most animals possess a chevron of light color at about the level of the flippers, and some may have a second chevron further back. The flippers and flukes are dark gray. Light-colored, concave healed scars (possibly caused by cookie-cutter sharks) are common on the body.

The baleen plates in this species number about 213–230 in each side of the upper jaw. They are up to 68 cm long and are very flexible and tough. The color of the plates is yellowish-white. The fringes are the finest and densest among all whales. The blow ranges from thin and columnar, to shorter and puffier, and may even be inconspicuous. This is the smallest baleen whale. The maximum length measured for a male is 6.1 m and that for a female is 6.5 m. Recent sightings off southern Australia suggest that they may reach lengths of up to 6.5 m quite regularly. They reach weights of at least 3,400 kg. At birth, pygmy right whales are about 2 m long.

Recognizable geographic forms None.

Can be confused with This species can easily be confused with common and Antarctic minke whales, but the differences in head shape (particularly the blunter rostrum and arched lower jaw of the pygmy right whale) and the white flipper bands present in dwarf minke whales will allow differentiation when specimens are seen clearly. From a distance, the back and dorsal fin could be confused with those of a beaked whale; however, beaked whales have very different head shapes.

60E 120E 180 120W 60W 0
60N 30N 0 30S 60S
Arctic Ocean
Pacific Ocean
Atlantic Ocean
Indian Ocean
2,500 NM
Primary Range
Secondary Range

Pygmy Right Whale
(Caperea marginata)
Global Distribution

The smallest of the baleen whales, the pygmy right whale is similar to a minke whale, but from the side it shows a noticeable arch to the lower jaw. Namibia. PHOTO: R. LEENEY

The rostrum of the pygmy right whale is downturned and much narrower than that of the otherwise very-similar minke whale. Namibia. PHOTO: R. LEENEY

The pygmy right whale is small enough that the prominent dorsal fin can be visible at the same time as the blowhole when the animal is at the surface; notice also the pale chevron just behind the blowhole. South of New Zealand. PHOTO: R. L. PITMAN

The sloping, narrow rostrum and arching lower lip of the pygmy right whale are clearly visible here; on the back are numerous cookie-cutter shark bite scars and a fairly conspicuous chevron. South of New Zealand. PHOTO: R. L. PITMAN

Distribution The pygmy right whale has a presumed circumpolar distribution in both coastal and oceanic waters, and is known only from a small number of records in the Southern Hemisphere, between about 30° and 55° (north of the Antarctic Convergence). There appear to be concentrations in plankton-rich waters around continents and islands, and also at the subtropical convergence in summer. There may be seasonal shifts in abundance, but this species does not appear to undergo a large-scale migration, as most of the larger baleen whales do.

Ecology and behavior This is the least known of all the baleen whales, as it is rarely sighted and has never been a primary target of commercial whalers. Groups of up to 14 individuals have been seen, but singles or pairs are most common. Recently, large aggregations of about 80–100 whales were observed in the southeast Indian Ocean and off Victoria, Australia, but groups this large may be uncommon. They are sometimes seen associated with other species of whales and dolphins. The inconspicuous small blow and quick, shallow surfacings of the pygmy right whale make it difficult to spot and observe at sea. The head and blowholes usually appear before the dorsal fin is exposed. Often, these animals bring their snout tips out of the water upon surfacing, making it possible to see the arched mouthline. They are capable of swimming

Pygmy right whales are not often seen at sea, but when a good look is obtained, they are quite identifiable. PHOTO: R. L. PITMAN

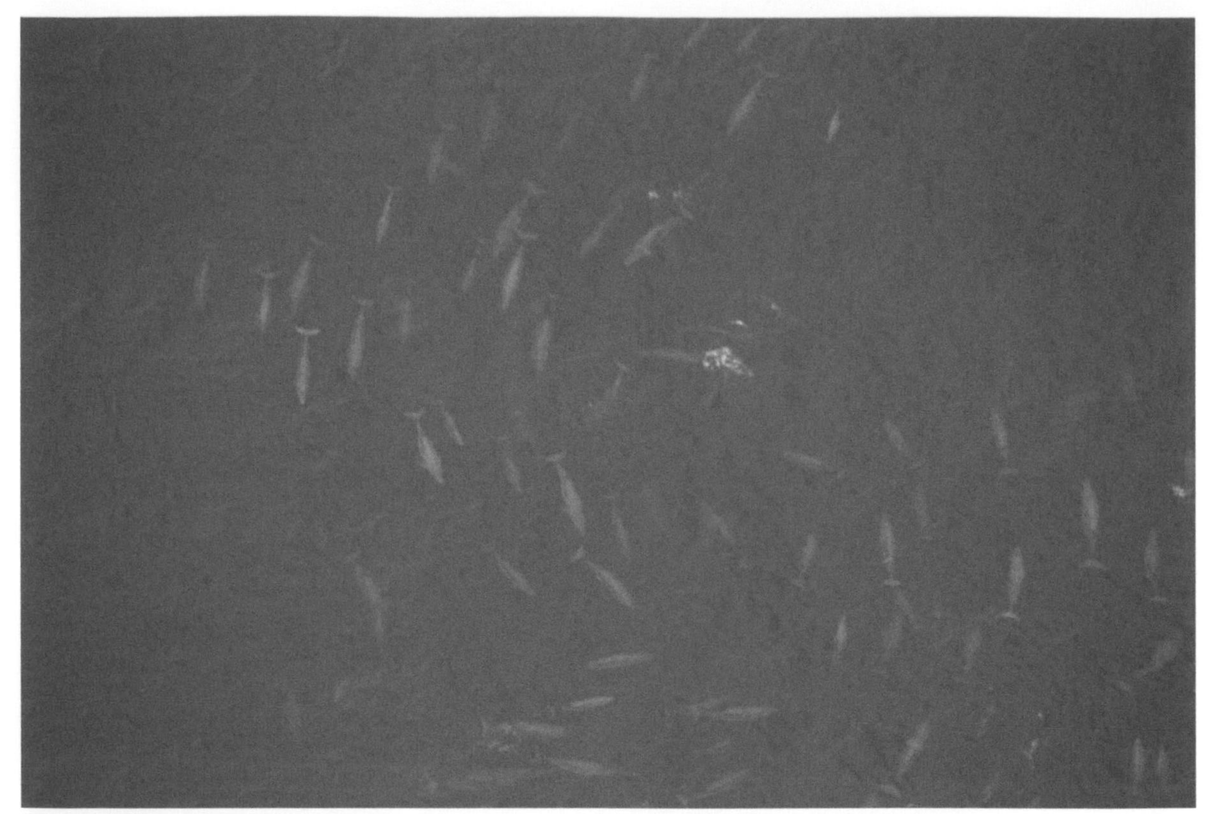

Although sightings of pygmy right whales at sea are extremely rare, they have sometimes been recorded in large groups. Here, 100+ individuals swirl in a vortex south of Australia—its purpose, whether for feeding or socializing, is unknown. PHOTO: BLUE WHALE STUDY

surprisingly fast (>6–8 knots without apparent effort). They typically dive for periods of about 4 min.

Very little is known about reproduction in this species, but the breeding season is thought to be protracted and may occur year-round. Sexual maturity is thought to occur at lengths greater than 5 m.

Feeding and prey Although there is little information on the diet, pygmy right whales are known to feed on calanoid and cyclopoid copepods, small euphausiids, and other small invertebrates (such as amphipods).

Threats and status This poorly-known species has never been commercially hunted (though a few opportunistic takes have been recorded). Although there is essentially nothing known of its status, there is no evidence that it is seriously threatened. It may be naturally rare, or perhaps its areas of concentration have not yet been discovered. There are no available estimates of abundance.

IUCN status Data Deficient.

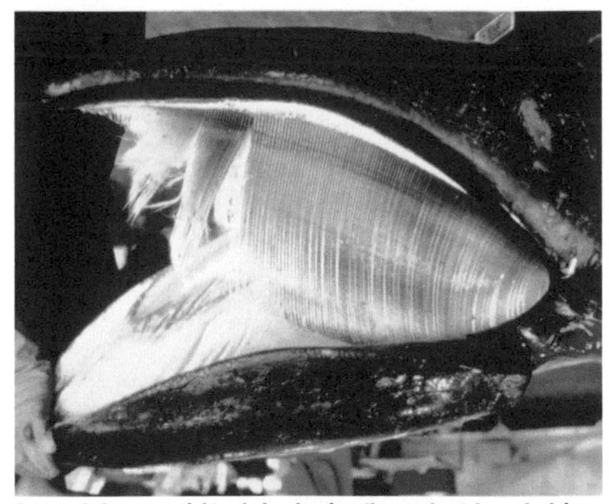

A stranded pygmy right whale showing the moderately arched jaws and exposed baleen plates. The baleen has extremely fine mesh for filtering very tiny invertebrates (mainly copepods). PHOTO: COURTESY OF S. LEATHERWOOD

Blue Whale—*Balaenoptera musculus*

(Linnaeus, 1758)

True Blue Whale

Pygmy Blue Whale

Recently-used synonyms *Sibbaldus musculus*.

Common names En.–blue whale; Sp.–*ballena azul*; Fr.–*rorqual bleu*.

Taxonomic information Cetartiodactyla, Cetacea, Mysticeti, Balaenopteridae. Five subspecies are tentatively recognized: the blue whale of the North Atlantic and North Pacific (*B. m. musculus*), the blue whale of the northern Indian Ocean (*B. m. indica*), the Antarctic blue whale (*B. m. intermedia*), the pygmy blue whale (*B. m. brevicauda*) of the Southern Hemisphere and northern Indian Ocean, and an unnamed subspecies found off Chile. The first three subspecies are collectively known as "true blue whales." The validity of *B. m. indica* is doubted by some. Hybrids between blue and fin whales have been described from nature.

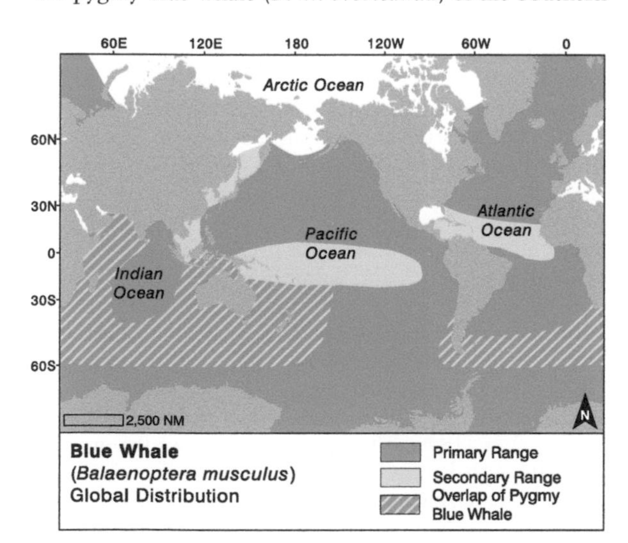

Species characteristics The blue whale is the largest animal ever known; however, size alone is not enough to distinguish it from other rorquals, as its size substantially overlaps with that of adult fin and sei whales. Like all rorquals, the blue whale is slender and streamlined. The head is broad and U-shaped (like a gothic arch), when viewed from above and relatively flat, when viewed from the side. Along the center of the rostrum, there is a single prominent ridge, which ends in an impressive "splash guard" around the blowholes. The flippers are long and pointed, and the dorsal fin is relatively small, variably shaped (sometimes reduced to just a nubbin), and placed about three-quarters of the way back from the rostrum tip. The broad, tapered flukes have a relatively smooth trailing edge and a prominent median notch. Of the five subspecies recognized, only the pygmy blue whale is known to have a significantly different body shape (see below).

Blue whales are bluish-gray dorsally and somewhat lighter underneath. The head is uniformly blue, but the back and sides are mottled blue and light gray. When viewed through the water surface they may appear dappled or uniformly light blue. There is light to extensive mottling on the sides, back, and belly, generally in the form of dark

The "galvanized" blue-gray coloration and nubbin dorsal fin are unique to the blue whale; the brownish coloration is due to diatoms. South Georgia. PHOTO: M. NOLAN

A cow and calf blue whale showing the long back of this species; the dorsal fins have *Xenobalanus* barnacles dangling from the tip and both fins have parallel tooth rake marks from killer whale encounters. Eastern North Pacific. PHOTO: J. TOWERS

Blue whales quite often show little or no dorsal fin; this one was perhaps removed by a killer whale earlier in its life. Eastern North Pacific. PHOTO: C. OEDEKOVEN/SWFSC

A diving blue whale with a *Xenobalanus* barnacle on the tip of its dorsal fin; small remoras clustered below the fin often move around on the body between surfacings. The corrugated back profile indicates a thin whale. Gulf of California. PHOTO: M. NOLAN

spots on a lighter surface, but sometimes the reverse. These mottling patterns are used by researchers to identify individuals. Diatom films on the ventral surface may be seen as an orangish-brown or yellow film, a characteristic which gave rise to the alternative name "sulphur-bottom" whale. There is generally a thin, white border around the flippers, on both the dorsal and ventral surfaces. The tail flukes often have light striations ventrally.

On the throat, there are 60–88 long ventral pleats extending to or near the navel. The mouth contains 260–400 pairs of black, broad-based baleen plates, each <1 m long. The bristles are quite coarse. The blow is tall and slender, reaching 10–12 m in height. Most adults of the Northern Hemisphere subspecies (*B. m. musculus*) are 23–27 m long (with females growing larger than males). The Antarctic blue whale (*B. m. intermedia*) is the largest, and generally measures up to 29 m, although a specimen over 33 m was once taken by whalers. Newborn blue whales are about 7–8 m long. Adults can weigh up to 180,000 kg, but most adults weigh between 72,000 and 135,000 kg.

Recognizable geographic forms

True blue whale (*B. m. musculus/indica/intermedia*)—Three of the five subspecies of blue whales share a similar external form (but differ from the pygmy blue, described below). They occur in the Pacific, Atlantic, Southern, and portions of the Indian Ocean. They are blue steel-gray in color. They are also somewhat longer than the pygmy blue, reaching lengths of up to 33 m. Compared to pygmy blues, these animals possess a relatively smaller head (i.e., a longer tail stock), slightly more ventral grooves, and longer baleen. The shape of the body has been described as "torpedo-shaped" (with a relatively narrower head than the pygmy blue).

Pygmy blue whale (*B. m. brevicauda*)—The pygmy blue whale is found in the Indian and southeastern South Atlantic oceans (suggestions that it occurs also off Baja California, Mexico, are erroneous). It is more silvery gray than the blue steel-gray of true blue whales. It has a relatively larger head (i.e., a shorter tail stock), slightly fewer ventral grooves, and shorter baleen. The shape of the body has been described as relatively "tadpole-shaped" (reflecting the relatively wider head). These differences are subtle. If a good view is obtained, it might be possible to distinguish pygmy blue whales from other blue whales at sea, but only by highly-experienced observers (who may be able to also use differences in surfacing behavior and prevalence of skin lesions as clues). Pygmy blue whales are smaller than true blue whales, with a maximum length of only about 25 m. However, they tend to be heavier than true blue whales.

Can be confused with When seen well, blue whales are easy to distinguish from other large whale species. At a distance, however, blue whales can be confused with the other large rorquals: primarily fin and sei whales. Although

A feeding blue whale rolls at the surface with throat pleats fully distended. Australia. PHOTO: P. GILL

A blue whale lunges at the surface on its right side; the white underside of its left pectoral flipper is visible as the blue-gray throat pleats just begin to billow out. South Africa. PHOTO: L. EDWARDS

The left lower jaw of a lunge-feeding blue whale arches high out of the water; this kind of surface-feeding during the daytime is uncommon. Notice that the top of the flipper is gray. South Africa. PHOTO: L. EDWARDS

When pressed, a blue whale can move remarkably fast, throwing up a large rooster-tail as it goes. South Orkney Islands. PHOTO: M. NOLAN

The blue whale is the only balaenopterid that regularly raises its flukes when it dives, often exposing its very deep caudal peduncle (tail stock). California coast. PHOTO: M. CARWARDINE

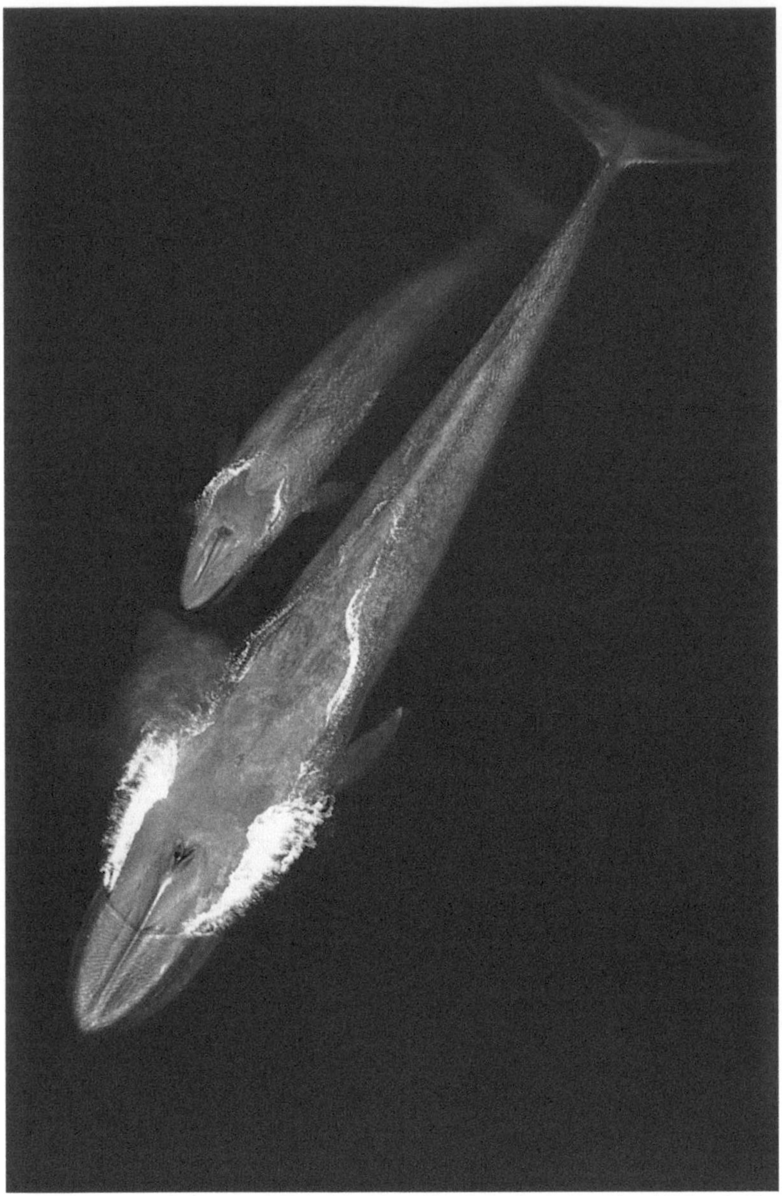

From the air, the pale blue-gray color of the blue whale is unmistakable; this adult female blue whale looks thin, probably from nursing her calf. Gulf of California. PHOTO: M. CARWARDINE

the great size of blue whale adults may aid in identification, the best clues for differentiating blue whales from fin or sei whales are careful attention to color pattern, head shape, and dorsal fin shape and position. Blow shape and height can also be useful, but the variability within species must be kept in mind.

Distribution Blue whales tend to be open-ocean animals, but they do come close to shore to feed, and possibly to breed, in some areas. Blue whales can be seen from tropical waters to the pack ice edges in both hemispheres, but they tend to avoid most equatorial waters. Some blue whales are resident, others are migratory, moving poleward to feed in the summer and toward the tropics to breed in the winter. The breeding grounds do not appear to be as well defined as are those of humpback, right, and gray whales. The exact distribution of the pygmy blue whale is not completely known, although it occurs in the northern Indian Ocean and throughout much of the Southern Hemisphere. In the Antarctic, the pygmy blue tends to prefer more northern waters than the true blue whale does. The two forms do overlap in distribution in some areas.

Ecology and behavior Blue whales are usually seen alone or in pairs. However, scattered aggregations of a dozen or more may develop on prime feeding grounds. Although shorter dives are most common, dives of up to 30 min, generally interspersed with long series of shorter surfacings (at 15–20 sec intervals), have been recorded (the longest dive documented was 36 min). Fluking-up is not uncommon upon beginning a dive, although not all blue whales are "flukers." Remarkably, some blue whales have been observed breaching, bringing nearly all of their bulk out of the water. Some researchers believe that pygmy blue whales have a tendency to surface without showing the dorsal fin or keel (unlike the standard blue whale, which usually does). However, this is probably not a very reliable character to use for identification. Their major natural predator is the killer whale.

Calves are born in winter, apparently in tropical/subtropical breeding areas (the specific locations of which are not known for most populations). Nursing calves may gain 90 kg per day, and appear to be weaned by the age of about 8 months. Not much is known of the mating system of the blue whale, but they are known occasionally to hybridize with other large whale species (such as fin and humpback whales). Blue whales are thought to live for over 80–90 years. The role that acoustics play in the mating system is not known,

The blow of the blue whale is columnar and extremely tall – it can reach as high as 12 meters! Monterey Bay. PHOTO: T. A. JEFFERSON

Blue whales regularly fluke up on their terminal surfacing before a deep dive. The undersides of the flukes are usually streaked gray. PHOTO: J. CALAMBOKIDIS

The blue whale has a broad, rounded rostrum; the blowholes of this animal are still pinched closed as it surfaces. PHOTO: P. OLSON

This blue whale flukes up as it dives near La Paz, Baja California, Mexico. The flukes of this species are tapered, with a straight trailing edge. PHOTO: R. W. BAIRD

but blue whales make some of the loudest sounds known in nature.

Feeding and prey Various species of krill (euphausiids) form the major part of the blue whale's diet, and on their feeding grounds blue whales can be observed lunging, often on their sides or upside-down, through great clouds of these invertebrates. The throat pleats become greatly distended when the whales lunge-feed, giving them a "tadpole" appearance.

Threats and status Blue whales were heavily hunted after the invention of the exploding harpoon and steam catcher boats made them accessible to whalers in the early part of the 20th century. This was the first species of rorqual to be depleted by whalers; it was highly desirable due to its large size. Blue whales were hunted in the North Pacific, North Atlantic and Southern oceans. There were illegal catches by the Soviets in the northern Indian Ocean, and possibly elsewhere. The Soviets took over 10,000 pygmy blues in the 1960s. Nearly all populations were severely depleted and the species was once thought to be on its way to extinction. This species is currently protected (since 1966), with a zero catch limit imposed by the International Whaling Commission. While some populations were drastically reduced or even wiped-out, at least some blue whale stocks (such as those in the eastern North Pacific) appear to be showing signs of recovery. The major concerns now are vessel collisions (especially from large cargo ships and tankers) and noise disturbance by human activities (in particular offshore oil and gas exploration and extraction). The blue whale is currently thought to number about 10,000–25,000 individuals globally, which is only about 3–11% of the population size in the early 1900s. There were thought to be about 4,000 pygmy blue whales in the early 1970s, but more recent estimates are not available.

IUCN status Endangered (overall species), Critically Endangered (Antarctic stock), Data Deficient (pygmy blue whale).

Fin Whale—*Balaenoptera physalus*

(Linnaeus, 1758)

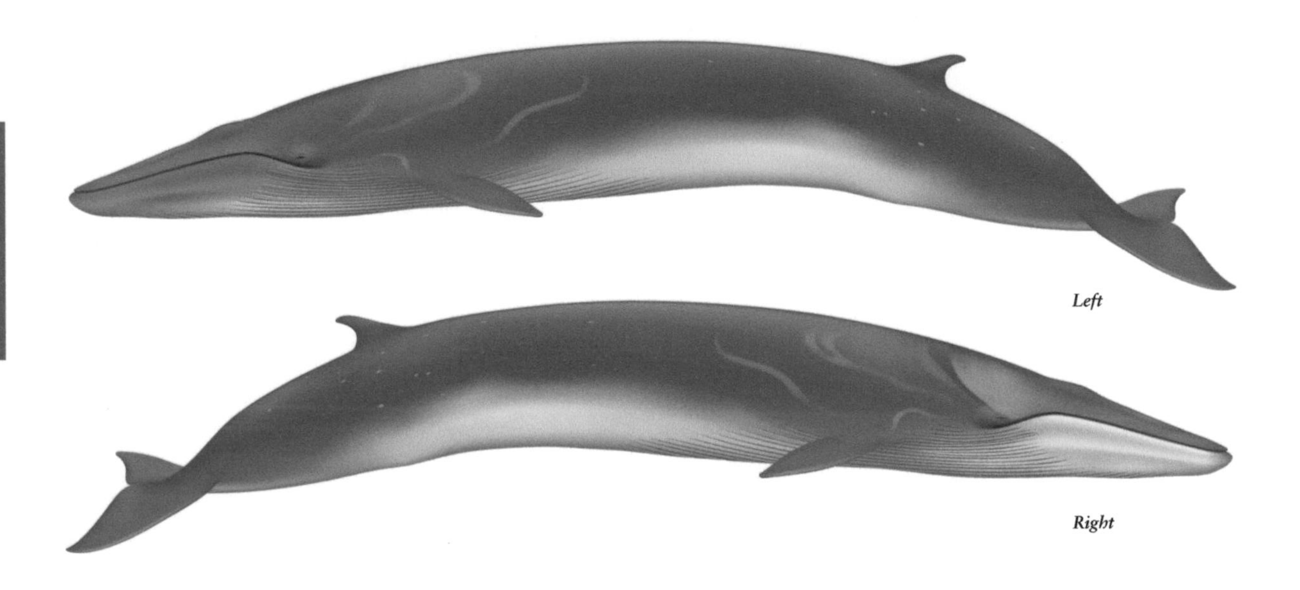

Recently-used synonyms None.
Common names En.-fin whale; Sp.-*rorcual común*, *ballena de aleta*; Fr.-*rorqual commun*.
Taxonomic information Cetartiodactyla, Cetacea, Mysticeti, Balaenopteridae. The Society for Marine Mammalogy recognizes three subspecies, one in the Northern Hemisphere (*B. p. physalus*), one in the Southern (*B. p. quoyi*), and the pygmy fin whale (*B. p. patachonica*). However, these subspecies designations are not widely used. A recently-described new species of rorqual, *Balaenoptera omurai*, may be closely related to the fin whale. There are reported hybrids between blue and fin whales.

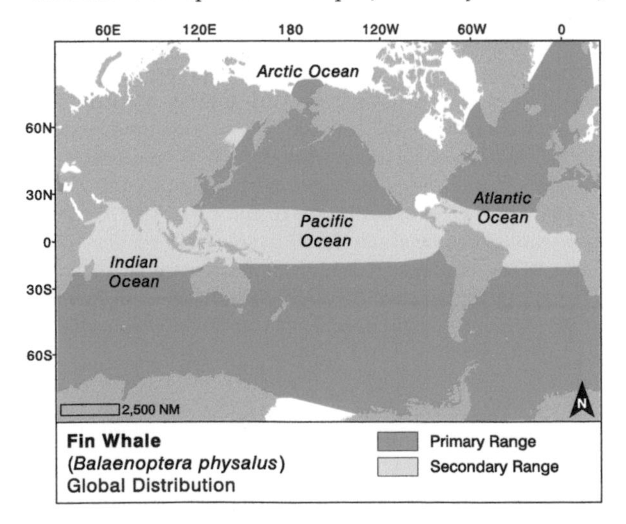

Species characteristics Fin whales are large (the second largest species of whale), but very sleek and streamlined. From above, the head is more V-shaped and pointed than that of the blue whale, and there is a single medial ridge on the upper surface of the rostrum. The dorsal fin is taller, more falcate, and set farther forward on the tail stock than in the blue whale The dorsal fin rises at a shallow angle from the animal's back (unlike in the other rorquals with prominent fins), and there is a prominent ridge along the tail stock between the dorsal fin and the flukes. The broad, tapered flukes have a smooth trailing edge and a prominent median notch. The flippers are long and tapered.

The most distinctive feature of the fin whale, however, is its coloration. The body is black or dark brownish-gray above and on the sides, shading to white below, but the head color is asymmetrical. The left lower jaw is mostly dark, while the right jaw is largely white. There tend to be several light-gray V-shaped "chevrons" (pointing forward) on the back behind the head. There are also typically other streaks or swirls of light coloring extending up from the belly (in particular over the right flipper), and from the eye. The undersides of the flukes are white, but have a gray border. The posterior portion of the body often has white, circular scars caused by bites of lampreys or cookie-cutter sharks. The bellies of fin whales may have a yellowish diatom film.

Fin whales often appear dark with silvery sides and without the mottling of blue whales; they regularly forage in groups. Eastern Canada. PHOTO: L. BOUVERET, OMMAG

Fin whales (and blue whales) are long enough that they often have a sinusoidal profile—when diving, the forward part of the back is convex and the rear part is still concave. British Columbia. PHOTO: J. TOWERS

An extremely long and dark back, with a conspicuous pale chevron distinguish this fin whale. Ligurian Sea. PHOTO: T. STENTON

Fin whales (like other rorquals) have a raised area around the blowholes, which is informally known as the splashguard. PHOTO: J. TOWERS

The underside of the fin whale's lower jaw shows the pigmentation asymmetry that results in its diagnostic white, right lower jaw and a dark left jaw when viewed at the surface. PHOTO: J. TOWERS

A white right lip (discolored with yellowish diatoms) and prominent black right eye stripe identify this as a fin whale. Southern Ocean. PHOTO: F. RITTER

Although rarely seen swimming fast, fin whales throw up considerable whitewater when they really get moving. Near South Georgia. PHOTO: M. NOLAN.

As a fin whale is biopsy darted in the Northeast Pacific, it offers up a clear view of the white lip, dark eye stripe (right side only), and a pale wavy pattern streaming from the ear hole. PHOTO: J. COTTON, SWFSC

In addition to white right lip and black right eye stripe, the swirling pigmentation patterning above the pectoral fin is clearly in evidence here. Antarctica. PHOTO: D. SINCLAIR

Fin whales have 260–480 baleen plates per side; the plates are dark gray to black, often striated with bands of gray and fringed with horizontal lines of yellowish-white to olive-green. Usually, the front one-half to one-third of the right side plates have much more light color than those on the left. The 50–100 throat pleats are long, and reach to the umbilicus. The slender blow is 4–6 m tall. Length at birth is 6–6.5 m. Adults can reach a maximum of 27 m in the Southern Hemisphere, but most Northern Hemisphere adults are less than 24 m long. Females are about 5–10% longer than males. Large animals may attain reported weights of up to 120,000 kg, but most probably weigh much less (<90,000 kg). Body weight varies seasonally.

Recognizable geographic forms None.

Can be confused with The other species of medium to large balaenopterids (blue, sei, Bryde's, and Omura's whales) are likely to be confused with fin whales at a distance. Size alone may be helpful for large fin whales, but care must be taken to distinguish small blue whales. Careful attention to color pattern, head shape, and dorsal fin shape and position will help to distinguish the various species. For instance, the color pattern of the fin whale (dark back, asymmetrical lower jaw, and light chevron and streaks in the thoracic area) is different from that of all other species, except the newly-discovered Omura's whale. Head shape of fin whales is much more slender and streamlined (and more pointed from above) than in blue whales, but is otherwise similar to the other species. Fin whales do not have accessory rostral ridges, as Bryde's whales do. Dorsal fin shape is one of the best clues—fin whales have a prominent dorsal fin that is usually falcate and rises out of the back at a very shallow angle (blue whales tend to have much smaller nubbin-like dorsal fins, and the other species generally have fins with a steeper rise from the back). The size and shape of the blow can also be helpful (generally shorter than in blue whales, but taller and more prominent than in the other species), but it should not be used for identification without other information. Clearly, a multitude of characters will be required to confirm an identification as a fin whale.

Distribution Surprisingly, for such a large and once commercially-important animal, the overall range and distribution of the fin whale is not well known. Fin whales inhabit primarily oceanic waters of both hemispheres. They are cosmopolitan, inhabiting all major oceans. When they are seen near shore, it is most commonly where deep water approaches the coast. Fin whales can be seen in temperate, and polar zones of all oceans, but are rare in tropical seas. Most populations are apparently migratory (a general poleward shift for feeding in the summer and a shift towards the subtropics for breeding in the winter),

Flippers splayed as it surfaces, this fin whale shows a blow pattern seen sometimes on rorquals when they first start to exhale: a single vertical blow with a lateral branch on either side. PHOTO: M. CARWARDINE

but their movements are complex and do not follow a simple pattern. There is little evidence for distinct calving areas. In fact, there appear to be resident groups in some areas (such as the Gulf of California, the East China Sea, off Japan, and perhaps the Mediterranean Sea, where they also show limited north/south movements).

Ecology and behavior Fin whales are capable of attaining high speeds, possibly as high as 37 km/h, making them one of the fastest great whales. They rarely raise their flukes on a dive, but they do occasionally breach. When fin whales surface, typically the blowholes appear briefly before the dorsal fin shows. Fin whales tend to be more social than other rorquals, sometimes gathering in pods of 2–7 whales, or more. In the North Atlantic, fin whales are often seen in large feeding aggregations, generally with humpback whales, minke whales, and Atlantic white-sided dolphins. Fin whales also sometimes associate with blue whales, and hybrids have been documented. Individual fin whales are identified by the pattern of the light streaks and chevrons on the back, along with the size and shape of the dorsal fin.

Not much is known of the social and mating system of fin whales. However, they appear to be typical of other baleen whales, in that long-term bonds between individuals are rare. Calving does not appear to take place in distinct nearshore areas (as it does, for instance, in gray and humpback whales). Young fin whales are born on dispersed breeding grounds in subtropical areas in midwinter. Gestation is 11–12 months. Sexual maturity occurs at ages of 6–10 years in males and 7–12 years in females, depending on the population and time period. Fin whales can live to be up to 80–90 years of age. The only known natural predator is the killer whale. Fin whales make a variety of low-frequency vocalizations, including the famous 20-Hz calls, which can be detected hundreds of kilometers away. Although these calls are produced throughout the year, they are emitted in patterns during the breeding season, suggesting that they may be used in breeding displays.

The undersides of a fin whale's flukes are white with a dark fringe. However, they are not often seen, as fin whales usually do not fluke up on a dive. PHOTO: B. WITEVEEN

Fin whales in echelon formation and feeding with throats distended; the pale chevron shows up very well from above. Notice also: black eye stripe on right side and asymmetric pale patch ahead of it. New England. PHOTO: NOAA, NMFS, NEFSC

Feeding and prey Fin whales feed on small invertebrates (euphausiids and copepods), schooling fishes (such as capelin, herring, mackerel, sandlance, and blue whiting), and squid. They are active lunge feeders (often lunging on their sides), gulping in huge volumes of food and water with the aid of their distended throat pleats (giving them a "guppy" appearance), and then using the baleen to filter the food particles out.

Threats and status The introduction of the explosive harpoon and steam-powered catcher boat in 1864 meant doom for large rorqual whales, which had previously been largely unobtainable for whalers. After depleting the blue whale, the fin whale was next in the line of fire, and this species was hunted relentlessly in all major oceans between the 1930s and 1960s. Tens of thousands of animals were taken in the Northern Hemisphere, and 725,000 were killed in the Southern Hemisphere. Small numbers of fin whales have been killed in recent years by Iceland in the North Atlantic. Fin whales occasionally get caught in fishing nets in the North Atlantic, but vessel strikes and disturbance by human-caused noise sources are more of a concern. Despite their history of heavy exploitation, there may be as many as 140,000 fin whales alive today. At least 600 or so apparently occur in the Gulf of California, Mexico, and well over 3,000 occur off the US west coast, where they are increasing rapidly. Fin whales are considered to be relatively abundant in both the North Pacific and North Atlantic. Southern Hemisphere stocks are not as healthy.

IUCN status Endangered (overall species), Vulnerable (Mediterranean stock).

When viewed from the air, balaenopterids have a surprisingly long and narrow shape—greyhounds of the sea. A sliver of the right white lip is visible on this fin whale. Northwest Atlantic. PHOTO: P. DULEY

Sei Whale—*Balaenoptera borealis*

Lesson, 1828

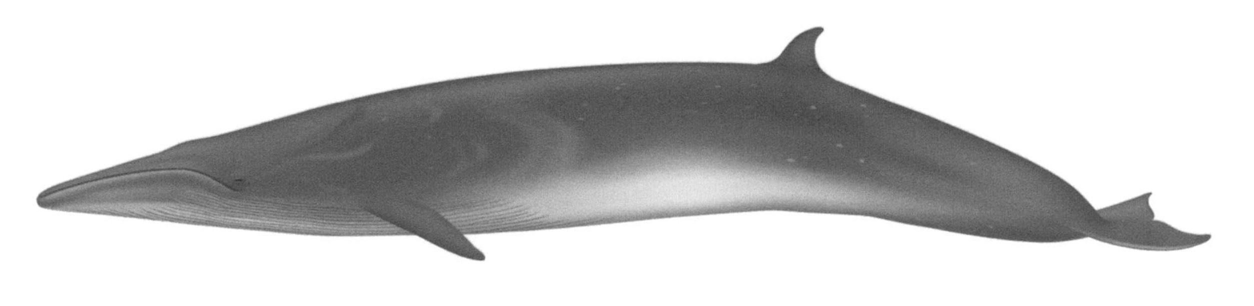

Recently-used synonyms None.

Common names En.-sei whale; Sp.-*rorcual sei, ballena sei*; Fr.-*rorqual de Rudolphi.*

Taxonomic information Cetartiodactyla, Cetacea, Mysticeti, Balaenopteridae. The Society for Marine Mammalogy recognizes two subspecies of sei whales, one in the Northern Hemisphere (*B. b. borealis*) and one in the Southern (*B. b. schlegellii*). However, these subspecies designations are not widely used.

Species characteristics Sei whales are similar in external appearance to fin, Omura's, and Bryde's whales, both of which also have a prominent falcate dorsal fin. All four have typical rorqual body shapes, which means they are very sleek and streamlined. In sei whales, the dorsal fin rises at a steep angle from the back. Compared to the similar dorsal fins of the Bryde's and Omura's whales, sei whale dorsal fins often have a "jointed" axis, with a strong backward angle about halfway up. Sei whales have a fairly pointed rostrum (when viewed from above) and only a single prominent longitudinal ridge on the rostrum (Bryde's whales have three), and a slightly arched head with a downturned tip (when viewed from the side). Unless the head can be seen at close quarters, however, Bryde's and sei whales can be especially difficult to distinguish at sea, and this has impeded our knowledge of the biology of this species.

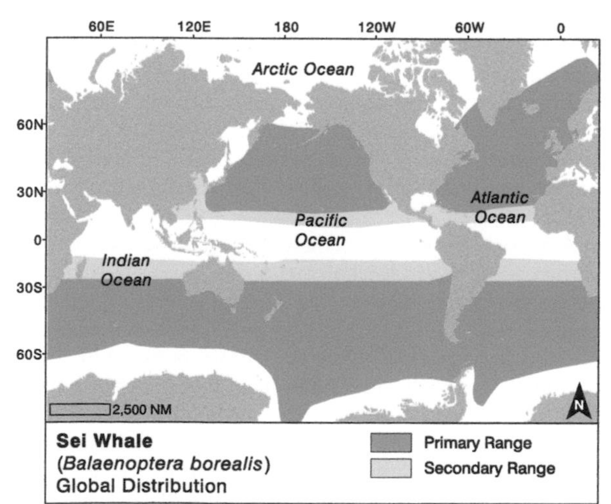

Sei Whale (*Balaenoptera borealis*) Global Distribution

Coloration is mostly dark gray or brown (it can be almost black), except for a whitish area on the belly. The back is often mottled with scars (probably from cookie-cutter shark and lamprey bites), and the skin surface often resembles galvanized metal.

The 32–65 (average about 50) ventral pleats are short for rorquals, ending far ahead of the navel. The 219–402 (average about 350) baleen plates on each side are black with very-fine fringes of light smoky-gray to white. They are less than 80 cm long, and tend to be narrower than in the other large rorquals. A few nearly white plates may occur. Sei whales produce a blow up to 3 m tall, which is high and columnar. Adult sei whales can be up to 18 m in length, although 12–17 m is a more typical length for adults. Females are slightly larger than males. Large adults may weigh 45,000 kg. At birth, sei whales are 4.5–4.8 m long.

Recognizable geographic forms None.

Can be confused with Sei whales are most likely to be confused with Bryde's whales, and less likely with fin and Omura's whales. Careful attention to body size, dorsal fin shape and position, head shape, and color pattern will help to distinguish among the four. The three head ridges of Bryde's whales (sei and fin whales always appear to have just one medial ridge), and larger size and asymmetrical head coloration of fin whales will help make them distinguishable from seis. Also, pay particular attention to the tip of the upper jaw, which has a strong tendency to be downturned in sei whales (but not in the other three). Despite the differences, sei

The sei whale usually shows its dorsal fin and blowhole at the same time at the surface; the dorsal fin juts up and then back. Azores. PHOTO: M. VAN DER LINDE

The rostrum of the sei whale is relatively narrow and often noticeably arched. Azores. PHOTO: M. VAN DER LINDE

Classic sei whale features: relatively arrow rostrum with a single median ridge, and brush mark on the side, mid-body. The dorsal fin is erect and swept back at the top and visible at the same time as the blowholes. Chile. PHOTO: R. HUCKE-GAETE

When a view such as this is obtained, sei whales can generally be distinguished from Bryde's whales by their absence of accessory rostral ridges. Near the Mariana Islands, western Pacific. PHOTO: J. COTTON

and Bryde's whales are still frequently confused, and in fact the two species can be frustratingly difficult to tell apart in typical situations. The classic character is the number of head ridges—three in Bryde's whales and only one in sei whales; but this can be very difficult to see on distant sightings and care must be taken to avoid getting confused by rippling water and glare on top of the head. Much of the literature on the two species still to this day contains identification errors. Finally, until the biology of the Omura's whale is better known, one should be very aware of the possibility of this species in the Indo-Pacific. Clearly, information on a multitude of characters will be required to confirm an identification as a sei whale.

Distribution The distribution of the sei whale is poorly understood for a number of reasons, not the least of which is their long-standing confusion with Bryde's whales. Another reason is that the species appears to occur largely in unpredictable patterns, being seen regularly in an area for several years, after which it may largely disappear. This type of pattern has been called "irruptive." Sei whales are largely open ocean whales, not often seen near the coast. They occur from the tropics to polar zones in both hemispheres and in all three major oceans, but are more restricted to mid-latitude temperate zones than are other rorquals. They do undergo seasonal migrations, although they apparently are not as extensive as those of some other large whales. They are very rare in the Mediterranean.

Ecology and behavior The sei whale is one of the least-known of the rorqual species. Groups of two to five individuals are most commonly seen, although not much is known of their social behavior. Sei whales are fast swimmers (reaching speeds of well over 25 km/hr), one of the fastest of all cetaceans. They tend to arch their backs less than other rorquals when surfacing, and often sink below the surface. When slow moving, sei whales surface with their blowholes and dorsal fin often visible above the water at the same time. Feeding sei whales tend to dive and surface in very predictable series, often remaining visible just below the surface between breaths. Sei whales virtually never lift their flukes upon diving. They apparently prefer to feed near dawn.

Calving occurs in midwinter, in low latitude portions of the species' range. Gestation lasts 10–12 months, and weaning is thought to occur at about 6–9 months.

Large sei whales from the Southern Hemisphere showing the conspicuous blow in this species and erect dorsal fin swept back toward the top. The animal at the back is a fin whale. Chile. PHOTO: R. HUCKE-GAETE

The pale brush mark on the side and erect, bent-back dorsal fin identify this sei whale feeding in the Beagle Channel. Argentina. PHOTO: V. BEAVER

Sei whales may hybridize with fin whales on occasion. They produce sounds similar to other rorquals, but there is much less known about their vocal behavior than for most other species in the family. Sei whales probably live to be more than 50 years of age. Their primary natural predator is the killer whale.

Feeding and prey Sei whales have more flexibility in their feeding behavior than other baleen whales. They generally skim copepods and other small prey types, rather than lunging and gulping, as other rorquals do. This may largely explain the relative fineness of the baleen fringes and the shortness of the throat pleats in this species. They do lunge feed on occasion, however. Other prey types taken when lunge feeding include krill, cephalopods, sardines, and anchovies.

Threats and status After blue and fin whales, the sei whale was next in the line of fire of the modern whaler's harpoon. It appears to be the preferred species among many whale-eating cultures. The heaviest period of exploitation of this fast species of whale was between the 1950s and 1970s. Whaling took place in the North Pacific (where at least 74,000 were taken) and North Atlantic (14,000

The brush marks are not visible on this low-swimming sei whale, but the pointed dorsal fin clearly shows the characteristic "up-and-back" shape of a sei whale. Azores. PHOTO: M. VAN DER LINDE

The relatively large dorsal fin juts up and then bends back toward the top; notice also the dark coloration and pale, forward-directed brush mark mid-body. New England. PHOTO: M. WHITE

Sei whales, like all members of the genus *Balaenoptera*, are slender and streamlined—built for speed. PHOTO: P. DULEY, NOAA, NMFS, NEFSC

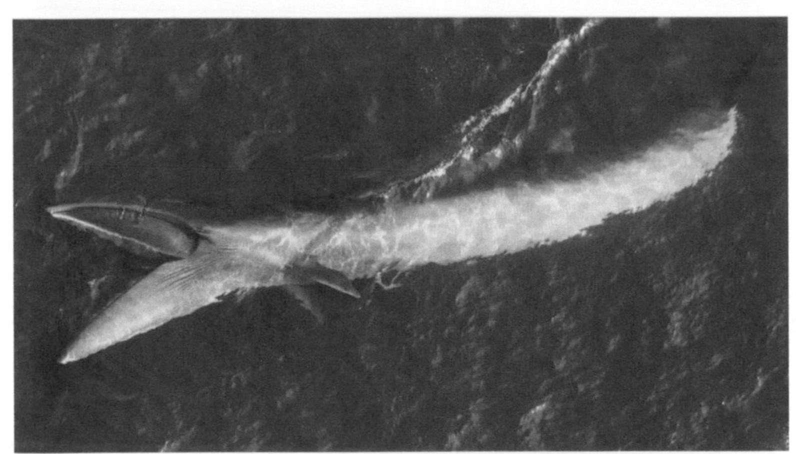

Viewed from above, rorquals such as this lunging sei whale can appear almost serpentine. New England. PHOTO: C. KAHN, NOAA, NMFS, NEFSC

The throat pleats of sei whales do not extend as far back on the body (terminating anterior to the navel) or expand as much as they do in larger species. New England. PHOTO: P. DULEY, NOAA, NMFS, NEFSC

whales taken) oceans, but most hunting was in the Southern Hemisphere (about 200,000, with at least 110,000 taken in the 1960s alone). Although fully protected by the International Whaling Commission since 1985, a few were taken in the North Atlantic by Iceland in the last few decades of the 20th century (70 whales from 1986–1988). Sources of mortality other than direct exploitation include probable vessel strikes and some entanglement in fishing gear, but in general these do not appear to be severe problems for sei whales. Populations in all areas were depleted by commercial whaling, although they currently appear to be doing quite well in the Northern Hemisphere. Current global abundance of the sei whale is not well-known, but is considered to be at least 80,000 individuals. Reliable recent data on trends for most populations are not available.

IUCN status Endangered.

Bryde's Whale—*Balaenoptera edeni*

Anderson, 1879

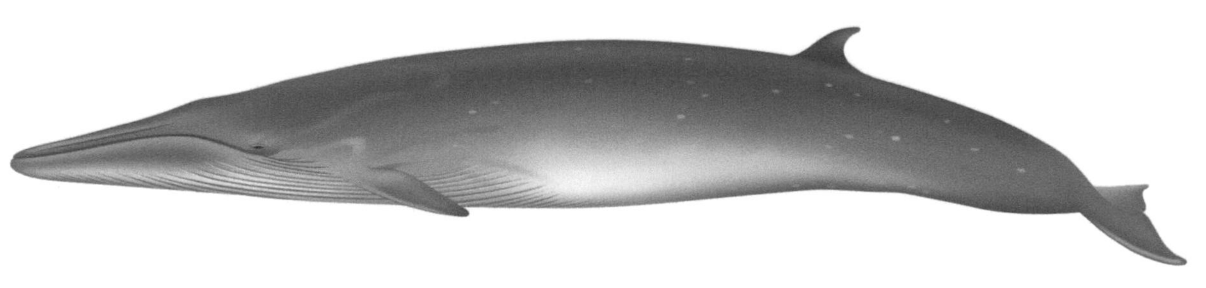

Recently-used synonyms *Balaenoptera brydei* (taxonomy is still unresolved).

Common names En.-Bryde's whale; Sp.-*rorcual tropical, ballena de Bryde*; Fr.-*rorqual de Bryde*.

Taxonomic information Cetartiodactyla, Cetacea, Mysticeti, Balaenopteridae. Until recently, all medium-sized rorquals were thought to be members of one of two species, *B. edeni* (Bryde's whale) or *B. borealis* (sei whale). However, we now know that there are at least 3–4 genetically-distinct types of these whales, including what in the past have been called pygmy or dwarf Bryde's whales. The species-level taxonomy remains unresolved. Their formal redescription awaits further taxonomic work. Provisionally, two subspecies are recognized: the offshore Bryde's whale (*B. e. edeni*) and Eden's whale (*B. e. brydei*). The International Whaling Commission continues to use the name *B. edeni* for the *edeni/brydei* complex whales, despite recognizing that there may be two species. To make matters even more complicated, in the past, the term "pygmy Bryde's whale" had also erroneously been used for specimens of what are now known to be a separate species (*Balaenoptera omurai*), which was only described in 2003.

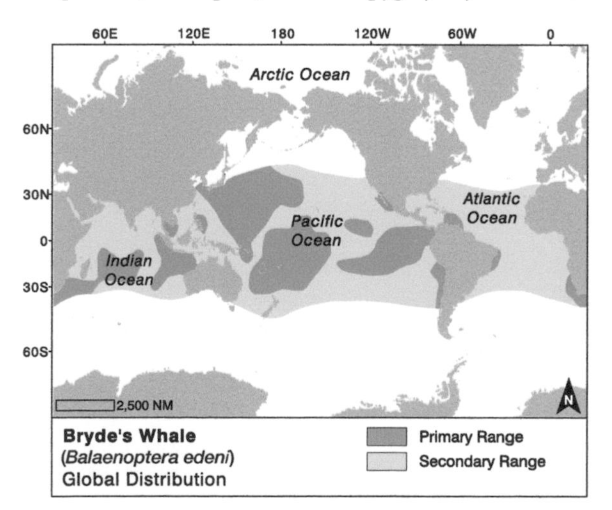

Species characteristics For many years, whalers and field observers did not distinguish between Bryde's and sei whales in their records. In the past few decades, however, whales of the two species have been able to be distinguished, even at sea. Bryde's whales have a very streamlined and sleek body shape. Most Bryde's whales have three prominent ridges on the rostrum (other rorquals generally have only one). This is perhaps the best characteristic to use for identification, although since some whales have poorly-developed ridges, and the ridges may be difficult to see on others, one is best advised to consider information on other characters as well. The shape of the head is somewhat pointed, when viewed from above. The head makes up about 25% of the body length. The Bryde's whale's dorsal fin is tall and falcate and generally rises abruptly out of the back, a feature that will help distinguish this species (and sei whales) from fin whales, in which the dorsal fin rises at a relatively shallow angle from the back. The dorsal fin is often notched on the trailing edge. The flukes are broad with a relatively straight trailing edge. Bryde's whales often exhale underwater, then surface with little or no visible blow. When visible, the blow can be either columnar or bushy. The height of the blow is variable.

Bryde's whales have a relatively simple, countershaded color pattern and are dark gray dorsally and lighter ventrally. There is often a pinkish tinge on the lower white areas, and the body may be covered in oval scars made by cookiecutter sharks. The upper jaw and lips are typically uniformly dark. The 40–70 throat pleats reach to or past the navel. The 250–370 pairs of gray baleen plates have light gray fringes, which are relatively coarse. The longest plates reach about 40 cm. Adults of this species can be up to 15.0 m long in males and 16.5 m long in females

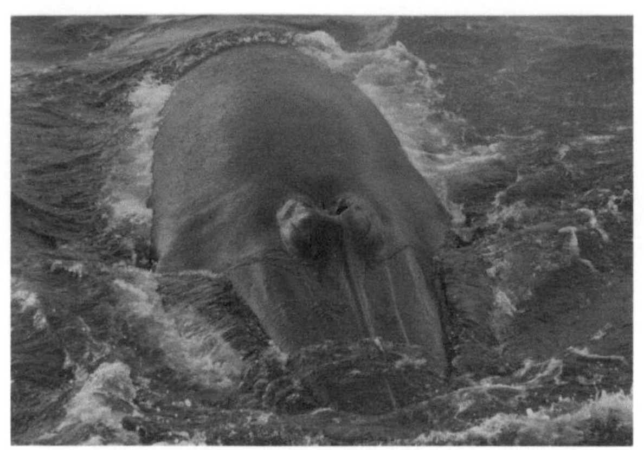

Although not always easy to discern, especially when water is washing over the rostrum, the accessory rostral ridge on either side of the median ridge is diagnostic in Bryde's whale. South Africa. PHOTO: L. EDWARDS

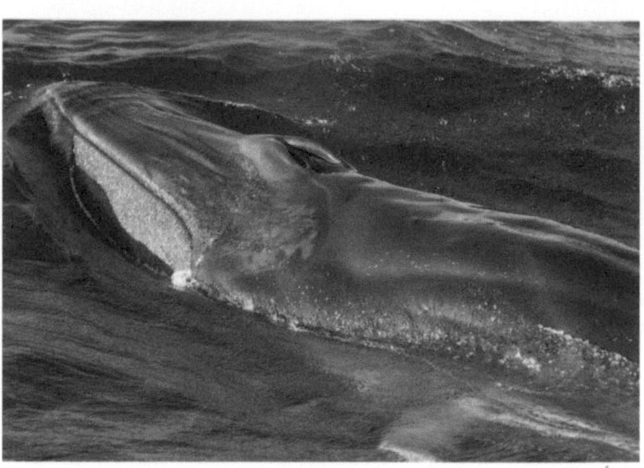

Some color pattern elements not often seen on Bryde's whales, but evident on this animal from Japan include a pale chevron and a pale "smoke plume" pattern coming out of the blowhole. Japan. PHOTO: K. ODA

A lolling or curious Bryde's whale spyhopping just offshore of Angola and showing the top of its rostrum. PHOTO: C. WEIR

This animal was genetically identified as "Eden's whale" a smaller, more neritic form (possibly a distinct subspecies or species) of Bryde's whale. Bay of Bengal. PHOTO: R. MANSUR

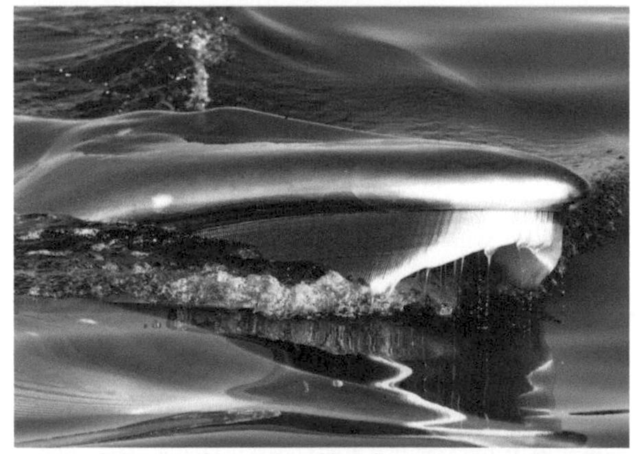

Bryde's whales sometimes swim with their mouth open at the surface, using their very fine baleen to skim small prey, much like right whales do. Gulf of California. PHOTO: M. NOLAN

Bryde's whales typically have a conspicuous blow, but they can make it almost invisible if they are approached too close by vessels or killer whales; sometimes they may even blow underwater. PHOTO: G. CRESSWELL

(females do reach greater lengths than males); newborns are about 4 m long. Maximum weight is about 40,000 kg.

Recognizable geographic forms There are one or more small forms of this species (for instance, off South Africa, southwestern Japan, Hong Kong/Macau, and Australia), which at this time cannot be reliably distinguished from common Bryde's whales in the field. These forms, generally referred to the subspecies *B. e. edeni* or variously to the species *B. edeni,* have a number of slight differences in their morphology and ecology, but have not been adequately described.

A Bryde's whale in the Gulf of California, Mexico: the dorsal fin is variable, but usually prominent, very falcate and often pointed at the tip. PHOTO: D. FERTL

Dorsal fin shape of the Bryde's whale is highly variable, but tends to be fairly prominent and often quite pointed. The white spots on this animal are cookie-cutter shark bites. PHOTO: I. VISSER

Can be confused with Bryde's whales are most likely to be confused with sei whales, and less likely with fin and Omura's whales. Careful attention to body size, dorsal fin shape and position, head shape, and color pattern will help to distinguish among the four. The three head ridges of Bryde's whales (sei and fin whales always appear to have just one medial ridge) and much larger size and asymmetrical head coloration of fin whales will help make them distinguishable from Bryde's. Also, pay particular attention to the tip of the upper jaw, which is generally quite flat in Bryde's whales (but has a strong tendency to be downturned in sei whales). Despite the differences, sei and Bryde's whales are still frequently confused, and in fact the two species can be frustratingly difficult to tell apart in typical situations. The classic character is the number of head ridges—three in Bryde's whales and only one in sei whales; but this can be very difficult to see on distant sightings and care must be taken to avoid getting confused by rippling water and glare on top of the head. Much of the literature on the two species still to this day contains identification errors. Omura's whale is similar, and one should be very aware of the possibility of this species in the Indo-Pacific. Sizes of Bryde's and Omura's whales are similar, but Omura's whales appear to have an asymmetrical head color pattern very similar to that of the fin whale. Clearly, a multitude of characters will be required to confirm an identification as a Bryde's whale.

Distribution This species has a circumglobal distribution and is found in the Atlantic, Pacific, and Indian oceans. Bryde's whales are largely creatures of the tropical and subtropical zones (inhabiting waters about 16°C or warmer) and generally do not move poleward of 40° in either hemisphere. They are found both offshore and near the coast in many areas, and tend to inhabit areas of unusually high productivity. They extend into some semi-enclosed seas, such as the Persian Gulf and Red Sea. Whales of this species are not known to make extensive north/south migrations, as do other species of baleen whale, although short migrations have been documented in some areas (especially among the offshore forms). Resident populations occur in certain areas, such as the Gulf of California and Gulf of Thailand.

Ecology and behavior Not much is known of the ecology of the Bryde's whale. Although generally seen alone or in pairs, Bryde's whales do aggregate into groups of up to 10–20 on feeding grounds. Unlike other rorquals, Bryde's whales often do not produce a visible blow. Breaching occurs occasionally. Bryde's whales generally arch their backs when diving, but virtually never fluke up on a dive. They are capable of swimming at speeds up to 25 km/hr, and can dive to about 300 m.

Bryde's whales are active lunge feeders. They often feed at the surface, which can attract seabirds and other pelagic predators. South Africa. PHOTO: L. EDWARDS

A Bryde's whale lunges vertically through a school of baitfish in the Caribbean. PHOTO: M. GREENFELDER

Sexual maturity occurs at lengths generally greater than 11.2 m in the common form, but the small forms mature at much shorter lengths. Unlike other rorquals, the warm-water Bryde's whale does not have a well-defined breeding season in many areas, and births can occur throughout the year. Bryde's whales produce low-frequency moans, the structure of which is just beginning to be investigated.

Feeding and prey Bryde's whales are primarily schooling-fish eaters (common prey species include pilchard, anchovy, sardine, mackerel, and herring), but they also take squid, krill, pelagic red crabs, and other invertebrates. They are very active lunge feeders, often changing direction abruptly when going after mobile fish prey. In the Gulf of Thailand, they often feed by coming up diagonally through a prey patch, with the mouth agape and the lower and upper jaws at an angle of 90° or greater. They have also been observed using bubble nets to corral prey.

Threats and status Bryde's whales were never hunted as heavily as their larger cousins, the blue, fin, and sei whales. Fewer than 8,000 were killed in the Southern Hemisphere in the 1900s. Due to this fact, most populations of the Bryde's whale have not been seriously depleted. In recent years, some Bryde's whales have been taken by the Japanese in the North Pacific, and low numbers of small whales have been killed by artisanal whalers from villages in Indonesia (although at least some of these may be Omura's whales—see below). There are no estimates of global abundance, although there are probably about 20,000–30,000 Bryde's whales in the North Pacific, and about 10,000 in the eastern tropical Pacific. The species is not considered to be endangered or threatened, and at least the western North Pacific stock is thought to be increasing. Habitat modification and noise disturbance may be additional human-caused threats.

IUCN status Data Deficient.

Omura's Whale—*Balaenoptera omurai*

Wada, Oishi, and Yamada, 2003

Recently-used synonyms None.

Common names En.-Omura's whale; Sp.-*ballena de Omura*; Fr.-*rorqual de Omura*.

Taxonomic information Cetartiodactyla, Cetacea, Mysticeti, Balaenopteridae. This species was described in 2003, and there is now abundant evidence from molecular genetic studies to confirm that it is a valid species, different from and not closely related to the Bryde's whale. In the past, some specimens of this species were included among "pygmy/dwarf Bryde's whales" in various studies, but we now know Omura's whale to be an early offshoot of the rorqual lineage, perhaps more closely related to the blue whale.

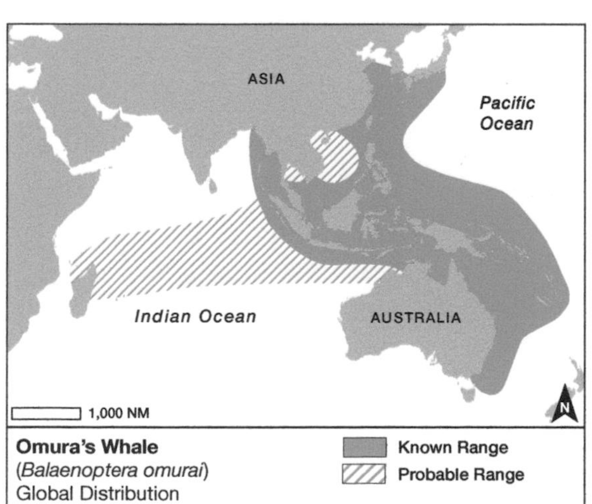

Species characteristics Omura's whales have a streamlined and sleek body shape. Although their external appearance is not yet fully documented, they apparently only have one prominent ridge on the rostrum (most Bryde's whales have three). The shape of the dorsal fin is generally like that of Bryde's and sei whales, i.e., tall and falcate and rising abruptly out of the back. However, there is a tendency for the dorsal fin to be more strongly falcate in Omura's whale. The flukes are broad, with a relatively straight trailing edge.

The color pattern of this species is not completely known. They are counter-shaded, lighter below, and darker above. The coloration most closely resembles that of the fin whale—with an asymmetrical lower jaw (white on the right and dark on the left). Also, at least some animals have light streaks and blazes that extend up from the light ventral side onto the darker back. The anterior edges and inner surfaces of the flippers are white, as is the ventral surface of the flukes, which have a black margin.

The 80–90 throat pleats reach to about level of the navel. The 180–210 pairs of baleen plates (less than in other species in the genus) are short and broad. They are yellowish-white at the front and black in the rear, with the middle ones intermediate (some may be two-tone). Adults of Omura's whales are generally no longer than

The right side of the lower jaw in Omura's whale is light in color, much like that of the fin whale. Near Madagascar. PHOTO: S. CERCHIO/WCS

Although the lower part of the left lower jaw may be light, the lip area is dark in Omura's whale. Near Madagascar. PHOTO: S. CERCHIO/WCS

A smallish dorsal fin may be characteristic of Omura's whale; notice also the conspicuous pale chevron and a fresh cookie-cutter shark bite in front of the dorsal fin. New Caledonia. PHOTO: C. GARRIGUE

This presumed Omura's whale is heavily pock-marked with cookie-cutter shark bites giving it a mottled appearance; the rippled profile in front of the dorsal fin indicates a skinny whale. New Caledonia. PHOTO: C. GARRIGUE

The pale spots on this Omura's whale are healed wounds from cookie-cutter sharks—a common scourge of tropical cetaceans. The species is small enough that the blowhole and dorsal fin are usually visible at the same time when the whale surfaces. The pale chevron is clearly visible. New Caledonia. PHOTO: C. GARRIGUE

More fin whale color pattern similarity: the pale area above the dark, right eye stripe loops further back on the right side than on the left. The identification features described on this page will require further fieldwork (including biopsy sampling) to confirm their validity. New Caledonia. PHOTO: C. GARRIGUE

about 11.5 m (although some may reach to 12.0 m). Females presumably grow slightly larger than males, as in other rorquals. Physical maturity may occur at lengths as short as 9.0 m. Maximum weight is not known (but is probably not over 20,000 kg). Newborns are about 3.5–4 m long.

Recognizable geographic forms None.

Can be confused with Omura's whale has only recently been described, and its external appearance is not yet well documented. For these reasons, special care must be taken in identifying this species and in ruling out other small to medium-sized balaenopterids (small fin, sei, Bryde's, and minke whales). When observed clearly, the complex color pattern of Omura's whale (with an asymmetrical lower jaw and light streaks and chevrons on the back) should make it relatively easy to distinguish these animals from all but small fin whales in the field. The dorsal fin may help here, as Omura's whales appear to have a very hooked fin rising at a steep angle, and fin whales almost always have fins rising at a shallow angle. The presence of three head ridges has for many years been taken to confirm a whale's identity as a Bryde's whale, but there is some suggestion that some Omura's whales may also have accessory head ridges (and be aware that rippling water on the head of other species can be mistaken for accessory head ridges). Both species of minke whales can also cause some confusion, but minke whales are generally slightly smaller, with a much sharper point to the head when viewed from above, and white bands/patches on the flippers are diagnostic of common minkes. Also, minkes have symmetrical head coloration, unlike that of Omura's whale. Clearly, a number of characters will be required to confirm an identification as an Omura's whale. Genetic samples may be required in some cases for confirmation.

Distribution This species has previously been considered to be restricted to the western Pacific and eastern Indian oceans, although the exact limits of its range are not well established. Recent probable records from Madagascar have been discovered, suggesting it may occur throughout the Indian Ocean. It is apparently restricted to tropical and subtropical waters, and mostly occurs over the continental shelf in relatively nearshore waters. The records available are mostly from the eastern Indian Ocean (off the Cocos Islands), Indonesian waters, the Philippines, Sea of Japan, and the Solomon Islands.

Ecology and behavior Very little is known of the ecology of this species. Although generally seen alone or in pairs, they may well aggregate into larger groups on feeding grounds. Based on the few confirmed sightings of these animals, they generally do not lift their flukes upon a dive.

Unlike most other rorquals, it is suspected that the warm-water Omura's whale probably does not have a well-defined breeding season. Virtually nothing is known of its reproductive biology. They live to ages of at least 38 years.

One can clearly see the dark left lower jaw of this Omura's whale, which is an example of bilateral asymmetry—the right lower jaw is light in color. Near Madagascar. PHOTO: S. CERCHIO/WCS

While the small size of Omura's whale may sometimes cause confusion with the similarly-sized minke whales, a view of the rostrum from above will show that it is not so acutely pointed as in minke whales. Near Madagascar. PHOTO: S. CERCHIO/WCS

Feeding and prey While there is little information on specific prey species, Omura's whales are probably primarily schooling-fish eaters. They are lunge feeders, like most other rorquals.

Threats and status Omura's whales were probably never hunted as heavily as their larger cousins, the blue, fin, sei, and Bryde's whales. Due to this fact, Omura's whale has probably not been seriously depleted, except possibly in the Philippines. They have been hunted during "scientific whaling" by the Japanese in the Solomon Sea, and near the Cocos Islands in the Indian Ocean. Omura's whales have also been killed by artisanal whalers from villages in the Philippines (and probably Indonesia as well). There are no estimates of abundance available for this species. Habitat modification and noise disturbance are probably among the human-caused threats.

IUCN status Data Deficient.

Common Minke Whale—*Balaenoptera acutorostrata*

Lacépède, 1804

Standard Minke Whale

Dwarf Minke Whale

Recently-used synonyms *Balaenoptera davidsoni*
Common names En.-common minke whale; Sp.-*rorcual enano, ballena minke*; Fr.-*petit rorqual.*
Taxonomic information Cetartiodactyla, Cetacea, Mysticeti, Balaenopteridae. The minke whales have long been considered to comprise only one species (*B. acutorostrata*), but in recent years the distinctness of the Antarctic minke whale (*B. bonaerensis*) became clear, and it has now been split-off as a separate species (see below). Three subspecies of the common minke whale are currently recognized: *B. a. acutorostrata* (in the North Atlantic), *B. a. scammoni* (in the North Pacific), and a third unnamed subspecies (the dwarf minke whale, mostly occurring in the Southern Hemisphere).

Minkes, including this white-shouldered form from Tonga, are quite curious at times and will sometimes allow a close approach. PHOTO: R. KIEFNER

Species characteristics Common minke whales are generally easy to distinguish from the larger rorquals. The body is quite sleek. The head is sharply pointed and V-shaped, viewed both from the side and from above, and the median head ridge is very prominent. The dorsal fin is tall (for a rorqual), recurved, and located about two-thirds of the way back from the snout tip. The flippers are narrow with pointed tips. The tapered flukes have a smooth trailing edge and a prominent notch. There are 50–70 moderately-short throat pleats (often extending just past the flippers) and 231–285 pairs of white to cream-colored baleen plates (however, some may be dark in color).

The minke whale's coloration is distinctive. It is dark gray dorsally and white ventrally, with streaks and/or lobes of intermediate shades on the sides. Some of the streaks may extend onto the back behind the head. The most distinctive light marking is a brilliant white patch on each flipper, which varies in extent between the dwarf and standard-form minkes (the white is not usually present on Antarctic minke whales). This band is generally visible through the surface when animals are swimming near the surface.

The blow tends to be diffuse and is often not visible at all. Adult length ranges from about 6.5–8.8 m. No significant external differences between the two Northern Hemisphere subspecies are known, but the dwarf minke is somewhat smaller. Females are longer than males in all forms. Length at birth for common minke whales is 2.0–2.8 m. Maximum body weight is about 9,200 kg.

Recognizable geographic forms

Standard minke whale (B. a. acutorostrata/scammoni)—Standard-form minke whales occur throughout the North Pacific and North Atlantic oceans. They may show varying amounts of mottling on the sides, belly, and back. The flippers are completely dark gray to black, except for a brilliant white band that runs across the middle half of each one. The shoulder area just above the flipper is generally dark. There may be a light gray rostral "saddle" in some individuals. The standard form of minke whale averages about 8.0 m for adult males and 8.5–8.8 m for adult females. Maximum body weight is about 9,200 kg.

Dwarf minke whale (B. a. subsp.)—The dwarf minke whale (an unnamed subspecies) occurs in the Southern Hemisphere, but may sometimes move north of the equator. It is distinguished from the standard-form minke mainly by coloration differences. Some of the streaks on the sides typically extend onto the back behind the head, and form a chevron pattern. The white blaze on the flipper extends up onto the shoulder area just above the flipper, and extends back to join light patches on the belly and thorax. The axilla is dark. There is also a well-developed light gray rostral "saddle." The baleen plates of the dwarf minke whale tend to have a narrow, dark fringe, and about half of the plates may be dark in color. The dorsal fin is slightly larger and the tail is slightly longer than in the common minke whale. This subspecies is somewhat smaller than the Northern Hemisphere ones, averaging about 6.5–7.0 m as adults. They reach a known maximum of 7.8 m and weights of up to only about 6,400 kg.

From above, the pointed head, small size, and conspicuous flipper band make common minkes easy to identify. Eastern North Pacific. PHOTO: NOAA FISHERIES

Can be confused with When seen well, common minkes are among the easiest to distinguish of the whales of the genus *Balaenoptera*, by a combination of their small size, usual absence of a visible blow, unique head shape, and distinctive color patterns (especially the white flipper bands). The Antarctic minke whale overlaps the dwarf minke in distribution at least in the southern summer,

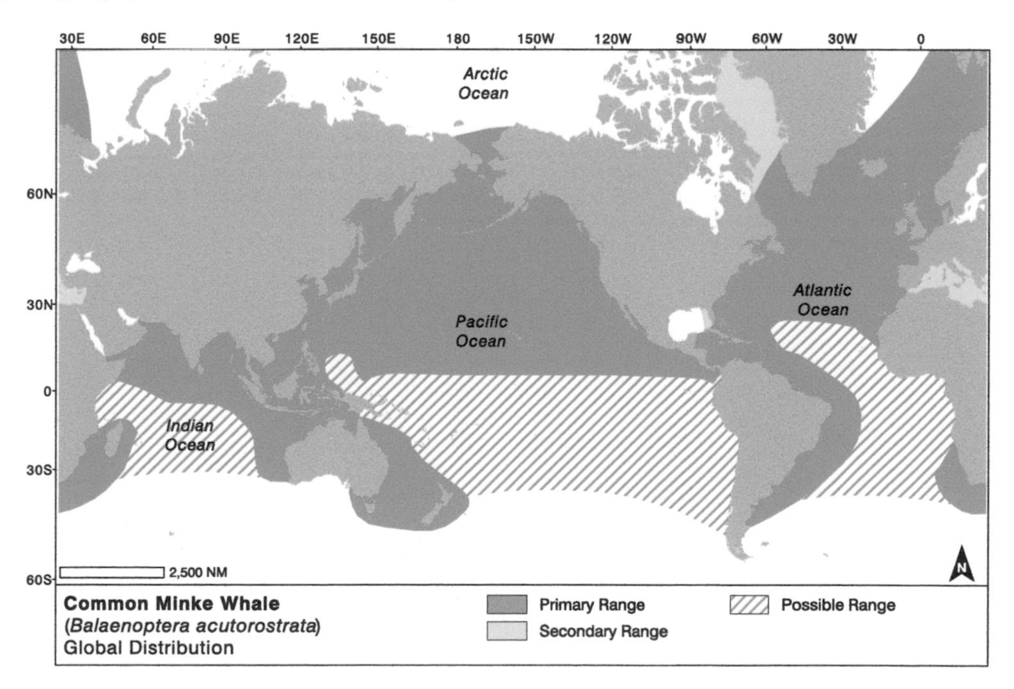

A common minke lunges on its side at the surface, tossing bait fish into the air as rhinoceros auklets look on; notice the white arm band. British Columbia. PHOTO: J. TOWERS

The narrow pointed rostrum and short, yellow baleen is evident on a common minke engulfing a shoal of sandlance off British Columbia. PHOTO: J. TOWERS

The distinctive pale patterning on the sides of common minke whales is often difficult to see unless they roll up fairly high out of the water—this is typically all you see of a minke. Iceland. PHOTO: C. WEIR

When frightened, common minkes can swim remarkably fast and will sometimes throw up a long series of large rooster-tails, until they disappear from sight. British Columbia. PHOTO: J. TOWERS

Clearly built for speed, this belly-view of a common minke shows its pointed head; the snowy white underparts will sometimes blush pink during periods of activity, probably as a measure to dump heat. PHOTO: K. ROBINSON

A breaching common minke shows off its snowy white belly and white flipper band. PHOTO: M. O. KAYA

A common minke whale displays its diagnostic white arm band visible as it hovers briefly over Canadian waters. British Columbia. PHOTO: J. TOWERS

A dwarf, or white-shouldered, minke also has a white flipper band but the white from the belly continues up above the insertion of the flipper from behind, somewhat obscuring the flipper. PHOTO: J.-P. SYLVESTRE

A dwarf minke off the Great Barrier Reef, Australia; notice the white area on the body at the insertion of the flipper on both animals, as well as the black collar that dips down on to the throat pleats between the eye and the flipper. PHOTO: W. OSBORN

and can be difficult to distinguish. The best way to distinguish them is by the presence/absence of the white flipper patch (Antarctic minkes generally do not have the patch). The pygmy right whale may also present identification problems from afar, but a good look at the head and color pattern will allow accurate identifications. Sei and Bryde's whales, and some beaked whales, may also present confusion, but generally only if the animals are seen at a distance.

Distribution Common minke whales are widely distributed from the tropics and subtropics to the ice edges in both hemispheres. The specific distribution, especially in the Southern Hemisphere, is not well known, due to lumping of all minke whales into one species in the past. Although they can be seen offshore as well, common minkes are more often seen in coastal and inshore areas. Minke whales are very rare in some tropical oceanic areas, such as the eastern tropical Pacific. They occur at least occasionally in the Mediterranean Sea. At least some populations migrate from high latitude summer feeding grounds to lower latitude winter breeding areas. Dwarf minke whales are found only in the Southern Hemisphere, as far south as 69°S, and some may overwinter in the Antarctic. They may have a circumpolar distribution, although this is not well-established.

Ecology and behavior Common minke whales sometimes aggregate for feeding in coastal and inshore areas of cold temperate to polar seas. Group sizes are generally small (singles, pairs, and trios), although larger groups of animals may aggregate on productive feeding grounds. They are often segregated by age, sex, and/or reproductive class, and appear to have a complex social structure. Minke whales do not fluke-up on a dive, but they do sometimes breach and perform other aerial behaviors. They often approach and swim around stationary vessels. Upon surfacing, the dorsal fin generally appears simultaneously with the blowholes. Minkes in Greater Puget Sound, Washington, have exclusive, adjoining home ranges, something that may be unique to baleen whales.

Calving probably occurs in dispersed low latitude areas (although the migrations of minkes are not as well defined as those of larger rorquals) in winter months. Some females may give birth annually, and gestation lasts about 10–11 months. Minke whales have recently been found to make a number of unique vocalization types (with colorful names like boings and "Star Wars"), although the functions of these are not well known. Longevity is unknown, but probably reaches about 50 years. Killer whales appear to be a significant predator of this small rorqual.

Feeding and prey The diversity of prey types of common minke whales primarily consist of small invertebrates (euphausiids and copepods) and small schooling fishes (sand lance, sand eel, salmon, capelin, mackerel, cod, coal

Dwarf minkes are individually-identifiable based on color pattern variation. Notice the white shoulder of this animal as it peers at the photographer. Great Barrier Reef, Australia. PHOTO: W. OSBORN

fish, whiting, sprat, pollack, haddock, herring, anchovy, saury, walleye pollack, and lanternfish), which they capture by lunging into large prey aggregations. They also take larger fish species (such as wolfish and dogfish) on occasion.

Threats and status Until the latter half of the 1900s, common minke whales were often considered too small to hunt, and were generally passed up. However, with the depletion of the larger rorquals, common minke whales have been heavily hunted in recent years (>100,000 in the North Atlantic alone), and this is now the species of whale most commonly taken by commercial and "scientific" whalers (at least in the Northern Hemisphere). It is still hunted by Norway in the North Atlantic and by Japan and Korea in the North Pacific and Antarctic. Despite this exploitation, the common minke whale generally remains abundant in most areas of its range (over 180,000 are thought to occur in the Northern Hemisphere, although accurate numbers are hard to come by). Some are caught in fishing gear and suffer from vessel strikes and habitat disturbance.

IUCN status Least Concern.

Antarctic Minke Whale—*Balaenoptera bonaerensis*

Burmeister, 1867

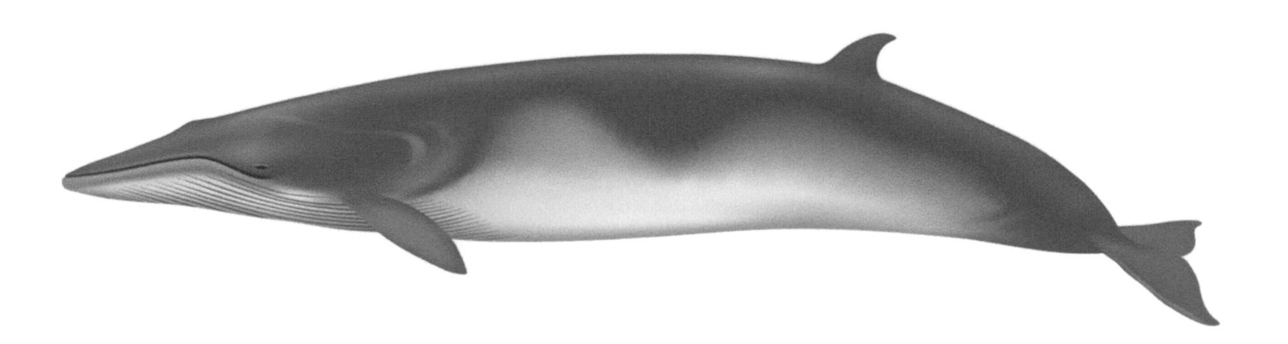

Recently-used synonyms None.

Common names En.-Antarctic minke whale; Sp.-*rorcual enano*, *ballena minke*; Fr.-*petit rorqual*.

Taxonomic information Cetartiodactyla, Cetacea, Mysticeti, Balaenopteridae. Until recently, all minke whales were classified as *B. acutorostrata*, but the distinctiveness of the Antarctic minke whale is now well established, and it has been designated its own species as of the late 1990s. This species is actually more closely related to the sei and Bryde's whales than to the common minke whale. Due to its tangled history with the common minke whale, our knowledge of the Antarctic minke's biology is limited. Probable hybrids between the two minke whale species have been discovered.

Species characteristics As in the common minke whale, the body is sleek and streamlined, the head is pointed, and there is a single prominent median ridge on top of the rostrum. The dorsal fin, situated about two-thirds of the way back from the snout tip, is quite tall and falcate (for a rorqual). The tapered flukes have a smooth trailing edge and a prominent notch. There are 22–38 short throat pleats, which extend just past the flippers (almost to the umbilicus) and 200–300 pairs of baleen plates, most of which are black. The color of the baleen is asymmetrical, and some of the anterior plates are white (more so on the right side).

Antarctic minke whales are dark gray dorsally and white ventrally, with streaks and/or lobes of grayish color on the lateral surface. Some of the streaks may even extend onto the back behind the head, forming a chevron. The distinctive white band on the flipper of Northern Hemisphere and dwarf minke whales is not present on Antarctic minke whales, and flippers are relatively uniformly colored. However, the flippers are not particularly dark in color—they are usually light gray, often with a distinct dark border.

The short blow is generally conspicuous in Antarctic waters, but can be more faint at times, and may even be invisible. Adult Antarctic minke whales average about 8.5–9.0 m, and can reach 10.7 m in length. Length at birth is about 2.8 m. Maximum body weight is about 9,100 kg.

Recognizable geographic forms None.

Can be confused with Antarctic minke whales are generally easy to distinguish from the larger rorquals that occur in the Antarctic (blue, fin, and sei whales), due to their small size, pointed head, specifics of the color pattern, and quicker movements. Although they are generally about 2 m longer, they can be quite difficult to distinguish from dwarf minke whales. When the flippers are seen, the best feature is the lack of distinct white flipper patches on Antarctic minkes. Pygmy right whales and some beaked whales may also cause some confusion at a distance, but a good look at the head and color pattern should make distinction possible.

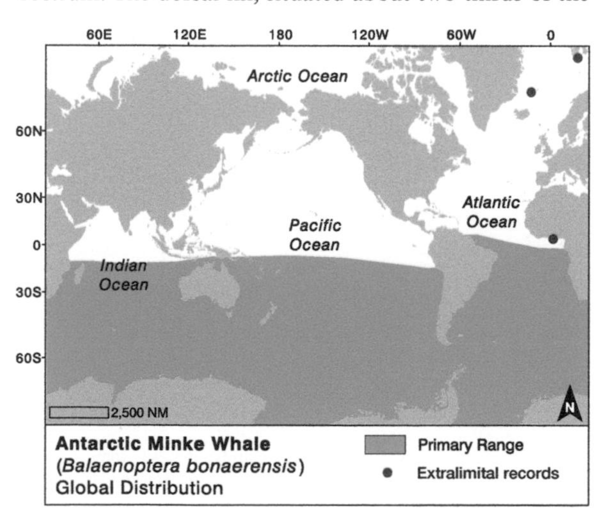

Antarctic Minke Whale (*Balaenoptera bonaerensis*) Global Distribution

Antarctic minkes can surface showing little or no blow; they often surface once or twice and then are not seen again. Antarctica. PHOTO: M. NOLAN

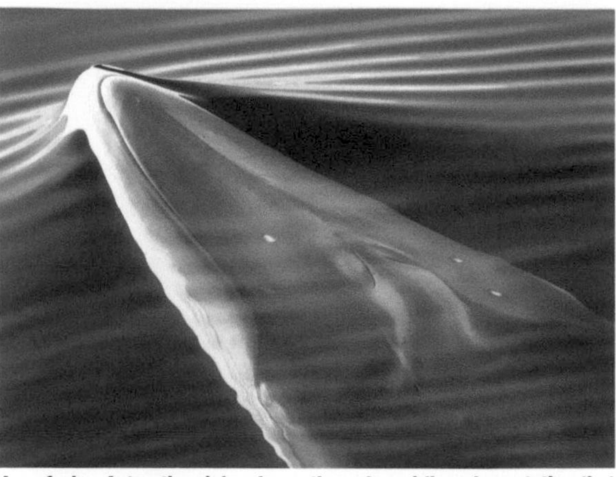

A surfacing Antarctic minke shows the pale swirling pigmentation that looks like smoke coming out of the blowholes. Antarctic Peninsula. PHOTO: S. HANSEN

Rolling over and exposing its throat pleats, the snowy white coloration of this Antarctic minke indicates that it has probably only recently arrived in Antarctic waters from its tropical breeding grounds. Antarctic Peninsula. PHOTO: S. HANSEN

After an extended stay in Antarctic waters, most whales, including this Antarctic minke, acquire a coating of diatoms giving them a yellowish, "sulfur-bottom" appearance. Drake Passage. PHOTO: J. DURBAN

A view of lateral color patterning on an Antarctic minke whale; this animal is arching for a dive and showing more than is typically visible. Antarctica. PHOTO: P. OLSON

Dorsal fin shape is quite variable in Antarctic minkes; some fins are sharply swept back toward the top, which may be age- or sex-related. Antarctica. PHOTO: M. NOLAN

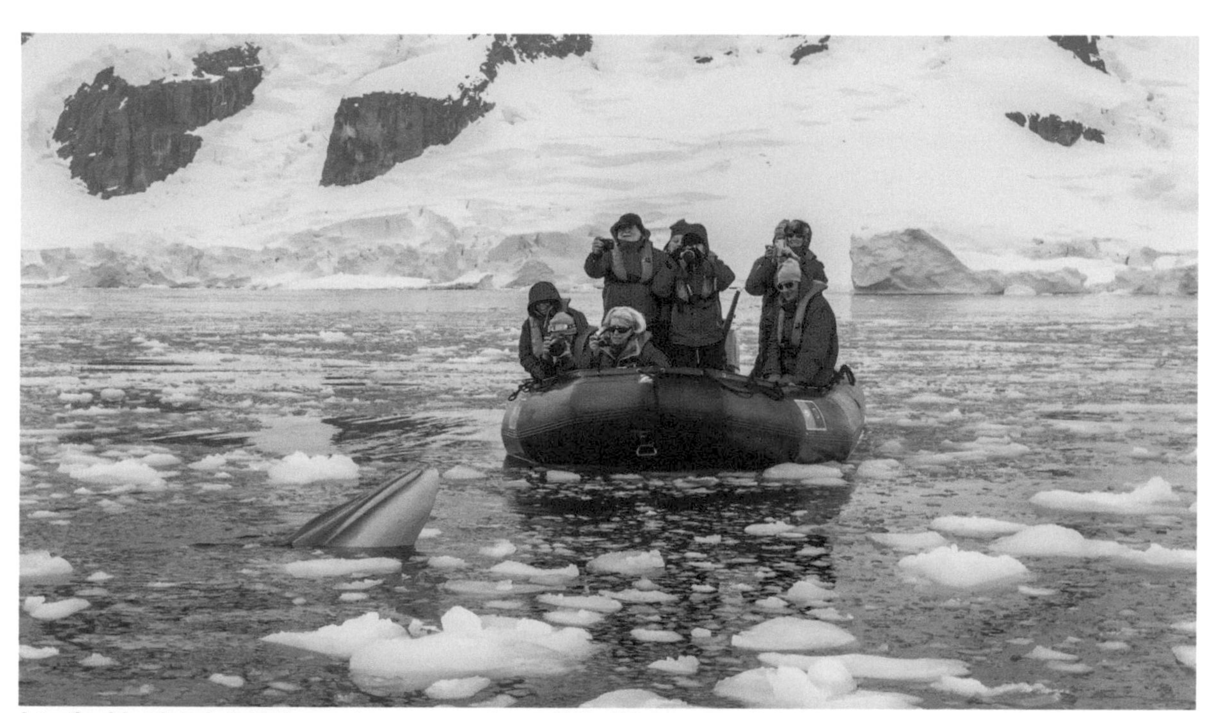

Antarctic minke whales in Antarctica seem to grow more curious and "friendly" every year—this individual seems to be mugging for the tourists' cameras. Antarctic Peninsula. PHOTO: M. NOLAN

Distribution Antarctic minke whales occur widely in coastal and offshore areas of the Southern Hemisphere, and are found from at least 7°S to the ice edges. Some may cross the equator into the Northern Hemisphere. There is a recent capture record for Togo in the Gulf of Guinea (6°N), and recently one Antarctic minke was documented from the arctic North Atlantic (this latter record is undoubtedly extralimital). The range is thought to be circumpolar, and in general they are more oceanic than dwarf minke whales. Although there is a general shift northward for breeding in the winter months (known breeding grounds are off Australia, South Africa, and Brazil), not all Antarctic minke whales migrate. They tend to be more polar than common minke whales, and most spend their summers in waters around the Antarctic continent. Some may even overwinter in the Antarctic.

Ecology and behavior The biology of the Antarctic minke whale is not well understood, at least partially due to a long-standing lack of distinction from common minke whales. Singles and pairs are not uncommon. Although groups elsewhere are generally much smaller, feeding aggregations in the Antarctic may contain hundreds of animals. Antarctic minke whales migrate to the Antarctic for the summer for feeding, and to more moderate climates in the winter for breeding. Although they are not as aerially active as some other whales, Antarctic minke whales do occasionally breach and spyhop, and sometimes even porpoise out of the water like dolphins when being chased. They often approach vessels, and are easily approachable when feeding. They do not fluke-up upon diving. Antarctic minke whales are often seen among the ice floes in the Antarctic summer months.

Antarctic minke whales exhibit similar life history parameters to those of Northern Hemisphere minkes. Annual and biennial breeding may be the norm, and peak calving occurs in July and August after a gestation of about 10 months. Attainment of sexual maturity occurs at average ages of 7–8 years for females and 8 years in males. Whales may segregate by age, sex, and reproductive status during migrations. Unique, species-specific vocalizations have been described. Killer whales appear to be important predators of this species.

Feeding and prey Antarctic minke whales eat mostly krill (which is very abundant in the Antarctic), although they do occasionally feed on small schooling fishes. They are lunge feeders, often actively plunging through patches of euphausiids with throat pleats distended.

Threats and status Large numbers of minke whales of both species (>100,000, mostly Antarctic minkes) have been killed in Antarctic waters in the past century. Japan's so-called "scientific whaling" focuses mainly on this species, which is still quite abundant in Southern Hemisphere waters. Japan recently has taken several hundred Antarctic minkes each year under special scientific permit, but the

The unbanded flippers, pointed rostrum and relatively small size identify this Antarctic minke; notice also the pale chevron pattern and the "smoke" pattern from the blowholes. The white spots by the dorsal fin are from recent cookiecutter shark bites acquired in tropical waters. Reunion Island. PHOTO: L. MOUYSSET

meat is marketed, and many conservationists see this program as simply a ploy to get around the International Whaling Commission's moratorium on commercial whaling. There is much controversy about their current abundance, but they clearly number several hundred thousand. A previous estimate of 670,000 is thought to be outdated, and current numbers are probably lower than this.
IUCN status Data Deficient.

Antarctic minke whales commonly inhabit dense pack ice in Antarctica, where due to confined space they often have to stick their head out of the water to breath. McMurdo Sound. PHOTO: S. KIM

Humpback Whale—*Megaptera novaeangliae*

(Borowski, 1781)

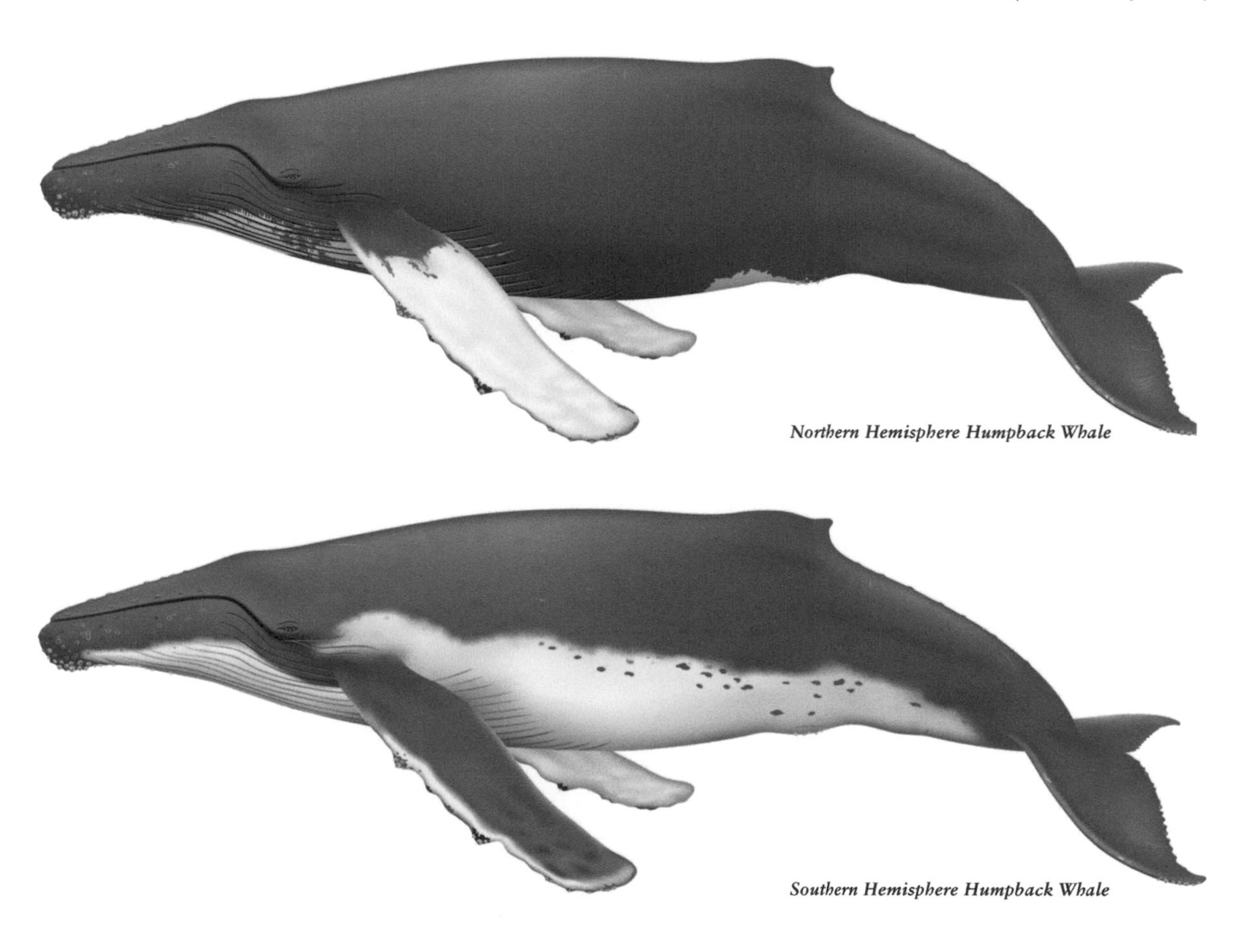

Northern Hemisphere Humpback Whale

Southern Hemisphere Humpback Whale

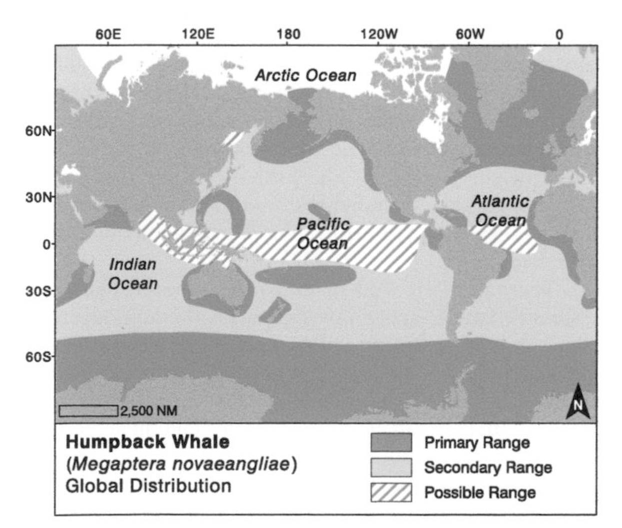

Humpback Whale
(*Megaptera novaeangliae*)
Global Distribution

Recently-used synonyms *Megaptera nodosa*.

Common names En.–humpback whale; Sp.–*rorcual jorobado, ballena jorobada*; Fr.–*baleine à bosse, mégaptère, jubarte*.

Taxonomic information Cetartiodactyla, Cetacea, Mysticeti, Balaenopteridae. Recently, three subspecies have been recognized: *Megaptera novaeangliae novaeangliae* in the North Atlantic, *M. n. kuzira* in the North Pacific, and *M. n. australis* in the Southern Hemisphere.

Species characteristics The humpback whale differs substantially from the general rorqual body plan. The body is more robust; the flippers are extremely long (up to one-third of the body length) with a series of bumps (known as tubercles), including two more prominent ones in consistent positions on the leading edge, which more-or-less divide the margin into thirds. The tail flukes have a deep notch and a serrated trailing edge. The dorsal fin is located approximately 2/3 of the way back from the

The lower jaws of the humpback can expand outward to make an even wider basket for catching prey during lunge-feeding. PHOTO: A. FRIEDLANDER

When hunting schooling fish, groups of humpbacks often coordinate their attacks, increasing individual efficiency to the benefit of all. Southeast Alaska. PHOTO: M. NOLAN

Calves of all large whales are surprisingly small when first born; humpback flippers are up to 5 m long, can weigh up to 1 ton and may be useful for defending the calf from predators. Baja California. M. LYNN, NOAA FISHERIES, SWFSC

The humpback in the foreground rests, belly up; its white flippers have barnacle-encrusted knobs, perhaps making them even more formidable weapons. Eastern Canada. PHOTO: L. BOUVERET, OMMAG

rostrum tip, and is low and broad-based (often sitting on a raised hump of tissue, which is most obvious when whales arch their backs to dive). It is highly variable in shape, ranging from a small nubbin to a prominent falcate fin—researchers use these variations to aid in identifying individual humpbacks. The head has a single low median ridge, which is lined with a series of tubercles. The anterior portion of the head also has several tubercles (each containing a single sensory hair). The tubercles are most prominent near the lips and on the chin.

The body is black or dark gray dorsally and generally has significant amounts of white ventrally (although largely black animals are known). The borderline between dark and light is highly variable and seems to differ by population (the white extends up onto the sides and back in many Southern Hemisphere humpbacks). The flippers are white on the ventral side and vary from all-white to mostly black on the dorsal surface—again this is variable among different populations. The ventral side of the flukes also varies from all black to all-white, with individually-distinctive patterns of white and black. The proportion of black and white on the flukes also varies geographically. This patterning, along with the serrations on the trailing edges of the flukes, are the primary means that biologists use to identify individual humpback whales. A few anomalously-white (possibly albino) individuals have been observed.

There are 270–400 black to olive baleen plates on each side of the upper jaw (although the anterior-most plates may be lighter in color), and 14–35 ventral pleats extending back to the navel or beyond. The blow is often rather low and bushy, reaching only 3 m; however, in large animals it can be quite tall and columnar. It may sometimes appear V-shaped.

Adult humpback whales are 11–17 m long (females are about 1–1.5 m longer than males) and newborns are

The smallish dorsal fin of the humpback whale sits on a step or hump, giving it a very distinct profile; this animal has propeller scars just below the dorsal fin. Southeast Alaska. PHOTO: M. NOLAN

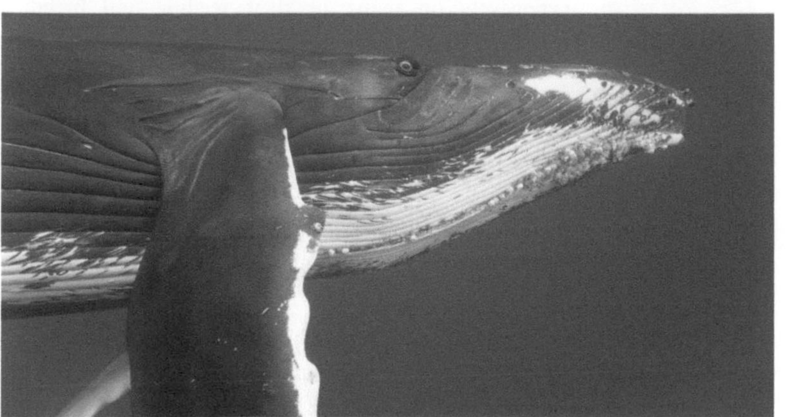

When confronted with something unusual in the environment, all whales, including humpbacks, will often open their eyes wide to get a better look. Hawai'i. PHOTO: M. NOLAN

The undersides of the flukes of humpback whales can range from virtually all-black, to combinations of black and white, to all-white. The sculpted trailing edge of this species is very distinctive. PHOTOS: C. OEDEKOVEN; H. FEARNBACH BOTH NOAA FISHERIES/SWFSC; F. UGARTE; R. L. PITMAN

The blows of humpbacks have more diverse shapes than those of any other large whale species, including an occasional V-shaped blow as shown here. PHOTO: P. OLSON

More so than any of the other large whales, humpbacks can be very demonstrative with their flukes and flippers, sometimes pounding the surface dozens of times for no apparent reason. Southeast Alaska. PHOTO: M. NOLAN

Southern Hemisphere humpbacks often have white bellies and the white often extends far up on their sides, unlike anything seen in the Northern Hemisphere. Ningaloo Reef, Australia. PHOTO: G. CHAN

Humpbacks often raise their flukes, which usually have pointed tips and an irregular trailing edge. Guadeloupe. PHOTO: L. BOUVERET, OMMAG

Cow and calf humpbacks are often accompanied by an "escort," usually an adult male, that is presumably interested in breeding with the female, but will also help defend the calf from predators. Baja California. PHOTO: M. CARWARDINE

A very distinctively-patterned Southern Hemisphere humpback strikes a pose in clear tropical waters. Tonga. PHOTO: R. KIEFNER

about 4.3 m in length and weigh about 680 kg. Weights of at least 40,000 kg are attained by adults.

Recognizable geographic forms Distinct geographic forms of the humpback whale are not widely recognized. However, Southern Hemisphere humpback whale populations generally have more extensive white on the lower part of the body, which may even extend up onto the back. They also typically have more extensive white on the flippers than their Northern Hemisphere counterparts.

Can be confused with At close range, the humpback is very distinctive, and is one of the easiest whales to identify. At a distance, however, there can be some confusion with other large whales, especially right, gray, and sperm whales. This is largely due to the similarity in their bushy blows and the absence of a large dorsal fin. When a closer look is obtained, humpbacks are generally easily distinguished by the combination of head and body shape, dorsal fin shape, and size and shape of the flippers.

Distribution The humpback whale is a cosmopolitan species. The only marine regions where they are clearly absent are in some equatorial regions, a few enclosed seas, and some parts of the high Arctic. Humpbacks do most of their feeding and breeding in coastal waters over continental shelves of all the continents, often near human population centers. This helps make them one of the most familiar of the large whales. They migrate from winter grounds in the tropics (breeding areas) to temperate and polar summer grounds, reaching the ice edges in both hemispheres (feeding areas). Their migrations often take them through oceanic zones, but in general, their migratory routes are not well-known. Humpbacks congregate in certain well-known "grounds" (many of these were first discovered by whalers), but they may also occur at lower densities in other areas outside these regions of concentration. Humpback whales in the Arabian Sea are unusual, in that they appear to remain there year-round. They are rare in the Mediterranean, and the few records from the Persian Gulf are considered extralimital.

Ecology and behavior Humpback whales have been studied extensively on many of their feeding and breeding grounds, and as a result they are one of the best known of all the large whales. Although they generally occur singly or in small groups, larger aggregations develop in feeding and breeding areas. Sometimes humpbacks gather into coordinated groups of up to 20 or more whales, which appear to work together to herd and capture prey (this is relatively common in Southeast Alaska, for instance). The only long-term bonds that are known are those of mothers and calves, although some whales do associate more frequently on feeding grounds. Humpbacks are probably the most acrobatic of all the great whales, sometimes performing full breaches that bring their entire bulk out of the water. Flipper and fluke slapping behavior is also common.

Humpback whale migrations are among the longest known for any mammal species, and can reach 8,000 km one-way. The specific reasons for the migration are often debated by scientists, but it is generally agreed that one reason is to take advantage of both the highly productive summer blooms of high latitudes and the energy-conserving properties of warm tropical areas in winter. Warmer waters also typically have lower densities of killer whales, their main predator. On the breeding grounds, males compete for access to estrous females, apparently using their well-known, complex songs as part of the breeding display. Males also compete physically, and (sometimes violent) male/male aggression is commonly observed on the wintering grounds. Calves are born on wintering grounds in tropical and subtropical regions, and most calves are weaned and independent from their mothers by the time they are one year old.

When diving, humpback whales frequently lift their flukes high out of the water, exposing the lower surface. Individual humpback whales can be identified using photographs of the distinctive markings on the undersides of their flukes. Such photos can be of great help in defining movements and migrations of this and other species.

A largely-black female humpback tends to her newborn calf; the dorsal fin of the calf is still bent over from when it was confined in the uterus. Eastern North Pacific. PHOTO: E. HIGUERA

Humpback whales are heavily parasitized, and the external surfaces of their bodies are often dotted with whale lice and barnacles (the latter especially tend to attach to the tubercles). Killer whales (and occasionally false killer whales) are known predators, and it is not uncommon to see killer whale tooth rake scars on their bodies and appendages. Longevity is not well known, but probably exceeds 50 years.

Feeding and prey Humpback whales have a diverse diet for a baleen whale, feeding largely on krill and a wide variety of small schooling fish (such as herring, sand lance, mackerel, sardines, anchovies, and capelin). They are adaptable lunge-feeders, and in some areas use bubble nets, bubble clouds, tail flicks, and other techniques to help concentrate prey for easier feeding. They are also one of the few species of baleen whales that appear to use cooperative feeding techniques.

Threats and status Humpback whales were hunted by commercial whalers in all major oceans, and many stocks have been seriously depleted by whaling that took place mostly in the 1900s. They were taken both by coastal and oceanic whaling operations, and some have continued to be taken by artisanal whalers in the Caribbean and the South Pacific in the past few decades. About 200,000 were killed in the Southern Hemisphere in the 20th century. The species has been fully protected by the International Whaling Commission since 1965, but the Soviets illegally killed almost 50,000 humpbacks until 1973. Additional threats include entanglement in fishing gear, vessel collisions, disturbance by human-caused noise and traffic, coastal habitat destruction, and climate change effects. Globally, there may be 90,000-100,000 humpback whales, including at least 12,000 in the North Atlantic, at least 20,000 in the North Pacific, over 59,000 in the Southern Hemisphere, and at least several hundred in the northern Indian Ocean. While some stocks are still depleted, most are showing evidence of recovery, in some cases very strongly (such as the North Pacific, which is increasing at 5–7%/year). As a species, the humpback whale clearly is in no immediate danger of extinction.

IUCN status Least Concern (overall species), Endangered (Arabian Sea and Oceania stocks).

Gray Whale—*Eschrichtius robustus*

(Lilljeborg, 1861)

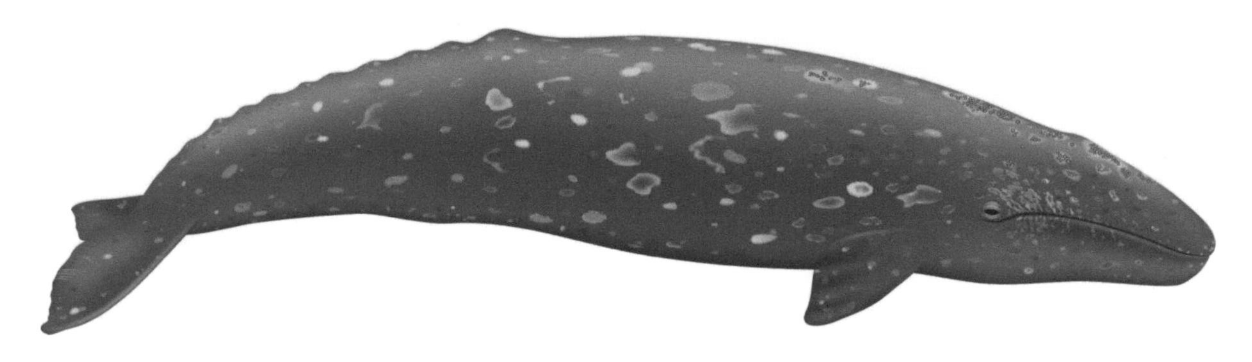

Recently-used synonyms *Eschrichtius gibbosus*, *Rhachianectes (Eschrichtius) glaucus*.

Common names En.–gray whale; Sp.–*ballena grise;* Fr.–*baleine grise*.

Taxonomic information Cetartiodactyla, Cetacea, Mysticeti, Eschrichtiidae.

Species characteristics Gray whales are easy to distinguish from other whale species. They are intermediate in robustness between right whales and rorquals. The upper jaw is moderately arched, and the head is somewhat triangular in top view and slopes downward toward the tip when viewed from the side. The small eyes are located just above the corners of the mouth. The flippers are broad and paddle shaped, with pointed tips. The flukes are wide (>3 m in adults) and deep, and have smooth, slightly-S-shaped trailing edges, with a deep median notch. There is a dorsal hump about two-thirds of the way back from the snout tip, followed by a series of 8–14 smaller "knuckles" on the dorsal ridge of the tail stock.

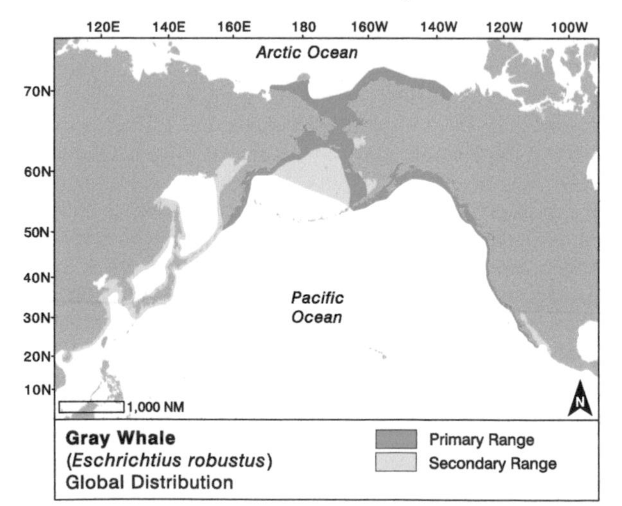

There may be several (generally 2–7) short, but deep, creases on the throat that allow some expansion of the throat during suction feeding.

Although young calves are dark charcoal gray, all other gray whales are brownish-gray to light gray in color. They are nearly covered with light-colored blotches and white to orangish patches consisting of whale lice and barnacles, especially on the head and tail. These patches of ectoparasites are very helpful in distinguishing this species, and gray whales may appear quite light-colored when viewed through the water's surface. The flippers and flukes of gray whales are often scarred by marks from killer whale bites.

The mouth contains 130–180 pairs of short (5–40 cm) yellowish baleen plates, with very coarse bristles. The blow is bushy, heart-shaped when viewed from ahead or behind, and rises less than 3–4 m. In some conditions it can appear a bit more columnar. At birth, gray whales are about 4.6–4.9 m long and weigh about 920 kg; adults are 11–15 m in length, with females slightly larger than males. Maximum body weight is about 45,000 kg.

Recognizable geographic forms None.

Can be confused with Gray whales are unique in body shape (somewhat torpedo-shaped, when viewed from above) and patterning (irregular blotches of white on gray), and there is usually no problem with identification at close range. From a great distance, however, they can sometimes be confused with other whales lacking a prominent dorsal fin (e.g., right, bowhead, sperm, or humpback whales). A closer look at the body shape and coloration will generally allow them to be distinguished.

Distribution Gray whales are today found only in the North Pacific Ocean and adjacent seas. There are two populations: a large eastern Pacific stock and a very small, remnant western North Pacific stock. They are primarily bottom feeders and are thus restricted to shallow continental shelf waters for feeding. In fact,

Gray whales inhabit shallow coastal waters, almost always within sight of land; here one spyhops in a kelp bed off British Columbia, Canada. PHOTO: J. TOWERS

Gray whales have two short throat pleats that allow the throat to expand and create suction to draw in water for feeding. Baja California. PHOTO: M. NOLAN

In addition to the mottled gray/white coloration, gray whales often have yellow/orangish patches from clusters of whale lice and barnacles that are found only on this species. Baja California. PHOTO: M. NOLAN

This gray whale has tooth rake marks on its head and flipper from a killer whale attack; the end of the pectoral fin was bitten off. Killer whale tooth rake marks are common on many large whales. Sakhalin Island, Russia. PHOTO: D. WELLER

Posterior to the dorsal hump, gray whales have a series of "knuckles" on the dorsal ridge that are exposed when the animal rolls over. Baja California. PHOTO: T. STENTON

The blow of the gray whale is conspicuous and normally bushy, but some have V-shaped blows, like this individual. Baja California. PHOTO: T. A. JEFFERSON

Gray whales have short, yellow baleen as seen in this very young calf. Baja California. PHOTO: M. NOLAN

In the breeding lagoons of Baja California, male gray whales often engage in sexual behavior, and show their penises. PHOTO: M. NOLAN

A closer look at the baleen of a gray whale—the frayed inner edges create a mesh to filter prey from the mud and water that the whale sucks in. Baja California. PHOTO: M. CARWARDINE

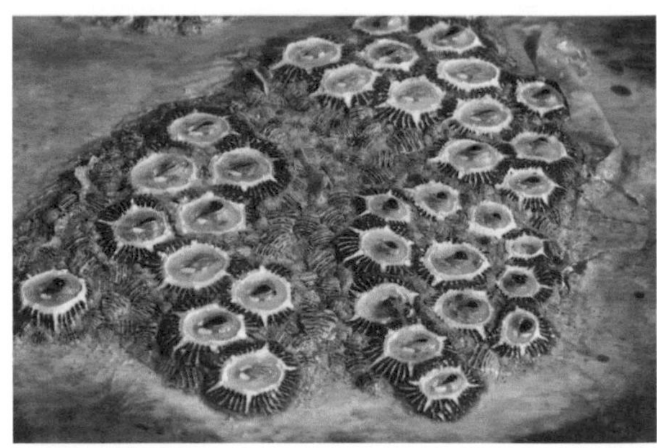

A close-up look at a patch of ecto-commensals on a gray whale's skin: shown are barnacles with whale lice crowded into the spaces in between. Baja California. PHOTO: T. A. JEFFERSON

they are the most coastal of all great whales, living much of their lives within a few tens of kilometers of shore (although they do feed great distances from shore on the shallow flats of the Bering and Chukchi seas). The eastern Pacific stock ranges from the southern Gulf of California, Mexico, to the Beaufort and Chukchi seas in the Arctic. The western Pacific stock ranges from the coast of southern China to the Sea of Okhotsk (however, only a portion of this historic range is currently occupied by the remnant population). Gray whales (referred to as "scrag whales" once occurred in the North Atlantic as well, but were wiped out there by whalers centuries ago. Occasional extralimital strays show up in Atlantic and even Southern Hemisphere waters (one was recently seen off southern Africa!).

Ecology and behavior The behavior and ecology of the gray whale has been relatively well-studied, especially on the breeding lagoons in Baja California, Mexico. Most gray whale groups are small, often with no more than three individuals, but gray whales do sometimes migrate in "pods" of up to 16. Large aggregations are common on the feeding and breeding grounds. Although they are generally slow-moving, this is a fairly aerially-active species. Breaching, spy-hopping, and other aerial behaviors are common, especially during migration, and in and near the breeding lagoons of Baja California. The annual migrations from winter breeding grounds in Mexico to summer feeding grounds in the Bering, Chukchi, and occasionally Beaufort seas is witnessed by tens of thousands of people each year along the west coast of North America. This is one of the longest migrations known for any mammal—some 15,000–20,000 km round trip. During migration, they generally swim at speeds of about 5–9 km/hour and dive for 3–5 minutes at a time. When feeding, gray whales roll on their sides to suck up epibenthic fauna in bottom sediments, and then surface with mud streaming out of the sides of their mouths. The unique feeding method leaves distinctive feeding pits on the shallow sea floor.

Peak calving of eastern Pacific grays occurs in mid-January. Conception occurs in late November to early December, during the southbound migration (before animals have reached the lagoons of Baja California). Sperm competition is thought to be important in the mating system. Apparent mating behavior has been seen throughout the year. The location of the western gray whale's wintering grounds is a mystery, although there is evidence that the southern coast of China was used in the past. Gray whales were known as "devilfish" by American whalers, due to their tendency to ferociously fight back when attacked. This is in contrast to the gentleness (termed "friendly" behavior) that they usually show to modern-day "whale-huggers" who approach and pet them in the wintering lagoons. Gray whales may routinely live to be over 40 years old, and one was aged at 75–80 years.

Gray whales typically feed on the bottom and often create plumes of filtered mud in areas of high prey density. Chukchi Sea, Alaska. PHOTO: L. MORSE

Feeding and prey Gray whales are suckers, using suction to take in food, water, and sediments in shallow waters—then expelling the water and sediment, while trapping the prey on the inside of their baleen plates. They feed primarily on swarming mysids, amphipods, and polychaete tube worms in the northern parts of their range, but are also known to opportunistically take pelagic red crabs, baitfish, and other food (crab larvae, mobile amphipods, herring eggs and larvae, cephalopods, and megalops).

Threats and status Gray whales were one of the first of the great whale species to be targeted by whalers in the Pacific. They are relatively large and slow-moving, and nearly their entire lives are spent in nearshore areas, with easy access from the shoreline. In these respects, they were ideal targets, however gray whales can also be aggressive when attacked (earning them the nickname "devilfish"), and many whalers lost their lives at the hands of a thrashing gray whale. The North Atlantic population was wiped-out by whalers, apparently sometime in the late 1600s or 1700s. Until recently, the western North Pacific (Korean-Okhotsk) stock was thought to have followed the North Atlantic group into extirpation. However, in the late 1900s a small remnant group (about 140 whales) was found to be spending its summers feeding off Sakhalin Island, Russia. It is not known if there are other such groups, but at least the population appears to be surviving. The eastern North Pacific (California-Chukchi) stock was twice reduced to near extinction in the last 150 years. Since its protection after World War II, the "California" gray whale has rebounded to numbers thought to be near pre-exploitation levels. In 1997/98, there were about 27,000 whales, but the numbers have dropped recently to under 20,000, as the population reaches an equilibrium with its environment. Climate change (especially warming of arctic waters) may have severe ramifications for food resource availability. Exploration and extraction of petroleum reserves may also have important implications for the viability of the species, especially in the Okhotsk Sea, where a large effort is underway to extract oil from an area within the only known feeding area of the western gray whale.

In the breeding lagoons of Baja California, gray whale mothers seem to encourage their calves to interact with delighted boatloads of whale-watchers—the advantage to the whales is not so clear. PHOTO: M. NOLAN

IUCN status Least Concern (overall species), Critically Endangered (western Pacific stock).

Sperm Whale—*Physeter macrocephalus*

Linnaeus, 1758

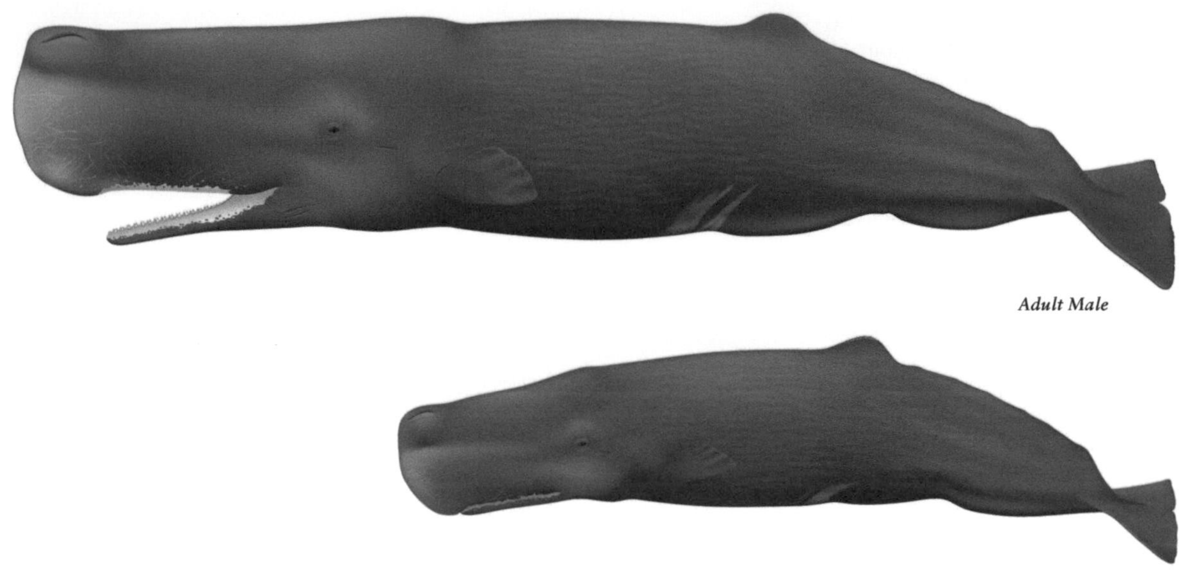

Adult Male

Adult Female

Recently-used synonyms *Physeter catodon*.

Common names En.–sperm whale; Sp.–*cachalote*; Fr. –*cachalot*.

Taxonomic information Cetartiodactyla, Cetacea, Odontoceti, Physeteridae. Great controversy existed in past years about whether *Physeter macrocephalus* or *P. catodon* is the correct scientific name for the species. The former won-out, and this name is now almost universally used. There was also some disagreement about whether the sperm whale is most closely related to other toothed whales or to baleen whales. Genetic data supporting a relationship with baleen whales were misinterpreted, and sperm whales are now considered to be perfectly-good odontocetes.

Species characteristics As the largest toothed cetacean, the sperm whale is unlikely to be confused with any other species. The body is somewhat laterally compressed and the head is huge (one-quarter to one-third of the total length, and an even greater proportion of the total bulk) and squarish, when viewed from the side. The lower jaw is much narrower than the upper, and is underslung, making it barely visible from the side. There are 2–10 short, deep throat grooves. The single S-shaped blowhole is set at the front of the head and is offset to the left. The flippers are short, but wide and spatulate. The flukes are broad and triangular with a nearly straight trailing edge (often with various nicks and notches), rounded tips, and a deep notch. The caudal peduncle is very deep, and may have a post-anal keel. There is a thick, low, rounded dorsal fin and a series of bumps, or crenulations, on the dorsal ridge of the tailstock. The body surface tends to be wrinkled behind the head and on the sides. Many (about 75%) adult females and some young males (about 30%) have white to yellowish calluses on the dorsal hump, whereas bulls almost never have them.

Sperm whales are predominantly black to brownish-gray, with variable white areas around the mouth and often on the belly. The white areas are of highly variable extent and pattern. White scratches and scars are common on the bodies (especially the heads) of some large adults. These are presumably caused mostly by other sperm whales and the beaks of their cephalopod prey. Anomalously-white individuals have been observed (in at least one case, this was an albino).

Functional teeth (18–26 pairs that fit into sockets in the upper jaw) are present in the lower jaw only. The teeth are large and may range from sharply pointed in young animals to rounded stumps in old individuals. The bushy blow projects up to 5 m and, because of the position of the blowhole, is directed forward and to the left. On windless days, such an angled blow originating from the most forward part of the head is diagnostic.

Newborn sperm whales are 3.5–4.5 m long and weigh about 1,000 kg. Adult females are up to 12.5 m and adult males are up to 19.2 m in length. Weights of up to 57,000 kg have been recorded. Sperm whales are

the most sexually dimorphic of all cetacean species, with males weighing nearly three times as much as females, and sometimes much more.
Calf—Length <½ of adult length, color pattern somewhat faint, mouthline relatively short, flukes very broad (anterior-posterior).
Subadult—Medium-sized (slightly smaller than adult females), dorsal fin callus may be present in some individuals.
Adult female—Relatively large size, pale dorsal fin callus usually present (although sometimes difficult to see), dorsal fin more anterior than in adult males.
Adult male—Very large body size (up to about 50% larger than females), often with extensive scarring on head, head proportionately large (up to ⅓ body length), upper jaw tends to overhang tip of lower jaw, dorsal fin more posterior than in adult females, callus rarely present on dorsal fin, trailing edges of flukes somewhat convex.

A young sperm whale, probably left alone at the surface while the adults forage far below. At least a dozen remoras (suckerfish) are attached to it; note also that the pectoral fin presses flat against the body, perhaps to prevent predators such as killer whales from grabbing onto it. Ogasawara Islands, Japan. PHOTO: H. MINAKUCHI

Recognizable geographic forms None.
Can be confused with Sperm whales are generally easy to distinguish from other large whales at sea, even at a great distance. The uniquely-angled blow is diagnostic, but one must be careful to take into account the effects of wind on a whale's blow. Only humpbacks, and possibly gray whales, would be likely to be confused with sperm whales, and only at a great distance. Once a closer look is obtained, the unique body shape, and exhalation patterns of sperm whales will make them distinguishable.
Distribution Perhaps only killer whales and humans are more widely distributed than the sperm whale. Sperm whales are somewhat migratory and are distributed in a cosmopolitan fashion from the tropics to the pack ice

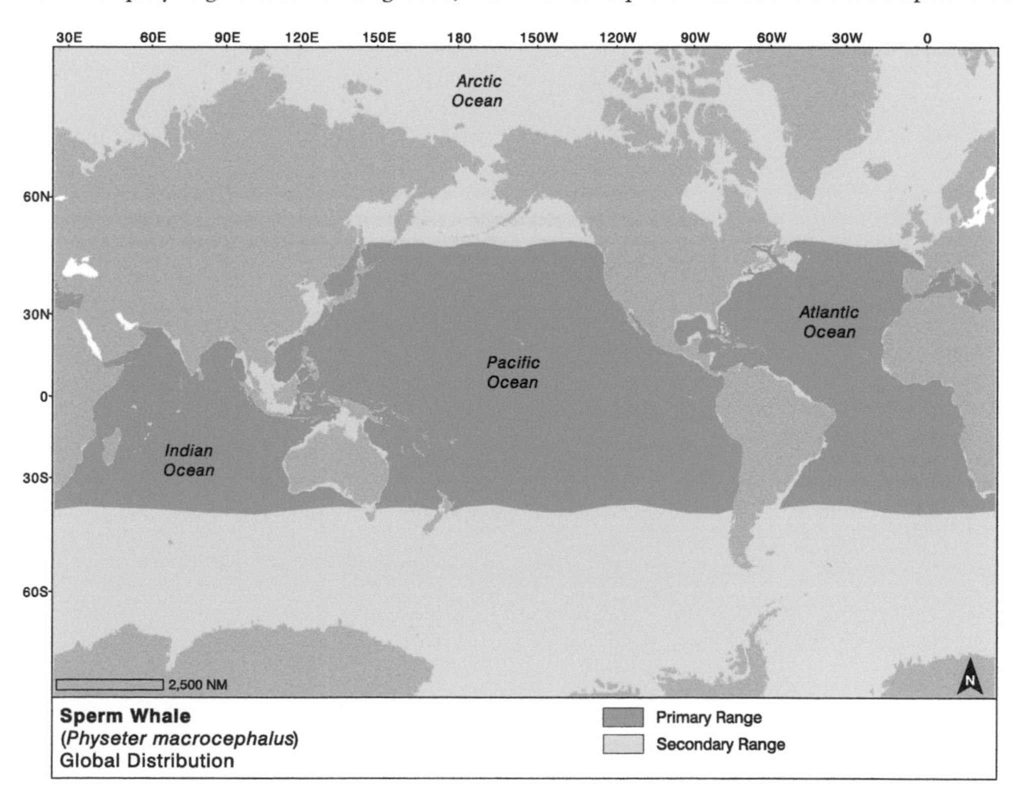

Sperm Whale
(Physeter macrocephalus)
Global Distribution

Sperm whales have teeth only in the lower jaw and in the upper jaw have just sockets that the lower teeth fit into; the mouth has a white border. Gulf of California. PHOTO: M. NOLAN

A young sperm whale raises up to take a look at the photographer—note the underslung jaw, wrinkled skin posterior to the head, and the square profile. Gulf of California. PHOTO: M. NOLAN

A sperm whale rolls away from the viewer and prepares to fluke up; in addition to the heavily wrinkled back there is a series of "knuckles" on the dorsal ridge behind the dorsal hump. Gulf of California. PHOTO: M. NOLAN

The blowhole of the sperm whale is positioned all the way forward and on the left side; the bump in the middle of the back is the rear edge of the enormous skull. The prominent dorsal hump shown here may indicate an adult male. Gulf of California. PHOTO: M. NOLAN

A pod of sperm whales resting at the surface; these are usually nursery herds comprised of females and young, but roving males sometimes join temporarily to look for receptive females. Gulf of California. PHOTO: M. NOLAN

A lone resting sperm whale—identified by its low, triangular dorsal hump on the left, and a low, puffy, forward-canted blow. Gulf of California. PHOTO: M. NOLAN

A large sperm whale lays sideways at the surface mouthing a prey item; note the square head and conspicuously-white, underslung lower jaw. PHOTO: M. CARWARDINE

Sperm whales fluke on nearly every deep dive, often raising the tail high and often vertically out of the water. Guadeloupe. PHOTO: L. BOUVERET, OMMAG

Sperm whales can be individually identified with photographs, based on the irregularities along the trailing margin of the flukes; tooth rake marks shown here indicate that killer whales may be responsible for much of that variation. Guadeloupe. PHOTO: D. MOUSSA

A researcher sits on a bowsprit and prepares to take a skin biopsy sample from a sperm whale using a crossbow. This type of sampling underpins much of the current scientific and conservation research done on cetaceans. Gulf of Mexico. PHOTO: V. BEAVER

The broad, almost-triangular flukes of the sperm whale with rounded tips and straight trailing edges are immediately distinguishable from those of all other large whale species. Gulf of California. PHOTO: M. NOLAN

edges in both hemispheres, although generally only large males venture to the extreme northern and southern portions of the range (poleward of about 40–50° latitude). Deep divers, sperm whales tend to inhabit continental slope and oceanic waters deeper than about 1,000 m, but some (especially adult males) do come closer to shore, especially where submarine canyons or other physical features bring deep water near the coast. They may even occasionally be seen over the continental shelf in specific areas (such as in the Gulf of California and off Long Island, New York). Their migrations are not as clear-cut as they are in most baleen whales, and in some tropical areas sperm whales appear to be largely resident. The distribution includes semi-enclosed seas with deep entrances, like the Gulf of Mexico, Caribbean Sea, Gulf of California, Sea of Japan, and Mediterranean Sea. However, it excludes seas with shallow, narrow entrances, like the Black Sea and Persian Gulf (there are also some records from the Red Sea, although probably extralimital). Sperm whales occur in higher densities in certain areas of high productivity, often near steep drop-offs and areas with strong currents. Many of the areas where sperm whales concentrate were discovered by Yankee whalers and are thus called the "grounds."

Sperm whales are intensely social and tactile animals and are often in direct physical contact with other group members. Azores. PHOTO: W. OSBORN

Ecology and behavior Although bulls are sometimes seen singly (especially above 40° latitude), sperm whales are more often found in medium to large groups of 20–30, consisting of females and their dependent young, but can occur in groups of up to 50 or more. In the past 20 years or so, the social system of sperm whales has been relatively well-studied. They are polygynous; adult males seem to employ a "roving" strategy for mating, associating with nursery groups of the much more social adult females and their offspring for only short periods of time. The female groups are comprised of units characterized by long-term stability. Sperm whales range widely, and adult males may move across and even between ocean basins. Although fights between males have seldom been observed, the evidence indicates that males compete for access to receptive females. Sexually-mature, but non-breeding, males that have left their natal groups may form bachelor herds. Most births occur in summer and fall, but reproductive rates are very low. Whales of this species can live to be at least 70 years old, and possibly much older. Sperm whales are sometimes attacked by killer whales (and very occasionally other odontocetes), but these rarely appear to be fatal. When attacked by killer whales or other predators, sperm whales may gather into a circle, with heads pointing inwards, and calves protected in the middle. This is known as a "marguerite" formation.

The two principal activities are foraging (in which whales gather in small clusters, fluking-up and diving consistently) and socializing (in which they gather into larger surface-active groups). Sperm whales may raft at the surface between bouts of diving. They are extremely deep and long divers, apparently capable of reaching depths of 3,200 m or more for well over two hours. Most commonly, during foraging they dive to about 400 m and for 30–45 minutes. Groups of females may spread out over distances of more than 1 km when foraging, but adult males tend to be solitary when foraging. Some dives of bulls, which are longer than those of the smaller cows, last as long as two hours. Fluking-up is common before a long dive, and rafting (whales lying nearly motionless at the surface) is common afterwards. The most common aerial behaviors are breaching and fluke-slapping.

Low-frequency, stereotyped, clicked vocalizations, some of which are termed "codas," are apparently distinct to groups of sperm whales and may act as acoustic signatures. Clicks are also probably used in echolocation, to help them locate prey. In addition, sperm whales also produce occasional "squeals" and other atypical sounds. Some scientists studying sperm whales argue quite convincingly that their complex social organization and behavior patterns fit the definition of "culture."

Feeding and prey An amazing variety of cephalopods and other invertebrates, deep-sea fish, and non-food items have been found in the stomachs of sperm whales from around the world. Cephalopods (squid and octopuses), however, are considered to be the major prey items. Primary prey species include squids of the genera *Architeuthis* (the giant squid), *Moroteuthis*, *Gonatopsis*, *Histioteuthis*, and *Galiteuthis*, as well as fishes like lumpsuckers and redfishes. Like all odontocetes, they seize individual prey items. Males eat larger prey than females and young. In some areas, sperm whales take fish from longlines.

Threats and status Sperm whales were the primary targets of Yankee whalers, based on the northeast coast of the US from the mid-18th to the mid-19th centuries. Their primary goal was to obtain oil (extracted from boiled down blubber and from the inside of the head), which was a prime commodity at the time. In addition to oil, there was the even more valuable ambergris, prized for its unique scent and properties as a fixative of aromatics. Later, a second major phase of exploitation occurred, and

A young (i.e., relatively unmarked) and curious sperm whale comes over to inspect a diver; notice the white lips. Azores. PHOTO: W. OSBORN

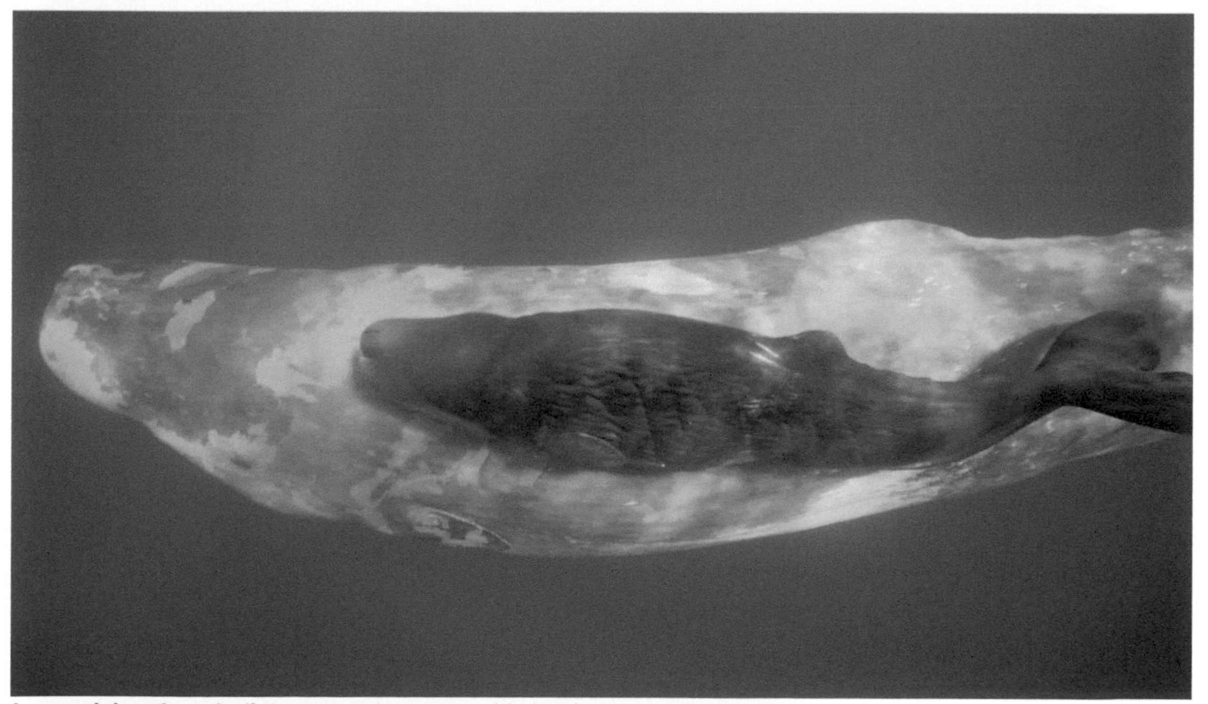

A sperm whale mother and calf; the corrugated appearance of the body is characteristic of all age and sex classes; the flippers lie flush to the body—perhaps as protection from predators. North Atlantic. PHOTO: D. PERRINE

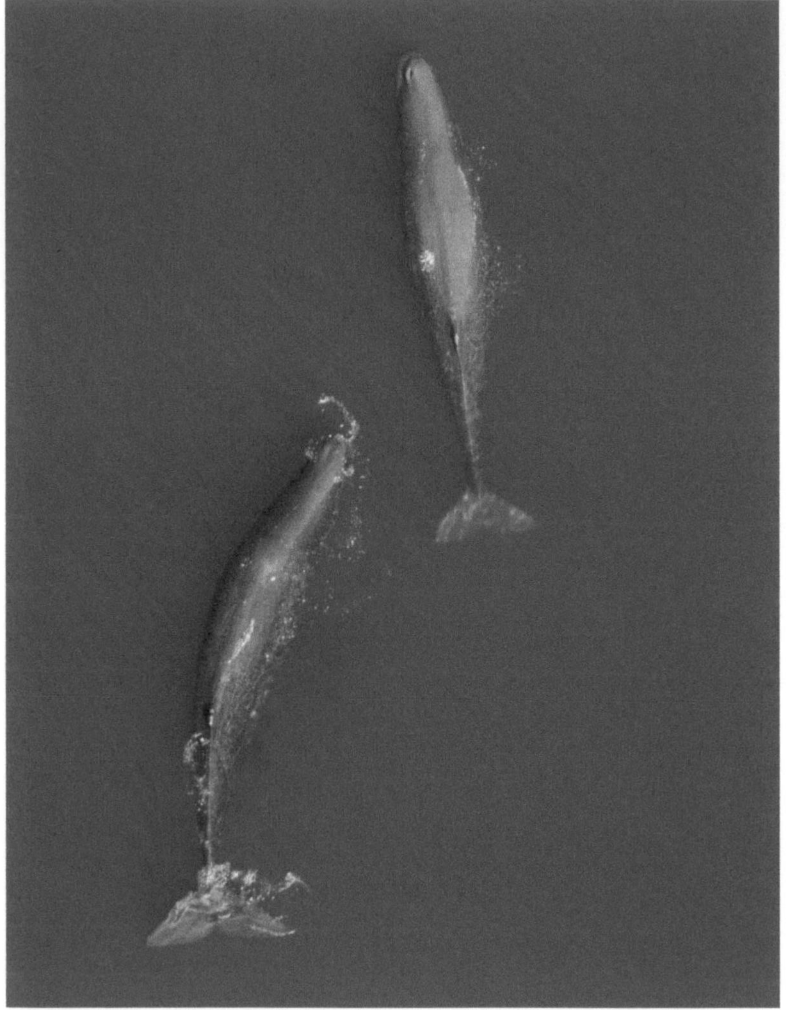
From the air, the blunt head, wrinkled back and sides, and triangular flukes of the sperm whale are unmistakable. Gulf of California. PHOTO: M. CARWARDINE

sperm whales were taken by pelagic and land-based whaling operations in many parts of the world, mostly from 1945 to 1980. This dramatically decreased over recent decades, although sperm whales are still occasionally killed, primarily in native, non-commercial hunts. The sperm whale has not been depleted to the level of some of the baleen whales (it is estimated that current abundance is about 1/3 of original levels), nevertheless some populations were severely reduced. In 1982, the International Whaling Commission banned the hunting of sperm whales (though it was not effective until 1988), but some may still be taken (generally in small numbers) by artisanal whalers in the Caribbean and in some of the eastern islands of Indonesia. In addition, Japan has recently hunted some under its so-called "scientific" whaling program. While sperm whaling today is but a small remnant of what it once was, the effects of whaling appear to be continuing. Especially in the southeast Pacific, populations are not showing strong signs of recovery. In other areas, sperm whales are relatively numerous, and these animals are very commonly seen in many tropical zones. There are at least 14,000 in the North Atlantic, 80,000 in the North Pacific (including at least 4,000 in the eastern tropical Pacific), and 9,500 in the Antarctic. Global abundance is not known, but is broadly estimated to be about 360,000, making it one of the most abundant of all the great whales. Today the threat from commercial whaling has largely been replaced by concerns about impacts of vessel strikes, fishing gear interactions, human-induced noise, chemical pollution and climate change.

IUCN status Vulnerable.

Pygmy Sperm Whale—*Kogia breviceps*

(Blainville, 1838)

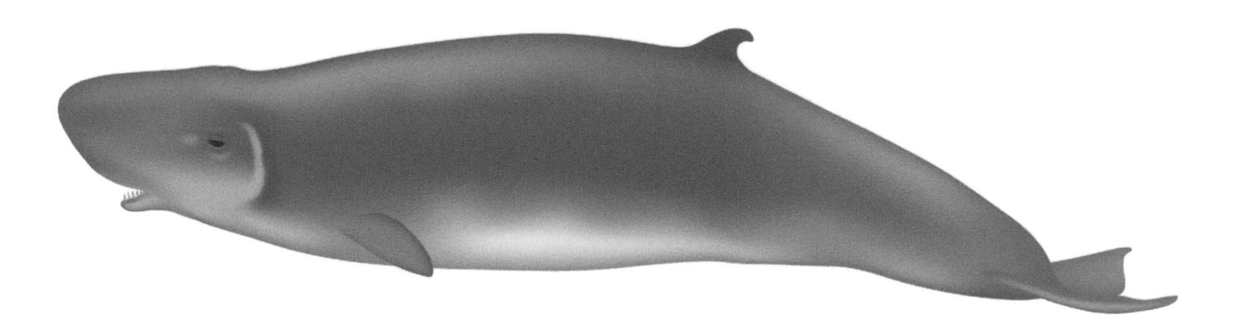

Recently-used synonyms None.
Common names En.–pygmy sperm whale; Sp.–*cachalote pigmeo*; Fr.–*cachalot pygmée.*
Taxonomic information Cetartiodactyla, Cetacea, Odontoceti, Kogiidae. Only one species *(K. breviceps)* was recognized in this genus until 1966, when studies clearly showed the pygmy and dwarf sperm whales to be distinct species.
Species characteristics Pygmy sperm whales have a very unusual body shape, which in some ways resembles that of a small sperm whale. They are quite robust, and are not as streamlined as most other odontocetes. Pygmy sperm whales have a shark-like head with a narrow underslung lower jaw. The head becomes more squarish in older individuals. The small flippers are set far forward on the sides near the head. The small dorsal fin (<5% of the body length) is usually set well behind the midpoint of the back. The dorsal fin is typically strongly falcate, with the tip well below the highest point. There is often the appearance of a hump on the back (between the blowhole and dorsal fin) in this species, especially when the animals are rafting at the surface. In pygmy sperm whales, the blowhole is positioned >10% of the way back from the snout tip. The ratio of the height of the dorsal fin, divided by the distance from the tip of the forehead to the anterior insertion of the dorsal fin is usually <8%.

Pygmy sperm whales are countershaded, ranging from dark brownish gray or black on the back to white below. Usually a somewhat darker patch encircles the eye. Often the belly has a pinkish tone. There is a light colored bracket mark, dubbed the "false gill," along each side of the head between the eye and the flipper. This is thought to be an adaptation related to mimicry of their shark predators.

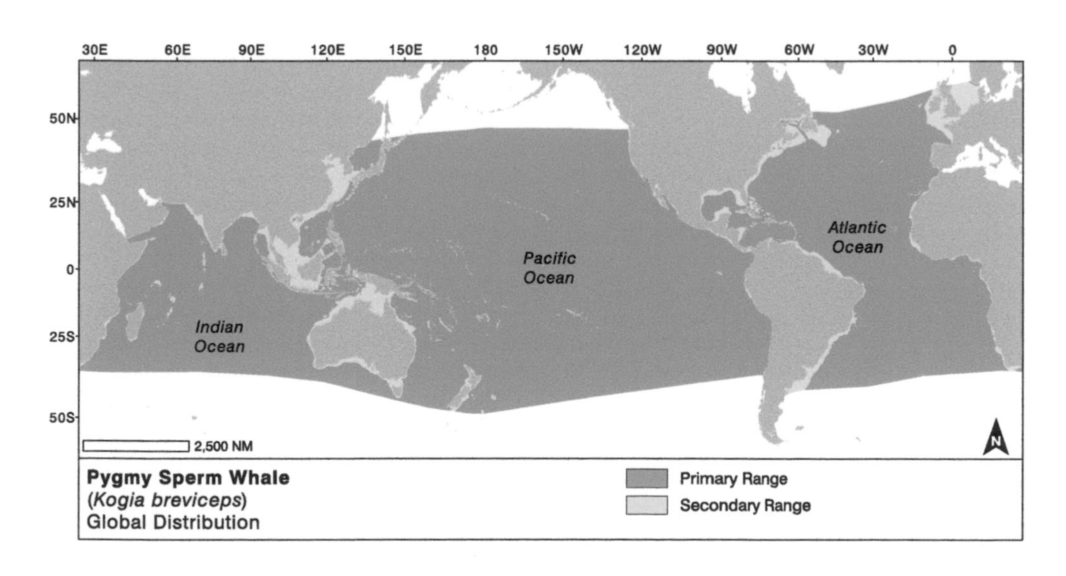

A captive pygmy sperm whale in Japan shows the squarish, slightly-overhanging head and the "false gill" pigmentation feature just in front of the flipper. The very small dorsal fin is indicative of *breviceps* versus *sima*. PHOTO: H. MINAKUCHI

The lower jaw contains 12–16 (sometimes 10 or 11) pairs of long, sharp, fang-like teeth that fit into sockets in the upper jaw. There are usually no teeth in the upper jaw. Adult pygmy sperm whales are 2.7–3.8 m long. Females grow somewhat larger than males. Adults may weigh as much as 450 kg. Newborns are about 1.2 m and weigh around 53 kg.

Recognizable geographic forms None.

Can be confused with Pygmy and dwarf sperm whales can be rather difficult to distinguish at sea. Pygmy sperm whales grow to somewhat greater total lengths, and have smaller, more rounded dorsal fins, generally set farther back on the body. The position of the blowhole is another good indicator (>10% of the way back from the snout tip in pygmy sperm whales). Also, look for the hump on the back that is often present in this species (but not in the dwarf sperm whale). There is some degree of overlap in most characteristics of these two species, and identifications must be made cautiously, even with a specimen "in hand." It is probably best to leave confirmed identifications to experts who have a great deal of experience with both species in the genus. For stranded animals that are not in fresh condition, genetic or biochemical analyses may be needed to distinguish the two species of *Kogia*.

Distribution Pygmy sperm whales are known from deep waters (outer continental shelf and beyond) in tropical to warm temperate zones of all oceans. They appear to be more common over and near the continental slope, although they also occur in very deep oceanic regions. This species appears to prefer somewhat more temperate waters than does the dwarf sperm whale. The frequency with which they strand in some areas (such as the southeastern United States [especially Florida] and South Africa) suggests that they may not always be as uncommon as sightings would suggest.

Ecology and behavior Pygmy sperm whales are very difficult to detect, except in extremely calm seas. Most sightings are of small groups of less than five or six individuals. Almost nothing is known of the behavior and ecology of this species, other than what has been learned from brief sightings during research cruises. They are generally not commonly seen alive at sea, but they are among the most frequently-stranded small whales in some areas. This may be due to the fact that they are easily missed

The dorsal fin of the pygmy sperm whale (shown here) is low and rounded compared to that of the dwarf sperm whale, which has a larger, more erect, and pointier fin. With some experience they are usually readily separable at sea. Delaware. PHOTO: M. BARAN

A pygmy sperm whale mother and calf lie motionless at the surface off Baja California. This logging behavior is typical of both species of *Kogia*; the rounded arch in the back and the low, rounded dorsal fin distinguish this species. PHOTO: M. W. NEWCOMER

The shark-like appearance of *Kogia* has led some to suggest that they could be shark mimics; features include an underslung jaw with shark-like teeth, tapered head, and "false gill" pattern; all evident in this fresh-stranded pygmy sperm whale. New South Wales, Australia. PHOTO: S. MOORE

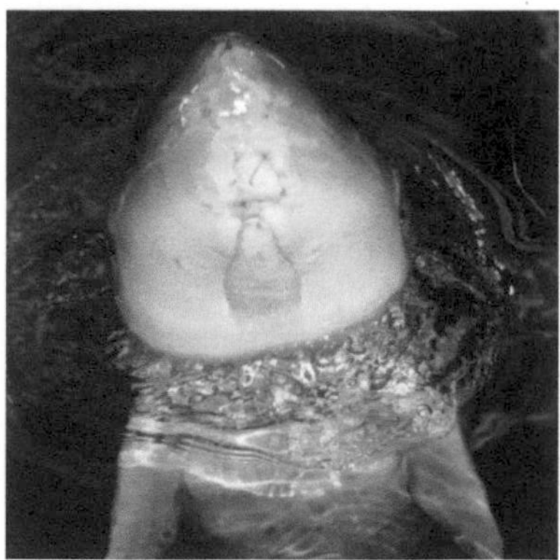

An underside view of a captive pygmy sperm whale showing its tapered head and the outline of its tiny underslung jaw. PHOTO: J.-P. SYLVESTRE

Pygmy sperm whales are not often seen at sea, so much of what we know about the natural history of these animals comes from strandings. PHOTO: C. WEIR

at sea, and are rarely seen in anything except calm seas. When seen at sea, they usually appear slow and sluggish, and often raft motionless at the surface with no visible blow. They may simply sink out of sight, or may roll to dive (especially if startled). However, captive animals have sometimes shown bursts of activity. Pygmy sperm whales may also emit a reddish-brown fluid from the anal area when startled. They do not show their flukes on diving. They are at least occasionally preyed upon by sharks and killer whales.

The vast majority of information available on this species comes from strandings, of both dead and living individuals. Very little is known of the reproductive biology of the pygmy sperm whale. Sexual maturity in South Africa appears to be reached at ages between 2.5 and 5 years. Births mostly occur from March to August, and females may reproduce annually. Some evidence suggests that sperm competition may be important in male reproductive success. From studies in South Africa, it appears that these are not long-lived animals, and maximum known longevity is only 23 years. Recent genetic studies suggest that there is some gene flow between the Atlantic and Indian oceans.

Feeding and prey Studies of feeding habits, based on stomach contents of stranded animals, suggest that this species feeds in deep water, primarily on cephalopods and, less often, on deep-sea fishes and shrimps. In South Africa, they take at least 67 different prey species, and appear to feed in deeper waters than do dwarf sperm whales.

Threats and status Although they have not been known to be taken in large numbers, either directly or incidentally in fisheries, pygmy sperm whales have been occasional victims of dolphin and small whale fisheries, as well as gillnet and purse seine operations. Other potential threats include plastic debris ingestion (and associated gut-blockage) and ship strikes. There are no estimates of global abundance, but there are thought to be at least 3,000 of them off California.

IUCN status Data Deficient.

Dwarf Sperm Whale—*Kogia sima*

(Owen, 1866)

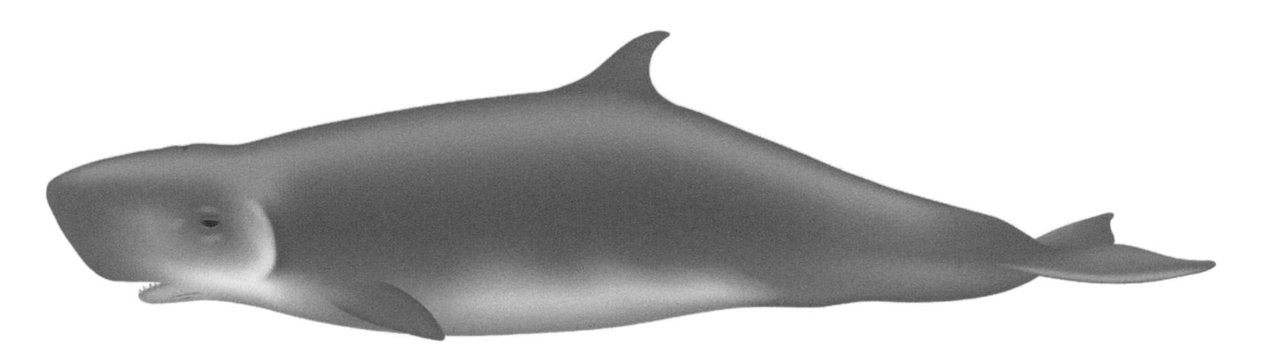

Recently-used synonyms *Kogia simus.*
Common names En.–dwarf sperm whale; Sp.–*cachalote enano*; Fr.–*cachalot nain.*
Taxonomic information Cetartiodactyla, Cetacea, Odontoceti, Kogiidae. Because this species was not generally recognized until the mid-1960s, there is still some confusion about which species is represented in the older literature. The specific name, *simus*, was changed to *sima* to reflect proper nomenclatural gender. A recent study of molecular genetic variation suggests that there may be two separate species of dwarf sperm whales, one in the Atlantic and one in the Indo-Pacific.
Species characteristics The dwarf sperm whale is similar in appearance to the pygmy sperm whale, with a robust body that tapers rapidly behind the dorsal fin. It has a rather triangular or squarish (more so in older animals) head profile and a narrow, underslung lower jaw. The head shape and light-colored "false gill slit" give the animal a somewhat shark-like appearance, like its congener (and this may be an example of adaptive mimicry). However, it has a larger dorsal fin (>5% of the body length), with the tip usually at the highest point. The dorsal fin is generally set near the middle of the back. The ratio of the height of the dorsal fin, divided by the distance from the tip of the forehead to the anterior insertion of the dorsal fin is usually >11%. The flippers are small, with somewhat blunt tips, and are positioned near the head. The dwarf sperm whale's blowhole is also positioned further forward (generally <10% of the way back from the snout tip). Generally, a pair of short grooves, similar to those in beaked whales, is present on the throat.

The overall appearance of this species is more dolphin-like than in the pygmy sperm whale. Like its congener, the dwarf sperm whale is countershaded, with a

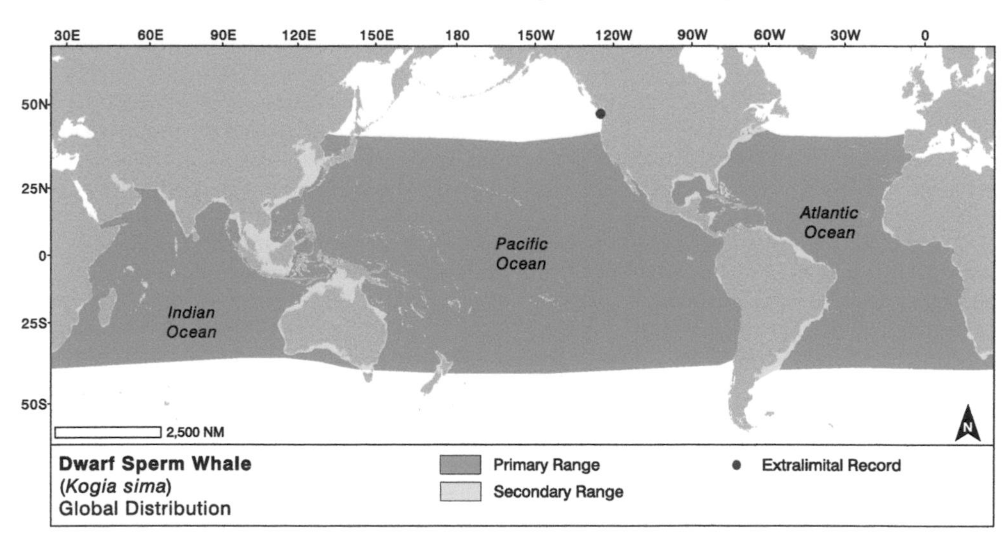

A live-stranded dwarf sperm whale showing the blowhole location to the left of center. North Carolina. PHOTO: K. RITTMASTER

A cow and calf dwarf sperm whale—this female has a slight bulge in the back, but it is not as pronounced as in the pygmy sperm whale. Gulf of California. PHOTO: C. HOWELL

A rare view of an approaching dwarf sperm whale showing its tapered head. Baja California. PHOTO: T. STENTON

A dwarf sperm whale breaches off West Africa—a very unusual sight for this normally cryptic species. PHOTO: M. ALDEN

Dwarf sperm whales are easily overlooked in anything but calm waters. Philippines. PHOTO: E. SABATER

brownish-gray (dorsal) to white (ventral) coloration. There is also a white bracket marking shaped like a shark's gill slit on the side of its head. A ring of darker color surrounds each eye. The belly often has a pinkish tinge. There are 7–12 (rarely up to 13) pairs of teeth in the lower jaw; sometimes up to three pairs of teeth are present in the upper jaw as well. The teeth are extremely sharp and fang-like. There is generally no visible blow.

Adults of this species are up to 2.7 m long (although at least in the northwestern North Atlantic they generally do not exceed 2.5 m) and may weigh up to 272 kg. Males may be slightly larger than females. Length and weight at birth are about 1 m and 14 kg.

Recognizable geographic forms None.

Can be confused with Dwarf sperm whales are most likely to be confused with pygmy sperm whales, which are very similar in appearance. Besides reaching smaller maximum lengths, dwarf sperm whales have taller, more dolphin-like dorsal fins, usually set more toward the middle of the back. Also, the position of the blowhole can help (>10% of the way back from the snout tip in pygmy sperm whales, and <10% in dwarf sperm whales). However, because sizes overlap and dorsal fins are variable in size and position, many at-sea sightings of *Kogia* whales may not be identifiable to species. Generally, it is difficult for non-experts to reliably distinguish the two species, especially in sightings as sea. For stranded specimens, genetic or biochemical means may be needed to confirm identifications.

Distribution The dwarf sperm whale, like the pygmy sperm whale, is known mostly from strandings. It is generally not a commonly-seen species at sea, although this may have more to do with its cryptic appearance than actual rarity. It appears to be distributed widely in tropical to warm-temperate zones, apparently largely offshore. Its distribution shows somewhat more of a preference for warmer waters than that of the pygmy sperm whale, and this species probably does not range as far into high-latitude waters. There is no evidence of migrations. There are records for some enclosed seas, such as the Persian Gulf. A single record exists for the Mediterranean, where the species is considered extralimital.

Ecology and behavior There is very little known of the ecology of this species. Much of what is known comes from records of dead- and live-strandings (one animal survived well over a year in rehabilitation). Group sizes tend to be small, most often less than about six individuals (although groups of up to 16 have been recorded). This species (like the pygmy sperm whale) is also typically shy and undemonstrative when observed at sea. They don't generally lift their flukes when they dive. Aerial behavior

***Kogia* spp. mostly remain motionless at the surface; they are rarely seen swimming and are easily overlooked in anything but flat waters; the short, straight back and relatively large, fairly erect dorsal fin of *K. sima* distinguishes it from *K. breviceps*.**
PHOTOS: (TOP LEFT TO RIGHT) BAJA CALIFORNIA, T. STENTON; GULF OF CALIFORNIA, M. NOLAN; NORTH CAROLINA, K. RITTMASTER; BAJA CALIFORNIA, C. HOWELL

is very rarely seen, although they do sometimes breach. They often float motionless at the surface, and may be mistaken for a piece of driftwood or flotsam at a distance. They tend to be difficult to approach, and may sink slowly below the surface or arch their backs to begin a dive. When startled, dwarf sperm whales may leave a large rust-colored cloud of fecal material behind as they dive.

In at least one area where detailed studies have been done (South Africa), there appears to be a calving peak in summer (December to March). Females may give birth annually, after a gestation period of about one year. Longevity does not appear to be long; the oldest known-age individual was just 22 years old. Sexual maturity is apparently attained at 2.5–5.0 years of age. Rather than direct male/male aggression, males may compete largely by sperm competition, although this has not been observed.

Feeding and prey Dwarf sperm whales appear to feed primarily on deep-water cephalopods, but also take other prey types. About 38 different prey species are known from South African waters, where this species is thought to feed in shallower water than does *K. breviceps*. They may also do some of their foraging near the sea bottom.

Threats and status Although never hunted commercially, these animals were sometimes harpooned by 19th century whalers. Dwarf sperm whales are sometimes killed in directed fisheries in the Caribbean and Indo-Pacific regions, and a few are known to have died incidentally in fisheries throughout their range. Dwarf sperm whales have a habit of eating plastic discarded or lost at sea, and their tendency to lay quietly at the surface may make them vulnerable to vessel strikes. In general, there are not known to be any serious human impacts, and populations are probably relatively less affected by human activities than are those of most other cetaceans. No estimates of total abundance exist, but about 11,000 occur in the eastern tropical Pacific, and a few hundred exist in the northern Gulf of Mexico.

IUCN status Data Deficient.

Baird's Beaked Whale—*Berardius bairdii*

Stejneger, 1883

Recently-used synonyms None.

Common names En.–Baird's beaked whale; Sp.–*zifio de Baird, ballena picuda de Baird*; Fr.–*baleine à bec de Baird, bérardien de Baird.*

Taxonomic information Cetartiodactyla, Cetacea, Odontoceti, Ziphiidae. There has been some suggestion that Baird's and Arnoux's beaked whales may actually be the same species, but this does not appear to be the case. Off Hokkaido, Japan, two geographic forms of Baird's beaked whale have recently been identified, and though their taxonomic status is currently unresolved, there is some evidence suggesting they may represent separate species.

Species characteristics Baird's beaked whales are the largest whales in the ziphiid family. They are slender (maximum girth is only 50–60% of total length), with a small head. They have a long, well-defined, tube-like beak and a sloping, rounded forehead. When viewed from above, the beak protrudes forward from the forehead. The body, however, is relatively more slender than that of the otherwise-similar bottlenose whales. There is often a visible depression that runs the length of the back between the blowhole and the dorsal fin. The small, but prominent, triangular dorsal fin is located about two-thirds of the way along the back and is generally rounded on top, especially on older animals. The tip is often curled under. The small, blunt-tipped flippers fit into depressions along the sides. There is the usual V-shaped pair of throat grooves characteristic of beaked whales, and some animals have accessory grooves. Though some animals have a median notch on the flukes, most have no notch, and some a slight convexity. The ends of the crescent-shaped blowhole in *Berardius* point backward (therefore the "hinge" is at back), instead of forward as in all the other beaked whale genera.

There are two pairs of visible teeth near the tip of the lower jaw, which erupt at about the time of sexual maturity in both sexes (rather unusual for beaked whales). The larger, forward pair of teeth in adults is visible at the tip of the protruding lower jaw, even when the mouth is closed. In older animals, the tips of the teeth may be rounded from heavy wear and on some individuals the teeth are heavily infested with barnacles.

Baird's beaked whales are dark brownish-gray to black, with a slightly lighter belly. Larger animals are usually heavily scarred with pale, linear, often parallel scratches (tooth rake marks) or splotches on the back and, often, on the undersides. Many of the scratches appear to be made by conspecifics, and may be quite extensive on adult males. On some animals, the forehead appears largely white. Whale lice and diatoms may be found on the skin surface and appendages, the latter giving the body a greenish-brown tinge.

Baird's beaked whales reach lengths of 10.7 m (males) and 11.1 m (females), and weigh up to 12,000 kg. There is less sexual size dimorphism in this species than in any other in the family Ziphiidae, with females only slightly larger than males. They are about 4.6 m long at birth. The conspicuous blow is low and bushy, and is often given in rapid succession.

Recognizable geographic forms Recently, morphological and molecular genetic evidence has been used to distinguish two forms, or possibly species, of Baird's beaked whales in the western North Pacific.

Slate-gray form—slate gray coloration, adults 10+ m long; more southerly distribution and probably more widespread.

Black form—black coloration, adults <7 m long; northern Hokkaido and Sea of Okhotsk.

Can be confused with Several of the other beaked whales (Cuvier's beaked whale and some species of mesoplodont) are found within the geographic range of Baird's beaked whale, but the larger size and unique head and dorsal fin shape, conspicuous blow and large, often densely-packed schools distinguish them. In *Berardius*, the beak is longer than the melon is high, which differs from other larger beaked whales (*Hyperoodon*, *Indopacetus*, and *Tasmacetus*). Minke whales could, in some circumstances, be confused with Baird's beaked whales. This might only occur at a great distance, but when a good look is obtained, differences in dorsal fin shape, head shape, group size, behavior at the surface and coloration make the two easily distinguishable.

A herd of Baird's beaked whales resting at the surface with their beaks pointed toward the photographer; the diagnostic flat or often concave back of this genus is clear from this view. Aleutian Islands, Alaska. PHOTO: P. WADE

Distribution Baird's beaked whales are found in deep oceanic waters and continental shelf edges of the North Pacific Ocean and the Japan, Okhotsk, and Bering seas. Their range extends from the southern Gulf of California in the eastern Pacific, and to the island of Kyushu, Japan, in the western Pacific. Though they may be seen close to shore where deep water approaches the coast, their primary habitats appear to be over or near the continental slope and near oceanic seamounts. There are seasonal inshore/offshore shifts in the eastern North Pacific. They may occur in the vicinity of drift ice in the northern Sea of Okhotsk. Off the Pacific coast of Japan, they migrate into waters over the continental slope from May to October,

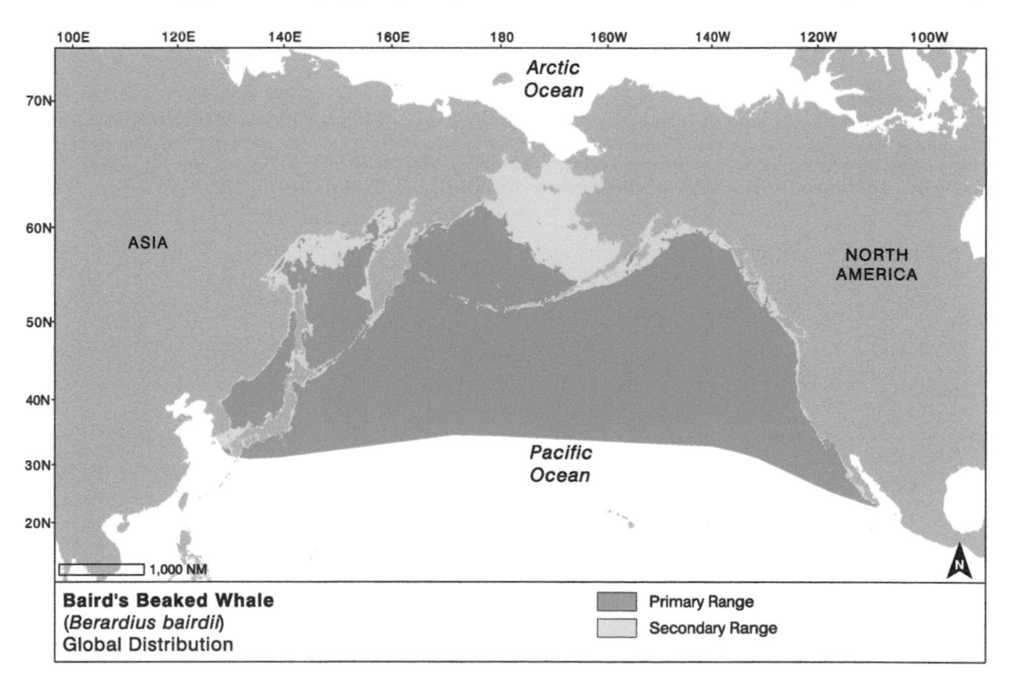

Baird's beaked whales have a low, often-rounded dorsal fin, placed far back on the back. The brownish coloration is typical and probably due to diatoms settling on the skin. Monterey Bay, California. PHOTO: J. POKLEN.

***Berardius* has a long, heavy beak and a relatively small melon. A pair of teeth erupt from the tip of the lower, underslung jaw in both sexes and one tooth is just visible here. Aleutian Islands, Alaska.** PHOTO: R. L. PITMAN

This is presumably a juvenile Baird's beaked whale—it has few scratches on it and the teeth do not appear to have erupted from the tip of the lower jaw yet. Aleutian Islands, Alaska. PHOTO: P. WADE

The animal in the foreground has tooth rake marks on its back from a killer whale—a likely predator of this species. Aleutian Islands, Alaska. PHOTO: R. L. PITMAN

All beaked whales have crescent-shaped blowholes, but only in the genus *Berardius* do the tips point toward the rear—this animal is pointing to the left. Aleutian Islands, Alaska. PHOTO: P. WADE

but where they go in winter is not known. Their distributional limits in oceanic waters of the mid-Pacific are also not well known.

Ecology and behavior Baird's beaked whales live in pods of 5–20 whales (average about six individuals), although groups of up to 50 are occasionally seen. They often assemble in tight groups drifting along at the surface. At such times, beaks are often seen as animals slide over one another's backs. They are deep divers, capable of staying down for over an hour (67 minutes is the current record). After a dive, they stay at the surface for up to 14 minutes, and may blow almost continuously after a long dive. They breach and perform other aerial behaviors (spyhops, fluke- and flipper-slaps) on occasion (more so than most other beaked whales).

This is an extremely long-lived species that apparently has an unusual life history pattern. From Japanese whaling data, it appears that males have lower mortality rates and live longer than females (up to about 84 vs. 54 years) and that females have no post-reproductive stage. There is a calving peak in March and April. The length of the gestation period is not known with any certainty, but is thought to be around 17 months.

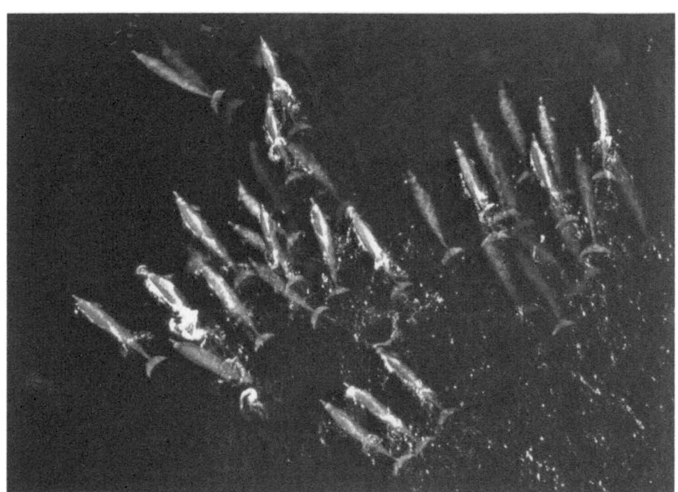

Baird's beaked whales often occur in large, dense schools, with animals often in physical contact with each other. Baja California, Mexico. PHOTO: M. LYNN, NOAA FISHERIES, SWFSC

Unlike other beaked whales, male and female *Berardius* both have teeth that erupt from the tip of the lower jaw, and adults of both sexes tend to be heavily scarred from tooth raking; juveniles (above, showing dorsal fin) are mostly unmarked. Pacific coast of Baja California. PHOTO: R. L. PITMAN

Three stocks are recognized in the western North Pacific (Sea of Japan, Okhotsk Sea, and Pacific Ocean), where these whales have been exploited for centuries. Tooth rake marks from killer whales are common on the bodies of Baird's beaked whales, as are cookie-cutter shark bites.

Feeding and prey Baird's beaked whales feed mainly on deepwater and bottom-dwelling gadiform fishes, cephalopods, and crustaceans. The also feed on some pelagic fishes, such as mackerel, sardines, and sauries. The diet off the Pacific coast of Japan consists of 82% fish and 18% cephalopods, while in the southern Sea of Okhotsk the proportions are 13% and 87%, respectively. They may do much of their feeding at depths of 800–1,200 m.

Threats and status Baird's beaked whales are one of the few species of ziphiids to be commercially hunted. Although small numbers were hunted by the Soviets, Canadians and Americans, major hunts occur only in Japan. The Japanese fishery started in the early 1600s and underwent several expansions and declines. At its peak, after World War II, over 300 whales were killed annually. Now the Japanese annual quotas add up to slightly over 60 whales. Other than hunting, and possibly occasional captures in fishing gear (especially incidental captures in drift nets), no other significant threats to the species are known. There are an estimated 1,100 Baird's beaked whales in the eastern North Pacific, including about 400 off the US west coast. Abundance in Japanese waters has been estimated at about 7,000 individuals (about 5,000 off the Pacific coast, 1,300 in the eastern Sea of Japan, and 660 in the southern Okhotsk Sea). The consensus is that the hunt is sustainable, and there is little immediate concern over the future of this species.

IUCN status Data Deficient.

Arnoux's Beaked Whale—*Berardius arnuxii*

Duvernoy, 1851

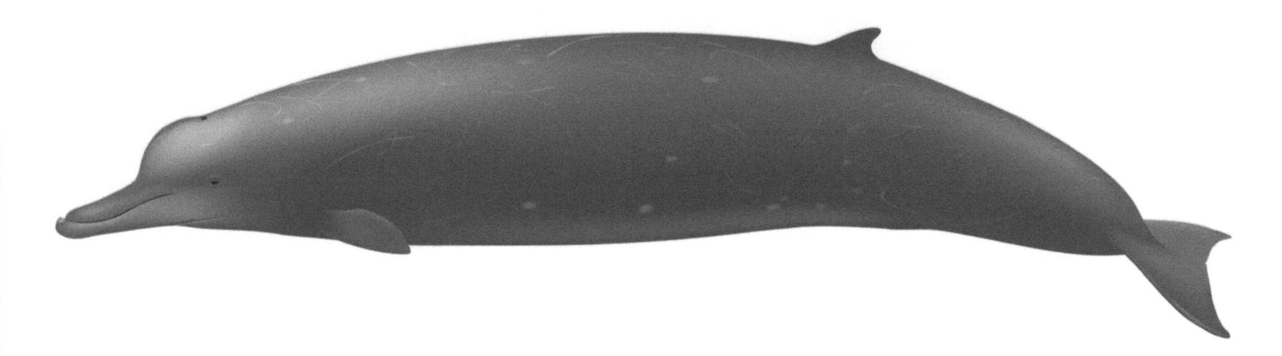

Recently-used synonyms None.

Common names En.–Arnoux's beaked whale; Sp.–*ballenato de Arnoux*; Fr.–*béradien d'Arnoux*.

Taxonomic information Cetartiodactyla, Cetacea, Odontoceti, Ziphiidae. Some researchers question whether Baird's and Arnoux's beaked whales are really separate species; however, the lack of known morphological differences may have more to do with lack of specimens than anything else. Their distributions are widely separated.

Species characteristics Similar in appearance to Baird's beaked whale (in fact, there are no known body shape differences between the two species), Arnoux's beaked whale is rather slender. This species has a small head, with a long, tube-like beak, moderately steep bulbous forehead, small rounded flippers (which fit into flipper pockets), short slightly falcate dorsal fin, and (usually) un-notched flukes. The dorsal fin is set about ⅔ of the way back from the beak tip, and is generally very rounded at the tip. A pair of V-shaped throat grooves is present, with accessory grooves present in some individuals. The blowhole concavity points forward, and therefore the "hinge" is at back. Only *Berardius* has this orientation.

Arnoux's beaked whales are slate gray to light brown in color; the head region is generally lighter than the rest of the body. The body is typically heavily scarred and scratched, and the underside tends to be lighter, and

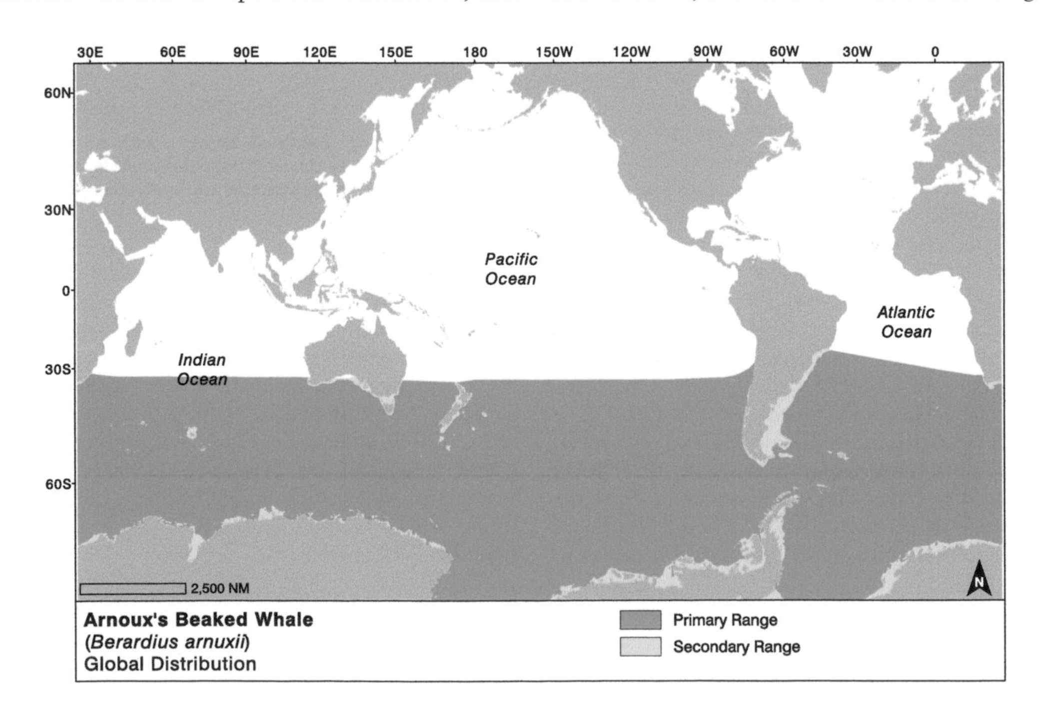

Swimming shoulder-to-shoulder in the prop-wash of a helicopter, these Arnoux's beaked whales can be identified by their long backs, bluff melons, long beaks, and prominent blows; they also have a flat back or a mid-dorsal groove down the middle of the back, a distinct feature of *Berardius*. Ross Sea, Antarctica. PHOTO: E. HEARD

is sometimes covered with white blotches. Scars caused by conspecifics (generally long scratches, most prominent on males), killer whales (shorter scratches and scrapes) and cookie-cutter sharks (round or oval, white scars) may be present on the body as well.

Two pairs of triangular teeth are present at the tip of the lower jaw; they erupt in both sexes and the forward pair is visible outside the closed mouth. The rear pair is somewhat smaller. Barnacles may attach to the teeth.

Arnoux's beaked whales reach a maximum known size of 9.3 m; females are probably a bit larger than males, as is generally true in beaked whales. Length at birth is unknown, but is probably around 4 m.

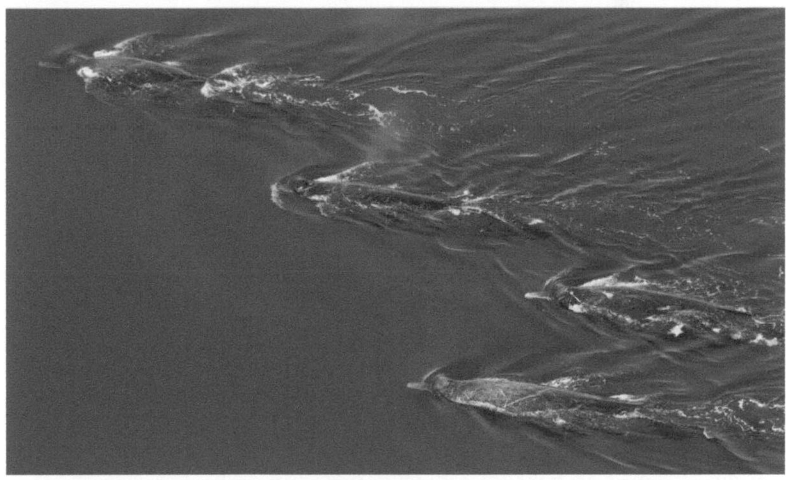

Arnoux's beaked whale is large with a conspicuous blow; it has a long, heavy, tubular beak, a distinct but relatively small melon, and heavy tooth raking on all of the adults (including females). In sunlight, they often appear brownish. Weddell Sea, Antarctica. PHOTO: M. SCHEIDAT

Recognizable geographic forms None.

Can be confused with Arnoux's beaked whales can be easily confused with southern bottlenose whales, which share much of their range and have a broadly similar body shape. Differences in coloration, head shape, dorsal fin shape, and tooth size and position should be sufficient to distinguish them, when clearly seen. Individuals of some species of *Mesoplodon* and *Indopacetus* could also be confused with this species at a distance, but they are generally much smaller. Minke whales may also be confused, at a distance, but at close range will present no identification problems.

Distribution Although this species probably has a vast circumpolar distribution in deep, cold, temperate and subpolar waters of the Southern Hemisphere, most records are from the southeast coast of South America, near the Antarctic Peninsula, South Africa, and southern Australia and New Zealand. Although the exact northern limits are not known, they generally do not reach the tropical zones. Although most records are south of 40°S, in some areas they may reach about 34°S (and there are even some records to as far north as 24°S).

Ecology and behavior Not much is known of the biology of this species, as it lives in a part of the world where little marine mammal research has been done, and it has never been hunted. Most groups number between 6–10 individuals, but some as large as 80 whales have been seen. Arnoux's beaked whales are reportedly shy of boats and

Arnoux's beaked whale is a large beaked whale, with a small, rounded dorsal fin and bulbous head. *Berardius* often appear brownish, especially in strong sunlight, and adults of both sexes are typically heavily scarred with tooth rake marks. PHOTO: M. SCHEIDAT

Arnoux's beaked whale is the only Southern Hemisphere species with apical teeth, a long beak, bluff melon, and either a flat back or a mid-dorsal groove. South Africa. PHOTO: M. MEYER

Arnoux's beaked surfaces at a breathing hole in the Antarctic ice; notice the massive beak, underslung jaw and tusks protruding from the tip of the lower jaw. PHOTO: BRITISH ANTARCTIC SURVEY

can dive for over an hour (maximum known dive of 70 min), although most dives probably last less than about 25 min. One group that was being followed by researchers in the Antarctic traveled more than 6 km underwater before resurfacing. Their elusive behavior and deep-diving capabilities can make observation difficult.

This species' reproductive biology is poorly known. It is unknown if it shares the unusual traits of its northern cousin, Baird's beaked whale (i.e., males living longer than females, and no post-reproductive stage in females). Some individuals have been trapped in ice and forced to spend the winter in the Antarctic. This may be a significant cause of mortality. Killer whales are also known to attack Arnoux's beaked whales. Little is known of their longevity, but their Northern Hemisphere cousins can live to be at least 84 years old.

Feeding and prey Very little is known about the feeding habits of this species, other than that they feed on squid (based on stomach contents). The feeding habits of Arnoux's beaked whales are assumed to be similar to those of their Northern Hemisphere relatives, Baird's beaked whales, thus consisting mostly of deepwater benthic and pelagic fishes and cephalopods.

Threats and status This species has never been hunted to any significant degree, and other threats are not known at this point. No abundance estimates are available, but Arnoux's beaked whale appears to be less common than the southern bottlenose whale, which shares much of its range. Although they may be naturally rare, there is no reason to believe that Arnoux's beaked whale is facing any serious threats to its survival.

IUCN status Data Deficient.

Cuvier's Beaked Whale—*Ziphius cavirostris*

G. Cuvier, 1823

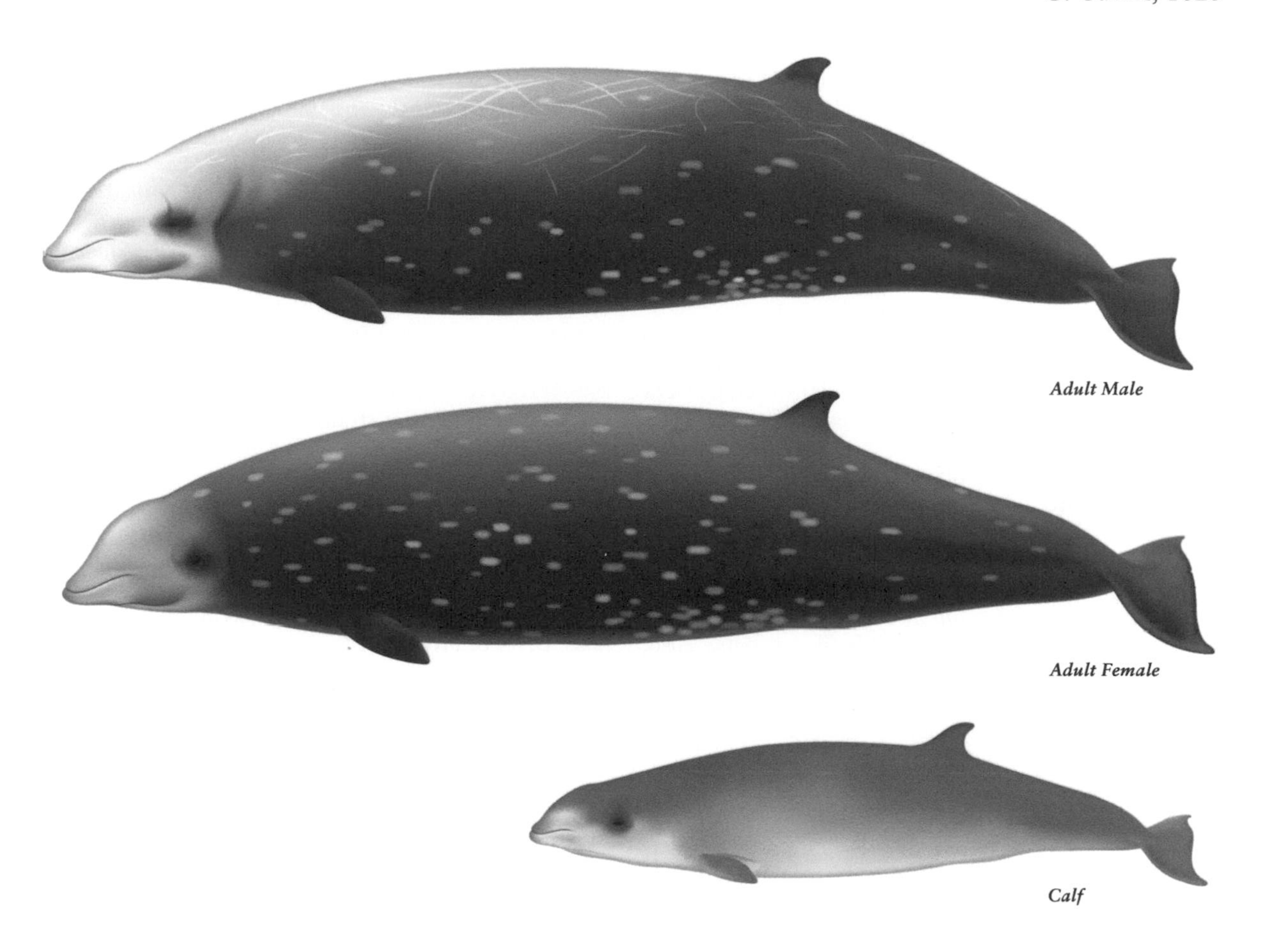

Recently-used synonyms None.

Common names En.–Cuvier's beaked whale, goose-beaked whale; Sp.–*zifio de Cuvier*, *ballena picuda de Cuvier*; Fr.–*ziphius*, *baleine à bec de Cuvier*.

Taxonomic information Cetartiodactyla, Cetacea, Odontoceti, Ziphiidae.

Species characteristics Cuvier's beaked whales have a body shape broadly similar to that of other species in the family, but are relatively robust, as beaked whales go. They have a short, poorly-defined beak, a smoothly-sloping forehead (although the melon appears to become more prominent, even bulbous, in adult males), and a mouthline that is curved along most of its length, with an upturn at the rear. A pair of V-shaped throat grooves is present. A diagnostic feature is the slight concavity on the top of the head, which increases in detectability in older animals. The blowhole is oriented with the opening at the back, therefore the "hinge" is at the front. The flukes are relatively large, and a fluke notch is only sometimes present. The dorsal fin is small and falcate, and is set about two-thirds of the way back from the snout tip. There are "flipper pockets," slight depressions into which the small, rounded flippers can be tucked to be kept flush with the body. Barnacles may be found on the flukes and dorsal fin.

The body is dark gray to light rusty brown, with lighter areas around the head and belly, although the ventral area darkens somewhat with age. Lighter color develops around the head and upper thorax with age in both sexes. The head and much of the upper back of adult males can be completely white. The eyes are usually surrounded by dark coloration, and there may be light crescent-shaped streaks around the eye area. Generally, adults are covered with white linear scratches and circular or oval marks. The latter are thought to be caused by bites from lampreys or cookie-cutter sharks. There may be orangish-yellow films (from algae or diatoms) on the body. Calves are more simply countershaded, dark above and lighter below.

There is a single pair of forward-pointing conical teeth at the tip of the lower jaw; they generally erupt only in adult males and are exposed outside the closed mouth in large bulls. The teeth may be infested with stalked barnacles. The blow is typically low and diffuse. It is often directed slightly forward. Although lengths of up to 8.5 m (females) and 9.8 m (males) have been recorded for Cuvier's beaked whales, total length measurements over 7.0 m are considered suspect. Maximum recorded weight is nearly 3,000 kg. Newborns are about 2.7 m long and weigh about 250–300 kg.

Recognizable geographic forms None.

Can be confused with All beaked whales are prone to identification problems, but when adult males are present this species is easier to identify than most. However, Cuvier's beaked whales are likely to be confused with other beaked whales, mainly various species of *Mesoplodon*. The very robust body (often visible above the surface as a broader back), blunt head (with only a very stubby beak), and lighter coloration (especially around the head, especially in adult males) may be sufficient to distinguish Cuvier's beaked whales, if visible. Whales of the genera *Hyperoodon* and *Berardius* are larger and have more bulbous foreheads and long tube-like snouts. Also, Longman's beaked whales may cause some confusion, but attention to coloration and head shape should allow distinction, if seen reasonably well.

Distribution Cuvier's beaked whales are widely distributed in offshore waters of all oceans, from the tropics to the polar regions in both hemispheres. Their range covers most marine waters of the world, with the exception of shallow water areas, and very high-latitude polar regions. They are found in many enclosed seas, such as the Gulf of California, Gulf of Mexico, Caribbean Sea, Mediterranean Sea (in fact, this is the only beaked whale species that regularly occurs in the Mediterranean), and the Sea of Okhotsk. They have the most cosmopolitan range of any beaked whale species. Although they can be found nearly anywhere in deep (>200 m) waters, they seem to prefer deeper waters over and near the continental slope, especially those with a steep bottom.

Ecology and behavior Due to their widespread distribution and relatively frequent stranding, Cuvier's beaked whale may be one of the most familiar of the beaked whales. They are found mostly in small groups of 2–7, but are not uncommonly seen alone. Their behavior tends to be rather elusive, and they can be difficult to approach. However, they have been seen breaching on a number of occasions. Cuvier's beaked whales are deep divers—currently the record-holder for deep diving among mammals. Dives of up to 2,992 m and lasting up to 138 minutes have been documented!

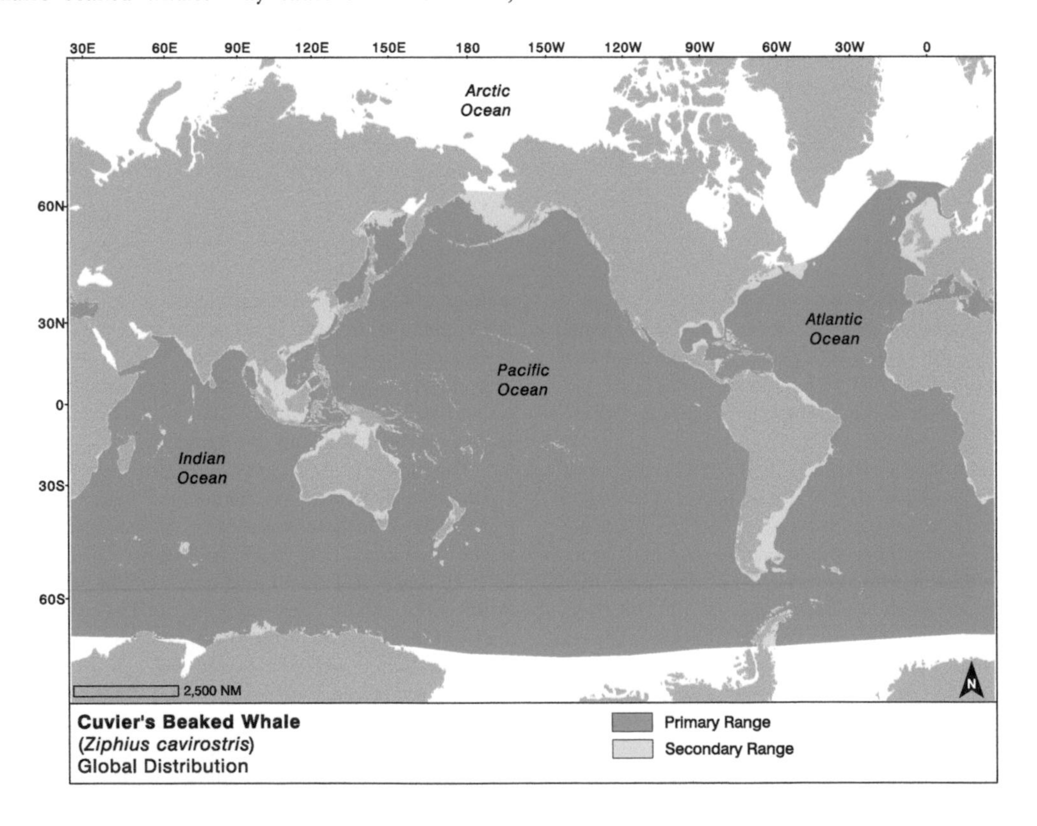

A young adult male *Ziphius*—note that the white is confined only to the head area, there are relatively few tooth rake scars, and although the teeth have erupted, they are still quite sharp (presumably unused); most of his battles are ahead of him. Gulf of Biscay. PHOTO: G. OCIO

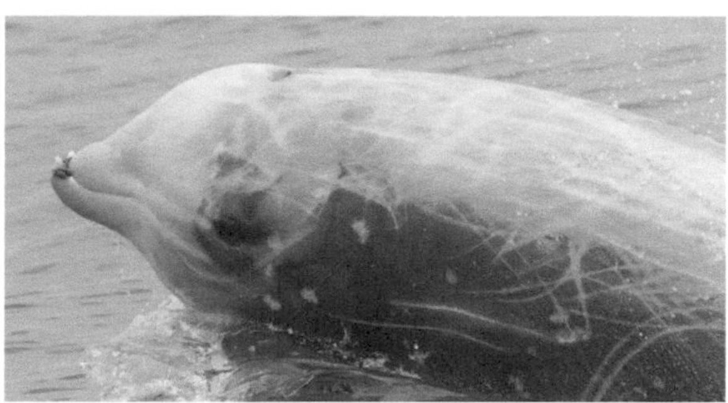

The chunky body and stubby, sloping beak of Cuvier's beaked whale is unmistakable; barnacles have started growing at the base of the teeth of this older male specimen. Bahamas. PHOTO: C. DUNN

The white on the head of adult male Cuvier's beaked whales extends further along the back with age so that older males can appear almost all-white, as in this older, battle-tested male. Bahamas. PHOTO: J. DURBAN

This heavily-raked male *Ziphius* has seen many clashes with other males; barely visible in the lower left is a biopsy dart about to hit its mark. Bahamas PHOTO: J. DURBAN

The numerous oval cookie-cutter shark bite scars coupled with only a single parallel tooth-rake track indicates that this *Ziphius* is probably an adult female; the brownish patches are diatoms. Bahamas. PHOTO: J. DURBAN

Young *Ziphius* have little or no scarring, a light gray or brownish body, and a pale head; as in all beaked whales, the dorsal fin is placed far back on the body. Diatom patches on the skin often give them an orangish or reddish appearance. Cape Hatteras, North Carolina. PHOTO: G. HOWELL

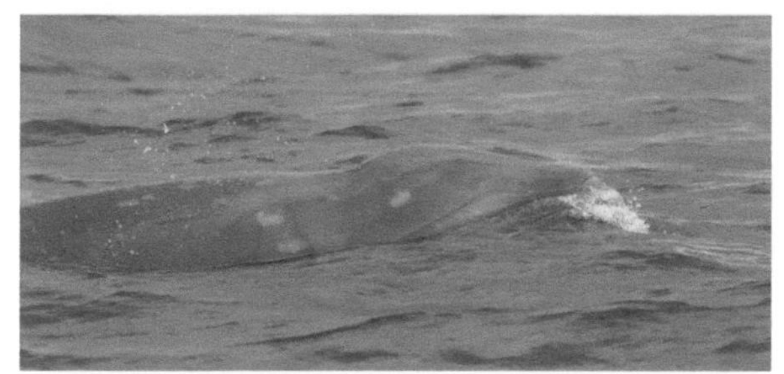

The same animal from the photo above, showing that females also have stubby beaks, but with less prominent lower jaws than adult males. Bahamas. PHOTO: J. DURBAN

Ziphius has a chunky body (compared to the slimmer mesoplodonts), a stubby beak, and a unique "smile;" this is probably an adult female: lots of cookie-cutter shark bites, but no parallel rake marks. PHOTO: R. BAIRD

A battle-scarred adult male Cuvier's beaked whale; its apical teeth, clearly visible at the tip of the lower jaw, have apparently been put into service many times. Gulf of Biscay. PHOTO: G. OCIO

In addition to linear scarring on the males, both male and female *Ziphius* acquire circular scars from cookie-cutter shark bites. This is especially true for populations in central and western ocean basins, such as this adult male off Japan. PHOTO: T. TAKAHASHI

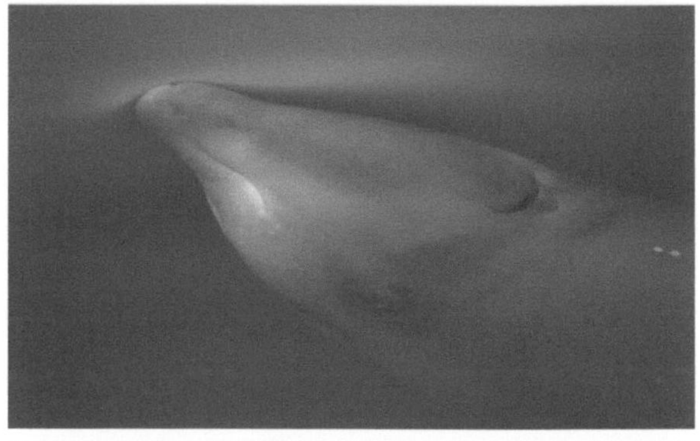

The short, stout beak of *Ziphius* (aka "goose-beaked whale"), with its underslung lower jaw, sets it apart from all other beaked whales. PHOTO: F. PONGIGLIONE, COURTESY T. PUSSER

***Ziphius* often appears light gray-brown and typically has white on the head; this animal has few cookie-cutter shark bites and no tooth rake scars—clear signs of a young animal.** PHOTO: J. DURBAN

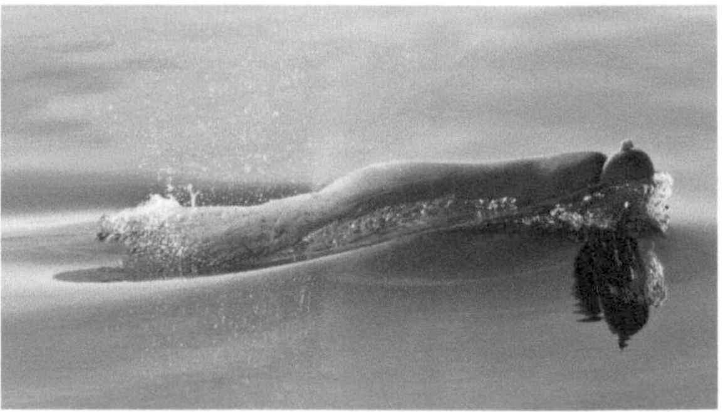

The underslung lower jaw of this adult male *Ziphius* strategically positions its pair of teeth for use as tusks; the teeth in this animal are worn from encounters with other males. Mediterranean Sea. PHOTO: M. ROSSO

An adult male *Ziphius* lunges, probably in preparation for a terminal dive; the beak is a little longer than typical for this species (the melon appears flatter than normal), and the lighting shows the deep furrowing associated with tooth rake marks. Off Dominica. PHOTO: L. BOUVERET

Seasonality of calving is not known in this species. In general, Cuvier's beaked whale life history is very poorly-known. Sexual maturity occurs at around 6.2 m. It is assumed that males fight for access to females (resulting in the long scratches seen on the bodies of most adult males), but this has never been witnessed. Strandings are relatively common (at least for beaked whales) in some areas, and this suggests that Cuvier's beaked whales are not as rare as was once thought. They sometimes come ashore and strand in groups. Killer whales are probable predators.

This is the only widely-distributed beaked whale species for which a global assessment of genetic diversity has been conducted. The results of this study suggest that there is probably little movement of Cuvier's beaked whales among different ocean basins, and that there may even be a distinct population in the Mediterranean Sea.

Cuvier's beaked whale is always huskier-looking than any of the mesoplodonts; the greenish-brown color of this animal is probably due to diatoms growing on the skin. Canary Islands. PHOTO: N. AGUILAR DE SOTO

Feeding and prey Cuvier's beaked whales, like all beaked whales, appear to prefer deep waters for feeding. Although few stomach contents have been examined, they appear to feed mostly on deep-sea squid, but also sometimes take fish and some crustaceans. They apparently feed both near the bottom and in the water column. Suction appears to be used to draw prey items into the mouth at close range. They may sometimes ingest non-food items.

Threats and status Never the main target of commercial whalers, Cuvier's beaked whales have sometimes been taken in other fisheries, however, such as those in the Caribbean islands, Indonesia, Taiwan, Peru, and Chile. A few (3–35 per year) were taken in past years in the Baird's beaked whale fishery off the coast of Japan. Some are occasionally taken in deep water drift gillnets. The only other threat that is known is the mortality of Cuvier's beaked whales that has apparently resulted from naval sonar exercises, such as those occurring recently in the Bahamas, Caribbean, Canary Islands, and the Mediterranean. This species appears to be particularly vulnerable to such events, the exact causes of which are not known. However, the sonar may result in changes in diving behavior or physiology that somehow result in an increase in bubbles in the blood (causing decompression sickness, or the bends) when returning to the surface from a deep dive. About 20,000 Cuvier's beaked whales are estimated to inhabit the eastern tropical Pacific. In the eastern North Pacific, over 90,000 are thought to occur, including about 1,600 in California and over 15,000 in Hawaiian waters. Off the US northeast coast, there are probably several hundred, and at least 95 are thought to inhabit the northern Gulf of Mexico. Although the accuracy of many of these estimates is questionable, Cuvier's beaked whales are among the most common and abundant of all the beaked whales.

IUCN status Least Concern.

Northern Bottlenose Whale—*Hyperoodon ampullatus*

(Forster, 1770)

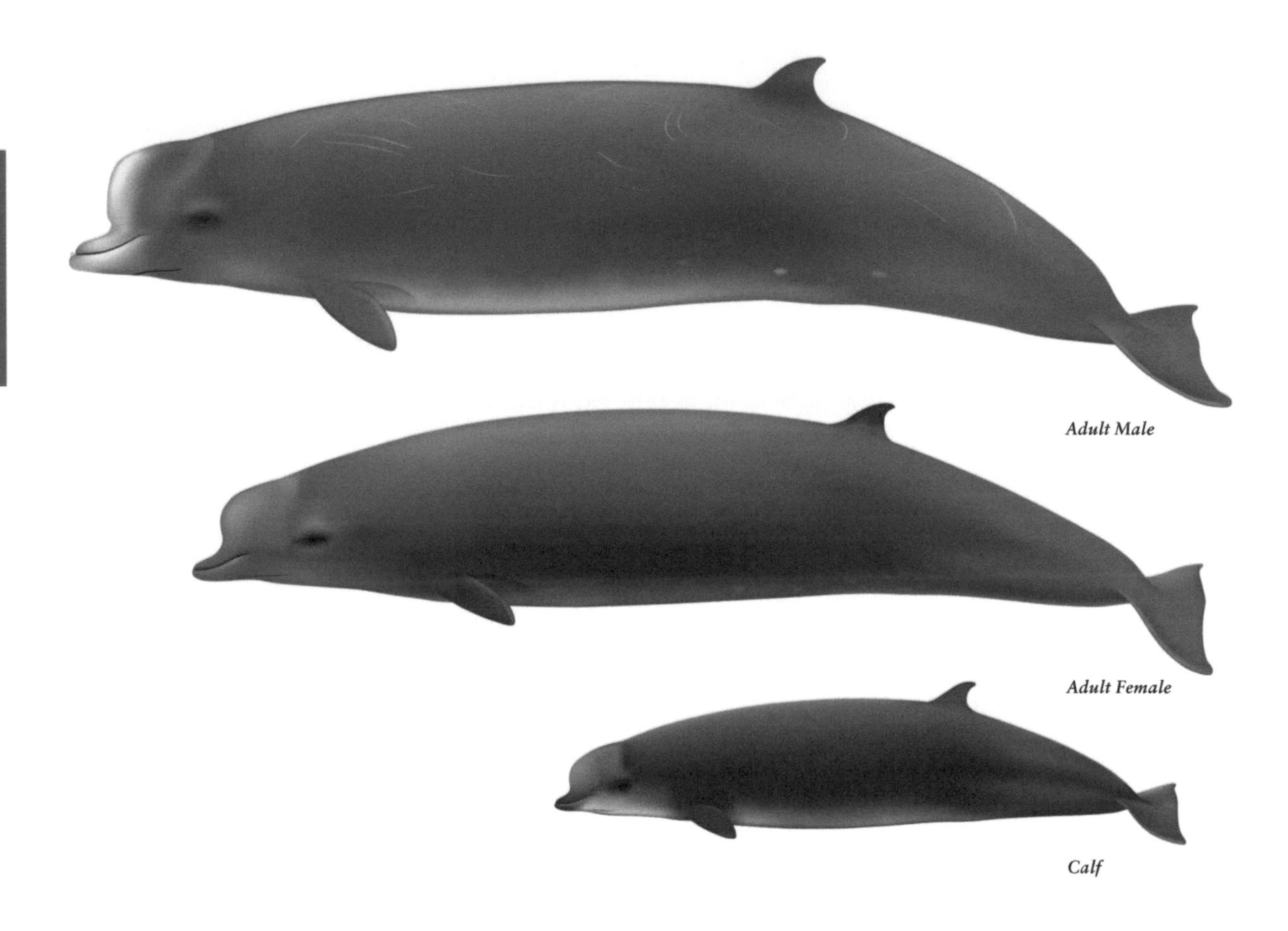

Adult Male

Adult Female

Calf

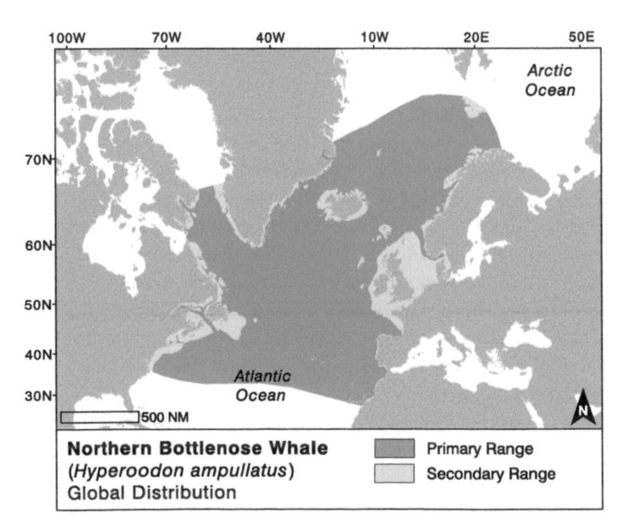

Recently-used synonyms *Hyperoodon rostratus.*

Common names En.–northern bottlenose whale; Sp.–*ballena nariz de botella del norte;* Fr.–*hyperoodon boréal.*

Taxonomic information Cetartiodactyla, Cetacea, Odontoceti, Ziphiidae.

Species characteristics Northern bottlenose whales are robust animals. They are appropriately named, having a moderately-long, tube-like beak that is distinct from the melon (reminiscent of the beak of some dolphins). In young animals and females, the rounded forehead slopes upward from the beak, but in adult males the forehead becomes very steep, with a nearly-squarish profile. The front of the melon may become very flat in some old bulls. A pair of forward-pointing grooves is found on the throat. The prominent dorsal fin (up to 30 cm tall) is falcate and pointed at the tip; it is located far back on the body. The flippers are small and blunt at the tips, and the flukes generally lack a median notch. The blowhole is oriented with

From the air *H. ampullatus* can look initially look like sperm whales, but notice the pale melon, prominent dorsal fin, and stubby beak when viewed from overhead. Northeast Atlantic. PHOTO: M. NIEMEYER

When northern bottlenose whales are resting at the surface they sometimes hang in the water at a 45° angle, exposing their bulbous melon and short beak. The Gully, North Atlantic. PHOTO: HAL WHITEHEAD'S LAB

***Hyperoodon* (both species) have a brownish coloration and a prominent, often pointed dorsal fin. The Gully, North Atlantic.** PHOTO: H. MOORS-MURPHY

Young northern bottlenose whales, in particular, have a pale melon, which extends a short way beyond the blowhole. The Gully, Northeast Atlantic. PHOTO: H. MOORS-MURPHY

The melon is vertical or slightly overhanging in some *H. ampullatus*, giving it a "bottlenose" shape in profile. The Gully, North Atlantic. PHOTO: HAL WHITEHEAD'S LAB

Very young bottlenose whales have a noticeably shorter beak than the adults. The Gully, North Atlantic. PHOTO: HAL WHITEHEAD'S LAB

From underneath, the pair of expandable throat grooves is evident, which allow for suction feeding. The Gully, North Atlantic. PHOTO: HAL WHITEHEAD'S LAB

A young and curious *H. ampullatus* shows off its rounded melon and slightly underslung lower jaw. The Gully, North Atlantic. PHOTO: M. MILLIGAN

Adult male northern bottlenose whales resort to head-butting to settle mating disputes, which presumably accounts for the flat head seen only in the males of this species. The Gully, North Atlantic. PHOTO: HAL WHITEHEAD'S LAB

Beak length is moderate in this species, though variable, with this individual representing the longer end of the range. The Gully, North Atlantic. PHOTO: HAL WHITEHEAD'S LAB

Both species of *Hyperoodon* have a forward-canted blow, although most of the time it just appears to be low and bushy. The Gully, North Atlantic. PHOTO: HAL WHITEHEAD'S LAB

the opening at the back, therefore the "hinge" is at the front.

Calves are generally brownish-gray on the body, with complex light coloration on the head. There is some disagreement as to whether young animals are countershaded. Adults are dark grayish to chocolate brown above and somewhat lighter below. The brownish tinge is often enhanced by a covering of diatoms. Some individuals are mottled with white to yellowish splotches and oval scars, which increase in number with age. Much of the melon and face may be light gray, or in adult males, nearly white. Older females often have a white band around the neck.

At the tip of the lower jaw are two conical teeth that erupt only in bulls, and are not always visible outside the closed mouth. They lean slightly forward, and sometimes have stalked barnacles attached to them. A second pair of teeth is sometimes buried in the gums behind the first, and 10–20 additional vestigial teeth may be found in the gums of both upper and lower jaws. Adult females are up to 8.6 m and adult males up to 10.0 m in length (perhaps up to 11.2 m). This species is unusual for beaked whales in having males growing significantly larger than females. They can weigh up to 7,500 kg. At birth calves are about 3.0–3.5 m.

In the foreground is a flat-headed adult male *H. ampullatus*; apparently this species has replaced tooth raking with head butting during aggressive encounters, because northern bottlenose whales do not have the extensive tooth rake marks of the southern species. The Gully, North Atlantic. PHOTO: H. MOORS-MURPHY

As in all beaked whales, *Hyperoodon* flukes, although rarely seen in the wild, typically lack a central notch. The Gully, North Atlantic. PHOTO: HAL WHITEHEAD'S LAB

Calf—Less than about ½ adult size, melon sloping and not bulbous. Dark eye patches and a light-colored forehead are present, body color darker than adults.

Female/subadult male—Full- or nearly adult-size, melon bulbous and gray in color, with a whitish neck collar behind the blowhole.

Adult male—Body size very large; melon extremely bulbous and often overhanging, with a flattened front (giving the melon a somewhat squared-off appearance), most of melon and head back to the eyes white.

Recognizable geographic forms None.

Can be confused with This is the largest beaked whale species in the North Atlantic. A good look will allow Cuvier's beaked whales to be distinguished from bottlenose whales by differences in head shape and body color. Mesoplodonts (only the Sowerby's beaked whale has a broadly-overlapping distribution) are distinguishable by their smaller size and more cone-shaped head. Northern bottlenose whales look similar to Longman's beaked whales (in fact, sightings of these animals in the past were attributed to the genus *Hyperoodon)*. However, Longman's beaked whales are only found in the Indo-Pacific and thus do not overlap northern bottlenose whales in range.

Distribution Northern bottlenose whales are found only in the North Atlantic, from New England to Baffin Island and southern Greenland in the west and from the Strait of Gibraltar to Svalbard in the east. The most famous population occurs in the "Gully" (a large submarine canyon off the coast of Nova Scotia, Canada). However, there have been strandings at least as far south as North Carolina in the western Atlantic. The pelagic distribution extends from the ice edges south to approximately 30°N. These cold temperate to subarctic whales are found in deep waters, mostly seaward of the continental

The large size (often with clearly visible blows), pale gray-brown coloration, pale, bulbous melon, and short but conspicuous beak readily identify *H. ampullatus* from the air in the North Atlantic.
PHOTO: K. O'BRIEN

shelf (and generally over 500–1,000 m deep) and near submarine canyons. There is some evidence from whaling records to suggest a north/south migration, but it is not clear-cut. They inhabit the most northerly waters of the Barents and Greenland seas in summer (May to August). Strandings have been recorded in the Baltic Sea, but this area is probably too shallow to be regularly inhabited by this species. The northern bottlenose whale forms an antitropical species pair with the southern bottlenose whale.

Ecology and behavior The behavior and ecology of northern bottlenose whales have been better studied than that of any other species in the family Ziphiidae. Most groups contain at least four whales, sometimes with as many as 20, and there is some segregation by age and sex. There is a well-studied, resident population in the "Gully." Much of their movement there is defined by short-term residence in ranges of about 25 km^2. Most dives last less than 10 minutes. However, these deep divers can remain submerged for an hour, possibly as long as two, and can reach depths of well over 1,400 m. These whales may be curious about vessels and have been known to approach and swim around stationary boats for some time. They are also known for their habit of "standing by" injured companions, which whalers used to facilitate hunting of an entire group. They can travel over long distances of over 1,000 km.

Northern bottlenose whale males may use their large heads to "butt" each other in male/male combat. They have a peak in calving in spring to early summer (April–June). Lengths of the gestation and lactation periods are both probably over one year. Sexual maturity is reached at about nine years in females, slightly later in males. Individuals in the Gully population can be identified by natural markings, and several long-term studies have been conducted on them. Their social organization is one of fission-fusion, with mostly very short-term associations (much like that of most dolphin species). Adult males, however, form long-term companionships with other males, the functions of which are unknown. Groups may be segregated by age and sex. Longevity is at least 37 years, possibly much longer.

Feeding and prey Although primarily adapted to feeding on squid (especially *Gonatus* spp.), these whales also eat fish (such as herring and redfish), sea cucumbers, starfish, and prawns. They apparently do much of their feeding on or near the bottom in very deep water (>800 m, and as deep as 1,400 m).

Threats and status The northern bottlenose was sought after for its oil (including a form of spermaceti oil in the head) and later for pet food. This is one of only a few species of beaked whales to be hunted commercially on a large scale, largely by Canada and Norway. Whalers took advantage of the habits of whales of this species to approach vessels. Hunts occurred from the 1850s to the 1970s, and over 80,000 whales were killed (with many more struck, but lost). They have also been hunted in a drive fishery in the Faroe Islands, with over 800 taken there. Current numbers are largely unknown, although there are an estimated 5,000+ in the waters around Iceland and the Faroe Islands. An estimated 40,000 occur in the eastern North Atlantic. In the Gully, the current resident population numbers only about 130 (with about 230 different individuals spending some time there). Most populations are probably still depleted, due to these large kills in the past. The Gully population may be further threatened by oil and gas activities, fishing, and shipping in the region.

IUCN status Data Deficient.

Southern Bottlenose Whale—*Hyperoodon planifrons*

Flower, 1882

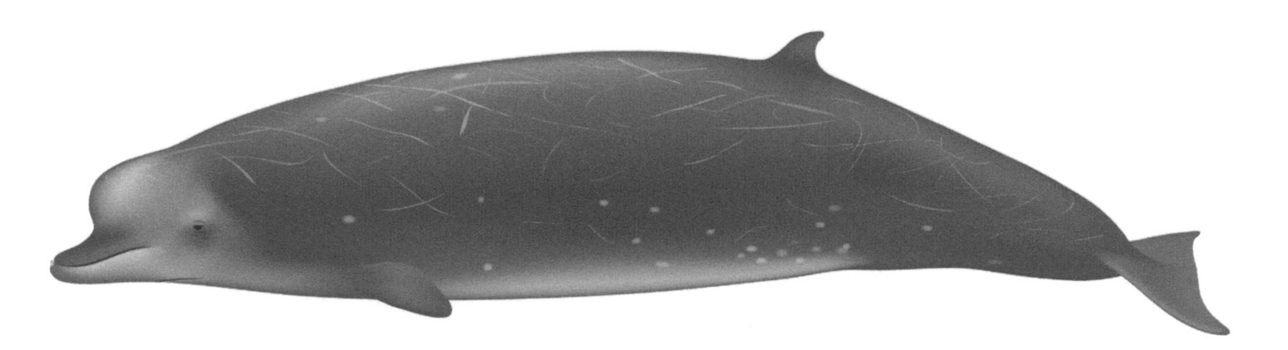

Recently-used synonyms None.
Common names En.–southern bottlenose whale; Sp.–*ballena nariz de botella del sur;* Fr.–*hyperoodon austral.*
Taxonomic information Cetartiodactyla, Cetacea, Odontoceti, Ziphiidae. For many years, there was speculation that a species of "tropical bottlenose whale" that had been repeatedly observed at sea in the Indo-Pacific may have been a representative of this species. However, it is now known to be Longman's beaked whale *Indopacetus pacificus* (see below).
Species characteristics This species resembles the northern bottlenose whale, with a robust body. The southern bottlenose whale also has a steep forehead, bulbous melon (it may be bluff and squared-off in adult males), and a tube-like, well-demarcated beak. There is a single pair of throat grooves. Along the rear half of the back is a small, but prominent, falcate dorsal fin. The small blunt flippers fit into "flipper pockets" along the sides of the animal. The flukes are wide, typically with no notch (or only a shallow one).

Southern bottlenose whales are light grayish-brown to dull yellow in color (diatoms have been said to be the cause of the yellowish tinge, but this is not confirmed). The belly and often much of the head are lighter. Large animals can be covered with light splotches, scratches, and scars. The color pattern of young calves and juveniles

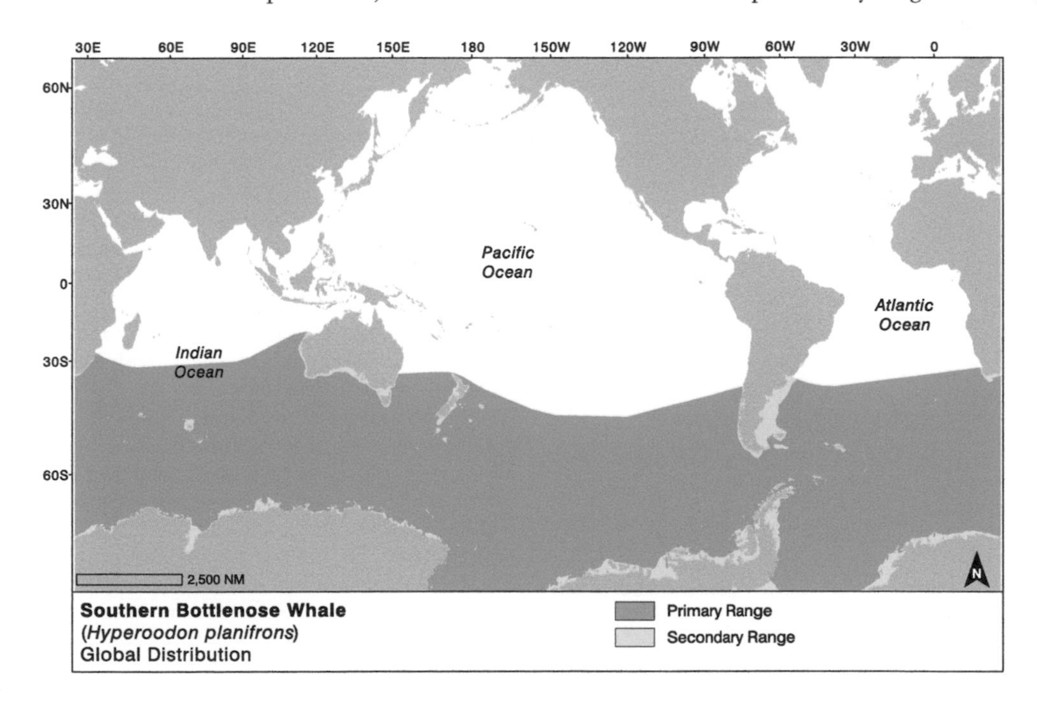

Southern Bottlenose Whale
(*Hyperoodon planifrons*)
Global Distribution

The blow of this southern bottlenose whale is conspicuous, but low and puffy; at close range like this it can be seen to be forward-canted also. Antarctica. PHOTO: K. SEKIGUCHI

Unlike northern bottlenose whales, when males of the southern species clash, they seem to rely on their teeth rather than head-butting; the result is that southern males are often heavily scratched up. PHOTO: P. OLSON

This *H. planifrons* is probably an adult female—there are lots of circular scars from cookie-cutter sharks, etc., but no linear scars. Southern Ocean. PHOTO: R. L. PITMAN

is bold and distinct (see below). The head (often back to behind the blowhole) of younger individuals is white, and distinctly demarcated from the darker back. Some adult males are very light in color along the entire back.

There is a single pair of conical teeth at the tip of the lower jaw, which erupt only in adult males, and are not visible outside the closed mouth. There is sometimes a smaller second pair. Rarely, several sets of vestigial (unerupted) teeth may be present in either jaw. The blow is short and bushy.

Maximum confirmed sizes are 7.5 m for females and about 7.0 m for males. However, specimens of up to 7.8 m have been reported. If females are, in fact, larger than males, this species would differ from its northern counterpart. However, the disparity is more likely a result of the small sample size of measured animals. Length at birth appears to be around 2–3 m.

Calf/juvenile—Less than about ⅔ adult size, melon sloping and not bulbous. Dark eye patches often present. Top and sides of head white to cream colored, with a dark stripe extending from the blowhole to the end of the beak. Spinal field relatively dark compared to adults.

Female/subadult male—Full- or nearly adult-size, melon bulbous and darker in color than in juveniles, with some remnants of the whitish head patch often still present. Spinal field lighter than in calves/juveniles.

Adult male—Body size large; melon bulbous and may be overhanging, with a flattened front. The spinal field and much of the back may be very light (almost white) in color. Usually they are heavily scarred.

Recognizable geographic forms None.

Can be confused with Southern bottlenose whales are most likely to be confused with Arnoux's beaked whales, and there is wide overlap in the distribution of these two species. They can be distinguished by differences in dorsal fin and head shape, as well as coloration differences. Cuvier's beaked whales and various mesoplodonts may cause some identification problems from a distance. They can be differentiated primarily by attention to size, head shape, and body patterning. Southern bottlenose whales may also be confused with Longman's beaked whale, and detailed views of the animals may be needed to distinguish the species. The overlap in distribution between these two species is minimal, however, with the Longman's preferring more tropical climates.

Distribution Southern bottlenose whales have a circumpolar distribution in the Southern Hemisphere, south of about 30°S. Most sightings are from about 57°S to 70°S. There are known areas of concentration between 58°S and 62°S in the Atlantic and eastern Indian Ocean sectors of their range. They apparently migrate, and are found in Antarctic waters during the summer, where they tend to occur within about 120 km of the ice edge (sometimes reaching the edge itself). Like other beaked whales, these

These *H. planifrons* are apparently a cow (foreground) and calf pair—the calf is unmarked and has a more contrasting color pattern. Southern Ocean. PHOTO: P. ENSOR

deep-water oceanic animals do not often stray over the continental shelf.

Ecology and behavior There is much less known about southern bottlenose whales than there is for their Northern Hemisphere counterparts. Groups of less than ten are most common, but groups of up to 25 have been seen. Nothing is known about their social organization. They are deep divers that can remain underwater for over an hour. Southern bottlenose whales do breach and perform other aerial behaviors on occasion and they may "porpoise" away from vessels. Based on what is known of their northern cousins, they are probably capable of diving to well over 1,000 m. They may undertake long seasonal movements of over 1,000 km.

Due to the fact that it has never been commercially hunted, and there have been no ecological studies, there is virtually nothing known of the reproductive biology of this species. Sparse information suggests that births off South Africa take place in spring to early summer. Longevity is not known.

Feeding and prey Southern bottlenose whales are thought to feed primarily on squid (over 200 squid beaks were found in the stomach of one specimen stranded in New Zealand), but also eat fish (including Patagonian toothfish) and possibly some crustaceans. They may compete with sperm whales for some of the same prey species, but bottlenose whales probably take smaller individuals.

Threats and status No significant exploitation of southern bottlenose whales is known, and they have never been hunted on a large scale. Some have been incidentally killed in driftnets. These animals are not rare by any means. They are the most common beaked whales sighted in Antarctic waters, and are clearly quite abundant there. One estimate puts their numbers south of the Antarctic Convergence at around 500,000 individuals.

IUCN status Least Concern.

A young (i.e., unmarked) *H. planifrons* breaching south of Madagascar in the Indian Ocean; the pale melon can be seen to extend beyond the constriction of the blowhole. Both species of *Hyperoodon* are rather chunky compared to most other beaked whales. PHOTO: R. L. PITMAN

Shepherd's Beaked Whale—*Tasmacetus shepherdi*

Oliver, 1937

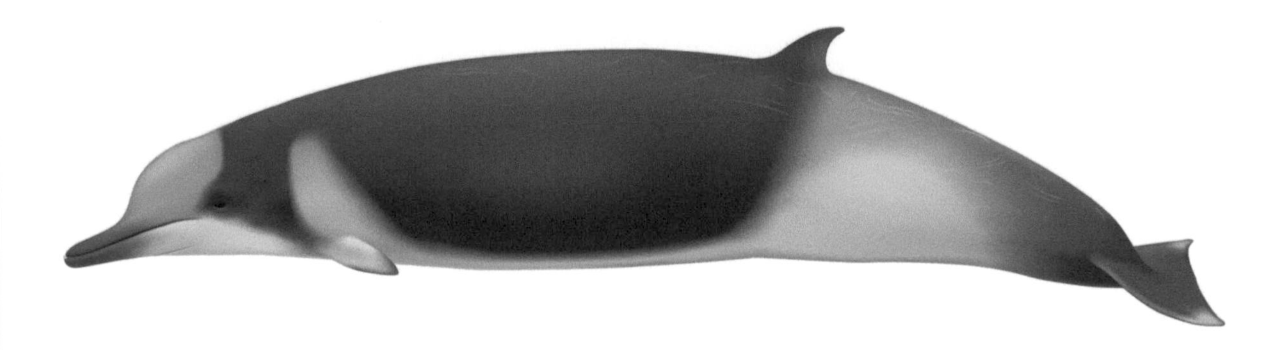

Recently-used synonyms None.
Common names En.–Shepherd's beaked whale; *Sp.–ballena picuda de Shepherd; Fr.–tasmacete.*
Taxonomic information Cetartiodactyla, Cetacea, Odontoceti, Ziphiidae.
Species characteristics Shepherd's beaked whales have spindle-shaped bodies (generally similar to those of the mesoplodonts), and long beaks that taper to a point, which are distinctly set off from their relatively steep foreheads. However, they have rather pronounced melons, which are unusual in that they are somewhat narrow and flattened laterally. There is no evidence that the melon gets larger with age. There is a shallow pair of throat creases. The flippers are small and tapered at the tips. The dorsal fin, set far back on the body, is short (about 30–35 cm high) and falcate. Generally, the notch between the flukes (characteristic of most cetaceans) is absent. The blowhole is oriented with the opening at the back, therefore the "hinge" is at the front.

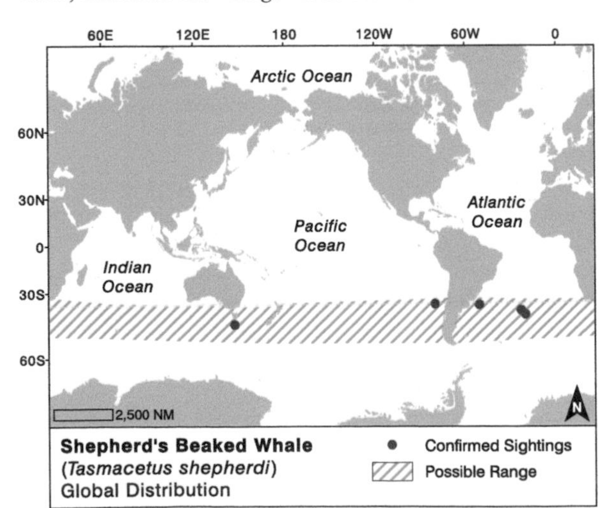

Shepherd's beaked whales have a distinctive color pattern, perhaps the most diagnostic of all the beaked whales. It appears to be present on all age and sex classes (including calves). The back and sides are mostly dark brownish ("chocolate") gray, and the belly is white or light gray. The light ventral field extends up onto the sides in three areas: 1) on the side of the face to just below the gape and eye, 2) in a rounded patch posterior to the flippers to just past midway up the side ("shoulder blaze"), and 3) on the tail stock posterior to the dorsal fin to just past midway up the sides. There is usually strong contrast between the light and dark areas, but the line separating them is generally not sharp. The dorsal fin, flukes, and beak are mostly dark, but there is also a pale "creamy" colored patch on the melon that extends onto the dorsal side of the beak. The light melon patch appears to be very prominent, when seen from the air. There are darkened eye patches. The color of the flippers appears to be variable, but the dark dorsal field extends onto the upper surface. Faint, paired scars are sometimes visible on what are presumed to be males.

Unique to beaked whales, this species normally has a mouthful of sharp, functional teeth. There are 17–21 teeth per row in the upper jaw, and 18–28 in the lower. At the tip of the lower jaw is a pair of enlarged terminal teeth, which erupt only in adult males. The blow observed from vessels has been inconspicuous, although it appears that it may be more prominent when viewed from the air. It is about 1–1.5 m tall and appears spiraled. Few specimens have been measured accurately. Lengths of up to 6.6 m (females) and 7.0 m (males) have been recorded, and measurements over 7 m are considered suspect. Length at birth is unknown, but is presumed to be around 3 m (the only known calf was 3.4 m long).
Recognizable geographic forms None.
Can be confused with When seen well, Shepherd's beaked whales should be quite easy to identify, due to their

Shepherd's is one of the easiest beaked whales to identify—the division between the darker cape and the lighter flank rises vertically, right below the dorsal fin, and is evident in both the cow and calf shown here. PHOTO: P. OLSON

A side view of a *Tasmacetus* calf showing the darker cape extending up through the middle of the dorsal fin. PHOTO: P. OLSON

unique color pattern. They can be confused with other beaked whales, especially mesoplodonts at a distance. However, they appear to be somewhat larger than most mesoplodonts, and have a sharper beak (from above) and more pronounced melon. They may also be confused with southern bottlenose and Longman's beaked whales, both of which have similarities in the shape of the head and beak, but careful attention to details of head shape and color pattern will help separate them.

Distribution Shepherd's beaked whales are primarily known from a few dozen strandings, all south of 30° S, around New Zealand, southern Australia, southern South America, the Juan Fernandez Islands, and Tristan de Cunha. There have been only a few sightings reported in the literature and the validity of some of those is suspect (or clearly erroneous). The confirmed sightings have been from northern New Zealand, southern Australia (including south of Tasmania), near Tristan da Cunha and in oceanic waters of the South Atlantic. It is presumed that they have a circumpolar distribution in cold temperate waters of the Southern Hemisphere. Like other members of the family, these are probably almost exclusively oceanic, deep-water animals, although they appear to concentrate near the shelf break.

Ecology and behavior Very little is known of the natural history of this species, making it one of the least-known of all cetaceans. There are no more than a few dozen records. Group sizes reported for four reliable sightings have been moderate (for beaked whales), ranging from 3–12 individuals. They have sometimes been observed to lift their beaks out of the water upon surfacing (and the apical teeth of bulls may be visible at such times), but little else is known of their behavior.

Many of the confirmed records are at least partially decomposed strandings. Much of what we know of this species' biology comes from strandings. There are only

Tasmacetus **is large enough to have a visible blow, which like other beaked whales, tends to cant forward when seen well. Southern Australia.** PHOTO: D. DONNELLY

Even from the air, *Tasmacetus* is readily identifiable by its pale melon and the clear boundary between the darker cape and the lighter flanks at the dorsal fin. Tristan da Cunha. PHOTO: P. BEST

The combination of a pale, well-developed melon, long beak, and (especially) the top of the pale shoulder blaze above the pectoral fin readily identify this animal as Shepherd's beaked whale. Southern Australia. PHOTO: D. DONNELLY

Even at a distance, the pale shoulder blaze and the cape rising up directly below the dorsal fin indicate that this is a Shepherd's beaked whale. Southern Australia. PHOTO: D. MANTLE

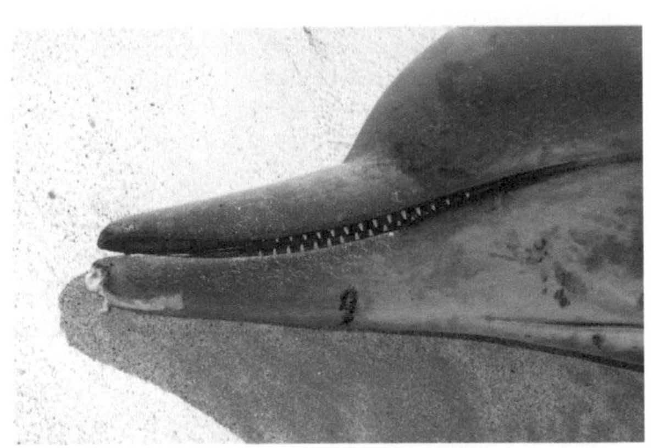

As can be seen in this stranded specimen, the dentition, along with beak and melon shape in *Tasmacetus* is much more dolphin-like than in any of the other known beaked whales. Western Australia. PHOTO: D. HOLLEY

a handful of fresh specimens and confirmed sighting records, although these are becoming more frequent.

Feeding and prey Shepherd's beaked whales are known to feed on several species of fish (primarily eel-pouts), as well as squid and crabs, possibly near the bottom in deep waters. This seems somewhat unusual, as most beaked whales appear to feed almost exclusively on cephalopods.

Threats and status As is true for most of the beaked whales, this species has never been hunted (at least that we know of), and fisheries interactions are not known. They may sometimes ingest plastic debris, which can result in death. Shepherd's beaked whales appear to be relatively rare, but there are no estimates of abundance available.

IUCN status Data Deficient.

Blainville's Beaked Whale—*Mesoplodon densirostris*

(Blainville, 1817)

Recently-used synonyms None.
Common names En.–Blainville's beaked whale, dense-beaked whale; Sp.–*zifio de Blainville, ballena picuda de Blainville;* Fr.–*baleine à bec de Blainville.*
Taxonomic information Cetartiodactyla, Cetacea, Odontoceti, Ziphiidae.
Species characteristics Blainville's beaked whales (like other species in the genus) are characterized by a spindle-shaped body, with a small head, small dorsal fin located about two-thirds of the way back from the snout tip, small and narrow flippers, and tapered flukes with no median notch. There is also a single pair of shallow throat grooves, and the blowhole is a crescent with the ends pointing forward, therefore the "hinge" is at the front. The beak is moderately long in adults, but much shorter and stubbier in younger animals.

This is one of the most distinctive of the mesoplodont beaked whales. The posterior half of the lower jaw of this species is highly arched, even in females and calves. In adult males, the arches are massive and very wide; flattened tusks erupt from the top of these arches (angled forward at about 45°) and the tips extend above the top of the upper jaw. In some individuals, the tusks may be covered by a tassel of barnacles. The "cheeks" of the lower jaw arches rise up along the melon, and the melon looks relatively flat in comparison. The "cheeks" are quite wide, especially when viewed from above, and they may have a "squared-off" top.

Coloration is relatively non-descript in this species. Adult Blainville's beaked whales are brownish-gray above and lighter below. Coloration of young has not been well described, but they appear to be generally countershaded, with a dark patch around the eye. The dark areas of larger animals tend to have round or oval white scars and widely separated, paired scratches are common on what are (often presumed to be) adult males. This species generally

A young male *M. densirostris* has fully-formed lower jaw arches, but the teeth have not erupted yet; he has, however, acquired a few rake marks from other males and numerous cookie-cutter shark bites. Bahamas. PHOTO: J. DURBAN

An adult male Blainville's beaked whale lunges and shows a rare, visible (forward-canted) blow on what is undoubtedly the start of a terminal dive. The teeth are raised above the top of the head where they may function almost as horns in clashes with other males. Western Pacific. PHOTO: A. Ü

shows more white scratches and scarring in adult males than in most other *Mesoplodon* species. There is often a yellowish-orange sheen (probably resulting from diatom films) on the head and anterior part of the thorax.

Maximum known size for males appears to be around 4.7 m, and for females also about 4.7 m (but on average, females are larger than males). Weights of up to 1,033 kg have been recorded. Length at birth is presumed to be between 2.0 and 2.5 m.

Calf—Less than about ½ adult size, arch of lower jaw present, but not exaggerated. Dark eye patches present, body color darker than adults.

Female/subadult male—Full or nearly adult size, back dark brownish-gray in color, lower jaw arched more strongly than in calf.

Adult male—Full or adult size, back dark brownish-gray in color, with numerous white parallel scratches on back, lower jaw with massive arches, and a forward-leaning triangular tusk sitting atop each arch.

Recognizable geographic forms None.

Can be confused with Generally, at sea adult males of this species will be distinguishable from other mesoplodonts. The high arching mouthline and massive flattened tusks (often covered in barnacles) that extend above the upper jaw may allow identification of bulls, if they are observed well, but some other species (such as *M. stejnegeri*) have similar features. The massive size of the arches in Blainville's beaked whales and relatively flat-looking melon are the best clues to separate this species from other, similar-appearing ones in the genus. Groups without bulls will require a closer look, but the arches in the jawline of this species are helpful. Coloration and head-shape differences, along with a more peaked dorsal surface, will help to distinguish this

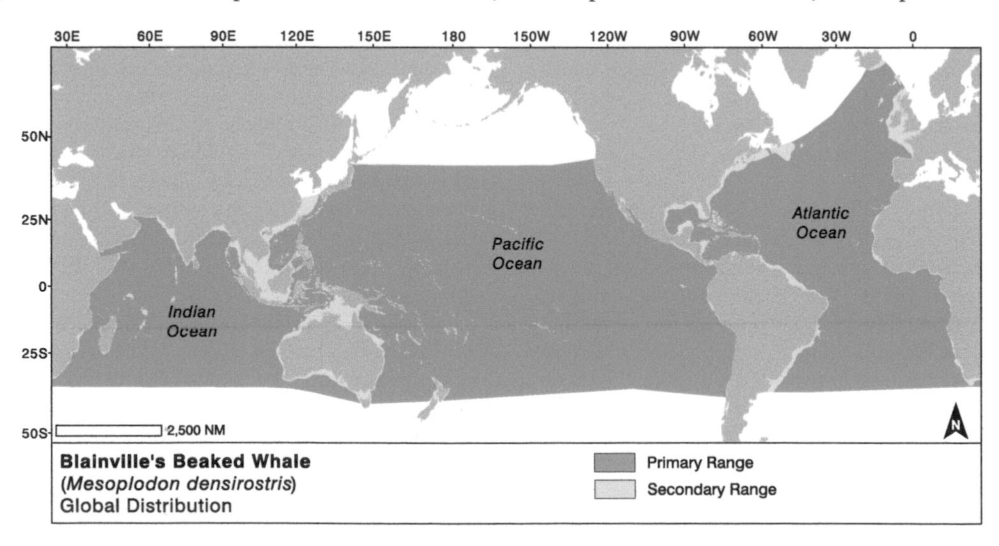

With its teeth raised well above the beak on thick, bony arches from the lower jaw, this adult male *M. densirostris* is a formidable opponent; stalked barnacles conceal both teeth, although the tooth on the far side may be broken off or worn down. PHOTO: M. G. VELASCO

Quite often, the teeth of *M. densirostris* (and other species of beaked whales) are obscured by tufts of stalked barnacles (*Conchoderma*); these "tassels" are useful for identifying the location of the teeth, which can aid in species identification. Hawai'i. PHOTO: R. BAIRD

An adult male Blainville's beaked whale with his tusks raised menacingly over his head; this species is often brownish in color. PHOTO: A. FRIEDLAENDER

Another adult male *M. densirostris* with one tooth exposed and the other covered with *Conchoderma* stalked barnacles. Typically there is only one male in a group and finding him can allow for identifying which species is present. Bahamas. PHOTO: C. DUNN

A very young calf *M. densirostris* (identified by the adults with it) still with fetal folds; it looks very similar to many other mesoplodonts at that age. Bahamas. PHOTO: J. DURBAN

This is probably a cow (foreground) and juvenile *M. densirostris*; the strongly-arched lower jaw separates these from other species of mesoplodonts that have longer beaks or straighter gapes. Bahamas. PHOTO: C. DUNN

This *M. densirostris* is probably a young male: although his massive lower jaw arch is elevated to the top of the head, his teeth apparently have not erupted yet, and he does not appear to have any linear tooth rake marks from other males. PHOTO: M. NOLAN

Females and a young (foreground) Blainville's beaked whales, all heavily pocked and marked with cookie-cutter shark bites; this is the most commonly-observed *Mesoplodon* in tropical waters around the world. Ogasawara Islands, Japan. PHOTO: T. TAKAHASHI

species from Cuvier's beaked whale, which overlaps its distribution extensively.

Distribution Blainville's beaked whales occur in temperate and tropical waters of all oceans. This species has the most extensive distribution of any species of the genus *Mesoplodon,* and is also one of the most tropical. Like other beaked whales, they are found mostly offshore in deep waters. A detailed analysis of habitat preferences in the Bahamas, where this species is commonly encountered, indicated that Blainville's beaked whales were found preferentially in waters of intermediate depth gradients and depths between 200 and 1,000 m (continental slope waters). These may be areas of increased prey availability caused by interactions of currents and local topography. Sightings are also common at some areas in the Hawaiian and Society Islands. They occur in many enclosed seas with deep water, such as the Gulf of Mexico, Caribbean Sea, and the Sea of Japan. However, there are only rare records of this species occurring in the Mediterranean.

Ecology and behavior There is more information available on the behavior and ecology of Blainville's beaked whale than for any other species of *Mesoplodon.* There is a population of this species that is being studied in detail in the Bahamas. Individual whales have been identified, based on natural marks, and this represents only the second case (in addition to the northern bottlenose whales in the Gully) that a beaked whale population has undergone long-term behavioral and ecological study. Groups of three to seven Blainville's beaked whales have been recorded, although singles or pairs are most common. In the Bahamas, adults are generally grouped into what appear to be harems, with a single adult male accompanying several adult females. Subadults appear to stay in separate groups. The harems tend to occur in more productive waters over the continental shelf waters or canyon walls, while subadults tend to occur in less productive waters inshore and offshore of these areas.

Dives of up to 1,400 m and over 54 minutes have been recorded in Hawaiian waters. However, they also appear to spend prolonged periods in the upper layer of the water column (<50 m deep), and this may be an adaptation to compensate for pushing the body's physiological limits on long, deep dives. Animals often lift their beaks out of the water upon surfacing, and sometimes slap them down on the surface again. Blainville's beaked whales do breach and perform other aerial behaviors on occasion.

There are almost no data on sexual maturity for any mesoplodont, but a nine-year-old female of this species had just reached maturity.

Feeding and prey Squid are apparently the main food items, but some deepwater fish may be taken as well. Like other mesoplodonts, these are thought to be suction feeders.

Threats and status Some Blainville's beaked whales have been taken incidentally by Japanese tuna boats off the Seychelles and western Australia, as well as directly by small cetacean hunters in various areas. They are sometimes injured or killed by naval operations using mid- and low-frequency sonar, which cause multiple animals to strand over an extensive area of coastline. Subadult whales, found in more offshore waters, appear to be most vulnerable. Some may also die from ingestion of marine debris. Overall, the species appears to be fairly common in most tropical seas. Estimates of abundance are generally not available, but there are thought to be about 2,100 in the waters of Hawai'i.

IUCN status Data Deficient.

A juvenile *M. densirostris* (identified by its associates); the short beak and strongly arched lower jaw are helpful for identifying young mesoplodonts, or at least eliminating some candidates. Hawai'i. PHOTO: R. BAIRD

A female or young male *M. densirostris*—the lack of marks suggests it is a young animal. The brownish coloration is probably due to diatoms on the skin. PHOTO: M. G. VELASCO

Note on Beaked Whales of the Genus *Mesoplodon*

The species of the genus *Mesoplodon* (collectively called mesoplodonts, a sort-of slang technical term) are very poorly known, and are in fact probably the most poorly known of all the genera of large mammals. Several of the species are known only, or primarily, from study of skeletal material or a few stranded (often rotten) carcasses. Because the external appearance and behavior of most species are so poorly documented and apparently similar among species, it is often very difficult, even for experts, to identify whales of the genus to species from sightings at sea. Even with a specimen in hand, museum preparation or genetic testing is often required for positive identification (except for adult males of some species). Useful field characters are overall size, beak length and shape, shape of mouthline, location and size of teeth, coloration and scarring patterns, and to a lesser extent, size and shape of the dorsal fin.

All species have the same basic body plan, with a laterally-compressed body. The blowhole is oriented with the opening at the back, therefore the "hinge" is at the front (the opposite of the situation in *Berardius*). Because of the similarity and poor documentation of external morphology and coloration among species, genetic techniques have recently become invaluable in distinguishing species. In fact, a new species (*M. perrini*—see below) was recently discovered, and several others were "resurrected" as legitimate species (*M. traversii* and *M. hotaula*) from genetic data. Although data are now rapidly building up, the distribution of most species is also poorly documented, and is still mostly known from stranding records for a few. More information is available from the eastern tropical Pacific than for most other areas, because of the extensive survey effort associated with the tuna fisheries there. Even for this area, however, the picture is very blurry.

There are 15 species of *Mesoplodon* currently recognized. The newest of these (*M. perrini*) was only described in 2002, and it is possible (perhaps even likely), that other undescribed species exist. All mesoplodonts have low, inconspicuous (usually invisible) blows and a small, triangular to slightly-falcate dorsal fin set about two-thirds of the way back from the beak tip. Most species are very similar in appearance, and in some cases only adult males are likely to be identifiable to species at sea.

Coloration is usually brownish-gray to olive, often with extensive white spots and scarring, especially in adult males. It should be noted, however, that the color pattern (and especially its age- and sex-related variation) is not well known for many species. The patterns described below are subject to change as our knowledge increases. In fact, our knowledge has expanded greatly in the last few years, as more fresh specimens are examined, and the first studies of the ecology of these mysterious whales begin to take shape. For instance, we have found that several species (e.g., True's, Gray's, Gervais', strap-toothed, pygmy, and Stejneger's beaked whales) have fairly complex color patterns, often with significant amounts of geographic, developmental, and sexual variation.

The single pair of teeth that grow outside the mouth—the tusks—do not appear to be used in feeding, and are apparently used in male/male combat. It has been suggested that the variation in their size, shape, and placement may be at least partially related to species recognition among multiple, sympatric species. The pattern of scarring is caused by intraspecific closed-mouth fighting among adult males, and has been suggested to act as an indicator of male "quality" in aggressive intraspecific interactions. The exact pattern of scars may be useful in narrowing identifications to several species. For example, sets of closely-paired scratches suggest two teeth located near the tip of the lower jaw (such as in True's, Perrin's, and Hector's beaked whales), while more widely spaced parallel scratches may implicate species with more elevated, and widely-separated teeth set farther back in the jaws (e.g., Blainville's, pygmy, Stejneger's, Andrews', and Hubbs' beaked whales). On the other hand, Gervais', Sowerby's, Gray's, ginkgo-toothed (and probably Deraniyagala's) beaked whales have teeth that are somewhat removed from the tip of the jaw, and which do not project above the upper jaw. Thus, these species would not be expected to produce paired scars, and any scratches on their bodies would likely be single. The strap-toothed whale is a bit enigmatic. Although it has long tusks that rise over the middle of the upper jaw, they curve inwards and backwards and allow only slight opening of the mouth, thus making it less likely that they would produce widely-separated paired scars. The newly-discovered spade-toothed beaked whale may be similar.

In general, mesoplodonts appear rather slow and sluggish. Most sightings are brief, as these whales do not spend much time on the surface. They are known to pursue mostly squid at great depths. Groups are usually small, almost always seven whales or less. Until recently, very little was known of their behavior and social organization. Recent studies on Blainville's beaked whales in the Bahamas suggest that adults may occur in harem-like groups, and other species may have similar social organization. Mesoplodonts are generally seen surfacing briefly, before disappearing on a long dive that may go down to 1,400 m or more. Breaching and other aerial behavior appears to be rare (but that may be partially due to the brevity of most sightings).

Generally, mesoplodonts are difficult to observe for more than short periods, due to their deep diving habits and apparent shyness of vessels. However, in recent years, sightings of more species at sea have been confirmed, and detailed long-term studies of populations of Blainville's beaked whales in the Bahamas and Hawai'i have been undertaken and are ongoing. As opportunities for observing these mysterious animals accumulate, we will be in a position to review our knowledge over the next several years. We may find that much of what we know (or think we know) needs to be revised.

Adult males of the various species of *Mesoplodon*, showing the shape of the jaw-line and shape and placement of the tusks:

A) Blainville's beaked whale PHOTO: N. B. BARROS
B) Gray's beaked whale PHOTO: G. LENTO, COURTESY M. DALEBOUT
C) Ginkgo-toothed beaked whale PHOTO: M. AIZAWA, COURTESY T. YAMADA
D) Deraniyagala's beaked whale PHOTO: L. THOMPSON
E) Hector's beaked whale PHOTO: E. P. MARTINE, COURTESY T. PUSSER
F) Perrin's beaked whale PHOTO: J. G. MEAD
G) Hubbs' beaked whale PHOTO: K.TOKUTAKE/HAKEIJIMA SEA PARADISE
H) Pygmy beaked whale PHOTO: J. URBAN
I) Sowerby's beaked whale PHOTO: N. TREGENZA
J) Gervais' beaked whale PHOTO: C. MANIRE
K) True's beaked whale PHOTO: J. G. MEAD
L) Strap-toothed beaked whale PHOTO: K. WESTERSKOV
M) Andrews' beaked whale PHOTO: AUSTRALIAN DEPARTMENT OF ENVIRONMENT AND CONSERVATION
N) Stejneger's beaked whale PHOTO: COURTESY T. YAMADA
O) Spade-toothed beaked whale (not adult male) PHOTO: P. LIVINGSTONE

A
B
C
D
E
F
G
H
I
J
K
L
M
N
O

Gray's Beaked Whale—*Mesoplodon grayi*

von Haast, 1876

Adult Male

Juvenile

Recently-used synonyms None.
Common names En.–Gray's beaked whale, Scamperdown whale; Sp.–*zifio de Gray;* Fr.–*baleine à bec de Gray.*
Taxonomic information Cetartiodactyla, Cetacea, Odontoceti, Ziphiidae.
Species characteristics Gray's beaked whales are characterized by a spindle-shaped body, with a small head, small dorsal fin located about two-thirds of the way back from the snout tip, small and narrow flippers, and tapered flukes typically with no median notch. There is also a single pair of shallow throat grooves, and the blowhole is a crescent with the ends pointing forward. They are relatively slender, with extremely long, narrow beaks, the longest of any of the mesoplodonts. The mouthline is straight, even in bulls. It is reminiscent of the beak of the rough-toothed dolphin. The melon bulges slightly between the beak and the blowhole.

This species has a distinctive color pattern. Although mostly gray, sometimes with a tan tinge, white patches are found in the genital region and on the beak (both upper and lower jaws) and much of the front of the forehead also becomes white in adults. Young animals appear to have much darker beaks and lighter bellies than adults. There may be dark patches around the eyes in young individuals. Cookie-cutter shark bite scars have been observed on the bodies of some individuals, and adult males may have a number of long linear scars on the body.

The long white beak of Gray's beaked whale is often visible underwater, as seen here. The dark head and pale area behind it is found on juveniles, but the scratching on this animal suggests it may occur on older animals also. Southern Ocean. PHOTO: R. L. PITMAN

There are two triangular teeth set in the middle of the lower jaw, which erupt only in bulls, and 17–22 pairs of small teeth in the posterior half of the upper jaw. The tusks of males are located midway along the length of the mouthline, and are mostly covered in gum tissue, with only the tips exposed. They may be

Young Gray's beaked whales have a dark beak with small teeth still present in the posterior part of the upper jaw. New Zealand. PHOTO: K. STOCKIN

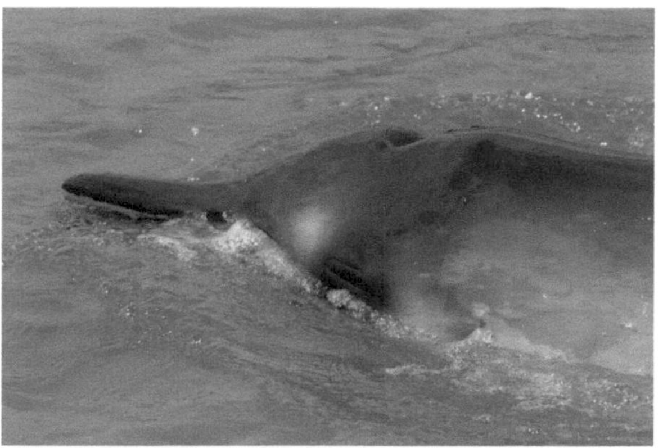

The same animal as left, showing the dark eye patch with white spot in front of it. New Zealand. PHOTO: K. STOCKIN

narrow or quite wide (up to 10 cm). A low and diffuse blow has been reported for some sightings.

Maximum known sizes are 5.6 m for males and 5.3 m for females. These animals are known to reach weights of at least 1,100 kg. Length at birth is estimated to be between 2.1 and 2.2 m.

Recognizable geographic forms None.

Can be confused with When seen from a distance, this species may be confused with any of the other species of the genus that overlap it in distribution (e.g., Andrews', Hector's, Blainville's, and strap-toothed beaked whales). The extremely long white beak of adults (often stuck up out of the water as the animal surfaces) and straight mouthline should allow Gray's beaked whales to be distinguished from other mesoplodonts, if a good look at these features is obtained.

Distribution Gray's beaked whale is primarily a Southern Hemisphere cool temperate species, which is apparently circumantarctic in occurrence. It typically occurs in deep waters beyond the edge of the continental shelf, with most records south of 30°S. There are many sighting records from Antarctic and subantarctic waters, and in summer months they appear near the Antarctic Peninsula and along the shores of the continent (sometimes among sea ice). Many of the stranding records are from New Zealand and southern Australia, South Africa, Argentina, Chile, and Peru. The area between the South Island of New Zealand and the Chatham Islands has been suggested to

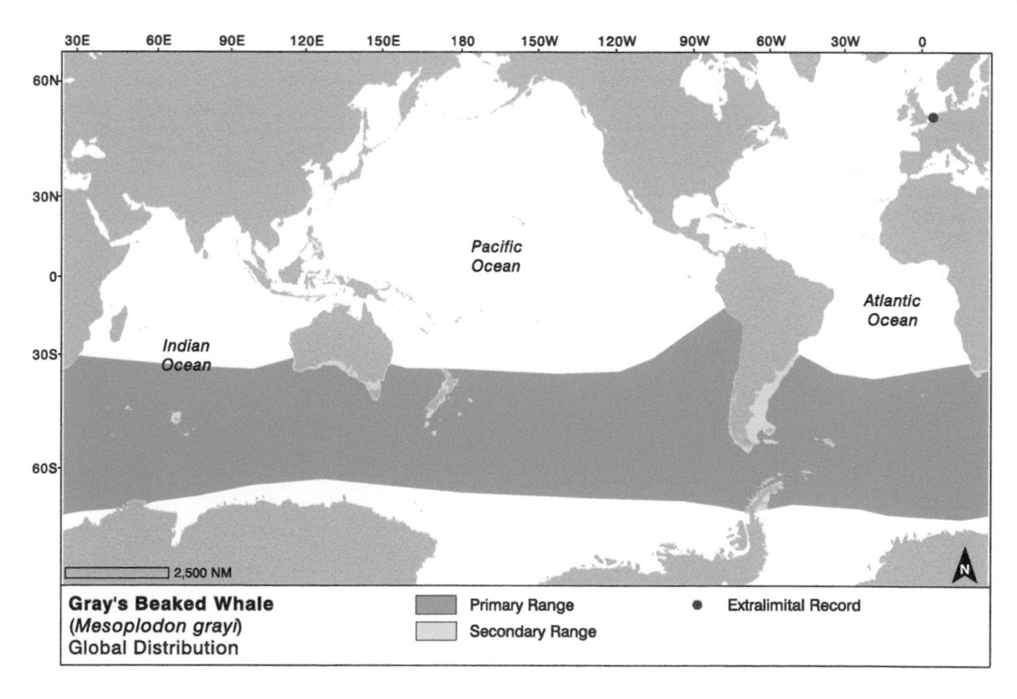

An adult male *M. grayi* showing heavy, parallel tooth rake marks and the long, white narrow beak of this species. PHOTO: M. NOLAN

When *M. grayi* surfaces, the long narrow beak often projects at a 45° angle. This is probably a younger animal because the beak has not turned completely white yet; the reddish coloration on the beak is from diatoms. PHOTO: P. OLSON

The farther animal with a dark head and white shoulder (and no markings) is probably the calf of the Gray's beaked whale in the foreground. PHOTO: P. OLSON

An adult male Gray's beaked whale showing the relatively-straight gape, as well as the size, location and orientation of the teeth in lower jaw, and extent of white on upper and lower beak. New Zealand. PHOTO: M. DALEBOUT

A young Gray's beaked whale showing a dark eye patch; the dark from the eye goes up over the head giving it a dark-capped appearance with a pale area behind it. The beak is long, but has not turned white yet. Also, small teeth are just visible in the upper jaw. New Zealand. PHOTO: NEW ZEALAND DEPARTMENT OF CONSERVATION

be a "hot spot" for sightings of this species. There is one record of a Gray's beaked whale straying into the Northern Hemisphere, a stranding record in the Netherlands. This was undoubtedly an extralimital occurrence, however.

Ecology and behavior Not a great deal is known of the biology of this species. Gray's beaked whales are seen mostly as singles or pairs; however, there is one record of a mass stranding of 25–28 of these whales (and other mass strandings are known). They are relatively frequently stranded, especially in New Zealand. Gray's beaked whales generally raise their long beaks out of the water when surfacing. Breaching, spyhopping, flipper-slapping, and fluke-slapping behavior has been observed, and they may almost "porpoise" out of the water when moving fast. They typically do not lift their flukes out of the water upon commencing a dive.

A female with a calf that was observed in Mahurangi Harbor, New Zealand, had a series of corrugated scars behind the dorsal fin suggesting that the animal had been struck by a vessel. Observations in such shallow waters are very unusual for this species.

Feeding and prey Like other mesoplodonts, they are thought to feed mainly on cephalopods in deep waters.

Threats and status This species may not be as rare as some other species of the genus *Mesoplodon,* based on the number of records. In particular, they seem to be fairly common around New Zealand. No significant exploitation is known, and there are no known estimates of abundance.

IUCN status Data Deficient.

Ginkgo-toothed Beaked Whale—*Mesoplodon ginkgodens*

Nishiwaki and Kamiya, 1958

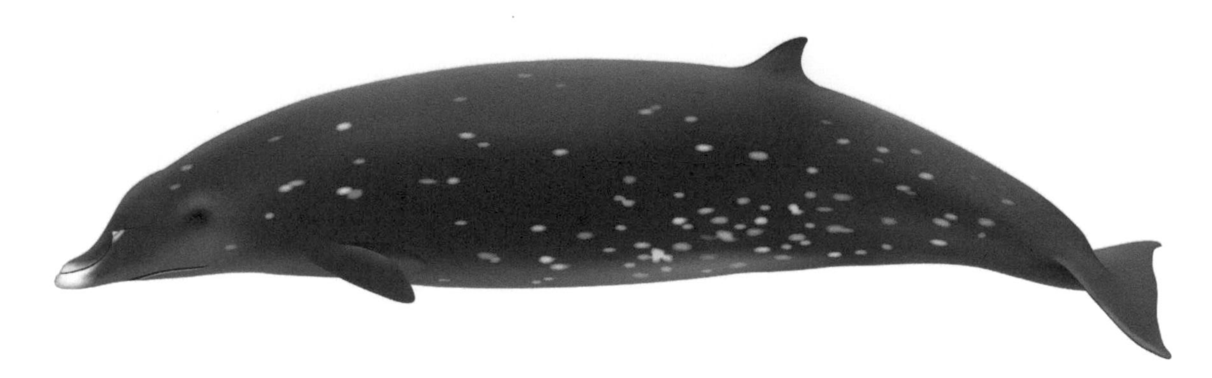

Recently-used synonyms None.
Common names En.–ginkgo-toothed beaked whale; Sp. –*zifio Japonés, ballena picuda de dientes de ginkgo*; Fr. –*baleine à bec de Nishiwaki.*
Taxonomic information Cetartiodactyla, Cetacea, Odontoceti, Ziphiidae. The newly-recognized species *M. hotaula* has recently been found to be valid and split off as a separate species.
Species characteristics Ginkgo-toothed beaked whales are characterized by a spindle-shaped body, with a small head, small dorsal fin located about two-thirds of the way back from the snout tip, small and narrow flippers, and tapered flukes with no median notch. There is also a single pair of shallow throat grooves, and the blowhole is a crescent with the ends pointing forward. The beak is moderate in length and the mouthline has a slight arch in adult males. The forehead has a shallow rise to it.

The color pattern is poorly known (few fresh specimens have been examined), but appears to consist largely of basic countershading. Adult male ginkgo-toothed beaked whales are dark gray, and the anterior portion of the beak is white. Females are reported to be somewhat lighter than males. It is unclear whether females share the white beak, but there is some evidence for at least a paler lower jaw in females. Light round or oval scars on the back and ventral surface of the tail stock also appear in adults. These white scars are considered to be caused by parasites (lampreys or cookie-cutter sharks). Unlike the situation in most other beaked whales, males generally have few or no white scratches (this may be due to the small amount of tooth that erupts from the gums).

Bulls have wide (up to 10 cm), round or elliptical, flattened tusks with an S-shaped outline sloping down from the point, which sit atop small arches slightly behind the middle of the lower jaw. The tusks are as wide or wider than they are tall. They barely break the gumline, and most of the tusks are buried in gum tissue. Teeth do not erupt from the gums in females. In juveniles (but not adults),

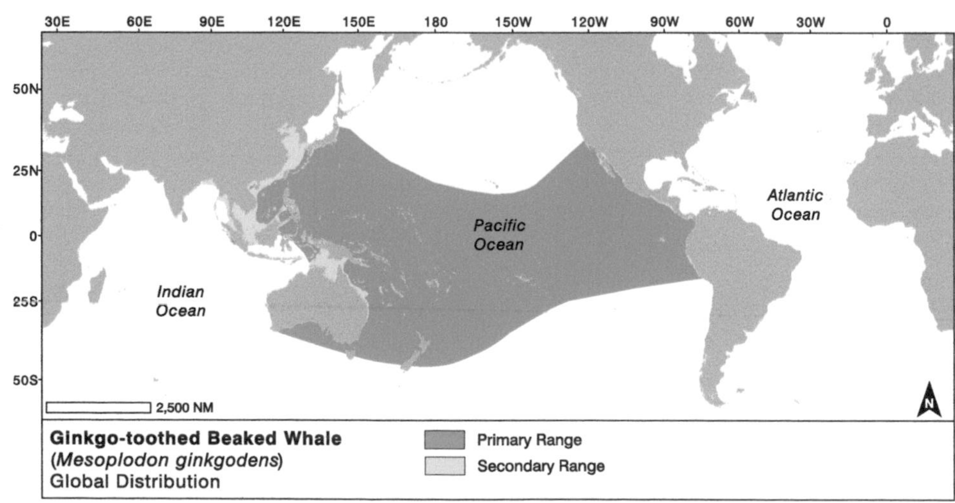

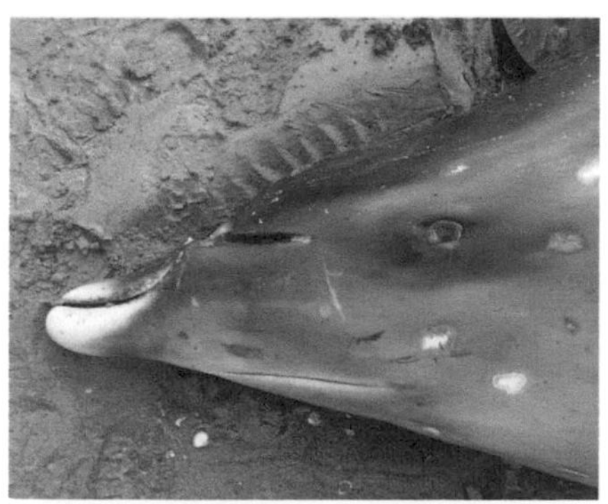
A young male ginkgo-toothed beaked whale; the teeth have apparently just recently erupted because there are no visible rake marks from other adult males on the body. New Zealand. PHOTO: NEW ZEALAND DEPT. OF CONSERVATION

The same animal as in the image to the left; ventral view—the white spots are healed cookie-cutter bite wounds, with an interesting concentration in the genital area of the animal. New Zealand. PHOTO: NEW ZEALAND DEPT. OF CONSERVATION

There are no confirmed photographs or sightings of live ginkgo-toothed beaked whales. The identity of this whale cannot be confirmed, but its visible features seem to match those of the ginkgo-toothed beaked whale. Baja California. PHOTO: F. LYMAN

This appears to be a young male: there is a prominent arch to the lower jaw, but no tooth is visible and there are no linear tooth rake scars. In addition to color patterns, the size of the melon, beak length, size and location of the arch in the lower jaw are all important features for identifying mesoplodonts. New Zealand. PHOTO: NEW ZEALAND DEPT. OF CONSERVATION

they resemble the leaves of the ginkgo tree. Maximum known body length is 5.3 m (for both males and females), although females appear to be larger on average. At birth, gingko-toothed beaked whales are thought to be about 2.0–2.5 m.

Recognizable geographic forms None.

Can be confused with The uniform dark pigmentation, small posteriorly-placed teeth, and lack of characteristic ziphiid scars make it very challenging to identify these animals at sea, even when adult males are seen. In particular, females of this species may be virtually indistinguishable from other mesoplodonts that share their range (e.g., Deraniyagala's, Hubbs', Blainville's, and Stejneger's beaked whales).

Distribution The ginkgo-toothed beaked whale is known from a few dozen widely-scattered (albeit sparse) strandings and captures (no confirmed sightings) in temperate and tropical waters of the Pacific Ocean, from Japan/Taiwan east to the shores of North America and the Galapagos Islands. There have also been a few records from New Zealand and Southern Australia, indicating that this species also inhabits the southwestern Pacific. Most records are from the seas around Japan. It is generally hypothesized that the range is continuous across the Pacific and possibly into the Indian Ocean, but until the species is reliably identified at sea, its true distribution will probably remain unknown.

Ecology and behavior Almost nothing is known of the ecology and biology of the ginkgo-toothed beaked whale, short of what little has been gleaned from stranded specimens. It has never been reliably identified alive in the wild, although several probable sightings have been reported.

Feeding and prey Ginkgo-toothed beaked whales are presumed to be primarily squid eaters, but may also take some fish.

Threats and status Ginkgo-toothed beaked whales have occasionally been taken by Japanese and Taiwanese whalers, and some have been caught in deepwater drift gillnets. There are no estimates of abundance, but the species does not appear to be very common.

IUCN status Data Deficient.

Deraniyagala's Beaked Whale—*Mesoplodon hotaula*

Deraniyagala, 1963

Recently-used synonyms None.
Common names En.–Deraniyagala's beaked whale; Sp.–*zifio de Deraniyagala*; Fr.–*baleine à bec de Deraniyagala*.
Taxonomic information Cetartiodactyla, Cetacea, Odontoceti, Ziphiidae. Although originally described in 1963 and quickly synonymized with *M. ginkgodens*, this species was only recently (2014) recognized as being valid (past records were generally lumped with *M. ginkgodens*).
Species characteristics Like other members of the genus *Mesoplodon*, Deraniyagala's beaked whales are characterized by a spindle-shaped body, with a small head, small dorsal fin located about two-thirds of the way back from the snout tip, small and narrow flippers, and un-notched flukes. There is also a single pair of shallow throat grooves, and the blowhole is a crescent with the ends pointing forward. The beak is moderate in length and the mouthline has a slight arch in adult males. The forehead has a shallow rise to it.

The color pattern is poorly-known (only a handful specimens have so far been examined), but appears to consist largely of basic countershading. Adult male Deraniyagala's beaked whales are dark gray, and apparently the anterior portion of the beak is lighter. It is unclear whether females share the white beak. Light spots on the back and ventral surface of the tailstock may also appear in adults. These white scars are probably caused by parasites (lampreys or cookie-cutter sharks). Specimens examined so far have not had the long, linear scars that are common on males of most other *Mesoplodon* species.

Bulls have tall, slightly recurved, flattened tusks with an S-shaped outline sloping down from the point, which emerge from atop small arches in the middle of the lower jaw. They are taller than wide. Most of the tusk's length is buried in gum tissue, and they may be broken off or worn. Teeth do not erupt from the gums in females. Maximum known adult length is 4.8 m, although they

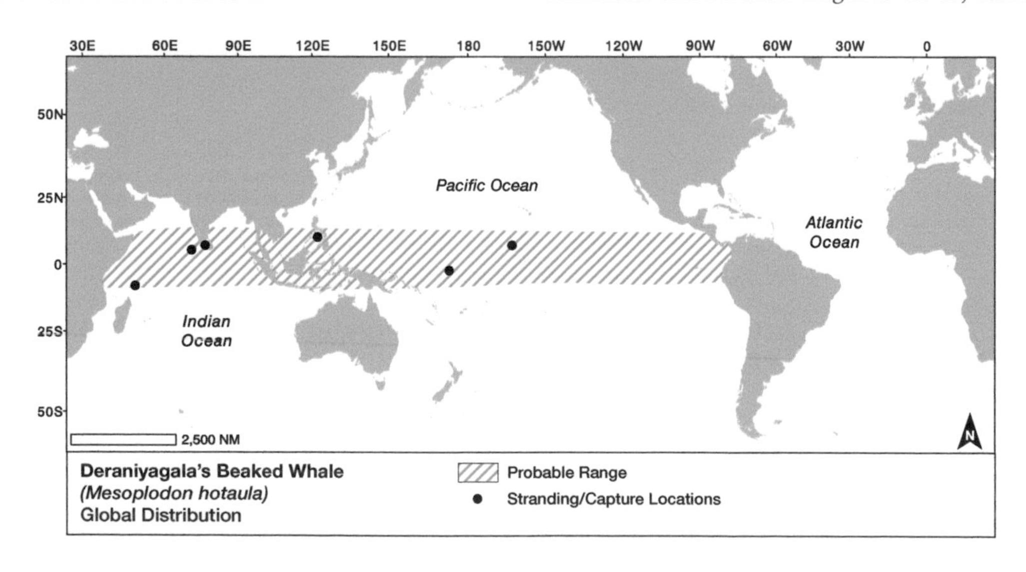

These photos of a stranding in the Seychelles are, to our knowledge, the only confirmed images of Deraniyagala's beaked whale. The top view of the head shows an erupted tooth indicating it was an adult male; the tooth was broken and stalked barnacles (*Conchoderma* spp.) are growing on it, giving it a "tasseled" appearance in life. It is unknown if the pale pattern on the beak, shown in the side view of the head, will be useful for identifying this species at sea. Seychelles. PHOTOS: L. THOMPSON

A cow and calf mesoplodont at Palmyra Island, central Pacific; the cow on the right is heavily pocked with healed cookie-cutter shark bites while the calf is unmarked. The scars on the cow have healed the same color as the surrounding skin so that, perhaps significantly, there is no white spotting evident on the upper surface. Their identity is unconfirmed, but probably *M. hotaula*. PHOTO: R. L. PITMAN

probably grow larger, as only seven specimens (3.9-4.8 m long) have so far been studied. Females appear to be larger than males on average in all mesoplodonts. Size at birth is not known, but is probably around 2 m.

Recognizable geographic forms None.

Can be confused with The uniform dark pigmentation, small posteriorly-placed teeth, lack of characteristic ziphiid scars, and poorly-known characters make it nearly impossible to identify these animals at sea, even when adult males are seen. In many cases, this species may be virtually indistinguishable from other mesoplodonts that may share its range (especially ginkgo-toothed beaked whales).

Distribution Deraniyagala's beaked whale is currently known from only seven widely-scattered strandings and captures (no confirmed sightings) in the tropical waters of the Indo-Pacific Ocean, from Sri Lanka, Kiribati, Line Islands, Philippines, Maldives, and Seychelles. It may span the entire tropical Indo-Pacific, but until the species can be identified at sea, its true distribution will probably remain unknown.

Ecology and behavior Essentially nothing is known of the ecology and biology of Deraniyagala's beaked whale, short of what little has been gleaned from the few stranded specimens. It has never been reliably identified alive in the wild. The broken teeth of the specimens examined so far suggest that males do use their teeth as weapons in male/male combat.

Feeding and prey Food habits have not been well studied. Deraniyagala's beaked whales are presumed to be primarily squid eaters, but may also take some fish.

Threats and status Nothing is known about the status of this species. There are no estimates of abundance, but based on the small number of records so far found, the species does not appear to be very common.

IUCN status Not listed.

Hector's Beaked Whale—*Mesoplodon hectori*

(Gray, 1871)

Recently-used synonyms None.
Common names En.–Hector's beaked whale; Sp.–*zifio de Hector, ballena picuda de Hector; Fr.–baleine à bec de Hector.*
Taxonomic information Cetartiodactyla, Cetacea, Odontoceti, Ziphiidae. Several sightings and strandings of beaked whales originally thought to be this species occurred in southern and central California in the 1970s. However, these are now known to have been Perrin's beaked whales (see below).
Species characteristics Hector's beaked whales are characterized by a spindle-shaped body, with a small head, small dorsal fin located about two-thirds of the way back from the snout tip, small and narrow flippers, and tapered flukes with no median notch. There is also a single pair of shallow throat grooves, and the blowhole is a crescent with the ends pointing forward. In adults, the beak is quite long, and the mouthline is relatively straight.

Body color of Hector's beaked whales was not well known until recently (2006), as few fresh specimens had been examined. However, it is now known that adult males are dark gray-brown above and slightly lighter below. Single and closely-paired scratches, as well as white cookie-cutter shark scars, may be present on the body. The most distinctive feature of the adult male is its white beak and anterior part of the head (although this description is based on only one specimen). Females and young, in contrast, appear to be relatively non-descript. In at least some individuals, probably younger animals, there is a dark patch surrounding the eye and running forward to connect to the melon and upper beak. The lower jaw may be light in color.

The single pair of flattened, triangular tusks are moderate in size and they are located near the tip of the lower jaw. They erupt only in bulls, and may be angled slightly outwards. Specimens of up to 4.3 m have been reported, although few have been measured. Females are probably slightly larger than males. Size of newborns is unknown, but is presumably about 1.9–2.0 m.
Recognizable geographic forms None.

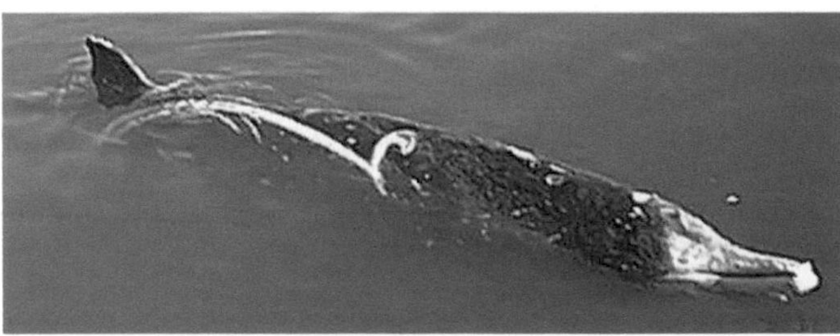

This is one of the only known photographs of an adult male Hector's beaked whale—from a stranding in Argentina. The long, all-white rostrum is very similar to that of Gray's beaked whale, but the tooth placement differs: instead of in the middle of the lower jaw, Hector's beaked whale teeth erupt at the tip of the lower jaw. These apical teeth are just visible in the photo. PHOTO: H. L. CAPPOZZO

Can be confused with If adult males are not present, it may be nearly impossible to identify these animals at sea. For bulls, the placement of the flattened teeth at the tip of the lower jaw, in combination with their white beaks and closely-paired scratches, should allow them to be distinguished from other species of *Mesoplodon* that are sympatric, if the head is seen quite well. Of the overlapping species, Gray's beaked whale can be similar in appearance. For adult males, note the presence (Hector's) or absence (Gray's) of paired scratches, and placement of the tusks at the tip (Hector's) or middle (Gray's) of the lower jaw.

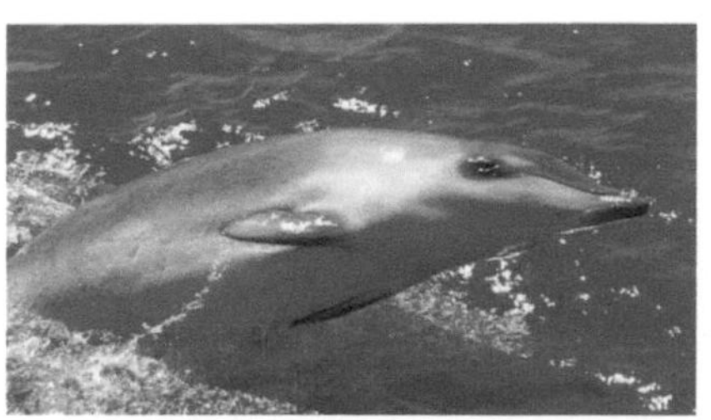

This juvenile Hector's beaked whale was identified from a genetic sample; otherwise it appears to be virtually indistinguishable from several other species. Western Australia. PHOTO: N. GALES

Adult male Hector's beaked whales have tusks that erupt near the tip of the lower jaw. PHOTO: H. L. CAPPOZZO

Distribution Hector's beaked whale is considered to be a Southern Hemisphere cool temperate species. The known records (mostly strandings) are from southern South America, South Africa, southern Australia, and New Zealand. The single sighting record is from southwest Australia. It has been speculated that the species has a continuous distribution in the Atlantic and Indian oceans at least from South America to New Zealand. Although there are no current records from the central and eastern Pacific Ocean, the range may prove to be circumpolar. They may be relatively common around New Zealand.

Ecology and behavior Almost nothing is known of the biology of Hector's beaked whale. Most of what had previously been attributed to this species was later found to be referable to Perrin's beaked whale (when that species was discovered and described in 2002). Even the external appearance of Hector's beaked whale was not adequately described until recently.

In the only known confirmed identification of this species alive at sea, a single individual was observed nearshore in western Australia—almost definitely atypical for the species. The animal breached several times near a research vessel. It dove for periods of up to 4 minutes, but these are clearly not representative of the diving capabilities of this deep-diving species.

Feeding and prey Little is known of the diet, but Hector's beaked whales are known to feed on squid, like most other beaked whales.

Threats and status There are no known threats, and no estimates of abundance. It is possible that this species is naturally rare; however, the paucity of records may have much to do with the challenge of identifying it at sea.

IUCN Status Data Deficient.

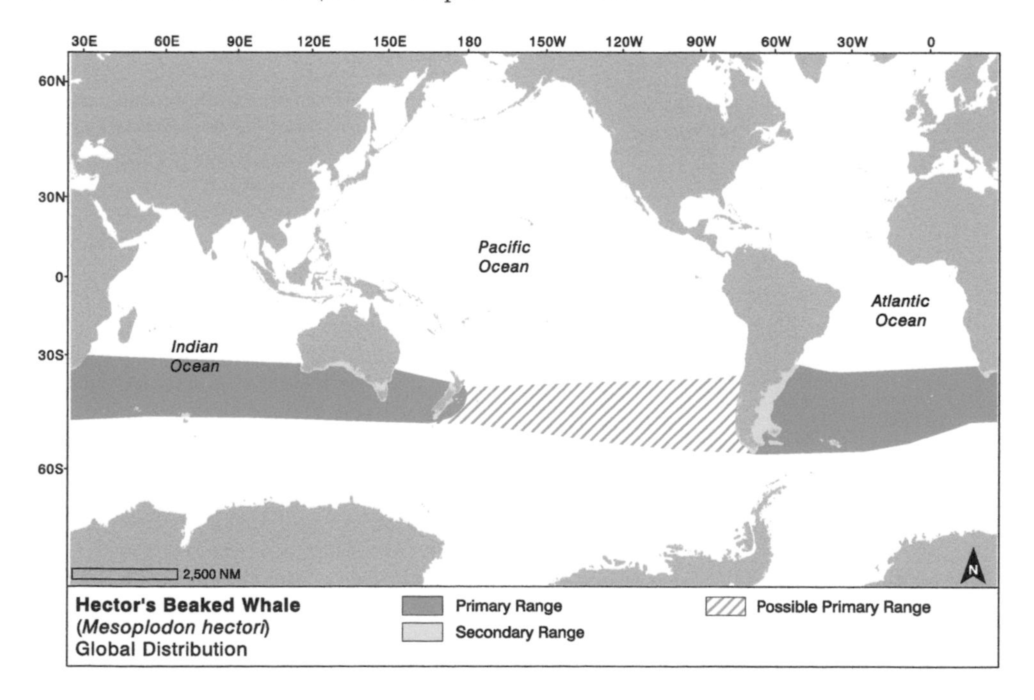

Ziphiidae

Hector's Beaked Whale

Perrin's Beaked Whale—*Mesoplodon perrini*

Dalebout, Mead, Baker, and van Helden, 2002

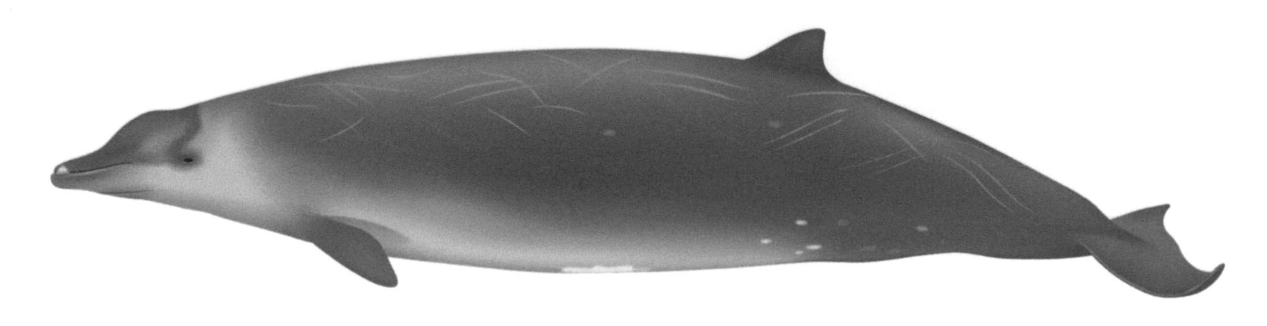

Recently-used synonyms None.
Common names En.–Perrin's beaked whale; Sp.–*zifio de Perrin;* Fr.–*baleine à bec de Perrin.*
Taxonomic information Cetartiodactyla, Cetacea, Odontoceti, Ziphiidae. In the 1970s, several sightings and strandings of beaked whales thought to be Hector's beaked whales occurred in California. Genetic evidence indicated that a new species was actually involved, and these are now known to have been Perrin's beaked whales.
Species characteristics Externally, this species appears very similar to Hector's beaked whale, and in fact, the two were thought to be the same until *M. perrini's* description in 2002. It has a spindle-shaped body, with a small head and a slight bulge to the melon, small dorsal fin located about two-thirds of the way back from the snout tip, small and narrow flippers, and tapered flukes with no median notch. There is also a single pair of shallow throat grooves, and the blowhole is a crescent with the ends pointing forward. The beak is shorter than in any other *Mesoplodon,* except the pygmy beaked whale; it is even shorter in calves.

Apparently, these animals are countershaded, with a dark gray back and whitish belly. Adult males, at least, have a white patch around the umbilicus, and a mask of dark gray color extending from near the gape to the eye and from there dorsally. The lower jaw and throat are light in color. The ventral surface of the flukes exhibits white striations. Any tooth scrapes on adult males tend to be singular, not paired. Cookie-cutter shark scars may be present on the body.

The flattened, triangular tusks of adult males are situated just behind the tip of the lower jaw, and the mouthline is relatively straight. The tusks are splayed outwards at an angle of about 15° and barnacles may infest the teeth. Females may have similar teeth, but they do not erupt.

Only five specimens have been examined, but these indicate that the animals can grow to at least 4.4 m (females) and 3.9 m (males).
Recognizable geographic forms None.
Can be confused with Only adult males will likely be identifiable from their external appearance. Perrin's beaked whale is most likely to be confused with Hector's beaked whale; however, the two probably do not overlap in distribution. A specimen of unknown origin will require expert examination of the skull or genetic analyses to distinguish between the two with certainty.
Distribution All five specimens so far examined have been from southern and central California (between 32° and 37°N), and it is likely that this species is endemic to the North Pacific Ocean (possibly even the eastern North Pacific). Like other members of the genus, it presumably prefers oceanic habitats with waters >1,000 m deep, and sightings thought to be of this species have occurred in deep waters.

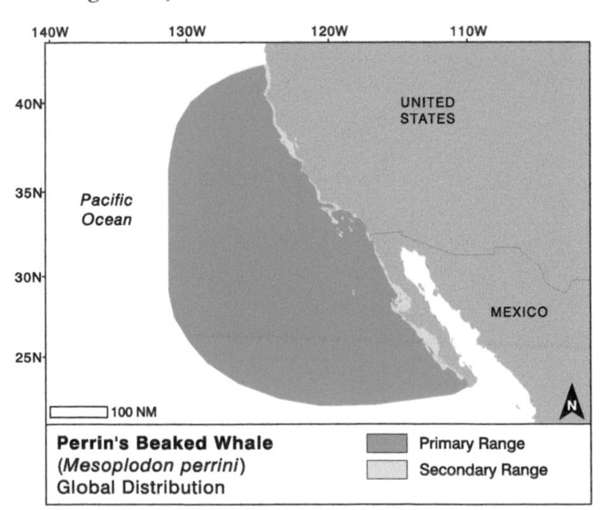

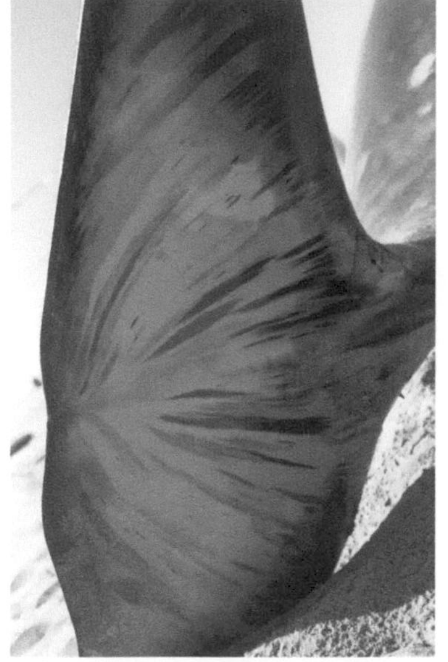

Various views of a stranded female Perrin's beaked whale from Southern California. The body shape and coloration is very similar to numerous other *Mesoplodon* species. However, the short beak has a relatively straight gape (typical for species with apical teeth), and this may allow for distinguishing from the other North Pacific mesoplodonts. The "starburst" color pattern on the underside of the flukes also occurs on Stejneger's beaked whale, but will probably never be seen in the wild. PHOTOS: A. SCHULMAN-JANIGER

Not much is known about the color pattern of Perrin's beaked whale, but the adult male has a large, triangular tooth near the tip of the lower jaw and a short beak, which should allow for positive identification at sea. Southern California. PHOTO: J. G. MEAD

Ecology and behavior This is one of the most recently-discovered species of marine mammal; its original description was only published in July 2002. Almost nothing is known of its biology or behavior. The specimens that are currently known were mostly erroneously thought to have been Hector's beaked whales in the past. The only possible sightings are of two groups off southern California, and they are not confirmed to be of this species. Calves may be weaned by the time they reach about 2.5 m in length.

Feeding and prey Based on a limited sample of stomach contents, Perrin's beaked whale probably feeds mainly on squids (including *Octopoteuthis* sp.). The remains of an unidentified invertebrate have been found in the stomach of an animal stranded in California.

Threats and status All records have been of strandings, so nothing is known of potential threats. No estimates of abundance exist.

IUCN status Data Deficient.

Hubbs' Beaked Whale—*Mesoplodon carlhubbsi*

Moore, 1963

Recently-used synonyms None.

Common names En.–Hubbs' beaked whale; Sp.–*zifio de Hubbs, ballena picuda de Hubbs; Fr.–baleine à bec de Hubbs.*

Taxonomic information Cetartiodactyla, Cetacea, Odontoceti, Ziphiidae. In the past, there was some suggestion that this species may simply be a subspecies of *M. bowdoini* (and some previous records were erroneously attributed to that species), but recent genetic studies confirm its specific distinctness.

Species characteristics Hubbs' beaked whales are characterized by a spindle-shaped body, with a small head, small dorsal fin (generally 22–23 cm high in adults) located about two-thirds of the way back from the snout tip, small and narrow flippers, and tapered flukes with no median notch. There is also a single pair of shallow throat grooves, and the blowhole is a crescent with the ends pointing forward. The beak is fairly short and there is a strong arch to the lower jaw of adult males, although the arch is more subdued in females. The length of the arched area is much smaller than in the Blainville's beaked whale (<1/3 of the lower jaw vs. well over 1/2). There is a distinct bulge on the melon.

Head of a 5.2 m female *M. carlhubbsi* that stranded in Oregon. The white upper and lower jaw separates it from other mesoplodonts within its North Pacific range. PHOTO: M. GRAYBILL

Males have massive flattened tusks erupting from the middle of narrow arches in each side of the lower jaw. The arch is actually mostly gum tissue surrounding the tusk, and only the tips of the teeth are exposed. The tusks protrude above the level of the upper jaw.

The basic body color is dark gray to black. Adult males of this species are more readily identifiable than individuals of most other species of mesoplodont. They have a white beak and a brilliant white "cap" or "beanie" on the melon ahead of the blowhole. The combination of these two features is diagnostic. Females and young are more generic, being more-or-less countershaded, but they may also have a light beak. Adult males are generally covered with long scratches, often arranged in pairs.

Maximum known size for both sexes is about 5.3–5.4 m. Weights of over 1,500 kg are attained. Newborns are thought to be about 2.5 m long.

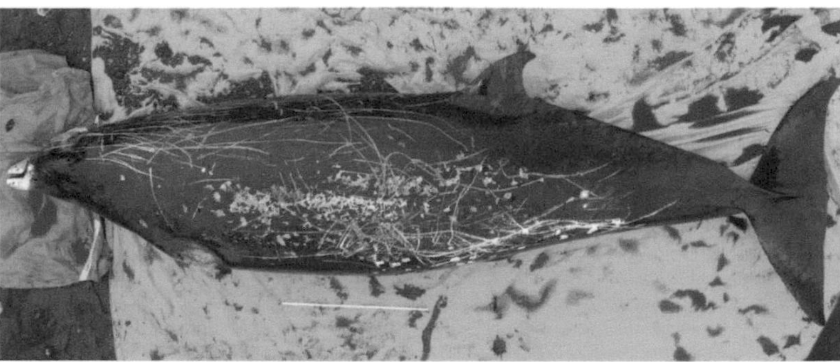
An adult male Hubbs' beaked whale with linear scarring from rakes from aggressive encounters with other males; interestingly there appears to be a concentration in the genital area. Although this species is apparently quite rare, the stout, white-tipped beak and white "beanie" (top of melon, not visible here), and the large tooth on a prominent arch on each side of the lower jaw makes this fairly easy to identify, if encountered at sea in the North Pacific. Japan. PHOTO: COURTESY T. YAMADA

Recognizable geographic forms None.

Can be confused with If a good look is obtained of the head, the white "beanie" and beak tip, and large tusks may allow bulls of this species to be distinguished from other species of *Mesoplodon*. Other age classes will be nearly impossible to distinguish from congeners, based solely on their external appearance. Genetic testing or cranial examination may be required.

Distribution Apparently limited to the North Pacific Ocean, Hubbs' beaked whale is known from central British Columbia to southern California in the east, and from Japan in the west. Sightings have been made off the coast of Oregon. It is an oceanic species, and the range is thought to be continuous across the North Pacific, although this is not confirmed.

Ecology and behavior Very little is known about the biology of this species, as only a few reliable sightings at sea have been made. This is somewhat surprising, as this is one of the more distinctive species of the genus (at least for adult males). The long, white, parallel scratches on the bodies of males are thought to be caused by closed-mouth fighting in this and other mesoplodonts. Hubbs' beaked whale has particularly dense bone in the rostrum as an adaptation to strengthen the upper jaw for such combat. Not much is known of the reproductive biology of this species, but calving may occur mainly in summer months.

Feeding and prey Hubbs' beaked whales feed on squid (including the genera *Gonatus, Onychoteuthis, Octopoteuthis, Histioteuthis*, and *Mastigoteuthis)* and some deepwater fishes.

Threats and status Hubbs' beaked whale has occasionally been taken by Japanese whalers in several small cetacean fisheries. Incidental catches in drift gillnets occur off the coast of California. There are no estimates of abundance.

IUCN status Data Deficient.

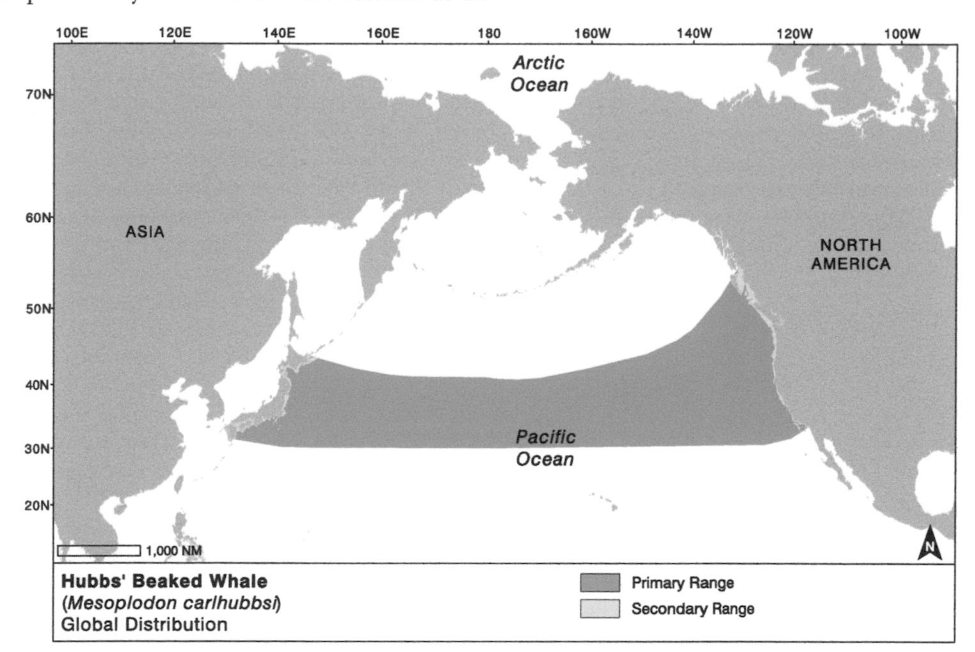

Zipiidae

Hubbs' Beaked Whale

Pygmy Beaked Whale—*Mesoplodon peruvianus*

Reyes, Mead, and Van Waerebeek, 1991

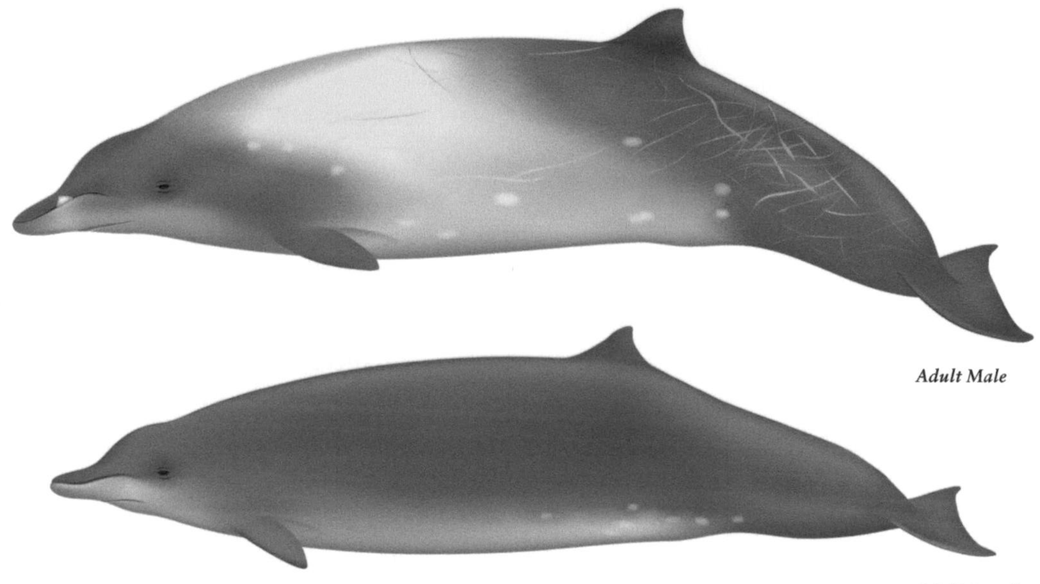

Recently-used synonyms None.

Common names En.-pygmy beaked whale, Peruvian beaked whale; Sp.- *ballena picuda; Fr.-baleine à bec pygmée.*

Taxonomic information Cetartiodactyla, Cetacea, Odontoceti, Ziphiidae. The pygmy beaked whale was described as a new species only in 1991. Whales now known to be of this species had previously been sighted in the eastern tropical Pacific. Since it was unknown to what species they belonged, they were listed in some previous field guides as "*Mesoplodon* sp. A."

Species characteristics The pygmy beaked whale is one of the most recently-described members of the genus, and appears to be the smallest of the species of *Mesoplodon.* These whales are characterized by a spindle-shaped body, with a small head with a slightly bulging melon, short beak, dorsal fin located about two-thirds of the way back from the snout tip, small and narrow flippers, and tapered flukes with no median notch. There is also a single pair of shallow throat grooves, and the blowhole is a crescent with the ends pointing forward. These animals have fairly short beaks, and slightly arched mouthlines. Their small, triangular, wide-based dorsal fins are shaped quite like those of harbor porpoises, and their tail stocks are deepened.

Two morphs exist, a scarred black-and-white form with a white swath across the back that is easily identified in the field (adult males), and another one that is largely uniformly colored on the back (females and subadults). The latter tend to be dark brownish-gray above and lighter below, with relatively little scarring.

The most distinctive characteristic is the teeth, however, which erupt from the middle of the lower jaw in

An aerial view of a heavily scarred *M. peruvianus*; a pair of erupted to teeth are visible. The head appears brownish, which may be due to diatoms on the skin. Eastern tropical Pacific.
PHOTO: NOAA FISHERIES, SWFSC

Adult males of *M. peruvianus* have a large white swath on the body that starts in the middle of the back in front of the dorsal fin and continues down the sides and toward the rear, forming a very broad band. This feature is apparently unique to this species and makes it easy to identify at sea. Michoacan, Mexico.
PHOTO: COURTESY J. URBAN

A pair of female or juvenile *M. peruvianus* surface in the rain in the eastern tropical Pacific—females and juveniles lack the white band of the adult males. This species has a very distinctive low, triangular dorsal fin, reminiscent of a harbor porpoise. The brown on the head is probably due to diatoms. PHOTOS: NOAA FISHERIES, SWFSC

adult males. They are quite small and oval or egg-shaped in cross section (although generally they are not visible in at-sea sightings). They lean forward at an angle of 20–40°. The mouthline has a slight to moderate arch.

This is the smallest known species of *Mesoplodon;* maximum known length is only about 3.7–3.9 m in both sexes. At birth, these animals are probably about 1.6 m long.

Calf—Approximately 1/2–2/3 the length of adults, coloration simply countershaded.

Female/subadult—Adult size, lower jaw with slight arch, no erupted teeth, back grayish-brown with lighter belly, scarring minimal or absent.

Adult male—Adult size, body may be especially robust, most of upper body dark brown to black, prominent white or cream-colored blaze extending from ventral area forward over back to just behind head (may present appearance of a chevron), often white beak tip, light patches on lower jaw near teeth, white scratches moderate to extensive, lower jaw moderately arched, with protruding tusks.

Recognizable geographic forms None.

Can be confused with Adult male pygmy beaked whales may be easily distinguishable from other mesoplodonts in their range (several species, but only M. *densirostris* appears to be common in the area) by the presence of a broad light swath that runs from the head and down the sides, on the otherwise dark brown or black body. Females and immatures are not easily distinguishable from other mesoplodonts. It is likely that osteological or genetic methods will need to be used to identify anything other than adult males.

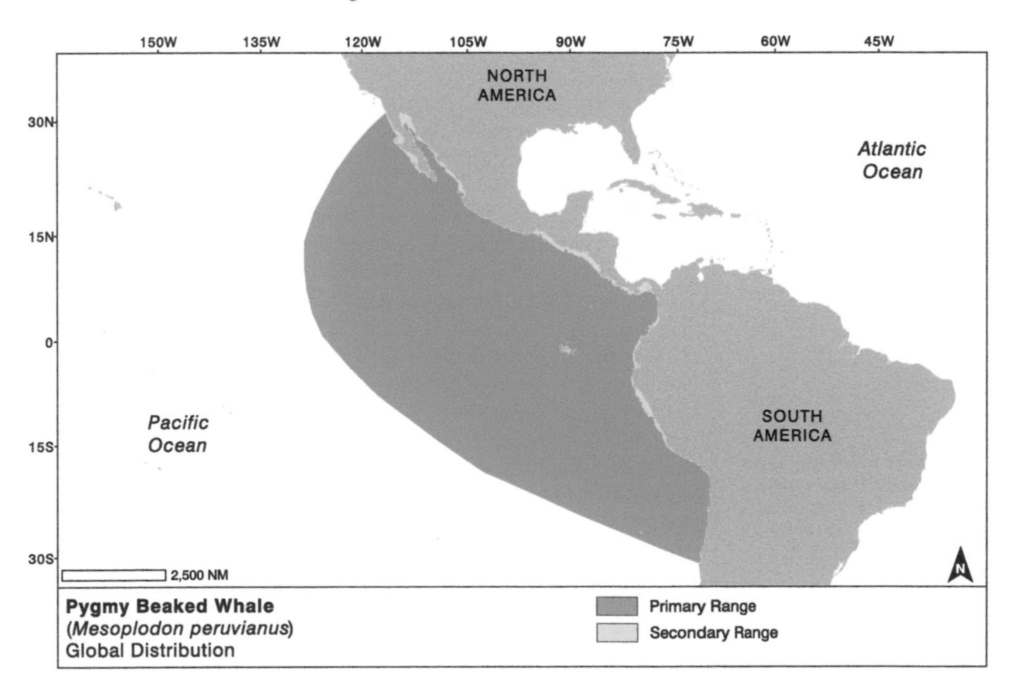

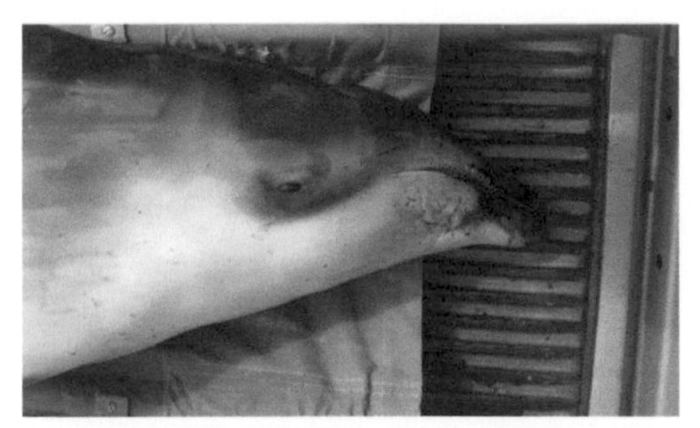

A calf *M. peruvianus* that stranded in California and was identified using genetics; calves of most mesoplodonts are largely indistinguishable at sea. PHOTO: NOAA FISHERIES, SWFSC

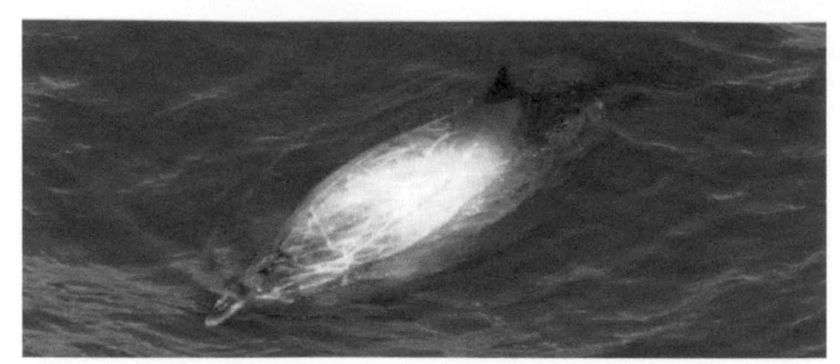

Adult male pygmy beaked whale photographed from the air in the eastern tropical Pacific. Note the white markings on the back, and the many tooth scars from other males. These features identify this as an adult male, even if the tusks cannot be seen. PHOTO: NOAA FISHERIES, SWFSC

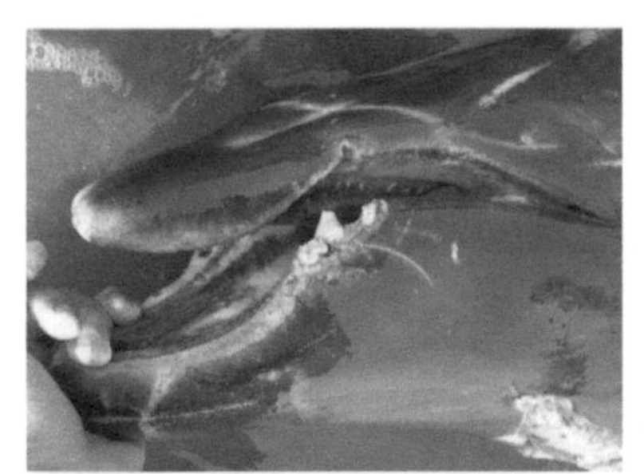

The tooth is placed on top of the arch in the lower jaw to increase its effectiveness and may have been broken in combat with another male. Michoacan, Mexico. PHOTO: COURTESY J. URBAN

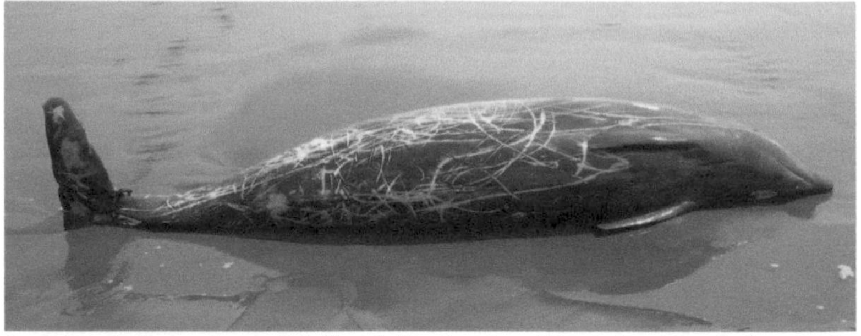

An adult male *M. peruvianus*—its belly and sides have long, linear tooth rake marks from tussles with other adult males; visible also are the paired throat grooves, used in creating suction for feeding. Michoacan, Mexico. PHOTO: COURTESY J. URBAN

Adult males of *M. peruvianus* have a large white swath on the body that starts in the middle of the back in front of the dorsal fin and continues down the sides and toward the rear, forming a very broad band. PHOTO: COURTESY J. URBAN

Distribution The pygmy beaked whale is known from a handful of specimens and several dozen sightings from the eastern tropical/warm temperate Pacific, including the Gulf of California. These records extend from about 30° to 28° and suggest that the species may be an eastern Pacific endemic. However, there is a single record of a stranding in New Zealand, possibly suggesting that this species may have a more extensive distribution than previously believed. Alternatively, the New Zealand record may be an extralimital wandering. This is the most frequently-sighted *Mesoplodon* whale in the eastern tropical Pacific (ETP). Like other members of the genus, it occurs in deep waters beyond the continental shelf.

Ecology and behavior Most groups have been of about two animals (average is 2.3 individuals), but have ranged up to five. The behavior of these animals appears to be similar to that of other species of mesplodonts. Groups may include mixed sex and age classes. During a sighting of a single male in the ETP, the animal breached three times.

Virtually nothing is known of the reproductive biology of this species. Small calves are generally not seen in the ETP late in the year, when effort there is concentrated, which suggests either a calving peak early in the year or that females segregate from adult males during the calving season.

Feeding and prey The diet consists of small mid-water fishes, oceanic squids, and shrimps. Presumably, these are taken at moderate to great depths.

Threats and status Little is known of the pygmy beaked whale's status or threats. Some have been killed in drift gillnets for sharks off Peru. Although there are no estimates of abundance available, they are seen quite frequently in the ETP, and they do not seem to be rare there.

IUCN status Data Deficient.

Sowerby's Beaked Whale—*Mesoplodon bidens*

(Sowerby, 1804)

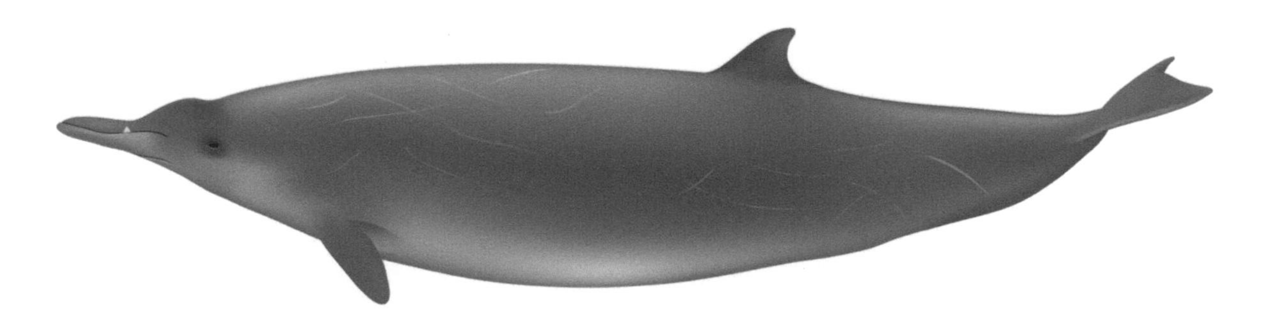

Recently-used synonyms None.
Common names En.–Sowerby's beaked whale, North Sea beaked whale; *Sp.–zifio de Sowerby; Fr.–baleine à bec de Sowerby.*
Taxonomic information Cetartiodactyla, Cetacea, Odontoceti, Ziphiidae.
Species characteristics Sowerby's beaked whales are characterized by a spindle-shaped body, with a small head, small dorsal fin located about two-thirds of the way back from the snout tip, small and narrow flippers, and tapered flukes with no median notch. There is also a single pair of shallow throat grooves, and the blowhole is a crescent with the ends pointing forward. Although there is some variation, Sowerby's beaked whales have a long and thin (for mesoplodonts) beak and often a prominent bulge on the forehead. There may be a very slight arch to the lower jaw, but typically the mouthline is straight.

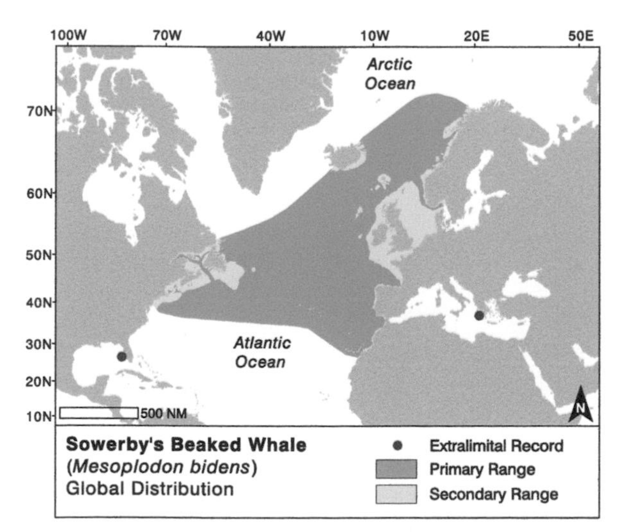

Coloration is not well known, but generally does not appear to be distinctive, with a charcoal gray back and somewhat lighter belly. White or light gray spots are common on the bodies of adults; however, young animals appear to have less spotting. Scarring of adult males is generally in the shoulder region and lower flanks. The scars generally consist of single scratches (not paired).

The two flattened, triangular tusks of adult males erupt from the lower jaw, about two-thirds of the way back from the beak tip. They are visible outside the closed mouth, although they are not particularly large. Males reach lengths of at least 5.5 m and females, 5.1 m. They attain weights of at least 1,300 kg. Newborns average about 2.4 m.
Recognizable geographic forms None.
Can be confused with Sowerby's beaked whales can be confused with other North Atlantic species of *Mesoplodon* (e.g., True's, Gervais', and Blainville's beaked whales); without a close look, even bulls would be difficult to distinguish from related species at sea. However, the limited distribution and presence of a long beak (and posteriorly-set tusks in males) may help to make an identification, if the head is seen well. This species may also appear somewhat darker than others in its range.
Distribution Sowerby's beaked whales are known almost exclusively from the colder waters of the North Atlantic, from at least Massachusetts to Labrador in the west, and from Iceland to Norway in the east. This is the most northerly distributed of the Atlantic species of *Mesoplodon,* and most records are north of 30°. The range is known to include the Baltic Sea, but only rarely (possibly as a stray) and the Mediterranean. The species appears to be more common in the eastern North Atlantic than in the western. Based on strandings, northern Europe appears to be the center of abundance. There is a single record from Florida in the Gulf of Mexico, but this

Long-beaked *Mesoplodon* species often surface with their beaks protruding at a 45° angle; these male Sowerby's beaked whales are further identified by their small pair of teeth set far back in the lower jaw, on a relatively small arch. The head and beak are blackish (the brown is probably from diatoms), with a lighter gray back starting behind the blowhole. North Atlantic. PHOTOS: (CLOCKWISE FROM TOP LEFT) H. WHITEHEAD LAB; AZORES. J. HART; THE GULLY. M. MILLIGAN; THE GULLY. M. MILLIGAN

An adult male Sowerby's beaked whale showing its dark head area and lighter gray back and sides. Because the teeth of male Sowerby's are relatively small, the linear scarring from other males appears to be fairly superficial compared to some other beaked whales. Northwest Atlantic. PHOTO: H. WHITEHEAD'S LAB

A rake-marked adult male Sowerby's—because the teeth do not rise above the level of the upper rostrum, the rake scars are mostly single lines; other species such as *M. mirus* tend to leave parallel tracks. Notice also that the pale back extends only as far back as just in front of the dorsal fin. Azores. PHOTO: J. HART

A young Sowerby's (identified by the cow it was with) strikes an odd pose with its out-stretched flippers; the beak length will increase with age. Azores. PHOTO: M. VAN DER LINDE

Sowerby's beaked whale is easy to identify, because it is normally the only long-beaked mesoplodont found in the North Atlantic; shown here is a rare behavior for any species of *Mesoplodon*. PHOTO: M. VAN DER LINDE

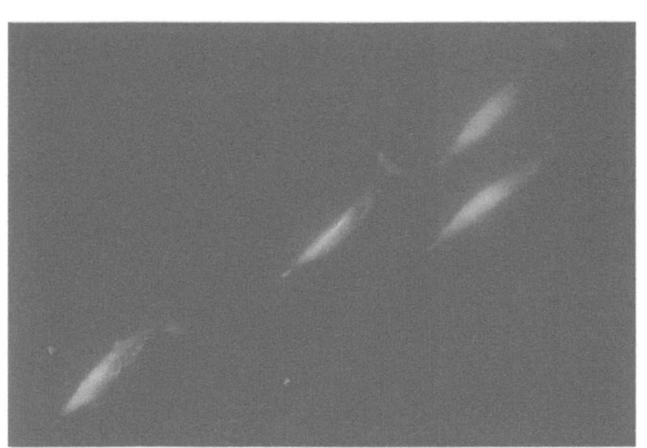

Photographed off New England, these beaked whales are identified as Sowerby's by their exceptionally long beaks, and as adult males by their dark heads, pale gray forward back and sides, and dark tail stocks and flukes; tooth rake marks are also visible. Four adult males traveling together is something unusual for normally-aggressive male mesoplodonts. PHOTO: R. DIGIOVANNI, NOAA FISHERIES, NEFSC

An adult male Sowerby's beaked whale that stranded in Europe, showing the location of the tusks. PHOTO: N. TREGENZA

***Mesoplodon bidens* brandishes its exceptionally-long beak In the North Atlantic. This is an adult female or a subadult male because it does not have an erupted tooth or the color patterning of an adult male.** PHOTO: H. WHITEHEAD LAB

Three female and/or young Sowerby's beaked whales in the North Atlantic, identified by their exceptionally-long beaks. Females and young apparently do not have the dark head and lighter back of the adult males. Azores. PHOTO: J. HART

appears to represent an extralimital wandering. As with other members of the genus, it occurs in deep waters past the continental shelf edge.

Ecology and behavior Very little is known of the natural history of this species beyond what has been learned from occasional sightings and strandings, which have generally involved singles and pairs. However, in several sightings observed at sea off Nova Scotia, groups ranged in size from 3–10 and some were of mixed composition. Mass strandings of up to six individuals of this species have been recorded. Recorded dives lasted 12–28 minutes. These whales often bring their heads up out of the water at a 45° angle when surfacing, and often arch their backs relatively high when diving. Breaching, spy-hopping, and fluke-slapping have been observed in sightings at sea. There have also been some instances in which the animals approached vessels.

The breeding season for Sowerby's beaked whale appears to be late winter to spring. Males possess tusks, which are presumably used for male/male combat over access to females (as in other species of the genus). As they reach maturity, structural changes in the tusks and lower jaw occur, which strengthen and reduce the prospects of injury to them. Essentially nothing else is known about the reproductive biology of this species.

Feeding and prey Sowerby's beaked whales feed on squid and small fish, including Atlantic cod.

Threats and status There is little specific information on the status or threats to this species, and no abundance estimates exist. However, some are known to have been incidentally killed by whalers in Newfoundland, Iceland, and in the Barents Sea. A few entanglements in fishing gear (e.g., driftnets) have been documented. Sowerby's beaked whale is not thought to be endangered.

IUCN status Data Deficient.

Gervais' Beaked Whale—*Mesoplodon europaeus*

(Gervais, 1855)

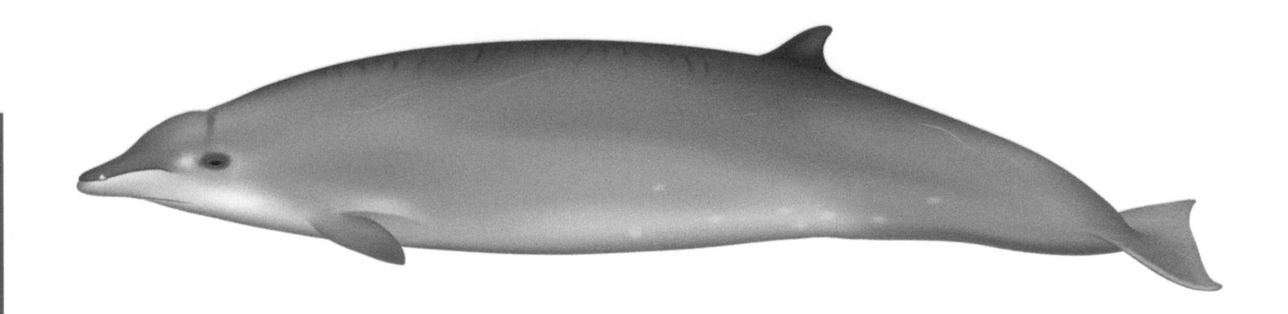

Recently-used synonyms *Mesoplodon gervaisi.*

Common names En.–Gervais' beaked whale; Sp.–*zifio de Gervais; Fr.–baleine à bec de Gervais.*

Taxonomic information Cetartiodactyla, Cetacea, Odontoceti, Ziphiidae.

Species characteristics Gervais' beaked whales are characterized by a spindle-shaped body, with an especially small head, small dorsal fin located about two-thirds of the way back from the snout tip, small and narrow flippers, and tapered flukes with no median notch. There is also a single pair of shallow throat grooves, and the blowhole is a crescent with the ends pointing forward. The beak is fairly short, and generally blends smoothly into the forehead.

The triangular teeth of adult males are found one-third of the distance from the tip of the lower jaw. Although they are mostly buried in gum tissue, the exposed tips are angled slightly outwards, and are visible outside the closed mouth. They may fit into grooves in the gums of the upper jaw. The mouthline tends to be relatively straight (with only a slightly raised area around the tusks).

In the Caribbean, a mesoplodont with a short beak and light gray sides with a darker gray dorsal band is almost certainly going to be a Gervais' beaked whale. Martinique. PHOTO: D. MOUSSA

At first look, Gervais' beaked whales may appear to have a relatively non-descript mesoplodont color pattern. They are moderate gray above and lighter gray below. However, females and young have a lighter back with a dark dorsal centerline and a series of faint, wavy stripes extending down a short distance from the centerline. In bulls, the darkening of the back appears to obscure these stripes. In young animals, the belly is white, but it appears to darken with age. In females, there is often a white patch in the genital area. Scarring is generally not heavy, even on adult males, and any scars present are generally single lines. There is often a dark patch around the eye, which appears to be more pronounced in this species than in some other mesoplodonts.

Gervais' beaked whale males attain lengths of at least 4.6 m, and adult females reach at least 4.8 m. Females appear to be larger, on average. Weights of at least 1,200 kg are attained. Newborns are about 2.1 m in length.

Recognizable geographic forms None.

Can be confused with Due to their relatively generic color pattern and non-distinctive tusks, Gervais' beaked whales are somewhat difficult to distinguish from overlapping mesoplodonts at sea. In the North Atlantic, it may be challenging to distinguish them from True's beaked whales (if no adult males are present). A very detailed view of the head of an adult male would allow identification, and that would be based largely on tusk shape and position. However, if the lighting is good, the wiggly dorsal stripes of some Gervais' beaked whales may be visible. Genetic study or details of the morphology of the skull will generally be required for positive identification.

Distribution Although sometimes depicted as a North Atlantic endemic, this species is probably continuously distributed in deep waters across the tropical and temper-

Gervais' beaked whale—the dark dorsal band and vertical striping are clearly visible even from the air. Cape Hatteras, North Carolina. PHOTO: A. FRIEDLAENDER

A rare underwater image of *M. europaeus*; notice the short beak and relatively straight gape. It has few marks on it and is probably a young animal. Bahamas. PHOTO: O. SHIPLEY

ate Atlantic Ocean, both north and south of the equator. Most records are from the east and Gulf coasts of North America, from New York to Texas, but Gervais' beaked whales are also known from several of the Caribbean islands. This is the most commonly stranded beaked whale in the southeastern United States. In the eastern Atlantic, they are known from the British Isles to Guinea-Bissau in West Africa. There is only one record of this species from the Mediterranean. There are also strandings at Ascension Island in the central South Atlantic, and along the coast of Brazil. There is speculation that its Southern Hemisphere distribution could extend to Uruguay and Angola.

Ecology and behavior The favored habitat of Gervais' beaked whales appears to be warm temperate and tropical waters. Little else is known of their biology. They have only sometimes been reliably identified alive in the wild, mostly in the eastern Atlantic, although many *Mesoplodon* sightings in the Gulf of Mexico are thought to have been of this species. Around the Canary Islands, they sometimes lift their heads out of the water upon surfacing. Live-stranded individuals have been held in captivity for short periods of time. Growth layer group counts in the teeth of one specimen suggest they live to at least 48 years of age. Other aspects of their life history are not known.

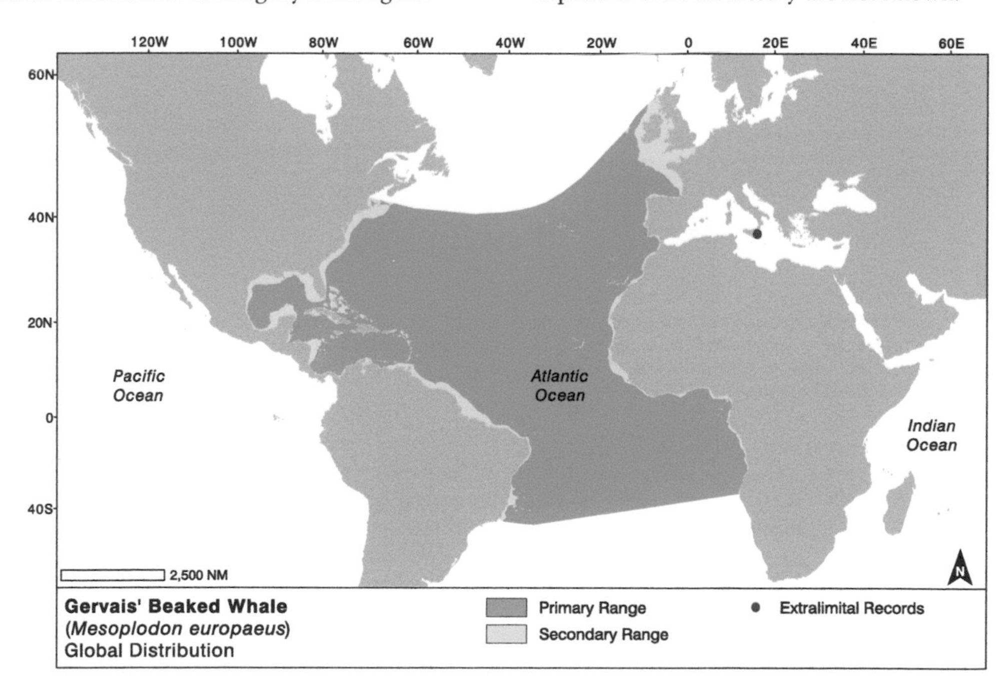

The erupted tooth of a male Gervais' beaked whale is small, only slightly elevated above the lower jaw and more than half-way to the tip of the jaw. Bahamas. PHOTO: K. MELILLO SWEETING

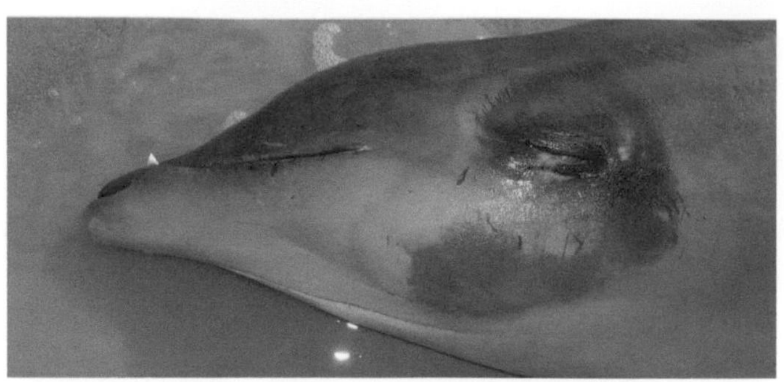

A stranded male *M. europaeus*: it has a short beak with a small tooth on a small arch. It is likely a young animal: the tooth is still sharp (not yet worn down), and there are no evident rake marks on the body. Florida. PHOTO: COURTESY T. PUSSER

An adult male Gervais' beaked whale surfaces, showing the location and size of the teeth. Guadeloupe. PHOTO: M. MEUNIER

In addition to the dark dorsal band, adult female Gervias' beaked whales have a dark eye patch with a small pale patch in front of it, another light area behind it, and the two are separated by a dark band that travels from the front of the eye, up and back behind the blowhole. Guadeloupe. PHOTO: C. MILLON

The dark dorsal ridge band and associated vertical striping appears to be diagnostic of *M. europaeus* females and juveniles. The biopsy dart (just striking the whale forward of the dorsal fin) collected a skin sample that genetically confirmed the identification. Bahamas. PHOTO: C. DUNN

Male *M. europaeus* (this individual had visible teeth) are slightly darker than the females and the dark band on the dorsal ridge is less conspicuous. Guadeloupe. PHOTO: M. MEUNIER

Feeding and prey Like other members of the genus, Gervais' beaked whales are known to feed primarily on squid, although some fish may be taken as well. There is also a record of a mysid shrimp found in the stomach of a stranded specimen.

Threats and status Specimens of Gervais' beaked whale have been entangled and killed in pound nets off New Jersey. Based on the frequency with which they strand, they are presumed to be relatively common in waters along the east coast of North America. No specific estimates of abundance exist; however, about 100 *Mesoplodon* whales occur in the northern Gulf of Mexico (most of which are probably of this species).

IUCN status Data Deficient.

True's Beaked Whale—*Mesoplodon mirus*

True, 1913

Recently-used synonyms None.

Common names En.–True's beaked whale; *Sp.–zifio de True; Fr.–baleine à bec de True.*

Taxonomic information Cetartiodactyla, Cetacea, Odontoceti, Ziphiidae. The existence of two apparently widely-separated and morphologically-distinct forms suggests possible subspecific (or even specific) differentiation. However, this remains to be confirmed.

Species characteristics

True's beaked whales are characterized by a spindle-shaped body, with a small head, small dorsal fin located about two-thirds of the way back from the snout tip, small and narrow flippers, and tapered flukes with no median notch. There is also a single pair of shallow throat grooves, and the blowhole is a crescent with the ends pointing forward. The forehead transitions into a moderately-short beak, with a straight or slightly-curved mouthline. The melon tends to be rounded and prominent, reminiscent of that of the bottlenose dolphin. Some animals may have a distinct indentation behind the blowhole.

The overall color pattern appears to be one of basic countershading, with a light belly and darker sides and back. However, there is significant geographic variation (see below). There is generally a dark ring around the eye, which often connects by a narrow line to the darker color on top of the head. There may be dark flecking on the throat and lower jaw, which probably increases with age. There is also a white urogenital patch in some animals. Calves may have a more simple countershaded pattern. These animals usually show little or no scarring. However, any scarring that exists is generally in the form of closely-spaced parallel scratches.

These beaked whales are characterized by the position of the mandibular teeth at the very tip of the lower jaw. The small tusks are oval in cross-section (i.e., acorn-shaped), lean forward, and are visible outside the closed mouth of adult males. The blow is usually bushy or indistinct. Both sexes are known to reach lengths of up to about 5.3–5.4 m (with females presumably slightly larger). Weights of up to 1,400 kg have been recorded. Newborns are probably between 2.0 and 2.5 m long.

Recognizable geographic forms

North Atlantic form—In the Northern Hemisphere, adult True's beaked whales possess a brownish-gray dorsal fin and upper tail stock, and the belly is light gray to white. However, on the entire dorsal surface they may be significantly lighter in color than other members of the genus within their North Atlantic range. The anterior half of the beak is dark gray to black, fading to a lighter gray behind.

Southern Hemisphere form—In most (or all) Southern Hemisphere adults, the white or light gray ventrum extends back to encompass the tail stock and underside of the flukes, and another extension goes forward to surround the dorsal fin (this light patch appears to fade rapidly postmortem, however). The light ventral surface of the flukes develop dark streaks radiating out from the

True's beaked whale has apical teeth: a single pair erupts from the tip of the lower jaw in adult males. This is a young adult male: it has no linear scarring and its teeth are still very sharp because they haven't seen much action. Mozambique. PHOTO: N. AYLIFFE

This is the same young male *M. mirus.* Mesoplodonts in general have relatively flat heads, and a dorsal fin placed far back. Most Southern Hemisphere True's beaked whales have a white caudal peduncle, fin and flukes, probably due to post-mortem darkening, this individual has only a hint of that color pattern. Mozambique. PHOTO: N. AYLIFFE

center of the trailing edge; the dorsal surface of the flukes is dark, however. The anterior half of the upper jaw is dark, while the lower jaw is nearly all light gray to white.

Can be confused with At sea, True's beaked whales are difficult to distinguish from other mesoplodonts that share their range. In the North Atlantic, it may be nearly impossible to distinguish them from Gervais' beaked whales (if no adult males are present). The unique color pattern (white patch on the tail stock and dorsal fin) may allow identification in at least the Southern Hemisphere. Otherwise, if the teeth of adult males are seen, they may be identifiable. The only other species in which males have oval teeth at the tip of the lower jaw is Longman's beaked whale. However, the latter species is generally much larger, and the forehead is much more steeply-rising than in True's beaked whale. Also, Longman's is a more tropical species, which does not occur in the North Atlantic.

Distribution True's beaked whales appear to have a disjunct, antitropical distribution. In the Northern Hemisphere, they are known only in the North Atlantic, from records in eastern North America (Nova Scotia to Florida), Bermuda, and Europe to the Canary Islands, the Bay of Biscay, and the Azores. They also occur at least in the southern Indian and Atlantic oceans, from South Africa, Madagascar, southern Australia, and Brazil. There is a recent stranding record also from the South Island of New Zealand, which may be considered extralimital. This peculiar disjunct pattern suggests that there may actually be separate species or subspecies in the Northern

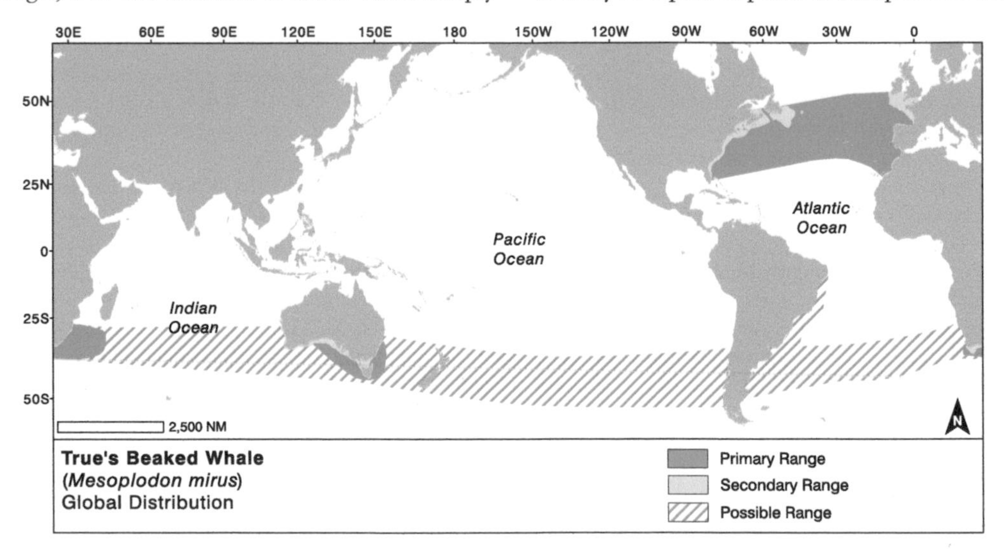

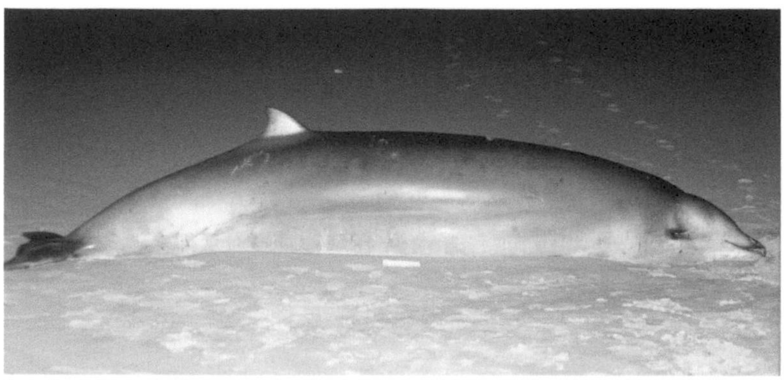

A northern True's beaked whale—probably a juvenile because the beak is short and there are relatively few markings on the body. As in all beaked whales, the fin is positioned toward the rear. PHOTO: COURTESY S. LEATHERWOOD

A ventral view of a stranded southern True's beaked whale showing its conspicuously white caudal peduncle and flukes. South America. PHOTO: S. SEBASTIAS

and Southern Hemispheres, and also that the southern form may extend into the South Pacific.

Ecology and behavior Since True's beaked whale has rarely been identified in the wild until recently, and is not one of the more commonly-stranded species, there is not much information available on the natural history of this species of beaked whale. Groups observed at sea have consisted of up to three individuals. They may show their beaks when surfacing. Energetic breaching behavior, up to 17 times in a row, has been observed in what is thought to be this species in the Bay of Biscay.

Essentially nothing is known of the life history of True's beaked whale, other than a record of a female that was simultaneously pregnant and lactating. Their unusual antitropical distribution and apparently different color morphs north and south of the equator lead one to believe that multiple species or subspecies may be involved, but this question will have to await further morphometric or genetic work.

Feeding and prey Like other members of the genus, stranded animals have had squid (mostly *Loligo* spp.) in their stomachs. They may also take fish, at least occasionally.

Threats and status Almost no information is available on the threats or status of this species. It appears never to have been hunted. True's beaked whales are not commonly identified at sea, and there are no estimates of abundance.

IUCN status Data Deficient.

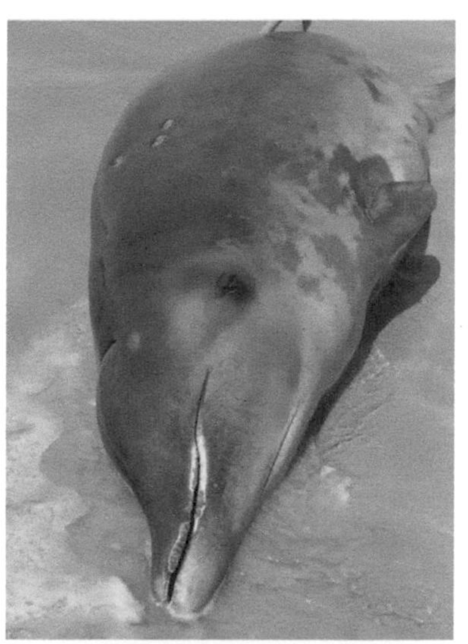

An adult female True's beaked whale from the Southern Hemisphere—when identifying it to species, the relatively straight gape and short beak would eliminate a number of mesoplodonts from consideration. Northern Australia. PHOTO: D. COUGHRAN

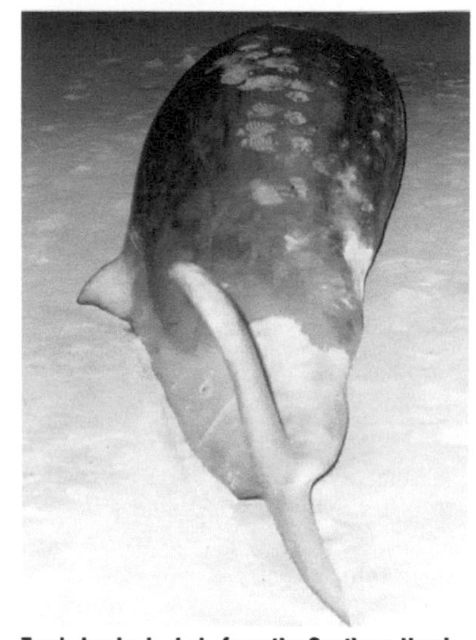

True's beaked whale from the Southern Hemisphere (perhaps a separate subspecies from the Northern Hemisphere form) has a conspicuous and unique color pattern—the flukes and caudal peduncle are white, and the white extends forward along the back to just in front of the dorsal fin, which is also white. South Africa. PHOTO: COURTESY PORT ELIZABETH MUSEUM

Strap-toothed Beaked Whale—*Mesoplodon layardii*

(Gray, 1865)

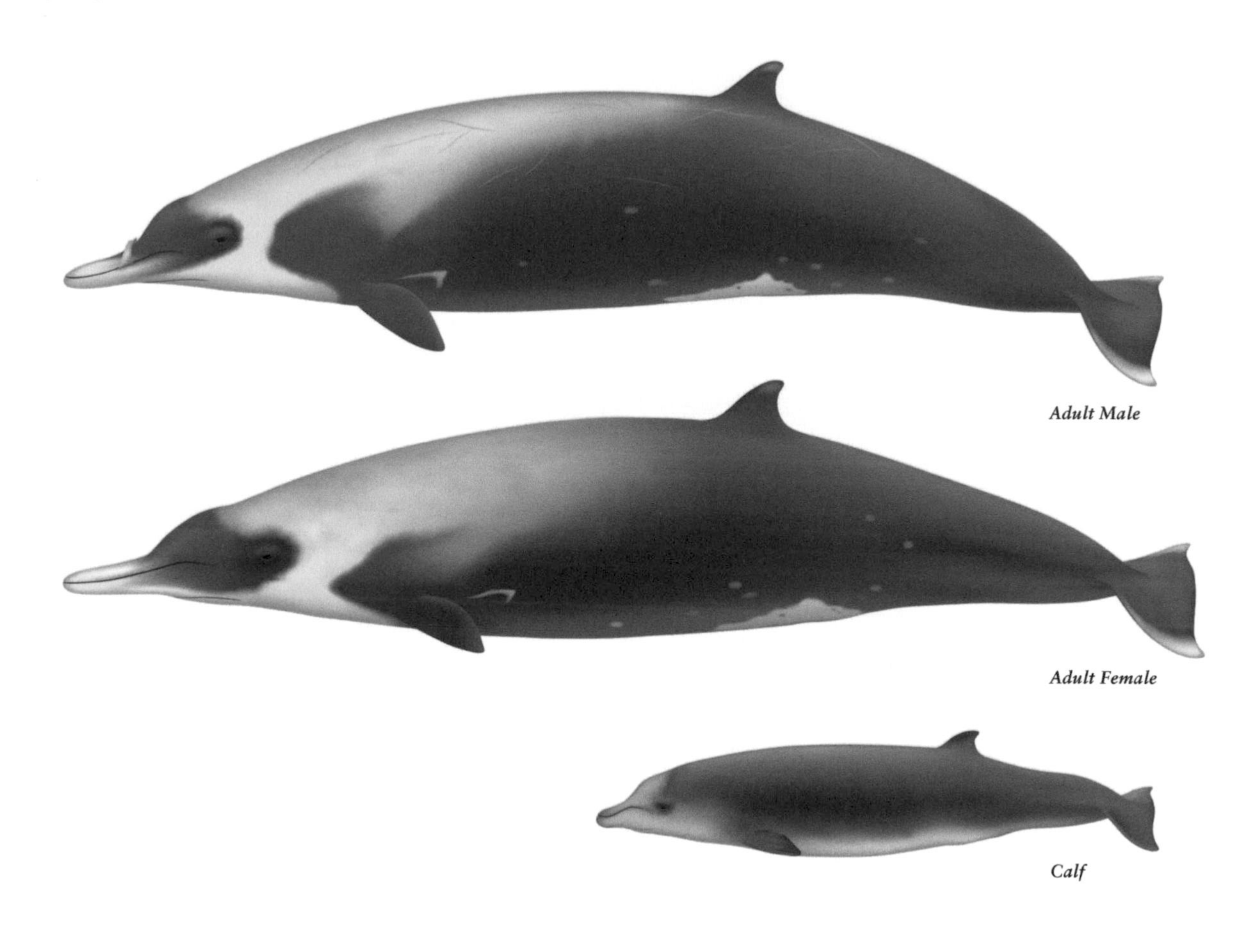

Adult Male

Adult Female

Calf

Recently-used synonyms None.

Common names En.–strap-toothed beaked whale; *Sp.–zifio de Layard; Fr.–baleine à bec de Layard.*

Taxonomic information Cetartiodactyla, Cetacea, Odontoceti, Ziphiidae.

Species characteristics Strap-toothed beaked whales are characterized by a spindle-shaped body, with a small head, small dorsal fin located about two-thirds of the way back from the snout tip, small and narrow flippers, and tapered flukes with no median notch. There is also a single pair of shallow throat grooves, and the blowhole is a crescent with the ends pointing forward. The beak is quite long, among the longest in the genus. The forehead may appear to have a slightly steeper rise than in most other mesoplodonts.

The complex adult color pattern is distinctive and is better known than that of most other mesoplodonts, as this species is documented from quite a large number of specimens. Subadults and calves do not appear to possess the distinctive coloration of adults, and they appear more generic, with a general countershaded pattern. However, the belly does darken with age. In larger animals, the body is mostly gray or black, sometimes with a purple or brown tinge. Distinct white to light gray patches develop in the thoracic area, and these are connected to a light throat patch, which extends to cover most of the lower jaw. The top of the head and the area around the eye remain dark, however. The melon darkens and the rostrum becomes white in adults. A white or light gray patch is found around the urogenital area. The result of all this is a rather complex adult color pattern of contrasting dark and white patches.

The dentition of adult males is very distinctive, one of the most bizarre sets of teeth of any mammal. The long (up to 30 cm) tusks emerge from near the middle of the lower jaw and curl backward at about 45° and inward, extending over the upper jaw, often preventing it from opening more than a few centimeters. There is a small denticle at the tip of each tooth, and the teeth often possess stalks of barnacles. How the animals eat with such an arrangement is unknown, although like other beaked whales, they do seem to use suction to bring prey into the mouth.

Adult females reach lengths of at least 6.2 m and males reach about 6.1 m, making this the largest of the mesoplodonts. They probably reach weights of well over 1,300 kg. Length at birth is unknown, but may be close to 3.0 m.

Recognizable geographic forms None.

Can be confused with The unique tusks of adult males of this species will make them easily identifiable, if seen. The primary confusion would be with the spade-toothed beaked whale, which has similar tooth morphology. Adult females may be identifiable if a good view of the unique color pattern is obtained. However, since nothing is known of the external appearance of the spade-toothed beaked whale, it will be virtually impossible to fully exclude this species in sightings at sea. Calves and juveniles, however, will likely be nearly impossible to distinguish from other mesoplodonts.

The unique color patterning of an adult male strap-toothed beaked whale on display out over the Southern Ocean. PHOTO: S. N. G. HOWELL

Distribution Strap-toothed beaked whales apparently have a continuous distribution in cold temperate waters of the Southern Hemisphere, mostly between 35° and 60°. There have been strandings in South Africa, Australia, Tasmania, New Zealand, the Kerguelen Islands, Heard Island, Argentina, Uruguay, Brazil, and the Falkland Islands. The seasonality of strandings suggests that this species may migrate. Like all beaked whales, they occur mostly in deep waters beyond the edge of the continental shelf. There is some evidence of sexual segregation in distribution.

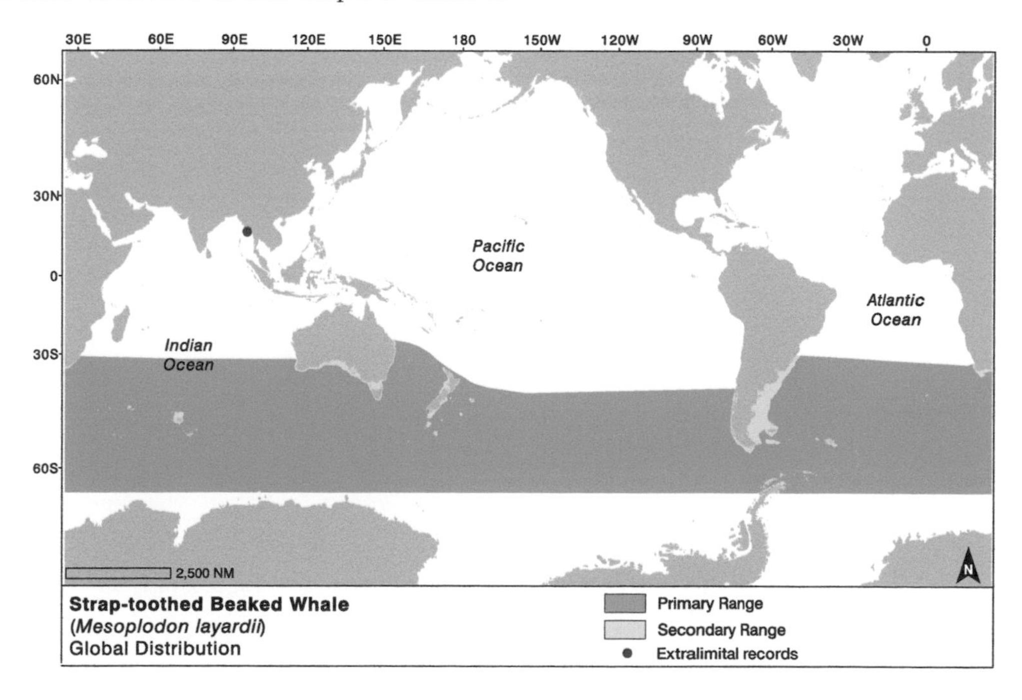

Ziphiidae

Strap-toothed Beaked Whale

Adult male strap-toothed beaked whales have very distinctive color pattern: the head is black but the distal half of the beak is white. The forward part of the back is pale gray and black beyond that. This animal has some ochre-colored diatom patches on it as well—not a regular feature. South Atlantic. PHOTO: N. GROVES

The dorsal fin and coloration of a female or young *M. layardii* (identified by adult male with group); it has patches of orange diatoms on its back. PHOTO: M. JENNER

A juvenile strap-toothed whale—the pale gray of the head and beak extend back to the blowhole and the rest of the body is a slightly darker shade of gray. PHOTO: M. JENNER

This strap-toothed whale is presumably a young male—most of the adult color patterning is evident, but more muted than in full adults. PHOTO: A. JONGENEEL

A juvenile strap-toothed beaked whale showing extent of paleness on the melon. PHOTO: M. JENNER

This adult male strap-toothed beaked whale stranded in Myanmar: the only known record for this species in the Northern Hemisphere. Notice the black head, white-tipped beak and the right tooth arching up over the top of the rostrum. PHOTO: COURTESY B. D. SMITH

Ecology and behavior Groups of up to three strap-toothed beaked whales have been seen. These animals are difficult to approach. Strap-toothed whales are commonly stranded, in fact more commonly than most other species of *Mesoplodon*. However, little has been learned from the few sightings of live animals. These animals often bring their head up out of the water at a 45° angle when surfacing. In adult males, the teeth will show in such surfacings. They have been observed to respond to ships by slowly sinking below the surface.

The bizarre teeth of adult males are thought to be used in male/male combat for female access, although this has never been observed directly. This may be a good example of strong sexual selection forces, as some males can only open their mouths about 3–4 cm. Calving appears to occur in the southern spring to summer months.

Feeding and prey Strap-toothed beaked whales eat primarily squid, although they may also take fish and crustaceans. Males appear to be restricted to eating only very small, slender squid (mostly <16 cm and 100 g), because of their unusual dental arrangement.

An adult male strap-toothed beaked whale stranded in Australia. Notice the extremely long tusks, covered in barnacles. PHOTO: K. WESTERSKOV

Threats and status There is little information available on the status of the strap-toothed whale, but based on the number of strandings, it is probably not a rare species. Whales of this species have not been hunted (that we know of), and there are no estimates of abundance.

IUCN status Data Deficient.

Andrews' Beaked Whale—*Mesoplodon bowdoini*

Andrews, 1908

Recently-used synonyms None.

Common names

En.–Andrews' beaked whale; Sp.–*zifio de Andrews;* Fr.–baleine *à bec de Bowdoin.*

Taxonomic information Cetartiodactyla, Cetacea, Odontoceti, Ziphiidae. Some researchers have suggested that this species and Hubbs' beaked whale may represent subspecies of the same species. However, recent genetic and morphological studies have supported the distinctness of the two species.

Species characteristics The external appearance of Andrews' beaked whale has only recently become known. We now know that this, along with its osteology, are similar to that of Hubbs' beaked whale. It has the basic *Mesoplodon* body plan, which consists of a spindle-shaped body, with a small head, small, somewhat triangular dorsal fin located about two-thirds of the way back from the snout tip, small and narrow flippers (fitting into flipper pockets), and tapered flukes with no median notch (sometimes a prominence is present instead). The beak is relatively short and can be quite stubby, and the forehead rises at a shallow angle. The mouthline of females is slightly arched, and in adult males the arch is much more pronounced. A pair of throat grooves is present, converging toward the tip of the jaw.

The basic color pattern appears to be dark bluish-gray, like that of most mesoplodonts. The anterior half of the

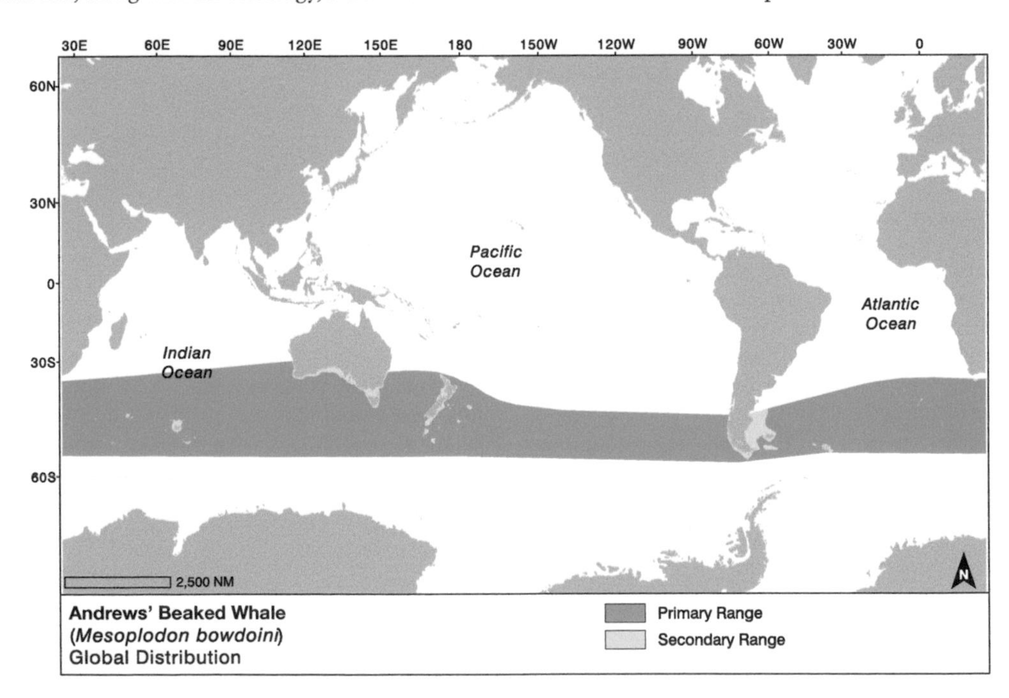

beak to just past the teeth is white in adult males (and maturing ones). Males may also be covered in white scratches, some of which may be paired. Females and juveniles are basically countershaded, apparently with a white lower jaw and dark upper jaw, and light patches in front of each eye. Cookie-cutter shark scars may be present on the body.

The laterally-flattened tusks of males emerge from the middle of the lower jaw on raised arches of gum tissue, and the tips are exposed, protruding slightly above the upper jaw and splaying outwards. The tusks are very reminiscent of those of the Hubbs' beaked whale (and for some time these two species were thought to be closely related). There is a denticle at the tip of each tooth, and stalked barnacles may attach to the tusks.

Only a few specimens have been measured, but both male and female Andrews' beaked whales reach at least 4.4 m in length. Length at birth is estimated to be about 2.2 m.

Recognizable geographic forms None.

Can be confused with The short, white beak tip may help to rule-out some other species of *Mesoplodon* that overlap in range (e.g, *M. layardii, M. hectori, and M. grayi,* which have much longer beaks). The tusks of bulls are fairly unique (only Hubbs' beaked whale has similar ones, and that species is a North Pacific endemic). If seen well, *M. bowdoini* tusks may allow these animals to be distinguished from most other mesoplodonts. Among those mesoplodonts that share their range, they are perhaps most likely to be mistaken for Blainville's beaked whales, but the lower jaw arch is much more massive and wider in that species, and it lacks the white beak.

Distribution To date, Andrews' beaked whale is known only from a few dozen stranding records between 32° and 55°S most of these have come from the South Pacific and Indian oceans (well over half are from New Zealand). Strandings have occurred in southern Australia, New Zealand, Tasmania, Tristan de Cunha, the Falkland Islands, Macquarie Island, and Argentina. The overall range may be circumpolar in the Southern Hemisphere; however, there is a gap in the known distribution between Chatham Island and the east coast of South America. It is presumably a creature of deep, offshore waters.

An adult male Andrew's beaked whale—the tip of the erupted tooth is barely visible. The white spotting is healed bite wounds from cookie-cutter sharks. Although the short beak with a white tip should make this species easy to identify at sea, we are aware of only one reliable at-sea sighting. New South Wales, Australia. PHOTO: M. JARMAN, NSW OFFICE OF ENVIRONMENT AND HERITAGE

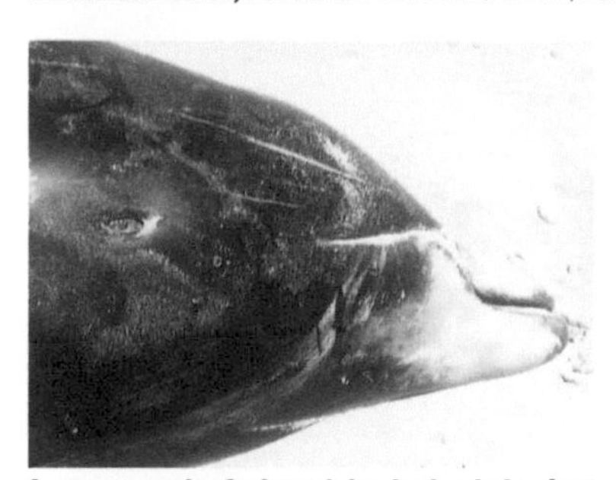

A young male Andrews' beaked whale, just beginning to develop the large jaw arches and tusks that are characteristic of the species. PHOTO: C. KEMPER

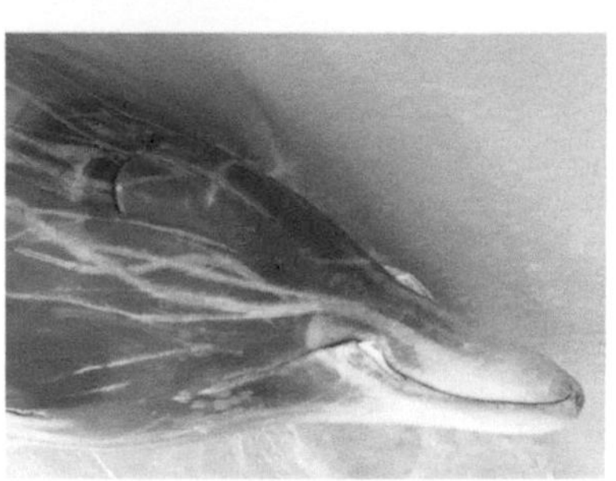

The numerous rake marks over the top of the head of this *M. bowdoini* and the worn tooth tips indicate that this adult male was a veteran of many fights with other males. Australia. PHOTO: D. COUGHRAN

Ecology and behavior Virtually nothing is known of the biology of this species, other than the few facts that have been gleaned from stranded individuals. This species has never been identified alive in the wild, so there is nothing known of its social organization and behavior.

Not much is known of the life history of this species, other than evidence for a summer-autumn breeding season, at least in the seas around New Zealand. A male specimen of 430 cm length that stranded in Uruguay appeared to be sexually immature.

Feeding and prey Andrews' beaked whales are assumed to feed primarily on cephalopods, like other members of the genus.

Threats and status Virtually nothing is known of the population status of Andrews' beaked whale. At least around New Zealand, the frequency of stranding records suggest that it may not be all that rare. No estimates of abundance exist.

IUCN status Data Deficient.

Stejneger's Beaked Whale—*Mesoplodon stejnegeri*

True, 1885

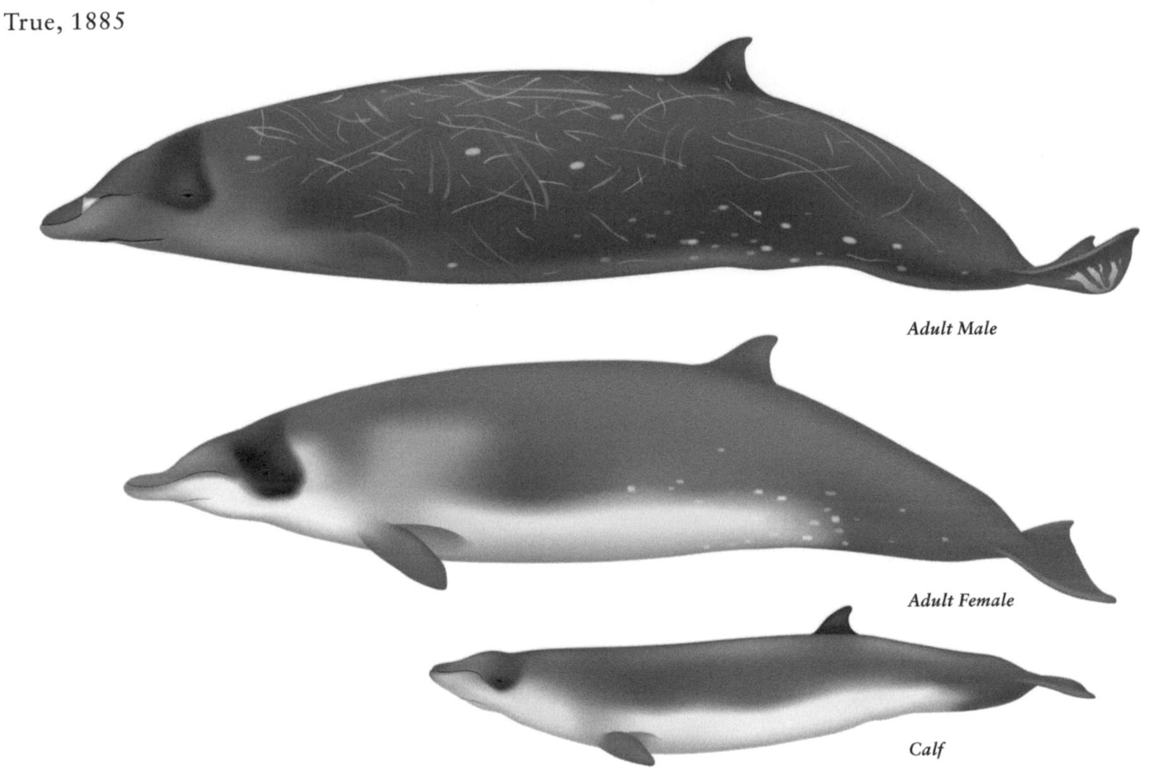

Recently-used synonyms None.
Common names En.–Stejneger's beaked whale; Sp.–*zifio de Stejneger*; Fr.–*baleine à bec de Stejneger*.
Taxonomic information Cetartiodactyla, Cetacea, Odontoceti, Ziphiidae.
Species characteristics Stejneger's beaked whales are characterized by a spindle-shaped, laterally-compressed body, with a small head, small dorsal fin located about two-thirds of the way back from the snout tip, small and narrow flippers, and tapered flukes with no median notch. There is also a single pair of shallow throat grooves, and the blowhole is a crescent with the ends pointing forward. The beak is relatively short, and the lower jaw contains a moderate arch in all age classes. The flipper pockets typical of mesoplodonts are present, and these are more darkly pigmented than the surrounding areas.

The color pattern on the head of this unidentified juvenile *Mesoplodon* is very similar to *M. stejnegeri*. Also, the flippers are pulled forward exposing the flipper pockets that allow the flippers to lie flush to the body and perhaps less accessible to predators like killer whales. Western North Pacific. PHOTO: J. O'HARA, COURTESY T. YAMADA

A breaching *Mesoplodon* from the western Pacific; although not identifiable to species, it does have the same pattern on the underside of the flukes as the Stejneger's beaked whale in the photograph on page 166. PHOTO: J. O'HARA, COURTESY T. YAMADA.

Apparently, adults of both sexes are uniformly brownishgray to nearly black in color, although females may be somewhat lighter on the back. The body is often covered with white mottling, especially on the ventral surface; these may represent healed scars from cookie-cutter shark bites (more common on older individuals). Adult males often have prominent scratches on the back, many of which may be arranged in paired patterns. Younger age classes appear to have a dark upper jaw and cranial cap, extending back to about the level of the blowhole, with extensions ventrally to surround the eyes. This gives the overall appearance of a "hood," and it appears to be more prominent on juveniles than adults. The area behind this appears to be somewhat lighter in color. The ventral surface of the flukes of adults has a series of gray or white concentric lines, radiating from the midpoint of the trailing edge. Such markings can be found in other mesoplodonts, but appear to be more exaggerated in adults of this species.

A live stranding of a Stejneger's beaked whale in the Aleutian Islands, Alaska. The dark area between the eyes and extending over the top of the head may be a useful feature for identifying this species at sea. PHOTO: E. D. ASH, COURTESY SMITHSONIAN INSTITUTION

The massive flattened tusks of males are situated near the middle of the lower jaw, and point forward and slightly inward. They are located on broad raised prominences or arches, so that the crowns extend above the upper surface of the beak. They may somewhat restrict the opening of the jaws. The size and shape of the teeth and arches are somewhat reminiscent of those in Blainville's beaked whale.

Stejneger's beaked whale females may be larger than males on average. The largest male measured was 5.7 m long. They reach weights of at least 1,600 kg. Newborns are assumed to be about 2.2 m in length. They may weigh about 80 kg at birth.

Recognizable geographic forms None.

Can be confused with When seen well, adult males will be distinguishable from most other mesoplodonts by tooth shape and position, and the shape of the lower jaw arch. Within the range of Stejneger's beaked whales, both Hubbs' and Blainville's beaked whale males have broadly similar teeth, but attention to details of the color pattern may allow an identification. Also, these other species are more common in the warm temperate to tropical waters of the North Pacific.

Distribution Stejneger's beaked whales are found in continental slope and oceanic waters of the North Pacific Basin, from central California, north to the Bering Sea, and south to the Sea of Japan (including the southern Okhotsk Sea). This appears to be primarily a cold temperate and subarctic species, and this is probably the only species of the genus common in Alaskan waters. It

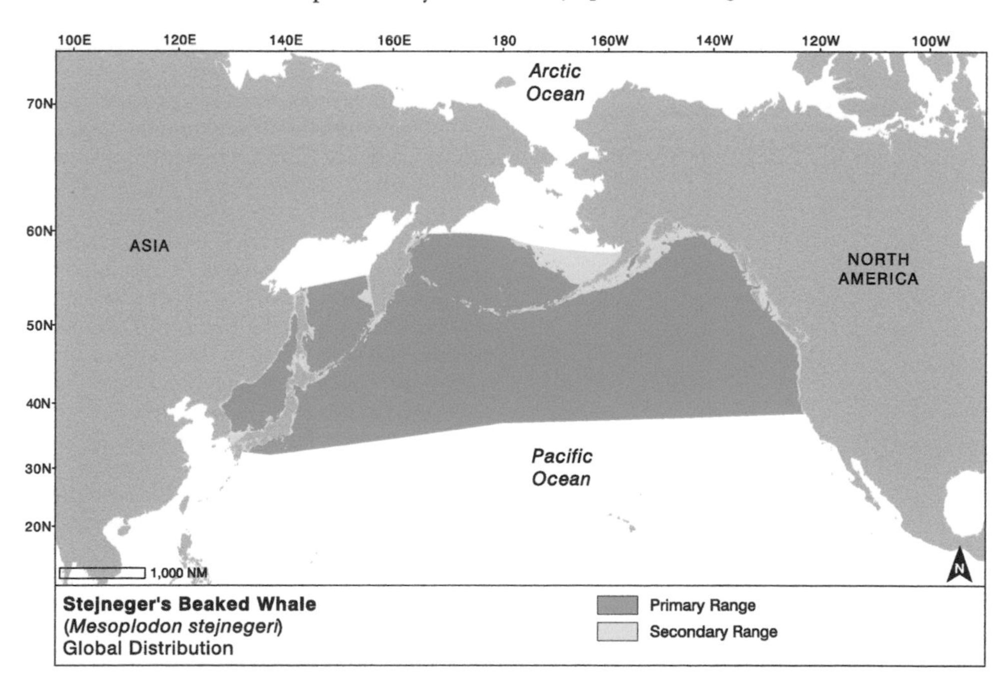

A stranding of four healthy-looking Stejneger's beaked whales in the Aleutian Islands, Alaska. Many of the white spots are from cookie-cutter shark bites, but some of the smaller scars may be from lamprey bites. Note also the radiating pattern on the underside of the flukes of the near animal. PHOTO: B. HANSON

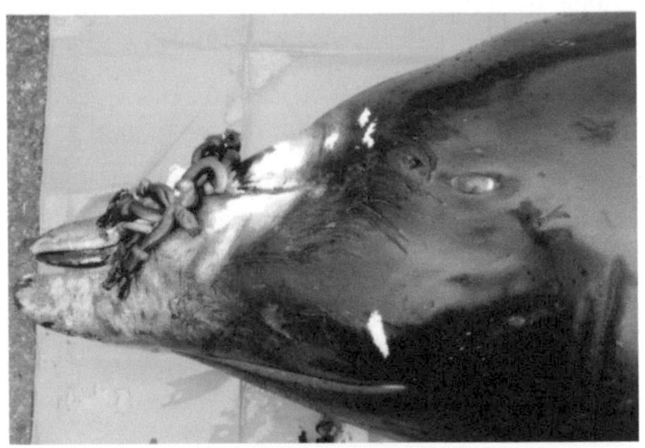

When beaked whales break their teeth (presumably from hitting the teeth of other males), stalked barnacles (*Conchoderma* sp.) often find a home, as on this adult male Stejneger's beaked whale. Japan. PHOTO: COURTESY T. YAMADA

This female or juvenile *M. stejnegeri* is very similar to females of many other mesoplodonts, which are often impossible to identify in the field. Clues can be found, however, in the length of the beak (short in this case) and the relative curvature of the gape (i.e., straight versus arched). Japan. PHOTO: T. YAMADA

With these two large, pointed teeth (tusks) raised prominently over the top of the rostrum, this adult male *M. stejnegeri* was prepared to do battle with other males. Haida Gwaii, British Columbia. PHOTO: C. TARVER

is most commonly stranded in Alaska, especially along the Aleutian Islands. Also, there are a large number of strandings (at least 34) from along the Sea of Japan coast of Japan, and much fewer along the Pacific coast. The large peak in strandings in this area in winter and spring suggests the species may migrate north in summer.

Ecology and behavior A great deal has been learned of the biology of this species in recent years, based on strandings that have occurred in the Aleutian Islands and along the west coast of Japan. In Stejneger's beaked whales, groups of 5–15 individuals have been observed, often containing animals of mixed sizes. Groups may be tightly bunched at the surface. Stejneger's beaked whales are presumably deep divers, feeding in the mesopelagic and bathypelagic zones. It has been hypothesized that there may be a resident population in the Sea of Japan and southern Okhotsk Sea. Based on a number of fetuses that have recently been examined, calving in the North Pacific appears to occur mainly from spring to early autumn, although the season may be protracted. Sexual maturity apparently occurs by about 4.5 m. Longevity is known to be at least 35 years, based on the aging of seven specimens.

Feeding and prey Stejneger's beaked whales are known to feed primarily on squids of the families Gonatidae and Cranchiidae, mostly in mesopelagic to bathypelagic depths. They also take some fish as prey. A pelagic tunicate was also found in the stomach of one stranded specimen. Stomachs of some specimens have contained non-food items, such as plastic bags and string.

Threats and status Not much is known about the status of Stejneger's beaked whale, but in the past some were taken in the Japanese salmon driftnet fishery in the Sea of Japan and in driftnets off the west coast of North America. They were also hunted to a certain degree in a Japanese whale fishery, along with Cuvier's beaked whales. There are no available estimates of abundance, but the species may not be rare in the northern North Pacific and Sea of Japan.

IUCN status Data Deficient.

Spade-toothed Beaked Whale—*Mesoplodon traversii*

(Gray, 1874)

Recently-used synonyms *Mesoplodon bahamondi*.
Common names En.–spade-toothed beaked whale; *Sp.–zifio de Travers;* Fr.–*baleine à bec de Travers*.
Taxonomic information Cetartiodactyla, Cetacea, Odontoceti, Ziphiidae. In 1995, a new species of beaked whale was described as *Mesoplodon bahamondi*. However, further study showed that it was actually the same as a previously-described beaked whale that had long been considered synonymous with the strap-toothed beaked whale. *Mesoplodon traversii* was found to be the valid scientific name of the "new" species.
Species characteristics Because only two specimens of this whale have ever been seen in the flesh (a decomposed cow and calf pair), little is known of the external appearance. Based on the limited information, spade-toothed beaked whales, like all mesoplodonts, are characterized by a spindle-shaped body, with a small head, small dorsal fin located about two-thirds of the way back from the snout tip, small and narrow flippers, tapered flukes with no median notch, and a single pair of shallow throat grooves. An important feature is the long, narrow beak, similar to that of Gray's and strap-toothed beaked whales. However, the melon appears to be more pronounced that in Gray's. The mouthline is relatively straight.

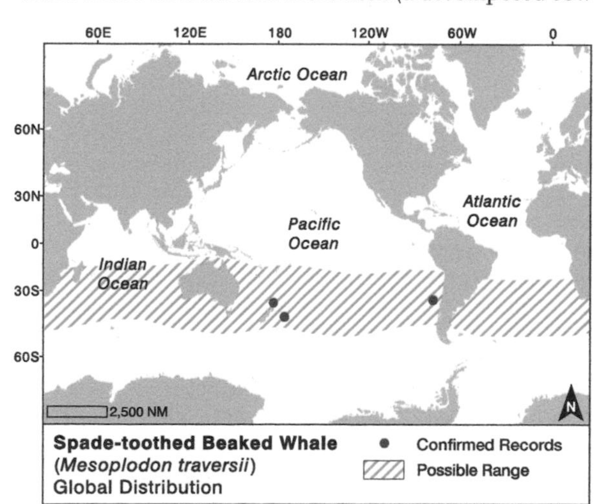

Coloration of the adults is not known for certain. Long-beaked mesoplodonts have been recorded at sea a few times in the eastern tropical Pacific just north of the equator. A single animal and a cow/calf pair all had a pale rostrum and a white lower jaw, and all three had a white patch behind the eye. These features all appear to be evident on the cow/calf pair that stranded in New Zealand. Adult males, however, could have quite different color patterning.

Adult males have a large tusk erupting from the middle of each side of the relatively straight lower jaw. It is presumed that the tusks extend above the rostrum, but it is unknown whether the lower part of the tooth is surrounded by gum tissue (as, for instance, in Hubbs' and Andrews' beaked whales) or whether it is largely exposed (as in the strap-toothed whale). The tusks lean backwards at an angle of about 45° and are wide and "spade-shaped." There is a large denticle at the tip, which juts up and serves as the cutting edge of the tooth. On older males, this denticle can be worn down smooth, apparently

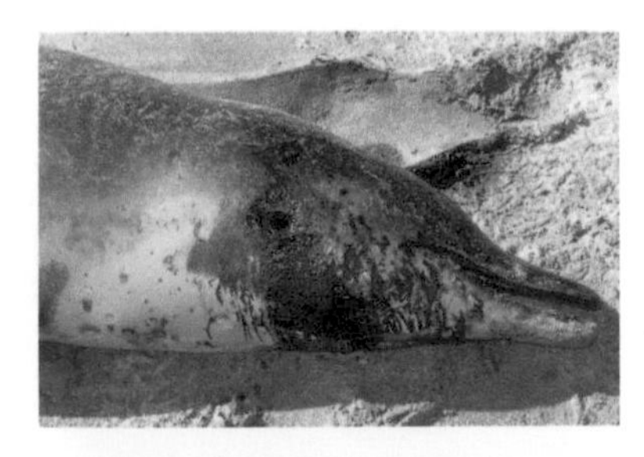

These are the only known photographs of *M. traversii* in the flesh—a cow (lower left and above) and calf (upper left) that stranded in New Zealand; the color pattern of this species is still unknown. Importantly, these photos indicate that it is a long-beaked species of *Mesoplodon. M. traversii* probably shares at least some of its range with two other long-beaked mesoplodonts, *M. grayi* and *M. layardii*; both of those are very distinctly patterned and likely distinguishable at sea from spade-toothed beaked whales. PHOTOS: P. LIVINGSTON

Confirmed sightings of the spade-toothed beaked whale have never been made at sea. The identity of this whale is not known, but it might be a member of this species. PHOTO: R. L. PITMAN

Another view of the calf above—it is not clear if the patterning shown here is real or a postmortem artifact. New Zealand. PHOTO: P. LIVINGSTON

from aggressive raking contact with other males. The tooth appears to wrap partially around the upper jaw.

Size is known only from two specimens, an adult female of 5.3 m and a juvenile male of 3.5 m.

Recognizable geographic forms None.

Can be confused with Until the external appearance of this species is better known, at-sea identification will be nearly impossible, and will most likely require expert examination. Among the other long-beaked mesoplodonts in the Southern Hemisphere, adult male strap-toothed whales have an easily-recognizable color pattern with a black and white beak, a black skull cap, and a white shawl over the forward back and sides; Gray's beaked whales have an all-white beak—it is likely that spade-toothed whales will have an entirely different coloration than either of these. The tooth is in the middle of the lower jaw, but is wide and arches backwards—similar but shorter than strap-toothed and much larger than Gray's. Hector's beaked whale also has a long, all-white beak, but the teeth erupt from the tip of the lower jaw.

Distribution Only about a half-dozen specimens are known to science: from northern Australia, White Island and the nearby Bay of Plenty (New Zealand), Pitt Island (Chatham Islands–holotype), and Robinson Crusoe Island (Juan Fernández Archipelago), off Chile. These indicate an at-sea distribution in subtropical waters of at least the South Pacific, with a possibility of ranging much further, especially to the west (Indian Ocean) and south.

Ecology and behavior The spade-toothed beaked whale may be the most poorly-known of all mammal species. Nothing is known of its biology and behavior, which is known only from two stranded specimens described above, a few skulls, and the mandibles and teeth from another. It is assumed to be similar to the other *Mesoplodon* species in general behavior and habits, though this can not be confirmed until there are sightings at sea attributable to this species.

Feeding and prey Nothing is known of the diet, other than an assumption that squid are the main prey.

Threats and status Nothing is known of the status of the species. It is probable that the species is relatively rare, given the small number of records available.

IUCN status Data Deficient.

Longman's Beaked Whale—*Indopacetus pacificus*

(Longman, 1926)

Recently-used synonyms *Mesoplodon pacificus.*

Common names En.–Longman's beaked whale, Indo-Pacific beaked whale; *Sp.–zifio de Longman;* Fr.–*baleine à bec de Longman.*

Taxonomic information Order Cetacea, Suborder Odontoceti, Family Ziphiidae. Some marine mammal scientists believe this species should be in the genus *Mesoplodon,* but recent genetic studies have suggested its generic distinctness. Until just a few years ago, this species was known only from two skulls. Sightings of what are now known to be this species in tropical waters were often mistakenly attributed to a whale of the genus *Hyperoodon* in the past.

Species characteristics Longman's beaked whales have moderately steep, bulging foreheads and moderate, tube-like beaks. Adults generally have a crease between the melon and beak. As in most other beaked whales, the dorsal fin is located behind the midpoint of the back, but it is a bit larger than in other beaked whales. The fin is falcate and shaped like a dolphin's dorsal fin. Typically, there is no notch on the trailing edge of the flukes. There is generally a pair of V-shaped grooves on the throat. The small, blunt flippers fit into "flipper pockets" on the sides. When seen, the blowhole is oriented with the ends pointing anteriorly, the opposite of the situation in Baird's and Arnoux's beaked whales, species with which it can be confused.

Coloration ranges from umber-brown to bluish-gray above, generally with light areas on the sides and around the head. These light areas are often separated by a diffuse dark band just behind the blowhole, which extends down to the flipper and to surround the eye. The dark band is continuous with the dark upper back. The above color pattern features appear to be most pronounced in young animals, and may fade with age. The white of the melon extends back only as far as the blowhole. Some

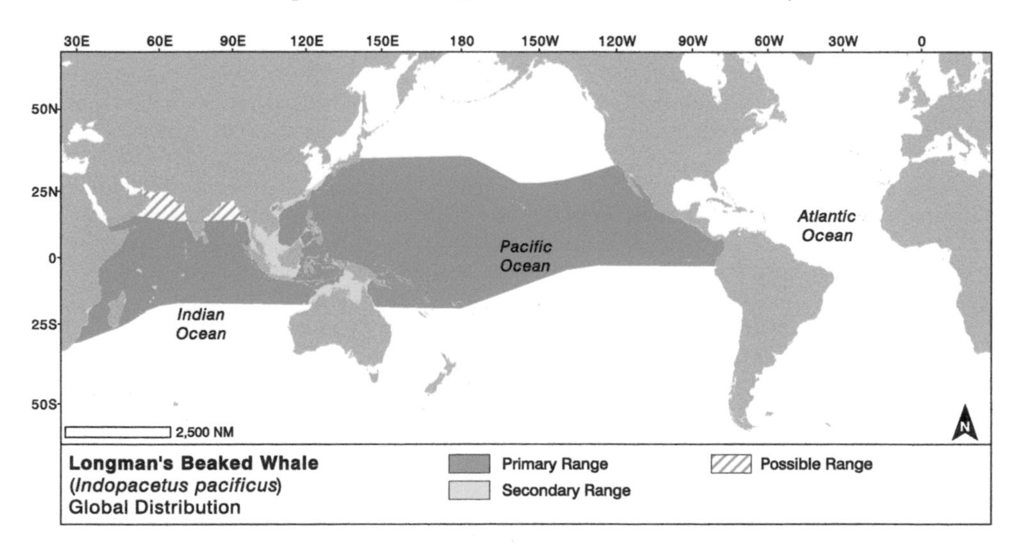

With numerous white cookie-cutter shark bite scars and almost no linear scarring, this Longman's beaked whale is almost certainly an adult female. The paleness of the melon extends to just behind the blowhole. Central Tropical Pacific. PHOTO: S. WEBB

***Indopacetus* is gray-brown with a noticeably large dorsal fin for a beaked whale, and most individuals have a conspicuously pale melon. It often travels in large groups of 10 or more, and unlike other beaked whales, it sometimes swims aggressively at the surface.** PHOTO: R. C. ANDERSON

Longman's is the largest tropical beaked whale: it has a conspicuous blow and a large, rounded melon that drops down almost perpendicular to the prominent beak (i.e., a "bottlenose" whale). The paleness of the melon extends back to the rear edge of the blowhole and down to just in front of the eye. Central Pacific. PHOTO: S. WEBB

A cow and calf Longman's beaked whale—calves always have a conspicuous pale melon; some adults do as well, but they may lose it with age. Maldives. PHOTO: C. JOHNSON

individuals have had scratches on the back, and white circular scars (probable cookie-cutter shark bites) are not uncommon. The upper surface of the flippers is dark and the underside is light. The upper surfaces of the flukes are dark (nearly black), and the lower surfaces generally have a series of light streaks.

There is a single pair of oval teeth at the tip of the lower jaw, which may be embedded in the gums. During at-sea sightings, the teeth have not been visible outside the closed mouth. The blow is low and bushy, but fairly conspicuous, and it may angle forward.

This is a relatively large beaked whale; size estimates from at-sea sightings have been in the range of 4–9 m. Two presumably adult females that stranded recently in Japan and the Maldives measured about 6.0 and 6.5 m, and a specimen live-stranded in the Philippines was 5.7 m long. Length at birth is not known, but a neonate measured 2.9 m long.

Recognizable geographic forms None.

Can be confused with The large size and relatively steep forehead of this species should rule out confusion with most species of beaked whales. However, the southern bottlenose and Baird's and Arnoux's beaked whales may cause some confusion. However, there is little overlap in distribution with these high latitude species. In shape and position, the teeth most closely resemble those of True's beaked whale. Examination of the skull by experts or genetic analyses may be required for positive identification. Longman's can especially be mistaken for southern bottlenose whales, but attention to details of head and body shape and coloration should allow distinction. Longman's are a bit more slender than southern bottlenose whales.

Distribution There have been many sightings at widespread locations in the tropical Pacific and Indian oceans of what we now know to be Longman's beaked whale.

The distribution of this species is not fully known, but it appears to be limited to the Indo-Pacific region. The available specimens are from Australia, Somalia, South Africa, the Maldives, Kenya, Sri Lanka, the Philippines, and Japan. The sightings come from scattered locations, many in deep oceanic waters, in the tropical to subtropical Indo-Pacific. Sightings have occurred in areas with surface water temperatures of 21–31°C These beaked whales are relatively infrequently seen in the eastern tropical Pacific and Hawaii, and may be more common in the western Pacific and western Indian Ocean. They also appear to be especially common around the Maldives archipelago.

Ecology and behavior Large, coordinated herds appear to be characteristic of Longman's beaked whales. Herds sighted in the Pacific Ocean have contained from one to an estimated 100 individuals, with many groups of 10 or greater (this is much larger than for Cuvier's beaked whale or the various *Mesoplodon* species). Groups in the western Indian Ocean have generally been smaller. They often swim in tight groups, and have been seen associated with pilot whales and bottlenose and spinner dolphins. Breaching has been observed, and these whales swim aggressively when moving along at the surface. Dive times may be quite long and can range up to at least 33 minutes.

All aspects of the biology of Longman's beaked whale are very poorly known—the species is known from only a few dozen specimens (as of 2014). It was known from only two skulls until as recently as 2003, when it was redescribed from several fresh strandings. A live-stranding occurred in the Philippines in 2004, and sightings have become more common in the western Indian Ocean in recent years. There is virtually nothing known of its reproductive biology. Probable predators of Longman's beaked whales are killer whales and large sharks.

Feeding and prey Nothing is known of its feeding habits, except for the stomach contents of a single specimen from Japan. These suggested that whales of this species may feed primarily on cephalopods, like other beaked whales.

Threats and status There is no known human exploitation of this species. Most records are of strandings or sightings, although some animals have been by-caught in Sri Lankan waters. Gillnet fisheries in the western Indian Ocean may be a threat. While certainly not the rarest of beaked whales, the paucity of recent sightings indicate that it is not particularly common either. The only estimate of abundance available is of about 760 individuals in the waters around Hawaii.

IUCN status Data Deficient.

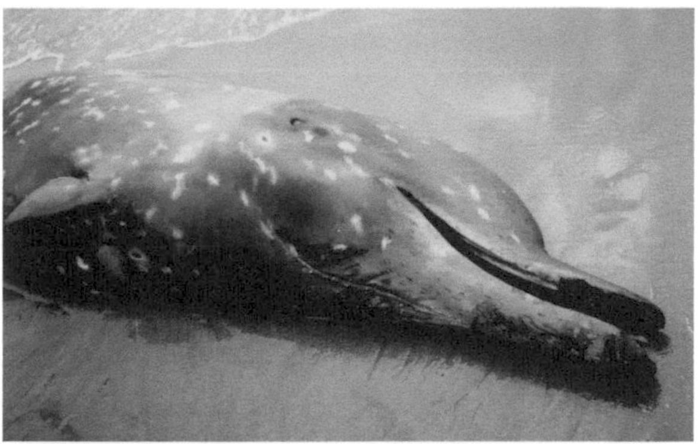

An emaciated *Indopacetus* that stranded in Japan; the white spots are cookie-cutter shark bites. PHOTO: COURTESY. T. YAMADA

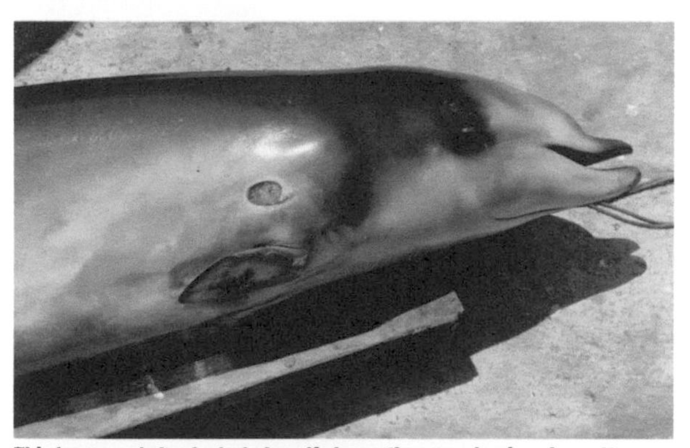

This Longman's beaked whale calf shows the same basic color pattern as the adult: a dark area behind the pale melon forms part of a dark patch around the eye, and also forms a band to the insertion of the flipper; behind the band a large pale area comes up from the belly, high onto the sides. South Africa. PHOTO: G. J. B. ROSS

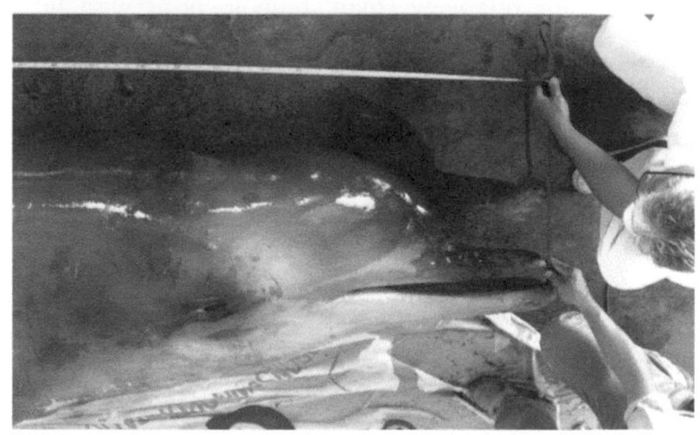

This image gives a sense of the large size of *Indopacetus* with its bulbous melon and prominent beak with a straight gape; the beak is generally dark above and lighter below. Philippines. PHOTO: J. M. ACEBES

Narwhal—*Monodon monoceros*

Linnaeus, 1758

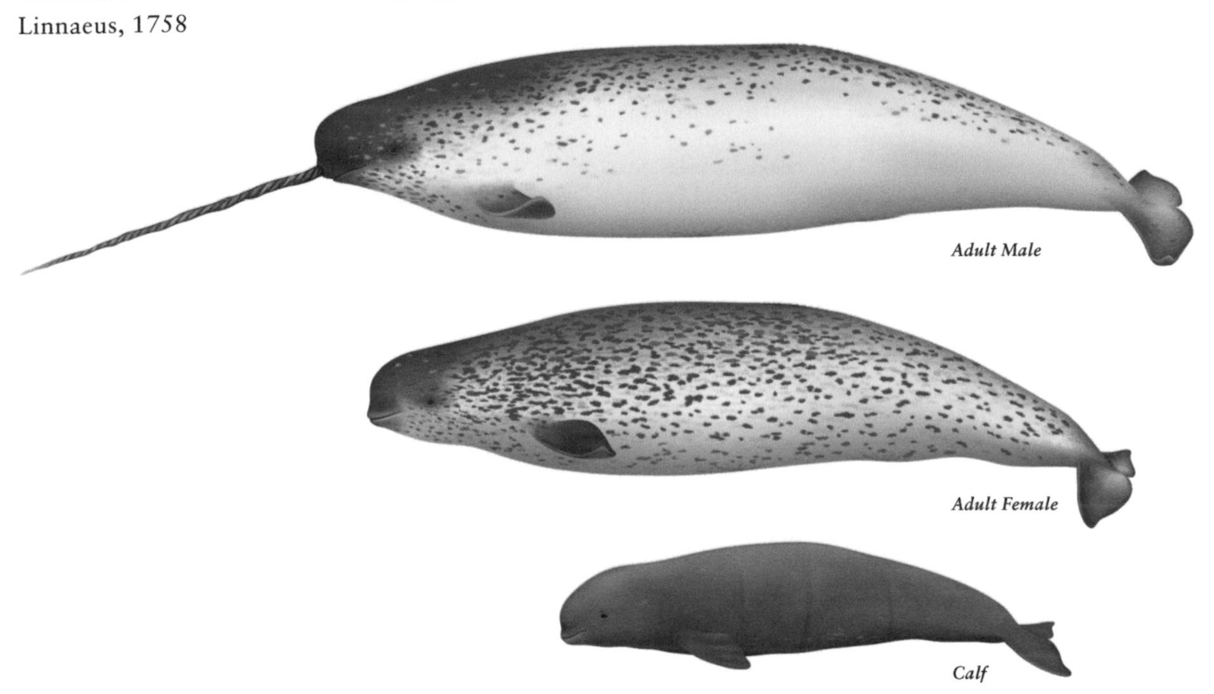

Recently-used synonyms None.

Common names En.–narwhal; Sp.–*narval*; Fr.–*narval*.

Taxonomic information Cetartiodactyla, Cetacea, Odontoceti, Monodontidae. Specimens that appear to be hybrids between narwhals and belugas have been described.

Species characteristics Narwhals are characterized by a robust body, relatively small, bulbous head with little or no beak, short blunt flippers that curl up at the tips in adults, absence of a dorsal fin (however, a low fleshy dorsal ridge is present on the posterior half of the back), and unusually-shaped flukes. The flukes of adults become straight to concave on the leading edge, and convex on the trailing edge. They are deeply notched and the tips tend to curl upwards, especially in older animals. This tendency is exaggerated in adult males. The mouthline is short and upturned.

Pacific Ocean

Arctic Ocean

Atlantic Ocean

1,000 NM

Narwhal
(*Monodon monoceros*)
Global Distribution

● Extralimital Record
Primary Range
Secondary Range

Young narwhals are uniformly gray to brownish-gray. As the animals age, they darken to all black and then white mottling develops. At this stage they appear spotted and the belly becomes light gray to white (with some dark mottling). Lightening continues as the animal ages. Older animals often appear nearly white, with some black mottling still remaining on the upper surfaces (especially the top and front of the head and along the dorsal ridge) and appendages (often as dark borders on the flukes and flippers).

The narwhal's most remarkable feature, however, is its teeth. There are only two teeth, both in the upper jaw. In females, these almost always remain embedded in the upper jawbone, but in males the left tooth normally grows out through the front of the head (starting at the age of about 2–3 years) and becomes a left-hand spiraled tusk up to 2.7 m long. Occasionally, females with a tusk or males with two developed tusks (double tuskers) are seen. Narwhals are the only cetaceans with such a tusk protruding from the front of the head, and the myth of the unicorn is thought to have derived from

This young male narwhal has a short tusk that will grow to be nearly 3 m long when it reaches its full development in adulthood. Baffin Island. PHOTO: M. CARWARDINE

Narwhals often raise their tusks out of the water and they regularly "cross swords." The tusk is rather brittle and broken tips are common, as seen here. PHOTO: J. K. B. FORD

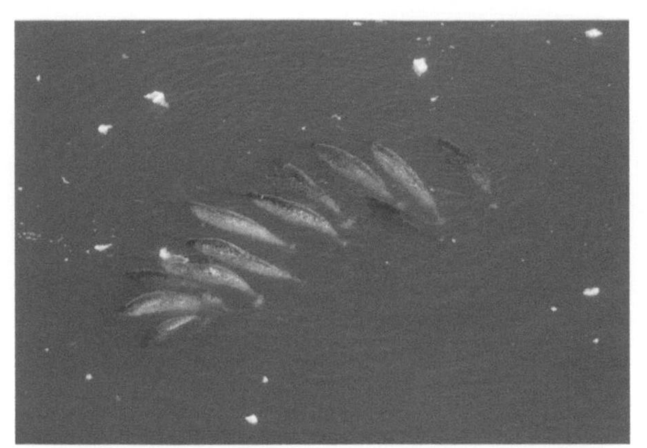

Narwhals in groups like this can be easily distinguished from belugas, even when no males are present, by the presence of individuals with extensive dark spotting on the body. PHOTO: K. LAIDRE

This mixed group of narwhals includes several different age classes, including adult males (light colored bodies and long tusks), and juveniles (darker bodies and no tusks). PHOTO: K. LAIDRE

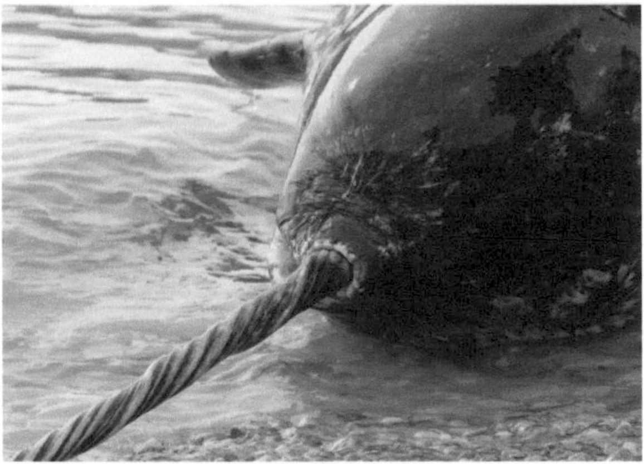

The upper left tooth grows with age to form the tusk of the narwhal, which is unique in the cetacean world. The lower jaw is slightly agape. PHOTO: K. LAIDRE

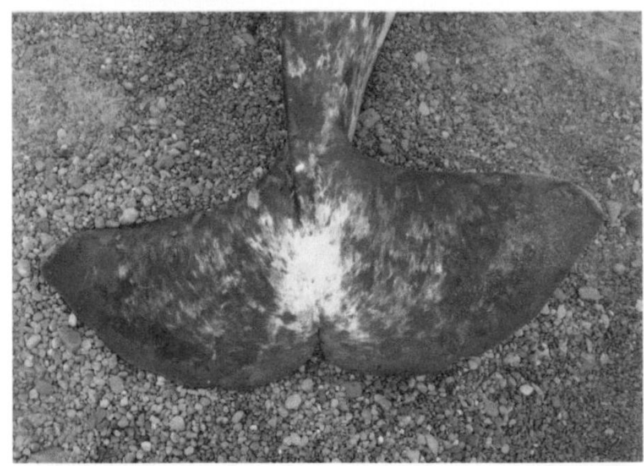

The flukes of adult narwhals become increasingly convex on the trailing edge, so much so that some individuals may almost look like the flukes are stuck-on backwards. PHOTO: K. LAIDRE

stories woven around narwhal tusks. Adult females can be up to 4.2 m and males up to 4.8 m long (not including the tusk). Large male narwhals can reach weights of over 1,600 kg, although females generally don't get over 1,000 kg. Narwhals are about 1.6 m long and 80 kg at birth.

Calf—Length <1/2 of adult length, color pattern uniform gray to nearly black with no spots, slight beak may be visible, no tusk.

Juvenile—Medium-sized (slightly smaller than adult females), body color dark gray to black with white spots on belly and sides, small tusk may be present in males.

Adult female—Relatively large size, ventral side largely white with varying amounts of black spotting (sometimes heavy) on sides and back, forehead bulbous, usually no tusk present.

Adult male—Large body size, lower body mostly white with varying amounts of black spotting especially in area around head and dorsal ridge, flipper tips generally curled up, flukes with convex trailing edge, head bulbous and often with overhanging melon, tusk present. Scarring on the head is fairly common.

Recognizable geographic forms None

Can be confused with Because of their arctic distribution and unique appearance, the narwhal is likely to be confused only with the beluga whale. Young belugas, especially, can look like narwhals, because of the gray coloration of youngsters of both species and the absence of a tusk in young narwhals. The absence of blotching or spotting on beluga whales is probably the best guide. Male narwhals can be easily distinguished by their tusks (no other cetaceans have a long forward-projecting tooth).

Distribution This is a strictly-arctic species; it is found mostly above the Arctic Circle year-round. It has a largely discontinuous range, separated by the island of Greenland. Narwhals basically inhabit the Atlantic sector of the Arctic, and there are few records for the Pacific segment. The principal distribution of the narwhal is from the central Canadian Arctic (Peel Sound and northern

Hudson Bay), eastward to Greenland and thence to the eastern Russian Arctic (around 180°W). They are rarely observed in the far eastern Russian Arctic, Alaska, or the western Canadian Arctic. There are annual migrations, primarily to open water in fall and back to inshore waters in spring. In summer, they follow the ice to more coastal areas. In winter, they remain in the pack ice, generally using cracks or holes. Five stocks are recognized on the basis of distribution and migration patterns.

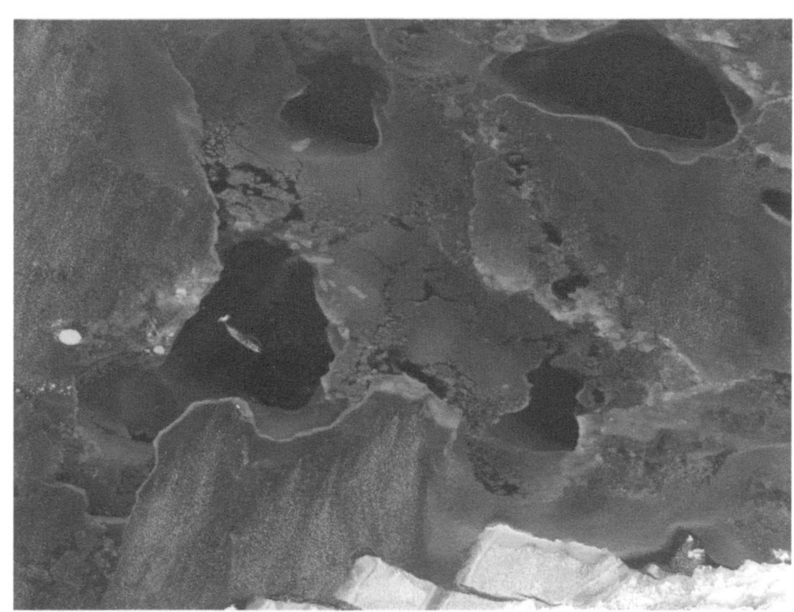

Narwhals are creatures of the Arctic ice. As such they are more likely to be negatively influenced by global climate change than most other species of cetaceans. PHOTO: K. LAIDRE

Ecology and behavior The ecology of the narwhal has been relatively well studied, due to the fact that it is such an important species in the culture and commerce of many northern peoples. In particular, many studies have recently been conducted on movements and diving, using satellite-linked transmitters attached to the dorsal ridge. Most pods of narwhals consist of 2–10 individuals, but there is some evidence that these groups are often parts of large dispersed herds of hundreds or even thousands of individuals, especially in summer. They scatter into smaller groups in winter, and at least in some areas, whales follow specific migratory routes (defined by sea ice formations). There is some age and sex segregation of narwhal groups, and all-male groups, as well as nursery groups, are common. Although belugas are sometimes seen in the same area as narwhals, they generally do not form mixed herds. The average swimming speed of migrating narwhals is about 5 km/h. They occasionally lift their flukes upon diving.

The tusk of male narwhals has long been a source of scientific controversy. It is now generally agreed that the tusk is not necessary for survival. It appears to be used in male-male competition for females, perhaps primarily as a display, although male narwhals have been seen "sparring" with their tusks above water. Adult males also are often seen with broken tusks and scarring on the head, providing further evidence of the aggressive use of tusks.

Young narwhals are born mainly in summer, from July through August, after a gestation of about 13–16 months. Nursing lasts for at least a year. They live to be at least 25 years old, and some may live to 50. Killer whales and polar bears are at least occasional predators.

Feeding and prey Fish, squid, and shrimp make up most of the narwhal's diet. A major component of their diet is made up of medium to large-size Arctic fish species, such as turbot, Arctic cod, and polar cod (the latter of which are often associated with undersides of ice). Narwhals feed at times in deep water and possibly at or near the bottom. Dives of up to nearly 1,200 m and 25 min are known, and there are some seasonal differences in the depth and intensity of diving. They use suction to bring prey items into the mouth.

Threats and status Although never the targets of large-scale commercial hunting, narwhals have been hunted by native peoples of the Arctic for their valuable tusks (thought to be the source of the unicorn legend) and highly sought-after flesh, which is eaten and used for dog food. Their skin (known as "muktuk") and blubber are highly prized by Arctic native peoples. In the Middle Ages in particular, there was a large profit to be made from selling narwhal tusks as unicorn horns to naïve buyers. Hunting continues to be the major threat to these animals. Narwhals are also prone to mortality from ice-entrapments (generally worsened by wind-driven or fast-forming ice), which may be happening more frequently these days, at least partially due to human-induced climate changes. When live, entrapped whales are discovered by Inuit hunters, they normally take advantage of the event by killing the animals (this is called a *savssat*). The tissues of narwhals contain high levels of environmental contaminants, although the specific effects of these are generally not known. While not endangered on a global scale, some populations are clearly depleted, and continue to be subjected to unsustainable exploitation. There are probably over 50,000 narwhals throughout their range (including at least 35,000 in northern Davis Strait and Baffin Bay, 1,400 in Hudson Strait, and 300 in Scoresby Sound).

IUCN status Near Threatened.

Beluga Whale—*Delphinapterus leucas*

(Pallas, 1776)

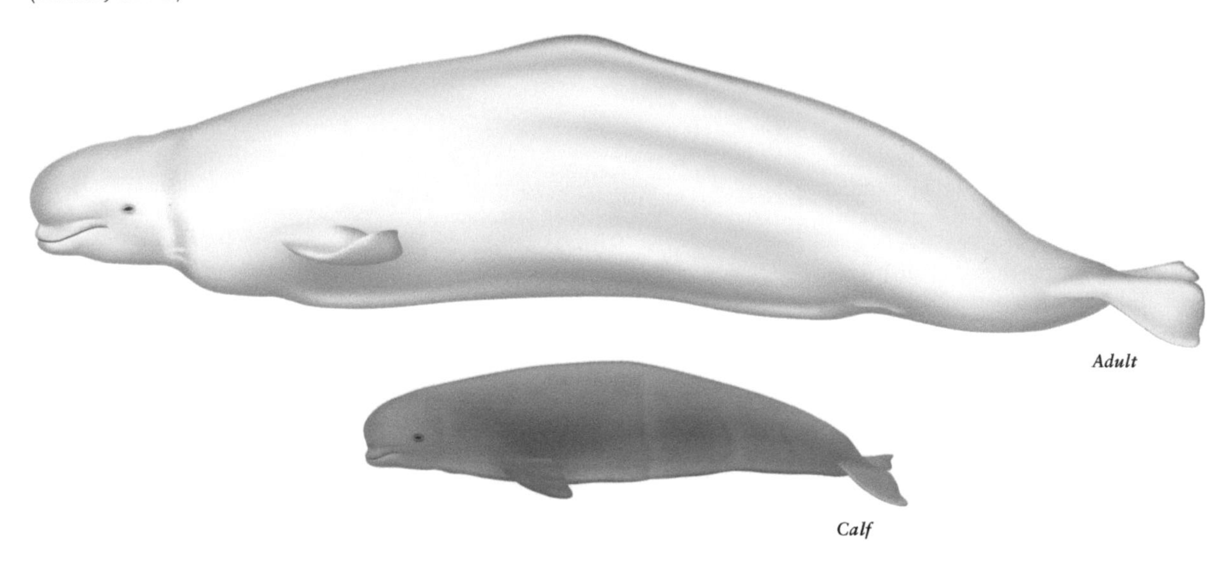

Recently-used synonyms *Delphinapterus freimani, Delphinapterus dorofeevi.*
Common names En.–beluga whale, white whale; Sp.–*beluga*; Fr.–*bélouga.*
Taxonomic information Cetartiodactyla, Cetacea, Odontoceti, Monodontidae. Specimens that appear to be hybrids between narwhals and belugas have been described.
Species characteristics The beluga whale, or white whale, is a robust animal—some individuals may be truly rotund, with folds of fat along the belly and sides. Its basic body shape is much like that of the narwhal; it has a small bulbous head with only a very short beak, and a cleft upper lip. There is no dorsal fin (instead, a shallow transversely-nicked ridge runs along the midline of the back) and small rounded flippers (with curled tips in adult males). The flukes are small, but relatively deep, and often have a convex trailing edge. Belugas are "blubbery"—the blubber layer may be up to 15 cm thick. Their bodies are supple and often wrinkled. There is often a visible neck, which is unusual for cetaceans. Because the cervical (neck) vertebrae are not fused, white whales can move their heads more than most other cetaceans and even have the ability to turn their heads sideways (which is very uncommon in whales and dolphins). The face and melon are also very supple (even the lips can move), and this species is capable of more "facial expressions" than any other species of whale or dolphin.

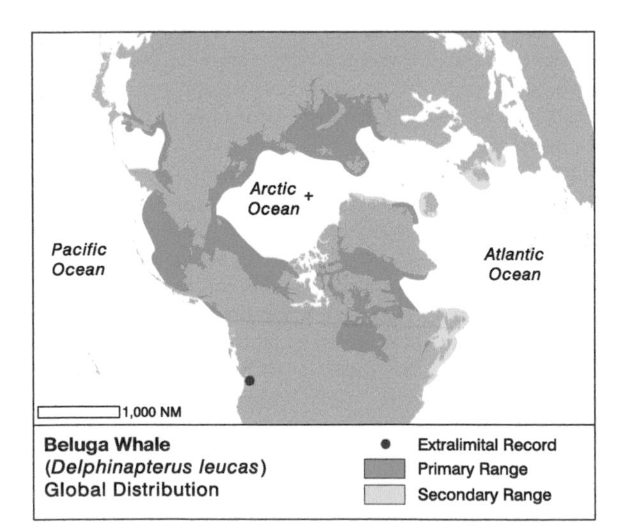

At birth, white whales are a creamy pale gray, and they rapidly turn dark gray to brownish-gray. They whiten increasingly as they age, reaching the pure white stage between 5 and 12 years of age. Some adults may have a yellowish tinge, especially when they congregate in estuaries in summer months.

The mouth generally contains 8–9 teeth, often heavily worn, in each row of the upper and lower jaws. The teeth may be worn down to the gums in some older animals. Most beluga whales are less than 5.5 m (males) or 4.3 m (females) long, and males are about 25% longer than females. Large animals may weigh up to 1,600 kg. Calves average about 1.6 m and 80–100 kg at birth.
Calf—Length 1/3–2/3 of adult length, color pattern uniform dark gray (may range from brownish to nearly black), slight beak may be visible.
Juvenile—Length 2/3–3/4 length of adults, color uniform gray (varies from dark to light gray).

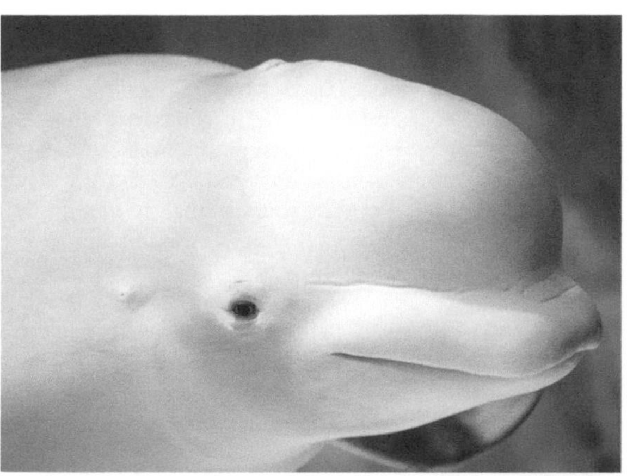

The head shape of a beluga whale is rather unusual. The beak is very short, but very wide. Both the forehead and lips are highly mobile, which is not the typical situation in cetaceans. Captivity. PHOTO: R. KIEFNER

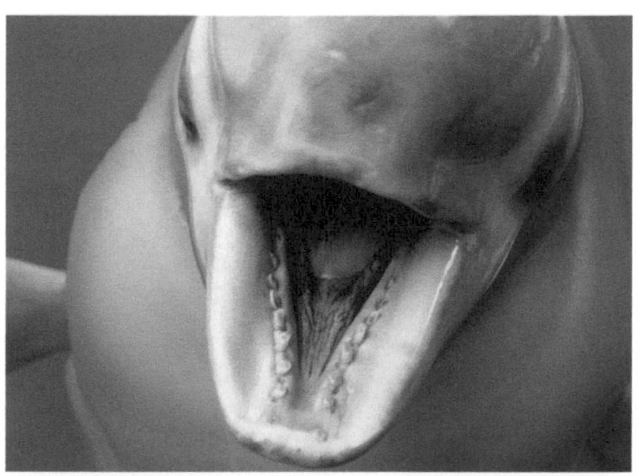

In older belugas, the teeth may be quite worn down. Notice also the very thick lips that characterize this species. Captivity. PHOTO: R. KIEFNER

When seen through the surface, the ghostly glow of a white whale is hard to mistake for any other species. However, it should be remembered that some older narwhals can appear very light in color as well. PHOTO: J.-P.. SYLVESTRE

There is no dorsal fin in the beluga, but instead there is a low dorsal ridge, which may sometimes be dark in color. Captivity. PHOTO: R. KIEFNER

Belugas have very flexible lips, which is quite unusual for a cetacean. PHOTO: J.-P.. SYLVESTRE

They don't often breach, but belugas do perform other aerial behaviors, such as spyhops, quite frequently. PHOTO: R. KIEFNER

They need a thick blubber layer to keep warm in the cold waters they inhabit, and some belugas can be so rotund that they develop rolls of fat along their sides. These fat pads are unique and are used as control surfaces in the absence of a dorsal fin. Captivity. PHOTO: R. KIEFNER

This unusual view of a beluga in a tank shows the diamond-shaped flippers quite well. Captivity. PHOTO: R. KIEFNER

Adult—Body very robust, color very light gray to white, with only some dark color on dorsal ridge and borders of appendages, flukes may be strongly convex on trailing edge, flipper tips may be curled upwards.

Recognizable geographic forms None.

Can be confused with Belugas can be confused with narwhals, which overlap in some parts of their range. The black and white spotting/blotching of narwhals, and the tusks of males of this species, should permit accurate identification in most situations. Particular care must be taken to distinguish calves of the two species, as both can be dark gray in color (and the narwhal's tusk is not yet developed). However, calves will usually be with adults, which should not be difficult to distinguish.

Distribution White whales have an almost panarctic distribution, and are found only in high latitudes of the Northern Hemisphere (mainly from 50–80°N) They are widely distributed throughout the arctic and subarctic regions, from the west coast of Greenland, west to eastern Scandinavia and Svalbard. They occur seasonally (mainly in summer) in coastal waters as shallow as 1–3 m deep; however, they often occur in deep (>800 m), offshore waters as well. Belugas enter estuaries (they are common in the lower St. Lawrence River estuary of eastern Canada), and even sometimes rivers; there are a few records of solitary individuals ranging thousands of kilometers up various rivers. The International Whaling Commission recognizes 29 management stocks of beluga whales, based on morphological, genetic, and distribution differences.

Ecology and behavior Belugas are well-studied, and much has been learned of their ecology and behavior from studies of carcasses and satellite tracking. The highly gregarious beluga is most often found in groups of up to about 15 individuals, but it is sometimes seen in aggregations of thousands (which periodically gather in shallow estuaries). Groups are often segregated by age and sex; all-male groups and mixed aggregations, including females and young, are known. Group structure is largely fluid.

In general, white whales are not showy at the surface and they do not often leap. These animals generally swim slowly, rolling at the surface. During the summer, when they aggregate in large numbers in shallow estuaries, they can be very active, and may do quite a bit of spyhopping, tail waving or fluke-slapping. In some areas, they may have distinct foraging and resting periods throughout the day. Their extreme loquaciousness, which was heard through the hulls of old-time whaling ships, earned them the nickname "sea canary." Vocalizations can be divided into whistles and pulsed calls. Some of the descriptive names given to their vocalizations include groans, buzzes, trills, and roars.

Dives may last up to 25 minutes, and can reach depths of >800 m. It is possible that they can dive to depths of 1,000 m or more, and they regularly appear to dive to the sea bottom. The beluga has an annual molt—something unusual for cetaceans. During this time, they may rub on the bottom to facilitate sloughing of dead skin and stimulate epidermal regrowth.

Sexual maturity in belugas is reached at ages of about 5 years for females and 8 for males. Calves are born in spring to summer, between April and September, depending on the population. Gestation probably lasts about

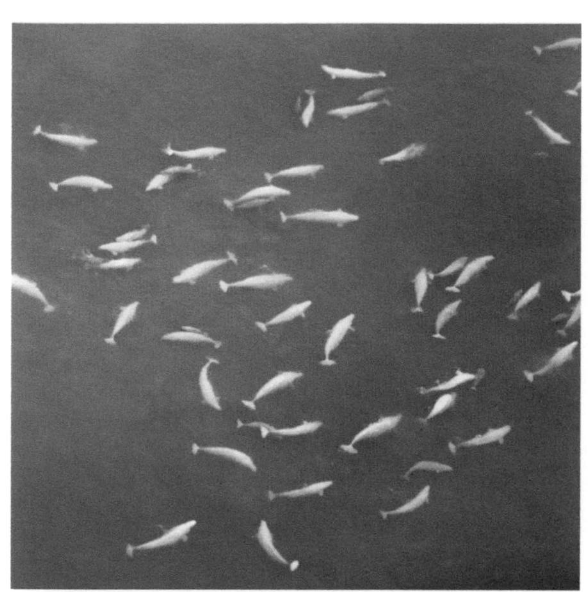

Very large aggregations of beluga whales may develop for feeding, mating/calving, and for molting—sometimes in very shallow waters. PHOTO: R. KIEFNER

A group of belugas, including both large white adults, and smaller grayish calves and juveniles, moves along through the arctic ice floes. PHOTO: V. BEAVER

12–15 months, and lactation occurs for about 2 years. Longevity is not well-known, but is expected to be at least 40 years. Killer whales and polar bears are known predators. Polar bears may wait at breathing holes in the ice, and pull whales from the water as they surface to breathe.

Feeding and prey The beluga has a diverse diet, which varies greatly from area to area. Although various species of fish are considered to be the primary prey items (including salmon, herring, and Arctic cod), beluga whales also feed on a wide variety of mollusks (such as squid and octopus) and benthic crustaceans (shrimps and crabs). They may even eat marine worms and some forms of zooplankton at times. Based on stomach contents, belugas are thought to feed mostly on or near the bottom (in waters up to 300 m deep). Belugas mostly hunt individually. They use their flexible lips to create suction and pull in prey items.

Threats and status Belugas have been hunted commercially in the past, and exploitation by native Arctic peoples continues today in a number of areas. Some stocks have been badly depleted, if not extirpated, by past hunting. Their predictable migration patterns and habit of aggregating in shallow water make them relatively easy to kill, and hunters have taken advantage of these facts. Entanglement in gillnet and other fisheries causes some mortality, and several hundred individuals have been captured alive for display and research over the past 150 years or so (in fact, this was one of the first species of cetacean to be held captive, beginning in New York in 1861). In addition, habitat degradation, behavioral disturbance (mainly by oil and gas exploration and extraction activities, as well as by hunters and fishermen), and pollution may be causing problems for these animals. The population in the St. Lawrence River is recovering more slowly than expected from past overhunting. The delay has been hypothesized to be related to the damaging effects of organochlorines and other pollutants on the health and reproduction of the animals. The status of some other populations is also being affected by environmental contamination. Abundance has been estimated for a number of the 29 recognized stocks. There is a small isolated population of approximately 350–400 belugas that occurs in Alaska's Cook Inlet—this population is seriously threatened. The beluga is certainly not endangered at the species level, but some populations are in serious trouble. There are considered to be well over 150,000 belugas throughout their range, including about 40,000 in the Beaufort Sea, 18,000 in the Bering Sea and other Alaskan waters, 28,000 in Baffin Bay, Canada, 25,000 in western Hudson Bay, about 10,000 in other waters of eastern Canada and west Greenland, and 21,000–30,000 in waters of the former Soviet Union.

IUCN status Near Threatened (overall species), Critically Endangered (Cook Inlet population).

Australian Snubfin Dolphin—*Orcaella heinsohni*

Beasley, Robertson, and Arnold, 2005

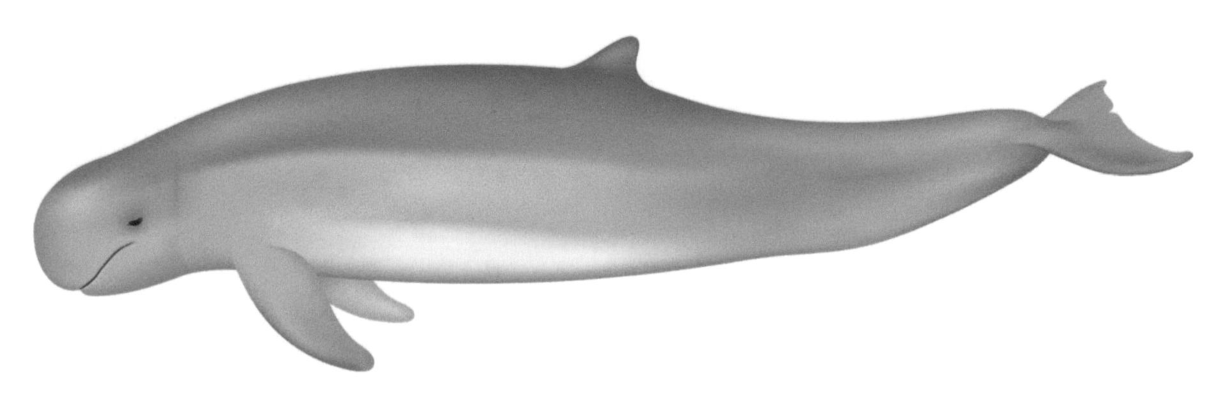

Recently-used synonyms None.
Common names En.–Australian snubfin dolphin; *Sp.–delfín del Heinsohn; Fr. –orcelle d' Australie.*
Taxonomic information Cetartiodactyla, Cetacea, Odontoceti, Delphinidae. For many years, only a single species was recognized in the genus. Recent studies have found evidence of two species, and *O. heinsohni* has now been split-off as a second species in the genus. This species occasionally hybridizes with the Australian humpback dolphin.
Species characteristics The Australian snubfin dolphin resembles the Irrawaddy dolphin and the finless porpoise, but unlike the latter species, it has a dorsal fin. The fin is small (averaging 4.4% of total length) and rounded, and is set slightly behind mid-back. The large flippers (width averaging 6.6% of total length) have curved leading edges and rounded tips. The head is blunt and can be bulbous, with no beak; the mouthline is straight; and there is a distinct neck crease. The dorsal groove characteristic of the Irrawaddy dolphin is absent in this species.

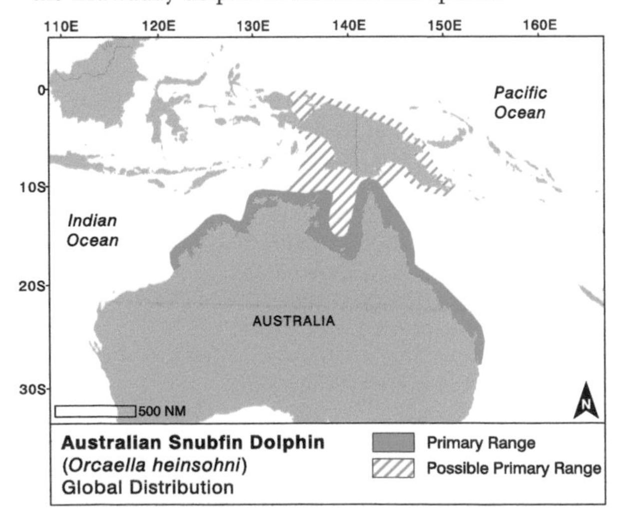

Coloration of the Australian snubfin dolphin is slightly different than in the Irrawaddy dolphin. These animals have a tripartite pattern, with a distinct dark grayish-brown cape, lighter sides, and whitish belly. In certain lighting it can take a reddish tinge. The margin of the cape is fairly straight and extends somewhat low on the sides. Tooth counts in each row are 11–22, averaging 18 (upper); and 14–19, averaging 17 (lower). The teeth have slightly expanded crowns. This is a relatively small dolphin, although it is a bit larger than the Irrawaddy dolphin. Adults range from 1.86 to 2.70 m. The average adult size is 2.18 m. They are known to reach weights of at least 130 kg. Scant evidence indicates that the length at birth is about 1 m.
Recognizable geographic forms None.
Can be confused with Australian snubfin dolphins can be confused with dugongs in most parts of their range. When a clear view is obtained, snubfin dolphins are easily distinguishable, because dugongs do not have a dorsal fin. They can also be confused with Australian humpback dolphins, but a good look at head and/or dorsal fin shape will clear up any problems. The possibility of hybrids should be kept in mind.
Distribution Australian snubfin dolphins are inhabitants of the shallow Sahul Shelf, and inhabit coastal, brackish waters of the tropical and subtropical zones of Australia. Their occurrence in Papua New Guinea has not yet been confirmed, but it is likely that the species occurs there. In Australia, they occur from Broome, Western Australia, to Port Alma/Fitzroy River, Queensland. Occasional sightings have been confirmed as far south as the Dampier Peninsula on the west coast and Brisbane River on the east coast of Australia. They most often occur near river and creek mouths, generally in waters <15 m deep (with a preference in some areas for waters <2 m deep).
Ecology and behavior Very little is known of the ecology of these animals, largely because they were only recently

The rounded forehead and often-visible neck of the Australian snubfin dolphin will assist in identification. Western Australia. PHOTO: S. ALLEN

split-off as a separate species from the Irrawaddy dolphin. Group sizes typically range from 1 to 10, but they sometimes occur in aggregations of up to 20 dolphins. Average group size is 5.3 individuals in northeast Queensland. Groups are somewhat fluid in structure, with dolphins having a series of constant companions and also casual acquaintances. There are numerous strong associations in at least Queensland.

Australian snubfin dolphins are commonly seen in the same areas used by Australian humpback dolphins. They are the subject of aggressive interactions from that species, and in Cleveland Bay, snubfin dolphins are sometimes aggressively chased and "harassed" by humpback dolphins. Individual dolphins have been identified by scars and marks on the back and dorsal fin. They are not especially active, but do make low leaps on occasion. They are not known to ride the bow waves of vessels.

Almost nothing is known of the life history of this species. A few dolphins that were aged from tooth growth layers reached adult size at 4–6 years, and some lived as long as 30 years. Longevity may be greater than this, however.

Feeding and prey Australian snubfin dolphins appear to be generalist feeders, taking a wide variety of fishes (including anchovies, sardines, eels, halibut, breams, grunters, and

Although they occur in murky waters in many parts of their range, snubfin dolphins in some parts of Western Australia are found in clear waters. PHOTO: A. BROWN

Australian snubfin dolphins will be quite easy to distinguish from the other small cetaceans that overlap them in distribution, which all have long beaks. Western Australia. PHOTO: J. SMITH, COURTESY OF A. BROWN

Australian snubfin and Irrawaddy dolphins are atypical for dolphins, in terms of having free neck vertebrae and the ability to bend their necks. PHOTO: C. HARRINGTON

Australian snubfin dolphins are found throughout the warmer, tropical waters of northern Australia, and apparently also occur in some areas off the island of New Guinea. Western Australia. PHOTO: A. BROWN

The Australian snubfin dolphin in front is clearly an older individual, as evidenced by the extensive scarring and wrinkled back. Western Australia. PHOTO: D. CLARKE, COURTESY OF A. BROWN

other estuarine species). They also eat cephalopods (squid, cuttlefish, and octopus), and crustaceans (shrimps and isopods, although the latter may be incidental).

Threats and status There are four reliable abundance estimates available for the Australian snubfin dolphin: <100 each in Cleveland Bay and Port Alma/Fitzroy River (both Queensland), 136–222 in Port Essington Harbour (Northern Territory) and 137 in Roebuck Bay (Western Australia). The nearshore occurrence of this species makes it vulnerable to detrimental human activities. Most of the range of the species in northern Australia has not yet been severely degraded, however, there is extensive development of port facilities in some areas that may impact Australian snubfin dolphins. Animals are known to have died from fishing and anti-shark net entanglement. Habitat loss, prey depletion and environmental contamination may be other, long-term threats. Vessel collisions appear to be a potential issue in some areas, such as Roebuck Bay, Western Australia. The designation of a series of marine parks in the northern waters of Australia is seen as a necessary prerequisite for the long-term survival of this species, although some populations need still greater protection.

IUCN status Near Threatened.

Irrawaddy Dolphin—*Orcaella brevirostris*

(Owen *in* Gray, 1866)

Recently-used synonyms *Orcaella fluminalis, Orcella brevirostris.*

Common names En.–Irrawaddy dolphin; *Sp.–delfín de Irawaddy; Fr.–orcelle.*

Taxonomic information Cetartiodactyla, Cetacea, Odontoceti, Delphinidae. Although there has long been considered to be only one species in the genus, recent studies have demonstrated there are in fact two: *O. brevirostris* and *O. heinsohni.*

Species characteristics The Irrawaddy dolphin resembles the finless porpoise, but unlike that species, it has a dorsal fin. It is a moderately-robust species. The dorsal fin is small (averaging 3% of total length) and rounded, and is set just behind mid-back. It is variably shaped. The large paddle-shaped flippers (width averaging 7.3% of total length) have curved leading edges and rounded tips. The head is blunt and bulbous, with no beak; the mouthline is straight, and there may be an indistinct neck crease.

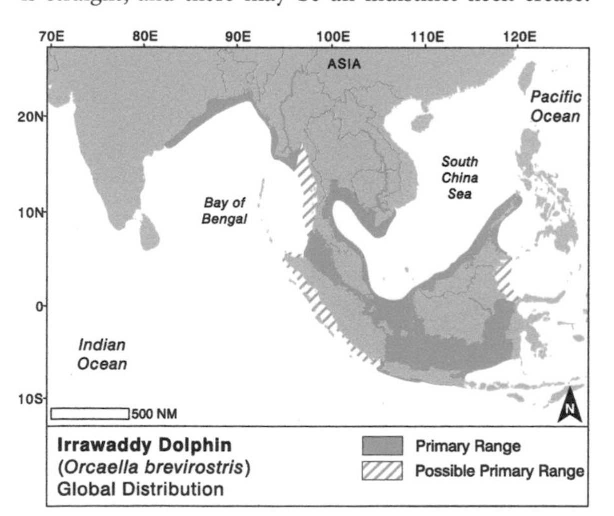

Irrawaddy Dolphin (*Orcaella brevirostris*) Global Distribution

There is generally a distinct dorsal groove running along the back from the neck region to just before the dorsal fin. The U-shaped blowhole is open toward the front, which is typical of most small odontocetes.

Coloration is somewhat variable, and animals kept captive indoors may lose much of their pigmentation, becoming almost white in color. In the wild, these animals are two-tone—the back and sides are gray to bluish-gray; the belly from the lower chin to the anus is somewhat lighter. The light ventral coloration extends up as an "elbow" onto the underside of the flippers.

Tooth counts in each row range from 8–19 and average about 15 (upper); and range from 11–18 and average about 13–14 (lower). The teeth have slightly expanded crowns. In dolphins from the Mahakam River population, the teeth may not erupt.

This is a relatively small dolphin; adults range from 1.73 to 2.75 m and average about 2.05 m. Males are somewhat larger than females. Average weight is about 115–130 kg. Scant evidence indicates that the length at birth is about 1 m, and weight is about 10–12 kg.

Recognizable geographic forms While there are clearly many populations of this species, and freshwater populations are clearly distinct (from each other, as well as from marine stocks), recognizable geographic forms have not yet been described.

Can be confused with Irrawaddy dolphins can be confused with finless porpoises or dugongs throughout most of their range, where these species overlap in distribution. When a clear view is obtained, Irrawaddy dolphins are easily distinguishable, because neither of the other species has a dorsal fin. The two species of *Orcaella* are not thought to be sympatric. They can be distinguished by coloration (two-tone in *O. brevirostris* vs. three-tone in *O. heinsohni*), and external morphology (*O. brevirostris* has a dorsal groove and *O. heinsohni* has a more distinct neck crease).

The bulbous head shape will be useful to distinguish Irrawaddy dolphins from the other small delphinids that share their coastal habitat, all of which have longer beaks. PHOTO: S. SAINO

Cetaceans with round heads rarely porpoise out of the water—a rare photograph of a leaping Irrawaddy dolphin from Malampaya Sound in the Philippines. PHOTO: M. MATILLANO

Although the head shape is similar, Irrawaddy dolphins have a more curved mouthline compared to that of finless porpoises, which is more straight. India. PHOTO: D. SUTARIA

The Irrawaddy dolphin is one of only a few species of cetaceans that have the lip mobility needed to spit water. India. PHOTO: D. SUTARIA

Although not considered a particularly-acrobatic species, Irrawaddy dolphins do perform a number of leaps and breaches at various times. Mekong River. PHOTO: D. SUTARIA

Irrawaddy dolphins often have a visible depression running longitudinally along their backs. Mekong River. PHOTO: R. L. PITMAN

Distribution Irrawaddy dolphins inhabit coastal, brackish, and fresh waters of the tropical and subtropical Indo-Pacific. They occur from Borneo and the central islands of the Indonesian archipelago, north to Palawan in the Philippines, and west to the Bay of Bengal, including the Gulf of Thailand. There are freshwater populations in the Ayeyarwady or Irrawaddy—(up to 1,500 km upstream), Mahakam (up to 560 km upstream), and Mekong (up to 690 km up-stream) rivers, and Songkhla Lake and Chilika Lagoon. The range is poorly documented in much of Southeast Asia. These animals appear to favor shallow estuaries throughout most of their coastal range.

Ecology and behavior Groups of fewer than six individuals are most common, but sometimes up to 15 Irrawaddy dolphins are seen together. Aggregations of up to 25 dolphins can be found in deepwater pools of the Mekong River during the dry season. Irrawaddy dolphins have been seen in the same areas as bottlenose and Indo-Pacific humpback dolphins, as well as finless porpoises. Irrawaddy dolphins are not especially active, but they do

Small groups of just a handful are more the norm for the Irrawaddy dolphin. However, sometimes larger groups do develop, such as this one in the Gulf of Thailand. PHOTO: L. MORSE

make low leaps, breaches, and spyhops on occasion. They are not known to bowride. These dolphins have been observed cooperating with fishermen to herd fish into nets in the Ayeyarwady River of Myanmar. Also, their mobile lips allow them to spit water, and there is some evidence suggesting that this is used as a feeding technique. A maximum dive time of 12 minutes has been recorded, but most dives are less than 3 minutes long. They will occasionally lift their flukes out of the water upon diving. The lack of fusion of most neck vertebrae means that Irrawaddy dolphins can move their head and bend their neck much more so than most other dolphin species.

The calving season is not well known. Some calves appear to have been born in June-August, but one captive female from the Mahakam River gave birth in December. Gestation has been estimated at 14 months. There is probably a great deal of geographic variation in life history parameters, though this is not yet documented. Maximum known longevity is about 30 years.

Feeding and prey Irrawaddy dolphins appear to be generalist feeders, taking a wide variety of fishes (including freshwater species, such as catfish), and cephalopods (squid, cuttlefish, and octopus). Dolphins from freshwater populations sometimes spit water while feeding, apparently to assist in capturing fish.

Threats and status The Irrawaddy dolphin may not be immediately threatened with extinction on a global scale, but certain populations appear to be reduced and are in serious danger of local extinction. This is especially true for freshwater stocks, such as those in the Mahakam, Ayeyarwady, and Mekong rivers, as well as that in Songkhla Lake, which each number less than 100 individuals. The nearshore and freshwater occurrence of this species makes it particularly vulnerable to threats from human modification and degradation of the coastal and freshwater environments. These result from fishing net entanglement, habitat loss, dam and waterway construction, prey depletion, environmental contamination, and vessel strikes. Live captures for aquaria display have also been a conservation concern in some areas (e.g., Thailand and Vietnam). Irrawaddy dolphins have been hunted directly in the past, at least in Cambodia, but are also revered by local people in many areas of Asia. Abundance estimates are available only for certain portions of the range: 77 animals in Malampaya Sound, Philippines; <93 in the Mekong River; 33–55 in the Mahakam River, Indonesia; 55–70 in the Ayeyarwady River, Myanmar, and 5,400 in Bangladesh. There are no quantitative measurements of trends in abundance, but there have been strong suggestions of population declines in some areas, and there is no doubt that the Mekong River population is small and declining.

Irrawaddy dolphins have the lumpy bodies, rounded heads, and low rounded dorsal fins characteristic of slow swimmers. Chilika Lake, India. PHOTO: D. SUTARIA

IUCN status Vulnerable (overall species), Critically Endangered (Mahakam River, Ayeyarwady River, Malampaya Sound, Mekong River, and Songkhla Lake populations).

Killer Whale — *Orcinus orca*

(Linnaeus, 1758)

Type A Killer Whale

Adult Female

Type B—large

Adult Female

Type B—small

Adult Female

Type C Killer Whale

Adult Female

Type D Killer Whale

Adult Female

Two type A killer whales flank an Antarctic minke whale that they and their group killed and ate; their classic black and white color pattern and medium-sized eye patch easily separate them from other Antarctic killer whale types. Western Antarctic Peninsula. PHOTO: R. L. PITMAN

A type C killer whale ("Ross Sea killer whale") works along a fast ice lead in McMurdo Sound, Antarctica; a fish-eater, it has a narrow, slanted eye patch and a less conspicuous cape than type B killer whales. At 6 m, it is also the smallest known killer whale. PHOTO: R. L. PITMAN

A small type B killer whale ("Gerlache killer whale") off the western Antarctic Peninsula with a large eye patch and prominent cape. All killer whales acquire a coating of yellowish diatoms in Antarctica, but when they return to the tropics, they shed the diatoms and turn a cleaner-looking, two-toned gray and white. PHOTO: R. L. PITMAN

Only recently described from field observations, type D killer whale ("sub-antarctic killer whale") has just a sliver of a white eye patch and a larger, more-rounded melon than the other forms; nothing is known about its feeding habits, except that it probably poaches toothfish off longlines. Scotia Sea, Antarctica. PHOTO: J.-P. SYLVESTRE

Recently-used synonyms *Grampus (Orcinus) rectipinna, Orcinus nanus, Orcinus glacialis.*
Common names En.–killer whale, orca; Sp.–*orca*; Fr.–*orque, épaulard*
Taxonomic information Cetartiodactyla, Cetacea, Odontoceti, Delphinidae. Although a single species of killer whale has been recognized for many decades, current genetic, morphological, and ecological evidence suggests that there are at least eight distinct forms of killer whales, including several with overlapping geographic ranges. The taxonomy of this genus is clearly in need of revision, and it is very likely that *O. orca* will be split into several species and/or subspecies over the next few years. Provisionally, the Society for Marine Mammalogy recognizes resident and transient (Bigg's) killer whales as un-named subspecies.

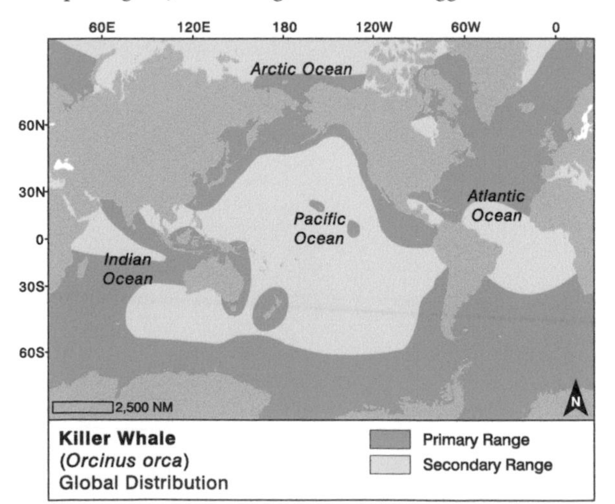

Species characteristics The killer whale is the largest dolphin and the most readily-identifiable of all cetaceans. The body is robust, but somewhat spindle-shaped; the very large dorsal fin is nearly as distinctive as the color pattern, reaching 0.9 m high in females and >2 m in males. The dorsal fins of females and young males are falcate and generally pointed or slightly rounded at the tip. Adult male dorsal fins are erect and triangular or even cant forward in some older individuals. The flippers are large (especially in older bulls) and rounded, and grow to lengths of up to 2 m (about 11–17% of total length for adult females, and 18–22% for adult males). The flukes are broad, with a straight or slightly-convex trailing edge. The fluke tips may be curled down in some large adults (especially bulls). Killer whales have blunt snouts, with short, poorly-defined beaks. The mouthline has a slight downward curve toward the corner of the gape.

The black-and-white color pattern of the killer whale is unique and unmistakable. The ventral surface from the

Fish-eating northern residents of the Pacific Northwest; the dorsal fin is more rounded at the tip than the mammal-eating form that occurs in much of the same waters. British Columbia. PHOTO: B. GISBORNE

Southern resident (fish-eating) killer whales are recognizable by their open saddles—the black incursion into the pale saddle patch behind and below the dorsal fin. San Juan Islands, Washington. PHOTO: D. ELLIFRIT

This adult male fish-eating (resident) killer whale shows off his absurdly large pectoral fins; a handicap for most activities, females may find these fins attractive. PHOTO: M. GREENFELDER

Adult male killer whales have massive dorsal fins, but they also have mammoth pectoral fins; this is a fish-eating ecotype from the central Aleutian Islands, Alaska. PHOTO: J. DURBAN

Fish-eating killer whales apparently feed on very small fish at times; this type C killer whale from McMurdo Sound, Antarctica, has a tiny ice fish in its mouth. PHOTO: R. L. PITMAN

tip of the lower jaw to the urogenital area and undersides of the flukes are white. White lobes extend up and back on the lower flanks behind the dorsal fin above, and behind each eye there is a white patch of variable shape and size depending on the population (see below). The rest of the body is black, except for a light gray "saddle patch" behind and below the dorsal fin; the saddle patch can be almost non-existent in tropical populations. The boundary between black and white is generally very distinct and sharp. Some populations are two-toned gray and white instead of black and white; for those, the pointed leading edge of the saddle travels forward as a thin line and forms the lower boundary to a dark gray dorsal cape that overlays paler gray sides. Very young animals of black-and-white populations sometimes also show a faint cape. Anomalously-white killer whales are occasionally reported.

Killer whales have, in each half of both jaws, 10–14 large, pointed, and slightly-recurved teeth, which are oval in cross section and up to 2.5 cm in diameter. In older animals, and throughout some populations, the teeth can be extensively worn and damaged by abscesses. Newborn killer whales are 2.1–2.6 m in length and weigh about 160–180 kg. Adult females are up to 8.5 m long and can weigh 7,500 kg; adult males can be up to 9.8 m and nearly 10,000 kg. The extent and significance of ecotypic variation among the different killer whale populations is only beginning to be adequately documented and understood (see below).

Calf—Much smaller than adults; the white-pigmented areas on neonates are tinged with a rusty-orange coloration that persists for 6–12 months or more before turning white. They also have a less conspicuous saddle patch.

Juvenile—Very similar to the adults, only smaller, with a falcate dorsal fin.

Adult female—Dorsal fin of moderate height (<1 m) and usually falcate; similar to juvenile males in all aspects.

Adult male—Noticeably larger than adult females—typically, ca. 1 m longer and can weigh over twice as much; dorsal fin extremely tall (>2 m in some cases) and erect or slightly canted forward; extremely large flippers, much larger than for females; flukes often convex, with tips curved downward.

Recognizable geographic forms Several geographic (and sometimes sympatric) forms of killer whales are field-identifiable, based mainly on the size, shape and orientation of the eye patch, and presence or absence of a dorsal cape (Southern Ocean forms); subtle differences in dorsal fin shape and saddle patch pigmentation (North Pacific), and relative angle of the eyepatch (North Atlantic). The taxonomic status of these forms is currently unresolved, but mounting evidence suggests there may be several species present.

A large type B killer whale ("pack ice killer whale") by a BBC film crew, dwarfing their inflatable boat; the dark gray cape and very large, white eye patch are clearly visible. Western Antarctic Peninsula. PHOTO: R. L. PITMAN

The range of the small type B killer whale overlaps considerably with large type B killer whales; besides being approximately 1 m smaller, it often occurs in large groups of dozens or more, and is not known to prey on marine mammals. It has a distinctive 2-tone gray-and-white color pattern, but is often yellowish with diatoms. Western Antarctic Peninsula. PHOTO: T. CHEESEMAN

North Pacific forms:

Resident killer whale—A medium-sized killer whale that often inhabits inland waterways and feeds only on fish (mainly salmon); medium-sized groups range from 1 to about 50 animals and averages about 12. An important distinguishing feature is the prevalence of an "open saddle"—a large dark pigmented area within the otherwise pale gray saddle.

Bigg's (transient) killer whale—Slightly larger than the fish-eating residents and usually occurs in smaller groups (usually 1–5 individuals); it has a more pointed dorsal fin overall, and never has the large black pigmented area in the dorsal saddle (aka "open saddle") common among residents. These are mammal-eaters that take mainly harbor seals in the Northeast Pacific but also minke whales, dolphins, porpoises, sea lions, etc.

Offshore killer whale—Smaller than either Bigg's or resident killer whales, offshores have a relatively smaller and more rounded dorsal fin. Their teeth are often worn down to the gum line, presumably from feeding on sharks; as implied by their name, they are often found out by the continental shelf edge. Individuals often tail-slap while swimming. Group size is very large—often 50–100 animals.

North Atlantic forms:

Type 1—A small (to 6.6 m), black-and-white form of killer whale with a parallel eye patch; feeds on small schooling fish (herring and mackerel), and worn teeth suggest possibly sharks also; takes seals on occasion.

Type 2—A large (to 8.4 m), black-and-white form; larger than type 1, and with an eye patch that slants downward in the rear; it may feed primarily (or entirely) on cetaceans.

Antarctic forms:

Type A—A large (to at least 9 m), black-and-white form with a medium-sized eye patch oriented parallel to the body axis; it normally has no visible dorsal cape. This form is circumpolar in Antarctic waters and is usually found in ice-free areas, where it preys mainly on Antarctic minke whales. This form migrates to lower latitudes and appears to spend more time there than B and C types.

Large Type B (Pack Ice killer whale)—A large (to at least 9 m), two-toned gray-and-white form with a darker gray dorsal cape and pale gray sides. Individuals acquire a coating of diatoms after some time in Antarctic waters, often giving them a dirty yellowish appearance. The eye patch is very large and oriented parallel to the body axis. Apparently circumpolar in Antarctic waters, this form feeds mainly on ice seals and is most often found around dense pack ice. It undertakes rapid, roundtrip migrations to the tropics, but apparently spends most of the year in Antarctic waters.

Small Type B (Gerlache killer whale)—This form is about 1 m shorter than the large type B and weighs perhaps only half as much. It is a two-toned gray-and-white form currently known only from (and possibly endemic to) the western Antarctic Peninsula and western Weddell Sea. It is light gray with a darker gray cape and has a large eye patch oriented parallel to the body axis. Individuals in Antarctic waters can acquire a thick yellowish coating of diatoms giving them a yellowish appearance. A deep diver, it is most often seen foraging in relatively ice-free waters where it appears to feed on fish or squid (and the occasional penguin). It undertakes rapid, roundtrip migrations to the tropics, but apparently spends most of its year in Antarctica.

Type C (Ross Sea killer whale)—A dwarf form of killer whale, it averages about 1–3 m shorter than the larger forms (adult females average about 5.2 m and adult males about 5.6 m, with a maximum of about 6.1 m). It is a two-toned gray-and-white form with a distinct cape and a narrow, slanted eye patch that is oriented at about a 45° angle to the long axis of the body. Individuals are often coated with diatoms that give them a dirty yellowish coloration; those just returning from the tropics appear cleaner, just gray and white. Type C killer whales are currently known only from eastern Antarctica, where they hunt for fish along the fast ice edge and deep into the leads. They undertake rapid, roundtrip migrations to the tropics, but apparently spend most of the year in Antarctica.

Type D killer whale—Only recently recognized, this form has a typical black-and-white killer whale color pattern, but with a tiny eye patch that lies parallel to the body axis. The melon appears to be more bulbous than in other forms and the dorsal fin may be more falcate and pointed. It is circumpolar in the Southern Ocean, but perhaps restricted to subantarctic waters.

A large group of Bigg's (or "transient") killer whales—mammal-eaters from the North Pacific. They tend to have a more pointed dorsal fin and larger body size than fish-eating "residents." Washington State. PHOTO: D. ELLIFRIT

Can be confused with Killer whales are easily recognizable due to the great size of the dorsal fin (especially adult males) and the unique black-and-white color pattern. At a distance, groups without adult males could be confused with Risso's dolphins or false killer whales.

Distribution The killer whale is the most cosmopolitan of all cetaceans, and may be the second-most wide-ranging vertebrate, after humans. Killer whales can literally be seen in any marine region, from the equator to the ice edges, and they have even been known to ascend rivers (although this is uncommon). They also occur in enclosed seas, such as the Mediterranean, Red Sea, Persian Gulf, and Gulf of California. They are generally most common in nearshore areas and at higher latitudes, where waters are most productive.

Compared to other killer whales, type D have tiny white eye patches, more bulbous heads, and they appear to be restricted to subantarctic waters around Antarctica. Scotia Sea, Antarctica. PHOTO: J.-P. SYLVESTRE

Ecology and behavior Killer whales are among the best-studied of all cetacean species, with a few long-term ecological studies spanning 3–4 decades. Studies in the eastern North Pacific, from Washington State to Alaska, have distinguished three ecotypes of killer whales, referred to as residents, Bigg's (transients), and offshores. They are distinguished by differences in acoustics, ecology, genetics, habitat and prey preferences, and social behavior, but with only subtle differences in coloration and external morphology. Residents are primarily fish-eaters, Bigg's killer whales eat mostly marine mammals, and offshores appear to be shark specialists. Studies suggest that while killer whales in other parts of the world also tend to form communities of reproductively-isolated prey specialists, phylogenetic relationships among the different types are complex (i.e., fish-eating did not evolve once and spread around the world). Current research suggests that Bigg's killer whale in the North Pacific is the most genetically divergent killer whale and the best candidate for separate species status.

Pods of resident killer whales in British Columbia and Washington represent one of the most stable societies known among non-human mammals; individuals stay in their natal pod throughout life. Pods are structured around matrilines (a mother and her offspring). Differences in vocal dialects among sympatric resident groups appear to help maintain pod discreteness and prevent inbreeding. Bigg's killer whales also have a matrilineal-based social structure, but there appears to be much more dispersal from the natal pod, which accounts for the smaller average group size.

Killer whales are among the most enthralling of all animals to observe in the wild. They do not typically ride bow waves, but this behavior has been seen in the Gulf of Mexico, and in some other areas. They may show great interest in vessels; at other times they may avoid them. They can be aerially active, and sometimes breach, spy-hop, flipper-slap and fluke-slap; they often perform these behaviors in bouts. The flukes are sometimes lifted out of the water when whales make a steep dive. When resting, they often travel in a line, abreast, and surface at regular intervals; males often travel separately from the females and young.

In the Northeast Pacific, calving occurs year-round, with a peak from October to March. Similarly, in the Northeast Atlantic, it occurs from late fall to mid-winter. Gestation appears to last for 15–18 months, which is extremely long for a mammal. The calving interval is about 5 years in the Northeast Pacific, and females have

A Bigg's killer whale ("transient") pounds a Dall's porpoise into submission; mammal-eating killer whales are one of the most spectacular predators on the planet. Pacific Northwest. PHOTO: J. TOWERS

Small type B killer whales probably feed mainly on fish (or squid), but they do snack on penguins occasionally (here, a gentoo); note the type B large eye patch. Western Antarctic Peninsula. PHOTO: H. FEARNBACH

A cow and calf large type B killer whale look over a crabeater (left) and Weddell seal—only the crabeater was allowed to get away. Very young calves of all killer whales are rusty orange for the first few months of their lives. Western Antarctic Peninsula. PHOTO: R. L. PITMAN

Fish-eating resident killer whales in the Pacific Northwest prey almost exclusively on salmon, as shown. Although they share coastal habitat with mammal-eating Bigg's killer whales, after years of observations, residents have never been observed to prey on marine mammals. Pacific Northwest. PHOTO: B. GISBORNE

a decades-long, post-reproductive period (menopause). The age at weaning is thought to be about 1–2 years, but may occur somewhat later. Sexual maturity is reached at ages of 10–15 years for females and about 15 years for males. These are extremely long-lived animals, with males living to 50–60 years and females 80–90 years, or even longer!

Feeding and prey Killer whales are known to prey on nearly every large animal species that swims in the sea, including nearly all groups of marine mammals, from sea otters to blue whales; various species of fish (including sharks and rays as well as bony fishes); and cephalopods. They also occasionally eat seabirds, including penguins, and marine turtles. They often use cooperative techniques to herd fish and to attack large prey. They have a great diversity of feeding strategies, including intentional beaching to take seals and sea lions in shallow waters, or creating waves to wash seals or penguins off ice floes. At least some populations are known to show extremely-narrow prey preferences. Thus, fish-eating residents in the Northeast Pacific are not only salmon specialists (at least during the summer months), but they take king (chinook) salmon almost to the exclusion of other salmon species, even when the latter are much more abundant. In Norwegian fjords, killer whales specialize on herring; in New Zealand, killer whales appear to feed mainly on sharks and other elasmobranchs, and type C killer whales in Antarctica are only known to feed on Antarctic toothfish. Among mammal-eating populations, Bigg's killer whales in Greater Puget Sound appear to preferentially take mainly harbor seals, and in Antarctic waters, type A killer whales specialize on Antarctic minke whales, and large type Bs eat mostly ice seals taken off ice floes.

Threats and status Killer whales have at one time or another been targets of both commercial and subsistence whalers, although never on a particularly large scale. Japanese coastal whaling operations have taken several hundred in the North Pacific, and Norwegian culling resulted in the taking of an average of 56/year in the mid-1900s. Russian whaling operations killed several thousand in Antarctica in the 1900s. They are still occasionally killed in small numbers in fisheries in Japan, Greenland, Indonesia, and the Caribbean islands. They have also been persecuted because of real or perceived interference with fisheries and their perceived threat to humans. Entanglements in fishing nets, vessel collisions, and live-captures for oceanaria have also contributed to their problems. Recently, the effects of high levels of environmental contaminants and acoustic "pollution" have become issues of concern, especially for Northeast Pacific populations. Although not considered to be highly abundant anywhere, killer whale populations in many areas appear to be healthy. There are thought to be on the order of 50,000 of these animals globally, with perhaps over half occurring in Antarctic waters alone. In the eastern tropical Pacific, there are estimated to be about 8,500 killer whales; there are an estimated 1,500 in coastal waters of western North America, from the Aleutian Islands to California, and about 2,000 in Japanese waters. The fish-eating population that inhabits the inland waters of Washington and southern British Columbia (southern resident killer whales) now numbers only about 88 individuals. It is considered to be endangered, and is listed as such under the US Endangered Species Act.

IUCN status Data Deficient.

Short-finned Pilot Whale—*Globicephala macrorhynchus*

Gray, 1846

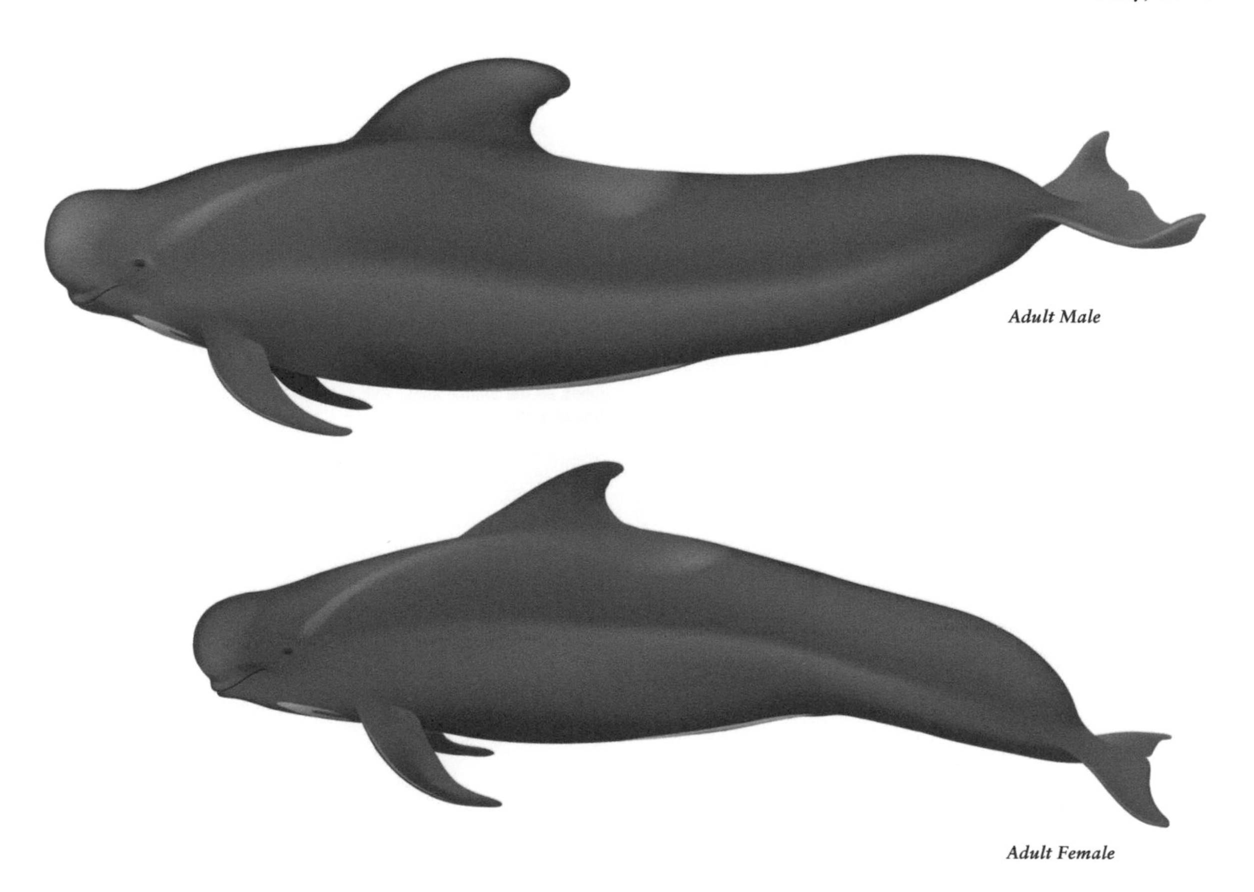

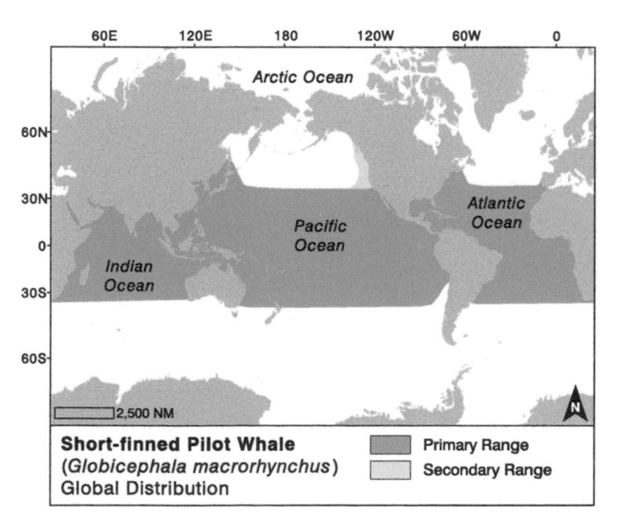

Short-finned Pilot Whale
(*Globicephala macrorhynchus*)
Global Distribution

Recently-used synonyms *Globicephala seiboldii, Globicephalus scammoni, Globicephala brachyptera.*

Common names En.–short-finned pilot whale; Sp.–*calderón de aletas cortas;* Fr.–*globicéphale tropical.*

Taxonomic information Cetartiodactyla, Cetacea, Odontoceti, Delphinidae. Hybrids with common bottlenose dolphins have been reported from captivity.

Species characteristics Short-finned pilot whales are large delphinids with long bodies, bulbous heads, dramatically upsloping mouthlines, and extremely short or nonexistent beaks. The shape of the head varies significantly with age and sex, becoming more globose in adult males. In some forms, the head of adult males may become flattened at the front and appear very squarish, when viewed from above or below. The dorsal fin, which is situated only about 1/3 of the way back from the head, is low and falcate, with a very wide base (it also varies with age and sex). The flippers are long and sickle-shaped,

Short-finned pilot whales have rounded heads, thick tailstocks, and an over-sized dorsal fin that occurs much further toward the head than in any other cetacean species. Hawai'i. PHOTO: M. NOLAN

A young short-finned pilot whale—the dorsal fin in younger animals is fairly normal-looking but becomes much larger, wider at the base and more lobed as the animal ages. This captive animal clearly has short pectoral fins as well. PHOTO: J.-P. SYLVESTRE

Short-finned pilot whales in the tropics typically look all black with little or no eye stripe or saddle patch. Their rounded melons suggest they do not travel very fast, so they rarely porpoise out of the water and do not ride bow waves. Malidves. PHOTO: C. WEIR

In the western and central North Pacific, some populations of short-finned pilot whales have conspicuously-flattened melons (do they butt heads?); the systematics of this form are not yet fully understood. Hawai'i. PHOTO: M. NOLAN

Short-finned pilot whales from the eastern North Atlantic are distinctively-patterned, with a well-defined and connected saddle patch and eye stripe. Azores. PHOTO: M. VAN DER LINDE

Short-finned pilot whales from the eastern North Atlantic often have a prominent saddle, which connects to the light eye blaze. Azores. PHOTO: M. VAN DER LINDE

A close-up view of an eastern Atlantic short-finned pilot whale; most populations worldwide are largely all black and show just a hint of this patterning. Canary Islands. PHOTO: F. RITTER

about 14–19% of the body length. The shape of the flippers is more curved in this species, than in the long-finned species. Adult males are significantly larger than females, with large, sometimes-squarish foreheads that may overhang the snout, very hooked dorsal fins with thickened leading edges, and deepened tail stocks with post-anal keels.

Except for a light gray, anchor-shaped patch on the chest, a gray post-dorsal fin saddle, and a pair of roughly-parallel bands high on the back that sometimes end as a light streak or teardrop above each eye, pilot whales are basically black to dark brownish-gray. This is the reason for one of their other common names, blackfish (the term blackfish is variously used, usually by fishermen, to refer to killer, false killer, pygmy killer, pilot, and melon-headed whales). However, a faint dorsal cape is often visible in good lighting. The characteristic delphinid "bridle" (consisting of an eye stripe and blowhole stripe) is present. Calves are paler than adults.

There are usually 7–9 short, sharply-pointed teeth in each tooth row. Short-finned pilot whales are about 1.4–1.9 m long at birth. Adults reach 5.5 m (females) and 7.2 m (males). Males may weigh nearly 3,600 kg.

Recognizable geographic forms Two geographic forms have been described from Japanese waters, differing in external and cranial morphology. These appear to represent separate subspecies, although they have not yet been formally described as such:

Japanese northern form—This form of pilot whale is found off the Pacific coast of Japan, north of 35°N. It is much larger than the southern form, with females attaining lengths of up to 5.1 m and males reaching 7.2 m. Newborns are about 1.85 m. There is a distinct, light gray saddle behind the dorsal fin. When viewed from above or below, the head of adult males is rounded.

Japanese southern form—This form of pilot whale is found off the Pacific coast of Japan, south of 39°N. It is much smaller than the northern form, with females reaching lengths of only 4.05 m and males reaching only 5.25 m. Newborns are about 1.4 m long, The saddle patch behind the dorsal fin is very faint and indistinct. When viewed from above or below, the head of adult males of this form is squarish.

Can be confused with In and near the areas of overlap (South Pacific; North Atlantic; off southern Africa; and off southern Brazil, Uruguay, and northern Argentina), the two species of pilot whales are difficult or impossible to distinguish at sea. Most sightings can be tentatively assigned to species, based on the area. Other smaller blackfish, such as false killer whales, and less commonly, pygmy killer and melon-headed whales, may be confused with pilot whales at a distance. Dorsal fin shape and position are the best clues to distinguishing pilot whales from these species.

Distribution Short-finned pilot whales are found in warm temperate to tropical waters of the world, generally in deep offshore areas. They do not usually range north of 50°N or south of 40°S There is some distributional overlap with their long-finned relatives *(G. melas)*, which appear to prefer cold temperate waters of the North Atlantic, Southern Hemisphere, and previously the western North Pacific. Only short-finned pilot whales are currently thought to inhabit the North Pacific, although distribution and taxonomy of pilot whales in this area are still largely unresolved. There are two geographic forms of short-finned pilot whale off Japan. They also occur in the Red Sea, but not the Mediterranean.

The short pectoral fins of these diving pilot whales and their tropical setting (Caribbean region) identify these as short-finned pilot whales. Guadeloupe. PHOTO: L. BOUVERET

Short-finned pilot whales from the northeastern Atlantic (e.g., Canaries, Azores) are perhaps the most strikingly patterned pilot whales anywhere; what appears to be an enlarged saddle sweeps up in front of the dorsal fin, then arcs down to contribute to an eye patch. Azores. PHOTO: W. OSBORN

Ecology and behavior In the eastern Pacific, short-finned pilot whales are commonly associated with other species (such as bottlenose, Pacific white-sided, common, and Risso's dolphins, and sperm whales). Pods of up to several hundred short-finned pilot whales are seen, and members of this highly-social species are almost never seen alone. Strong social bonds may partially explain why pilot whales are among the species of cetaceans that most frequently mass-strand. Although detailed studies of behavior have only begun recently, pilot whales appear to live in relatively-stable maternal groups.

In 1982–83, a strong El Niño event brought about major ecosystem changes off the southern California coast. Short-finned pilot whales, which were previously common especially around the California Channel Islands, essentially disappeared from the area (presumably due to the absence of spawning squid), and remain extremely rare there to this day. They may have been displaced by Risso's dolphins.

Sexual maturity occurs at around 8–9 years for females and 13–17 years for males. Females become post-reproductive at around 40 years, but may continue to suckle young for up to 15 additional years, suggesting a complex social structure in which older females may give their own or related calves a "reproductive edge" through prolonged suckling. Calving peaks occur in spring and fall in the Southern Hemisphere, and in fall and winter in most Northern Hemisphere populations. However, the southern form off the Pacific coast of Japan gives birth mostly in July and August. Longevity is at least 63 years.

Feeding and prey Although they also take fish, pilot whales are thought to be primarily adapted to feeding on squid. One of the main genera taken off the California coast is the market squid (*Doryteuthis* sp.). They show the tooth reduction typical of other squid-eating cetaceans.

Threats and status Short-finned pilot whales have been killed directly in drive fisheries in Japan and in harpoon fisheries in the Caribbean and Indonesia. This species is also taken as by-catch in several fisheries in the North Pacific, including driftnet fisheries for swordfish and sharks. At least in the past, they have also been incidentally taken in the squid purse seine fishery that operates off the California coast, and some have been shot by fishermen, who see them as competition for squid. Several short-finned pilot whales have been captured for public display and research in the US and Japan. There are no estimates of global abundance, but some estimates for specific areas do exist. Off the west coast of the US there are about 1,000, in the eastern tropical Pacific there are about 500,000, in the Sulu Sea of the Philippines there are about 7,700, and off Japan there are around 60,000 (53,000 of which are of the southern form). Clearly, the species is not endangered.

IUCN status Data Deficient.

Long-finned Pilot Whale — *Globicephala melas*

(Traill, 1809)

Recently-used synonyms *Giobicephala melaena, Globicephala edwardii.*

Common names En.–long-finned pilot whale; Sp.–*calderón común;* Fr. –*globicéphale commun.*

Taxonomic information Cetartiodactyla, Cetacea, Odontoceti, Delphinidae. Three subspecies are recognized in some classifications (including by the Society for Marine Mammalogy): *G. m. melas* in the North Atlantic, *G. m. edwardii* in the Southern Hemisphere, and an un-named subspecies in Japanese waters of the North Pacific (now likely extinct).

Species characteristics Externally, the long-finned pilot whale resembles its short-finned relative, *G. macrorhynchus.* The body is relatively robust, but the tail stock is long. The head is globose, with an upsloping mouthline, and there is only a slightly discernible beak (if any). The flippers are extremely long (18–27% of the body length) and slender, with pointed tips and a strongly-angled leading edge that forms an "elbow." The dorsal fin is about one-third of the way back from the snout tip, and is low, extremely wide based, and falcate. The tail stock is deepened (it remains of more-or-less uniform depth from the saddle patch to just ahead of the flukes). Compared to females, male pilot whales have larger, more bulbous heads (which may actually be rather squarish); larger, thicker dorsal fins; and deeper tail stocks.

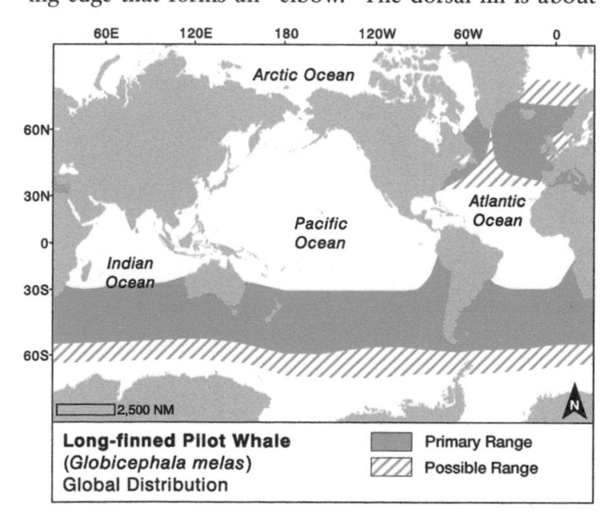

Predominantly dark grayish-brown to black in color, pilot whales have a white to light gray anchor-shaped patch on the chest, extending back to the urogenital area, a light gray post-dorsal fin saddle, and light gray "eyebrow" streaks. A light line between the forward extension of the saddle and the eyebrow streak may be present, forming a cape—this tends to be especially prominent in Southern Hemisphere animals (North Atlantic animals generally do not have it). Calves are significantly lighter than adults, and may have a distinct brownish tinge.

Inside the mouth are 8–13 pairs of sharp, pointed teeth in each jaw. These animals are extremely sexually dimorphic. Adults reach 6.7 m (males) and 5.7 m (females) in length. Females reach weights of up to 1,300 kg, and males up to 2,300 kg. Newborns are 1.7–1.8 m long, and weigh about 75 kg.

Recognizable geographic forms In the Southern Hemisphere, at least some pilot whales appear to have a more extensive "eyebrow" streak and more exaggerated post-dorsal fin saddle than in North Atlantic specimens. However, the extent and degree of potential overlap of these characters has not yet been adequately studied.

Can be confused with In some temperate waters (e.g., South Pacific; North Atlantic; off southern Africa; and off southern Brazil, Uruguay, and northern Argentina), long-finned and short-finned pilot whales overlap in distribution. In these areas, the two species are virtually impossible to distinguish at sea. Tooth counts and relative flipper lengths (both of which are generally not useful in most at-sea sightings) are the only reliable means

Pilot whales regularly occur in groups of dozens or hundreds. During the daytime they often just rest at the surface (logging), but they can also be found traveling at moderate speed. South Atlantic Ocean. PHOTO: M. NOLAN

In addition to white eye stripes, southern long-finned pilot whales have conspicuous white "saddles"—the pale patch behind the dorsal fin. PHOTO: P. OLSON

Pilot whales have a short stubby beak, which is often obscured by the large rounded melon; a faint eye stripe can also be seen angling up and back behind the eye of this *G. melas*. PHOTO: H. GOLDSTIEN

A neonate long-finned pilot whale (foreground) swims alongside its mother; the vertical lines are fetal folds—temporary crease lines from when the calf was folded up in utero. Western Australia. PHOTO: R. L. PITMAN

Adult male pilot whales often develop massive, lobed dorsal fins, as in this long-finned bull from Western Australia. PHOTO: R. L. PITMAN

Southern Hemisphere long-finned pilot whales are typically glossy black and have conspicuous white eye stripes. Scotia Sea. PHOTO: F. RITTER

Pilot whales rarely show their ventral pattern so obligingly, or offer such a clear view of the relative pectoral fin length, which clearly identifies this as a long-finned pilot whale. Norway. PHOTO: F. BROMS

of separating the two. In the lower latitude areas of its range, the long-finned pilot whale can be confused with false killer and less likely, pygmy killer and melon-headed whales; however, the differences in head shape and dorsal fin shape and position (in particular, the short, forward-placed, wide-based dorsal fin of the pilot whale) should permit correct identification.

Distribution Long-finned pilot whales occur in temperate and sub-polar zones. They are found in oceanic waters and some coastal waters of the North Atlantic Ocean, including the western Mediterranean Sea and North Sea. In the western North Atlantic, they occur in high densities over the continental slope in winter and spring months. In summer and autumn months, they move over the shelf. Long-finned pilot whales were previously found in the western North Pacific, but appear to be absent there today. The circumantarctic population(s) in the Southern Hemisphere occur as far south as the Antarctic Convergence, sometimes to 68°S. They are isolated from those of the Northern Hemisphere.

Ecology and behavior Pilot whales are highly social; they are generally found in pods of about 20–100, but some groups contain over 1,000 individuals. These large pods are generally dispersed in smaller subgroups of 10–20. Based on photo-identification and genetic work, pilot whales appear to live in relatively stable, maternally-based pods like those of killer whales, and not in fluid groups characteristic of many smaller dolphins.

The mating system is thought to be polygynous; this is consistent with the observed sexual dimorphism and adult sex ratio. Pilot whales are apparently deep divers. Groups often forage in broad ranks, sometimes with other species. Although they sometimes are aerially active, pilot whales are often seen rafting in groups at the surface, apparently resting. They often spyhop, but breaching is much less common.

This is one of the species most often involved in mass strandings. Strandings are fairly frequent, for instance, on Cape Cod (Massachusetts, USA) beaches from October to January. Their tight social structure also makes pilot whales vulnerable to herding, and this has been taken advantage of by whalers in drive fisheries off Newfoundland, the Faroe Islands, and elsewhere. Pilot whales frequently associate with other marine mammal species, including several species of dolphins and large whales.

Breeding of long-finned pilot whales can apparently occur at any time of the year, but peaks occur in summer in both hemispheres. Mating occurs primarily in spring to summer. Sexual maturity occurs at about 8 years for females and about 12 for males. Longevity is about 35–45 years for males and more than 60 for females.

Feeding and prey Primarily squid eaters, pilot whales will also take small to medium-sized fish, such as mackerel, when available. Other fish species taken include cod, turbot, herring, hake, and dogfish. They will sometimes also ingest shrimp. Most feeding appears to take place at depths of 200–500 m.

Threats and status Long-finned pilot whales have been taken directly in several large-scale drive fisheries in the North Atlantic Ocean. The most famous of these occurred previously in Newfoundland, and another one still operates in the Faroe Islands. The annual catches in the Faroes were about 1,000–1,500 in the 1990s. Other such drive fisheries used to occur in the USA (Cape Cod), Norway, Iceland, Greenland, Ireland, and Scotland (Orkney and Hebrides Islands). A drive fishery in the Falkland Islands has taken whales from the southern population(s). In addition, there are incidental catches in several fisheries, especially trawls, driftnets, and longlines. Other threats that have been identified include live-captures for captive display and the effects of environmental contaminants. There are estimated to be about 200,000 long-finned pilot whales in summer south of the Antarctic Convergence in the Southern Hemisphere; approximately 10,000–20,000 in the western North Atlantic; and about 780,000 in the eastern North Atlantic. The species is clearly not endangered.

IUCN status Data Deficient.

False Killer Whale—*Pseudorca crassidens*

(Owen, 1846)

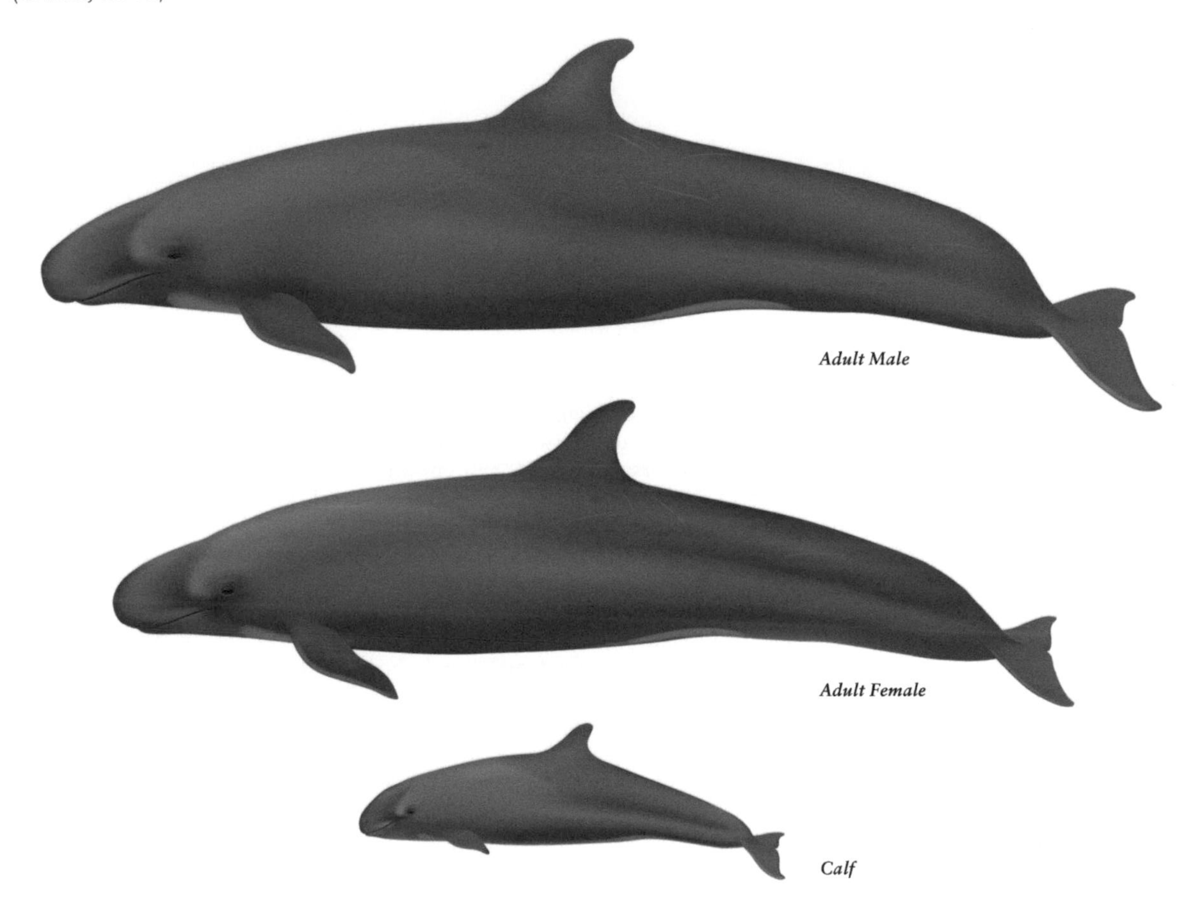

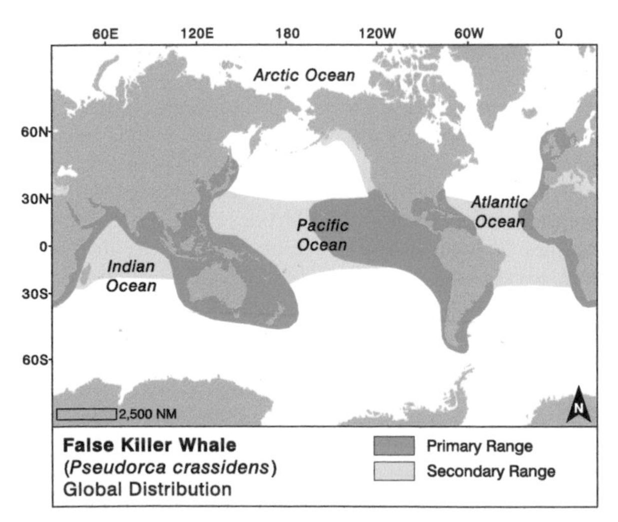

Recently-used synonyms None.

Common names En.–false killer whale; *Sp.–orca falsa; Fr.–faux-orque.*

Taxonomic information Cetartiodactyla, Cetacea, Odontoceti, Delphinidae. Hybrids between this species and common bottlenose dolphins have been reported from captivity.

Species characteristics The false killer whale has a long, slender, cigar-shaped body, a rounded (but narrow when viewed from above or below) overhanging melon (more so in males than in females), and no discernible beak. The dorsal fin, although variable in shape, tends to be falcate and slender along most of its height, and is generally somewhat rounded at the tip (however, there is a great deal of within-species variation). It is located near the midpoint of the back and is relatively small in relation to the length of the back. The flippers have rounded tips and a characteristic hump on the leading edge, probably the

False killer whales can be widely scattered over kilometers when they are foraging or intensely social, as the occasion demands. The do not normally porpoise out of the water when traveling fast, and will only occasionally bowride if a vessel is traveling at a very slow speed. Hawai'i. PHOTO: S. WEBB

Some larger *Pseudorca* individuals have a flattened tip to their rostrum; the cause for this is unknown. PHOTO: P. OLSON

The melon of the false killer whale is blunt, but tapered and slightly overhangs the lower jaw. Guadeloupe. PHOTO: C. MILLON

Pseudorca normally look glossy black, but in bright light they can appear dark slate gray. Guadeloupe. PHOTO: C. MILLON

Peering into the maw of a captive *Pseudorca* with its impressive array of stout teeth in the upper and lower jaws. Japan. PHOTO: H. MINAKUCHI

False killer whales are known to feed mainly on fish, but how they catch incredibly-fast fish like wahoo (shown here) is unknown; in some places, however, they have learned to take fish off fishing lines. Hawai'i. PHOTO: R. BAIRD

A *Pseudorca* rolls on its side in the eastern tropical Pacific, waving its uniquely-shaped pectoral fin in the air; the *Pseudorca* flipper appears to have a bulge on the leading edge. PHOTO: R. L. PITMAN

The underside of *Pseudorca* normally appears all dark, although a subtle, lighter gray pattern can be seen in good lighting. Eastern tropical Pacific. PHOTO: NOAA FISHERIES, SWFSC

False killer whales occasionally launch out of the water—mostly during feeding events. Hawai'i. PHOTO: NOAA FISHERIES, SWFSC

species' most diagnostic character. This gives the flippers an S-shaped appearance. The flukes are also relatively slender and there is a median notch.

The false killer whale has a very simple color pattern. It is one of several species of delphinids that some people call "blackfish." It is a large, dark gray to black dolphin, with a faint light-gray patch on the chest and belly, and sometimes light gray areas on the head. Any indication of a cape is generally very faint and indistinct. The characteristic delphinid "bridle" (consisting of an eye stripe and blowhole stripe) is present.

Each jaw contains 7–12 pairs of large conical teeth, which are round in cross-section. Adult false killer whales are up to 6 m (males) or 5 m (females) long; newborns are 1.5–2.1 m. Large males may weigh up to 2,000 kg.

Recognizable geographic forms Skulls of false killer whales from Australia, South Africa, and Scotland have been shown to differ, and this suggests the existence of different populations in these areas. However, distinctive geographical forms have not been described in terms of external appearance.

Can be confused with False killer whales are most commonly confused with other "blackfish," especially pygmy killer and melon-headed whales, and less commonly, pilot whales. The differences in head shape (bulbous in pilot whale adults and rounded, but more slender, in false killers) and dorsal fin shape and position (very wide-based and located on the forward part of the back in pilot whales, and much more slender and positioned mid-back in false killers) will make it easy to separate pilot whales. Size will be helpful with the other two species, as false killer whale adults are much larger than pygmy killer and melon-headed whales. Small false killer whales will present the most difficulty, as they may overlap in size with these two. Shape of the head (more elongated and narrow from above in false killers), dorsal fin (more rounded at the tip in false killers), relative size of dorsal fin (small in relation to amount of back showing in false killers), and flippers (S-shaped in false killers, with the diagnostic flipper hump) will be the best characters to use in distinguishing them. Ideally, a multitude of characters should be used to confirm an identity as a false killer whale.

Distribution False killer whales are found in tropical to warm temperate zones, generally in relatively deep, offshore waters of all three major oceans. In addition to deep, oceanic areas, they do sometimes occur over the continental shelf and appear to move into very shallow waters on occasion. They do not generally range poleward of about 50° in either hemisphere. However, some animals may move into shallow and more high-latitude waters, on occasion. They are found in many semi-enclosed seas and bays, but only occasionally occur in the Mediterranean Sea. There are also records from the Persian Gulf and Red Sea.

Ecology and behavior As is the case for most of the tropical oceanic delphinids, this species is less studied than the more temperate species. In some areas, false killer whales take fish from longlines and therefore are not popular with fishermen. This species is considered to be extremely social. Groups of 10–60 are typical, though much larger groups are known. Some groups are very tight-knit, while others may have many subgroups spread over a kilometer

False killer whales often have a distinct dorsal fin shape, as in the top-right animal: the leading and trailing edges appear parallel, with the top somewhat bent back. Eastern tropical Pacific. PHOTO: E. HIGUERA

or more of ocean. Individual false killer whales have been photo-identified in central America, New Zealand, and Hawai'i. Some site fidelity has been noted, although there is a considerable amount of inter-island movement in Hawai'i. The Costa Rica study also supported the idea that false killer whales may exhibit stable associations (like their larger cousin, the killer whale).

This is one of the most common species involved in cetacean mass strandings. In one case, over 800 stranded together. The false killer whale is a lively, fast-swimming cetacean, which behaves more like the spritely smaller dolphins than other mid-sized cetaceans (helpful in identification). It occasionally rides bow waves of vessels. False killer whales are known to behave aggressively toward other small cetaceans, and have even been seen chasing and attacking dolphins and large whales. However, these aggressive incidents appear uncommon, and they are often seen interacting peacefully with other species in Hawai'i.

No dramatic seasonality in breeding is known for the false killer whale, although one population has an apparent peak in late winter. Surprisingly (considering how commonly it strands), life history of this species has not been well-studied. Females reach sexual maturity at between 8–11 years of age, while males reach sexual maturity at 11–18 years of age. The calving interval has been estimated at 6–7 years. Females become reproductively senescent in later years. Longevity may reach 57 years for males and 62 for females.

Feeding and prey Although false killer whales eat primarily fish and cephalopods, they also have been known to attack other cetaceans, and on at least one occasion, even sperm whales! They eat some large species of fish, such as mahi mahi, wahoo, billfish, and tunas.

Threats and status False killer whales have a reputation for taking fish from longlines and other hook and line fisheries. Because of this habit, they have been shot, harpooned, and otherwise persecuted. In Japan, they have been targets of drive fisheries, largely because of perceived competition with fishermen. Bycatch levels in the US longline fishery in the central tropical Pacific are above those thought to be sustainable, and the US longline fleet represents only about 5% of total fishing effort. Incidental catches in other fishing gear, such as driftnets, and purse seines occur at least occasionally. Live captures (mostly in the US and Japan) and damaging effects from pollutants are among the other potential threats to this species. There are few estimates of abundance, and none for total population size, but in general most false killer whale populations are not thought to be in particular trouble from human activities. There are three known populations in Hawaiian waters: two island-associated populations and a pelagic population. The island-associated population around the main Hawaiian Islands was listed as Endangered on the US Endangered Species List in 2012 and is thought to number between 150 and 200 individuals.

IUCN status Data Deficient.

Pygmy Killer Whale—*Feresa attenuata*

Gray, 1874

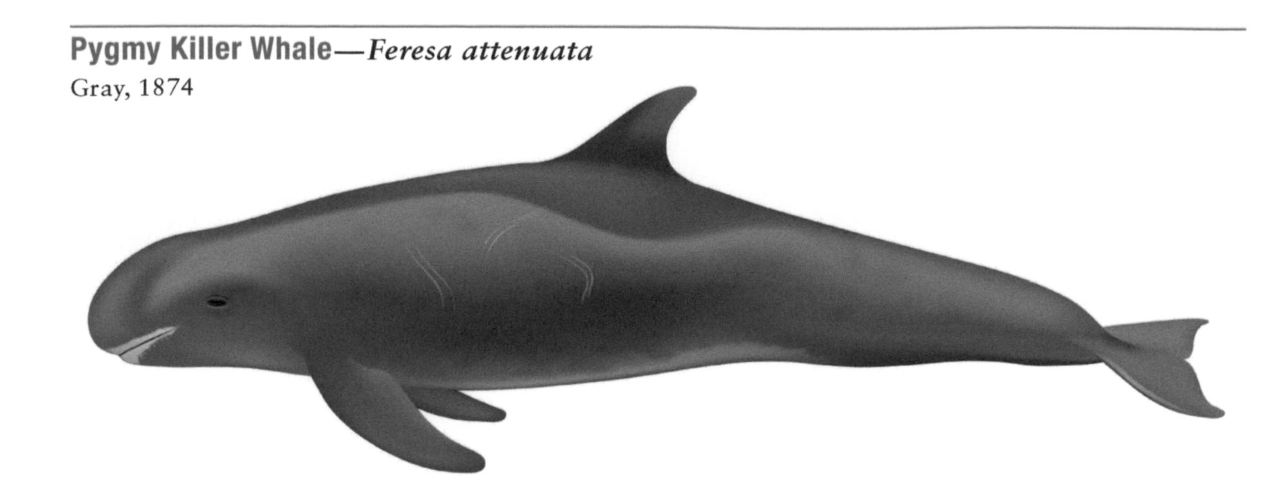

Recently-used synonyms *Feresa intermedia.*
Common names En.–pygmy killer whale; *Sp.–orca pigmea*; *Fr.–orque pygmée.*
Taxonomic information Cetartiodactyla, Cetacea, Odontoceti, Delphinidae.
Species characteristics The body of the pygmy killer whale is slender to moderately robust, narrowing in depth significantly posterior to the dorsal fin. The head is rounded (even bulbous) in profile and from above, and has no beak (except in some young animals). The head also appears rounded (as opposed to triangular) from above. The dorsal fin is tall and slightly falcate, usually rising at a relatively shallow angle from the back. The flippers have rounded tips, with a convex leading edge and concave trailing edge. The pygmy killer whale is often confused with the false killer whale and melon-headed whale, which are both very similar in general appearance.

The color of the pygmy killer whale's body is dark gray to black, with a fairly-prominent narrow cape that dips only slightly below the dorsal fin, and a white to light gray ventral band that widens around the genitals. Also, the lips and beak tip are often white, especially in adults. Occasionally, this white color may extend to the entire front of the head (apparently in older animals). There is a dark patch on the crown, extending from behind the blowhole to the mouth. The characteristic delphinid "bridle" (consisting of an eye stripe and blowhole stripe) is present. White cookie-cutter shark scars are quite common on the bodies of many larger pygmy killer whales, and some animals may also have white, widely-spaced rake scars.

The upper jaw contains 8–11 pairs of teeth, and the lower jaw has 11–13 pairs. The teeth are relatively large and heavy.

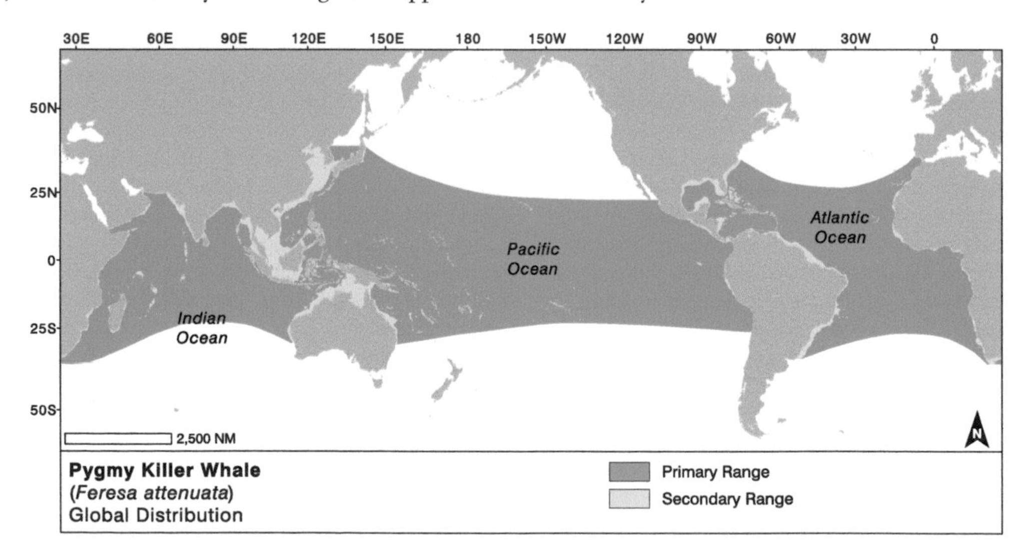

Pygmy killer whales interacting with a humpback whale in Hawai'i, showing their rounded heads, rounded tips of the flippers, large white ventral patches, and narrow capes. PHOTO: H. MINAKUCHI

A rather thin-looking *Feresa* displays its narrow cape and round head; the white cookie-cutter shark bite scars on the throat and belly are tinged pink, probably from blood pumped to the skin surface to dissipate body heat. Indonesia. PHOTO: D. KREB

Feresa often swim shoulder-to-shoulder in "chorus-lines," but they don't travel fast enough to porpoise out of the water like dolphins do. Eastern tropical Pacific. PHOTO: G. FRIEDRICHSEN

The extensive white around the face and lips, and the white blotching on the chest and belly, distinguish *Feresa* from the similar-looking *Peponocephala*. The white spots on the chest are healed cookie-cutter shark bites. Hawai'i. PHOTO: R. BAIRD

Newborns are about 80 cm long, adults up to 2.6 m. Males are apparently slightly larger than females. Maximum known weight is 225 kg.

Recognizable geographic forms None.

Can be confused with Pygmy killer whales are most easily confused with melon-headed whales, and less commonly with false killer whales. Flipper shape (rounded tips in the pygmy killer whale, pointed tips in the melon-headed whale, and with a prominent hump in the false killer whale) and the contour of the cape (relatively distinct and straight in the pygmy killer whale, indistinct and dipping low below the dorsal fin in the melon-headed whale, and virtually absent in the false killer whale) are the best features for distinguishing these three species. In good lighting conditions the clear demarcation of the cape is the easiest trait to distinguish this species. Also, the shape of head when viewed from above (rounded in this species and in pilot whales, more triangular in the melon-headed whale) may be useful.

Distribution This is a tropical/subtropical species that inhabits oceanic waters around the globe (Atlantic, Pacific and Indian oceans), generally not ranging north of 40°N or south of 35°S. They are rarely seen in nearshore waters, but may occur relatively close to shore around oceanic islands, where the water is deep and clear.

Ecology and behavior There is little known of the biology of the pygmy killer whale. It is rarely encountered in most areas, making it one of the least-known of all the delphinids. Groups generally contain about 12–50 individuals, although herds of up to several hundred have been seen. In Hawaiian waters, pygmy killer whales show high fidelity to specific islands, and association patterns are strong and stable. Their movements tend to be slow and lethargic compared to the similar-appearing melon-headed whale, which is a more spritely species. It generally does not bow ride, but has been known to on occasion. Leaps and spyhops are sometimes seen. Most of the few animals that have been held in captivity have been very aggressive.

Some key *Feresa* features visible here include large, blunt-tipped flippers; extensive white on chin (rear animal), and conspicuous white belly patch on each animal. Guadeloupe. PHOTO: L. BOUVERET

Feresa will occasionally approach the bow wave of a very slow moving vessel, but they are not generally bowriders. They have the blunt head and flexible neck of a slow swimmer; notice also the relatively narrow cape. Eastern tropical Pacific. PHOTO: R. L. PITMAN

Not much is known of the reproductive biology of this species. Reliable estimates of length and age at sexual maturity are not available, but it is thought that these animals probably reach adulthood when over 2.0 m.

Feeding and prey Pygmy killer whales eat mostly fish and squid, although they occasionally have been known to attack other dolphins, at least when the dolphins are involved in tuna fishery interactions in the eastern tropical Pacific. However, this kind of behavior has not been reported in other parts of the species' range, where they apparently restrict their diet to fish and squid. At least in Hawai'i, most feeding appears to occur at night.

Threats and status Pygmy killer whales have been killed both directly and opportunistically in harpoon and driftnet fisheries (Caribbean islands, Sri Lanka, Philippines, Taiwan, and Indonesia) and incidentally in various types of fishing gear (most areas of the species' range). The only abundance estimate available for a large region is of about 39,000 in the eastern tropical Pacific. Although few serious conservation problems are known, this species does not appear to be particularly abundant anywhere that it has been sighted. Although widespread, it has been suggested that it may be a naturally rare species. In Hawai'i, populations appear to be small, and this, along with their limited movements, suggests the species may be particularly vulnerable to human impacts.

IUCN status Data Deficient.

Melon-headed Whale—*Peponocephala electra*

(Gray, 1846)

Recently-used synonyms *Lagenorhynchus asia, Lagenorhynchus electra, Electra electra.*
Common names En.–melon-headed whale; *Sp.–calderón pequeño, delfín cabeza de melón*; *Fr.–péponocéphale.*
Taxonomic information Cetartiodactyla, Cetacea, Odontoceti, Delphinidae.
Species characteristics Melon-headed whales are moderately-robust small delphinids. The dorsal fin is tall and slightly falcate, and is located near the middle of the back. The head shape is somewhat triangular, especially when viewed from above or below. The head of adults can be bulbous, and the melon can be overhanging in adult males. However, younger animals have a more sloping forehead, although there is generally little hint of a beak, except in small calves. The flippers are sickle-shaped and have pointed tips. Compared to females, males have more bulbous melons, longer flippers, taller dorsal fins, and tail stocks with pronounced post-anal humps (often called keels).

Although often described as simply black with a few light patches in older literature, the melon-headed whale actually has an interesting color pattern. The body is generally charcoal gray to dark gray (young are lighter gray), with a white urogenital patch with irregular margins. On many individuals, there is also an anchor-shaped patch of light color on the underside of the head, just ahead of the flippers. On larger animals, the black triangular "mask" on the face of melon-headed whales distinguishes them from the somewhat more uniformly-colored pygmy killer whale. A darker stripe runs forward from the eye to the beak tip, and there is a pale, streaked blowhole stripe, which widens as it runs toward the tip of the upper jaw. Much of this facial coloration is subtle and can only be seen when the lighting is favorable. The lips and tip of the lower jaw of larger animals are light gray or white. Melon-headed whales also have a rounded cape that dips much lower below the dorsal fin than that of pygmy killer whales,

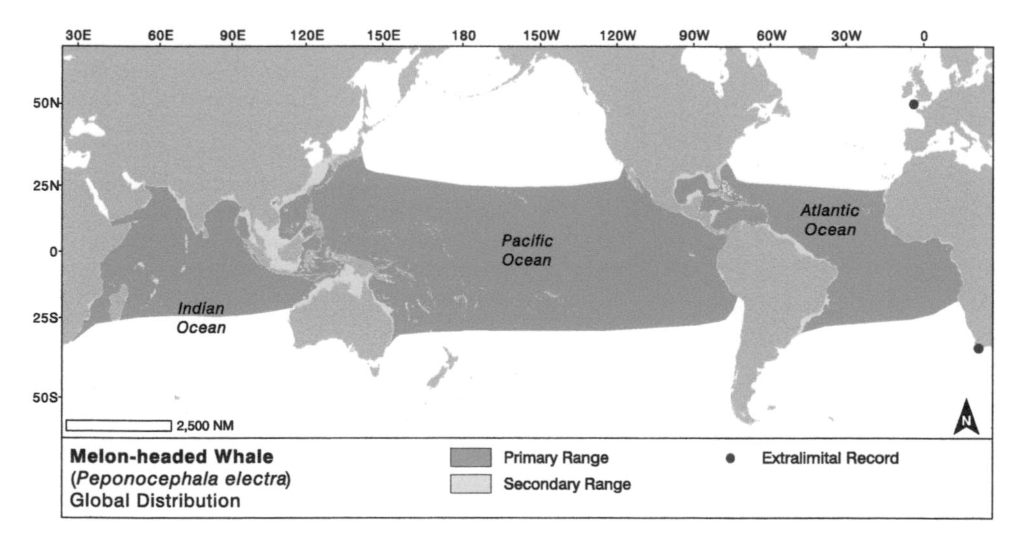

When traveling at high speeds, melon-headed "whales" porpoise clear of the water, similar to the smaller dolphins. Central Pacific. PHOTO: J. COTTON

A very young *Peponocephala* calf swims with its mother—it has the lumpy body and fetal folds of a neonate. Bahamas. PHOTO: R. L. PITMAN

Peponocephala form large, dense, fast-swimming schools, although they can also be found resting or milling during the daytime. The characteristic thin white lips are not present on younger animals. Central Pacific. PHOTO: J. COTTON

although its margin is often faint and there is little contrast between it and the lighter flanks. The cape may not be visible in sightings at sea, unless the lighting is good.

These animals have 20–25 small slender teeth in each tooth row. The teeth are much more similar to those of the smaller dolphins than they are to those of other blackfish species. Melon-headed whales reach a maximum of about 2.78 m, with males reaching somewhat greater lengths than females. Length at birth is thought to be about 1 m or less. Maximum known weight is about 275 kg.

Recognizable geographic forms None.

Can be confused with Melon-headed whales can be difficult to distinguish from pygmy killer whales, especially when seen at sea from a distance. Head shape, flipper shape, and the shape and distinctness of the cape can be useful in identification. Pygmy killer whales have rounded (as opposed to pointed) flipper tips and only 8–13 pairs of robust teeth (as opposed to 20–25 pairs of slender ones). Also, melon-headed whales tend to have a more triangular head shape (when viewed from above or below), and females and young have a beak, albeit very short and poorly defined. False killer whales may also be confused with this species at a distance, but size and close attention to body shape and coloration will allow them to be distinguished.

Distribution The range of the melon-headed whale coincides almost exactly with that of the pygmy killer whale in tropical/subtropical oceanic waters between about 40°N and 35°S. They are rarely found nearshore, unless the water is very deep. The few high-latitude strandings are thought to be extralimital records, and are generally associated with incursions of warm water.

Ecology and behavior Melon-headed whales are highly social, and are known to occur usually in pods of 100–500 (with a known maximum of about 2,000 individuals). They are often seen swimming with other species, especially Fraser's dolphins in the Gulf of Mexico, eastern tropical Pacific, and Philippines. Melon-headed whales often move at high speed, porpoising out of the water regularly, and are eager bowriders in many areas, often displacing other species from the bow wave. In the calm waters of the Philippines and some other tropical archipelagos, this species is often seen in large schools of rafting individuals in resting formation.

This is a species that is sometimes involved in mass strandings, generally containing many dozens or even hundreds of animals. The largest ones are known have included up to 250 animals. Photo-identification studies in Hawaiian waters have shown that there is some long-term residency lasting at least two decades.

There is some scant evidence to indicate a calving peak in July and August, but this is inconclusive. In Japanese waters, females reach sexual maturity at about 11.5 years and 2.35 m, and males at 16.5 years

Melon-headed whales lack the distinct white ventral patches and spotting that are common on the otherwise similar pygmy killer whales. Palmyra Atoll, Central Pacific. PHOTO: R. L. PITMAN

The dark "bandit mask" of *Peponocephala* is visible only in good lighting; notice the thin white lips and the pale pattern on the throat. Bahamas. PHOTO: R. L. PITMAN

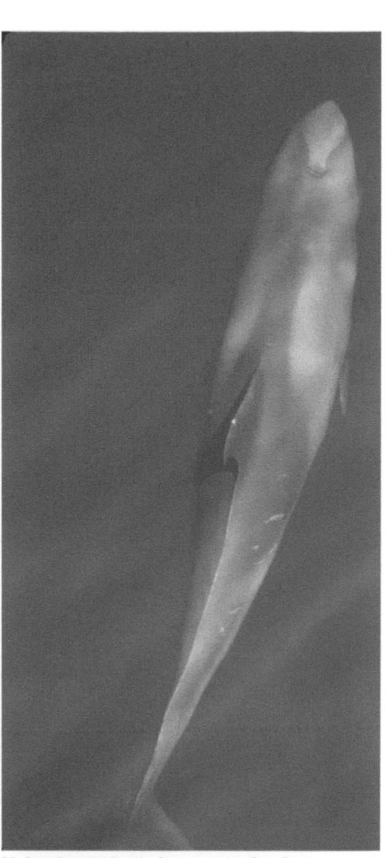

Melon-headed whales can swim fast enough that they sometimes bowride; from above notice the unique, indistinct cape pattern: narrow behind the blowhole and dipping down along the sides below the dorsal fin. Their build and behavior is very dolphin-like—giving rise to an old common name: the "Electra dolphin." Eastern tropical Pacific. PHOTO: R. PITMAN

***Peponocephala* have all-gray underparts with a slightly paler "blackfish" anchor patch on the throat and belly; there are cookie-cutter shark bite scars on its ventral surface, but they do not leave white spots as they do on *Feresa*. Bahamas.** PHOTO: R. L. PITMAN

Seen here are the pointed fins and beak of a fairly fast swimmer; note also the black face and thin white lips. Bahamas. PHOTO: R. L. PITMAN

and 2.44 m. Little else is known of the species' life history or reproductive biology.

Feeding and prey Melon-headed whales are known to feed on several species of squid and small fish. They appear to feed mainly deep in the water column. Some shrimps have also been found in the stomachs of melon-headed whales.

Threats and status Although no regular, large hunts are known, melon-headed whales have been taken in various fisheries. This includes direct killing in drive fisheries in Japan and harpoon/driftnet fisheries in the Caribbean, Sri Lanka, the Philippines, and Indonesia, as well as incidental catches in tuna purse seines in the eastern tropical Pacific. They may be among the species that are susceptible to acoustic impacts from naval sonar operations. This species is relatively common in some areas of its range, such as parts of the Philippines and around some archipelagos in the western tropical Pacific. The only abundance estimates available for large regions are of about 45,000 animals in the eastern tropical Pacific, and 2,000 in the northern Gulf of Mexico. In Hawai'i, there are two known populations. A small population resident of the Big Island numbers about 450, and a larger population that ranges among all the main islands numbers an estimated 5,800 individuals. On a global scale, the melon-headed whale is not considered to be threatened or endangered.

IUCN status Least Concern.

Risso's Dolphin—*Grampus griseus*

(G. Cuvier, 1812)

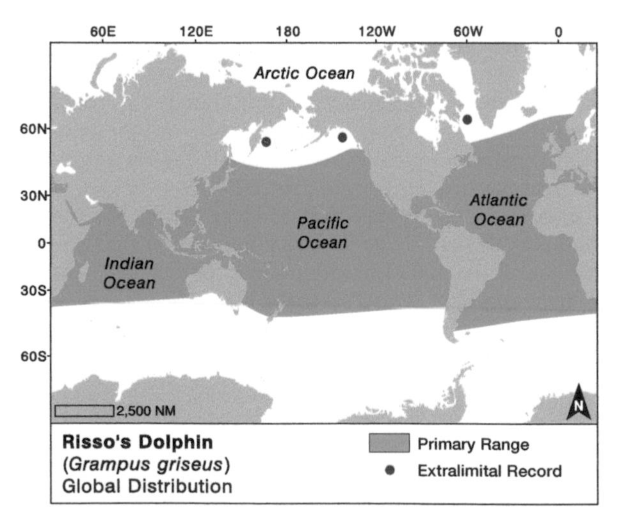

Recently-used synonyms *Grampidelphis griseus.*

Common names En.–Risso's dolphin; *Sp.–delfín de Risso;* Fr.–*grampus, dauphin de Risso.*

Taxonomic information Cetartiodactyla, Cetacea, Odontoceti, Delphinidae. Hybrids with common bottlenose dolphins have been reported from captivity and in the wild.

Species characteristics Risso's dolphins have a rather unique body shape. They are robust, blunt-headed animals without distinct beaks. The forehead is bulging, but with a more squarish profile than the rounded melon in some other small cetaceans. One of the most distinctive features is a shallow vertical crease on the front of the melon. The flippers are long, pointed, and recurved. The dorsal fin is very tall and slender, somewhat falcate (though there is extensive individual variation in shape), and most often pointed at the tip. Risso's dolphins have mouthlines that slope upward. The tail stock is generally very shallow, especially directly in front of the flukes.

In higher latitudes, adult *Grampus* are often ghostly white with dark eye patches and very tall, dark dorsal fins; in the distance, they are often identifiable by the size of the fin alone. Monterey Bay, California. PHOTO: T. A. JEFFERSON

In tropical waters, Risso's dolphins are generally much less white, so that their darker gray cape is clearly discernible. Eastern tropical Pacific. PHOTO: NOAA FISHERIES, SWFSC

***Grampus* have a reduced number of teeth, and they are found only in the lower jaw—these features are usually associated with suction feeding and a diet of squid. Captivity, Japan** PHOTO: H. MINAKUCHI

This recently-born *Grampus* is wholly unmarked and still has fetal folds. Mediterranean Sea. PHOTO: NOAA FISHERIES, SWFSC

A juvenile *Grampus* just starting to acquire some marks—the dorsal fin and around its base have already begun to turn black. California. PHOTO: C. OEDEKOVEN

Although largely obliterated here, the blackfish "anchor pattern" is still visible on the chest above the flippers. Risso's dolphins get whiter with age, due to lost pigmentation and from scarring due to various traumas. Baja California. PHOTO: T. A. JEFFERSON

With their large appendages and blunt heads, Risso's dolphins usually travel too slowly to porpoise out of the water or bowride—normally they plow through the water, throwing up spray. But, when excited they can be quite active, leaping and breaching. California. PHOTO: K. WHITAKER

Before age-related whitening obscures the patterning, younger *Grampus* show a narrow cape, and white patches on the belly and chest. Monterey Bay, California. PHOTO: R. L. PITMAN

A quaint old name for Risso's dolphin was "bosom-headed whale" in reference to the cleavage in the melon, a feature unique to this species. California. PHOTO: C. OEDEKOVEN

This gives the animal the appearance of having most of its bulk ahead of the dorsal fin.

At sea, the best identification characteristic is the coloration and scarring. Adult Risso's dolphins range from dark gray to nearly white, but are typically covered with white scratches, spots, and blotches. Many of these are thought to result from the beaks and suckers of squid, their major prey, but others may be caused by the teeth of other Risso's dolphins. In fact, this species is the most heavily-scarred of all the dolphins. Higher-latitude animals are much lighter overall than animals from the tropics. The chest has a whitish anchor-shaped patch just ahead of the flippers, and there is another white patch of variable extent around the urogenital area. The appendages tend to be darker than the rest of the body. Young animals range from light gray to dark brownish-gray and are relatively unmarked. The scars and scratches are highly unique, and can be used in conjunction with dorsal fin scars, to identify individuals. In older individuals, the lips often contrast strongly with the surrounding area. There is generally no sign of the typical delphinid "bridle" (i.e., eye and blowhole stripe complex) in adults. Calves and juveniles are more of a brownish-gray color, and have few or no scratches and scars. Young animals also have a distinct cape that extends down over the eye, and dips slightly below the dorsal fin.

The teeth of this species are also unique; there are 2–7 pairs of stout, pointed teeth in the front of the lower jaw, and usually none (but occasionally 1–2 pairs) in the upper jaw. Some or all of the teeth may be worn-down in, or missing from, older adults.

Newborns are 1.1–1.5 m long and adults range up to at least 3.8 m long. Some evidence suggests geographic variation in body size. There appears to be little or no sexual dimorphism in the length of adults. Weights of up to 400 kg have been recorded, and the maximum may be near 500 kg.

Recognizable geographic forms None.

Can be confused Risso's dolphins are generally easy to identify when seen at close range, as they are the only medium-sized, blunt-headed cetaceans that are typically light in color. However, from a distance they may be confused with other large delphinids with a tall dorsal fin (such as bottlenose dolphins, false killer whales, and even killer whales). When visible, the light, extensively-scarred bodies and squarish heads of Risso's dolphins make them unmistakable. If the vertical crease on the forehead is observed, there can be no doubt about the identity of the specimen.

Distribution This is a widely-distributed species, inhabiting primarily deeper waters of the continental slope and outer shelf (especially with steep bottom topography) from the tropics through the temperate regions in both hemispheres, from about 64°N to 46°S. They also occur in oceanic areas, beyond the slope, such as in the eastern tropical Pacific, though generally at lower densities. The have a strong preference for temperate waters between about 30° and 45° latitude. They are found from Newfoundland, Norway, the Kamchatka Peninsula, and Gulf of Alaska in the north to the tips of South America and South Africa, southern Australia, and southern New Zealand in the south. Their range includes many semi-enclosed bodies of water, such as the Gulf of Mexico, Red Sea, Sea of Japan, and Mediterranean Sea.

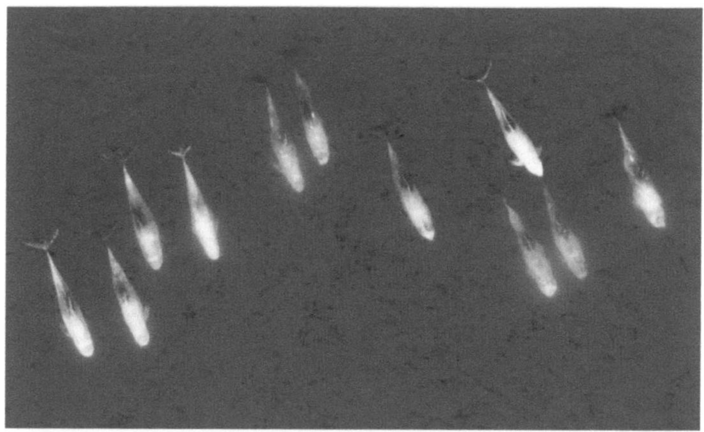

From the air, *Grampus* in higher latitudes appear mostly white, with a conspicuous dark area around the base of the dorsal fin. Eastern North Pacific. PHOTO: NOAA FISHERIES, SWFSC

Ecology and behavior These large dolphins are often seen surfacing slowly and somewhat lethargically, although they can be energetic, sometimes breaching, spyhopping, or porpoising, and occasionally bowriding. Most herds tend to be small to moderate in size (10–100), but groups of up to an estimated 4,000 have been reported. Social organization has not been well studied in most areas; however, it appears that young animals remain with their natal school until around puberty. Adult females gather together, and males move among schools. Group composition is relatively fluid (although with some elements of stable subgroups), at least in Monterey Bay, California, and the Azores. Risso's dolphins commonly associate with other species of cetaceans (especially Pacific white-sided and northern right whale dolphins off the California coast), including large whales such as gray and sperm whales. The have been seen behaving aggressively toward these other species at times. Dive times of up to 30 min. have been reported.

There have only been a few studies on the life history parameters of Risso's dolphins. In at least the North Atlantic, there appears to be a summer calving peak. Calving peaks appear to differ in different parts of the North Pacific—summer/fall off Japan, and fall/winter off California. Sexual maturity appears to occur at lengths of about 260–277 cm for both sexes, with some apparent regional differences. It has been suggested that the extensive scarring on this species may have evolved, at least partly as a mechanism to allow animals to gauge the "quality" of other group members during social interactions. Longevity in this species is at least 35 years.

The most famous Risso's dolphin was undoubtedly "Pelorous Jack," who accompanied steamers and ferries into and out of Admiralty Bay in New Zealand for many years in the early part of the 20th century. Long-term changes in the occurrence of Risso's dolphins in some areas (e.g., off Catalina Island and in central California) have been linked to oceanographic conditions and movements of spawning squid.

Feeding and prey Risso's dolphins feed on crustaceans and cephalopods, but seem to prefer squid and octopus. Squid bites may be the cause of at least some of the scars found on the bodies of these animals. In the few areas where feeding habits have been studied, they appear to feed mainly at night.

Threats and status Occasional direct killing of Risso's dolphins has occurred. This is generally as a result of the dolphins stealing fish from longlines, or in multispecies small cetacean fisheries, such as those that occur in Sri Lanka, the Caribbean islands, and Indonesia. One regular hunt occurs in Japan, where about 250–500 are taken per year in a drive fishery. There are also records of incidental catches in several fisheries, in particular driftnet fisheries, but also in purse seines. Some Risso's dolphins have been captured for live display, although there are not many of them in oceanaria today. There are some abundance estimates for specific areas. There are estimated to be about 110,500 in the ETP, 6,300 off the western United States, 2,400 in Hawaiian waters, 20,500 off the eastern US coast, 1,600 in the northern Gulf of Mexico, 2,200 around the UK, 700 in the Northeast Atlantic area, 2,800 in parts of the Mediterranean, 83,300 off Japan, 1,500 in the Philippines, 400 off Taiwan, and 5,500–13,000 off Sri Lanka. There are no estimates of global abundance, but based on the above, this may number more than half a million animals. The species is not considered threatened or endangered.

IUCN status Least Concern.

Tucuxi—*Sotalia fluviatilis*

(Gervais and Deville *in* Gervais, 1853)

Recently-used synonyms *Sotalia pallidus, Sotalia tucuxi.*
Common names En.–tucuxi; *Sp.–bufeo gris, bufeo negro Fr.–sotalia.*
Taxonomic information Cetartiodactyla, Cetacea, Odontoceti, Delphinidae. Recently, it was found that the riverine and marine forms of *Sotalia* are two different species: *S. fluviatilis* in the Amazon River system and *S. guianensis* in coastal marine waters of central and South America. Those in the Orinoco River system may be a separate species.
Species characteristics This small dolphin resembles the bottlenose dolphin in general body shape. It is significantly smaller than its congener, the Guiana dolphin. It is rather stocky, with a moderately long and narrow beak, and broad flippers. The dorsal fin is shorter, more triangular and wide-based than in the bottlenose dolphin—it is sometimes recurved at the tip. The tucuxi has a rounded melon, which is not separated from the beak by a distinct crease (as it is in most other dolphins).

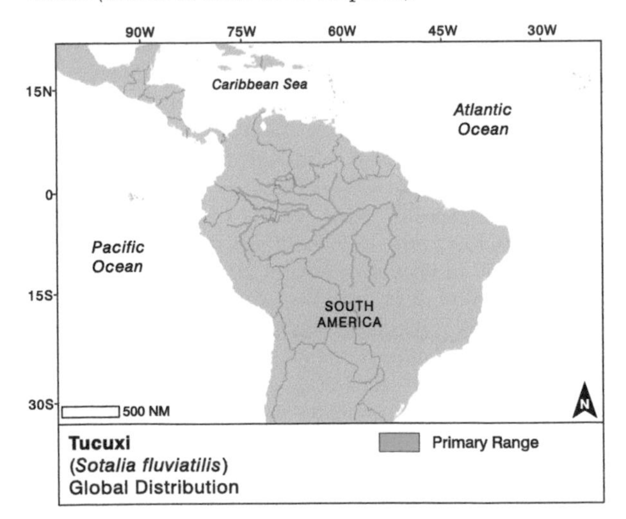

On the upper surface, tucuxis are dark-bluish or brownish-gray, fading to light gray or white (usually with a pinkish tinge) on the belly. There is a broad, somewhat poorly-defined stripe from the eye to the flipper, with a distinct lower boundary between the dark above and light below. The characteristic delphinid "bridle" (consisting of an eye stripe and blowhole stripe) is present. A faint lateral area of lighter gray is present behind the flipper, and another one laterally from mid-body to the anus. The flippers and flukes are dark gray on both surfaces.

The mouth contains 28–35 small, pointed teeth in each upper tooth row and 26–33 in each lower. Adult female dolphins of this species are only up to about 1.52 m in length, and males may be slightly smaller, with measurements so far reaching 1.49 m. Size at birth is poorly-known, but is estimated to be 0.7–0.8 m.
Recognizable geographic forms None.
Can be confused with In the Amazon River system, Amazon River dolphins (botos) are the only other cetaceans present, and it is usually easy to distinguish the two species. Differences in size, coloration, dorsal fin shape, head shape, and behavior are the best clues to distinguishing them. The tucuxi and Guiana dolphin may be sympatric in waters near the estuary of the Amazon River, and it would be nearly impossible to distinguish them "at sea."
Distribution Tucuxis are almost exclusively freshwater animals, and occur in the Amazon River and possibly the Orinoco system as well (although it is thought to be the Guiana dolphin or a new species that inhabits Orinoco waters, up to about 12 km upstream). The tucuxi is found throughout the Amazon drainage basin, as far inland as southern Peru, eastern Ecuador, and southeastern Colombia.
Ecology and behavior Dolphins of this species live mostly in small groups of four or fewer, although they may be found in groups of up to 30. They are social animals,

The tucuxi in some way resembles a small bottlenose dolphin, but with a shorter dorsal fin. Colombia. PHOTO: F. TRUJILLO

Tucuxis not only live in river mainstems, but also inhabit lakes and oxbows appended to the river. They may occur among floating seagrass. Colombia. PHOTO: F. TRUJILLO

This tucuxi is a bit darker than most; though apparent coloration can vary quite a bit and is always affected by lighting. Colombia. PHOTO: F. TRUJILLO

These two individuals show the tucuxi's distinctive features well—especially the long beak with a curved mouthline, and the short wide-based dorsal fin. Colombia. PHOTO: F. TRUJILLO

Although not considered among the most acrobatic of dolphins, tucuxis do porpoise and perform various types of leaps. Colombia. PHOTO: F. TRUJILLO

The belly of the tucuxi is white, but when animals are very active in warm waters, it may have a pinkish hue. PHOTO: F. TRUJILLO

and they often feed in groups, sometimes appearing to use cooperative techniques. Dive times are generally less than 1.5–2.0 minutes. They do not bowride, but they are sometimes quite active, with various types of leaps and other aerial behaviors seen.

They are generally shy and difficult to approach. During the flood season, animals may move into smaller tributaries, but apparently do not move into the inundated forest to feed (as Amazon River dolphins often do), staying mainly in the main river channels, tributaries and lakes. Tucuxis are largely sympatric with botos in the Amazon system, but usually do not interact with them (although they have been known to do so at times, and surprisingly *Sotalia* appear to be dominant).

Life history has been little studied; most of the studies involve the marine species. At least in Brazil, calving apparently occurs primarily during the low water period, October to November. Sexual maturity is reached at about 1.3–1.4 m in length. Lifespan is estimated to be at least 30–35 years.

Feeding and prey A wide variety of fish, mostly small schooling species, are eaten. Feeding occurs both individually and in large groups.

Threats and status There are no global estimates of abundance available, and only estimates for specific portions of the range have been made. The tucuxi has been taken in fishing gear, especially gillnets and seines, in many areas of its range, and may be used for fish bait in Peru. Damming of rivers has also contributed to problems for this species. Additional threats include the damaging effects of gold mining with mercury, habitat loss/destruction, vessel collisions, environmental contaminants, and behavioral disturbance. In the past, live captures have also resulted in the loss of some animals. The species is not uncommon, and in many parts of the Amazon River system it is actually quite abundant. It occurs in some of the highest densities known for any cetacean species, and receives some protection there from myths and legends that discourage killing.

IUCN status Data Deficient.

Guiana Dolphin—*Sotalia guianensis*

(Van Bénéden, 1864)

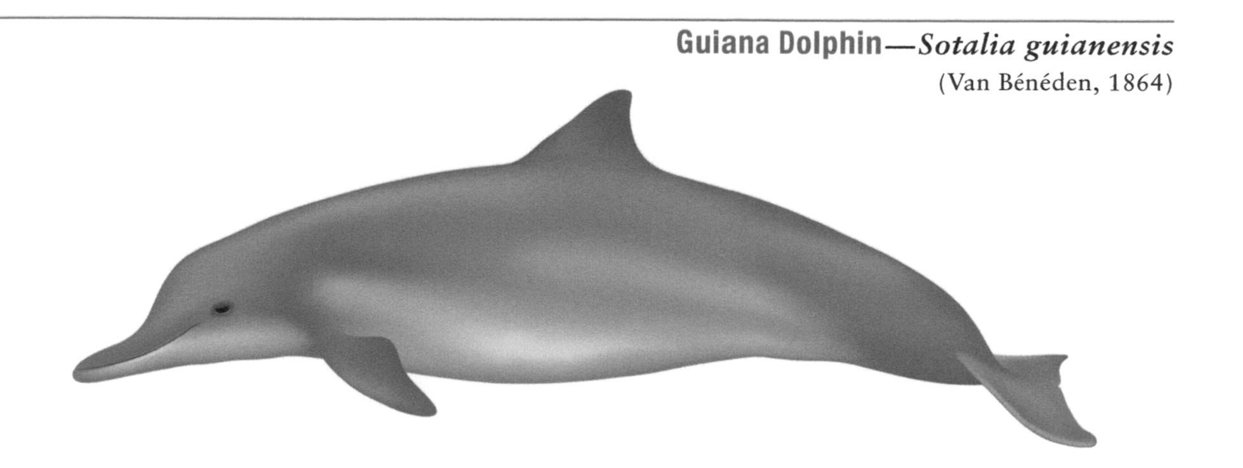

Recently-used synonyms *Sotalia brasiliensis.*
Common names En.–Guiana dolphin, costero; *Sp.–bufeo negro, tonina; Fr.–sotalia.* (Note: In Brazil, they are called *boto, boto común*, or *golfinho cinza.)*
Taxonomic information Cetartiodactyla, Cetacea, Odontoceti, Delphinidae. Recently, it was found that the riverine and marine forms of *Sotalia* are two different species: *S. fluviatilis* in the Amazon River system and *S. guianensis* in coastal marine waters of Central and South America.
Species characteristics This small dolphin resembles the bottlenose dolphin in general body shape. It is somewhat larger than its close relative, the tucuxi. It is rather chunky, with a moderately long and narrow beak, and broad flippers. The dorsal fin is shorter and more triangular and wide-based than in the bottlenose dolphin—it is often recurved at the tip. The Guiana dolphin has a rounded melon, which is not separated from the beak by a distinct crease (as it is in most other dolphins).

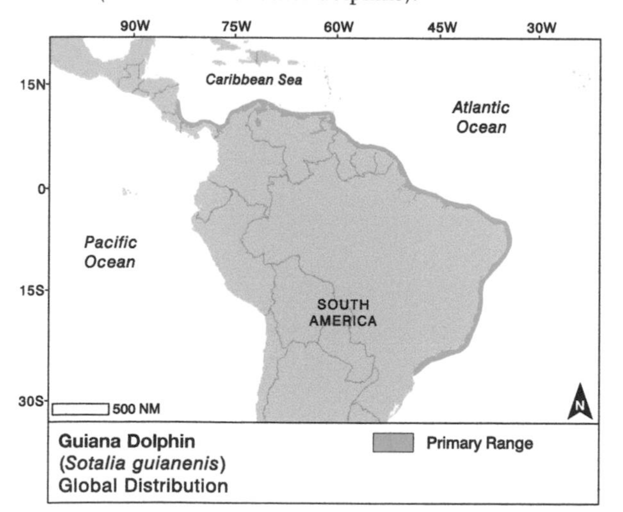

On the upper surface, dolphins of this species are dark-bluish or brownish-gray, fading to light gray (often with a pinkish tinge) on the belly. The belly tends to be darker than in the tucuxi. There is a broad, somewhat poorly-defined stripe from the eye to the flipper. They may have another light streak of gray on the sides of the tail stock. The characteristic delphinid "bridle" (consisting of an eye stripe and blowhole stripe) is present. The undersides of the flippers and flukes are gray.

The mouth contains 30–36 small, pointed teeth in each upper tooth row, and 28–34 in each lower row. This is slightly more than in the tucuxi.

Adult dolphins of this species can be up to nearly 2.2 m in length (the largest known male was 1.7 m and the largest known female 1.87 m). They reach weights of at least 121 kg. Size at birth is estimated to be 0.8–1.15 m.
Recognizable geographic forms None.
Can be confused with Bottlenose dolphins could be mistaken for Guiana dolphins, but they are much larger, with taller, more falcate dorsal fins. Franciscanas might also be difficult to distinguish from *Sotalia* in coastal waters, but only if not seen well. The franciscana generally has a smaller body, much longer beak, and squarish (rather than pointed) flippers. Also, the franciscana has a shorter, more rounded dorsal fin. Near the mouth of the Amazon, the Guiana dolphin may overlap with the tucuxi (unconfirmed), and distinguishing them would be nearly impossible without a specimen "in hand."
Distribution Guiana dolphins are found along the Atlantic coast of Central and South America, mostly in shallow, nearshore marine waters, and in estuaries. They occur from Nicaragua (perhaps even Honduras) to Florianopolis, southern Brazil. The species has also been reported around some Caribbean islands, such as Trinidad and Tobago, as well as the Abrolhos archipelago, off Brazil. There are at least three populations in the Caribbean,

Even when only a bit of the body is seen, Guiana dolphins are generally easy to identify—here the long, slender beak is evident. PHOTO: F. TRUJILLO

This Guiana dolphin calf still has a lumpy look and fetal folds, indicating it is still very young. Southeast Brazil. PHOTO: L. FLACH

These dolphins can be quite active; this Guiana dolphin is part of a large school in Bahia Norte, Brazil. PHOTO: T. A. JEFFERSON

Around Florianopolis, Brazil, Guiana dolphins reach the southern end of their range in Bahia Norte. Here a juvenile surfs alongside its mother. PHOTO: T. A. JEFFERSON

Guiana dolphins are inhabitants of coastal and estuarine waters, and as such, live nearly their entire lives close to land. Suriname. PHOTO: S. TSANG

Guiana dolphins can be quite aerially active, performing occasional breaches and various kinds of leaps. Suriname. PHOTO: W. MEIJLINK

including those in Lake Maracaibo, an estuarine system in Venezuela, as well as in the Morrosquillo Gulf of Colombia. Dolphins of the genus *Sotalia* also occur in the Orinoco River basin, as far as about 300 km upriver (to about Ciudad Bolivar), but it is thought that *S. guianensis* only reaches about 12 km upstream from the mouth.

In warm waters, and especially when animals are active, Guiana dolphins may show a pinkish hue to the belly and undersides. Suriname. PHOTO: M. DE BOER

Ecology and behavior Guiana dolphins live mostly in groups of four or fewer, although they are found in groups of up to 50 at times. They are social animals, and they often feed in groups, sometimes appearing to use cooperative techniques. Dive times are generally less than 1.5–2.0 minutes. They do not bowride, but they are sometimes quite active, with various types of leaps and other aerial behaviors seen. Long-term photo-identification studies have shown that at least some individuals display long-term residency. Guiana dolphins are easier to approach than are tucuxis. Along the coast of Brazil, there are apparently at least three populations. They are often associated with confluences of rivers in shallow waters.

Life history has been studied in some detail recently. Although births do occur throughout the year, there is some evidence of calving peaks in this species. There appears to be a 2-year calving interval. Sexual maturity is reached at about 1.6–1.8 m and 5–8 years. Lifespan is estimated to be at least 30–35 years.

Feeding and prey Guiana dolphins consume a wide variety of pelagic and demersal fishes (from at least 25 families) and cephalopods (from at least 5 families). Crustaceans are also sometimes taken. Feeding occurs both individually and in large groups.

Although they mostly live in murky waters with low visibility, Guiana dolphins do sometimes occur in clear waters, such as this. PHOTO: F. TRUJILLO

Threats and status There are no global estimates of abundance available for the Guiana dolphin, and only estimates for specific portions of the range have been made (e.g., about 420 animals in Guanabara Bay, 57–124 in Caravelas River Estuary, 245 in Babitonga Bay, and 182+ in Paranaguá Estuary, all in Brazil). They have been taken in fishing gear, especially gillnets and seines, in many areas of their range. There has been some direct killing for human consumption and for shark bait. Additional threats include habitat loss/destruction, vessel collisions, seismic operations, environmental contaminants, behavioral disturbance and hand-feeding by tourist boats. In the past, live captures may have also resulted in the loss of some animals. The species is relatively common throughout much of its range along the Central and South American coast.

IUCN status Data Deficient.

Rough-toothed Dolphin—*Steno bredanensis*

(G. Cuvier *in* Lesson, 1828)

Recently-used synonyms *Steno rostratus.*

Common names En.–rough-toothed dolphin; Sp.–*esteno, delfín de dientes rugosos*; *Fr.–sténo.*

Taxonomic information Cetartiodactyla, Cetacea, Odontoceti, Delphinidae. Hybrids with common bottlenose dolphins have been reported from captivity.

Species characteristics The rough-toothed dolphin is relatively robust, with a long, conical head and no demarcation between the gently-sloping melon and the moderately long beak. It has a somewhat-reptilean appearance, unique among small cetaceans. This species has large flippers (seemingly oversized for the animal), which are set farther back on the body than in most other dolphin species. The dorsal fin is prominent and slightly falcate. Adult males may have a post-anal hump of connective tissue on the underside of the tail stock. The shape of the flukes and flippers are unremarkable.

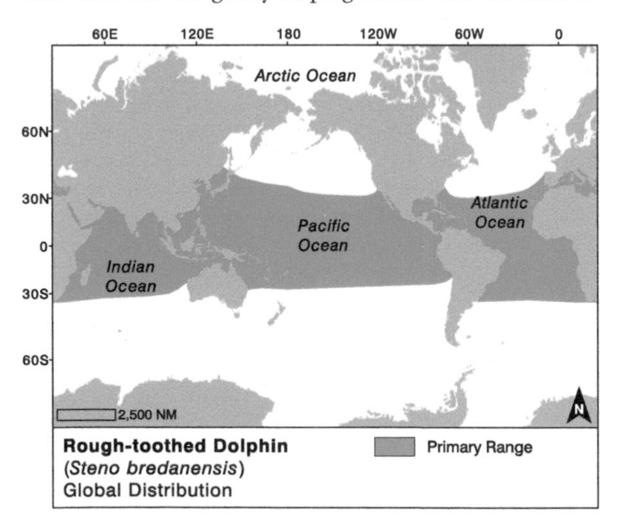

Rough-toothed Dolphin (*Steno bredanensis*) Global Distribution

The color pattern is generally three-tone. The body is dark gray, with a prominent, narrow dorsal cape that dips slightly down onto the side below the dorsal fin. The belly, lips, and much of the lower jaw are usually blotched with white patches, sometimes with a pinkish tinge. There may be a dark eye patch and the characteristic delphinid "bridle" (consisting of an eye stripe and blowhole stripe) is present. The sides are an intermediate shade of gray. White scratches and spots, apparently mostly caused by bites of cookie-cutter sharks and probably other rough-toothed dolphins, often are present on the bodies of older individuals. Young animals are lighter in color than adults, have a muted, less-contrasting color pattern, and generally lack the white spots.

The jaws contain 19–28 stout teeth in each row with subtle, but detectable, vertical wrinkles or ridges. These ridges give rise to the species' English common name. Adults are up to about 2.65 m long (with males slightly larger than females), and may occasionally reach 2.8 m.

Skimming along the surface is a behavior well known in rough-toothed dolphins. Eastern tropical Pacific. PHOTO: NOAA FISHERIES/SWFSC

These rough-toothed dolphins are lined up in a synchronized "chorus line," something that this species is well known for. Canary Islands. PHOTO: J. ZAERA

The color pattern of the rough-toothed dolphin consists of a dark back, lighter gray sides, and a white belly that is demarcated by a variable border. The pink color is from flushing of blood. Angola. PHOTO: C. WEIR

Rough-toothed dolphins sometimes appear quite lethargic and inactive, but they can also become quite energetic, leaping and porpoising out of the water. Bahamas. PHOTO: J. DURBAN

The cone-shaped head of the rough-toothed dolphin, with no demarcation between the melon and upper beak, is unmistakable—no other dolphin has such a head shape. Guadeloupe. PHOTO: L. BOUVERET

The rough-toothed dolphin in back may be a juvenile, but it still shows the species' main distinctive characters quite well—smooth forehead, high cape, and long beak. Hawai'i. PHOTO: K. YUI

Although it is difficult to tell in this photo, rough-toothed dolphins are one of only a few species of dolphin with longitudinal ridges or wrinkles on their teeth—a very useful diagnostic feature when seen. Canary Islands. PHOTO: F. RITTER

More than any other dolphin species, rough-toothed dolphins have a distinct tendency to associate with floating debris—wooden boards, coils of rope, tree trunks and branches, etc. Eastern tropical Pacific. PHOTO: NOAA FISHERIES/SWFSC

This leaping rough-toothed dolphin appears to have a dark belly, but in fact, most members of the species have a whitish belly. The dark, narrow cape shows well here. Costa Rica. PHOTO: D. PALACIOS-ALFARO

In rough-toothed dolphins, there is a dark, but very narrow dorsal cape. It shows well in this photo, and one can see the slight dip behind the dorsal fin. Guadeloupe. PHOTO: L. BOUVERET

The young rough-toothed dolphin on the left in this photo has not attained adult size yet, but it still shows the species' main distinctive characters quite well—smooth forehead, large flippers, light gray sides, and blotchy light belly. Guadeloupe. PHOTO: L. BOUVERET

Rough-toothed dolphins often surface in "chorus lines" with animals lined up abreast, although it would be hard to identify them from this photo alone. Eastern Pacific. PHOTO: NOAA FISHERIES/SWFSC

They are known to reach weights of up to 155 kg. Length at birth is unknown, but is likely close to 1 m.

Recognizable geographic forms None.

Can be confused with Because of their unique color pattern and head shape, rough-toothed dolphins are generally easy to identify, when seen at close range. However, they may be mistaken for bottlenose dolphins, if seen at a distance. The narrow cape and cone-shaped head will be the best clues to identification of rough-toothed dolphins in such situations.

Distribution The rough-toothed dolphin is a tropical to subtropical species, which generally inhabits deep, oceanic waters of all three major oceans, rarely ranging north of 40°N or south of 35°S. However, in some areas (such as off the coast of Brazil and West Africa), rough-toothed dolphins may occur in more shallow coastal waters. They are found in many semi-enclosed bodies of water (such as the Gulf of Mexico, Caribbean Sea, Red Sea, and Gulf of California), but only occasionally occur in the Mediterranean Sea.

Ecology and behavior Rough-toothed dolphins have been seen most commonly in groups of 10–20, although herds of over 100 have been reported. They may appear lethargic and slow-moving, especially in comparison to the smaller oceanic dolphins of the genera *Delphinus* and *Stenella*. At other times, they move at high speed with the chin and head just above the surface, in a distinctive skimming behavior, sometimes described as "surfing." They do bowride at times and aerial behavior, including spyhops and various leaps and slaps are not uncommon. They have been seen feeding opportunistically around trawlers.

Rough-toothed dolphins are thought to be capable of very deep-diving, and dives of up to 15 min have been recorded. In the eastern tropical Pacific, they tend to associate with floating objects and sometimes with other cetaceans (especially pilot whales and other dolphins).

Not much is known about the reproductive biology of this species. Studies in Japan suggest that males reach sexual maturity at about 14 years of age and 2.25 m, and females at 10 years and 2.1–2.2 m. Longevity may be around 36 years or more.

Feeding and prey Rough-toothed dolphins feed on a variety of cephalopods and fishes, including such large fish as *mahi mahi* (also called dorado or dolphinfish). Recently, it has been suggested that these dolphins may be adapted to be specialist feeders on *mahi mahi*.

Threats and status No fisheries are known to specifically target this species, but rough-toothed dolphins are one of several species killed in direct fisheries in Japan, Sri Lanka, Indonesia, the Caribbean, Papua New Guinea, the Solomon Islands, and West Africa. They are some-times taken as by-catch in purse seine fisheries for tuna in the eastern tropical Pacific, and in gillnet fisheries in Sri Lanka, Brazil, and the offshore North Pacific. Their offshore distribution in most areas should reduce potential problems of habitat loss and alteration. There are few estimates of abundance from large areas, except for the eastern tropical Pacific (146,000 animals) and northern Gulf of Mexico (450–850 animals). This species is generally not one of the more common species of tropical dolphins, but there is no reason to believe that it is threatened on a global scale.

IUCN status Least Concern.

Atlantic Humpback Dolphin—*Sousa teuszii*

(Kükenthal, 1892)

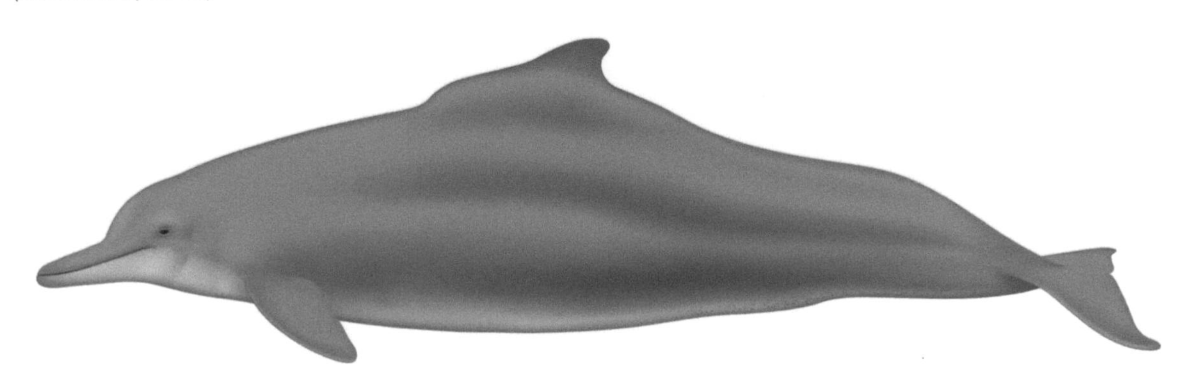

Recently-used synonyms None.

Common names En.-Atlantic humpback dolphin; *Sp.-delfín jorobado del Atlántico*; *Fr.-dauphin à bosse de l'Atlantique*.

Taxonomic information Cetartiodactyla, Cetacea, Odontoceti, Delphinidae. The validity of this species was questioned in the past, but recent studies on morphology and genetics have made it clear that the species is indeed valid and separate from humpback dolphins in the Indian and Pacific oceans.

Species characteristics Atlantic humpback dolphins have a robust body; long, slender beak (but with a very shallow crease between the beak and melon); broad flippers with rounded tips, and a moderately-deepened tail stock. The dorsal fin is variable in shape, but is generally small and falcate. It has a much more rounded tip than in the similar-appearing Indian Ocean humpback dolphin. The dorsal fin emerges from a wide hump or ridge of connective tissue on the back. All animals, except for possibly very young calves, appear to possess the hump. Although this species is poorly known, it is probably sexually dimorphic, like the Indian Ocean humpback dolphin.

Coloration is somewhat variable. Animals are typically slate gray on the sides and back, and lighter gray to whitish below. The coloration may appear brownish in some lighting. A faint delphinid "bridle" (consisting of an eye stripe and blowhole stripe) is present. Some individuals have dark spots or flecks on the tail stock and near the base of the dorsal fin, and whitish scarring on the dorsal fin and ridge.

Tooth counts are 27–32 per upper tooth row, and 26–31 per lower row.

Adults are up to about 2.8 m in length (probably with males reaching greater sizes than females), and weigh up to 284 kg. Length at birth is thought to be about 1 m, but there are few specimen records.

Recognizable geographic forms None.

Can be confused with The similar-looking common bottlenose dolphin also inhabits the inshore range of the Atlantic humpback dolphin. The two can be distinguished by differences in beak length, dorsal fin shape (including the presence of the hump in *Sousa)*, and coloration. This species is externally very similar to *Sousa plumbea* that occurs in the western Indian Ocean, but there should not be any identification problems, as the two do not overlap in distribution.

30W 20W 10W 0 10E 20E 30E 40E
30N 20N 10N 0 10S 20S
AFRICA
Atlantic Ocean
500 NM
N

Atlantic Humpback Dolphin
(Sousa teuszii)
Global Distribution
Primary Range

Distribution Atlantic humpback dolphins occur in nearshore waters off tropical to subtropical West Africa, from Western Sahara south to at least southern Angola. In some areas, they may occur as distinct populations, separated by areas of low or zero density. For instance, there are no contemporaneous records for Togo and Ghana in the Gulf of Guinea. They are found primarily in estuarine and shallow coastal waters. Humpback dolphins may occupy lower reaches of rivers to slightly beyond the tidal

The dorsal hump is prominent on all age classes of Atlantic humpback dolphins, even on young calves and juveniles. Congo. PHOTO: T. COLLINS

This is almost definitely an adult male, based on the very large dorsal hump present on the back. The tip of the dorsal fin has been damaged—it is usually rounded. Angola. PHOTO: C. WEIR

Atlantic humpback dolphins have a nearly-uniform dark gray body, and what is seen above the surface is generally all gray. Angola. PHOTO: C. WEIR

Atlantic humpback dolphins look very similar to their relatives, Indian Ocean humpbacks, but the dorsal fin tends to be more rounded. Angola. PHOTO: C. WEIR

Although they do not appear to be as aerially active as most other dolphin species, humpback dolphins do occasionally breach and perform other aerial behaviors, such as this aerial scan. Angola. PHOTO: C. WEIR

Atlantic humpback dolphins often "patrol" along the beach, cruising just offshore of the breaking waves. Angola. PHOTO: C. WEIR

Atlantic humpback dolphins most often occur in small groups of 2–10, and large aggregations are less frequently seen. Angola. PHOTO: C. WEIR

influence, but there is no evidence that there are separate freshwater populations. They frequent the surf zone just offshore of the breakers.

Ecology and behavior Groups of Atlantic humpback dolphins generally contain 2–10 individuals, occasionally up to 30 or 40 animals may gather in scattered subgroups. Groups generally feed very near shore, often within easy view of the beach, and sometimes just beyond the turbid surf zone. These animals do not bowride. Although they do breach and leap on occasion, they generally appear not to be very aerially active. Compared to some populations of the Indo-Pacific humpback dolphin, they can be shy and difficult to approach at times. Little else is known of the behavior of these animals. They sometimes form mixed-species groups with bottlenose dolphins. Tail-up dives are frequently exhibited by some groups.

Breeding has been documented in March and April, but the reproductive season may well be more protracted. Virtually nothing else is known of their life history.

Feeding and prey Atlantic humpback dolphins feed on nearshore schooling fishes such as mullet and bongo. Off the coast of Mauritania, fishermen have been documented to cooperate with bottlenose dolphins to capture mullet with beach seines, and at least occasionally Atlantic humpback dolphins may also be involved. They sometimes cooperatively forage as groups to herd mullet, but also often forage individually with tail-up dives suggesting hunting for non-schooling benthic and reef fish.

Threats and status Incidental catches in fishing nets are considered to be the greatest threat to these animals, and some Atlantic humpback dolphins may be taken directly by local people in West Africa. Habitat loss/degradation, vessel collisions, environmental contamination, anthropogenic noise, climate change, and the "marine bushmeat" trade may be additional threats. There are few statistically defensible estimates of abundance available for anywhere in the range (the only recent one is of about 10 animals in Namibe Province, Angola). Although there are thought to be at least several hundred in Guinea-Bissau, in most other areas abundance appears to be no more than about 100 individuals. This species has a restricted range, and it is likely that there are no more than a few thousand of these animals left, in more-or-less isolated populations. Due to the evidence of low densities in most areas of study, and significant threats from human activities, the Atlantic humpback dolphin is considered to be threatened with extinction on a global scale.

IUCN status Vulnerable.

Indo-Pacific Humpback Dolphin—*Sousa chinensis*

(Osbeck, 1765)

Recently-used synonyms *Sousa borneensis*.
Common names En.–Indo-Pacific humpback dolphin; Sp.–*delfín jorobado del Indo-Pacífico*; Fr.–*dauphin à bosse de l'Indo-Pacifique*.
Taxonomic information Cetartiodactyla, Cetacea, Odontoceti, Delphinidae. This species has only recently been recognized as separate from *S. plumbea* and *S. sahulensis*. A unique subspecies may exist in Taiwan.

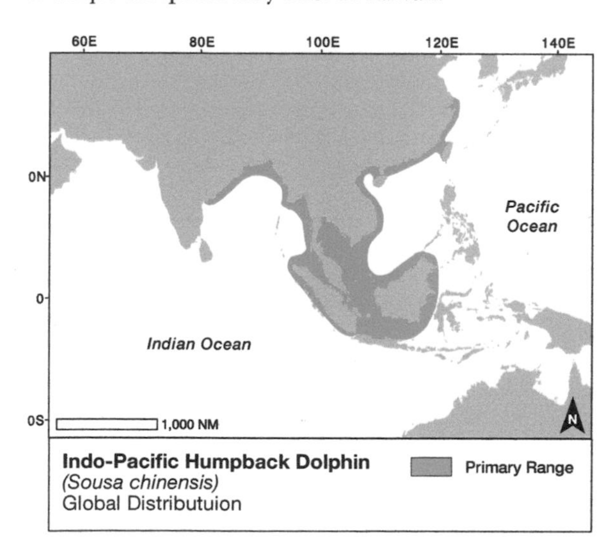

Species characteristics Dolphins of this highly-variable species are characterized by robust bodies with long, well-defined beaks. The beak is distinctly set-off from the rounded melon, but not by a deep crease (as it is in many other dolphin species). In Indo-Pacific humpback dolphins the dorsal ridge that is present on Indian Ocean humpback dolphins is absent. The dorsal fin is short, but very wide-based and ranges from nearly triangular to slightly falcate. The flippers and flukes are relatively broad with rounded tips. Old adults may be wrinkled and have corrugated skin, especially on their tail stocks.

The color pattern varies with age, sex, and area. In this species, dark gray calves lighten with age. Juveniles begin to lose the dark color in blotches, leaving the white ground color to show through. In most areas of Southeast Asia, adults become pinkish-white, often with dark spots and blotches concentrated on the dorsal side. In at least southern China, the spotting disappears in most old females, leaving them ghostly white (though often with a pinkish tinge). There is generally no sign of the typical delphinid "bridle" (i.e., eye and blowhole stripe complex).

There are 32–38 sturdy teeth in each upper tooth row, and 29–38 in each lower row. Maximum known body lengths are 2.7 m (males) and 2.6 m (females), but there appears to be very little (if any) sexual dimorphism in most populations. Weights of up to 240 kg have been recorded in Chinese waters. Newborns are around 1 m in length.

In Borneo, *Sousa* have exceedingly long-based and low dorsal fins. This has led some researchers to consider these animals to be a separate species, but they are not currently recognized as such. PHOTO: G. MINTON

In at least some parts of their range, the dorsal fin lightens well before the rest of the body does. There is much geographical variation in coloration in *S. chinensis*. Bangladesh. PHOTO: R. MANSUR MOWGLI

Younger Indo-Pacific humpback dolphins have much more dark pigment on the bodies than old adults; thus it is easy to recognize them. Hong Kong. PHOTO: HONG KONG DOLPHIN CONSERVATION SOCIETY

In Bangladesh, Indo-Pacific humpback dolphins lose their pigmentation, as they do in Chinese waters. However, even the larger adults generally retain large areas of dark coloration, and they do not appear to become pure white, as in China. PHOTO: R. MANSUR MOWGLI

Calf—About 1/3–1/2 of adult size; dark gray to black dorsal surface, with no spotting.
Juvenile—About 2/3–3/4 adult size; gray dorsal surface, often with some lightening of the back and dorsal fin.
Subadult—Close to adult size; dark ground color of the dorsal surface fades in blotches to leave extensive gray-black spotting/mottling.
Adult—Ground color white (often with pinkish tinge); spotting on dorsal surface may be extensive to absent.

Recognizable geographic forms

chinensis-*type*—Coloration of adults generally ranges from gray to pure white, often with moderate to heavy spotting on much of the body (including the dorsal fin and back). The dorsal fin is relatively short, and generally slightly falcate.

borneensis-*type*—Coloration of adults ranges from gray to white, generally with dark blotches and spotting on much of the body (including the dorsal fin and back). The dorsal fin is extremely short and wide-based, and more nearly triangular.

Can be confused with This species is likely to be confused with the Indian Ocean humpback dolphin, with which it overlaps in range in the Bay of Bengal area. The Indo-Pacific species does not have the dorsal hump of the Indian Ocean species, and often (though not always) has extensive areas of white color on the body. Humpback dolphins can also be confused with bottlenose dolphins (especially Indo-Pacific bottlenose dolphins), which share much of their range). Differences in dorsal fin shape, head shape, and color can be used to distinguish between the two. The surfacing behavior is also different, and may be useful once the observer has seen a number of groups of both species. Other species that share the nearshore range of this species (e.g., Irrawaddy dolphins, common dolphins, finless porpoises) should be easy to distinguish, based on size, general color pattern, head shape and dorsal fin shape.

Distribution Indo-Pacific humpback dolphins are found from central China (near the mouth of the Yangtze River) in the east, through the Indo-Malay Archipelago, and westward around the coastal rim of the Indian Ocean to as far west as the Orissa coast of India. They are inhabitants of tropical to warm temperate, coastal waters and they often enter river mouths, estuaries, and mangroves. The species' distribution extends to enclosed seas such as the Gulf of Thailand. They only rarely occur more than a few tens of kilometers from shore.

Ecology and behavior Groups tend to be small, containing fewer than 10 individuals, although up to 30–40 have been seen together on occasion in Chinese waters (usually following fishing boats). Group structure has been studied using photo-identification techniques. At least off Hong Kong, fluidity in group structure is the rule. But, in Taiwan there is much more stability in group structure. Ranging patterns appear to vary, depending on the type of coastline involved, but residency of at least

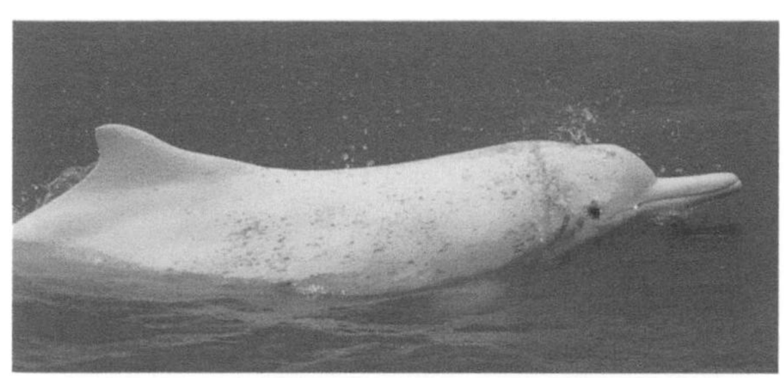

When surfacing to breathe, Indo-Pacific humpback dolphins often bring the upper part of their head and their eyes above the surface—perhaps to look at the photographer? Hong Kong. PHOTO: T. A. JEFFERSON

In the waters of Hong Kong, older humpback dolphins may appear completely white, though often with a pinkish tinge. The pink color is temporary and caused by blood flushing to the surface, and not by red pigments. PHOTO: T. A. JEFFERSON

some individuals appears to be typical. In Hong Kong, individuals have home range sizes that average about 135 km^2.

These dolphins are moderately acrobatic, frequently being seen spy-hopping, breaching, and performing other leaps out of the water. Groups in Hong Kong waters often used to feed behind active fishing trawlers (trawling is now illegal). In Hong Kong, they do not feed around reefs, but often consume fish in shallow, murky waters. Sometimes they are seen with mud on their bodies, suggesting they feed on the bottom. Bowriding behavior is extremely rare, although some individuals may occasionally ride swells or vessel wakes. In Hong Kong, dolphins are very accustomed to heavy vessel traffic, but in other areas (such as Borneo) they appear to be much more wary of vessels, even shy.

Little is known about reproductive biology throughout most of the range. Sexual maturity appears to be reached at around 9–10 years for females, and a few years later for males. Mating and calving occur all year, at least in China, but there appears to be a calving peak in late spring to summer. Newborns are about 1 m long. They can live to be at least 38 years old, and it is suspected that some live well into their 40s.

Feeding and prey Indo-Pacific humpback dolphins appear to be rather opportunistic feeders. Feeding is primarily on a very wide variety of nearshore, estuarine, and reef fishes. They may also eat cephalopods in some areas, although crustaceans appear to be rare in the diet.

Threats and status Indo-Pacific humpback dolphins are not known to be hunted directly anywhere in their range in significant numbers. However, they are often caught in fishing nets (such as gillnets and trawls). Because they are coastal in nature, habitat loss, vessel strikes, and pollution are additional important threats to these animals, and they are vulnerable to human threats. In particular, organochlorines (especially DDT and its derivatives) and trace metals have been found to be very high for several populations (such as those off Hong Kong). Coastal development is a major issue in many areas, especially in China, Taiwan, and Singapore. The species is not in any immediate danger of extinction, but certain populations are known to have been depleted. Most of the few abundance estimates obtained have been in the lower hundreds of dolphins, but there appear to be at least 2,000 in the Pearl River Estuary of southern China (this population includes Hong Kong). The subpopulation that occurs in Hong Kong's western waters is declining, and the small population off Taiwan's west coast (ca. 75 dolphins) also appears to be declining. Other population estimates are about 86 off Xiamen, 237 off the Leizhou Peninsula, and >98 for the Beibu Gulf. It is doubtful that there are more than 10,000 of these dolphins in the global population.

IUCN status Vulnerable (overall species), Critically Endangered (Eastern Taiwan Strait population).

(Left) Although Indo-Pacific humpback dolphins share the long beak of their Indian Ocean relatives, the adults are generally much lighter in coloration. (Right) This animal is not likely yet fully adult, as indicated by the extensive dark spotting on its body. Hong Kong. PHOTOS: T. A. JEFFERSON

Indian Ocean Humpback Dolphin—*Sousa plumbea*

(G. Cuvier, 1829)

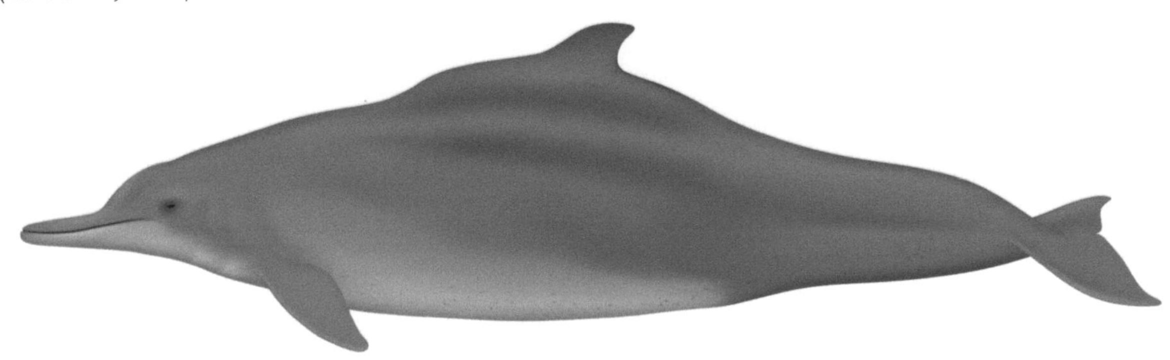

Recently-used synonyms *Sousa lentiginosa.*
Common names En.–Indian Ocean humpback dolphin; Sp.–*delfín jorobado del Océano Índico*; Fr.–*dauphin à bosse de l'océan Indien.*
Taxonomic information Cetartiodactyla, Cetacea, Odontoceti, Delphinidae. This species has only recently been split off from *Sousa chinensis* and *Sousa sahulensis*.
Species characteristics Indian Ocean humpback dolphins are characterized by robust bodies with long, well-defined beaks. The beak is distinctly set-off from the rounded melon, but not by a deep crease (as it is in many other dolphin species). The small, highly falcate and pointed dorsal fin sits atop a very wide hump, or ridge, in the middle of the animal's back. The hump is present in young animals, but gets proportionately larger in older animals, especially adult males. In some areas, there also appear to be well-developed dorsal and ventral ridges on the tailstock, which are also more exaggerated in males.

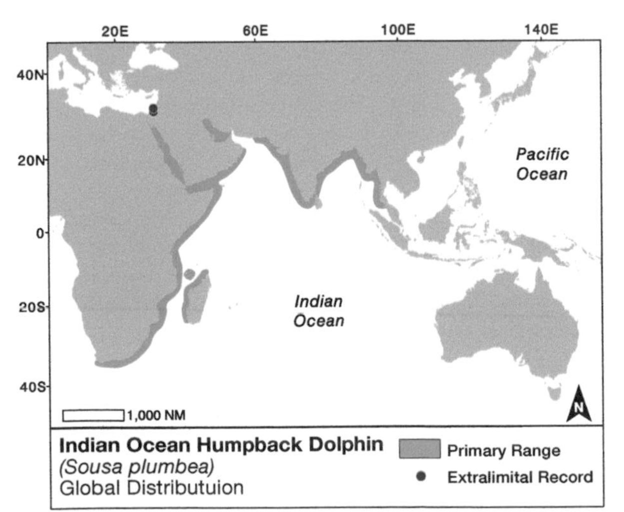

The flippers and flukes are relatively broad, generally with rounded tips.

The color pattern varies somewhat with age. These dolphins are mostly brownish-gray, with a slightly lighter belly. The characteristic delphinid "bridle" (consisting of an eye stripe and blowhole stripe) is present, though faint. Generally, lighter-colored calves darken with age to become dark lead gray above and light gray below as adults (sometimes with white scarring around the dorsal fin/hump and head). There is often a white margin to the dorsal fin and hump, especially in older animals. Spotting tends to be relatively sparse, if present at all.

There are 33–39 sturdy teeth in each upper tooth row, and 31–37 in each lower row.

Maximum known body lengths are 2.8 m (males) and 2.6 m (females). Indian specimens may grow larger than this, although reported lengths of 3.2 m are considered suspect. They are sexually dimorphic, with males reaching much greater sizes than females. Weights of up to 280 kg have been recorded. Newborns appear to be around 1 m in length.

Calf—About 1/3–1/2 of adult size; off-white to light gray coloration, with no spotting.
Juvenile/Subadult—Greater than about 2/3 adult size; uniform slate-gray dorsal surface.
Adult—Body robust, generally with prominent dorsal hump and deepened tail stock; dorsal surface and head often whitened, scarred, and sometimes spotted.

Recognizable geographic forms

plumbea-*type*—These animals are found throughout the range. Coloration is generally brownish-gray, with a lighter belly, and spotting is limited to small areas of the tail stock. The only white areas on the dorsal surface of the body are generally dorsal scarring in large adults.
lentiginosa-*type*—These animals appear to be mainly found around India and Sri Lanka. Coloration is much lighter than in the *plumbea*-type, with larger adults having

In waters of Sri Lanka and southeastern India, there is a geographic form of humpback dolphin with more extensive light color on the belly and sides. This has been called the *lentiginosa*-form. Sri Lanka. PHOTO: R. NAYAKARRA

In this photo, the uniform gray coloration that is typical of most Indian Ocean humpback dolphins is clearly visible. India. PHOTO: D. SUTARIA

While Indian Ocean humpback dolphins are generally all gray, there can be some lightening around the beak and dorsal fin in older animals. India. PHOTO: D. SUTARIA

Although younger animals are almost completely gray, older Indian Ocean humpback dolphins may develop spotting and scarring on the dorsal hump, and white on the beak. Iran. PHOTO: H. MOSHIRI, COURTESY T. COLLINS

The small dorsal fin, and large hump on the back of Indian Ocean humpbacks are unmistakable. No other dolphin species in their range has such an appearance. Iran. PHOTO: H. MOSHIRI, COURTESY T. COLLINS

Indian Ocean humpback dolphins breach and perform other aerial maneuvers, although not very often. South Africa. PHOTO: B. ATKINS

Humpback dolphins can be the hosts to several species of external parasites and commensals, including remoras. South Africa. PHOTO: B. ATKINS

This is likely to be a female or young male Indian Ocean humpback dolphin, since the dorsal hump is not very large. Oman. PHOTO: COURTESY T. COLLINS

a white ground color. Spotting may be extensive, but the overall appearance of most adults is more light than dark.

Can be confused with Indian Ocean humpback dolphins are most likely to be confused with Indo-Pacific humpback dolphins, which overlap in the Bay of Bengal area. The prominent dorsal hump, smaller dorsal fin, and much darker color of adults will help to distinguish this species. Bottlenose dolphins may also cause confusion. Differences in dorsal fin shape (including presence of the hump on humpback dolphins), head shape, and color can be used to distinguish between the two. The surfacing behavior is also different, and may be useful once the observer has seen a number of groups of both species. Other species that share the nearshore range of this species (e.g., common and spinner dolphins) should be easy to distinguish, based on size, general color pattern, head shape and dorsal fin shape.

Distribution Indian Ocean humpback dolphins are found in a nearshore strip from Cape Town, South Africa, to the Bay of Bengal, as far east as the Mergui Archipelago, Myanmar (aka Burma). They are inhabitants of tropical to warm temperate coastal waters and they concentrate around the mouths of rivers, estuaries, and mangroves. The species' distribution extends to enclosed seas, such as the Persian Gulf and Red Sea. There are a few extralimital records for the Mediterranean (apparent strays from the Red Sea). They rarely occur in waters deeper than 30 m.

Ecology and behavior Groups tend to be small, containing fewer than 10 individuals, although in Arabian waters, groups of over 100 are sometimes seen. Group structure has been studied using photo-identification techniques mainly in southern Africa. In Mozambique, groups have some long-term stability, though this appears unusual. Off South Africa (as in probably most of the range), fluidity in group structure is the rule. Ranging patterns appear to vary, depending on the type of coastline involved, but in southern Africa, most individuals have more-or-less linear ranges that do not extend far offshore. Residency of at least some individuals appears to be typical, but movements of several hundred linear kilometers are known.

Off South Africa, where the behavior of these dolphins has been most thoroughly studied, herds often patrol slowly parallel to shore and preferentially use sandy bays for resting and socializing, and use rocky

Indian Ocean humpback dolphins have a prominent dorsal hump even as young calves. This mother and calf were caught and killed in anti-shark nets off South Africa. PHOTO: NATAL SHARKS BOARD

coastline and larger estuarine areas for foraging. Some individuals have been seen to beach themselves temporarily while pursuing fish nearshore. These animals are moderately acrobatic. They feed nearshore and around reefs, but often consume fish near the bottom in shallow, murky waters. Bowriding behavior is extremely rare. They are quite shy of boats, at least in South Africa and near the mouth of the Persian Gulf.

Little is known about their reproductive biology, which has only been studied in detail in South Africa. Sexual maturity appears to be reached at around 10 years for females, and around 12–13 years for males. Mating and calving occur all year, at least in South Africa, but there appears to be a calving peak in summer months. Newborns are about 1.0–1.1 m long, and some individuals live to over 40 years.

Feeding and prey Indian Ocean humpback dolphins appear to be rather opportunistic feeders. Feeding is primarily on a very wide variety of nearshore, estuarine, and reef fishes. They may also eat cephalopods in some areas, although crustaceans appear to be rare in the diet.

Threats and status Indian Ocean humpback dolphins are not known to be hunted directly anywhere in their range in significant numbers, although some direct killing may occur in India and Madagascar. However, they are often caught in fishing nets (such as gillnets and trawls), and also in anti-shark nets off South Africa. Because they are coastal in nature, habitat loss, vessel strikes, oil spills, and pollution are additional important threats to these animals, and they appear to be especially vulnerable to human threats. In particular, organochlorine levels have been found to be very high for several populations (such as those off South Africa). Certain populations are considered to be depleted. Most of the few abundance estimates obtained have been in the lower hundreds of dolphins (450 in Algoa Bay, 170–244 in Richards Bay, 105 in Maputo Bay, ca. 60 in the Bazaruto Archipelago, and 58–65 off Zanzibar). There may only be a dozen or so left off Sri Lanka. It is doubtful that there would be more than about 10,000 of these animals in the entire global population, but as a species they are not in any immediate danger of extinction.

IUCN status Vulnerable.

Australian Humpback Dolphin—*Sousa sahulensis*

Jefferson and Rosenbaum, 2014

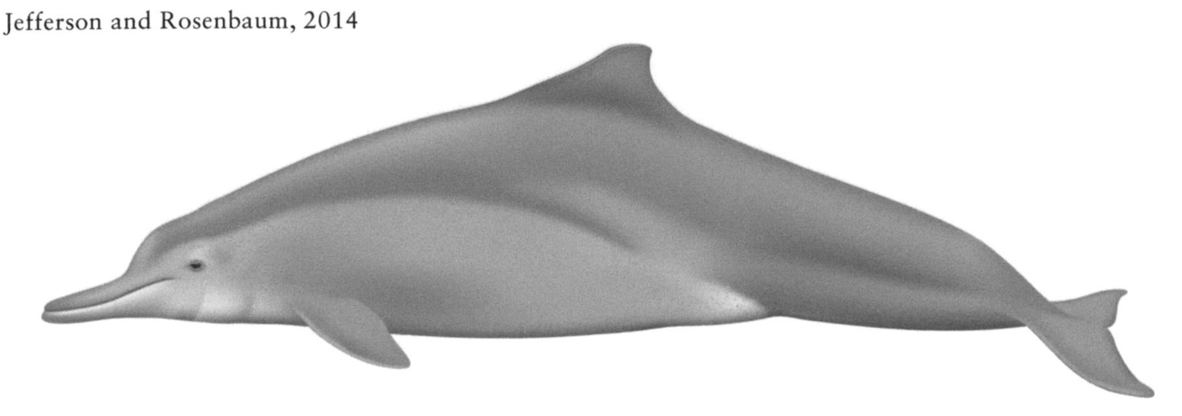

Recently-used synonyms *Sousa queenslandensis*.

Common names En.-Australian humpback dolphin; Sp.-*delfín jorobado Australiano*; Fr.-*dauphin à bosse d'Australie*.

Taxonomic information Cetartiodactyla, Cetacea, Odontoceti, Delphinidae. This species has just recently (2014) been split off from *Sousa chinensis* and recognized as a distinct species. As such, its biology is still poorly known. Hybrids with *O. heinsohni* have been documented.

Species characteristics Australian humpback dolphins are characterized by robust bodies with long, well-defined beaks. The beak is distinctly set-off from the rounded melon, but not by a deep crease (as it is in many other dolphin species). In Australian humpback dolphins the dorsal hump (which is prominent on *S. plumbea* and *S. teuszii*) is absent, and the dorsal fin is moderately short and quite wide-based. The flukes and flippers are relatively broad, and tend to have rounded tips.

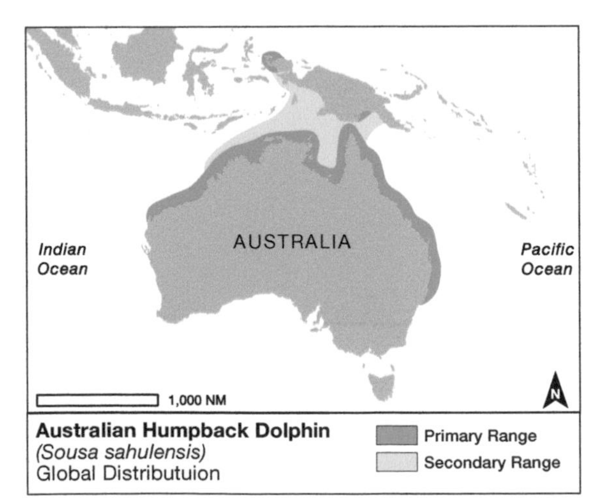

Australian Humpback Dolphin
(Sousa sahulensis)
Global Distributuion

The color pattern varies with age. Dark gray calves lighten with age, and some (apparently old) animals may appear very light in color. However, most subadults and adults still have extensive dark areas on the body, and variable amounts of dark flecking and scarring. There is a separation of the dark back and the lighter lower sides and belly by a slightly curved, diagonal cape line, with indistinct margins. The dark cape is often invaded by a faint light patch from below, just behind the dorsal fin. If there is a delphinid blowhole stripe it tends to be quite faint.

There are 31 to 35 sturdy teeth in each upper tooth row, and 31–34 in each lower row.

Maximum known body lengths are slightly over 2.6 m for both sexes, and there appears to be slight sexual dimorphism. Body lengths up to 2.7 m (sex unknown) have been reported. There are few data on body weight, but it is expected that they probably reach at least 240 kg. Newborns appear to be around 1 m in length.

Recognizable geographic forms None.

Can be confused with Australian humpback dolphins are most likely to be confused with bottlenose dolphins (especially Indo-Pacific bottlenoses), with which they share their entire distribution. Differences in dorsal fin shape, head shape, and coloration can be used to distinguish between the two. The surfacing behavior is also different, and may be useful once the observer has seen a number of groups of both species. Other species that share the nearshore range of this species (e.g., Australian snubfin dolphins and common dolphins) should be easy to distinguish, based on size, general color pattern, head shape, and dorsal fin shape.

Distribution Australian humpback dolphins are found in tropical/subtropical parts of Australia (from the Queensland/New South Wales border to Ningaloo Reef) and at least southern New Guinea (reliable records are from the Gulf of Papua, PNG). They are inhabitants

This breaching Australian humpback dolphin shows the diagnostic color pattern of this species, which includes a diagonal cape on the flank. Ningaloo Reef. PHOTO: R. L. PITMAN

The dorsal fin of the Australian humpback dolphin is very low and wide-based, with little or no hint of a hump. Western Australia. PHOTO: A. BROWN

While often appearing nearly-uniform gray, some Australian humpback dolphins have white areas on the dorsal fin. Western Australia. PHOTO: T. HUNT

This individual is suspected to be an old adult, as evidenced by the extensive white on the dorsal fin and beak. Western Australia. PHOTO: C. ACKERMAN

Tool use among wild dolphins does not appear to be common, but this is one species where it has been observed with some regularity. Dolphins use sponges in Western Australia, although it is not certain what they use them for. PHOTO: D. CLARKE

The distinctive dorsal cape and light belly of this species show well in this photo of two breaching individuals. Australia. PHOTO: I. BEASLEY

Australian humpback dolphins are denizens of shallow waters, and they sometimes strand themselves on shallow flats while pursuing prey. PHOTO: I. BEASLEY

of coastal waters and they often enter rivers, estuaries, and mangroves associated with the shallow Sahul Shelf. The Northwest Cape, Western Australia, appears to be a "hotspot" for the species.

Ecology and behavior Ecology and behavior of Australian humpback dolphins are not well known. Groups tend to be small, mostly containing fewer than 10 individuals. Group structure has been studied using photo-identification techniques mainly in Queensland. At least in that area, fluidity in group structure is the rule. In Queensland, some animals appear to be permanent residents, while others are more transitory. These dolphins use sponges as play items and perhaps as tools—a behavior almost unique to this species.

Australian humpback dolphin behavioral ecology has mostly been studied in detail in Queensland. There, they are moderately acrobatic. Groups in Moreton Bay often feed behind active trawlers, sometimes in association with Indo-Pacific bottlenose dolphins. They mostly consume fish near the bottom in shallow, murky waters. Bowriding behavior is essentially unheard of. They tend to be relatively shy of boats.

Almost nothing is known about their life history, but what little is known suggests that they are similar to *Sousa chinensis*. Mating and calving most likely occur all year, but with peaks at certain times of the year.

Feeding and prey Australian humpback dolphins appear to be rather opportunistic feeders. Feeding habits have not been studied from large samples, but they are known to feed primarily on a very wide variety of nearshore, estuarine, and reef-associated fishes. Cephalopods are not often consumed.

Threats and status Australian humpback dolphins are not known to be hunted directly in significant numbers. However, they are caught in fishing nets (such as gillnets and trawls), and also in anti-shark nets at least off Queensland. Because they are coastal in nature, habitat loss, wildlife ecotourism, vessel strikes, and pollution are additional potential threats to these animals. Apparently, the species is not in any immediate danger of extinction, but there are probably no more than a few thousand of these animals overall. Some populations may already have been depleted. Most of the available abundance estimates obtained have been in the low hundreds of dolphins (e.g., <100 in Cleveland Bay, 150 in two communities in the Great Sandy Strait Marine Park area). The northwestern part of Australia is undergoing rapid changes, with the recent development of oil and minerals extraction operations, therefore better impact assessment efforts must be directed toward the species there.

IUCN status Not listed.

Common Bottlenose Dolphin—*Tursiops truncatus*

(Montagu, 1821)

Offshore Adult

Coastal Adult

Coastal Calf

Recently-used synonyms *Tursiops gillii, Tursiops nuuanu, Tursiops gephyreus, Tursiops australis.*

Common names En.–common bottlenose dolphin; *Sp.–tursión, tonina;* Fr.–*grand dauphin.*

Taxonomic information Cetartiodactyla, Cetacea, Odontoceti, Delphinidae. All bottlenose dolphins around the world were previously recognized as *T. truncatus*, but the genus has been split into two species: *T. truncatus* and *T. aduncus* (the smaller Indo-Pacific bottlenose dolphin). Many other species of bottlenose dolphins have been described, but most of these are thought to be synonymous with *T. truncatus*. The taxonomy of bottlenose dolphins is still confused, however, due to the great extent of geographical variation, and it is very possible that additional species will be recognized in the future. For instance, more and more evidence is suggesting that coastal bottlenose dolphins in some areas of the Atlantic belong to a different species from those in offshore waters. Currently, the Society for Marine Mammalogy recognizes two subspecies: *T. t. ponticus* in the Black Sea, and *T. t. truncatus* elsewhere. Hybrids with several other delphinid species (e.g., rough-toothed, long-beaked common, and Risso's dolphins, and false killer and short-finned pilot whales) are known, both from captivity and in the wild.

Species characteristics The common bottlenose dolphin is probably the most familiar of the small cetaceans because of its coastal habits, prevalence in captivity worldwide, and frequent appearance on television and in advertising. It is a large, relatively-robust dolphin, with a short to moderate-length stocky beak that is distinctly set off from the melon by a crease. It has a gently-curved mouthline that dips down from the tip of the beak, then back up, and finally down again at the gape (many people feel the mouthline resembles a smile). The dorsal fin is tall and falcate, and set near the middle of the back. The flippers are recurved and somewhat pointed at the tips.

The flukes are wide and possess a median notch. The shape of the body is generic for a dolphin, but there is extensive geographical variation in specifics of body shape, appendages, and coloration.

The color pattern is the most generalized of all the marine dolphins, and varies from light gray to nearly black on the back and sides, fading to white (sometimes with a pinkish hue) on the belly. The belly and lower sides are rarely spotted, and when they are, the spots are generally small flecks. Although the back is generally dark gray, there is sometimes also a faint dorsal cape on the back, generally only visible at close range. Sometimes a faint spinal blaze may also be present. Often, there are brushings of gray on the body, especially on the face. A generalized delphinid "bridle," consisting of blowhole and eye stripes is present, but is variable in its intensity. A wide, but very faint, stripe often extends forward from the flipper and connects to the gape. A light patch on the side of the melon is often present, bordered by dark stripes from the blowhole to the apex of the melon. There may be a pale throat chevron and genital patch. Anomalously white individuals have been observed (in at least one case, this was a confirmed albino).

Bottlenose dolphins have 18–27 pairs of stout, pointed teeth in each jaw. They are generally smooth in contour. In older animals, many of these may be worn down or missing. Length at birth is about 1–1.3 m. Adults range from 1.9–3.8 m, with males tending to be somewhat larger than females in at least some populations. There is wide variation in size between different populations, with the largest known specimens coming from the eastern North Atlantic, around the United Kingdom. Maximum weight is at least 650 kg, although most common bottlenose dolphins are much smaller.

Recognizable geographic forms Many different geographical forms of the common bottlenose dolphin have been described for different areas of the range. However, in most cases there is some question as to their validity (since they are often based on small or inadequate samples). Even for those cases where their distinctness is well-recognized, it is often the case that reliable external morphological characters have not been provided and range limits are generally not known, which compromises our ability to distinguish them in the field. Hybridization may contribute to some of this confusion. In many areas of the world, such as the western North Atlantic, eastern North Pacific, and Peru, there appear to be two "ecotypes," a coastal ecotype and an offshore ecotype. These forms have been distinguished by various morphological, ecological, and physiological features, but differences are subtle and regionally-variable. Therefore, separation into ecotype should be left to experienced observers.

Burrunan bottlenose dolphin (australis-type)—This form was described as a distinct species, but this has not been recognized and we consider it here to be a geographic form of *T. truncatus*. Found only off southeastern Australia, this form has a distinct tripartite coloration pattern, with a light-colored face and intermediate light gray sides that dip below the dorsal fin. It reaches only about 2.8 m in length, much smaller than most other common bottlenose dolphin populations.

Can be confused with Because of their somewhat generic appearance, common bottlenose dolphins can be mistaken for several other species of dolphins, depending on the region. Identification of the species may be mostly a process of elimination of other potential species. They are most likely to be confused with Indo-Pacific bottlenose dolphins where the two species overlap in distribution. Overall size, head and beak shape, and subtleties of coloration are the best cues (for instance, common bottlenose dolphins are generally quite a bit larger, more strongly countershaded, and ventral spotting is rare). There can be confusion in the tropical Atlantic Ocean with young, as yet unspotted, Atlantic spotted dolphins, along the east coast of South America with dolphins of the genus *Sotalia*, and in the Indo-Pacific and off West Africa with humpback dolphins. When seen from a distance, they can also be confused with Risso's or rough-toothed dolphins. Such confusion will generally only occur when the animals are not seen well; in most situations, bottlenose dolphins can be identified by their various combinations of coloration, body shape and size, and head (especially beak) shape characteristics.

Distribution Common bottlenose dolphins are very widely-distributed. They are found most commonly in coastal and continental shelf waters of tropical and temperate regions of the world. They occur in most enclosed or semi-enclosed seas (e.g., the North Sea, Mediterranean Sea, Black Sea, Caribbean Sea, Gulf of Mexico, Red Sea, Persian Gulf, Sea of Japan, Gulf of California). They frequent bays, lagoons, channels, and even the mouths of rivers. Although they can also be found in very deep waters

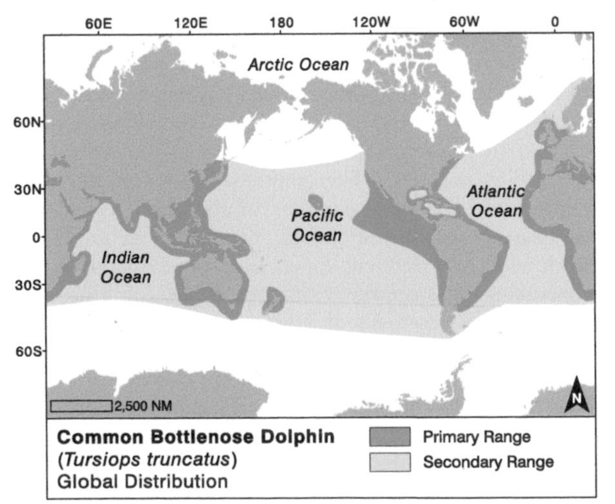

Common Bottlenose Dolphin
(*Tursiops truncatus*)
Global Distribution

In waters of the southeastern United States, common bottlenose dolphins feed on mullet, sometimes catching the fish in mid-air as they leap from the water. Florida. PHOTO: T. PUSSER

Common bottlenose dolphins use a wide variety of feeding techniques, and possibly the most interesting is strand feeding—where dolphins rush out of the water onto mud banks, grab a fish, and then slide back into the water. South Carolina. PHOTO: T. PUSSER

Prey species of the common bottlenose dolphin are highly variable. They eat cephalopods, shrimp, and fish ranging from small schooling species to types as large as salmon. Scotland. PHOTO: UNIVERSITY OF ABERDEEN

Remoras often attach to the bodies of dolphins in warm waters, and common bottlenose dolphins are no exception. Eastern tropical Pacific. PHOTO: R. L. PITMAN

Bottlenose dolphins have extremely diverse feeding habits, and they may sometimes try to eat things that are larger than their heads. Scotland. PHOTO: UNIVERSITY OF ABERDEEN

Common bottlenose dolphins usually do not have spotting on the belly, but some animals do, at least in the Atlantic Ocean. Captivity, Bahamas. PHOTO: R. L. PITMAN

These bottlenose dolphins leave no doubt as to their identity—the short, stubby beak, subtle color pattern with a series of facial stripes, and unmistakable "smile" are all visible here. Bahamas. PHOTO: H. MINAKUCHI

These intensively-socializing common bottlenose dolphins off Guadeloupe display the generic color pattern and body shape that are characteristic of this species. PHOTOS: L. BOUVERET

in some oceanic regions (e.g., the eastern tropical Pacific), population density always appears to be higher nearshore. Except for their occurrence around the United Kingdom, northern Europe, and southern New Zealand, they generally do not range poleward of 45° in either hemisphere.

Ecology and behavior More is known of the biology of this species than of any other dolphin. Its behavior has been studied, both in captivity and in a large number of different coastal areas throughout the range. Group size is commonly less than 20, but large herds of several hundred are sometimes seen in offshore areas (such as the eastern tropical Pacific and western Indian Ocean). Bottlenose dolphins are commonly associated with other cetaceans, including both large whales and other dolphin species (mixed schools with Indo-Pacific bottlenose dolphins have been found, for instance off China and Taiwan). They have been observed to attack and kill harbor porpoises, in both the Atlantic and Pacific oceans.

Based on a number of studies of nearshore populations, bottlenose dolphins seem to live in relatively fluid (fission/fusion) groupings with somewhat-closed societies. Mother/calf bonds and some other associations may be strong, but other individuals may be seen from day-to-day with a variety of different associates. In many inshore areas where they are not migratory, they maintain definable, long-term multi-generational home ranges; in others, they are migratory, generally ranging further. The behavior and social systems of these animals are highly adaptable and diverse.

The bottlenose dolphin is the most common species of dolphin held in captivity. It has proven highly adaptable and is easily trained. Much of what we know of the general biology of dolphins comes from studies of bottlenose dolphins, both in captivity and in the wild. Some solitary dolphins in the wild have become "friendly" and interacted with humans for extended periods of time. Bottlenose dolphins are avid bowriders, in both inshore and offshore waters and may perform acrobatic leaps while riding. They also sometimes ride the waves produced by the heads of large whales (something researchers call "snout-riding"). Bottlenose dolphins tend to be quite active (especially when feeding or socializing), often slapping the water with their flukes, leaping, and performing other aerial behaviors.

Life history has been studied in many areas, and there is wide geographic variation in reproductive parameters. Females mature at 5–13 years of age, and males at 9–13 years. Spring and summer or spring and fall calving peaks are known for most populations. Calves are typically suckled for 1.5–2 years, although nursing can last longer, especially for a last-born calf. Females may live to over 50 years of age and males to 40–45 years.

Feeding and prey Common bottlenose dolphins are generalist feeders, which eat a wide variety of prey species, mostly fish (with a tendency towards sciaenids, scombrids,

This common bottlenose dolphin appears to be completely black. Although the coloration is highly variable in this species, most individuals have a dark dorsal cape, lighter gray sides, and a variably-whitish belly. Gulf of California. PHOTO: M. NOLAN

A new species of bottlenose dolphin was described recently as *Tursiops australis.* While the species is not currently recognized as valid by most scientists, it does appear to at least be a distinct form with some unique coloration features. Southern Australia. PHOTO: D. DONNELLY

Typically, common bottlenose dolphins do not have dark spotting on the belly, as their congeners commonly do. However, there are exceptions, as can be seen in this very light animal from the Bahamas. PHOTO: H. FEARNBACH

The common bottlenose dolphin is a species that is known for being aerially active. Breaches, spyhops, porpoising and various types of leaps are common in these animals. Angola. PHOTO: C. WEIR

A mother and calf of the common bottlenose dolphin porpoise out of the water, showing the robust body shape and non-descript color pattern. Gulf of California. PHOTO: M. NOLAN

In some areas of the tropical Atlantic Ocean, common bottlenose dolphins possess some dark spotting on the belly. One wonders whether this may be a case of occasional mating with the sympatric Atlantic spotted dolphin. Guadeloupe. PHOTO: L. BOUVERET

Common bottlenose dolphins are avid bowriders, and will ride on bow waves of vessels ranging from small, outboard motor boats to large oceanic tankers and cargo vessels. Gulf of California.
PHOTO: M. NOLAN

and mugilids) and squid. They sometimes eat shrimps and other crustaceans. In some areas, they apparently take the most abundant or easiest prey at the time, but in other areas they demonstrate apparent preferences. Feeding behavior is varied, mostly involving individual capture of fish, but it can range from cooperative foraging on schooling fish, to chasing fish onto mud banks, to feeding behind shrimp trawlers.

Threats and status Common bottlenose dolphins have been hunted directly in several areas. The largest takes in recent years have been in the Black Sea, but large numbers have occasionally been taken in drive fisheries in Japan and Taiwan as well. Some animals have also been taken in small cetacean fisheries in the Caribbean, Peru, Sri Lanka, West Africa, and Indonesia. Bottlenose dolphins were deliberately hunted on the east coast of the US in the late 1800s and early 1900s. Finally, significant numbers have been taken in live-capture fisheries for display, research, and use by the military. They continue to be collected for captive display and research in some parts of the world. Common bottlenose dolphins interact with fisheries throughout their range, and are often seen following behind and feeding from shrimp trawlers. Incidental catches are also known from throughout the species' range. Catches have occurred in gillnets, driftnets, purse seines, trawls, and on hook-and-line gear. Mortality related to recreational fishing is also being documented with increasing frequency, in some cases exceeding commercial catches. Coastal bottlenose dolphins are susceptible to habitat destruction and degradation by human activities. Vessel collisions and the effects of environmental contaminants also result in some mortality. There have been several die-offs of bottlenose dolphins in recent years, most often linked to poisoning from biotoxins of natural origin. Anthropogenic contaminants are also a serious concern, especially in that they can affect the immune and reproductive systems. There are no estimates of overall abundance, although abundance has been estimated for some parts of the range. In the eastern tropical Pacific, there are about 243,500, in Japan 317,000, off the eastern US coast 10,000–13,000, in the northern Gulf of Mexico 35,000–45,000, off Natal, South Africa 900 (however, many of these may be *T. aduncus)*, and in the Black Sea 7,000. There are also at least minimum estimates (based on photo-identification) for many smaller study areas. Clearly, the species is not endangered, although some populations may be threatened by human activities.

IUCN status Least Concern (overall species), Critically Endangered (Fiordland population), Endangered (Black Sea population), Vulnerable (Mediterranean population).

Indo-Pacific Bottlenose Dolphin—*Tursiops aduncus*

(Ehrenberg, 1833)

Recently-used synonyms *Tursiops catalania.*
Common names En.–Indo-Pacific bottlenose dolphin; *Sp.–delfín mular del Océano Índico; Fr.–grand dauphin de l'océan Indien.*
Taxonomic information Cetartiodactyla, Cetacea, Odontoceti, Delphinidae. The validity of this species was in doubt for many years, and until the last few years, most marine mammal biologists classified all bottlenose dolphins as *T. truncatus*. Now, the Indo-Pacific bottlenose dolphin is known to be distinct, based on concordance in genetics, osteology, and external morphology. Recent genetic studies have even suggested that dolphins currently recognized as *T. aduncus* off southern Africa might be a third species in the genus *Tursiops*. Future studies will be needed to confirm this. In the past few years, some doubt as to the specific identity of the well-studied bottlenose dolphins in Shark Bay, Australia, has also been raised.
Species characteristics Indo-Pacific bottlenose dolphins look very similar to common bottlenose dolphins, with a relatively robust body, moderate-length beak, and tall falcate dorsal fin. However, they tend to be somewhat more slender than common bottlenose dolphins, and the beak is relatively longer and more slender. Also, the melon tends to be slightly less convex. The flippers are typically dolphin-shaped—recurved with acutely-rounded tips. The dorsal fin is tall and slightly falcate (generally less so than in the common bottlenose); it is relatively larger and more wide-based than in the common bottlenose. The flukes are wide and possess a median notch.

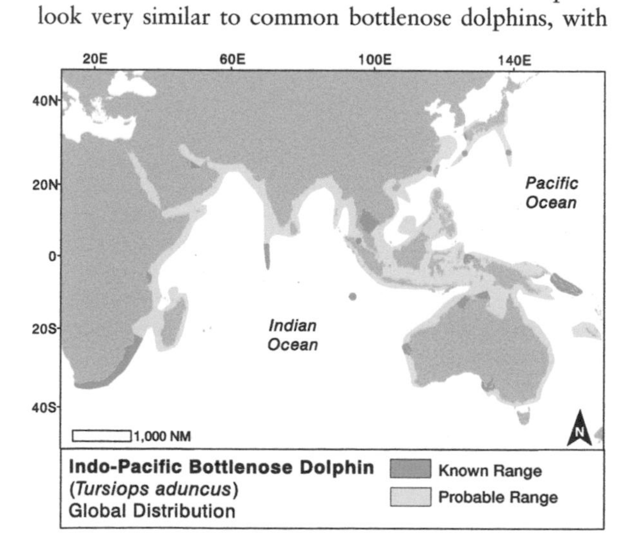

Coloration (although variable) tends to be somewhat lighter than in most common bottlenose dolphins. The belly is generally off-white to pale gray, and tends to grade smoothly to darker gray on the lateral and dorsal surfaces. The moderate to dark-gray cape is generally more distinct, and extends back onto the tail stock. There is usually a light spinal blaze extending to below the dorsal fin (this may be quite prominent in some animals). The most distinctive feature is generally the presence of prominent black spots or flecks on the bellies of adults of this species (these are very rarely present on common bottlenose dolphins). However, not all Indo-Pacific bottlenose dolphins necessarily have noticeable ventral spotting. In particular, young animals appear to lack ventral spots. There is also often a dark ring around the eye. The delphinid "bridle," consisting of a blowhole and eye stripe is also present.

The teeth number 21–29 in each half of the upper and lower jaws. The teeth are a bit more slender than those of common bottlenose dolphins. Although maximum size is geographically variable, Indo-Pacific bottlenose dolphins grow to lengths of about 2.7 m and 230 kg. Males tend to be somewhat larger than females. Length at birth is about 85–112 cm.
Recognizable geographic forms None.
Can be confused with This species can be difficult to distinguish from the larger common bottlenose dolphin in sightings at sea, and the two do overlap in distribution

In South Africa, the type locality for the species, Indo-Pacific bottlenose dolphins have been studied extensively. PHOTO: L. EDWARDS

Although they generally appear to be less acrobatic than their congeners, Indo-Pacific bottlenose dolphins nevertheless are capable of some impressive leaps. Bangladesh. PHOTO: R. MANSUR MOWGLI

The dorsal fin tends to be more wide-based and slightly less falcate in Indo-Pacific bottlenose dolphins than in common bottlenoses. Bangladesh. PHOTO: R. MANSUR MOWGLI

Indo-Pacific bottlenose dolphins can be challenging to distinguish from common bottlenose dolphins, and they do overlap in distribution in many areas, so a suite of characters may be needed to identify them. Bangladesh. PHOTO: R. MANSUR MOWGLI

quite extensively in the Indo-Pacific. Multiple characters will be required to distinguish them in areas of overlap. The best field characters to look for are overall size and robustness, the length and slenderness of the beak, extent of a spinal blaze, and the presence/absence of dark spots or flecks on the belly. Attention to the dorsal fin may also be useful to experienced observers (larger and wider-based in *aduncus* and relatively smaller and more slender in *truncatus*). The Indo-Pacific species may also have a slightly brownish tinge to its cape, when seen in bright sunlight.

Distribution Indo-Pacific bottlenose dolphins are found only in the warm-temperate to tropical Indo-Pacific, though they nearly reach the cool temperate region (for instance, off central Honshu, Japan). They range from South Africa in the west to eastern Australia and the Solomon Islands/New Caledonia in the east. They are found throughout the islands and peninsulas of the Indo-Malay archipelago, and the distribution extends into some enclosed bodies of water like the Gulf of Thailand, Red Sea, and Persian Gulf. They occur almost exclusively over the continental shelf, mostly in shallow coastal and inshore waters. There are also populations around some oceanic Indo-Pacific island groups (e.g., the Ryukyu and Amami Islands of southern Japan, the Maldives, and the Cocos (Keeling) Islands of the southern Indian Ocean).

Ecology and behavior Indo-Pacific bottlenose dolphins have not been studied as well as their congeners, but some detailed studies have been done in South Africa, Australia, and recently in Japanese waters. These animals occur in groups ranging in size up to the low hundreds, but groups of less than 20 are much more common. In most areas, they do not appear to be avid bowriders. They sometimes occur in mixed groups with common bottlenose dolphins and other delphinid species. Off western Australia, where Indo-Pacific bottlenose dolphins come into shallow beaches to be fed and touched by humans, this species' behavior has been intensively studied. There, male Indo-Pacific bottlenose dolphins often band together into "coalitions" to garner young females from the group. The social structure is characterized by a fission-fusion type of society, with great fluidity in group membership. The dol-

Indo-Pacific bottlenose dolphins reach the northern limits of their range in Japanese waters, but even such cooler-water populations show the species' ID characters. Ogasawara Island. PHOTO: H. MINAKUCHI

One of the most useful characteristics of the Indo-Pacific bottlenose dolphin is spotting on the belly (sometimes heavy, as in the animal in the foreground). Although not often visible on a normal surfacing, when seen, it can help confirm an ID. Ogasawara Island. PHOTO: H. MINAKUCHI

phins in Shark Bay have also been found to use sponges as foraging tools, and there appears to be cultural transmission of this behavior from mother to offspring.

Although reproductive activity occurs throughout the year, where it has been studied, breeding peaks in spring and summer months. The gestation period is about 12 months, and calves are weaned after 18–24 months of lactation. Ventral spotting develops at around the time of sexual maturity, which occurs for both sexes at around 2.30–2.35 m in length. Individuals may be 12 years or older at the time of sexual maturity.

Sharks often prey on these animals, at least in areas where they have been well studied, such as South Africa and eastern and western Australia.

Feeding and prey Feeding is on a large variety of schooling, demersal and reef fishes, as well as cephalopods. Most prey items are less than 20 cm in length. Off eastern Australia, dolphins feed behind trawlers, often in association with sympatric Australian humpback dolphins.

Threats and status Some Indo-Pacific bottlenose dolphins are taken in the small cetacean fisheries of Sri Lanka, the Solomon Islands, and possibly Indonesia as well. Live-captures for oceanarium display have taken place in Taiwanese, Solomon, and Indonesian waters in recent years, and the aquarium industry's preference for this as a captive display species makes it vulnerable to depletion from such catches. Until it was outlawed in 1990, this species was involved in a large-scale drive fishery in Taiwan's Penghu Islands. Incidental catches occur in a number of fisheries throughout the range, including gillnets and purse seines. The largest known of these takes up to 2,000 per year in the Taiwanese driftnet fishery operating in Indonesian waters (this fishery formerly operated in northern Australian waters). In addition, Indo-Pacific bottlenose dolphins are killed in anti-shark gillnets in South Africa and Australia. Since this is a coastal species, it is subjected to a number of other human threats—including habitat destruction/degradation, vessel collisions, and environmental contamination. Indo-Pacific bottlenose dolphins interact with fisheries throughout their range, and are often seen following behind and feeding on fish from trawlers. Few estimates of abundance have been made. Although the species is not considered rare, they do tend to occur in small, isolated populations. Undoubtedly, some of these populations have been depleted by human activities. There are estimated to be about 900 bottlenose dolphins off Natal, South Africa, and most of these are probably *T. aduncus* (the rest are *T. truncatus*). Other local estimates of abundance are: 380 in Japan; >600 in Shark Bay, Australia; 700–1,000 off Point Lookout, Australia; 334 in Moreton Bay, Australia; 136–179 in Zanzibar; >24 in southern Taiwan; and 44 in the northeastern Philippines.

IUCN status Data Deficient.

Pantropical Spotted Dolphin—*Stenella attenuata*

(Gray, 1846)

Offshore Adult

Coastal Adult

Recently-used synonyms *Stenella graffmani, Stenella dubia.*
Common names En.–pantropical spotted dolphin; Sp.–*estenela moteada, delfín manchado pantropical;* Fr.–*dauphin tacheté pantropical.*
Taxonomic information Cetartiodactyla, Cetacea, Odontoceti, Delphinidae. Two subspecies are recognized: *S. a. attenuata* in oceanic tropical waters worldwide, and *S. a. graffmani* in the coastal waters of the eastern tropical Pacific. Recent genetic work suggests that the genus *Stenella* is actually paraphyletic (not a natural grouping), and it is likely that it will be restructured in coming years. Dolphins that appear to be hybrids between pantropical spotted and spinner dolphins have been observed off the Fernando de Noronha archipelago of Brazil.

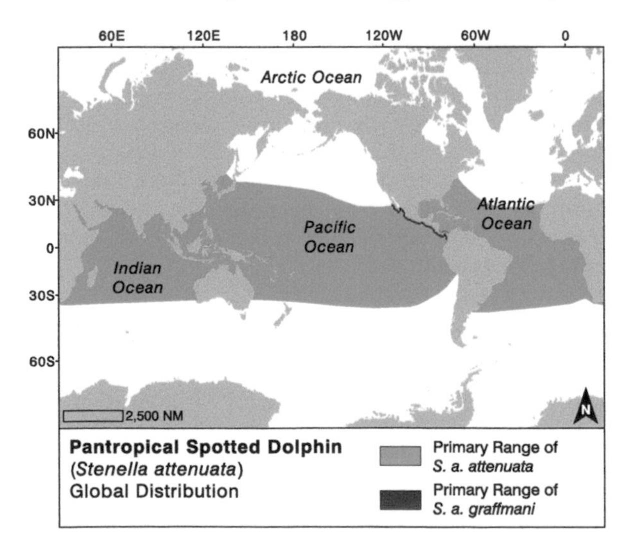

Species characteristics Pantropical spotted dolphins (sometimes simply called "spotters") are generally fairly slender, streamlined animals. However, the coastal form in the eastern tropical Pacific is moderately robust. They have a long, slender beak that is separated from the melon by a distinct crease. The dorsal fin is very narrow, falcate, and usually slightly rounded at the tip, and the flippers are very slender and strongly recurved. There may be a small post-anal protuberance (especially in adult males). The flukes are unremarkable.

The most distinctive color pattern component of this species is the dark dorsal cape, which is high above the flipper and sweeps very low on the side below the dorsal fin. Although unspotted at birth, by adulthood spotted dolphins have varying degrees of white mottling on the dark cape. The tail stock, posterior to the up-sweep of the cape, is often streaked with light extensions of the ventral field. The spotting ranges from very slight (or even non-existent) in offshore animals to heavy enough to obliterate the cape in coastal dolphins. The lower sides and belly of adults are gray and the lips and beak tip tend to be brilliant white. A dark gray band encircles the eye, and continues forward to the apex of the melon as a

Though the degree of spotting is highly variable in the pantropical spotted dolphin, younger animals usually have dark spots on the belly and lower sides of the head. Guadeloupe. PHOTO: L. BOUVERET

Larger adults of the pantropical spotted dolphin have white lips and beak tips, and these may flush pink in animals that are overheated. Guadeloupe. PHOTO: L. BOUVERET

Anomalously white individuals may appear in any species. This pantropical spotted dolphin is probably an albino, although there are other conditions that can cause such aberration. Eastern tropical Pacific. PHOTO: NOAA FISHERIES, SWFSC

The flippers are small and strongly curved and the dorsal fin tends to be quite falcate and narrow in the pantropical spotted dolphin. PHOTO: J. COTTON

A coastal pantropical spotted dolphin (*S. a. graffmani*), which has such extensive white dorsal spotting that the cape margin has been virtually obliterated. PHOTO: SWFSC/STAR

Among the most acrobatic of all dolphins, pantropical spotted dolphins are well known for their high leaps. India. PHOTO: D. SUTARIA

The best ID feature of the pantropical spotted dolphin is probably the dark cape, which dips very low on the side well ahead of the dorsal fin. That, especially when combined with the belly spotting and white lips, makes these animals unmistakable. Bangladesh. PHOTO: R. MANSUR MOWGLI

Not all pantropical spotted dolphin are spotted. In fact, some populations do not have extensive spotting, and the shape of the cape is a much better feature than spotting to use in identification. Guadeloupe. PHOTO: L. BOUVERET

Adult pantropical spotted dolphins have a grayish belly in most areas (dark spotting fuses and lightens to form the continuous gray). Notice the remora on the flipper of this dolphin. Eastern tropical Pacific. PHOTO: J. COTTON

Pantropical spotted dolphins have much variability in their color patterns, but the cape always dips low on the side just ahead of the dorsal fin. Gabon. PHOTO: C. WEIR

narrow eye stripe. This and the blowhole stripe complete the characteristic delphinid "bridle." There also is a dark gape-to-flipper stripe. Younger animals may have extensive dark ventral spotting as well.

In each tooth row are 34–48 slender, sharply-pointed teeth. At birth, pantropical spotted dolphins are about 80–85 cm long. Adults range in length from 1.6–2.4 m (females) or 1.6–2.6 m (males), depending on the population. Maximum recorded weight is 119 kg.

Neonate—Neonates are about 1/4–1/3 of adult size; with a two-part color pattern (dark gray above, white below), with no spotting.

Two-tone—Two-tone calves are about 1/3–3/4 adult size; with a two-tone gray color pattern and no spotting.

Speckled—Speckled juveniles are at least 3/4 of adult size; the two-tone color pattern is still evident, with dark ventral spots developing.

Mottled—Mottled adults/subadults have dark ventral spotting (beginning to converge and fuse) and light dorsal spotting.

Fused—The two-tone color pattern is still evident, usually with light dorsal spotting and the ventral spots, mostly fused into a solid gray belly; the beak tip and lips are variably white.

Recognizable geographic forms

***Offshore spotted dolphin* (S. a. attenuata)**—Offshore spotted dolphins are slightly smaller and more slender than the coastal form, and the beak tends to be more slender as well. Dorsal spotting is much less dense, and in some populations can be virtually nonexistent in adults. Adults range in length from 1.6 to 2.4 m long, and tend to weigh somewhat less than in the coastal form.

***Coastal spotted dolphin* (S. a. graffmani)**—In coastal waters of the eastern tropical Pacific, a coastal form of the pantropical spotted dolphin exists. It is larger and stockier, with a thicker beak than the offshore form. Spotting tends to

When viewed underwater like this, the bright white lips and beak tips of adult pantropical spotted dolphins may glow like a flashlight through the bluish water. Hawai'i. PHOTO: M. NOLAN

be much more extensive on the coastal form, and in some adults the white dorsal spots can be so dense as to completely obscure the cape. Coastal spotted dolphins range from 1.8–2.6 m long, and can weigh up to at least 119 kg.

Can be confused with Pantropical spotted dolphins can be confused with several other long-beaked oceanic dolphins. Spinner dolphins, which share much of the range, can be distinguished by differences in dorsal fin shape, beak length, and color pattern. Atlantic spotted dolphins can look similar, but attention to head shape, dorsal fin shape, and color pattern details will allow correct identification. In addition to Atlantic spotted dolphins, both bottlenose and humpback dolphins can also be spotted (generally most extensively on the belly), but will be distinguishable by attention to differences in body shape and size. In fact, presence/absence of spotting is not really a very good character to use in identification of *S. attenuata,* as the species can range from heavily spotted to unspotted. The shape of the dorsal cape is a better field character.

Distribution Pantropical spotted dolphins are found in the Pacific, Atlantic, and Indian oceans. They are also found in the Persian Gulf and Red Sea. These are mostly creatures of offshore tropical zones, although they do occur close to shore in some areas where deep water approaches the coast (e.g., off the west coast of central America and Mexico, around the Hawaiian islands, off some islands in the Caribbean, off Taiwan, and in the Philippines). As their name implies, these animals are pantropical, found in all oceans between about 40°N and 40°S, although they are much more abundant in the lower latitude portions of their range. In the eastern tropical Pacific, the offshore form primarily inhabits waters with a sharp, shallow thermocline and surface water temperatures of over 25°C.

Deep, blue, oceanic waters are the home of most pantropical spotted dolphin populations. Ogasawara Island, Japan. PHOTO: H. MINAKUCHI

Ecology and behavior Pantropical spotted dolphins are among the most abundant dolphins in the eastern tropical Pacific (ETP) and are the primary species involved in the tuna/dolphin interaction there. In at least the Pacific and Indian oceans, spotted dolphins associate with yellowfin tuna, spinner dolphins, and other oceanic predators; the fishermen take advantage of this association to help them locate and catch tuna more efficiently. They are also among the most common species of cetaceans in portions of the Atlantic and Indian oceans.

School sizes are generally less than 100 for the coastal form, but offshore herds may number in the thousands. They may form schools with some age and sex segregation. These gregarious animals are fast swimmers, often engaging in acrobatics (like breaches and side slaps, but they do not typically spin), and frequently bowriding (except on the tuna fishing grounds of the ETP, where they generally have learned to avoid boats). They may swim at speeds of at least 22 km/hr, and dives of up to 3.4 min in length have been recorded through time/depth recorder studies.

The life history of this species has been thoroughly studied, due to the large number of specimens available through kills in the ETP tuna fishery, as well as drive fisheries of Japan. Gestation lasts about 11.5 months. Sexual maturity is reached at ages of 9–11 years in females, and 12–15 in males. The typical calving interval is 2–3 years, and age at weaning varies strongly by population. Although pantropical spotted dolphins breed year-round, there are two calving peaks in the ETP, one in spring and one in fall.

Some of the developmental variation in the amount of dorsal spotting can be seen on these bowriding pantropical spotted dolphins. Such variation will be apparent in any school of moderate or large size. Guadeloupe. PHOTO: M. MEUNIER

Feeding and prey Offshore spotted dolphins feed largely on small epi- and mesopelagic fishes, squids, and crustaceans that associate with the deep scattering layer (DSL). In some areas, flyingfish are also important prey. The diet of the coastal form is poorly known, but is thought to consist mainly of larger and tougher fishes, perhaps mainly bottom-living species.

Threats and status The heaviest known mortality of this species has been in the ETP purse seine fishery for tuna. Since the interaction was first documented in the late 1960s, millions of spotted dolphins have died in the nets (from 1959 to 1972, about three million offshore spotters were killed). Current mortality in this fishery has been greatly reduced by years of modifications to the fishing practices, fleet changes, and US and international legislation. Current concerns are focused on the potential effects of fishery-related stress, and its role in preventing recovery of the populations. In addition, pantropical spotted dolphins are taken incidentally in a number of other purse seine, gillnet, and trawl fisheries throughout the range. Large direct kills occur sporadically in the Japanese small cetacean drive and harpoon fisheries, and much smaller direct kills have occurred in the dolphin fisheries of the Caribbean, Sri Lanka, Philippines, Indonesia, St. Helena, and the Laccadive and Solomon Islands. Most of these kills have not been adequately monitored and the effects on the populations are usually not known. They have been live-captured in some areas (e.g., Hawai'i), and some individuals have survived for a short time in captivity. Overall, the species is very abundant, perhaps one of the most abundant dolphins in the world. In the eastern tropical Pacific, there were estimated to be 228,000 coastal spotters in 2000. The northeastern offshore spotted dolphin (the stock most affected by the ETP tuna fishery) numbered about 647,000 in 2000. This stock is not showing clear signs of recovery, despite the dramatic decline in mortality in recent years. The western/southern offshore stock (which is less affected by the fishery) numbered about 800,000. About 440,000 inhabited Japanese waters in the early 1990s. There were estimated to be over 47,000 in the northern Gulf of Mexico. Clearly, the species is in no danger of extinction on a global scale.

IUCN status Least Concern.

Atlantic Spotted Dolphin—*Stenella frontalis*

(G. Cuvier, 1829)

Recently-used synonyms *Stenella plagiodon.*

Common names En.–Atlantic spotted dolphin; Sp.–*delfín pintado;* Fr.–*dauphin tacheté de l'Atlantique.*

Taxonomic information Cetartiodactyla, Cetacea, Odontoceti, Delphinidae. There is some indication that this species may be closely related to the Indo-Pacific bottlenose dolphin. If so, then the genus-level taxonomy of these two groups may need revision. There is known to be some geographic variation in this species, and evidence for a distinct geographic form in the more temperate waters of the North Atlantic is accumulating.

Species characteristics The Atlantic spotted dolphin, in many ways, resembles the Indo-Pacific bottlenose dolphin more than it does the pantropical spotted dolphin. In body shape, it is somewhat intermediate between the two, neither very robust, nor very slender. It has a moderately-long, but somewhat-thick beak (strongly reminiscent of that of the bottlenose dolphin). There is a distinct crease between the melon and beak. The dorsal fin is tall and falcate, and the flippers are recurved and of the typical dolphin shape. The flukes are unremarkable.

There is much developmental variation in the color pattern. Atlantic spotted dolphins begin life with unspotted background coloration. Young animals look much like slender bottlenose dolphins, with a dark cape, light gray sides, and white belly. The characteristic delphinid "bridle" (consisting of an eye stripe and blowhole stripe) is present. There is generally a distinct spinal blaze (which is variable in its development, but more strongly developed in older individuals). Development of larger spots on both dorsal and ventral surfaces progresses as the animal ages; some individuals become so heavily spotted that the cape margin and spinal blaze are totally obliterated. However, in some populations, adults are essentially unspotted

Two young Atlantic spotted dolphins leap off the bow of a research vessel in the western Atlantic. These dolphins are avid bowriders in most parts of their range. PHOTO: P. OLSON

(these are generally in offshore and/or more temperate areas). The white spots are generally larger than the small ones typical of the pantropical spotted dolphin.

There are 32–42 relatively-stout teeth in each upper tooth row and 30–40 in each lower row. Adults are up to 2.3 m long and 143 kg in weight. Newborn Atlantic spotted dolphins are 0.8–1.2 m long.

Neonate—Neonates are about 1/4–1/3 of adult size; with a two-tone color pattern (dark gray above, white below), with no spotting.

Calf—Calves are 1/3–3/4 adult size; with a two-tone color pattern, and possibly a few dark ventral spots.

Juvenile—Juveniles are at least 3/4 of adult size; the two-tone color pattern is still evident, with some dark ventral and light dorsal spots developing.

Adult—Adults have a mottled (dark ventral spotting and light dorsal spotting) to fused (same as above, with the ventral spots fusing) color pattern, often with a white-tipped beak.

Recognizable geographic forms Although two distinct geographic forms of the Atlantic spotted dolphin have been distinguished, primarily from osteological and color pattern data, the exact ranges and extent of overlap in their external appearance are not adequately known. The larger, more heavily-spotted form is mainly found over the continental shelf in warmer waters (the largest specimens known are from the continental shelf of North America), while a small, more lightly-spotted form occurs in more oceanic areas further north (and around offshore islands, like the Azores).

Can be confused with Atlantic spotted dolphins can be most easily confused with bottlenose dolphins (although they do not overlap in distribution with the Indo-Pacific species). The differences in size and robustness are good clues, but may require a trained eye to distinguish them in some at-sea sightings. Heavy spotting is generally a good characteristic for Atlantic spotted dolphins; however, some may be nearly unspotted, and some bottlenose dolphins may have spotting and blotches on the belly and sides. Pantropical spotted dolphins may also be difficult to distinguish from Atlantic spotted dolphins, but attention to body robustness, beak and dorsal fin shape, and color pattern differences will allow them to be separated.

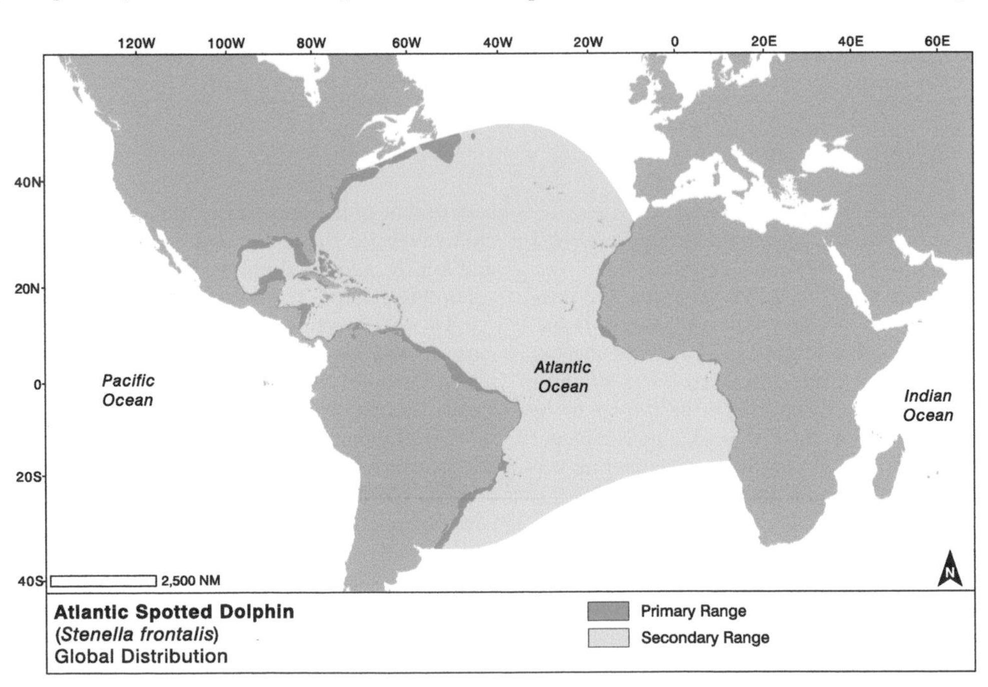

The body shape of the Atlantic spotted dolphin tends to be moderately robust and can appear similar to that of bottlenose dolphins, but the heavy spotting gives them away. Guadeloupe. PHOTO: L. BOUVERET

From this perspective, the resemblance of this species to the Indo-Pacific bottlenose dolphin is apparent. Guadeloupe. PHOTO: L. BOUVERET

This juvenile Atlantic spotted dolphin appears almost entirely unspotted, but shows the dorsal cape and spinal blaze that characterize the species. Canary Islands. PHOTO: D. GAULTIERI

Coloration varies dramatically among different populations of Atlantic spotted dolphins, with differences in spotting development, expression of the dorsal cape, etc. Canary Islands. PHOTO: D. GAULTIERI

This Atlantic spotted dolphin has a very well-developed spinal blaze, although its shape is rather unusual. PHOTO: P. OLSON

The Atlantic spotted dolphin may appear very similar to the Indo-Pacific bottlenose dolphin. Fortunately, the two occupy different ocean basins, so there will be little cause for confusion. Gulf of Mexico. PHOTO: V. BEAVER

This is an offshore form of the Atlantic spotted dolphin. In these animals, there is more exaggerated division of the color pattern into three parts, with a distinct whitish belly. PHOTO: L. STEINER/ WHALE WATCH AZORES

This Atlantic spotted dolphin has an anomalous piebald pattern, and might not have been identifiable to species were it not in a school with normal-colored individuals. Brazil. PHOTO: COURTESY M. BOROBIA

Atlantic spotted dolphins typically have a combination of dark spots on the lighter, lower parts of their body, and light spots on the dark dorsal cape. Bahamas. PHOTO: H. MINAKUCHI

Young Atlantic spotted dolphins may have only very slight spotting development. Bahamas. PHOTO: R. KIEFNER

Atlantic spotted dolphins often play with kelp or other objects in their environment. Bahamas. PHOTO: H. MINAKUCHI

Distribution This species is found only in the Atlantic Ocean, from southern Brazil to New England in the west, and to the coast of Africa in the east (the exact limits off West Africa are not well-known). They are not found in the Mediterranean Sea. Their tropical to warm-temperate distribution is mostly over the outer continental shelf and upper continental slope, but they also inhabit some deep oceanic waters. They occur near some oceanic island groups, such as the Azores. In the Bahamas, they spend much of their time over shallow (6–12 m) sand flats.

Ecology and behavior Small to moderate-sized groups, generally of less than 50 individuals, are characteristic of the Atlantic spotted dolphin. Coastal groups usually consist of 5–15 animals. These are acrobatic animals and they are known to be avid bowriders. They are mostly shallow divers (most dives apparently <10 m), but dives of 40–60 m and up to 6 min have been recorded. Dolphins in the Bahamas have been observed to capture fish hiding in the soft, sandy bottom by sticking their beaks into the sand.

Atlantic spotted dolphins in the clear, warm waters off the Bahamas allow people to swim with them, and this population has become a major tourist draw. People come from around the world to interact with these dolphins in their natural habitat. Studies there show that Atlantic spotted dolphins have a fluid group structure, like that of bottlenose and other small dolphins. Groups are often segregated by age and sex. They may interact with common bottlenose dolphins (where this has been observed, the bottlenose dolphins often act aggressively, harassing and attacking the smaller Atlantic spotters). Sharks are known predators.

There is not much known of the species' life history, but more tropical populations would be expected to have a protracted breeding season. Age at sexual maturity is thought to occur somewhere between 8 and 15 years in females (when females are in the mottled color phase). The calving interval ranges from 1–5 years, and averages about 3 years. Nursing of calves can last for up to 5 years. Longevity is not well known.

Feeding and prey A wide variety of epi- and meso-pelagic fishes and squids, as well as benthic invertebrates, are taken by this species. There are known to be some regional differences in the diet.

Threats and status Overall, this species is not known to suffer greatly at the hands of humans. Incidental catches in fisheries are known for several areas of the range (Brazil, the Caribbean, off the east coast of the United States, and Mauritania). No direct killing is known, other than occasional catches in the Caribbean Lesser Antillean dolphin fisheries, and maybe also off West Africa. The only abundance estimate available is for the northern Gulf of Mexico, of about 3,200 animals. The species is not in any extinction danger on a global scale.

IUCN status Data Deficient.

Spinner Dolphin—*Stenella longirostris*

(Gray, 1828)

This may be a dwarf spinner dolphin, a subspecies apparently found only in Southeast Asia and tropical waters of northern Australia. Bangladesh. PHOTO: R. MANSUR MOWGLI

Recently-used synonyms *Stenella roseiventris, Stenella microps.*
Common names En.–spinner dolphin; Sp.-*estenela giradora, delfín girador;* Fr.–*dauphin longirostre.*
Taxonomic information Cetartiodactyla, Cetacea, Odontoceti, Delphinidae. Four subspecies are currently recognized: *S. l. longirostris* in oceanic tropical waters worldwide, *S. l. orientalis* in the offshore eastern tropical Pacific, *S. l. centroamericana* in the coastal eastern tropical Pacific, and *S. l. roseiventris* in Southeast Asia and northern Australia. Recent genetic work suggests that the genus *Stenella* is actually paraphyletic, and it is likely that it will be restructured in coming years. Dolphins that appear to be hybrids between this species (with pantropical spotted and Clymene dolphins) have been observed off the Fernando de Noronha archipelago of Brazil.
Species characteristics The spinner dolphin is a slender species, with an extremely long, thin beak. Also, the head is very slender at the apex of the melon. The flippers are slender and recurved. The dorsal fin usually ranges from slightly falcate to erect and triangular. In adult males of some populations, however, the dorsal fin can be so canted forward that it looks as if it were stuck on backwards. Also, the tail stock may become very deepened, with an enlarged post-anal hump (some times called a keel) of connective tissue, and the tips of the flukes may curl upwards somewhat. These three features are correlated in their development in adult males.

Spinners generally have dark eye-to-flipper stripes and dark lips and beak tips. The characteristic delphinid "bridle" (consisting of an eye stripe and blowhole stripe) is visible in at least Gray's spinner dolphin. Individuals of most spinner dolphin populations have a tripartite (three-part) color pattern (dark gray cape, light gray sides, and white belly) and only minor differences in appearance of males and females. This includes Gray's spinner dolphins *(S. l. longirostri)* and dwarf spinners *(S. l. roseiventris).* In these animals, the upper beak is dark, and most of the lower jaw is white; the beak tip tends to be dark. In the eastern tropical Pacific (ETP), three other forms with

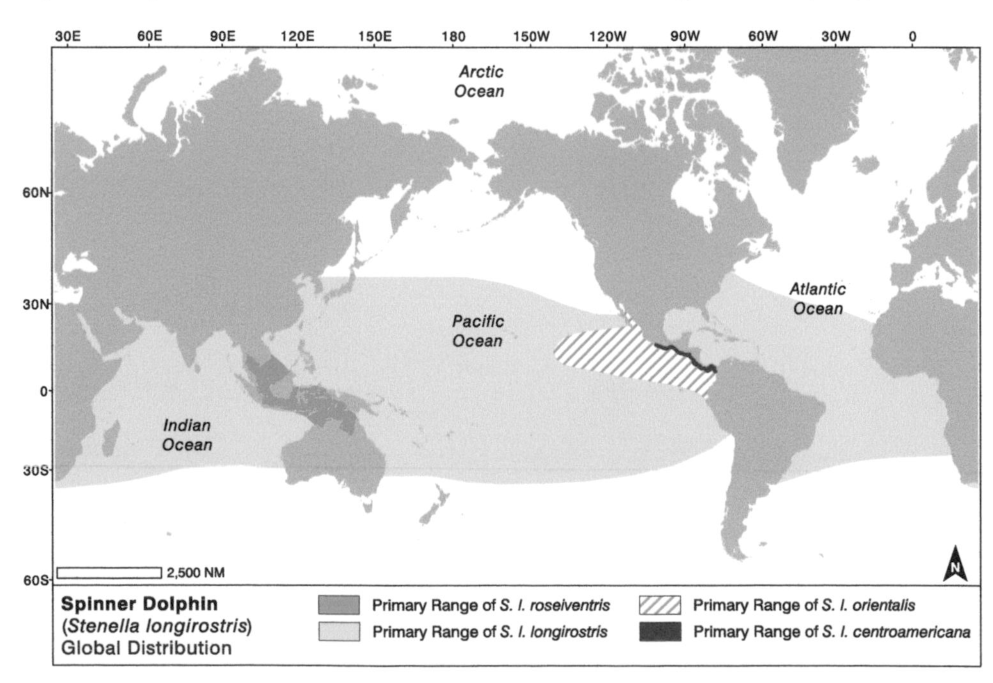

This is an eastern spinner dolphin, a form restricted to the eastern tropical Pacific, and identifiable by its monotone color pattern and extreme sexual dimorphism (in adult males, the dorsal fin is strongly forward-canted, the fluke tips are upturned, and the tail stock has a large ventral keel). PHOTO: R. L. PITMAN

very different color patterns are found (see Recognizable Geographic Forms below). Geographical variation in spinner dolphins has not been as well described for most areas outside the eastern Pacific.

In spinner dolphins, there are 40–62 pairs of very fine, pointed teeth in each jaw, with some differences between subspecies. This is more than in any other cetacean species, except the franciscana and some long-beaked common dolphins.

Newborn spinner dolphins are about 75-80 cm long; adults reach 2.0 m (females) and 2.35 m (males). There is much geographical variation; for instance dwarf spinners reach maximum lengths of only about 1.58 m. Spinners are known to reach weights of at least 82 kg (although dwarf spinners can be mature at 25 kg). Males appear to be somewhat larger than females in all forms.

Neonate—About 1/2 of adult length; color pattern muted.
Calf—Calves are generally less than 3/4 adult size; head relatively large; with a muted color pattern.
Subadult—Length generally greater than 3/4 adult size; body more slender than that of adults.
Adult female—Shows the typical color pattern for their respective geographical form (see below); dorsal fin more erect than in subadults.
Adult male—Adult males show the typical color pattern for their respective geographical form; they have more erect or canted dorsal fins, more deepened tail stocks, and enlarged post-anal humps (the differences may be slight [Gray's/dwarf] or moderate [whitebelly], and reach great extremes in eastern and central American spinners).

Recognizable geographic forms

Gray's spinner dolphin (S. l. longirostris)—This is the typical form of spinner dolphin found in most areas of the world (with the exception of some waters of tropical Asia and the eastern tropical Pacific). They have the tripartite color pattern described above, with a falcate to triangular dorsal fin, small to non-existent post-anal hump, and relatively small dorsal fin and flippers. Tooth counts are generally in the range of 40–60 per row. Adult females are about 1.39–2.04 m long, and males are 1.60–2.08 m.

In the eastern tropical Pacific, there are two endemic forms of spinner dolphins that do not have the typical tripartite color pattern, but instead are essentially monotone "battleship" gray. Central American spinner. PHOTO: C. OEDEKOVEN

Eastern spinner dolphin (S. l. orientalis)—Eastern spinner dolphins are found only in the waters of the eastern tropical Pacific Ocean, mostly offshore. They have a monotone steel gray color pattern, with white only as patches around the genital area and axillae. They have exaggerated sexual dimorphism, with males possessing strongly-canted dorsal fins, medium to large post-anal humps, and upturned fluke tips. The beak is relatively long. Eastern spinners are relatively small, with adult females 1.52–1.93 m (mean = 1.71 m) and adult males 1.60–1.99 m in length (mean = 1.76 m).

Central American spinner dolphin (S. l. centroamericana)—Central American spinner dolphins, previously called Costa Rican spinners, are poorly known. This subspecies is found only in coastal waters of the eastern tropical Pacific.

In Hawaiian waters, and elsewhere in the world, coastal spinner dolphins move into shallow, near-shore waters during the daytime to rest. PHOTO: H. MINAKUCHI

They have a monotone color pattern similar to that of the eastern spinner, although apparently lacking the white ventral patches. There is pronounced sexual dimorphism, with adult males possessing strongly-canted dorsal fins, large post-anal humps, and up-turned fluke tips. Central American spinners are the largest subspecies, with adult females 1.75–2.11 m and adult males reaching over 2.16 m in length. The beak is even longer than in eastern spinners.

Whitebelly spinner dolphin—A third type of spinner dolphin in the eastern tropical Pacific, called the whitebelly spinner, appears to be an intergrade between *S. l. longirostris* and *S. l. orientalis.* They occur in large schools in more offshore regions of the ETP. Whitebelly spinners are more robust, with a two-part color pattern and less exaggerated sexual dimorphism than the other forms in the ETP Adult females are 1.57–1.98 m in length (mean = 1.76 m), and adult males 1.60–2.35 m (mean = 1.80 m).

***Dwarf spinner dolphin* (S. l. roseiventris)**—There is a dwarf form of the spinner dolphin, which is found in waters of Southeast Asia and northern Australia. It has a tripartite color pattern, similar to that of Gray's spinner dolphin. There is an erect to falcate dorsal fin, which is proportionately large (9.5–13% of total length). The flippers are also relatively large. Sexual dimorphism is reduced, and tooth counts are much lower than in the other forms (about 41–52 teeth per row). Dwarf spinners (as the name implies) are much smaller than others of the species, with adult females ranging from 1.38–1.45 m and adult males 1.29–1.58 m (only 2/3–3/4 the length of other spinner subspecies).

A long, slender beak, tripartite color pattern, and relatively triangular dorsal fin characterize spinner dolphins in most areas. PHOTO: M. NOLAN

Can be confused with From a distance, other long-beaked oceanic dolphins can look like spinner dolphins. Spinners are most likely to be confused with Clymene dolphins in the Atlantic, but careful attention to color pattern differences and head and body shape differences will allow them to be distinguished. In any ocean, they may be confused with pantropical spotted, striped and common dolphins (especially the long-beaked species, *Delphinus capensis*). Although the general body shape is very similar, careful attention to relative beak length, dorsal fin shape, and color patterns should allow for accurate identification.

Distribution The range of the spinner dolphin is pantropical and nearly identical to that of the pantropical spotted dolphin, encompassing oceanic tropical and subtropical zones in both hemispheres. Limits are near 40°N and 40°S. Spinner dolphins are found in the Red Sea and Persian Gulf, but not the Mediterranean Sea. Unlike pantropical spotted dolphins, spinners often rest in shallow, coastal waters and may spend their days in sandy-bottomed bays of oceanic islands and coral atolls. However, much of their range is oceanic, especially in the eastern tropical Pacific. The dwarf spinner occurs almost exclusively in shallow waters in southeast Asia and northern Australia, and apparently feeds over shallow reefs (it is at least partially sympatric with Gray's spinner dolphin).

Ecology and behavior The spinner dolphin is named for its habit of leaping from the water and spinning up to seven times on its long axis, before falling back to the water. This behavior may be repeated many times in "bouts." This is one of the most aerial of all dolphins and these animals often perform breaches, side-slaps, fluke-slaps, flipper-slaps, and spins (on the long axis). In most areas they are active bowriders (the main exception is the ETP, where these dolphins have been harassed by fishermen, who encircle them to catch tuna swimming below). Herd sizes range from less than 50 up to several thousand. Associations with spotted dolphins are common in the ETP, and they occasionally associate with several other marine mammal species. They may dive to 600 m or deeper in pursuit of prey, although most feeding is probably done in shallower waters. In the ETP, at least, they range over vast distances of open ocean in search of suitable patches of prey.

The behavior of Hawaiian spinner dolphins has been quite well studied. Moderate-sized groups of dolphins move into shallow sandy bays to rest in the daytime and then move offshore in the late afternoon/evening for nighttime feeding (mostly near dusk and dawn) in nearby continental slope and oceanic waters. Dolphins are highly aerial during the ascent from rest to foraging and often engage in "zig-zag" swimming (characterized by a great deal of stereotyped aerial behavior), thought to test the readiness of the school to move offshore. Societies in the main Hawaiian Islands are characterized by a fission-fusion system, in which group associations are very fluid. Similar behavior patterns have been noted at other island groups in the Atlantic and Indian oceans. Interestingly, at Midway Atoll (and probably at other islands and atolls in the Northwestern Hawaiian Islands), spinners have a very different social structure, with stable groups of long-term associates.

Life history has been studied much more extensively than for most other dolphin species, largely due to the specimens available from the tuna fishery. Gestation lasts 10 months, nursing lasts 1–2 years, and the calving interval is about 3 years. Sexual maturity is attained at ages of about 4–7 for females and about 7–10 for males. Breeding is diffusely seasonal, and calving peaks in different populations range from late spring to fall. The

Spinner dolphins almost invariably have dark lips and beak tips, which helps to distinguish them from pantropical spotted dolphins, with which they overlap and often associate. Ningaloo Reef, Australia. PHOTO: R. L. PITMAN

A dwarf spinner dolphin from waters of the southern South China Sea. These animals are much smaller than other subspecies of spinner dolphins, and have relatively large appendages. PHOTO: T. AQUINO

Most spinner dolphins worldwide have a three-part color pattern of a dark gray cape, white belly, and intermediate gray sides. Ogasawara Island, Japan. PHOTO: H. MINAKUCHI

Although it varies quite a bit in shape, a relatively triangular dorsal fin is diagnostic of the spinner dolphin. Hawai'i. PHOTO: M. NOLAN

Some other dolphins also spin, but none do so with as much frequency and vigor as the spinner dolphin. Hawai'i. PHOTO: M. NOLAN

different subspecies have evolved various degrees of male/male competition for females (probably ranging from overt competition to sperm competition). As a result, the mating system apparently ranges from more polygynous in the eastern (and probably central American) spinner to more polygynandrous in whitebelly spinners.

Feeding and prey Most spinner dolphins feed predominantly at night, on small (<20 cm) midwater fishes of many different families (including myctophids), squids, and sergestid shrimps, and rest during much of the day. Their association with spotted dolphins and yellowfin tuna results in their entanglement in tuna purse seines in the ETP. Dwarf spinner dolphins are exceptional, however; they presumably feed during daylight hours on small, reef-associated organisms (benthic reef fishes and invertebrates).

Threats and status This is the second-most important species of dolphin involved in the ETP tuna purse seine fishery (after the pantropical spotted dolphin). The eastern spinner dolphin population is estimated to have been reduced to less than 1/3 of its original size by the tuna fishery kill. Although current mortality is greatly reduced, the population is not recovering at its expected potential, and fishery-related stress may be at least partially responsible for this. Other incidental kills occur throughout the range in a number of different fisheries, including driftnets, purse seines, and trawls. Dwarf spinners are caught incidentally in shrimp trawls in the Gulf of Thailand. In some cases, human use of by-caught dolphins has led to directed fisheries. Direct kills occur in several areas, including the Caribbean, Sri Lanka, the Philippines, Indonesia, and occasionally Japan. They may also be taken in West Africa. Their habit of resting in coastal waters leads to problems of harassment by dolphin-watching boats in a number of areas. Some have been captured for public display, but they generally do not do well in captivity. The spinner dolphin is one of the most abundant dolphins in the world. There were about 800,000 whitebelly spinners in the ETP in 2000. The eastern spinner dolphin (the spinner stock most heavily impacted by the ETP tuna fishery) numbered about 450,000 in 2000, and despite large reductions in the kill, does not appear to be showing evidence of recovery. There are estimated to be over 11,000 in the northern Gulf of Mexico. There are no abundance estimates for the dwarf subspecies, but its limited range and significant incidental kills may have resulted in conservation problems. So, while the species may not be in danger of extinction overall, some populations and subspecies clearly are threatened by human activities.

IUCN status Data Deficient (overall species), Vulnerable (eastern spinner dolphin).

Clymene Dolphin—*Stenella clymene*

(Gray, 1850)

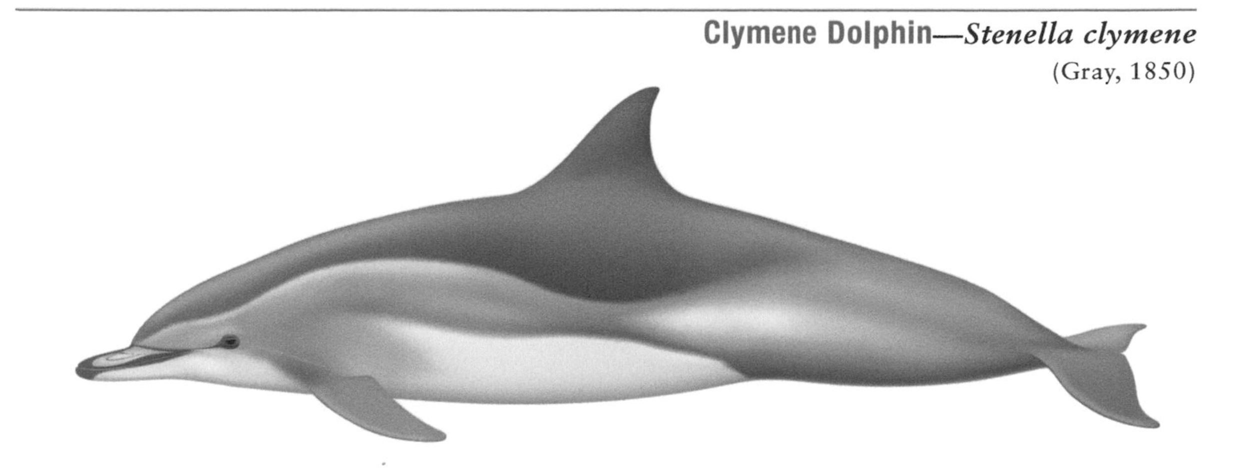

Recently-used synonyms None.

Common names En.–Clymene dolphin, short-snouted spinner dolphin; Sp.-*delfín Clymene;* Fr.-*dauphin de Clymène.*

Taxonomic information Cetartiodactyla, Cetacea, Odontoceti, Delphinidae. Studies using molecular techniques have recently shown that the Clymene dolphin evolved as a hybridization event between spinner and striped dolphins. Recent genetic work suggests that the genus *Stenella* is actually paraphyletic, and it is likely that it will be restructured in coming years. Although first described in 1846, the authorship of the scientific name is more correctly attributed to Gray (1850). Presumed hybrids between Clymene and spinner dolphins have been observed at sea off the Fernando de Noronha archipelago of Brazil.

Species characteristics The Clymene dolphin is externally similar to the spinner dolphin, but is smaller and generally more robust, with a much shorter and stockier beak. There is a distinct crease between the beak and melon. The dorsal fin is erect, and only slightly falcate. Despite these external similarities with the spinner dolphin, skull morphology belies the close relationship of the Clymene dolphin to the striped dolphin. The flippers are smoothly recurved, typical of those of other species in the genus. The flukes are unremarkable.

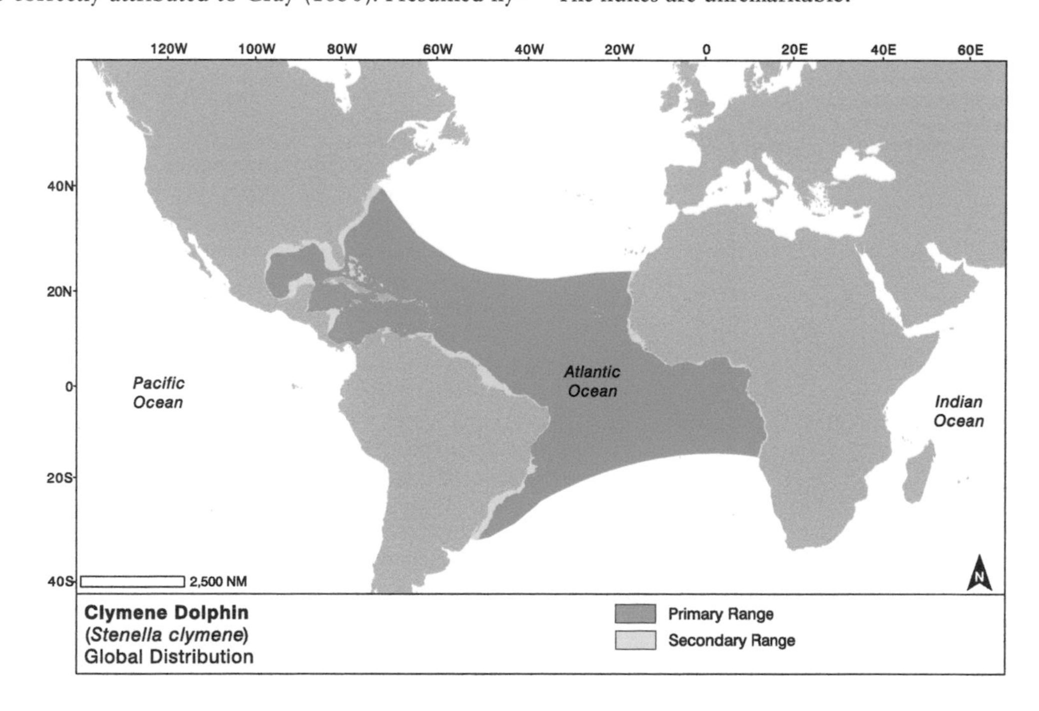

A Clymene dolphin from the Gulf of Mexico. The color pattern is very similar to Gray's spinner dolphin, but the body is shorter and stouter, with a much shorter beak. It generally has a three-part color pattern, but shows lots of individual variation, as evident here. PHOTO: R. L. PITMAN

The Clymene dolphin strongly resembles the spinner dolphin, but is more robust, has a shorter beak, bolder facial stripes, a stronger dip to the cape below the dorsal fin. The two are very closely related. US east coast. PHOTO: C. FAIRFIELD

The Clymene dolphin's range overlaps both those of the spinner and striped dolphins, and it is now understood that this species evolved through extensive hybridization of the latter two species. PHOTO: P. OLSON

A three-part color pattern, with a dark gray cape, light gray sides, and white belly, is characteristic of this species. The cape dips in two places: above the eye, and below the dorsal fin. The beak is mostly light gray, but the lips and beak tip are black. There is also a dark stripe on the top of the beak, from the tip to the apex of the melon, and typically a dark "moustache" marking on the middle of the top of the beak. The eye is also surrounded by black, and a dark gray stripe runs from the eye to the flipper. There may also be an indistinct dark band running diagonally along the flanks, from the anus forward.

Tooth counts are much lower than in spinners: 39–52 teeth per row. The teeth are slender and pointed. Clymene dolphins are the smallest dolphins in the genus *Stenella*. They are known to reach at least 1.97 m (males) and 1.90 m (females) in length and sexual maturity appears to be reached by about 1.7–1.8 m (however, this is based on a very small sample). Neonatal length is unknown, but is less than 1.2 m. This species reaches weights of at least 80 kg.

Recognizable geographic forms None.

Can be confused with Clymene dolphins are most easily confused with spinner dolphins, but are more robust, with shorter, stubbier beaks. Also, the color pattern is slightly different; the two dips in the cape and the dark "moustache" marking on top of the beak will allow Clymene dolphins to be distinguished. The body shape of Clymene dolphins also closely resembles that of the short-beaked common dolphin, as does the color pattern in a superficial way. Common dolphins can best be distinguished by their hourglass pattern, cape that forms a V below the dorsal fin, chin-to-flipper stripe, and absence of a "moustache" on the beak. The striped dolphin also has a similar body shape, but the color pattern is very different, and a decent look should make it easy to distinguish these two species.

Distribution The Clymene dolphin is found only in the tropical and subtropical Atlantic Ocean, including the Caribbean Sea and Gulf of Mexico. This species has a notable warm-water preference, although there are records as far north as New Jersey on the US east coast (associated with the northward extension of the warm Gulf Stream) and as far south as southern Brazil. The limits on the West African coast are not well-known, but extend from at least Angola north to Mauritania. The Clymene dolphin is not known to enter the Mediterranean Sea. This is a deep-water, oceanic species, not often seen near shore (except where deep, oceanic water approaches the coast).

Ecology and behavior Very little is known of the Clymene dolphin's natural history. This is partly due to the long-standing confusion between this and other similar long-beaked tropical delphinids, as well as the species' somewhat-restricted offshore habitat. Schools tend to be smaller than those of spinner dolphins, generally less than 200 animals. In the Gulf of Mexico, most schools are less than about 60–80 individuals. Schools often appear to be segregated by age and sex. They have been reported to associate with common dolphins off West Africa, and with spinner dolphins in the Caribbean.

These quick and agile dolphins are aerially active. They have been reported to spin up to 3–4 revolutions on the long axis before falling back to the water, but the spins are less elaborate and acrobatic than those of the spinner dolphin. Clymene dolphins are avid bowriders, and have been seen playing with seaweed while riding the bow wave of a research vessel in the Gulf of Mexico. Because so few specimens have been examined, almost nothing is known of the life history of this species. Both sexes appear to reach sexual maturity by the time they are about 180 cm in length.

Feeding and prey Very few stomachs have been examined, and even fewer observations of feeding behavior reported in the literature. Clymene dolphins apparently feed on small fish (including myctophids) and squid at moderate depths, presumably mainly at night.

Threats and status Although they are known to be taken by harpoon occasionally in dolphin fisheries in the Caribbean Lesser Antilles, and incidental captures in fishing nets do occur throughout much of the range, the Clymene dolphin is not known to suffer any heavy exploitation at present. The only possible exception may be off the coast of West Africa, where this species is possibly one of several taken in reasonably-large numbers in tuna purse seines in the Gulf of Guinea. The extent of that problem has been well hidden by fisheries officials, but it must be assumed that incidental kills are significant (as they have been in the eastern tropical Pacific). Clymene dolphin abundance has been estimated for the northern Gulf of Mexico, where about 12,000 are thought to occur.

IUCN status Data Deficient.

Three Clymene dolphins leap alongside a research vessel in the Gulf of Mexico, showing the species' diagnostic characters. In many ways, they appear almost intermediate between spinner and striped dolphins. PHOTO: T. PUSSER

Bowriding Clymene dolphins can be identified by the black-tipped beak with a wide, dark stripe on the mid-line of the rostrum, and diagnostic "moustache" marking on top of the beak. Gulf of Mexico. PHOTO: T. PUSSER

This Clymene dolphin that live-stranded in Texas shows very clearly the moustache marking and beak stripes that are diagnostic of the species. PHOTO: TEXAS MARINE MAMMAL STRANDING NETWORK

Striped Dolphin—*Stenella coeruleoalba*

(Meyen, 1833)

Recently-used synonyms *Stenella euphrosyne, Stenella styx.*

Common names En.–striped dolphin; Sp.–*estenela listada, delfín listado;* Fr.–*dauphin bleu et blanc.*

Taxonomic information Cetartiodactyla, Cetacea, Odontoceti, Delphinidae. Recent genetic work suggests that the genus *Stenella* is actually paraphyletic, and it is likely that it will be restructured in coming years.

Species characteristics The striped dolphin has the basic body shape typical of the *Stenella/Delphinus* group, although it is somewhat-more robust (especially in the thoracic region) than spinner and pantropical spotted dolphins. The beak is moderate in length, with a distinct crease between the melon and beak. The dorsal fin is tall and slightly falcate, and the flippers are recurved and pointed (typical dolphin shape). The flukes are not notably different from those of others in the genus (with slender blades and acutely-rounded tips).

The striped dolphin's color pattern is stunning: a white or pinkish belly and dark gray dorsal cape (although it may appear bluish-gray) are separated by a light gray thorax. A light gray spinal blaze (highly variable in extent and intensity) extends from the thoracic area into the cape, to just under the dorsal fin. The mostly-black beak sends back a dark stripe (the basis for the species' common name), which encircles the eye and

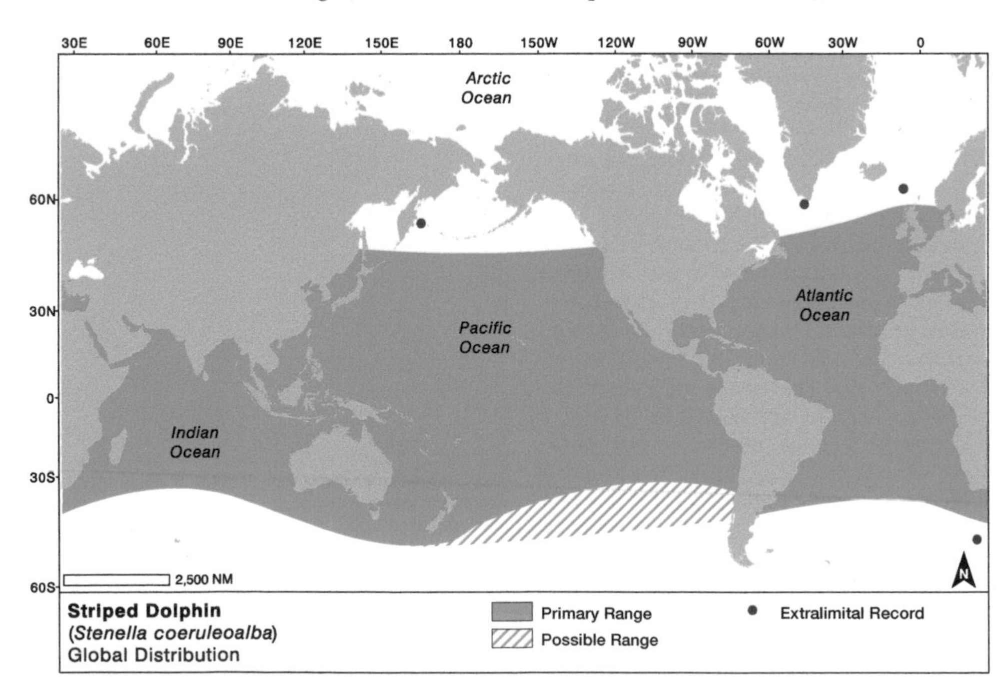

then widens and runs back to the anus. There is an eye-to-flipper stripe and usually a short accessory stripe between the other two. The characteristic delphinid "bridle" (consisting of an eye stripe and blowhole stripe) is present. The appendages are light gray to black.

The striped dolphin's mouth contains 40–55 slender, pointed teeth in each tooth row. Adult striped dolphins are up to 2.56 m long; males grow slightly larger than females. Newborns are about 93–100 cm in length. Maximum known weight is about 156 kg. There is geographical variation in the size of adults from different populations (for instance, dolphins from the southwestern Mediterranean are about 5–8 cm shorter than those from the adjacent northeastern Atlantic).

This striped dolphin has a long, thin spinal blaze, and a very slender lateral stripe, slightly unusual, but nonetheless well within the known range of variation in this species. Atlantic Ocean. PHOTO: P. OLSON

Recognizable geographic forms While significant geographic variation in osteological and body length characteristics has been documented, it is not currently possible to reliably recognize different geographical forms of this species in the field.

Can be confused with Although the body shape is similar to that of other species in the *Stenella/Delphinus* group, striped dolphins are generally easy to distinguish by their unique color pattern. They are most likely to be confused with common dolphins (*Delphinus* spp.), but a good look at the color pattern should clear up any problems. Fraser's dolphins also have an eye-to-anus stripe, and therefore may cause some confusion, but are much more robust, with much smaller appendages. When the color pattern is not visible, there may be some confusion with other species of the genus *Stenella*. However, when visible, attention to details of the color pattern will allow easy identification.

An anomalously-colored striped dolphin—though the basic elements of the striped dolphin color pattern can still be seen. Central Pacific. PHOTO: J. COTTON

Distribution This is a widely-distributed species, found in the Atlantic, Pacific, and Indian oceans, as well as many adjacent seas. Although also primarily a warm-water species, the range of the striped dolphin extends higher into temperate regions than do those of any other species in the genus (spotted and spinner/Clymene dolphins). For instance, this is the only species of the genus that routinely reaches northern European waters. Normal range limits are about 50°N and 40°S, although there are extralimital records from the Kamchatka Peninsula, southern Greenland, Scandinavia, the Faroe Islands, and the Prince Edward Islands. Striped dolphins also are generally restricted to oceanic regions and are seen close to shore only where deep water approaches the coast. In some areas (e.g., the eastern tropical Pacific), they are mostly associated with convergence zones and regions of upwelling. There

An eastern tropical Pacific striped dolphin performing a characteristic leap in which the tail is thrown high in the air. PHOTO: C. OEDEKOVEN/NOAA FISHERIES/SWFSC/CSCAPE

A young striped dolphin leaps beside a research vessel. The body of this species can range from quite robust to somewhat slender. Mediterranean Sea. PHOTO: C. JOHNSON

There is a great deal of geographical and individual variation in the color pattern of the striped dolphin, but the side stripe and spinal blaze usually make it easily identifiable, when seen well. PHOTO: T. PUSSER

The expression of the spinal blaze in the striped dolphin is highly variable, with some animals displaying extensive ones, and others (like this captive one) showing just a hint. PHOTO: J.-P. SYLVESTRE

is a well-known and well-studied population in the Mediterranean Sea. Off the west coast of North America, they occur far offshore, in waters influenced by the warm, northward-flowing Davidson Current.

Ecology and behavior Striped dolphins are fast swimmers, and appear to be more easily "spooked" than other tropical dolphins. This and their color pattern have prompted fishermen in the eastern tropical Pacific to call them "streakers." They are well-known in the ETP for their habit of running from vessels in low "splashy" leaps, at high speeds. They are very acrobatic, performing frequent breaches and other aerial maneuvers (including a behavior called roto-tailing, in which the dolphin whips its tail in a circle as it performs a high arcing leap out of the water). They often ride bow waves, except in the eastern tropical Pacific, where they tend to run from vessels.

Although most striped dolphin herds number between a few dozen and 500 individuals, these dolphins sometimes assemble into herds of thousands. At least off Japan, there appears to be some age/sex segregation of such herds, with individuals moving among juvenile, adult, and mixed schools. They are thought to be capable of diving to depths of 200–700 m to obtain prey. Off Japan, where the biology of this species has been best studied, there are two calving peaks: one in summer, another in winter. Sexual maturity occurs at ages of 7–15 years for males, and 5–13 for females, generally when specimens are about 2.1–2.2 m long. Density-dependent changes in life history parameters have been documented for some areas. Gestation lasts an estimated 12–13 months. The breeding system is thought to be polygynous. Striped dolphins are estimated to live for as long as 58 years.

Feeding and prey The diet of this species consists primarily of a wide variety of small, midwater and pelagic or benthopelagic fish, especially lanternfish and cod, and squids. Striped dolphins apparently feed in pelagic to benthopelagic zones, in continental slope or oceanic regions.

Threats and status Striped dolphins are the main delphinid species involved in small cetacean harpoon and drive fisheries in Japanese waters. Catch levels vary widely, but

In the Maldives, two striped dolphins porpoise alongside a research vessel, giving a good look at some of the variation in coloration. PHOTO: T. A. JEFFERSON

in some years were over 20,000 animals, although the takes were generally much lower (<1,000 per year) in recent years. Despite the recent reductions, the populations (more than one may be involved) are thought to be seriously depleted. This species has also been directly captured in smaller numbers in the Caribbean, Sri Lanka, and occasionally in the Mediterranean. Incidental catches occur throughout the range in various types of fishing gear, especially purse seines and driftnets. This species was formerly caught in large numbers in pelagic driftnets in the Mediterranean and in the North Pacific, and there are catches in pelagic trawls and driftnets in western Europe. A massive die-off in the Mediterranean Sea from 1990–1992 is thought to have been at least partly related to lowered immunity to disease (morbillivirus) from environmental contaminants (especially organochlorines like PCBs). In the 1980s, there were an estimated 820,000 striped dolphins around Japan; however, this number is probably reduced now. Off the west coast of North America, there are about 20,000. In the ETP, there are estimated to be around 1 million striped dolphins. Estimates in the western North Atlantic are about 62,000, and in the Gulf of Mexico about 4,400. There were about 225,000 in the Mediterranean in the mid-1990s (after the mass die-off of 1990–1992). Clearly, this is not an endangered species, but due to the potential impacts of various human activities, the status of certain populations is very much a matter of concern.

The striped dolphin has a lovely color pattern of streaks and stripes on its sides. The combination of the light spinal blaze and dark lateral stripes rule out all other dolphin species. Offshore Spain. PHOTO: C. OEDEKOVEN

IUCN status Least Concern.

Short-beaked Common Dolphin—*Delphinus delphis*

Linnaeus, 1758

Recently-used synonyms None.

Common names En.–short-beaked common dolphin; Sp.–*delfín común de rostro corto;* Fr.–*dauphin commun à petit bec.*

Taxonomic information Cetartiodactyla, Cetacea, Odontoceti, Delphinidae. Until 1994, all common dolphins were classified as a single species: *D. delphis*. However, it is now known that at least two species exist within the genus: the short-beaked (*D. delphis*) and long-beaked (*D. capensis*) common dolphins. There is a distinct short-beaked form in the Black Sea, which is currently recognized as a subspecies: *D. delphis ponticus*.

Species characteristics Common dolphins have only recently been split into two species, and there may be additional species not yet recognized. Short-beaked common dolphins have tall, slightly-falcate dorsal fins, with pointed tips. The short-beaked species is somewhat more robust than the long-beaked species, with a shorter (but still moderately long) beak, and a more rounded, bulging melon. There is a deep crease between the melon and beak. The flippers are the typical shape for dolphins: slender, recurved, and pointed at the tips. In some individuals (presumably mostly adult males), there is a prominent post-anal hump of connective tissue—forming a ventral "keel." The flukes are unremarkable.

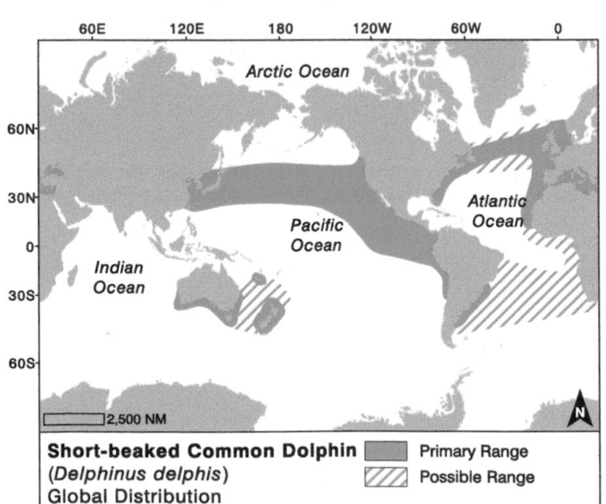

Short-beaked common dolphins are strikingly marked, with a dark brownish-gray back, white belly, and tan to ochre thoracic patch. This thoracic patch dips below the dorsal fin and combines with an area of streaked light gray on the tail stock to produce the common dolphins' most characteristic feature: an hourglass pattern on the side. In short-beaked common dolphins, the thoracic patch is relatively light, contrasting strongly with the dark cape. The flipper-to-anus stripe (generally well-developed in the long-beaked species) is weakly developed or absent. The flippers and dorsal fin can range from completely dark to almost entirely white, but most often have a small light patch in the center. The chin-to-flipper stripe does not closely approach the gape, and it narrows forward of the eye. The lips are black, and there is a dark, distinct stripe running from the apex of the melon to encircle the eye. The upper surface of the beak is generally light gray, with a black tip, and a dark line that runs from the tip to the apex of the melon. A light blowhole stripe runs from the apex to the blowhole. The color pattern is highly variable.

There is an uncommonly-occurring color morph (dark morph) that has been identified from most areas of the range of the species, in which the large light-colored thoracic patch appears not to develop (or to develop incompletely). This results in a dark gray lateral area, and gives the individual an appearance somewhat like that of a bottlenose or spinner dolphin. Otherwise, the color pattern appears normal. Anomalously all-white and all-black individuals have also been observed.

This short-beaked common dolphin shows an extreme at one end of the range of color pattern expression—the belly is brilliant white, and the thoracic patch very light ochre, both of which contrast strongly with the dark cape. US west coast. PHOTO: J. COTTON/NOAA FISHERIES/SWFSC

Short-beaked common dolphins in the cool temperate waters of the North Atlantic Ocean often appear to have a more dusky color pattern, though this pattern is anomalous. Azores. PHOTO: M. VAN DER LINDE

This is a more typical short-beaked common dolphin, with a highly contrasting thoracic patch, distinct dark cape, bold facial stripes, and light color on the flippers. New Zealand. PHOTO: M. NOLAN

This is an infrequently-occurring color pattern of the short-beaked common dolphin in which the dark dorsal overlay obliterates the light thoracic patch. Bay of Biscay, and US west coast. PHOTOS: (LEFT) R. WHITE, (RIGHT) NOAA FISHERIES/SWFSC

The short-beaked common dolphin is generally more chunky in appearance than its congener, with some dolphins being quite robust. Southern California. PHOTO: T. A. JEFFERSON

This short-beaked common dolphin is so dark as to render the species' main color pattern features barely discernible—in contrast to the bold patterns seen on more tropical populations. Angola. PHOTO: C. WEIR

There are 41–57 pairs of small, sharp, pointed teeth in each jaw. At birth, short-beaked common dolphins are about 80–90 cm long. Adult short-beaked common dolphins in the eastern Pacific are 1.6–2.2 m (females) or 1.7–2.3 m (males), with occasional specimens to 2.35 m. Those in the northeast Atlantic appear to grow much larger, up to about 2.50 m. Some specimens may reach up to nearly 2.70 m. Short-beaked common dolphins may weigh up to 200 kg.

Calf—1/3 to 1/2 adult size, color pattern muted and borders faint, light areas often with an ochre tinge, dorsal fin falcate.

Subadult—Nearly adult size, color pattern more developed, but not as bold as in adults, dorsal fin falcate.

Adult female—Full size, color pattern generally bold, often with light patches on the dorsal fin and flippers, dorsal fin relatively erect.

Adult male—Large size, color pattern generally bold, often with light patches on the dorsal fin and flippers, dorsal fin generally erect, and post-anal hump usually present.

Recognizable geographic forms While there is certainly extensive geographical variation in the coloration and external appearance of short-beaked common dolphins (for instance, a longer-beaked form in New Zealand and southern Australia), in most cases the variation has not yet been described and compared adequately for distinct geographic forms to be reliably recognized in practice. Only the following forms can be reliably recognized:

Standard-form common dolphin (D. d. delphis)—Found in many areas of the Pacific and Atlantic oceans, including the Mediterranean Sea. There is great geographical variation in total length, with the maximum approaching 2.70 m. Tooth counts are generally greater than in the Black Sea form, reaching up to 57 teeth in each tooth row. Coloration is highly variable among populations.

Black Sea common dolphin (D. d. ponticus)—Found only in the Black Sea, this is a dwarf form. Total length averages 1.62 m for males and 1.58 m for females, with the maximum known only 2.19 m. Tooth counts are low, and range from 41–52 teeth per row. There are slight color pattern differences, with the dark lateral band generally more prominent than in other short-beaked common dolphins.

Can be confused with Where they are sympatric (e.g., at least southern California, the west coast of Mexico, southern Japan, Korea, Peru and northern Chile, eastern South America, West Africa, and possibly southern Africa), the two species of common dolphins may be difficult to distinguish, especially in sightings at sea. The short-beaked species has a shorter beak, more steeply-rising melon, more robust body, a more contrasting color pattern, a narrower chin-to-flipper stripe, and a greater tendency to have white patches on the fins. In addition, the Clymene dolphin may cause confusion in the Atlantic Ocean, but that species has a "moustache" marking and a rounded (as opposed to V-shaped) margin to the cape.

Distribution The short-beaked common dolphin is an oceanic species that is widely distributed in tropical to cool temperate waters of the Atlantic and Pacific oceans. However, they apparently do not occur in the Gulf of Mexico, and most parts of the Caribbean Sea. Also, there are no confirmed records for the Indian Ocean, and it is doubtful that the species occurs there (with the possible exception of the areas around southeast Africa and southwest Australia). The short-beaked common dolphin occurs from nearshore waters to thousands of kilometers offshore, although separate populations apparently occur in some enclosed seas, such as the Mediterranean and Black seas. In most areas, this species appears to have a strong preference for upwelling-modified waters and areas with steep sea-bottoms (such as seamounts and escarpments).

Ecology and behavior Large, boisterous groups of common dolphins are often seen whipping the ocean's surface into a froth as they move along at high speed. Herds range in size from about ten to over 10,000. Schools are often segregated by age and sex. Associations with other marine mammal species (especially pilot whales) are not uncommon. They have been observed in the vicinity of schools of long-beaked common dolphins, but the two species do not appear to intermix often. Schools in the eastern tropical Pacific are sometimes associated with yellowfin tuna (although not nearly as often as are spotted and spinner dolphins), and have thus been impacted by the purse-seine fishery there.

Active and energetic bowriders (except in prime tuna fishing zones of the ETP), short-beaked common dolphins may be very familiar to most seagoers in low- to mid-latitudes. They also ride the "snout waves" of large whales at times. They are often aerially active, performing various breaches and leaps. Individuals are also highly vocal; sometimes their squeals can be heard above the surface as they bowride. Foraging dives as deep as 200 m have been recorded.

Short-beaked common dolphins have reported calving intervals of 1–3 years, and the reported age at sexual maturity varies greatly among populations (3–12 years for males and 2–7 years for females). Some of this variation may result from real differences, at least partly attributable to density-dependent responses to different histories of exploitation. Gestation apparently lasts 10–11 months. In the Black Sea, peak calving occurs in summer months (June to August). Longevity is at least 25 years.

Feeding and prey The prey of short-beaked common dolphins consists largely of small schooling fishes and squid. In some areas (e.g., southern California), common dolphins feed mostly at night on creatures associated with the deep scattering layer (DSL), which migrates toward the surface in the dark. In other areas, they feed mainly

Short-beaked common dolphins have a lot of variation in the expression of the facial stripes, but the chin-to-flipper stripe is virtually always slender. This is in contrast to the long-beaked species. California. PHOTO: S. WEBB

Fast-moving schools of short-beaked common dolphins may whip the ocean's surface into a froth as they cruise along. Some schools of this species may number over 10,000 individuals! California. PHOTO: S. WEBB

on epipelagic schooling fish. Squids may form a more important part of the diet of the short-beaked common dolphin than it does for the long-beaked species.

Threats and status Short-beaked common dolphins are taken in many fisheries worldwide. Huge direct catches formerly occurred in countries bordering the Black Sea. Common dolphin stocks there have declined and the fishery has not operated since 1983. Some direct mortality still occurs off Japan and in the Mediterranean. However, the recent common dolphin decline in the Mediterranean Sea appears to have been primarily caused by prey depletion and poor habitat quality. The eastern tropical Pacific tuna fishery incidentally takes common dolphins from several stocks, and some of these may have been depleted by past levels of mortality. There are also incidental catches in various fisheries, in particular gillnets and pelagic trawls, throughout the range. Globally, this is a very abundant species, with many available estimates for the various areas where it occurs: about 370,000 (western US coast), 3,000,000 (eastern tropical Pacific), 30,000 (eastern US coast), 96,000 (Black Sea/Turkish Straits System), 61,000 (eastern Atlantic continental shelf), 14,700 (Alboran Sea), and 75,000 (Celtic Sea shelf). Therefore, while certain populations are in danger (such as the one in the Mediterranean Sea),
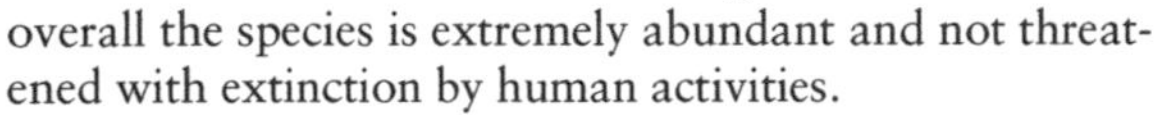
overall the species is extremely abundant and not threatened with extinction by human activities.

IUCN status Least Concern (overall species), Endangered (Mediterranean population), Vulnerable (Black Sea population).

There are two young calves among these short-beaked common dolphins; they are distinguishable by their muted color patterns, and very light beaks. Southern California. PHOTO: J. COTTON

In this subgroup of a school of short-beaked common dolphins, the variability in expression of the light patch on the dorsal fin can be seen— from completely absent to filling nearly the entire fin. US west coast. PHOTO: NOAA FISHERIES/SWFSC

Long-beaked Common Dolphin—*Delphinus capensis*

Gray, 1828

Recently-used synonyms *Delphinus bairdii, Delphinus tropicalis.*

Common names En.–long-beaked common dolphin; Sp.–*delfín común de rostro largo;* Fr.–*dauphin commun à long bec.*

Taxonomic information Cetartiodactyla, Cetacea, Odontoceti, Delphinidae. In the past, the geographic form in the Indo-Pacific with an exceptionally long beak was thought by some to be a separate species, *D. tropicalis.* However, morphometric studies suggested that it might be a subspecies of *D. capensis,* and it is currently recognized as *D. c. tropicalis.* More recent work using molecular genetic techniques indicates that it is actually more closely related to the short-beaked common dolphin, and further suggests that most long-beaked populations have evolved separately from a *D. delphis* ancestor. Hybrids with common bottlenose dolphins in captivity and dusky dolphins in the wild have been reported.

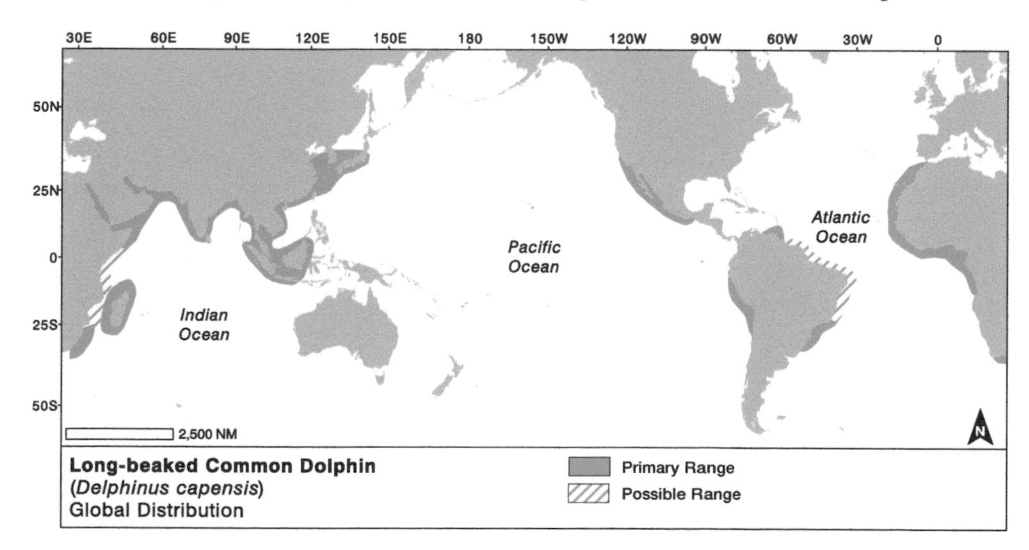

South Africa's Cape region is the type locality of the long-beaked common dolphin, and thus the name *D. capensis* will remain with these animals, even if other long-beaked forms are found to be different species. PHOTO: M. SIDWELL

Species characteristics Because of the recent discovery that common dolphins in the eastern Pacific represent two species (rather than only one, as was commonly thought), much of the older biological information available for dolphins of the genus *Delphinus* cannot be reliably applied to one or the other species. Long-beaked common dolphins have the same basic body morphology as the short-beaked species, but tend to be more slender. They have tall, slightly-falcate dorsal fins, and flippers that are recurved and pointed at the tips. Besides having longer beaks than short-beaked common dolphins (up to at least 9.7% of total length), long-beaked common dolphins have a somewhat more-flat appearance to the melon, which rises from the rostrum at a relatively low angle. There is a deep crease between the melon and beak. In some individuals (mostly adult males), there may be a prominent post-anal hump of connective tissue—forming a ventral "keel."

Long-beaked common dolphins are characterized by an hourglass pattern on the side, forming a V below the dorsal fin. In this species, the coloration generally appears somewhat muted, compared to the short-beaked species. The thoracic patch is relatively dark, contrasting less with the cape, and often the border between them appears "smeared." The flipper-to-anus stripe is generally moderately to strongly developed. The chin-to-flipper stripe fuses with the lip patch at or just anterior to the gape, and remains relatively wide ahead of the eye. The eye patch is imbedded in the flipper-to-anus stripe. The characteristic delphinid "bridle" (consisting of an eye stripe and blowhole stripe) is present. Light patches on the dorsal fin and flippers are only occasionally present, and are small or faint when they are.

This species has higher tooth counts than any other species of delphinid, although there is strong overlap with several other species. There are 47–67 sharp, pointed teeth in each tooth row. Most measured adults were 2.02–2.54 m (males) or 1.93–2.22 m (females) long, although they may reach lengths of over 2.60 m. Weights of at least 235 kg are attained.

Recognizable geographic forms

***Standard long-beaked common dolphin* (D. c. capensis)**—The nominate subspecies occurs in coastal areas of the Pacific and Atlantic oceans. It is slightly shorter than the other, with adult females of 1.93–2.22 m and males of 2.02–2.54 m. The beak is relatively shorter (ranging from about 6.9–7.6% of the total length). Tooth counts are significantly lower—47–60 teeth in each upper tooth row, and 48–57 in the lower row. There are reports of tooth counts up to 62, however.

***Indo-Pacific common dolphin* (D. c. tropicalis)**— The *tropicalis* form of common dolphin apparently only occurs in the Indian and far western Pacific oceans. This subspecies is apparently slightly longer than the nominate one, reaching lengths of at least 2.60 m. The beak is exceedingly narrow and long, the small amount of data available suggesting it reaches about 9.4–9.7% of total length. There are significantly higher tooth counts than in the nominal subspecies— 54–67 (upper) and 52–64 (lower) teeth per row.

Can be confused with The two species of common dolphins may be difficult to distinguish, especially in sightings at sea. The long-beaked species has a longer beak, less steeply-rising melon, more slender body, a more muted color pattern, a wider chin-to-flipper stripe (which often merges with the lip patch, making much of the lower jaw dark), and a reduced tendency to have white patches on the fins. Also, the spinner dolphin may easily be confused with this species, especially with the *tropicalis* subspecies in the Indo-Pacific. The body shape of the two species is nearly identical, but close attention to color patterning (especially the shape of the cape and facial stripes) should allow distinction.

Distribution Long-beaked common dolphins inhabit more nearshore and tropical waters than the short-beaked species, generally occurring within 180 km of the coast. They sometimes come into shallow water just a few meters deep very near shore. The *capensis* subspecies appears to occur in discrete areas, and populations are known from the east coast of South America, West Africa, southern Africa, southern Japan and Korea (and possibly northern China), central California to southern Mexico, and Peru and Chile. The *tropicalis* subspecies ranges around the rim of the Indo-Pacific, from at least the Red Sea/Somalia

These are classic long-beaked common dolphins–long and slender, with long beaks, dark "face masks," and dark lateral stripes. Gulf of California. PHOTO: M. NOLAN

In the eastern North Pacific, young long-beaked common dolphins tend to have muted color patterns, with less contrast and distinctness between light and dark. The chin-to-flipper stripe and lateral stripe are much less developed in these young animals, as well. Baja California. PHOTO: T. A. JEFFERSON

Indo-Pacific common dolphins (*D. c. tropicalis*) leap at the bow of a vessel in the Arabian Sea area. This subspecies can be recognized by its exceedingly long beak and very high tooth counts. PHOTO: R. BALDWIN

In the eastern North Pacific, adult long-beaked common dolphins tend to have more bold and contrasting color patterns, and in some cases the chin-to-flipper stripe and the dark lateral stripe can be so wide and dark as to merge and form a "bandit mask." Baja California. PHOTOS: T. A. JEFFERSON

Certain forms of long-beaked common dolphin have the highest tooth counts among any of the oceanic dolphins, even exceeding those of spinner dolphins. Gulf of California. PHOTO: M. NOLAN

Some long-beaked common dolphin populations in the Caribbean region have muddy color patterns, without the bold contrast between dark and light seen elsewhere. These animals from Suriname may comprise a unique subspecies or geographic form. PHOTO: M. DE BOER

In South Africa, long-beaked common dolphins migrate into nearshore waters every year to take advantage of the feast provided by the annual "sardine run." PHOTO: R. KIEFNER

(including the Persian Gulf, and possibly further south along the east African coast) to Taiwan and southern/central China.

Ecology and behavior The long-beaked common dolphin is highly gregarious. Herds of less than a dozen to several thousand are formed. Not much is known of the composition of schools, but it is suspected that there is some degree of age and sex segregation, as is known for other species of *Delphinus* and *Stenella*. They sometimes associate with other species of cetaceans.

Long-beaked common dolphins are swift and agile creatures, often porpoising out of the water, while moving at high speed in great splashing herds. These dolphins are capable and willing bowriders, and often exhibit a great deal of aerial activity, including breaches and various types of acrobatic leaps. They are highly vocal and can sometimes be heard whistling from above the water's surface.

One or the other species of common dolphin tends to predominate in the stranding record for southern California for a particular period of time. In the years following the 1982–83 El Niño event, the long-beaked form was most common, and it has been seen nearshore with increasing frequency in the last few years. This species occurs in some inshore waters, such as the Persian Gulf, Red Sea, Gulf of Thailand, and Gulf of California. In South African waters of Natal, they are associated with the annual "sardine run," which they follow north and inshore.

There is little information on the life history of populations of this species. What little data are available generally come from areas where the two species of *Delphinus* are sympatric and may thus be mixed with data from *D. delphis* populations (e.g., off West Africa and southern California). Sexual maturity probably occurs at lengths of around 2.0 m. Because this species is generally more tropical than its congener, it is thought that its reproduction may show less seasonality. Hybrids between this species and *T. truncatus* have occurred multiple times in captivity.

Feeding and prey A wide variety of small schooling fishes (e.g., pilchards, sardines, anchovies, and hake) and squids are taken as prey. Apparently, cooperative feeding techniques are sometimes used to herd fish schools (e.g., in the northern Gulf of California and off southern California). It is thought that most feeding is done in relatively shallow waters.

Threats and status This species is commonly taken in gillnets off southern California. Long-beaked common dolphins are only occasionally involved in the eastern tropical Pacific tuna fishery. They are sometimes taken in drive fisheries off Japan, and there is concern about the

The series of dark stripes in the facial region of the long-beaked common dolphin are generally more extensively developed than in the short-beaked species. PHOTO: M. NOLAN

When seen from below, the dark "face mask" and chin-to-flipper stripe of long-beaked common dolphins are obvious. Gulf of California. PHOTO: M. NOLAN

This is an unusual looking common dolphin from Angola, southwestern Africa. It seems to have some features of both long-beaked and short-beaked common dolphins, and its taxonomic status is uncertain. PHOTO: C. WEIR

growing problems of direct kills off Peru. There has been a large direct kill off northern Venezuela, but its current status is not known. In the Indian Ocean and Chinese waters, long-beaked common dolphins are among many species of small cetaceans taken in gillnets, trawls, and purse seines. Some dolphins of this species have been live-captured for display, but they do not do as well in captivity as the more coastal bottlenose dolphins. While not as numerous as the short-beaked species, overall, the species is relatively abundant and not in any danger of extinction on a global scale. There are few estimates of abundance, however. The stock that occurs off the west coast of the US and Mexico numbers about 279,000 and there are thought to be about 15,000–20,000 long-beaked common dolphins off South Africa.

IUCN status Data Deficient.

Fraser's Dolphin—*Lagenodelphis hosei*

Fraser, 1956

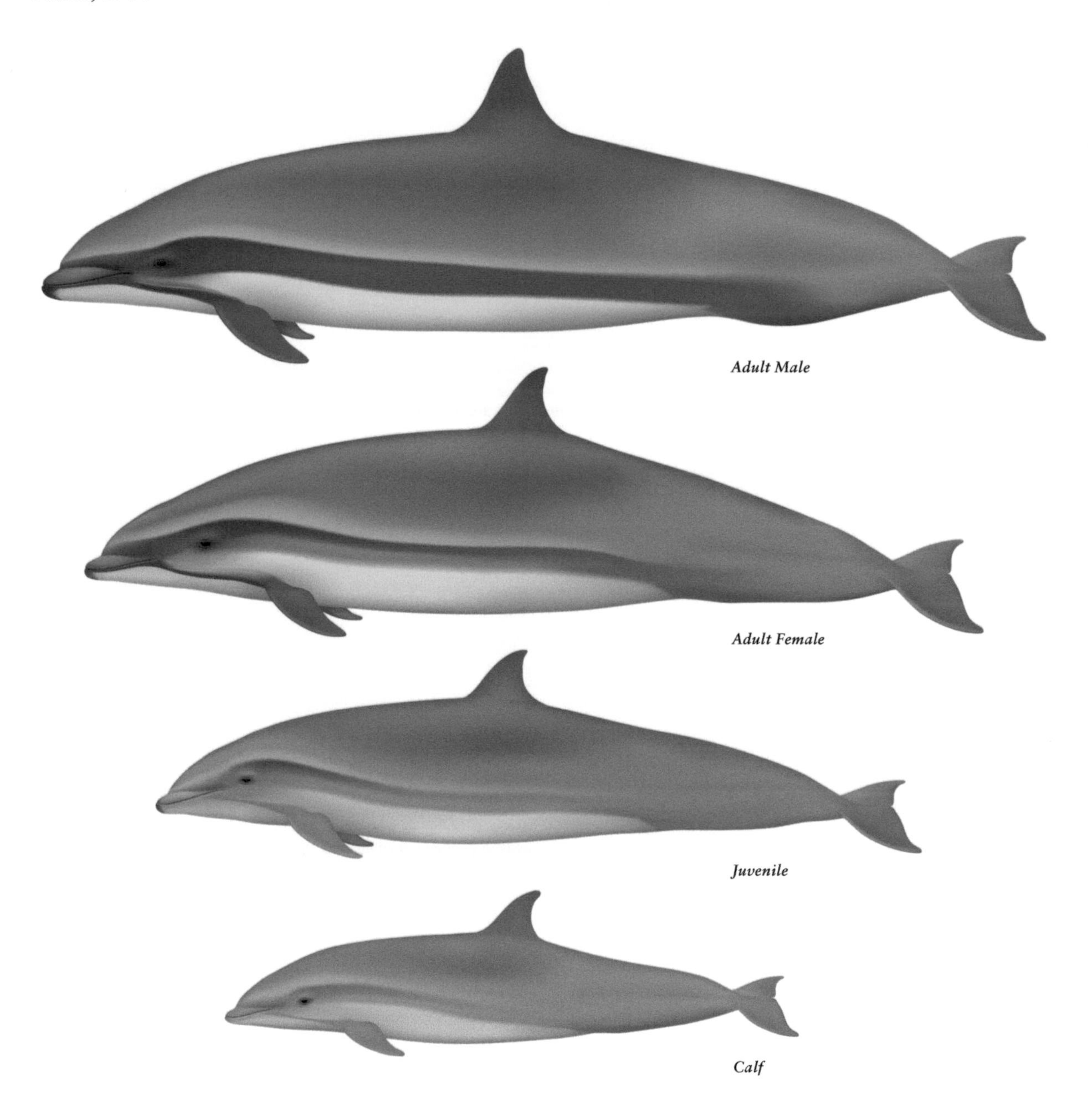

Recently-used synonyms None.
Common names En.–Fraser's dolphin; Sp.–*delfín de Fraser;* Fr.–*dauphin de Fraser.*
Taxonomic information Cetartiodactyla, Cetacea, Odontoceti, Delphinidae. For almost two decades after its discovery in 1956, this species was known only from skeletal material, until it was "rediscovered" in the early 1970s.
Species characteristics Fraser's dolphin is a very distinctive dolphin, with an extremely stocky body and very small appendages. The short dorsal fin (<9.5% of total length) is triangular or only slightly falcate, and tends to be more erect in adult males. The small flippers are pointed at the tips, and the flukes are concave on the trailing edge. There is a very short (<3% of total length) and stubby, but well-defined, beak. Adult males often have a large post-anal hump or keel, composed of connective tissue.

The color pattern can be striking; the most distinctive feature (when present) is a dark band of varying thickness, running from the face to the anus (in some regions, the band is indistinct). This band is scarcely apparent on young animals, and appears to widen and darken with age in some animals, especially adult males (but is geographically variable). There is also a flipper stripe that starts at midlength along the lower jaw (in some animals the lateral stripe is so wide that it merges with the flipper stripe, creating a dark face mask, sometimes informally called a "bandit mask"). Otherwise, the back is dark brownish-gray, the lower sides are cream-colored, and the belly is white or pink. Young animals may have particularly pinkish bellies. The tip of the beak and lips are dark, and there is a dark stripe from the tip of the upper jaw to the apex of the melon. The characteristic delphinid "bridle" (consisting of an eye stripe and blowhole stripe) is present.

There are 38–44 pairs of teeth in each jaw. Maximum size is at least 2.7 m (males) and 2.6 m (females). They

Most smaller tropical dolphins have long beaks, but Fraser's dolphin has a very short beak, more similar to that of the *Lagenorhynchus* dolphins than to the tropical dolphins it is more closely related to. Maldives. PHOTO: T. STENTON

Part of a large group of Fraser's dolphins leaps at the bow of a research vessel in the Maldives, showing the short beak, tiny appendages, and robust body of the species. When seen at such close range, there is no mistaking them. Outside the eastern Pacific Ocean, only adult male Fraser's dolphins appear to show the very dark, wide lateral stripe. PHOTO: T. A. JEFFERSON

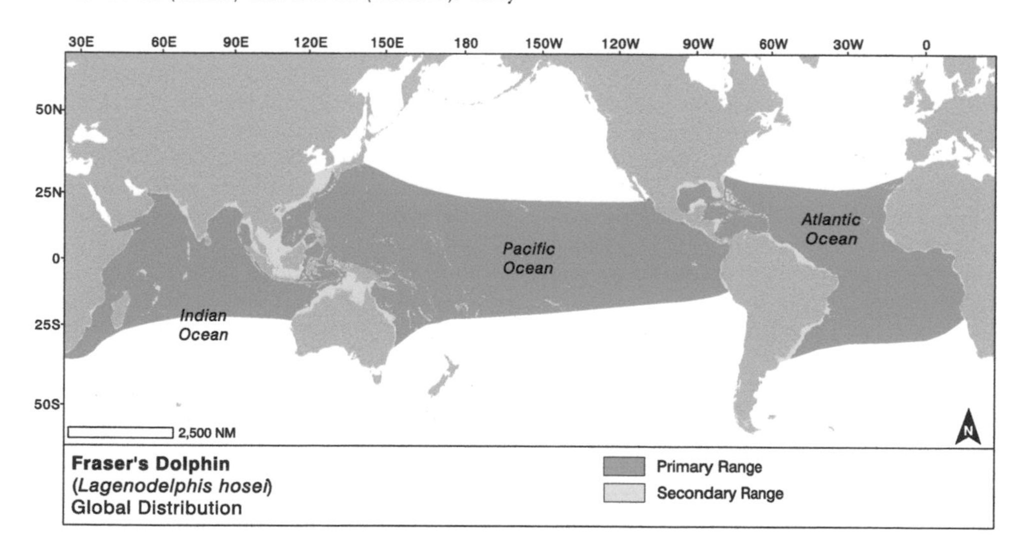

The appendages of Fraser's dolphins are tiny for the body size. The dorsal fin, flippers, and flukes just seem too small. Maldives. PHOTO: T. STENTON

The belly of a Fraser's dolphin is brilliant white, but when the animals need to dump heat (such as when very active and/or in very warm water), they often take on a pinkish tinge. Guadeloupe. PHOTO: L. BOUVERET

A Fraser's dolphin surfaces in the Indian Ocean. It can be identified as an adult male by the very dark and wide lateral stripe, bandit mask, and erect dorsal fin. PHOTO: J. KISZKA

may reach weights of over 210 kg. Newborns are estimated to be 1.0–1.1 m long.

Calf—1/3 to 1/2 adult body size, color pattern muted and lateral stripe faint or absent, dorsal fin falcate, and post-anal hump not present.

Subadult—Close to adult body size, color pattern not strongly contrasting and lateral stripe faint or absent, dorsal fin falcate, and post-anal hump not present or only slightly developed.

Adult female—Full body size, color pattern with lateral stripe faint to moderate, "bandit mask" may be present, dorsal fin slightly falcate, and post-anal hump not present or only slightly developed.

Adult male—Large body size, color pattern strongly contrasting with lateral stripe moderate to dark and broad, "bandit mask" generally present, dorsal fin erect and triangular, and post-anal hump moderate to large and well developed.

Recognizable geographic forms Although there is some evidence that Fraser's dolphins in the Atlantic are larger, with a more weakly-developed eye-to-anus stripe than those in the Pacific or Indian oceans, distinct geographic forms have not been adequately described for reliable identification.

Can be confused with When seen well, the unique body shape of Fraser's dolphin should rule out confusion with most other species, but striped dolphins, which can be robust and also have an eye-to-anus stripe, can be confused with Fraser's at a distance. Identification should be assisted by the fact that this is the only small, tropical species of oceanic dolphin with a short, well-defined beak.

Distribution Fraser's dolphin has a pantropical distribution, largely between 30°N and 30°S in all three major oceans. It is an oceanic species that prefers deep offshore waters, but it can be seen near shore in some areas where deep water approaches the coast (such as the Philippines, Taiwan, and some islands of the Caribbean and the Indo-Malay archipelago). Strandings outside the tropics (such as those in France and Uruguay) are generally considered to be extralimital, and are usually associated with anomalously-warm water temperatures.

Ecology and behavior Fraser's dolphins remain somewhat mysterious. There is still little known of the ecology of this oceanic species. Herds tend to be large, consisting of hundreds or even thousands of dolphins, often mixed with other species, especially melon-headed whales in the eastern tropical Pacific, Japan, and the Philippines; as well as pilot whales, and Risso's, spotted and spinner dolphins in the Philippines. Little is known of their social structure, but they do strand *en masse* on occasion.

These are active animals, generally seen moving along in dense schools at great speeds, and often whipping the sea surface into a froth with their low-angle leaps. Fast-moving herds create a great deal of white water. In some areas, they are considered shy and difficult to approach; in others they are more approachable. They do not generally bowride in the eastern tropical Pacific, but appear to

The lateral stripe in Fraser's dolphin is geographically variable with respect to its prevalence within groups, and with regard to its extent and intensity on individual animals. In the eastern Pacific, a heavy dark stripe can be found among animals of both sexes and all age classes, except perhaps the very youngest animals. Eastern tropical Pacific. PHOTO: R. L. PITMAN

In waters of the Pacific Ocean, some individuals of Fraser's dolphin schools possess very dark and wide lateral stripes. These are large adults, with males having the most exaggerated stripes. Central Pacific. PHOTO: C. OEDEKOVEN

do so in most other areas. In the Philippine Sulu Sea, this species is most likely to bowride on slow-moving vessels.

Not much is known of the life history of Fraser's dolphin. In Japan, the only place where life history has been studied in detail, sexual maturity is reached at 7–10 years and 2.2–2.3 m for males, and at 5–8 years and 2.1–2.2 m in females. Calving peaks in spring and autumn have been documented in Japan, and in South Africa calving appears to peak in summer. Longevity is at least 18 years.

Feeding and prey Fraser's dolphins appear to feed on midwater fish (especially myctophids), squid, and crustaceans. Physiological studies indicate that Fraser's are capable of quite deep diving (and it is thought that they do most of their feeding deep in the water column in waters up to 600 m deep), but they have been observed to feed near the surface as well.

Threats and status Direct killing of Fraser's dolphins has been documented in Japan, the Lesser Antilles, Sri Lanka, Indonesia and the Philippines. Incidental catches in purse seines (eastern tropical Pacific and the Philippines), gillnets and drift-nets (South Africa, Japan, the Philippines, and Sri Lanka), and trap nets (Japan) are also known. Their range in tropical, oceanic waters probably subjects them to less problems of habitat degradation than for most other dolphins. This species does not appear to be particularly abundant anywhere (with the possible exception of some areas in the Philippines), although there is no serious concern about its global conservation status. There are estimated to be about 289,000 Fraser's dolphins in the eastern tropical Pacific. Only a few hundred may be present in the northern Gulf of Mexico.

IUCN status Least Concern.

White-beaked Dolphin—*Lagenorhynchus albirostris*

(Gray, 1846)

Recently-used synonyms None.

Common names En.–white-beaked dolphin; Sp.–*delfín de hocico blanco;* Fr.–*dauphin à bec blanc.*

Taxonomic information Cetartiodactyla, Cetacea, Odontoceti, Delphinidae. Recent genetic studies suggest that this species is not closely-related to any of the others currently placed in the genus *Lagenorhynchus*. It is likely that future work will split out the other species, leaving this one as the only member of the genus *Lagenorhynchus*.

Species characteristics White-beaked dolphins are extremely robust dolphins. The beak is very short and thick (usually 5–8 cm long), but is set-off from the melon by a shallow crease. The dorsal fin is very tall and only slightly falcate, with a pointed tip. The flukes and dorsal fin get proportionately larger as the animal ages. The pointed flippers are moderately broad and long (up to 19% of the total length), with a falcate trailing edge. The tail stock is not particularly deepened, tapering gradually as it approaches the flukes.

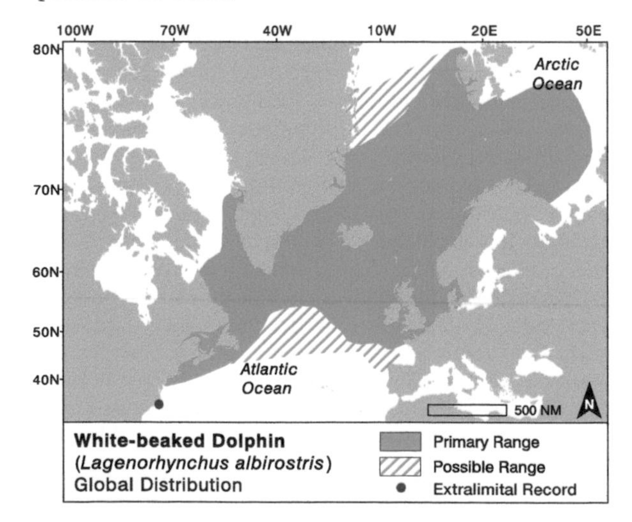

The color pattern is highly variable, but the animals are mostly black to dark gray on the upper sides and back. The beak and most of the belly are white to light gray, and (especially the beak) are often mottled, flecked, or ashy gray. The light color of the beak extends dorsally to cover the apex of the melon. The characteristic delphinid "bridle" (consisting of an eye stripe and blowhole stripe) is present. Light streaks may surround the eyes. An area of light gray with an indistinct border originating on the upper flank broadens to cover most of the tail stock—this is one of the species' most distinctive characteristics. There is generally a black thoracic patch, surrounded by lighter coloring. There is often dark or light flecking in the region between the eye and the flipper. The dorsal fin, flippers, and flukes are mostly dark.

Twenty-two to 28 sharp, but moderately-large (compared to other members of the genus) teeth line each half of each jaw. Adults are 2.4–3.1 m in length (males grow larger than females) and weigh between 180 and 350 kg. Newborns are between 1.1 and 1.2 m long, and weigh approximately 40 kg.

Recognizable geographic forms None.

Can be confused with White-beaked dolphins are most likely to be confused with Atlantic white-sided dolphins, which share an almost identical range. They can be distinguished by differences in coloration (which are very divergent) and beak shape. Also, in the few areas where they overlap with bottlenose dolphins, care must be taken to distinguish between these two when seen at a distance.

Distribution White-beaked dolphins inhabit cold temperate to subpolar waters of the North Atlantic Ocean, from Cape Cod and Portugal, north to central Davis Strait, southern Greenland, and Svalbard. The range includes Iceland, the United Kingdom, and most Scandinavian waters (but apparently not very far into the Baltic Sea). The species does not normally occur in the Mediterranean, but there are a few questionable records from at least the

Not all white-beaked dolphins have white beaks! Spitsbergen. PHOTO: M. NOLAN

A white dorsal ridge behind the large dorsal fin is a distinguishing feature of the white-beaked dolphin. Iceland. PHOTO: C. WEIR

***L. albirostris* is a chunky, slower-swimming dolphin with a large fin and distinctive white band on the side. Scotland.** PHOTO: C. WEIR

White-beaked dolphins tend not to porpoise cleanly out of the water when they travel fast, but skim just over the surface instead. Scotland. PHOTO: C. WEIR

Another view of the characteristic oblique white band on the side of the white-beaked dolphin; the stubby white beak is also visible here. Scotland. PHOTO: C. WEIR

A view of the white ventral surface and black side patch of the white-beaked dolphin. England. PHOTO: T. BRERETON

White-beaked dolphins have a large dark patch below the dorsal fin and another above and behind the flippers, separated by an oblique white band. England. PHOTO: T. BRERETON

Even from the air, the conspicuous, oblique white band and prominent dorsal fin of *L. albirostris* allows for easy identification. Norway. PHOTO: F. BROMS

White-beaked dolphin, like all *Lagenorhynchus* species, has a complex color pattern rarely seen in its entirety like this. Norway. PHOTO: R. SVENSEN

western Mediterranean. They inhabit continental shelf and offshore waters, although there is evidence suggesting that their primary habitat is in waters less than 200 m deep.

Ecology and behavior The behavior and ecology of white-beaked dolphins has received little detailed study. Groups of less than 30 white-beaked dolphins are most common, but herds of many hundreds (or even occasionally thousands) have been seen. These animals are active, often leaping and breaching, and they often approach vessels to ride bow or stern waves. However, they may be quite elusive in some areas. They sometimes associate, while feeding, with large whales (such as fin and humpback whales), and are known to form mixed groups with a number of other dolphin species (including common bottlenose and Atlantic white-sided dolphins). Cooperative feeding has been observed. Little is known of group structure or association patterns, but there is some evidence of age and sex class segregation among schools. Mass strandings are relatively rare.

There appears to be a protracted calving peak in summer to early fall (May to September), but not much else is known about reproduction in this species. Attainment of sexual maturity has not been well-studied, but males appear to become mature at around 2.5 m and females at around 2.4 m.

Feeding and prey White-beaked dolphins feed on a variety of small mesopelagic and schooling fishes (such as herring, cod, haddock, poor cod, bib, hake, and whiting), squids, and crustaceans.

Threats and status Although not a target of any commercial fisheries, some white-beaked dolphins are shot in Greenlandic waters, and they have been hunted opportunistically in other countries (e.g., Norway, Iceland, and Canada) as well. In addition, incidental catches in gillnets, cod traps, and trawl nets are also known from several areas of the species' range. Some dolphins may die from collisions with vessels, although there is little documentation of this. Some also die from getting entrapped in encroaching ice. The species is not considered to be threatened. Declines in abundance have occurred in some areas, such as the Gulf of Maine (however, this may have more to do with habitat shifts than changes in population size). At least 7,800 white-beaked dolphins are estimated to inhabit the North Sea and adjacent waters. There are few estimates of abundance, but it is clear that there are many tens of thousands (or even low hundreds of thousands) of these animals throughout their range.

IUCN status Least Concern.

Atlantic White-sided Dolphin—*Lagenorhynchus acutus*

(Gray, 1828)

Recently-used synonyms *Leucopleurus acutus.*
Common names En.–Atlantic white-sided dolphin; Sp.–*delfín de flancos blancos;* Fr.–*dauphin à flancs blancs de l'Atlantique.*
Taxonomic information Cetartiodactyla, Cetacea, Odontoceti, Delphinidae. The Atlantic white-sided dolphin is currently placed in the multi-species genus, *Lagenorhynchus.* However, molecular analyses suggest that it is not closely related to any of the other species, and it is possible that it will be split off from them into its own genus *(Leucopleurus)* in the next few years.
Species characteristics Atlantic white-sided dolphins are robust and deep-bodied (maximum girth can be up to 60% of total length), with a tail stock that is strongly deepened into "keels" above and below. The deepening of the tail stock is more pronounced in adult males than in other age/sex classes. The beak is very short, but is somewhat well-defined from the rounded melon. The dorsal fin is tall (more than 12% of the total length), slightly falcate, and typically pointed. The flippers are moderately large and pointed at the tips, and there may be tubercles on the leading edges. The flukes are unremarkable.

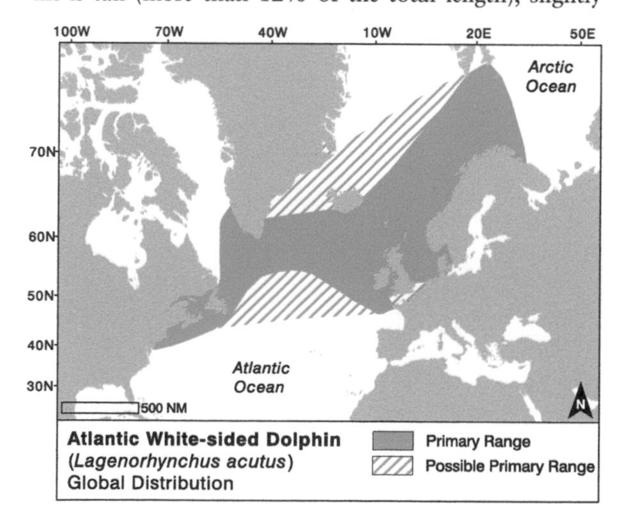

The color pattern is complex and striking. The back and upper sides, upper jaw, dorsal fin, flippers, and flukes are black or dark gray, and a thin, dark line runs backwards from the beak and meets a black patch around the eye. The characteristic delphinid "bridle" (consisting of an eye stripe and blowhole stripe) is present. There is a thin eye-to-flipper stripe. The lower jaw and belly, as far as the urogenital area, are white. In between, the sides from just ahead of the eyes to the base of the flukes are light gray. Along the upper margin of the gray side is a white patch from below the dorsal fin to midway along the tail stock. There is another narrow band, this one yellow to ochre in color, at the lower margin of the dark upper flank, from mid-tail stock to just in front of the flukes. Calves appear to have a muted pattern with less contrast than adults. Some individuals with unusual pigmentation patterns have been observed, most commonly, dolphins with more-extensive amounts of white on the body.

Each tooth row contains 30–40 teeth. The teeth are small and pointed. Adult Atlantic white-sided dolphins reach 2.8 m (males) or 2.5 m (females) in length and about 235 kg (males) or 182 kg (females)in weight. Newborns are 1.1–1.2 m.
Recognizable geographic forms None.
Can be confused with Confusion is most likely with the white-beaked dolphin, which shares a nearly identical range (however, Atlantic white-sided dolphins are more common in the southern parts of the range). The two can be distinguished most readily by size, color pattern, and subtle head shape differences. The Atlantic white-sided dolphin is smaller and more slender, and its color pattern is much more bold and striking, with more distinct flank patches. It also lacks the white tail stock of the white-beaked dolphin.

L. acutus **readily porpoises out of the water when traveling fast; the common name "white-sided dolphins" seems a bit of a misnomer. New England.** PHOTO: M. OKTAY KAYA

The narrow ochre patch on the tail stock of *L. acutus* is one of the few examples of true color on a cetacean, which are normally just shades of gray, black and white. New England. PHOTO: M. WHITE

With their regular cape pattern, Atlantic white-sided dolphins appear more similar to *Stenella* dolphins than to other *Lagenorhynchus* species, which have more complex patterning. New England. PHOTO: M. WHITE

Lagenorhynchus **dolphins all have stubby beaks and prominent dorsal fins in common.** PHOTO: M. OKTAY KAYA

An aberrantly-patterned Atlantic white-sided dolphin stranded in Scotland—the white and ochre pattern element is missing from the flank. PHOTO: J. BATTEN

The stubby beak and white and ochre color pattern on the flanks readily identify this as *L. acutus.* New England. PHOTO: M. WHITE

Distribution Atlantic white-sided dolphins are found in cold temperate to subpolar waters of the North Atlantic, from about 38°N (south of Cape Cod) in the west and the Brittany coast of France in the east, north to southern Greenland, Iceland, and southern Svalbard. The range includes the United Kingdom and the northern coasts of Scandinavia, although they rarely enter the Baltic Sea. They also sometimes move quite far up the Saint Lawrence River of eastern Canada. The preferred habitat appears to be deep waters of the outer continental shelf and slope, but these dolphins are found in relatively shallow and oceanic waters across the North Atlantic as well. Seasonal shifts in abundance have been noted in several areas of the range.

Ecology and behavior Much of what we know of this species' biology comes from examination of several mass-stranded herds. Herd size varies considerably, with most groups numbering less than 100 (the average in New England coastal waters is about 52). However, up to 2,000–4,000 are sometimes seen in a single school, and there may be age and sex segregation of herds. Older immature individuals are not generally found in reproductive herds of mature females and young. Little is known of long-term association patterns, but there is some evidence of some stable subgroups within large herds. They often strand in large groups. These dolphins associate and feed with large baleen whales (fin and humpback whales), and are known to form mixed groups with long-finned pilot whales and a number of other dolphin species (including bottlenose and white-beaked dolphins).

Atlantic white-sided dolphins are lively and acrobatic, often breaching and tail-slapping. Aerial behavior is more common in larger groups. Larger animals, especially males, are avid bowriders, and also will ride the stern wakes of vessels. These animals are not known to

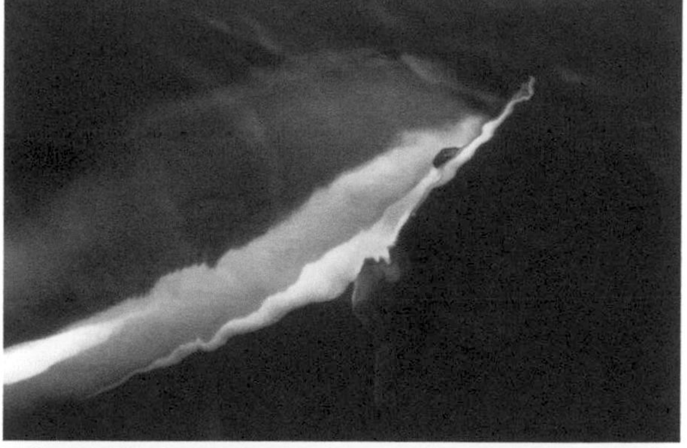

Unlike the other *Lagenorhynchus, L. acutus* has essentially a tripartite color pattern with dark gray cape, pale gray sides, and white belly. PHOTO: M. OKTAY KAYA

Except for the stubby beaks and larger dorsal fins, these Atlantic white-sided dolphins appear very similar to any of several *Stenella* dolphin species. New England. PHOTO: M. WHITE

The markedly deeper tail stock of the adult male *L. acutus* in the foreground distinguishes it from the female/juvenile ahead of it. Norway. PHOTO: F. BROMS

A view of the white ventral patterning of *L. acutus*; as in many cetaceans, the genital area is highlighted. Norway. PHOTO: F. BROMS

be long divers, with the longest recorded dive of a tagged dolphin being just about 4 min. Coordinated "ball" feeding on schools of sand lance has been observed off New England. Atlantic white-sided dolphins appear to have extensive movements, and there is currently no evidence for separate populations.

Life history is rather poorly known, despite the fact that this species live-strands in large herds. Calves are born over an extended period around the summer season, with apparent peaks in June and July. Sexual maturity occurs at around 2.01–2.22 m in females, and at larger sizes for males (different studies suggest either 2.3–2.5 m or 2.15–2.30 m). These correspond to ages of 6–12 years. Longevity can be at least 22 years.

Feeding and prey Atlantic white-sided dolphins feed mostly on small schooling fish (such as herring, mackerel, cod, smelt, hake, and sand lance), shrimp, and squid. They often feed in association with large whales and other species of dolphins.

Threats and status Atlantic white-sided dolphins have been hunted in drive fisheries in Norway, Newfoundland, Greenland, and the Faroe Islands. Incidental catches are known from many different areas of the species' range, and although they do get caught in gillnets, captures in mid-water trawls appear to be the most significant issue. Recent incidental catches in trawls off Ireland have been quite large. This species is quite abundant throughout its range, and the total numbers are estimated to be in the hundreds of thousands. There are thought to be over 52,000 Atlantic white-sided dolphins off the eastern North American shoreline, and about 96,000 off the west coast of Scotland. On a global scale, the species is not considered to be threatened by human activities.

IUCN status Least Concern.

Pacific White-sided Dolphin—*Lagenorhynchus obliquidens*

Gill, 1865

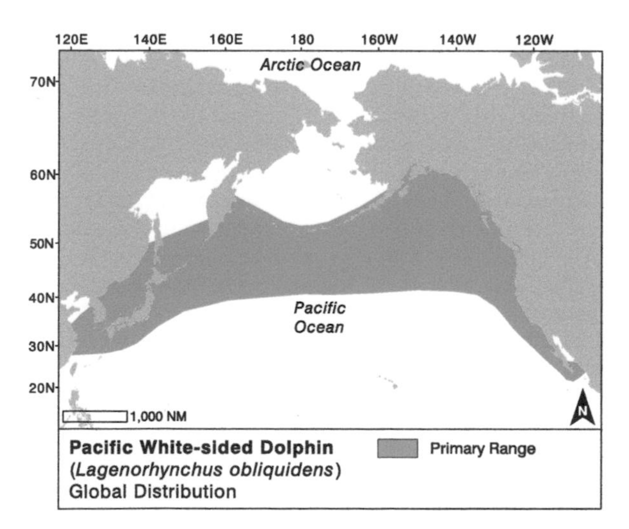

Recently-used synonyms *Lagenorhynchus ognevi.*

Common names En.–Pacific white-sided dolphin; Sp.–*delfín de costados blancos del Pacífico*; Fr.–*dauphin à flancs blancs du Pacifique.*

Taxonomic information Cetartiodactyla, Cetacea, Odontoceti, Delphinidae. Some biologists previously suggested that the Pacific white-sided dolphin may be a subspecies of the dusky dolphin, but systematic research has disproved this idea. This species has a complicated taxonomic position, and it may be split out into a separate genus (possibly *Sagmatias*) in the near future.

Species characteristics Pacific white-sided dolphins, like all members of the genus *Lagenorhynchus,* are stocky animals with very short, thick beaks. There is a shallow crease, which defines the beak from the melon. The large flippers are recurved, with slightly rounded tips. The dorsal fin, one of the most diagnostic features, is large and strongly recurved in some individuals. In some older

A close-up view of the facial patterning and bi-colored dorsal fin of *L. obliquidens.* This one is presumably an adult male, based on the very hooked dorsal fin. Monterey Bay. PHOTO: M. COTTER

Younger Pacific white-sided dolphins have a less lobed and prominently hooked dorsal fin. Vancouver Island. PHOTO: M. MALLESON

All cetaceans occasionally have aberrantly-colored individuals among their ranks; this leucistic ("partial albino") *L. obliquidens* would probably not be identifiable from its coloration alone. Monterey Bay. PHOTO: T. A. JEFFERSON

Pacific white-sided dolphins rarely pass up an opportunity to ride the bow of a passing vessel. California. PHOTO: C. OEDEKOVEN

animals (apparently adult males), the dorsal fin becomes so hooked (lobate) that the tip may curl down to a point below the midpoint of the dorsal fin height. It remains relatively wide nearly to the tip. The flukes can have a slightly concave trailing edge and possess a median notch

The color pattern is complex and gorgeous. The dark gray back and light gray sides are distinctly set-off from the white belly by a black border. Light gray streaks beginning on the sides of the melon above the eyes sweep back and then downwards gradually past the dorsal fin and expand into large gray flank patches. In the thoracic area they form light gray "suspender stripes." The lips and beak tip are black, while most of the lower jaw is white. The characteristic delphinid "bridle" (consisting of an eye stripe and blowhole stripe) is present. The anterior part of the dorsal fin is dark gray, and the posterior portion is light gray to white. The upper surfaces of the flippers may also contain light patches. The color of young calves is muted, much lighter than in adults, and newborns may have an orangish tinge.

There are several uncommonly-occurring morphs. The most common one (referred to as "Brownell type") has the suspender stripe above and just behind the eye greatly expanded and white in color, and a very dark thoracic patch. Another one is an animal with the regular color pattern, but with the gray and black shades replaced by an orangish-rust color. The third is largely white (but is not an albino), with some remnants of the normal color pattern (e.g., black beak tip, lips, and top of head, and dark gray fringing on appendages and sides). All-black individuals have also been reported. Several individuals with these unusual color morphs may be present in large schools.

Each tooth row contains 23–36 pairs of relatively-fine, sharply-pointed teeth. Eastern North Pacific adults of this species reach 2.5 m in length in males and 2.4 m in females. In the western Pacific, they reach up to 2.4 m (males) and 2.3 m (females). Maximum weight is about 198 kg. Length at birth is estimated to be 92–100 cm in the central Pacific.

Acrobatic Pacific white-sided dolphins launching out of the water, as if on cue. British Columbia. PHOTO: J. TOWERS

Recognizable geographic forms Two geographical forms have been described from the eastern North Pacific and two for the western Pacific. However, they are known to differ from each other mainly in modal length and cranial characteristics, and are generally not distinguishable in the field.

Can be confused with Among sympatric species, Pacific white-sided dolphins are most likely to be confused with both common dolphins, because all three species are found in large schools and have large light-colored flank patches. Beak length (much longer in common dolphins) and specifics of the color pattern (e.g., no suspender stripes on common dolphins) are the best keys to distinguishing them. Also, common dolphins generally have a much more erect dorsal fin.

Distribution Pacific white-sided dolphins inhabit cool temperate waters of the North Pacific and some adjacent seas (e.g., Sea of Japan, Okhotsk Sea, southern Bering Sea). Although they are widely distributed in deep offshore waters, they also extend onto the continental shelf and very near shore in some areas. They reach their southern limits at the mouth of the Gulf of California (and occasionally venture northward in the Gulf to or beyond La Paz) and southern Japan (records from Taiwan are considered to be misidentifications). They occur seasonally in some of the inshore waters of the Pacific Northwest (inland waters of Washington, British Columbia, and southeast Alaska). Seasonal inshore/offshore and north/south movements have been documented. On both eastern and western sides of the Pacific, separate stocks have been documented.

Pacific white-sided dolphins have a brilliant white belly, separated from the sides by a dark border. Eastern Pacific. PHOTO: M. NOLAN

Social and highly acrobatic, Pacific white-sided dolphins can be found in very nearshore waters or hundreds of miles offshore. California. PHOTO: S. WEBB

This is a recurring anomalous type of Pacific white-sided dolphin, often dubbed the "Brownell" type, after the biologist who first described it. California. PHOTO: S. WEBB

Color pattern variants of *L. obliquidens* are fairly common, often with a dark area above the flipper instead of the usual "white sides." Western Pacific. PHOTO: K. SEKIGUCHI

Ecology and behavior Often seen in large herds of hundreds or even thousands, these highly-gregarious dolphins are also commonly seen with a large variety of other marine mammal species, especially northern right whale dolphins and Risso's dolphins. Predation by killer whales has been observed. There is often segregation of schools, according to age and sex. They are highly acrobatic and playful, commonly bowriding, and often leaping, flipping, or somersaulting. However, there have been few detailed studies of the behavior of this fascinating species of dolphin. Large herds are often observed feeding in an apparently-cooperative manner on large schools of fish. Dives of over 6 min have been recorded. Males in the central North Pacific reach sexual maturity at about 10 years and 1.7–1.8 m, and females at 8–11 years of age and 1.75–1.86 m. Calving in some areas apparently occurs during a protracted summer breeding season, which extends into fall. In the central Pacific, calving takes place in late winter to spring.

Feeding and prey Pacific white-sided dolphins feed mostly in epipelagic and mesopelagic waters on small schooling fishes, such as lanternfish, anchovies, sauries, horse mackerel, and hake, as well as cephalopods. There is evidence that these dolphins feed mostly on deep scattering layer (DSL) organisms, possibly using cooperative foraging techniques.

Threats and status Pacific white-sided dolphins have never been the primary targets of Japanese drive fisheries, but some dolphins have been harpooned and taken in drives in Japanese waters. They are taken in a number of fisheries (mainly gillnet and driftnet fisheries, but also trawls and purse seines) in the eastern North Pacific, and in the 1980s and early 1990s, several thousand per year were killed in the now-much-reduced squid driftnet fisheries that operated in the offshore central Pacific by Japan, Taiwan, and Korea. Estimates of total Pacific white-sided dolphin abundance range from about 930,000 to 990,000, but both of these may be overestimates (due to unaccounted-for dolphin attraction to survey vessels). The species is not considered to be threatened.

IUCN status Least Concern.

Dusky Dolphin—*Lagenorhynchus obscurus*

(Gray, 1828)

Recently-used synonyms *Lagenorhynchus fitzroyi, Lagenorhynchus superciliosus.*

Common names En.–dusky dolphin; Sp.–*delfín obscuro;* Fr.–*dauphin sombre.*

Taxonomic information Cetartiodactyla, Cetacea, Odontoceti, Delphinidae. Some researchers used to believe that the Pacific white-sided dolphin was a subspecies of *L. obscurus,* but their specific distinction has now been established. This species has a complicated taxonomic position, and it may be split out into a separate genus (possibly *Sagmatias*) in the near future. Four subspecies are recognized: *L. o. obscurus* in southern Africa, *L. o. fitzroyi* in Argentina, *L. o. posidonia* in Peru/Chile, and an unnamed subspecies in New Zealand (suggested by some to be *L. o. superciliosus*). Hybrids with common dolphins and southern right whale dolphins have been reported from the wild off South America and New Zealand.

Species characteristics The dusky dolphin is a small, moderately-robust species. The rostrum is short and pointed, and clearly demarcated from the melon (forehead). The conspicuous dorsal fin is moderately falcate and pointed. The flippers are moderately curved on the leading edge, with bluntly rounded tips. The flukes are unremarkable.

The body coloration is complex, and is generally countershaded, dark gray to blue black above and white below. The sides are marked with blazes and patches of pale gray. In front of the dorsal fin, they bear a broad light gray thoracic patch that encompasses the face, most of the head, and thorax, tapering towards the belly. A separate crescent-shaped flank patch reaches the top of the tail stock, just in front of the flukes. At the front of this, the flank patch splits into two blazes, a shorter ventral and a longer dorsal one; this latter one narrows and stretches up onto the back, almost to the blowhole. The

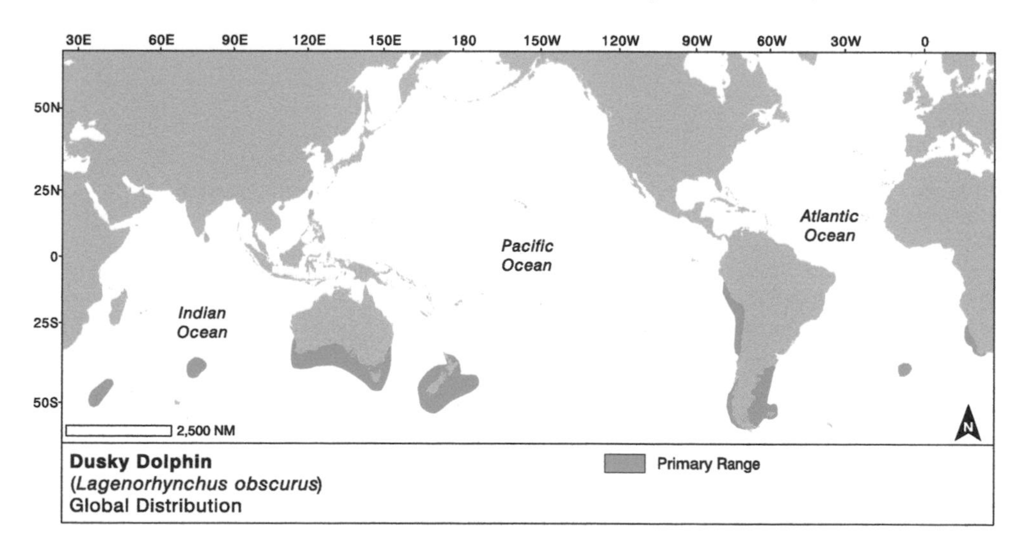

Dusky Dolphin
(*Lagenorhynchus obscurus*)
Global Distribution

Dusky dolphins are among the most acrobatic of all marine mammals, and schools often have multiple individuals launching out of the water, as if shot out from cannons. New Zealand. PHOTO: COURTESY B. WÜRSIG

One of the most acrobatic dolphin species, even among schools of milling, apparently-resting dusky dolphins, individuals will be performing aerial feats. New Zealand. PHOTO: M. NOLAN

rostrum is gray-black around the tip, tapering back to darken just the lips near the gape. The eye is set in a small patch of gray-black. The characteristic delphinid "bridle" (consisting of an eye stripe and blowhole stripe) is present. A variable crescent of pale gray contrasts the trailing half of the dorsal fin with the dark-colored front half, and the flippers are pale gray, but darken around the edges.

There are 27–36 small, pointed teeth on each side of each jaw. The maximum recorded length is 2.1 m. Most adults are less than 2 m long, and there is not known to be any significant difference in the size of males and females. Specimens from Peru appear to be larger than those from South Africa or New Zealand, but sample sizes are small. Healthy adults weigh 70–85 kg. Length at birth has been reported to be 80–100 cm.

Recognizable geographic forms Other than the apparent differences in size among dusky dolphins from different stocks, there are no known reliable differences in external appearance of dusky dolphins from different parts of the range.

Can be confused with At sea, dusky dolphins can be distinguished from the closely-related, but larger and more robust, Peale's dolphin, primarily by careful attention to

Like several other *Lagenorhynchus* species, dusky dolphins have a bi-colored dorsal fin and complex color patterning. New Zealand. PHOTO: H. MINAKUCHI

The lovely face of the "not-so-dusky" dusky dolphin. New Zealand. PHOTO: M. NOLAN

Dusky dolphins often occur in large, dense schools and are quick to approach passing vessels to ride the bow wave. Peru. NOAA FISHERIES, SWFSC

All fast-swimming dolphins, including duskies, regularly porpoise out of the water to save energy when they travel fast. New Zealand. PHOTO: J. Y. WANG

Dusky dolphins often make acrobatic leaps, but when traveling fast, they may porpoise or just slice along the surface. Chile. PHOTO: P. CACERES

The pale flank patch of *L. obscurus* is very distinctive: a white "smile" with a narrow band rising up and forward, along the back toward the head. Peru. PHOTO: NOAA FISHERIES, SWFSC

color pattern components (especially the much-darker beak of Peale's dolphin). They may also be confused with common dolphins, which share a light anterior thoracic patch. The much longer beak of the common dolphins will be the best feature to distinguish them.

Distribution Dusky dolphins are widespread in the Southern Hemisphere. They occur in apparently-disjunct populations in the waters off New Zealand (including the Chatham and Campbell islands), southern Australia, central and southern South America (including the Falkland Islands), and southwestern Africa. They also occur around some oceanic island groups (e.g., Tristan de Cunha/Gough, Prince Edward/Marion, Crozet, and Amsterdam/St. Paul islands). This is a more-or-less coastal species and is usually found over the continental shelf and slope. They do dive over deep waters in some areas (e.g., New Zealand), but always along continental slopes. Inshore/offshore shifts in abundance have been noted for Argentina and New Zealand.

Ecology and behavior The behavior of dusky dolphins has been studied in detail in both Argentina and New Zealand. They are highly social, gregarious animals. They sometimes form impressive herds of over 1,000 individuals, but are more likely to occur in groups of 20–500. These dolphins are known to gather into large cooperative groups for feeding on schooling fish. Group foraging is thought to be coordinated through use of a series of acrobatic leaps. Some stable subgroups exist, but the overall social structure appears to be one of fluid associations. Dusky dolphins are one of the most acrobatic of all the dolphins, frequently leaping high out of the water, at times tumbling through the air in a variety of different types of leaps. They readily approach vessels to engage in bowriding. Many species of cetaceans have been observed in association with dusky dolphins.

In Peru, calving is believed to peak in August to October (spring), while in Argentina, South Africa and New Zealand, calving appears to occur in summer months (November to February). Sexual maturity is reached at ages of about 4–6 years in females and 4–5 years in males (the lower values may reflect a density-dependent response to past exploitation).

Seasonal inshore/offshore movements are known for several parts of the range, including both Argentina and New Zealand. Individuals are highly adaptable and may change their feeding strategy to conform to local conditions.

Feeding and prey Dusky dolphins take a wide variety of prey, including southern anchovy near the surface in shallower waters, as well as mid-water and benthic prey, such as squid, hake, and lanternfishes. They may also engage in nocturnal feeding, in association with the deep scattering layer, such as occurs off Kaikoura Canyon in New Zealand. Off southwest Africa, they tend to move offshore during upwelling conditions.

Threats and status Dusky dolphins are known to be taken directly in the multi-species small cetacean fisheries of Peru and Chile, which make use of both driftnets and harpoons for capture. Although recent catches appear to have decreased, the Peruvian catch is still thought to be unsustainable. Dusky dolphins are also killed incidentally in gillnets in New Zealand (though it has decreased to insignificant levels in the past decade), and in mid-water trawl nets in Argentina. There are few statistically-defensible abundance estimates available. However, over 12,000 are thought to occur off New Zealand, and at least 6,600 off the Patagonia coast. Although the species is in no danger of extinction on a global scale, some populations are thought to have been seriously depleted by human activities (e.g., those off Peru).

IUCN status Data Deficient.

Hourglass Dolphin—*Lagenorhynchus cruciger*

(Quoy and Gaimard, 1824)

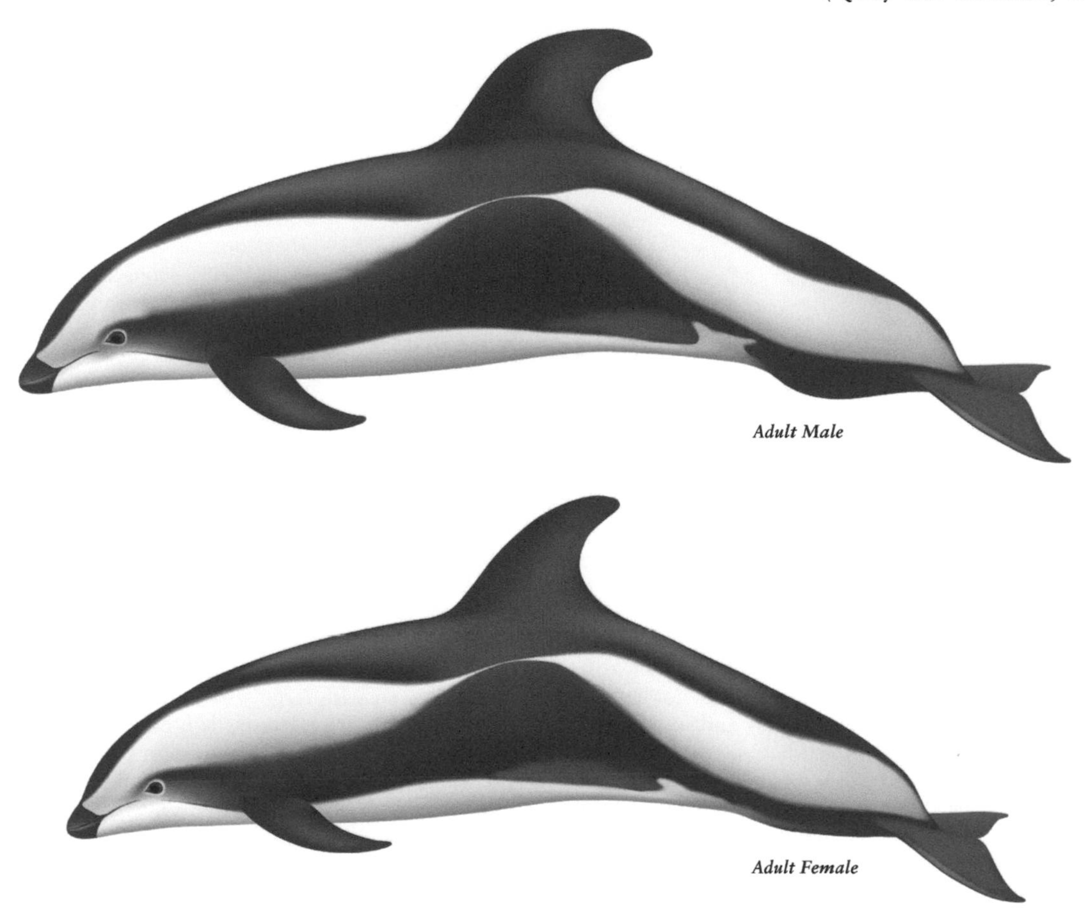

Adult Male

Adult Female

Recently-used synonyms *Lagenorhynchus wilsoni.*
Common names En.–hourglass dolphin; Sp.–*delfín cruzado;* Fr.–*dauphin crucigère.*
Taxonomic in formation Cetartiodactyla, Cetacea, Odontoceti, Delphinidae. This species has a complicated taxonomic position, and it may be split out into a separate genus (possibly *Sagmatias*) in the near future.
Species characteristics Hourglass dolphins are moderately to extremely robust, with extremely-short and stubby (but well-defined) beaks. The moderately-tall dorsal fin is set midway along the back. The markedly hooked dorsal fins seen on some individuals probably develop at the onset of maturity. In some animals (possibly adult males), the dorsal fin also possesses a sharp backward bend about halfway up, and remains very wide to near its tip, with a flattened top. The tail stock is generally deeply keeled in both sexes, but more so in males (which may also have a visible post-anal protuberance). The flippers are recurved, with a concave trailing edge, and the flukes are relatively unremarkable.

Hourglass dolphins are strikingly marked: distinctly black above and white below. The black sides are broken by a bold white flank patch that covers most of the tail stock in a wedge shape, tapering as it rises towards the fin. There, it meets the vertex of a white dorsal-spinal blaze that widens above the flippers, passes above the eyes to cover the sides of the face and finally converges at the gape with the white of the chest and throat. These white markings resemble an hourglass in shape and give the dolphin its common name. The rostrum, forehead, and top of the head are also black. The characteristic delphinid "bridle" (consisting of an eye stripe and blowhole stripe) is present. A white, hook-shaped mark curves up to the black side below the flank patch, near the genital

Hourglass dolphins riding the bow wave of a fin whale in the Southern Ocean; large whales usually seem perturbed by the attention. PHOTO: F. RITTER

Recognizable geographic forms None.

Distribution Hourglass dolphins are distributed in a circumpolar pattern in the higher latitudes of the southern oceans (no other small dolphin species occurs regularly at such high latitudes). They range to the ice edges in the south, but the northern limits are not well known (they are found to at least 45°S, although some occasionally reach 33°S). The most southerly sighting is from near 68°S, in the South Pacific. Hourglass dolphins appear to be largely oceanic, and they are mostly seen in deep offshore waters. However, some sightings have been made in waters of 200 m or less, near islands and banks. This is the only small dolphin species regularly found south of the Antarctic Convergence.

aperture. The flippers, dorsal fin, and flukes are all black. The coloration of calves and juveniles has not been described, but based on what we know about related species, is probably somewhat muted.

Tooth counts of 26–34 (upper) and 27–35 (lower) on each side have been recorded. The teeth are rather slender and pointed. Less than 20 specimens have been measured, so the cited measurements probably do not represent the full range. The maximum lengths and weights so far recorded were 1.9 m and 94 kg (males), and 1.8 m and 88 kg (females). Length at birth is assumed to be about 1 m.

Can be confused with As the only small, oceanic dolphin with a pointed dorsal fin in subantarctic and Antarctic waters, the strikingly-marked hourglass dolphin should not be confused with other species.

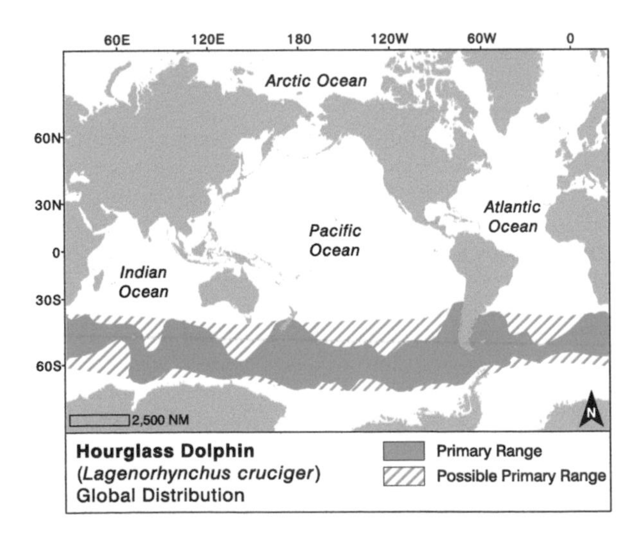

Ecology and behavior Very little is known about hourglass dolphin biology; in fact, this is one of the most poorly known of all the small cetaceans. Groups tend to be small, which is unusual for a small oceanic delphinid. Although herds of up to 60 have been seen, groups of 1–8 are more common. Hourglass dolphins have been encountered with several other species of cetaceans, and are often seen with large whales (especially fin whales). These dolphins are enthusiastic bowriders, often leaping as they race towards the bow or stern of a ship. They can also move rapidly without leaping, usually when avoiding a vessel; at such times they cause a highly visible "rooster tail" of spray. Stranding records of this species are quite rare, probably at least partially related to its oceanic, high-latitude range. Almost nothing is known of the life history of this dolphin species. Based on a handful of specimens, females may reach sexual maturity at around 1.80–1.85 m, and males appear to mature by 1.74 m. Calves have been observed in January and February.

Feeding and prey The stomach contents of the five specimens of hourglass dolphins that have been examined for prey contained small fish (including myctophids), squids, and crustaceans. They often feed in aggregations of seabirds and in plankton swarms.

Threats and status The only known exploitation of hourglass dolphins has been several individuals taken for scientific research, one taken in a Japanese experimental drift net in the South Pacific, and three specimens incidentally killed in a gillnet operation in the South Pacific Ocean. Because of its habitat in one of the most remote marine regions known, this species is not thought to be threatened by human activities. There are estimated to be 144,000 hourglass dolphins south of the Antarctic Convergence.

IUCN status Least Concern.

One of the most strikingly-patterned dolphin species and an avid bowrider, the hourglass dolphin is a welcome sight to mariners in the Southern Ocean. Drake Passage. PHOTO: V. BEAVER

The strongly hooked dorsal fin and prominent post-anal keel identify this as an adult male. PHOTO: D. SINCLAIR

Instead of just black and white, for unknown reasons some hourglass dolphins also show shades of brownish-gray. Southern Ocean. PHOTO: M. SCHUY

Individually-recognizable color pattern variation in hourglass dolphins is clearly evident to the human eye and almost certainly to other dolphins. Drake Passage. PHOTO: V. BEAVER

Hourglass dolphins typically travel in small groups of less than 10 animals and are easily overlooked as they slip in to ride the bow of a passing vessel. Drake Passage. PHOTO: D. GAULTIERI

One of the most beautiful dolphins, the striking black and white hourglass pattern of *L. cruciger* is unmistakable. PHOTO: D. SINCLAIR

Color pattern variants are common among *Lagenorhynchus* dolphins. The animal in the foreground is an anomalously-patterned *L. cruciger* from South Georgia. PHOTO: I. CUMMING

Peale's Dolphin—*Lagenorhynchus australis*

(Peale, 1848)

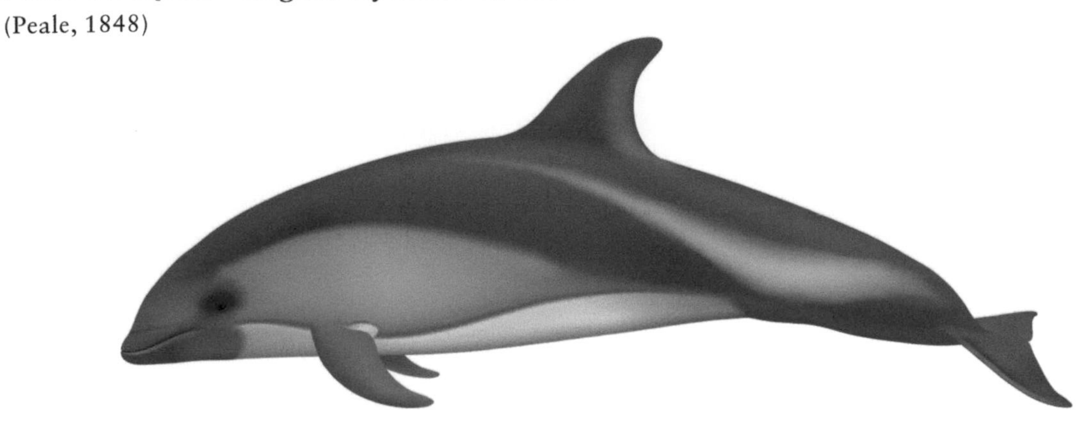

Recently-used synonyms *Sagmatias australis, Sagmatias amblodon.*

Common names En.–Peale's dolphin; Sp.–*delfín austral;* Fr.–*dauphin de Peale.*

Taxonomic information Cetartiodactyla, Cetacea, Odontoceti, Delphinidae. This species has a complicated taxonomic position, and it may be split out into a separate genus (possibly *Sagmatias*) in the near future.

Species characteristics The general body shape of Peale's dolphin is typical for dolphins of the genus *Lagenorhynchus*. Few specimens have been examined, but Peale's dolphins are fairly robust and stocky. They have a short, stubby beak, which is only moderately well-delineated from the melon. They have a tall, slightly falcate dorsal fin. The flippers are recurved and pointed, and the flukes are similar to those of other dolphins.

Peale's dolphins share coloration pattern components with both dusky and Pacific white-sided dolphins. Peale's dolphins are grayish-black above and white below. They have a curved flank patch of light gray with a single dorsal spinal blaze, or "suspender," fading into the black of the back near the blowhole. A large pale gray thoracic patch extends from the eye to mid-body; it is separated from the white below by a well-developed dark stripe. The stripe loops up above a small white patch under the flipper. The flippers are gray-black, and the dorsal fin is dark gray-black, with a thin crescent of light gray on the trailing margin. The most distinctive feature is that the beak tip, lips, and entire lower jaw are dark gray to black in color (this coloration is unique in the genus *Lagenorhynchus)*. The characteristic delphinid "bridle" (consisting of an eye stripe and blowhole stripe) is present. Calves have a lighter color pattern than adults.

Tooth counts for the few examined specimens were up to 37 (upper) and 34 (lower) on each side of the jaw. Only a few dozen specimens have been measured. The largest male recorded was 2.2 m long and the largest female was 2.1 m. Of only five specimens weighed, the heaviest (an adult female) weighed 115 kg. Length at birth is thought to be about 1 m.

Recognizable geographic forms None.

Can be confused with Peale's dolphins are most easily confused with dusky dolphins, which are broadly similar in body shape and general color pattern. The face, rostrum, melon and most of the chin of Peale's dolphins are dark gray-black, as if encased in a mask. This feature, plus the well-developed black stripe below the thoracic patch, helps to distinguish Peale's dolphins from dusky dolphins. At a distance, they may also be confused with hourglass dolphins, but with a better look, the two are easily distinguished.

Distribution Peale's dolphins are apparently confined to waters around South America, from the southern tip to about the latitudes of Santiago, Chile (33°S) and central Argentina (38°S). The distribution may extend south well into Drake Passage. They are regularly seen around the

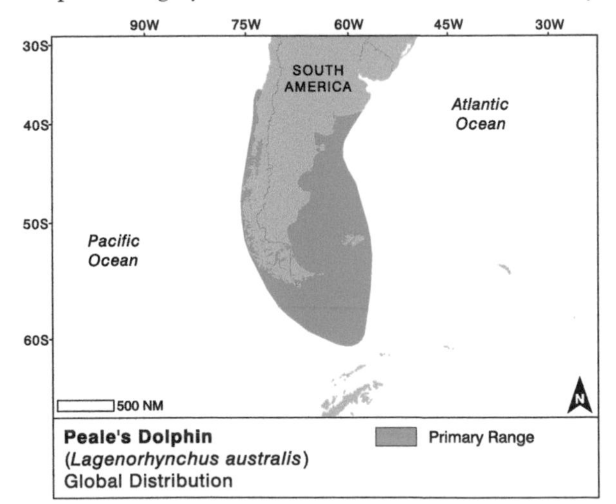

Peale's Dolphin
(*Lagenorhynchus australis*)
Global Distribution

Peale's dolphin is sometimes called "black-faced dolphin"—a feature that distinguishes it from the otherwise-similar dusky dolphin. Argentina. PHOTO: R. PITMAN

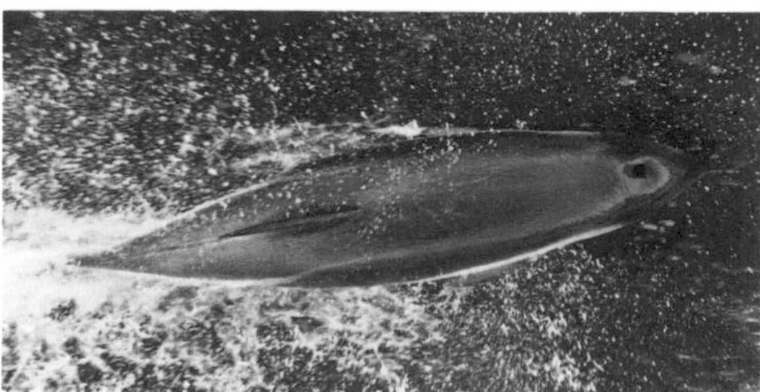

A bowriding Peale's dolphin seen from above, showing all black head and face, robust body, black and white *Lagenorhynchus* patterning, and pale "suspenders" on either side of the dorsal fin. Falkland Islands. PHOTO: M. NOLAN

Peale's dolphin has pale thoracic and caudal patches separated by a thick, oblique black swath; the black face is clearly evident here also. Falkland Islands. PHOTO: D. STILL

A chunky dolphin that occurs in small bands, Peale's is not an avid bow-rider; notice the bi-colored dorsal fin found in several *Lagenorhynchus* species. PHOTO: A. FRIEDLAENDER

Details of facial coloration and the stubby *Lagenorhynchus* beak of Peale's dolphin. PHOTO: G. PAOLO SANINO

Peale's is a chunky dolphin that typically inhabits green coastal waters, usually within sight of land. Falkland Islands. PHOTO: M. NOLAN

Falkland Islands. Peale's dolphins are relatively coastal animals, found in bays and inlets, around islands, and over the continental shelf. They are frequently seen close to shore (often in waters <20 m deep off southern Chile), even shoreward of kelp beds. Occasional sightings have been made in deeper waters, up to about 300 m deep. Inshore/offshore movements have been documented in some areas of the range. One apparently-extralimital sighting of what appears to be this species was reported from tropical waters of Palmerston Atoll (18°S). However, several other sightings of *Lagenorhynchus*-like dolphins in tropical/subtropical waters of the Indo-Pacific have since come to light. This brings up the possibility of a more tropical form of the Peale's dolphin or an undescribed species in this area.

Ecology and behavior Very little is known about the biology of this species. Peale's dolphins have been seen in small groups (5–30 are typical), but some groups of up to 100 have been observed. They often occur in the riptides at the entrances to fjords and channels, and are often seen swimming slowly in and around kelp beds, especially in the Falkland Islands. Dives are known to last up to nearly 3 min, with an average of about 28 seconds.

Aerial behavior, including breaching, spyhopping, and slapping the head, flukes, and flippers on the surface is not

Peale's dolphins are coastal animals and are often found around kelp beds in some parts of their range. Southern Chile. PHOTO: S. HEINRICH

The snowy white underparts contrast with the black face of Peale's dolphin in this ventral view. Chile. PHOTO: R. HUCKE-GAETE

Peale's dolphins sometimes engage in aerial antics while riding the bow waves of passing vessels. Beagle Channel. PHOTO: R. PITMAN

A Peale's dolphin shows its underbelly and chin coloration as it looks back at the photographer. Chile. PHOTO: J. ACEVEDO

The dark face of Peale's dolphin distinguishes it from all others in the genus. Beagle Channel. PHOTO: R. PITMAN

uncommon. Peale's dolphins frequently bowride, and will sprint to a ship's bow. At the bow, they often speed ahead, leap high into the air and fall back into the water on their sides, producing a large splash with a loud slapping noise. They produce a wide splash or wave when swimming near the surface. Peale's dolphins will associate with other cetacean species, especially Commerson's dolphins.

Virtually nothing is known about reproduction, with the exception of a 2.1 m mature female and immature females of 1.9 and 2.0 m. Calves have been reported from spring through autumn.

Feeding and prey Little is know of food and feeding habits in this species, with less than 20 stomachs having been examined. These contained mostly demersal fish, octopus, and squid species that occur in shallow waters and in kelp beds. Some shrimps have also been found in stomachs.

Threats and status A few Peale's dolphins have been taken for scientific research. More substantially, Peale's dolphins have been killed for crab bait in both Chile and Argentina. At times, this exploitation was heavy enough to consider the species at risk of local extirpation, but the mortality has apparently subsided in recent years. Some incidental catches are known in anti-predator nets in Chile, and in gillnets and trawls in Argentina. The only known estimate of abundance is of about 200 animals in southern Chile, but the species is considered to be fairly abundant, at least around the Falkland Islands.

IUCN status Data Deficient.

Northern Right Whale Dolphin—*Lissodelphis borealis*

(Peale, 1848)

Recently-used synonyms None.
Common names En.–northern right whale dolphin; Sp.–*delfín liso del norte;* Fr.–*dauphin à dos lisse boréal, dauphin aptère boréal.*
Taxonomic information Cetartiodactyla, Cetacea, Odontoceti, Delphinidae.
Species characteristics The northern right whale dolphin (along with its Southern Hemisphere relative) is the slenderest of all small cetaceans. The body is slim and torpedo-shaped, and the tail stock is extremely shallow (top to bottom). At close range, northern right whale dolphins are unmistakable; they are the only small cetaceans in their range with no dorsal fin. The flippers are small, narrow, and delicate-looking. The flukes are narrow, but deep, and possess a median notch. The beak is short, but well-defined, and is relatively broad at its base.

Northern right whale dolphins are primarily black, with a distinctly-bordered white band from the throat to the fluke notch that widens to cover the entire area between the flippers, and a white patch just behind the tip of the lower jaw. The trailing edges of the flukes have a crescent-shaped patch of light gray edging above and white below. The characteristic delphinid "bridle" (consisting of an eye stripe and blowhole stripe) is faint. Young animals have muted color patterns of dark creamy gray to light gray. There is an uncommonly occurring color morph (often called the "swirled" form, sometimes mistakenly considered a subspecies, *L. b. albiventris*), in which animals have a more extensive white field, which reaches onto the lower sides and often includes the lower sides of the face. In the thoracic area, this white field extends above the levels of the flippers and generally includes a white swirl at the base of the upper surface of the flippers.

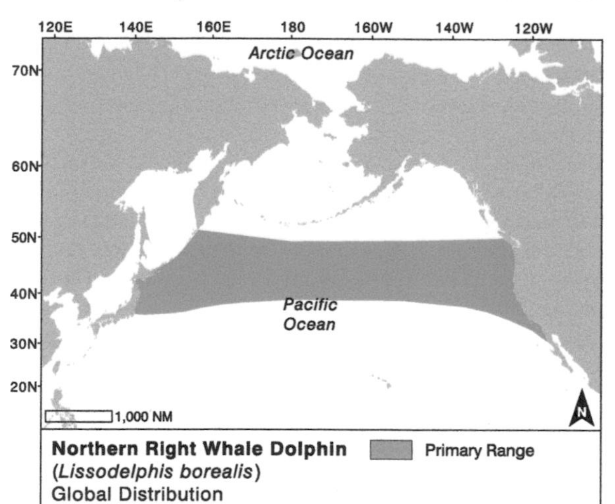

Northern Right Whale Dolphin (*Lissodelphis borealis*) Global Distribution

The mouth contains 37–54 pairs of sharp, slender teeth in each jaw. Measured adults have been up to 2.3 m (females) and 3.1 m (males). Length at birth is about 1 m. Maximum known weight is 115 kg.
Recognizable geographic forms None.
Can be confused with The slender, finless body should allow easy separation from all other sympatric North Pacific small cetaceans. However, porpoising California sea lions may cause some confusion at a distance.
Distribution The northern right whale dolphin is an oceanic species, inhabiting cool and warm temperate regions of the North Pacific only, between about 30°N and 50°N Although it generally does not penetrate as far north as the Bering Sea, there are some northern (probably extralimital) records from along the Aleutian Islands and in the Gulf of Alaska. Also, it is generally not thought to enter the Sea of Japan or Okhotsk Sea, and it is seen nearshore only where submarine canyons or other features bring deep water close to the coast. It forms an anti-tropical species pair with the southern right whale dolphin. The

303

Sprinting northern right whale dolphins—notice the extremely narrow tail stocks and tiny flukes. California. PHOTO: C. OEDEKOVEN

In the distance, the finless northern right whale dolphins can be confused with a traveling group of California sea lions. California. PHOTO: J. SCARFF

Northern right whale dolphins spyhopping, probably having a look around. The tiny beaks are an indication that they take very small prey. California. PHOTO: K. WHITAKER

Northern right whale dolphins are more likely to bowride when they are mixed with other species that more readily approach vessels, such as Pacific white-sided dolphins. California. PHOTO: K. WHITAKER

habitat of this species is characterized by deeper waters from the outer continental shelf to the oceanic regions.

Ecology and behavior Most schools number between 100 and 200 individuals, but groups of up to 3,000 have been seen. Some herds are very tightly packed. Groups of northern right whale dolphins mixed with other marine mammals, especially Pacific white-sided dolphins (with which they share a nearly identical range) and Risso's dolphins, are not uncommon.

Northern right whale dolphins are fast swimmers, sometimes creating a great surface disturbance with their low-angle leaps and belly flops. When moving more slowly, these animals may cruise along and barely break the surface to breathe. They bowride, especially when accompanied by other species of dolphins. Aerial displays are not uncommon, and they breach and perform side-slaps at times.

In the central North Pacific, life history has been well-studied, based on a large sample of specimens killed in the North Pacific squid driftnet fisheries. There appears to be a calving peak in late summer (July and August), and females generally have a 2-year calving interval. Gestation lasts about 12 months. Both sexes become sexually mature at about 10 years of age, when males are about 2.15 m, and females 2.00–2.01 m. They can live to be at least 42 years of age. Reproduction in other areas of the range is very poorly-known.

Feeding and prey Although market squid and lanternfish are the major prey items for northern right whale dolphins off southern California, a variety of other species are taken by this species throughout the range. These include various species of cephalopods, hake, sauries, and several species of surface and midwater fishes.

Threats and status Northern right whale dolphins have never been hunted extensively in a major fishery, although they have sometimes been taken in Japan's small cetacean fisheries. Incidental catches have occurred in Japanese and Russian purse seines, Japanese salmon driftnets, and American shark and swordfish driftnets. However, the largest takes in recent years were in the North Pacific

A wholly different dolphin shape—*Lissodelphis* spp., including this northern right whale dolphin, manage to navigate perfectly well without a dorsal fin; notice the very delicate beak. Monterey Bay. PHOTO: H. MINAKUCHI

Northern right whale dolphins don't often show off their strikingly white underparts like this, which include a chin spot, white chest with narrow ventral stripe, and white flukes. California. PHOTO: NOAA FISHERIES, SWFSC

Aberrant color patterning is fairly common among northern right whale dolphins—this one is beginning to resemble its southern counterpart. This is sometimes called the "swirled" variant. US West Coast. PHOTO: D. DOOLITTLE

Northern right whale dolphins riding in a ship's quarter-wake: the animal on the right has more white than usual on its chin and throat. California. PHOTO: S. WEBB

squid driftnet fishery, which was operated by vessels from Japan, Taiwan, and Korea. Nets from this fishery took many thousands of northern right whale dolphins annually in the central North Pacific, until they were shut down by international law in the early 1990s. There is some evidence to suggest that the large kills in the latter case may have caused serious depletion of the population(s). There are thought to be about 68,000 of these animals in the North Pacific, and slightly over 20,000 of these animals occur in US west coast waters of California, Oregon, and Washington. Now that the squid driftnet fishery is closed-down, this species is no longer thought to be threatened.

IUCN status Least Concern.

Southern Right Whale Dolphin—*Lissodelphis peronii*

(Lacépède, 1804)

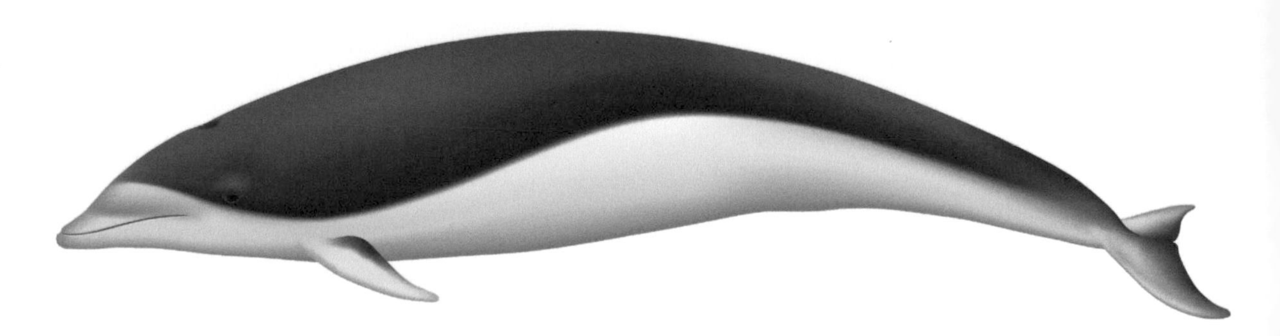

Recently-used synonyms None.

Common names En.–southern right whale dolphin; Sp.–*delfín liso austral;* Fr.–*dauphin aptère austral.*

Taxonomic information Cetartiodactyla, Cetacea, Odontoceti, Delphinidae. Hybrids with dusky dolphins have been reported in the wild.

Species characteristics Southern right whale dolphins, along with their Northern Hemisphere counterparts, are the most slender of all cetaceans (although this species appears to be a bit more robust than the Northern Hemisphere species). The body shape is very similar to the northern right whale dolphin (very slender and torpedo-shaped), with a short, well-demarcated beak (which is quite wide at the base), small recurved flippers, extremely shallow (top to bottom) tail stock, and no hint of a dorsal fin or ridge. The flukes are narrow, but deep, and possess a median notch.

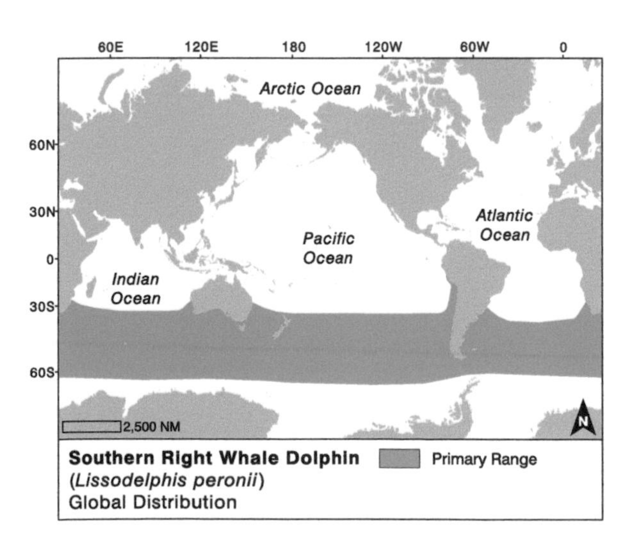

Southern Right Whale Dolphin
(Lissodelphis peronii)
Global Distribution

Southern right whale dolphins are strikingly black and white. The white coloration of the ventral area extends well up on the sides; the sharp line demarcating black above and white below runs from the tail stock forward, dips down to the flipper insertion, and then sweeps back up to cross the melon between the blowhole and beak crease. The flippers are mostly white, but the trailing edge has a black band. The flukes are white below, and above are dark gray, fading to white on the leading edge. Anomalous all-black southern right whale dolphins have been observed, as have dolphins thought to be hybrids between this species and the dusky dolphin. The presumed hybrids have been observed in Argentina, and they appear to be almost exactly intermediate between dusky and right whale dolphins.

The mouth is lined with 44–49 sharp, pointed teeth in each row. These dolphins reach lengths of at least 3.0 m and weights of 116 kg. Length at birth is very poorly-known, but is probably about 1 m.

Recognizable geographic forms None.

Can be confused with The unique body shape, coloration, and complete absence of a dorsal fin of this species should make it difficult to confuse with any other marine mammal species. Only porpoising groups of sea lions or fur seals would likely cause confusion, and that would only be at a distance.

Distribution Southern right whale dolphins are found only in cool temperate to subantarctic waters of the Southern Hemisphere, mostly between about 30°S and 65°S. The southern limit appears generally to be bounded by the Antarctic Convergence. The range extends furthest north along the west coast of continents, due to the cold counterclockwise currents of the Southern Hemisphere. The northernmost record is at 12°S in central Peru. This is an oceanic species, coming close to shore only in deepwater coastal areas.

Dolphins in high latitudes tend to be just black and white, and the southern right whale dolphin is no exception. Chile. PHOTO: P. CACERES

The elegant and unmistakable southern right whale dolphin. This animal appears to have *Xenobalanus* barnacles on its right flipper. Off southern Chile. PHOTO: R. L. PITMAN

Easy to identify, but not often seen, the southern right whale dolphin occurs in temperate waters throughout the Southern Hemisphere. Chile. PHOTO: P. CACERES.

Often in schools of hundreds, in the distance, the finless profiles of southern right whale dolphins can make them look like waves rolling at the surface. New Zealand. PHOTO: T. PUSSER

With its designer color patterning and elegant slender form, the southern right whale dolphin is arguably the most beautiful cetacean. Southern Ocean. PHOTO: L. MORSE

When traveling rapidly in large schools, southern right whale dolphins can roil the surface, like this group off South Africa. PHOTO: J. SMITH

Color pattern variation is common among southern right whale dolphins—including occasional, all-dark individuals as shown here on the left. Southwest of Australia. PHOTO: P. A. OLSON

Ecology and behavior This is one of the most poorly-known of all the dolphins. Large, active schools are characteristic of the southern right whale dolphin. Some estimates of group size range to over 1,000 animals. Associations with other marine mammal species are common, especially with dusky dolphins and long-finned pilot whales. Like their northern cousins, dolphins of this species are active, energetic swimmers, often coming out of the water in clean low-angle leaps, as they move at high speed. Swimming speeds of up to 22 km/hr have been reported, and dives of up to 6.4 min have been recorded. Fluke-slaps and other aerial displays are not uncommon. Southern right whale dolphins bowride occasionally. Virtually nothing is known of this species' reproductive biology.

Feeding and prey A variety of fish and squid have been reported as prey; lanternfish (myctophids) are especially common in southern right whale dolphin stomachs. They are probably capable of diving to depths in excess of 200 m in pursuit of prey.

Threats and status Southern right whale dolphins have been taken directly in recent years in Peru and Chile for crab bait and for human consumption. Bycatches have been reported in small numbers in some areas. The only incidental catch of any magnitude that is known is in the swordfish gillnet fishery off Chile. There are no estimates of abundance for the southern right whale dolphin, and virtually nothing is known of the status of any population of the species.

IUCN status Data Deficient.

Commerson's Dolphin—*Cephalorhynchus commersonii*

(Lacépède, 1804)

South American Form

Kerguelen Form

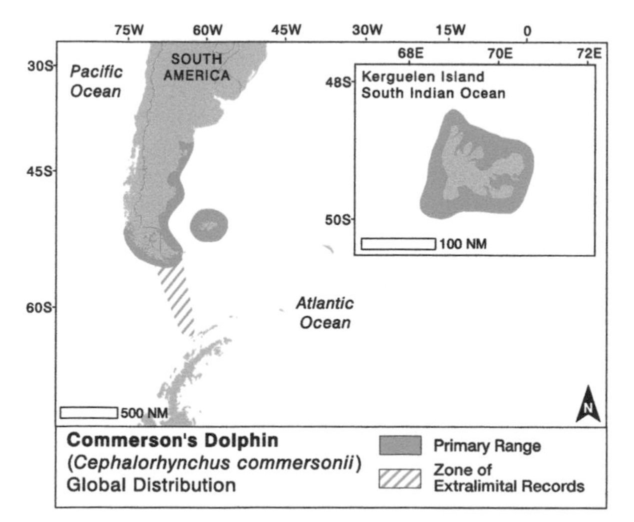

Commerson's Dolphin
(Cephalorhynchus commersonii)
Global Distribution

Recently-used synonyms None.

Common names En.–Commerson's dolphin; Sp.–*tonina overa*; Fr.–*dauphin de Commerson.*

Taxonomic information Cetartiodactyla, Cetacea, Odontoceti, Delphinidae. Two subspecies are recognized: *C. c. commersonii* in southern South America, and *C. c. kerguelenensis* in the Kerguelen Islands. The Kerguelen subspecies was apparently founded by a few individuals as recently as 10,000 years ago. There is growing evidence that the animals around the Falkland Islands may be a separate subspecies as well.

Species characteristics The robust Commerson's dolphin is similar in body shape to porpoises (phocoenids), as are other species of the genus *Cephalorhynchus*. The head is blunt, with only the slightest hint of a beak, and a relatively straight mouthline. The dorsal fin is small (compared to the other species in the genus) and rounded, rising at a

Seemingly more beautiful than they need to be, only in captivity do we get such a clear view of the stunning color pattern of Commerson's dolphin. The low, rounded dorsal fin is characteristic of this species. PHOTO: J.-P. SYLVESTRE

shallow angle from the back, and leaning slightly towards the tail. The flippers and flukes have rounded tips, and there are serrations on the leading edges of the flippers.

The coloration is strikingly contrasted dark and light. There is a light band that completely encircles the body dorsally from just behind the blowhole to in front of the dorsal fin, and ventrally from behind the flippers to behind the genital area. There is a large white patch on the throat, and a black oval to heart-shaped patch around the genitals that tends to vary in shape between males and females (females generally have the heart-shaped patch pointing forward, males pointing backward). The rest of the animal is dark gray or black, including the top of the head, flippers, dorsal fin, and flukes. Newborn animals and calves have a muted pattern of mostly gray tones (instead of black and white), as do many small cetaceans. In calves, the borders between the dark and light regions of the body are less distinct than in adults. Any evidence of the typical delphinid "bridle" (i.e., eye and blowhole stripe complex) is faint.

Twenty-eight to 35 small, pointed teeth line each tooth row. Length at birth ranges from 65–75 cm. Commerson's dolphin adults can reach sizes of up 1.8 m and 86 kg. Females are slightly larger than males in this species.

Recognizable geographic forms

South American subspecies (C. c. commersonii)—South American Commerson's dolphin adults are smaller than the Kerguelen subspecies, reaching only 1.5 m and 66 kg. The body of adults is black and white, with the border between them generally very distinct.

Kerguelen Islands subspecies (C. c. kerguelenensis)—Commerson's dolphins from the Kerguelen Islands are larger than their South American counterparts. These animals reach nearly 1.8 m in length and weights of 86 kg. The dark areas of the body tend to be a dark gray, rather than black. The light band that encircles the body is mostly light gray, but the ventral portion is white. The border between light and dark is generally less distinct than in adult *C. c. commersonii.*

Can be confused with The only other black and white small cetacean likely to be confused with this species is the spectacled porpoise, but the dorsal fin shape and color pattern differences should make these two easily discernible. In some areas, Commerson's dolphins may also be confused with Chilean dolphins, which have a similarly-rounded dorsal fin, but are mostly dark gray (with no distinct black and white patterns on the dorsal surface).

Distribution There are at least two disjunct populations of Commerson's dolphins, those off South America (Chile and Argentina) and the Falkland Islands, and those off the Kerguelen Islands (about 8,500 km away). Recently, a record of a sighting of a single individual south of Cape Town, in South African waters, was reported, although this should undoubtedly be considered extralimital. There are also unsubstantiated reports of this species at South Georgia, but these have been rejected by some recent workers. Commerson's dolphins occur in relatively shallow, coastal waters. They sometimes move very close to shore, even inside the breakers.

Ecology and behavior Small groups of less than 10 individuals are the norm for this species, although they do sometimes aggregate into groups of over 100. These are quick, active animals. They are known to engage in various types of leaps and other aerial maneuvers. They frequently ride bow waves, breakers, and oceanic swells. Commerson's dolphins often swim upside down, and spin on their long axes while darting around. Several individuals have been brought into captivity in the last few decades, and the species seems to do relatively well in captivity.

The breeding season is in the southern spring and summer, between October and March. Gestation lasts slightly less than a year. Sexual maturity occurs between 5 and 9 years. Not much else is known of their reproductive biology. Longevity is at least 18 years.

Feeding and prey South American Commerson's dolphins appear to be opportunistic, feeding primarily near the bottom. Feeding there is on a diverse diet of various

Commerson's dolphins from Kerguelen Island (shown here) comprise an isolated population that is considered a separate subspecies; they are distinguished by their somewhat-larger size and more grayish overall coloration. PHOTO: D. HATCH, COURTESY F. TODD

A rare view of the stunning ventral color pattern of a Commerson's dolphin off Chile. PHOTO: J. ACEVEDO

A coastal species, Commerson's dolphin is almost always seen in green water, within sight of land. Chile. PHOTO: J. ACEVEDO

Although not particularly fast, Commerson's dolphins will porpoise out of the water to get to the bow for a free ride from a passing vessel. PHOTO: D. SINCLAIR

Commerson's, like all dolphins, are very tactile and often in physical contact with each other. Captivity in Japan. PHOTO: H. MINAKUCHI

Commerson's dolphins have rounded appendages (fin and flippers) and a deep, relatively narrow body. Chile. PHOTO: J. ACEVEDO

Commerson's dolphin usually ply nearshore waters, here with a sandy bottom probably less than 1 m deep. PHOTO: H. MINAKUCHI

Dolphins are the original surfers. As a largely-coastal species, Commerson's dolphins are often seen in the surf zone, riding the waves. Falkland Islands. PHOTO: J.-P. SYLVESTRE

The dusky coloration of the animals on the left usually indicates they are younger, but this seems to be more prevalent among all Commerson's dolphins in the Falkland Islands (as here). PHOTO: M. NOLAN

species of fish, cephalopods, crustaceans, and benthic invertebrates. In the Kerguelen Islands, they have a more restricted diet, mostly consisting of fish.

Threats and status Commerson's dolphins have been hunted for food and crab bait in the southern parts of their range. In addition, this species is caught incidentally in several fisheries, primarily those using gillnets. Despite these threats, the species still appears to be reasonably numerous, and is not thought to be in danger of global extinction. There are few estimates of abundance for this species, but they are reportedly quite common in the Strait of Magellan. About 40,000 are thought to occur in Argentine waters. The Kerguelen population is restricted in range, and is therefore probably small and vulnerable to human impacts.

IUCN status Data Deficient.

Heaviside's Dolphin—*Cephalorhynchus heavisidii*

(Gray, 1828)

Recently-used synonyms None.

Common names En.–Heaviside's dolphin, Haviside's dolphin; Sp.–*delfín de Heaviside*; Fr.–*dauphin de Heaviside*. This species was named after Captain Haviside, but due to an error, it was thought that it was named after another person, Captain Heaviside. Thus, the common name should actually be Haviside's dolphin, but due to convention and habit, Heaviside's is still commonly used.

Taxonomic information Cetartiodactyla, Cetacea, Odontoceti, Delphinidae.

Species characteristics Heaviside's dolphin is one of the more poorly-known species of dolphins. The shape of the body is similar to that in other *Cephalorhynchus* dolphins: stocky, with a short blunt snout, and blunt-tipped flippers. The dorsal fin is basically triangular (very different from the rounded fins of the other genus members), but it is significantly taller than in most porpoises. The trailing edge of the dorsal fin often has a slight outward bulge. The flukes are similar in size and shape to those of other dolphins, with a notch, and can be slightly concave on the trailing edge.

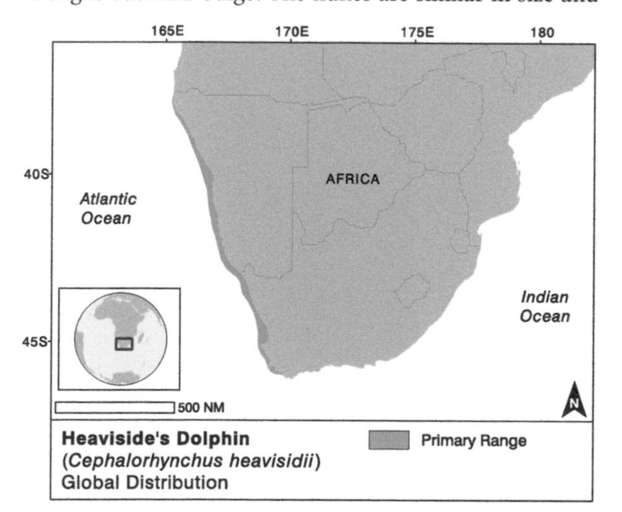

Heaviside's Dolphin (*Cephalorhynchus heavisidii*) Global Distribution

The body is predominantly various shades of gray, with a dark cape, which starts at the blowhole, remains extremely narrow in the thoracic region, and then widens to dip low on the sides below the dorsal fin. Most of the thoracic region is light gray. The area around the eye and much of the face is often a darker shade of gray. There is a white ventral patch that begins just behind the flippers, and splits into three arms behind the umbilicus. The middle arm encloses the urogenital area and the side arms extend only to below the midline (this is reminiscent of the pattern in killer whales). There is also a white diamond-shaped patch between the anterior insertions of the flippers, and separate white spots in the axillae. Several predominantly-white, possibly albino, individuals have been seen.

Heaviside's dolphins have 22–28 small, sharp teeth in each tooth row. Adults of this species can reach at least 1.7 m in length, and 75 kg in weight. Newborn size is unknown, but is likely to be 80–85 cm.

Recognizable geographic forms None.

Can be confused with The only other small cetaceans within this species' range are larger dolphins (e.g., dusky, common, and southern right whale dolphins). Their larger size, different color patterns, and falcate dorsal fins (or absence of a fin in the southern right whale dolphin) should make them easy to distinguish from Heaviside's dolphins.

Distribution This species of dolphin is restricted to southwest Africa (Namibia and South Africa), with records from about 17°S to the southern tip of Africa. It is commonly seen along the west coast of South Africa in the Cape Town region. As is true of the other species in the genus, it is a coastal animal that inhabits mostly shallow waters <100 m deep.

Ecology and behavior Much has been learned of the biology of this species in the past few years. They are

The rounded appendages and chunky body indicate that Heaviside's dolphin is not a particularly fast swimmer. It also has a distinctive cape pattern: narrow in the front and dropping low in front of and below the dorsal fin. South Africa. PHOTO: N. TAKANAWA

Perhaps as an example of convergent evolution or mimicry, the ventral color patterning of Heaviside's dolphin is somewhat similar to that of the killer whale. South Africa. PHOTO: T. PUSSER

Heaviside's dolphins are usually found in small, often-tight groups where they live intensely-social lives. Southern Africa. PHOTO: S. ELWEN

Heaviside's dolphins are not particularly shy, and will often perform various aerial behaviors around boats. PHOTO: S. ELWEN

A leaping Heaviside's dolphin shows the ventral color pattern that is somewhat reminiscent of that of the killer whale—a possible case of mimicry. South Africa. PHOTO: T. PUSSER

A a pair of Heaviside's dolphins launch themselves into the air, though briefly. The uniquely-triangular dorsal fin is characteristic of this species. Namibia. PHOTO: S. ELWEN

seen mostly in small groups of less than 10, with pairs and trios being most common. Heaviside's dolphins are generally active and sometimes boisterous; they are known to ride bow waves. The animals are quick and agile, often zooming around vessels and repeatedly leaping out of the water. They rest in nearshore waters during the day and forage offshore at night, mostly on hake.

Preliminary studies of genetic structure suggest that there might be a single population of this species, but further studies are needed to confirm if this is really true. Photo-identification studies show that Heaviside's dolphins have small home ranges of 876–1,990 km^2), and most alongshore movements are less than 88 km. Association patterns indicate that they have a fission-fusion type society, with essentially random mixing of individuals. Virtually nothing is known of the reproductive biology of Heaviside's dolphin. Some calves are born in the southern summer (November to January).

Feeding and prey The main diet of Heaviside's dolphin consists of demersal fish, such as hake. Some other pelagic schooling fishes and cephalopods (including octopus) are also taken.

Threats and status In general, Heaviside's dolphin appears to be facing fewer threats than the other members of its genus. Some animals are known to be taken incidentally in fisheries, such as in beach seines, purse seines, trawls, and gillnets—the latter are of particular concern. Some animals are apparently illegally taken (shot or harpooned) for human consumption as well. There is also concern about the effects of the growing whale watching industry of South Africa. The major concern has to do with the limited range of the population, and its vulnerable coastal habitat. The overall population size is likely to be over 6,400 individuals, and this bodes well for their long-term survival, in contrast to other species of the genus.

IUCN status Data Deficient.

In Heaviside's, the narrow portion of the dark cape extends only as far as the blowhole; note also the stubby beak common to all *Cephalorhynchus* species. Namibia. PHOTO: S. ELWEN

An obliging Heaviside's dolphin shows the distinctive color patterning on the chest of this species. Namibia. PHOTO: S. ELWEN

Hector's Dolphin—*Cephalorhynchus hectori*

(Van Beneden, 1881)

Recently-used synonyms *Cephalorhynchus albifrons.*
Common names En.–Hector's dolphin; Sp.–*delfín de Hector;* Fr.–*dauphin d'Hector.*
Taxonomic information Cetartiodactyla, Cetacea, Odontoceti, Delphinidae. There are two subspecies recognized: *C. hectori hectori* along the South Island of New Zealand, and *C. h. maui* along the west coast of the North Island.
Species characteristics The typical robust *Cephalorhynchus* body shape is evident in this species. The head is blunt, with only the slightest hint of a beak. The dorsal fin is low (although relatively broad) and rounded (some people say it reminds them of Mickey Mouse ears), with the leading edge rising at a shallow angle from the back. The flippers are rounded and paddle-shaped, with blunt tips. The flukes are unremarkable.

The predominant color of Hector's dolphin is light gray. The dorsal fin, flukes, flippers, area around the blowhole, and much of the face are dark gray to black. A dark "collar" extends from the area above the eyes to behind the blowhole. Ventrally, the animals are largely white. The lower part of the head, starting just behind the black lower jaw tip is white, as is the area from just behind the flippers to the urogenital area. Arms of white from this patch also extend part way up the sides. The white ventral patches can be invaded by black between the flippers, or can be completely separated by a continuous black area. There are also small, white axillary and dark-gray urogenital patches (the latter are smaller and not apparent in some females). The delphinid "bridle" (consisting of an eye stripe and blowhole stripe) is present, but the blowhole stripe is very wide, forming a "cap" on the head.

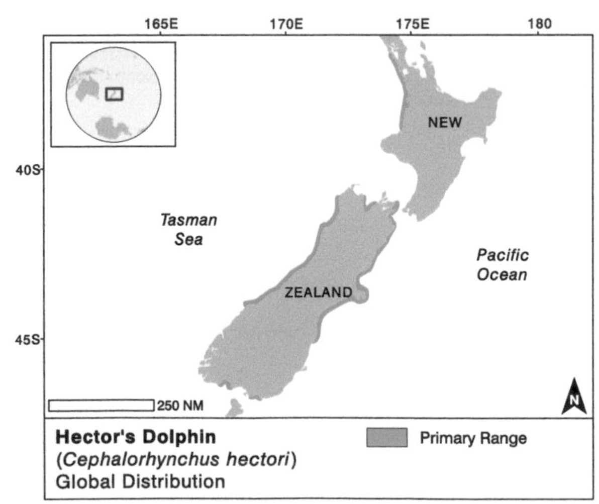

The mouths of Hector's dolphins contain 24–31 fine, pointed teeth in each row. Hector's dolphin adults reach lengths of 1.63 m (females) and 1.46 m (males)—females are somewhat larger than males. Newborns are about 60–70 cm long. Weights of up to 57 kg have been reported.

Recognizable geographic forms

South Island Hector's dolphin (C. h. hectori)—The nominal form of the species occurs in three populations around the South Island of New Zealand. It is slightly shorter than the North Island subspecies, with adults reaching 1.53 m (females) and 1.44 m (males). There is a dark, heart-shaped patch around the penile opening of males.

Maui's dolphin, sometimes called North Island Hector's dolphin (C. h. maui)—This genetically-separate form of Hector's dolphin occurs only off the North Island. It is slightly longer than the South Island subspecies, with adults reaching 1.63 m (females) and 1.46 m (males). Also, the dark, heart-shaped patch around the penile opening of males is either reduced or absent. Other external differences have not yet been described.

Can be confused with Other dolphins (common, dusky, bottlenose, southern right whale, etc.) are found around New Zealand, but should be easy to distinguish from the

The dorsal fin of Hector's dolphin is rounded and overhangs the posterior insertion; the dark at the base of the dorsal fin may be a remnant of what was once a cape. Tooth rake marks on the side and dorsal fin are from interactions with conspecifics. New Zealand. PHOTO: M. NOLAN

Hector's dolphins are quite aerially active, and can be highly acrobatic at times. PHOTO: I. VISSER

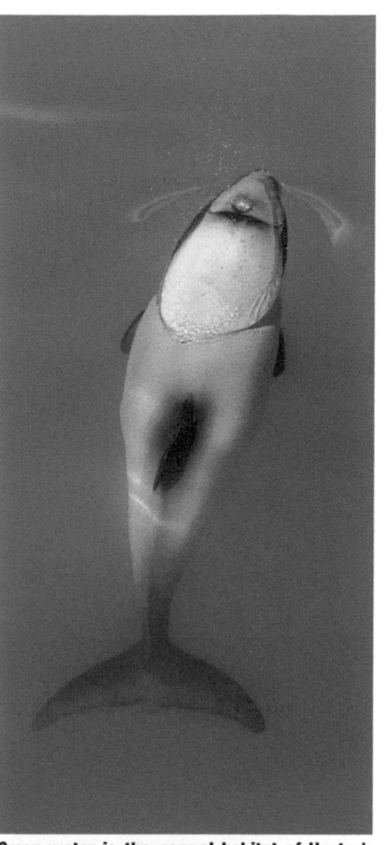
Green water is the normal habitat of Hector's dolphin, and they are almost always seen within sight of land. New Zealand. PHOTO: M. NOLAN

small Hector's dolphin, largely on the basis of dorsal fin shape. These other species mostly have pointed fins, as opposed to the round ones of Hector's dolphin. Hector's dolphin is also significantly smaller than other species of dolphins in New Zealand. Coloration in Hector's dolphin is also unique.

Distribution Hector's dolphin is endemic to New Zealand, giving it one of the most restricted distributions of any cetacean. The animals are found in shallow coastal waters, almost always within about 8 km of shore and <75 m deep. Hector's dolphins are strongly concentrated in shallow, turbid waters close to shore in summer months, and are more widely dispersed in winter. They are most common off the South Island and the west coast of the North Island. There are three genetically-separate populations in the South Island, and the single, small North Island population has recently been named as a separate subspecies *(C. h. maui)*.

Ecology and behavior The habits and biology of Hector's dolphin have been well studied in the last couple of decades, and this is undoubtedly the best-known species of the genus. Hector's dolphins live in groups of 2–8 individuals. Larger aggregations of up to 50 can be seen at times. These are generally the result of several smaller groups coalescing, but they usually do not last long. Hector's dolphins are active, sometimes acrobatic animals, on occasion leaping out of the water quite energetically. They are known to engage in bowriding activity. Photo-identification studies have demonstrated that at least some individuals are resident in small areas (about 30 km of coastline) year-round. No two sightings of an individual have been more than 106 km apart. The social organization is rather fluid, and long-term associations between individuals are rare. The sounds produced by this species are mostly high-frequency, narrow-band pulses, very similar to those made by the true porpoises (phocoenids).

Life history has been better-studied than for the other species in the genus. The mating and calving season is in the southern spring through summer. Gestation lasts 10–11 months. Sexual maturity is reached at ages between 5 and 9 years, and adult females give birth every 2–4 years. They can live to be at least 20 years of age.

Friendly and inquisitive, the diminutive Hector's dolphin has a beautiful color pattern when seen in good light; notice the lobe that extends up and back from the white ventrum. New Zealand. PHOTO: S. DAWSON

An obliging Hector's dolphin shows the distinctive color patterning on the sides of this species. New Zealand. PHOTO: S. DAWSON

Each of the four *Cephalorhynchus* dolphins has a species-specific shape to the dorsal fin; in Hector's it is very rounded and undercut. New Zealand. PHOTO: M. NOLAN

The typical view of Hector's dolphin is of a small group, slow-rolling in nearshore waters; this is Maui's dolphin, a critically endangered subspecies found off the North Island of New Zealand. PHOTO: D. DONNELLY

Feeding and prey Hector's dolphins engage in opportunistic feeding on several species of small fish and squid. The diet is more varied on the east coast of the South Island (eight species make up 80% of the diet) than on the west coast (only four species make up 80%).

Threats and status Recent surveys show that the South Island Hector's dolphin populations collectively number about 7,300 individuals, including about 900–1,100 animals at Banks Peninsula. Maui's dolphin is estimated to number no more than about 100 animals, and both numbers and range have apparently been declining rapidly over the past 30 years.

Hector's dolphin faces serious pressures from human activities. Like its congeners, gillnets are the major concern (with trawl fisheries also causing some mortality), but unlike the other species, the main problem in some areas is recreational gillnet fishing. Up to 57 individuals per year were known to have been caught in a small part of the range in the mid-1980s. That number has been reduced through legislation and the creation of two marine mammal sanctuaries (including the Banks Peninsula Marine Mammal Sanctuary). Despite these laudable efforts, the species is still facing serious problems and most populations are still thought to be in decline. Maui's dolphin is considered to be at serious risk of becoming extinct in the near future.

IUCN status Endangered (overall species), Critically Endangered *(C. h. maui)*.

Chilean Dolphin—*Cephalorhynchus eutropia*

(Gray, 1846)

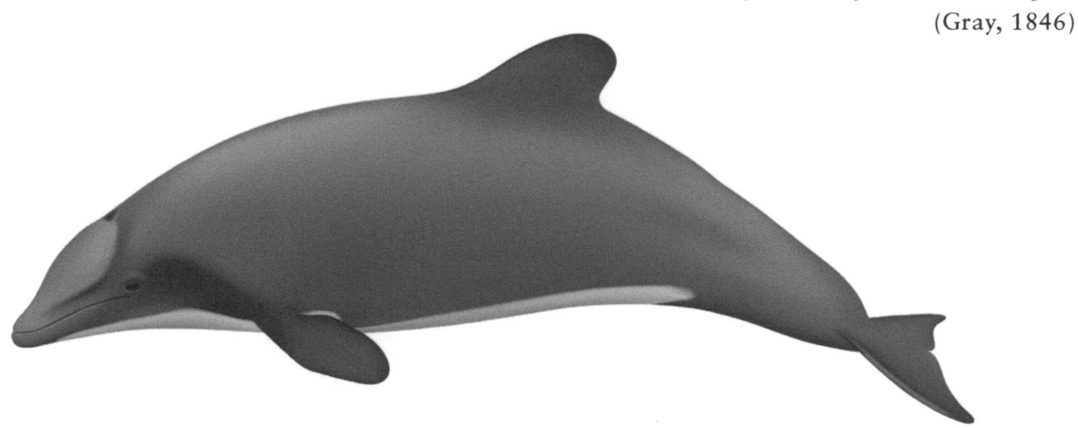

Recently-used synonyms *Cephalorhynchus albiventris.*
Common names En.–Chilean dolphin; Sp.–*delfín chileno;* Fr.–*dauphin noir du Chili.*
Taxonomic information Cetartiodactyla, Cetacea, Odontoceti, Delphinidae.
Species characteristics This little-known dolphin is robust, with a short, poorly-defined beak, like other members of the genus. The head is slightly more pointed than in other *Cephalorhynchus* dolphins. The dorsal fin is moderately low, wide based, and rounded. It is somewhat larger than in Commerson's dolphin, and leans slightly backwards. The flippers are paddle-shaped, with rounded tips, much like those of other members of the genus. The flukes are similar to those of most other dolphin species.

The color of the Chilean dolphin is mostly medium gray, with a darker gray band extending from the blowhole to above the eye. There is often darker gray on the sides of the face, and in a wide band from around the eye to the flipper. The delphinid "bridle" (consisting of an eye stripe and blowhole stripe) is present, but the blowhole stripe is very wide, forming a "cap" on the head. On the belly are large white patches from behind the flippers to the urogenital area, and from ahead of the flippers to the beak tip. These patches are separated by a dark gray band between the flippers. There are also small white patches in the axillae, and thin gray patches around the urogenital area (the latter are sexually-and individually-variable).

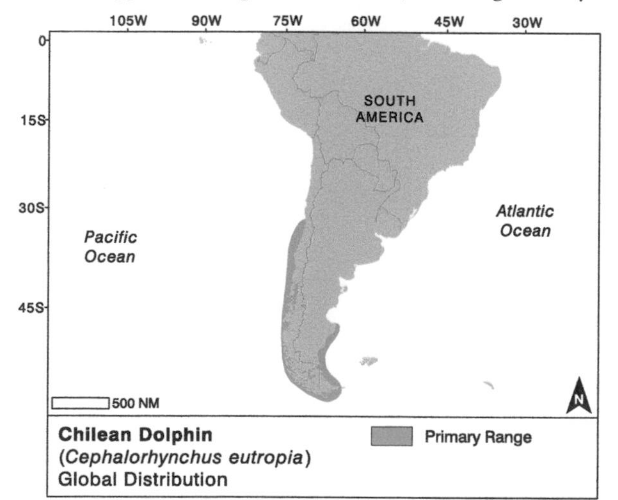

Chilean dolphins have 29–34 small pointed teeth in each row. Adults of this species reach at least 1.7 m (size at sexual maturity has not been sufficiently documented). Length at birth is unknown, but is probably somewhat less than 1 m. Chilean dolphins reach weights of at least 63 kg.
Recognizable geographic forms None.
Can be confused with Chilean dolphins can be confused with Commerson's dolphins in the areas where they overlap, around the southern tip of South America. The large white areas on South American Commerson's dolphins will be the best clue. Burmeister's porpoises may also be confused with this species in some cases. Here, dorsal fin shape will be the best character to distinguish them (rounded in Chilean dolphins, and concave, slender, and almost banana-shaped in Burmeister's porpoises).
Distribution The Chilean dolphin is found only along the Chilean and southern Argentine coast, from about 30°S to Cape Horn. As is true of other members of the genus, it is found in shallow coastal waters, and sometimes enters estuaries and rivers. In southern Chile it is found mostly within about 500 m from shore in waters <20 m deep. It occurs in the inshore channels and fjords of the west coast of Tierra del Fuego, such as the Strait of Magellan and Beagle Channel.
Ecology and behavior Until recently, there had been very few sightings of these animals by researchers, and this is the least-known member of the *Cephalorhynchus* genus. Groups tend to be small, between 2 and 15 members, but much larger aggregations have been recorded. They have

A young Chilean dolphin surfaces next to its mother, showing the characteristic wide-based, very rounded dorsal fin, rounded flipper, and stubby beak. Chile. PHOTO: R. MORAGA

Chilean dolphins show more subtle shades of gray compared to other *Cephalorhynchus* species, but have the same stubby beak as their congeners. Chile. PHOTO: R. HUCK-GAETE

Normally slow-swimmers, Chilean dolphins will sometimes race along the surface to approach a passing vessel. Chile. PHOTO: R. MORAGA

Chilean dolphins are normally seen just slow rolling (here, toward the left); the animal on the right is a young calf still that still has fetal folds. PHOTO: S. HEINRICH

The dorsal fins of Chilean dolphins are usually dark, with dark also around the base. Chile. PHOTO: R. HUCKE-GAETE

Chilean dolphins live in bays and shallow coastal waters; their small size and rounded dorsal fins, plus their lack of white areas on the sides or back, make them easy to identify. Chile. PHOTO: R. HUCKE-GAETE

been observed to form mixed groups with Peale's dolphins on several occasions. Although active, they tend to be shy and somewhat difficult to approach (possibly as a result of having been hunted), and generally appear to avoid boats. They only occasionally ride bow waves. Their movements appear to be quite limited, with most dolphins being resident to a small area. Individuals, which have been identified by markings on the back and dorsal fin, concentrate their activities in discrete bays and channels.

Most sightings of young Chilean dolphins have been from October to April. Sexual maturity may be reached at 5–9 years. Little else is known of their reproductive biology, and longevity is not known.

When traveling fast, the Chilean dolphin does not normally porpoise out of the water, but instead skims along the surface with the chin just out of the water. Chile. PHOTO: F. VIDDI

Feeding and prey Chilean dolphins feed on shallow-water fishes (like sardines and anchovies), cephalopods, and crustaceans. They have also been seen to eat young released salmon. Green algae have been found in stomachs, but this may have been ingested incidentally.

Threats and status Chilean dolphins have been hunted for many years for food and crab bait. Up to 1,300–1,500 per year were harpooned in the late 1970s and early 1980s. Such activities are now illegal, but enforcement is difficult. Gillnet catches also affect the species. Aquaculture farms for salmon and mussels may also result in some threats to Chilean dolphins, at least partly by restricting their movements and eliminating available habitat. The only reliable abundance estimate available is of 59 dolphins for an area of southern Chile. Although, the status of virtually all populations is unknown, some are almost certainly threatened with local extinction.

IUCN status Near Threatened.

Dall's Porpoise—*Phocoenoides dalli*

(True, 1885)

Adult Male, dalli-***type***

Adult Female, dalli-***type***

Calf

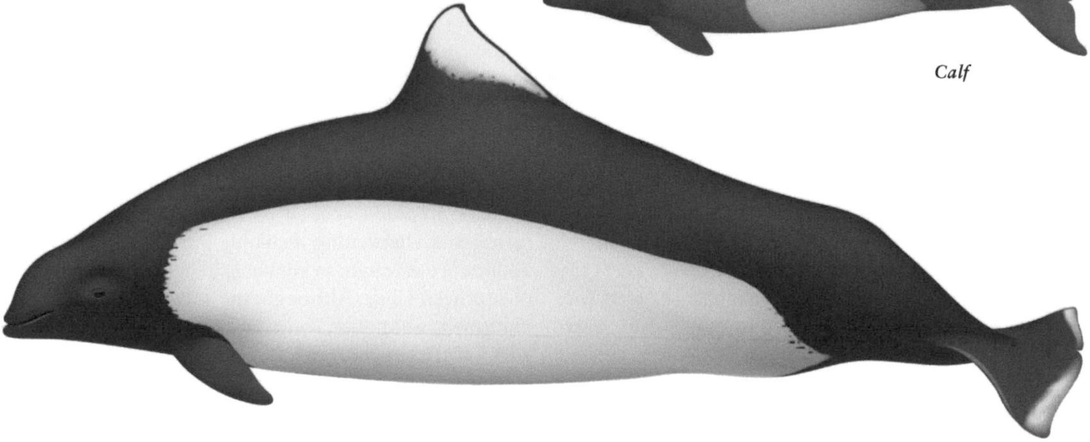

Adult Male, truei-***type***

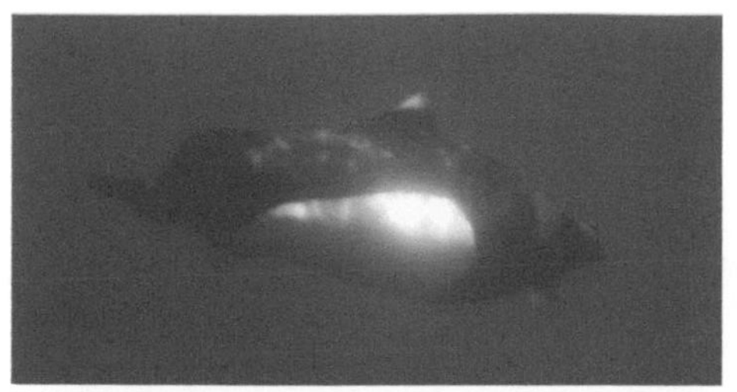

This atypical view of a Dall's porpoise, taken from the bow port of a research vessel, shows this to be an adult male. Notice the deepened tail stock, ventral keel, canted dorsal fin, and extreme thoracic girth. Southern California. PHOTO: R. L. PITMAN

Dall's porpoises are avid bowriders; in fact this is the only species of true porpoise that regularly rides the bow and stern waves of vessels. US west coast. PHOTO: C. OEDEKOVEN

Recently-used synonyms *Phocoenoides truei.*

Common names En.–Dall's porpoise; Sp.–*marsopa de Dall*; Fr.–*marsouin de Dall.*

Taxonomic information Cetartiodactyla, Cetacea, Odontoceti, Phocoenidae. Although there was some past controversy, the two major color forms currently represent separate subspecies: *P. d. dalli* throughout the northern North Pacific, and *P. d. truei* in the northwestern North Pacific. In the waters of the Northeast Pacific, hybrids between Dall's and harbor porpoises are relatively common, and these animals usually result from a Dall's mother and a harbor porpoise father. The hybrids appear to be intermediate between the two species in most aspects of morphology and behavior (though coloration is not). Such hybrids may occur in other areas as well.

Species characteristics Dall's porpoises are robust animals, with a wide-based, triangular dorsal fin (with a falcate tip), and small flippers placed quite near the head. The dorsal fin may be canted forward in some individuals, mostly strongly in large adult males. The flukes are relatively deep and in some older animals the trailing edge becomes convex and the tips very rounded. There is a prominent median notch. The small head has a short beak, with no clear demarcation from the melon. From above, the head appears nearly triangular and very small.

120E 140E 160E 180 160W 140W 120W
70N 60N 50N 40N 30N 20N
Arctic Ocean
Pacific Ocean
1,000 NM
N

Dall's Porpoise
(Phocoenoides dalli)
Global Distribution

Primary Range
Overlap of *truei*-type Primary Range

Dall's porpoises are strikingly marked, with a black body and bright white lateral patches that are continuous ventrally, although young animals have muted color patterns of dark and light gray instead of black and white. The flank patch extends up the sides about midway, from the urogenital area forward to around the level of the dorsal fin, or even to around the area of the flipper insertions. Some individuals may have flecks of black on the white flank patch (occasionally quite extensive). In addition, there is white to light gray "frosting" on the upper portion of the dorsal fin and the trailing edges of the flukes. There are two commonly-occurring color types, *dalli*-type (with a smaller flank patch) and *truei*-type (which has a larger flank patch). Some individuals may have a small area of white to gray flecking on the ventral part of the tail stock, just ahead of the fluke insertion. The characteristic delphinoid "bridle" (consisting of an eye stripe and blowhole stripe) is present. All-white (possibly albino) and all-black (melanistic) individuals have been observed in various places throughout the range. Rarely, porpoises intermediate between the *dalli*- and *truei*-types occur as well.

Dall's porpoise has the smallest teeth of any cetacean. There are 23–28 tiny, spade-shaped teeth in each tooth row. They look rather like grains of rice. Newborn Dall's porpoises are about 1 m long. Adults are about 1.7–2.2 m (females) or 1.8–2.4 m (males), with significant variation among different populations. The Sea of Japan population animals are the largest, and body length appears to decrease as one moves from west to east across the North Pacific. Maximum weight is about 200 kg.

Neonate—About 1/2 adult length; head relatively large; slate gray in color, with no frosting on the flukes or dorsal fin; flank patch light to moderate gray.

Juvenile—About 2/3–3/4 adult length; color pattern has developed to dark and light gray; light gray frosting beginning to develop on dorsal fin and flukes.

Subadult/Adult female—3/4 to full adult size; color pattern basically black and white, with prominent light gray to white frosting.

Younger Dall's porpoises may still have grayish tones, instead of full black and white, in particular around the head. British Columbia. PHOTO: R. MCDONELL

Slow rolling Dall's porpoises are easy to identify at close range, due to the white frosting on the dorsal fin. But at a great distance, they may be difficult to distinguish from harbor porpoises. Southeast Alaska. PHOTO: T. A. JEFFERSON

The deepened tail stock, canted dorsal fin, and thoracic hump on the Dall's porpoise on the right clearly identify it as an adult male. Alaska. PHOTO: J. NAHMENS

The *truei*-type Dall's porpoise is now recognized as a separate subspecies of Dall's porpoise. It is found in Japanese waters, though very rarely an individual will show up in the eastern Pacific. Notice the larger white flank patch. Japan. PHOTO: K. SEKIGUCHI

When moving along rapidly, Dall's porpoises do not porpoise out of the water. Instead, they slice along the surface, making a distinctive sloppy splash, called a roostertail splash. Southeast Alaska. PHOTO: T. A. JEFFERSON

Adult male—Large size; color pattern highly contrasting black and white, with extensive frosting; dorsal fin strongly canted; moderate to large post-anal hump; peduncle deepened; flukes generally have convex trailing edge; head often appears relatively small.

Recognizable geographic forms

Dalli-*type Dall's porpoise (*P. d. dalli*)*—The nominate subspecies occurs in several populations throughout the entire range in the North Pacific, Sea of Japan and Bering and Okhotsk seas. The white flank patch extends forward only to the level of midpoint of the dorsal fin (Sea of Japan population) or the level of the leading edge of the dorsal fin (other populations). Adult females of different populations in the western Pacific average 1.76–1.96 m and males 1.81–2.05 m. Maximum total length is about 2.39 m (for Sea of Japan/Okhotsk Sea specimens).

Truei-*type Dall's porpoise (*P. d. truei*)*—The single population of the *truei-type* is found only in the western part of the range, in the Okhotsk Sea, and along the Pacific coast of Japan, occasionally reaching as far north as the western Kamchatka Peninsula and western Aleutian Islands. In this form, the white flank patch extends much further forward than in the *dalli*-type—to about the level of the insertion of the flipper. Adult females average 1.89 m and males 2.02 m. Maximum total length is only about 2.20 m, significantly shorter than in the other subspecies.

Can be confused with Dall's porpoises are likely to be confused only with harbor porpoises, and even then, only if seen at a great distance. When seen well, the differences in color pattern and dorsal fin shape will be readily apparent. Highly-experienced biologists may be able to discern the two species at a distance based on their surfacing behavior, with Dall's porpoises lifting their tail stocks higher above the surface and rolling more "squarely." On the other hand, inexperienced observers sometimes mistake them for "baby killer whales." One must always keep in mind the possibility of hybrids, although they are usually found in groups with regular Dall's porpoises.

Distribution Dall's porpoises are found only in the North Pacific Ocean and adjacent Bering, Okhotsk, and Japan seas. They inhabit deep waters of the warm temperate through subarctic zones, between about 30°N and 62°N. They may occasionally occur as far south as about 28°N off the coast of Baja California, during unusually cold-water periods. They occur far offshore in oceanic zones, as well as very nearshore where deep water approaches the coast. They are even commonly seen in the inshore waters of Washington, British Columbia, and Alaska. There is apparently a single *truei*-type population that migrates between the Pacific coast of Japan and the Okhotsk Sea; *dalli*-types predominate in all other areas of the range.

Ecology and behavior This may be the fastest swimmer of all small cetaceans, at least for short bursts. When swimming rapidly, Dall's slice along the surface, producing a characteristic V-shaped "roostertail" of spray. At other times, the animals move slowly and roll at the surface with a distinctive squarish profile, creating little or no disturbance (i.e., "slow rolling"). They are avid bowriders, moving back and forth with jerky movements, and often coming from seemingly nowhere to appear at the bow of a fast-moving vessel. They tend to lose interest in vessels that are moving at only slow to moderate speeds. Breaching, porpoising, and other kinds of aerial behavior are extremely rare in this species. Dall's porpoises are found mostly in small groups of 2–12. Although aggregations of up to several hundred or more have been reported, these are extremely rare. Groups appear to be fluid, often forming and breaking up for feeding and playing.

The International Whaling Commission currently recognizes eight stocks of this species, based on pollutant loads, parasite faunas, and distribution patterns of cow/calf pairs. Other than color type information (*dalli*-type or *truei*-type), the stocks cannot be reliably distinguished by appearance at sea. Most Dall's porpoise calves are born in late spring and summer, and there is great geographic variation in age and length at sexual maturity among populations. Maturity occurs at lengths of 1.72–1.87 m and ages of 4–7 years for females, and 1.75–1.96 m and 3.5–8 years for males. Gestation lasts 10–12 months and lactation is thought to last less than one year. This species regularly interbreeds and hybridizes with harbor porpoises in the waters of the Northeast Pacific.

Feeding and prey Dall's porpoises are apparently opportunistic feeders, taking a wide range of surface and midwater fish and squid, especially soft-bodied species like lanternfish (myctophids) and gonatid squid. Occasional krill, decapods, and shrimps found in stomachs are not considered normal prey.

Threats and status Dall's porpoise is one of the primary species taken directly by Japanese fishermen for human consumption. Porpoises are harpooned as they ride the bow waves of catcher boats, and in the past few decades the annual catch has been as high as 40,853 (1988). In addition, Dall's porpoises have been taken incidentally in large numbers in several North Pacific driftnet fisheries, although several of these fisheries are now defunct. This was the primary species taken as marine mammal bycatch in the Japanese mothership and land-based salmon driftnet fisheries that operated in the North Pacific from the 1950s through 1990. Smaller incidental catches occur in several fisheries using gillnets and trawls in Russian, and US and Canadian west coast waters. Environmental contaminants are also thought to be a threat, and high levels of organochlorines may reduce testosterone levels in males and reduce calf survival, thereby influencing reproduction and survival. Dall's porpoises are common in many parts of the North Pacific and the density is high in many areas. The total abundance of the species is probably over 1.2 million individuals, and there are estimated to be about 104,000 along the Pacific coast of Japan, 554,000 in the Okhotsk Sea, 86,000 in Alaska, and 100,000 along the west coast of the US.

IUCN status Least Concern.

In the inshore waters of southern British Columbia and Washington State, Dall's porpoises and harbor porpoises frequently mate and hybridize. The hybrids generally have dark dorsal color patterns, with no evidence of the white flank patch. British Columbia. PHOTO: M. MALLESON

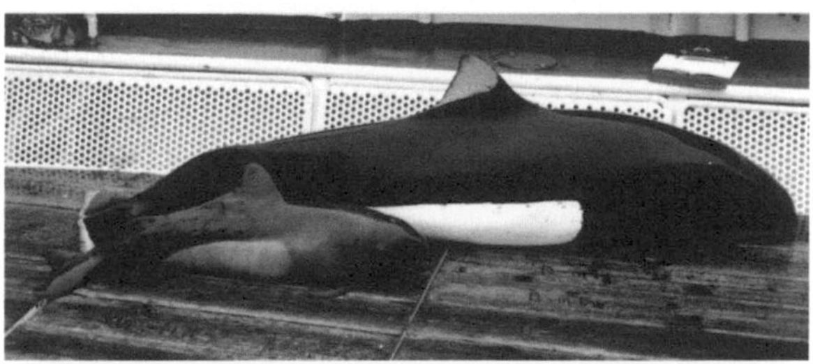

A Dall's porpoise mother and calf killed in a western North Pacific driftnet. The coloration of calves is muted in shades of gray, and the "frosting" on the dorsal fin and flukes doesn't develop until later in life. PHOTO: T. A. JEFFERSON

Three color morphs of Dall's porpoise. The *dalli*-type (top) is the typical pattern in most of the range, the black-type (middle) is seen rarely throughout, and the *truei*-type (bottom) is generally only found in the western North Pacific. PHOTO: T. C. NEWBY, COURTESY S. LEATHERWOOD

Harbor Porpoise—*Phocoena phocoena*

(Linnaeus, 1758)

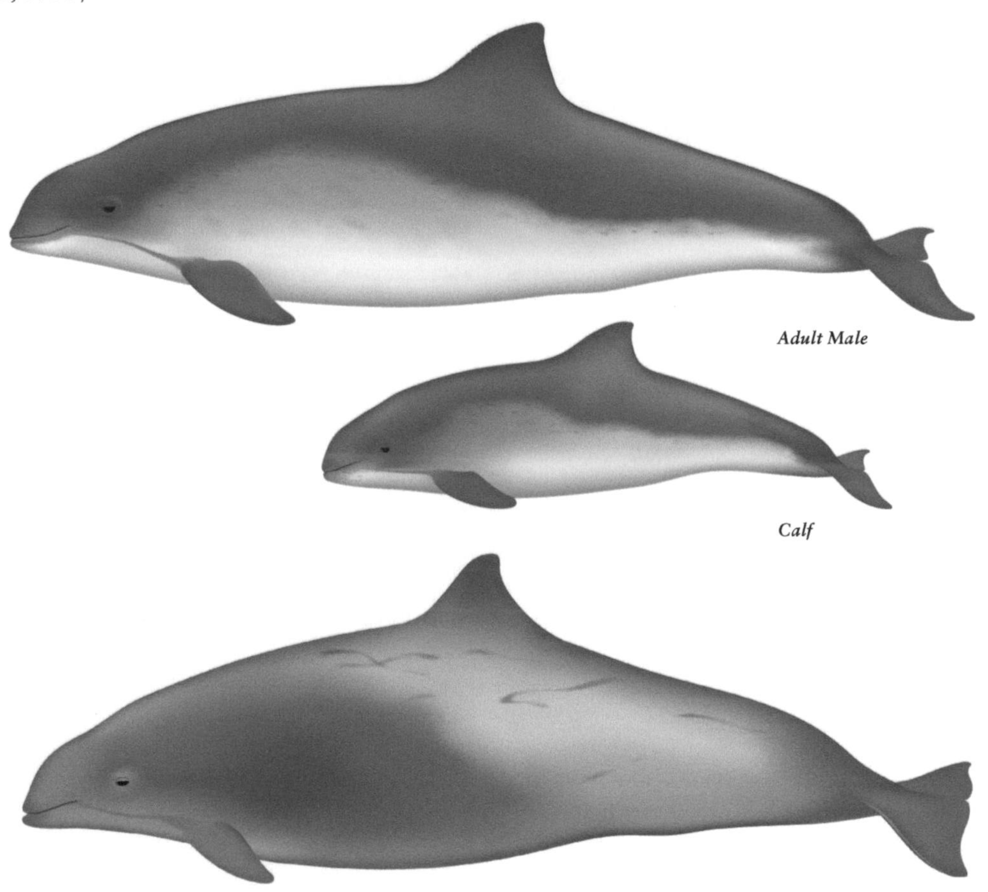

Adult Male

Calf

Phocoena X Phocoenoides *Hybrid*

Recently-used synonyms *Phocoena communis, Phocoena vomerina, Phocoena relicta.*

Common names En.–harbor porpoise; Sp.–*marsopa común*; Fr.–*marsouin commun.*

Taxonomic information Cetartiodactyla, Cetacea, Odontoceti, Phocoenidae. Four subspecies are recognized: *P p. phocoena* in the North Atlantic, *P p. vomerina* in the eastern North Pacific, *P p. relicta* in the Black Sea, and an unnamed subspecies in the western North Pacific. In the waters of the Northeast Pacific, hybrids between harbor and Dall's porpoises are relatively common, and these animals usually result from a Dall's mother and a harbor porpoise father. The hybrids appear to be intermediate between the two species in most aspects of morphology and behavior. Such hybrids may occur in other areas as well.

Species characteristics The harbor porpoise is a relatively stocky cetacean, with a blunt, short-beaked head. The body can be quite rotund, due to a large allocation of body mass to blubber (up to 37%, by weight, in calves). Placed about midway along the back is a short, wide-based, triangular dorsal fin, generally with small bumps (often called denticles or tubercles) on the leading edge. The flippers are small and somewhat rounded at the tips. The flukes have a concave trailing edge, and are divided by a prominent median notch; the tips are rounded. The straight mouthline slopes upward towards the eye.

The color pattern is relatively simple at first glance. Counter-shading is apparent in the harbor porpoise's color pattern; the animals are generally medium to dark gray on the back and white on the belly. The sides are intermediate, with the border area often splotched with various shades of gray. The flippers and lips are dark, and there is a dark gray gape-to-flipper stripe, which is variable in width. The throat often has a series of dark streaks that extend from the lip patch back to the area between the flippers. While the color pattern is actually quite variable (especially in the thoracic and facial areas), there does not seem to be any consistent difference between the sexes or among populations.

Nineteen to 28 small, spatulate, blunt teeth line each tooth row. Maximum known female length for various populations in the North Atlantic ranges from about 1.46–1.89 m, and for males about 1.32–1.78 m. Most adult harbor porpoises are less than 1.8 m long; maximum length is about 2.0 m. Females grow somewhat larger than males. Weights range from 45–70 kg for adults. Newborns are 70–90 cm long, depending on population.

Recognizable geographic forms There is significant geographic variation in various external features, and the Black Sea population of harbor porpoises (now considered a separate subspecies, *P. p. relicta)* is apparently genetically distinct from its nearest neighbors in the eastern North Atlantic (and those that occur occasionally in the Mediterranean Sea). The Black Sea porpoise appears to represent a dwarf form, not reaching lengths greater than about 1.50 m. However, for the most part, reliably-recognizable geographic forms have not been documented in this species.

Can be confused with Harbor porpoises, if seen clearly, should not be confused with any of the various species of dolphins that share their range. The only other porpoise that overlaps in the North Pacific, Dall's porpoise, can be confused with this species when backlit fins are seen at a distance. However, the black and white color pattern and slight difference in dorsal fin shape of Dall's porpoise will be identifiable when seen well. Also, the behavior of the two species tends to be different, with Dall's either roostertailing or bringing the tail stock higher out of the water when rolling. Hybrids are always a possibility, especially in inshore waters of the Pacific Northwest. No other porpoises occur in the North Atlantic, and there should be little problem with identification there.

Although more commonly seen in small groups, harbor porpoises do gather into larger groups to feed and socialize. San Francisco Bay. PHOTO: B. KEENER

Distribution Harbor porpoises are found in cool temperate to subpolar waters of the Northern Hemisphere. They are usually found in shallow waters, most often near shore, although they occasionally travel over deeper offshore waters. Their preferred habitats are characterized by a diversity of water depths (though generally <100 m), substrate types, and prey resources. In the North Pacific, they range from central California and northern Honshu, Japan, to the

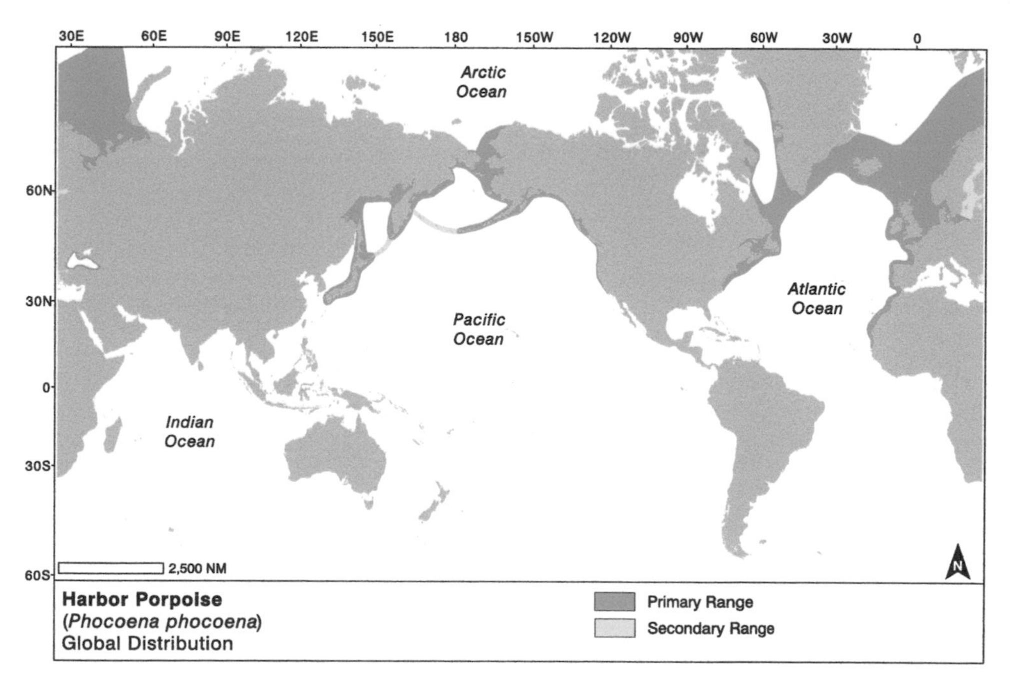

Though it is often difficult to see on live sightings of harbor porpoises, these animals have a complex color pattern, with much individual variation, especially with regard to the border between dark back and light undersides. Belgium. PHOTOS: J. HAELTERS

The dorsal fin of the harbor porpoise is low and subtriangular, with small tubercles or bumps along the leading edge. However, these will be difficult to see in most at-sea sightings. Monterey Bay. PHOTO: T. A. JEFFERSON

Harbor porpoises usually roll slowly at the surface, but sometimes they move more quickly, creating a "pop splash." PHOTO: A. FRIEDLANDER

A typical view of two harbor porpoises moving slowly in Monterey Bay. The low, broad-based dorsal fin and bland color pattern are visible. PHOTO: T. A. JEFFERSON

Anomalously-white harbor porpoises are seen periodically, and a photo-identification project in San Francisco Bay has observed one of these individuals. PHOTO: B. KEENER

southern Beaufort and Chukchi seas. In the North Atlantic, they are found from the southeastern United States to southern Baffin Island (they apparently do not enter Hudson Bay) in the west, and Senegal, West Africa, to Novaya Zemlya in the east. They also occur around southeast and western Greenland, Iceland, and the Faroe Islands. There is also a single stranding record from the Azores. There is a population (or possibly two) in the Black Sea and the Sea of Azov, but they do not regularly occur throughout most of the Mediterranean, except in the northern Aegean Sea. This is the only cetacean species regularly found in the Baltic Sea. Major populations in the North Pacific and North Atlantic are isolated from each another, and many provisional stocks have been recognized.

Ecology and behavior Most harbor porpoise groups are small, generally consisting of less than five or six individuals. They do, at times, aggregate into large, loose groups of 50 to several hundred animals, mostly for feeding or migration.

Behavior tends to be inconspicuous, compared to most dolphins and porpoises. Harbor porpoises almost never approach boats to ride bow waves (though they will ride stern wakes), and often actively avoid vessels. When moving slowly they tend to surface in a slow gentle roll. However, when moving fast, they may surface in a behavior often called "pop-splashing" (which looks somewhat different from the "roostertailing" of Dall's porpoises). Breaches and other leaps are rarely seen, although they do exhibit aerial behavior on occasion, especially when males are attempting to mate. Harbor porpoises sometimes lie nearly motionless at the surface for brief periods between submergences, although it is not known why they do this. They are capable of diving to at least 220 m, and for over 5 min. (although most dives last about 1 min.).

Reproductive biology has been quite well studied in some parts of the range. Sexual maturity is generally reached at 3–4 years of age (at lengths ranging from about 1.2–1.5 m), with some geographic and density-dependent variation. Most calves are born from late spring through midsummer (April to August). Gestation lasts 10–11 months and calves are weaned by 1 year of age. During the mating season, the testes of males grow much larger. Harbor porpoises have some of the largest testes known among any mammal (relative to body size), and the mating system is considered to be promiscuous, involving sperm competition. Some porpoises give birth annually, but others may have a calving interval of two or more years.

In the North Atlantic Ocean (including the Black and Azov seas), 14 populations are recognized, and in the North Pacific, at least 10 different stocks occur. Harbor porpoises regularly interbreed and hybridize with Dall's porpoises in the waters of Greater Puget Sound. They are very short-lived for cetaceans. Most harbor porpoises do not live much beyond 10 years, but some do reach at least 24 years.

Feeding and prey Harbor porpoises eat a wide variety of fish and cephalopods, and the main prey items appear to vary regionally. Small, non-spiny schooling fish (such as herring and mackerel) are the most common prey in many areas, and many prey species are benthic or demersal.

Threats and status The harbor porpoise faces many threats at the hands of humans. The species has been hunted in many areas of its range, and the major kills have been in the Bay of Fundy, Danish Baltic Sea, Black Sea, and Greenland. Today, the most significant threat in most areas is incidental catches in fishing nets, primarily various types of gillnets. Kills of over 1,000 per year have been documented for the Gulf of Maine, West Greenland, North Sea, and Celtic Shelf, but smaller kills occur in many other areas (probably in virtually every part of the species' range). In addition to gillnets, harbor porpoises are also taken in some areas in trawls, Japanese set nets, herring weirs, pound nets, and cod traps. Finally, other types of threats include vessel traffic, noise, and depletion of prey by overfishing. The effects of environmental contaminants may also pose a threat, especially in heavily-industrialized areas. Detrimental effects of exposure

Bottlenose dolphins in some parts of the world attack and kill harbor porpoises on a regular basis—they do not eat the carcasses, and the killing appears to be conducted mainly or exclusively by young males. Monterey Bay. PHOTO: M. COTTER

Harbor porpoises have a very interesting sex life based on sperm competition. The act of copulation appears to be very brief, lasting only a few seconds. San Francisco Bay. PHOTO: B. KEENER

to environmental contaminants have been documented. Globally, the species is not rare, nor in any immediate danger of extinction (although certain populations, such as that in the Black Sea, may very well be). Abundance has been estimated for selected portions of the species' range: 73,000 along the US west coast, 89,000 in Alaskan waters, 89,000 in the Gulf of Maine/Bay of Fundy, 27,000 in the Gulf of St. Lawrence, 28,000 in Iceland, 11,000 in Norway, 36,000 in the Kattegat and vicinity, 600 in the Baltic Sea, 268,000 in the North Sea area, 36,000 in Ireland and the western UK, and between 3,000 and 10,000 in the Black Sea/Sea of Azov. Abundance in the western North Pacific is not known, but is probably lower than for the other ocean basins. Taken together, these numbers suggest the global abundance of the harbor porpoise is probably greater than 675,000 individuals.

IUCN status Least Concern (overall species), Critically Endangered (Baltic Sea population), Endangered (Black Sea population).

Spectacled Porpoise—*Phocoena dioptrica*

Lahille, 1912

Adult Male

Adult Female

Calf

Recently-used synonyms *Australophocaena dioptrica, Phocoena obtusata.*

Common names En.–spectacled porpoise; Sp.–*marsopa de anteojos*; Fr.–*marsouin de Lahille.*

Taxonomic information Cetartiodactyla, Cetacea, Odontoceti, Phocoenidae. This species was briefly considered to be in its own genus, *Australophocaena,* but due to findings of recent genetic and morphometric studies, it is now again placed in the genus *Phocoena.*

Species characteristics The spectacled porpoises' overall body shape is typically porpoise-like, robust with a blunt head, showing only a hint of a beak. The dorsal fin is moderate-sized to large and rounded, with a convex trailing edge (which is uncommon in small cetaceans). In adult males, the dorsal fin becomes oval-shaped and extremely large—so much so as to look disproportionate to the body. The leading and trailing edges are both convex. The flippers are small, with rounded tips, and the flukes are unremarkable.

The two-tone color pattern of spectacled porpoises is very distinctive. Above a line that runs down the side at the level of the eye, they are dark gray to black (except that the line sweeps upward at the tail stock, just before the flukes). Below this line, they are white, with the exception of black lips and a dark gape-to-flipper stripe (the latter is apparently not present on all adults). There is a black patch, surrounded by a fine white line (the "spectacle") around the eye. There is a faint blowhole stripe extending to the apex of the melon. The flukes are gray above and white below; the flippers are variably colored, either all dark or grayish-white with gray edges. A pale saddle on the back around the dorsal fin has recently been confirmed on sightings of live animals at sea (this probably fades extremely rapidly after death).

Young animals have muted gray color patterns, and adult males appear to be darker on the back than females.

Inside the mouth are 17–23 (upper) or 17–20 (lower) spade-shaped teeth in each row. Although few specimens have been measured, adult male spectacled porpoises reach lengths of at least 2.24 m and adult females are up to about 2.04 m. They can reach weights of at least 115 kg. Newborns are probably about 1 m.

Calf—Much smaller than adults (about 1/2 size), with a muted color pattern of light and dark gray.

Juvenile—Slightly smaller than adult size (2/3 to 3/4), with a relatively small dorsal fin.

Adult female—Adult size (up to 2.1 m); dorsal fin with relatively shallow leading edge and <12 cm tall.

Adult male—Adult size (up to 2.3 m); dorsal fin with steeper leading edge and extremely large (up to 25 cm tall) and wide based.

Recognizable geographic forms None.

Can be confused with The spectacled porpoise is not likely to be confused with other species when seen well. But at a distance, there can be some confusion with Commerson's dolphin and Burmeister's porpoise, which both share portions of its range. These three species can best be distinguished by careful attention to dorsal fin shape and color pattern differences. Adult male spectacled porpoises, with their absurdly-large dorsal fins, should be easy to identify. The dorsal saddle of the spectacled porpoise, although faint and difficult to see, will provide a good diagnostic character, when observed.

Although rarely seen and not known to ride bow waves, occasionally spectacled porpoises approach vessels, affording a close look if the water is calm. Subantarctic waters. PHOTO: L. MORSE

Distribution Previously known primarily from the southern coast of eastern South America, from Uruguay to Tierra del Fuego, this species is apparently also found offshore in the Southern Hemisphere. There are records from the Falkland Islands, South Georgia, Kerguelen Islands, Heard Island, Macquarie Island, the Auckland Islands, and Tasmania. Although rarely seen at sea (there are only a few dozen live sightings), this information suggests that the spectacled porpoise may be circumpolar in the subantarctic zone (with water temperatures of at least 1–10°C). Sightings have occurred in oceanic waters and around oceanic

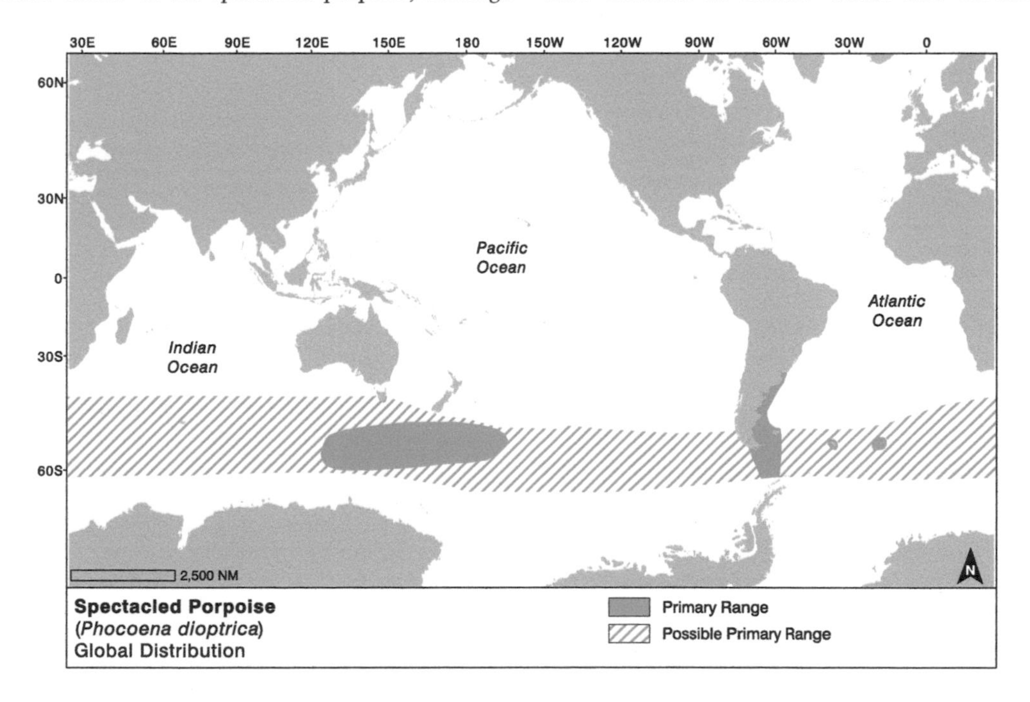

Spectacled Porpoise
(*Phocoena dioptrica*)
Global Distribution

The dorsal fin of the adult male spectacled porpoise is enormous, and seems to be too large for the size of the animal. Subantarctic waters. PHOTO: P. OLSON

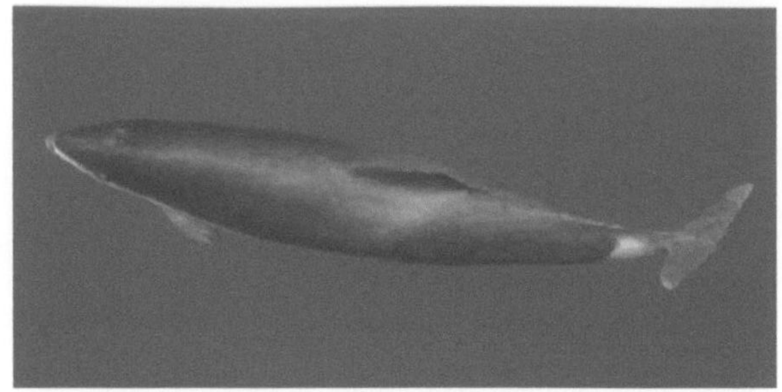

Spectacled porpoises have a very faint post-dorsal fin saddle, though it will be difficult to see in at-sea sightings, and fades quickly after death. Subantarctic waters. PHOTO: L. MORSE

Spectacled porpoise dorsal fins are more rounded at the tip than those of other species of porpoises. Subantarctic waters. PHOTO: L. MORSE

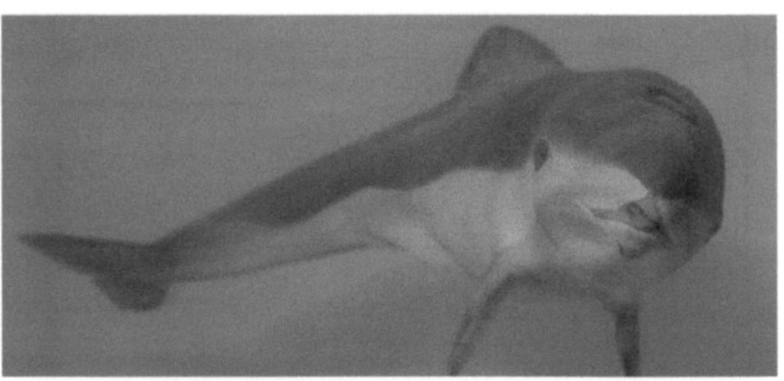

A young spectacled porpoise that was live-stranded in southern Australia swims in its tank, providing an unusual view. PHOTO: C. KEMPER

This mother and calf spectacled porpoise show the unique color pattern of the adults, as well as the muted pattern that is characteristic of young calves. Subantarctic waters. PHOTOS: P. OLSON

islands, as well as in some rivers and channels, but the at-sea ecology of these animals remains very poorly known. The southernmost sighting available is from 64°34′S.

Ecology and behavior This is one of the world's most poorly-known cetaceans. In the few known sightings, group sizes were small, mostly singles, pairs, or trios, with groups ranging up to five individuals (average about two). Cow/calf pairs generally are accompanied by a single adult male. These animals are very inconspicuous when surfacing slowly and do not generally ride bow waves. They are capable of fast swimming, and they do approach vessels on occasion. There is virtually nothing known about potential migrations or seasonal movements. From the small amount of information available, births appear to occur in the southern spring to summer. Essentially nothing else is known of this species' behavior and biology.

Feeding and prey Only four stomachs have been examined, and the contents included anchovies, stomatopods, and a small amount of algae (the latter probably ingested incidentally).

Threats and status Not much is known about the status of this species. However, like all phocoenids, spectacled porpoises are caught in gillnets. At least one record of a capture in a midwater trawl is also known. In the past, this species was apparently harpooned by Tierra del Fuegan natives, as well as fishermen and whalers. There are no estimates of abundance available for this species.

IUCN status Data Deficient.

Burmeister's Porpoise—*Phocoena spinipinnis*

Burmeister, 1865

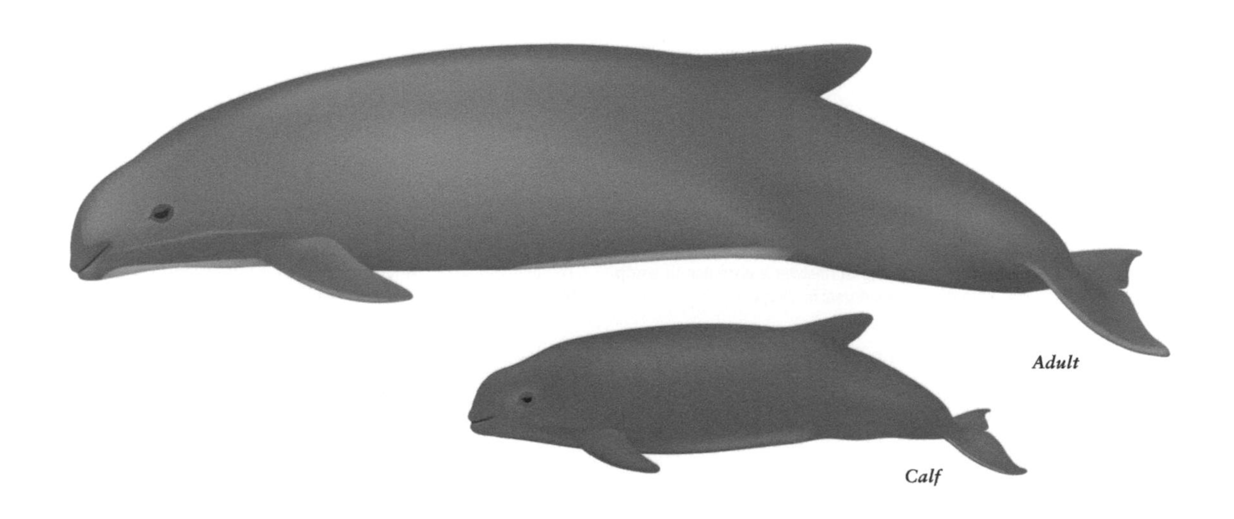

Recently-used synonyms None.

Common names En.–Burmeister's porpoise; Sp.–*marsopa espinosa*; Fr.–*marsouin de Burmeister.*

Taxonomic information Cetartiodactyla, Cetacea, Odontoceti, Phocoenidae.

Species characteristics The unique dorsal fin of Burmeister's porpoise rises at a very shallow angle from behind the midpoint of the back, and the trailing edge is straight to convex. Additionally, there are tubercles along the leading edge of the fin (this characteristic gave the species its scientific name). Other than this, the species has a rather typical robust phocoenid body form, with a blunt, nearly beakless head and broad-based flippers with rounded tips. The flukes are typical of those of other phocoenids.

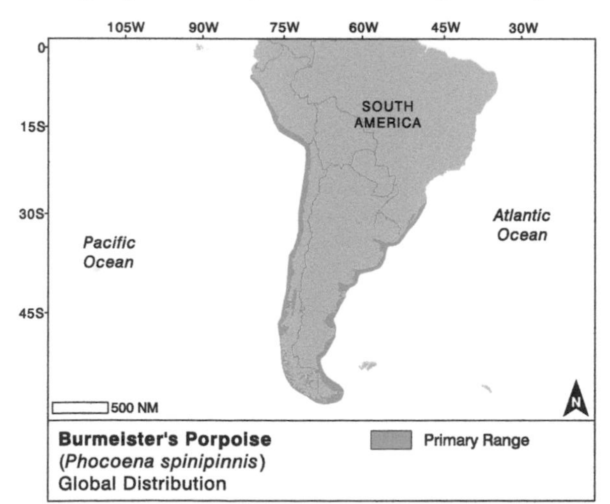

Coloration is charcoal to dark gray, with a light-gray to whitish abdominal field, including light streaks on the chin and belly. The characteristic delphinoid "bridle" (consisting of an eye stripe and blowhole stripe) is present. Burmeister's porpoises have well-defined dark eye patches, dark lips, and dark chin-to-flipper stripes (defined by lighter areas above and below). These flipper stripes are individually variable and asymmetrical; they are narrower and extend further forward on the right side.

Teeth number 10–23 in each upper tooth row and 14–23 in each lower row. As in other phocoenids, the teeth are spatulate. Most adults are up to 1.85 m in length, although animals from Uruguay up to 2.0 m have been recorded. Males grow larger than females. Newborns are 0.8–0.9 m. Maximum weight is about 85 kg.

Recognizable geographic forms None.

Can be confused with From a distance, Burmeister's porpoises can be confused with South American fur seals and South American sea lions, which often stick their fins in the air (these can look like Burmeister's porpoise dorsal fins). They can also be confused with Chilean dolphins off the west coast of South America. Dorsal fin shape will be the best clue to distinguish these two species of small cetaceans. Differences in coloration, dorsal fin shape, and

An exceptional view of a slow-surfacing Burmeister's porpoise in southern Chilean waters, showing the unique dorsal fin shape. PHOTO: S. HEINRICH

The peculiar back-swept dorsal fin of a Burmeister's porpoise, showing the raised tubercles, the function of which are unknown. Southern Chile. PHOTO: S. HENIRICH

A close-up of the dorsal fin of a Burmeister's porpoise captured in Peruvian waters. Notice the tubercles on the leading edge of the dorsal fin, a useful identification character, although not distinctive to this species. PHOTO: R. L. PITMAN

The dorsal fin of the Burmeister's porpoise is one of the most unusual among all cetaceans. It is banana-shaped, and has a series of large tubercles or bumps along the leading edge (which gives it its specific name). Chile. PHOTO: R. HUCK-GAETE

swimming style should allow Burmeister's porpoises to be distinguished easily from Commerson's dolphins and spectacled porpoises, and head and dorsal fin shape will be the best characteristics to allow distinction from franciscanas.

Distribution Burmeister's porpoises are distributed in shallow, coastal waters of South America, from southern Brazil (about 28°48′S), south to Cape Horn in Tierra del Fuego, and thence north to northern Peru (to about 5°01′S). It is unclear whether the distribution is continuous between the Atlantic and Pacific oceans. They occur in the inshore bays, channels, and fjords of Tierra del Fuego and southern Chile, and have even been seen upstream in some rivers. Although most sightings are made within view of shore, Burmeister's porpoises have been observed up to 50 km from the coast.

Ecology and behavior Very little is known about the natural history of this species. Most sightings are of less than six individuals, but aggregations of up to 70 have been reported. Behavior of Burmeister's porpoise is generally inconspicuous; they surface with little surface disturbance. They may swim quickly away from approaching vessels, and are not known to ride bow waves.

There appears to be a protracted summer calving peak; most births in Peru apparently occur in late summer to fall. Pregnancy lasts 11–12 months. In Peru, males reach sexual maturity at about 1.60 m and females at about 1.55 m in length. Recent genetic studies have indicated that porpoises in Peru form separate populations from those in southern Chile and in Argentina. The possibility of multiple populations in Peruvian waters is also considered likely.

Feeding and prey Feeding is on demersal and pelagic fish species, such as anchovies and hake, as well as squid and shrimps.

Threats and status Burmeister's porpoises are taken incidentally in fishing nets, especially gillnets throughout

The muted color patterns on the belly of a Burmeister's porpoise are probably never visible on a live animal at sea. Peru. PHOTO: R. L. PITMAN

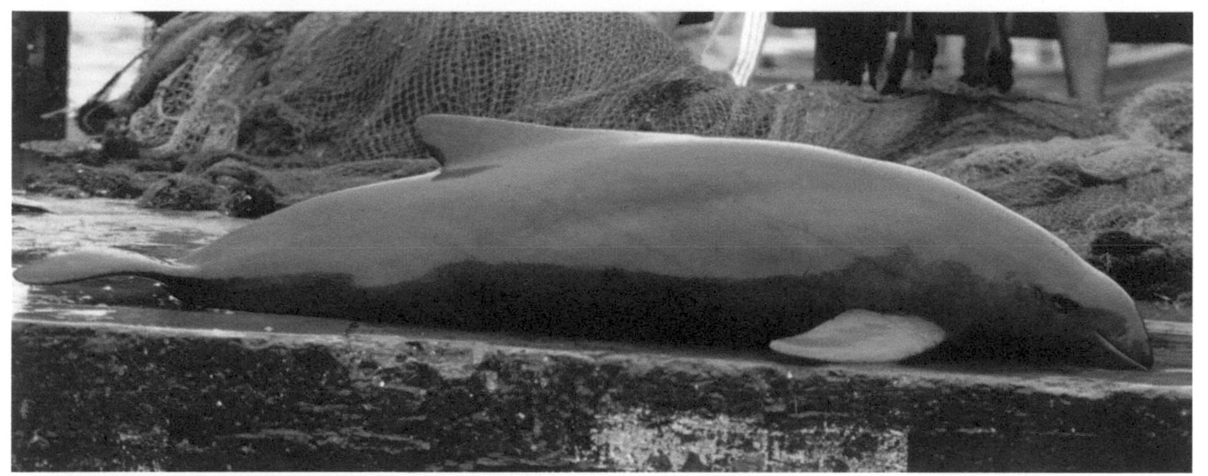

A Burmeister's porpoise caught in Peruvian waters. The typical phocoenid body shape and distinctive dorsal fin are visible. PHOTO: R. L. PITMAN

their range. In southern Chile, porpoises may be killed directly by harpooning. Most of these kills are not known to be large (although monitoring, in most cases, is sparse). However, in Peru, up to 2,000 porpoises per year have been taken, both incidentally and directly, and in addition to being used as shark bait, the meat may be used for human consumption. In addition, to these more immediate threats, it is likely that environmental contaminants are having detrimental impacts on at least some Burmeister's porpoise populations. There are no estimates of abundance or trends for this species, but at least the stock in Peruvian waters is likely threatened by the high levels of capture there.

IUCN status Data Deficient.

Vaquita—*Phocoena sinus*

Norris and McFarland, 1958

Recently-used synonyms None.
Common names En.–vaquita or Gulf of California (harbor) porpoise; Sp.–*vaquita marina*; Fr.–*marsouin du golfe de Californie, vaquita marina.*
Taxonomic information Cetartiodactyla, Cetacea, Odontoceti, Phocoenidae.
Species characteristics The vaquita is among the smallest of all marine cetaceans. Although morphologically it shares much in common with other phocoenids, it has a taller, more falcate dorsal fin and larger flippers. The dorsal fin is more like that of a dolphin, than a porpoise (reaching up to 11–13% of total length). There are tubercles on the leading edge of the dorsal fin. Like other porpoises, however, it is stocky, with a blunt beakless head. The melon has a distinct bulge to it.

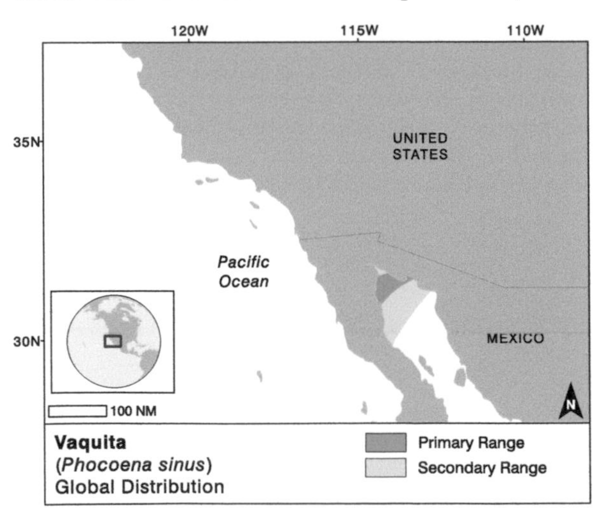

Vaquitas have unique color patterns. The characteristic delphinoid "bridle" (consisting of an eye stripe and blowhole stripe) is present, and there are often dark streaks on the underside of the head. There are distinct black to dark gray lip patches and eye rings; otherwise the body is light brownish-gray with a whitish belly. There is a chin-to-flipper stripe. A small black spot surrounds the external ear opening, and there may be a faint dark streak running from this forward to the eye. Calves tend to be somewhat darker than adults.

In the small number of specimens examined to date, there have been 16–22 pairs of teeth in the upper jaw and 17–20 pairs in the lower jaw. Known maximum length is 1.5 m (females) and 1.45 m (males), but few specimens have been examined. Females are larger than males, as is true in several species of the porpoise family. Newborns are around 70 cm in length.
Recognizable geographic forms None.
Can be confused with When seen at a distance, the tall dorsal fin of the vaquita must be carefully distinguished from those of bottlenose and long-beaked common dolphins,

both of which are regularly seen in the vaquita's range. However, the small group size, unique body shape (e.g., no prominent beak), and different behavior will generally allow the vaquita to be easily distinguished.

Distribution The restricted distribution of the vaquita appears to be defined by relatively murky, shallow (<40 m deep) waters in the northern section of the Gulf of California (with a core area of ca. 2,000 km^2, centered at Rocas Consag). This is near the estuary of the Colorado River. However, there are some suggestions that the range traditionally may have extended further south in the Gulf as well. The animals are most commonly seen in the western portion of the upper Gulf, between San Felipe Bay and Rocas Consag.

Ecology and behavior Very little is known of the biology of the vaquita. As is generally true for porpoises, they occur in small groups, most often of about two, although groups of up to 8–10 have been sighted. Sometimes, large numbers of such groups loosely aggregate in a very small area (only several hundred square meters). They are cryptic and relatively inconspicuous in their behavior, and they do not ride bow waves. Although they often avoid motor vessels underway, they may be curious and approach drifting vessels. Aerial behavior is rare.

The vaquita is small, shy, cryptic, and rare. Detecting these animals in anything but flat calm conditions is nearly impossible. Northern Gulf of California. PHOTO: T. A. JEFFERSON

A vaquita mother and calf surface in the northern Gulf of California. The relatively-tall (for a porpoise) dorsal fin can be seen quite clearly here. PHOTO: P. OLSON

Life history parameters appear to be similar to those of other, better-studied species of porpoises. Females reach sexual maturity sometime between 3 and 6 years of age, and males are assumed to be similar. Calving appears to be seasonal, and most calves are apparently born in the spring (i.e., around March/April). Gestation is about 10–11 months, and females appear to calve about every other year. The maximum known longevity is 21 years.

Feeding and prey Only a few dozen stomachs have been examined, and the contents indicated opportunistic feeding on a wide array of demersal and benthic fishes (e.g., grunts and croakers), squids, and crustaceans.

Threats and status The vaquita is considered the most endangered species of cetacean in the world (with the apparent extinction of the baiji in about 2006). The only proven (and without a doubt, the main) threat is incidental catches in fisheries, especially various types of gillnets. Estimated mortality from gillnet fishing for past years was at least 39–84 vaquitas, which is certainly unsustainable. There are also a number of other potential threats (e.g., habitat changes associated with reduction in freshwater flow from the Colorado River, inbreeding depression, and environmental contamination), though there is no evidence that these are affecting the population at this time. The most recent estimate of total abundance for the vaquita was 97 individuals (in 2014). With the population apparently declining by about 19%/year, there would be much fewer than 90 vaquitas alive in early 2015. The species is clearly in danger of extinction in the next couple of years.

Conservation of the vaquita is being attempted through the creation of a bio-sphere reserve and a vaquita refuge that includes waters outside the reserve in

These two vaquitas surfacing in calm waters show the unique facial coloration (dark eye rings, and lip patches) of the species. Northern Gulf of California. PHOTO: T. A. JEFFERSON

Although vaquitas don't typically bring their eyes out of the water, they may be curious enough to do so when idling boats are nearby. Northern Gulf of California. PHOTO: T. A. JEFFERSON

A photo of several vaquitas killed in totoaba gillnets near El Golfo de Santa Clara in the northern Gulf of California. The animal in front is a 70-cm female calf, and others are adult males (back) and females (middle), each around 135 cm in length. PHOTO: A. ROBLES, COURTESY OF L. ROJAS-BRACHO

Like other species in the genus *Phocoena*, vaquitas have tubercles or bumps along the leading edge of the dorsal fin. Northern Gulf of California. PHOTO: T. A. JEFFERSON

the upper Gulf of California. An international committee (CIRVA) convened by the Mexican government meets regularly to recommend protection measures. The best bet for the vaquita's survival is the switchover from gillnets to other (vaquita-safe) types of fishing gear, which is now being implemented by the Mexican government. However, compliance has been fraught with problems, and the effectiveness of these measures is still very much in doubt. Despite all this, the new Mexican president is giving serious attention to the vaquita, and the habitat of the species appears relatively healthy, so there is still a glimmer of hope for the vaquita. But time is quickly running out.

IUCN status Critically Endangered.

Indo-Pacific Finless Porpoise—*Neophocaena phocaenoides*

(G. Cuvier, 1829)

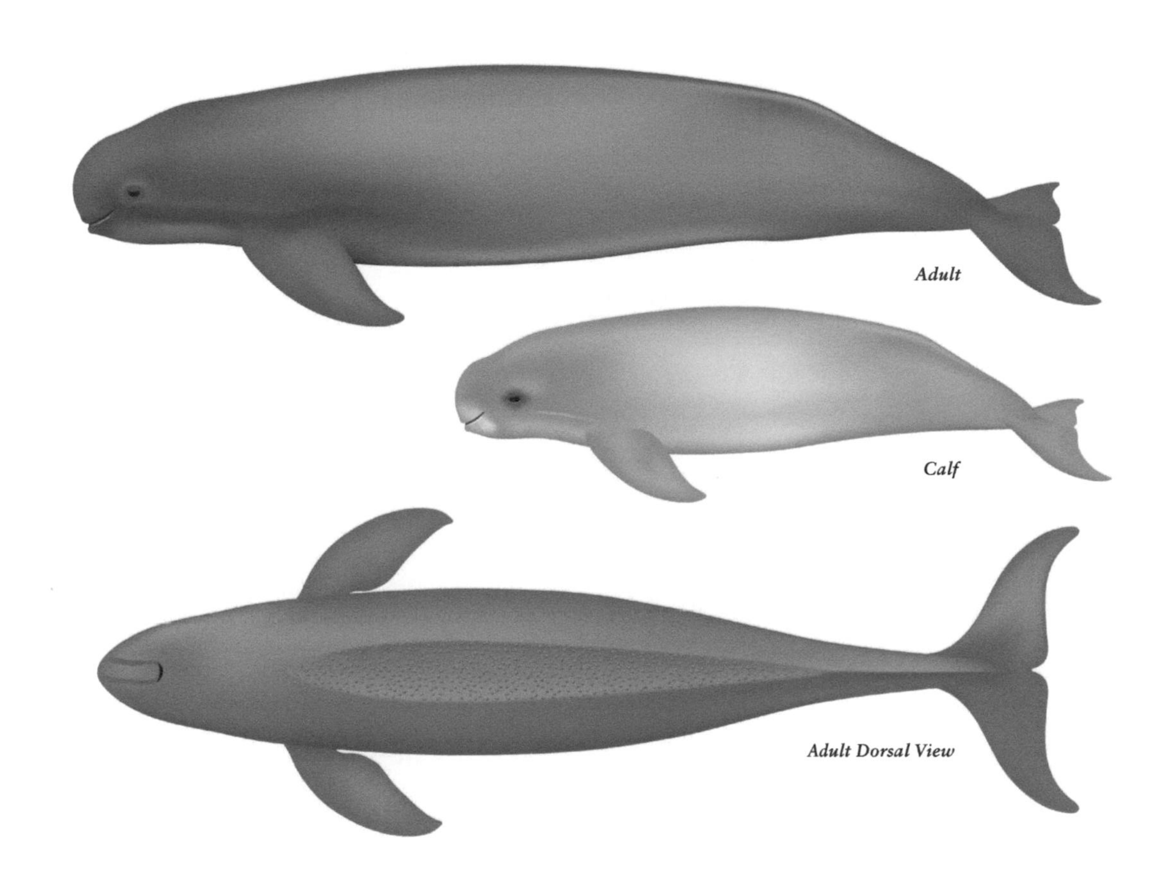

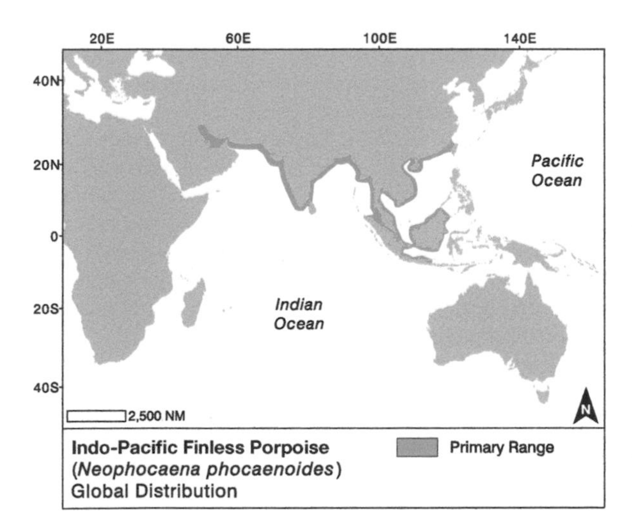

Recently-used synonyms *Neomeris phocaenoides*
Common names En.–Indo-Pacific finless porpoise; Sp.–*marsopa lisa o sin aleta*; Fr.–*marsouin aptère.*
Taxonomic information Cetartiodactyla, Cetacea, Odontoceti, Phocoenidae. It has only recently been recognized that the two major finless porpoise forms—i.e., the wide-ridge type (*N. phocaenoides*) and narrow-ridge type (*N. asiaeorientalis*) actually represent separate species.
Species characteristics As the name implies, the Indo-Pacific finless porpoise has no dorsal fin, and this is the most distinctive characteristic. In some ways, these animals resemble small, slender beluga whales. The head is beakless; the rounded forehead rises steeply from the snout tip. The body shape, in general, is more slender than in other porpoises. The bodies of at least some finless porpoises are soft and mushy, and the neck is very flexible. Instead of a dorsal fin, the Indo-Pacific finless porpoise has an area of small bumps or tubercles on its back, run-

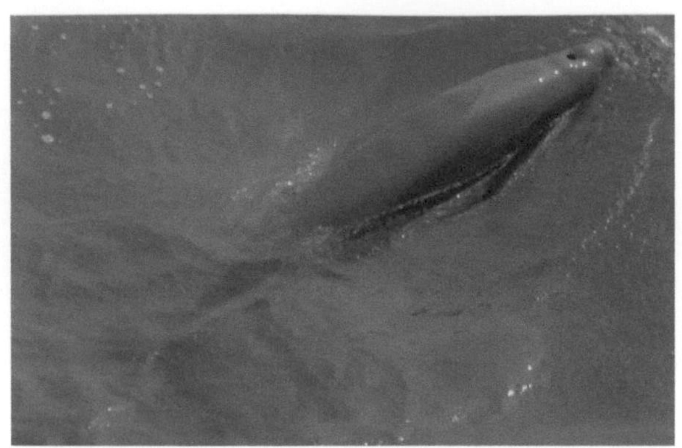

Although the wide dorsal structure of the Indo-Pacific finless porpoise will generally not be visible in distant sightings, the two finless porpoise species apparently only overlap in the Taiwan Strait area. Hong Kong. PHOTO: S. K. HUNG/HONG KONG DOLPHIN CONSERVATION SOCIETY

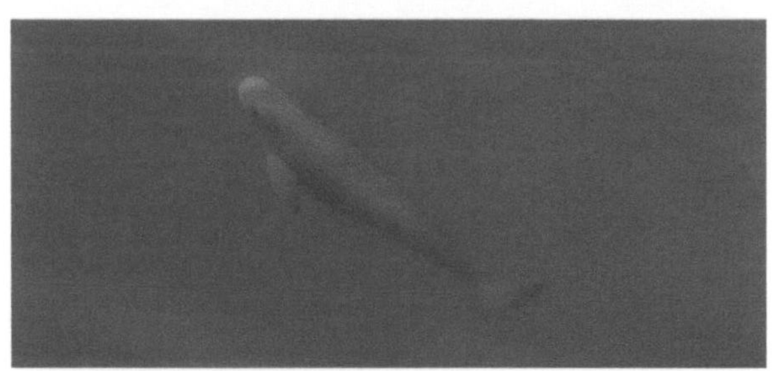

Indo-Pacific finless porpoises will be difficult to detect in anything but calm seas. Hong Kong. PHOTO: S. K. HUNG/HKDCS

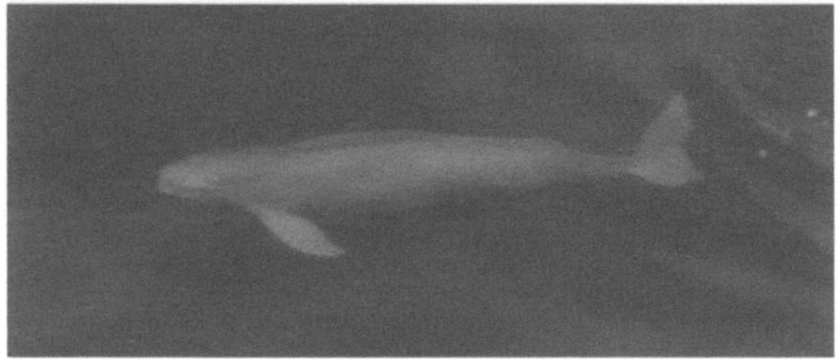

In some parts of their range, Indo-Pacific finless porpoises overlap in distribution with dugongs, and care may be needed to distinguish them. Hong Kong. PHOTO: S. K. HUNG/HKDCS

This is a more typical view of Indo-Pacific finless porpoises—several backs rolling at the surface. They can resemble car tires bobbing in the water, so look carefully. Hong Kong. PHOTO: S. K. HUNG/HKDCS

ning from just forward of mid-back to the tailstock. This "dorsal ridge" ranges in width from 4.8–12 cm, lined with 10–25 rows of tubercles. The trailing edge of the flukes is concave and the flippers are large, ending in rounded tips. Regional differences in body size and morphology have been documented, with separate stocks suggested, based mainly on skull morphology.

The common name that was used in the past, "finless black porpoise," apparently resulted from descriptions of dead animals, after post-mortem darkening. Indo-Pacific finless porpoise adults are actually dark gray in color, often with lighter areas on the throat and around the genitals. They are light gray at birth, and darken to dark gray as adults. In young animals, a dark gape-to-flipper stripe may be visible, as well as a light area around the mouth, and sometimes dark streaks on the throat. The characteristic delphinoid "bridle" (consisting of an eye stripe and blowhole stripe) is present.

Tooth counts range from 15–22 in each tooth row. The teeth are small and slender. Adults of this species can reach 1.7 m in length, although individuals from most populations are much smaller, rarely exceeding 1.5 m (males tend to be slightly larger than females). Indo-Pacific finless porpoises are about 75–85 cm at birth.

Recognizable geographic forms None.

Can be confused with The smooth back of the finless porpoise should make it easy to distinguish from other cetacean species, such as humpback, bottlenose, and Irrawaddy dolphins, and South Asian river dolphins, which share parts of its range. The two species of finless porpoise only overlap in the Taiwan Strait area, and careful attention to the width and tubercles of the dorsal structure can be used to distinguish them. Finless porpoises may be most likely to be confused with dugongs, where the two species overlap in tropical and subtropical waters. The double-nostrils on the snout tip and very different mouth shape should serve to distinguish dugongs.

Distribution The Indo-Pacific finless porpoise occurs in shallow, tropical to warm temperate waters of the Indo–Pacific region. It mostly is found in coastal waters, including shallow bays, mangrove swamps, and the estuaries of some large rivers. The range extends from the Taiwan Strait of central China, in a narrow "ribbon" south and west along the northern rim of the Indian Ocean to the Persian Gulf, including a few estuaries and the lower reaches of some large rivers in the Asian subcontinent.

Ecology and behavior Indo-Pacific finless porpoises are generally found as singles, pairs, or in groups of up to 20. However, groups of up to about 50 have been reported in Chinese coastal waters. When such large groups are seen, they are generally opportunistic aggregations of smaller subgroups gathering to take advantage of a productive food source.

Like other porpoises, their behavior tends to be less energetic and showy than that of dolphins. However, contrary to what their body shape might suggest, they are fast and agile swimmers, often making high-speed sharp turns and rolling on their long axes while chasing down prey. They do not ride bow waves, and in some areas appear to be quite shy of boats. In most areas, aerial behavior appears to be quite rare. When startled by a powered vessel, they may create splashes as they move quickly away from the boat. Most dives are relatively short, and the longest recorded dive time is only about four minutes. In Hong Kong waters, groups of finless porpoises spend about 60% of their time at or near the surface.

Reproduction has been studied in Chinese waters. Sexual maturity of both sexes occurs at 3–6 years of age, and 1.3–1.5 m (males) or 1.2–1.45 m (females) in length. Calving occurs at different times of the year in different regions (most often either spring/summer or winter). These animals appear to live somewhat longer than other species of porpoises, with some populations having a maximum known longevity of 18–25 years. A specimen from Hong Kong in southern China was found to have lived to 33 years, which is quite unusual for a phocoenid.

Feeding and prey Small fishes, squids, and crustaceans form the diet of finless porpoises. Some animals have also ingested some plant material, probably incidentally. Common prey species are fishes of the families Apogonidae, Carangidae, Clupeidae, Sparidae, and Engraulidae; cephalopods of the families Loligonidae, Sepiidae, and Octopodidae; and panaeid shrimps.

Threats and status Indo-Pacific finless porpoises are not known to be directly killed in large numbers anywhere, but they are often incidentally killed in fishing gear throughout their range. Gillnets appear to represent a particularly serious threat in nearly all areas of the range. However, they are also taken in trawls, beach seines, set nets, stow nets, and traps in certain areas. Significant numbers have also been live-captured in China and the Gulf of Thailand. This coastal species also suffers from serious problems of habitat loss and degradation, vessel strikes, and the effects of environmental contamination. Estimates of Indo-Pacific finless porpoise abundance have only been made for a few areas: Bangladesh (1,400) and Hong Kong and surrounding waters (>220). The species is not in immediate danger of extinction, and global abundance is probably over 10,000. However, certain populations may be seriously threatened by human activities.

IUCN status Vulnerable.

The dorsal "ridge" of the Indo-Pacific finless porpoise is wide, even in young calves, and it contains several lines of tubercles or bumps. Hong Kong. PHOTO: S. K. HUNG/HONG KONG DOLPHIN CONSERVATION SOCIETY

Two finless porpoises taken in fishing nets in the same area of Fujian Province, central China. The animal in front is of the *phocaenoides* species, with a wide dorsal ridge. The animal behind is of the *asiaeorientalis* species, with a narrow dorsal ridge. PHOTO: COURTESY OF S. LEATHERWOOD

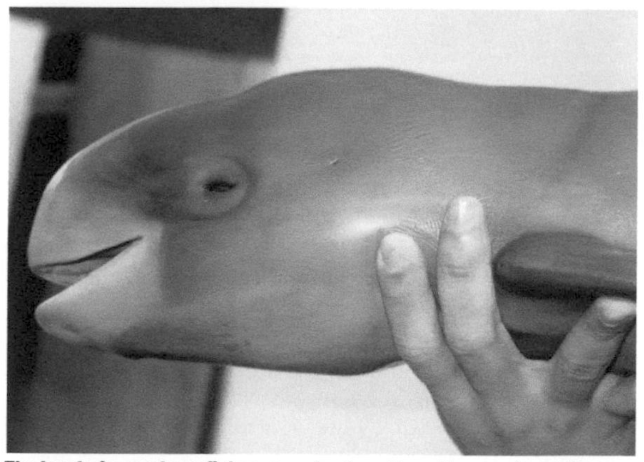

The head of a newborn finless porpoise from Hong Kong waters. The very slender body, light coloration, and whitish areas around the mouth are typical for neonates in this area. PHOTO: S. K. HUNG

Narrow-ridged Finless Porpoise—*Neophocaena asiaeorientalis*

(Pilleri and Gihr, 1972)

Recently-used synonyms *Neomeris (Meomeris) asiaeorientalis, Neophocaena sunameri.*

Common names En.–Narrow-ridged finless porpoise; Sp.–*marsopa lisa o sin aleta*; Fr.–*marsouin aptère*.

Taxonomic information Cetartiodactyla, Cetacea, Odontoceti, Phocoenidae. It has only recently been recognized that the two major finless porpoise forms—i.e., the wide-ridge type (*N. phocaenoides*) and narrow-ridge type (*N. asiaeorientalis*) actually represent separate species. Two subspecies are recognized: *N. a. asiaeorientalis* in the Yangtze River of China, and *N. a. sunameri* found throughout the rest of the range.

Species characteristics As the name implies, narrow-ridged finless porpoises have no dorsal fin, and this is their most distinctive characteristic. In some ways, they resemble small, slender beluga whales. The head is beakless; the rounded forehead rises steeply from the snout tip. The body shape, in general, is more slender than in other porpoises. The bodies of at least some narrow-ridged finless porpoises are soft and mushy, and the neck is very flexible. Instead of a dorsal fin, the finless porpoise has an area of small bumps or tubercles on its back, running from just forward of mid-back to the tailstock. This dorsal ridge ranges in width from 0.2 to 1.2 cm, with 1–10 rows of tubercles. The trailing edge of the flukes is concave and the flippers are relatively large, ending in rounded tips. Regional differences in body size and morphology have been documented, with separate forms recognized, based mainly on skull morphology and molecular data.

The common name that was used in the past, "finless black porpoise," apparently resulted from descriptions of dead animals, after post-mortem darkening. In most areas, narrow-ridged finless porpoises are gray in color, often with lighter areas on the throat and around the genitals. This species has the reverse pattern of its congener: young are dark gray, and lighten as they age. In at least Japan and northern China, the adults are a light creamy-gray.

Tooth counts range from 15–21 in each tooth row. The teeth are small and slender. Adults of this species can reach 2.27 m in length, although animals from most populations are much smaller, rarely exceeding 2.0 m (males tend to be slightly larger than females). Narrow-ridged finless porpoises are about 75–85 cm at birth.

Recognizable geographic forms

Yangtze finless porpoise (N. a. asiaeorientalis)—This subspecies occurs throughout the Yangtze River system (and possibly into the estuary). It has a narrow dorsal ridge (0.2–0.8 cm), which originates anterior to mid-length. The ridge rarely exceeds 1.5 cm high and has 1–5 rows of tubercles. Adults reach up to 1.77 m, and are uniform dark gray.

East Asian finless porpoise (N. a. sunameri)—This subspecies occurs in temperate waters of northern China (from the Taiwan Strait north), Korea and Japan. The dorsal ridge is narrow (0.2–1.2 cm) and high (1.2–5.5 cm), with 1–10 rows of tubercles. It is the largest form of finless porpoise, with adults ranging up to 2.27 m long. Coloration of adults is light gray to cream, with newborns much darker (nearly black).

Can be confused with The two species of finless porpoise are sympatric in the Taiwan Strait area, and careful attention to coloration and the shape and width of the dorsal ridge will be required to identify them correctly. The smooth back of the narrow-ridged finless porpoise should make it easy to distinguish from most other coastal cetacean species, such as bottlenose and common dolphins, which share parts of its range.

Distribution The narrow-ridged finless porpoise occurs in shallow, cool temperate waters of the western Pacific Ocean. It mostly is found in coastal waters, including shallow bays, mangrove swamps, estuaries, and some large rivers. However, it does occur quite far from shore (up to 240 km) in some areas with "shallow" water <200 m deep. The range extends from central Japan and northern China, south and west to the Taiwan Strait, including a few estuaries and rivers in the Asian subcontinent. One of the best-known populations (and the only known wholly-freshwater one) is found in the Yangtze River of China, where it ranges up to 1,600 km upstream from the mouth, to near the headwaters in the Three Gorges area.

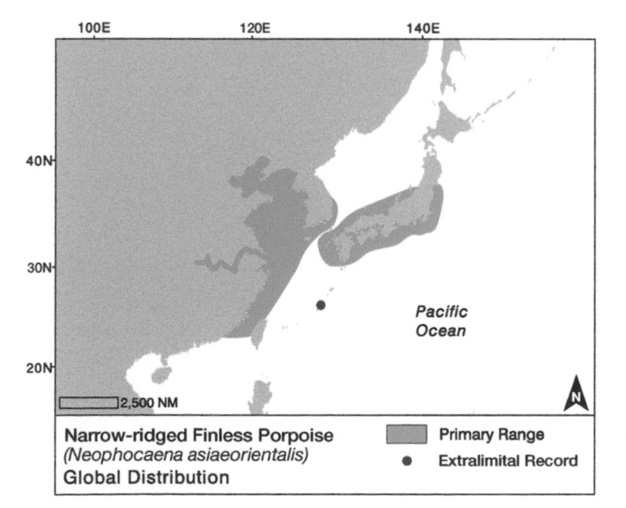

A Japanese finless porpoise. Notice the very light coloration of this adult animal, which is typical for porpoises around Japan. PHOTO: G. ABEL

Ecology and behavior Narrow-ridged finless porpoises are generally found as singles, pairs, or in groups of up to 20. However, groups of up to about 50 have been reported in Chinese coastal waters. When such large groups are seen, they are generally opportunistic aggregations of smaller subgroups gathering to take advantage of a productive food source.

Like other porpoises, their behavior tends to be less energetic and showy than that of dolphins. However, contrary to what their body shape might suggest, they are fast and agile swimmers, often making high-speed sharp turns and rolling on their long axes while chasing down prey. They do not ride bow waves, and in some areas appear to be shy of boats. Mothers have been reported to carry calves on the tubercled area on their backs, but this behavior has not been confirmed. In the Yangtze River, finless porpoises are known to leap from the water and perform "tail stands." However, in other areas, aerial behavior appears to be quite rare. When startled by a powered vessel, they may create splashes as they move quickly away from the boat. Most dives are relatively short, and the longest recorded dive time is only about four minutes.

Reproduction has been studied in Japanese and Chinese waters. Sexual maturity of both sexes occurs at 3–6 years of age, and 1.32–1.45 m in length. Calving occurs at different times of the year in different regions (most often either spring/summer or winter). These animals may live somewhat longer than other species of

Narrow-ridged finless porpoises are amazingly agile and flexible. Captivity. PHOTO: G. ABEL

The body of the narrow-ridged finless porpoise can be quite long and slender—surprisingly streamlined for a true porpoise. Captivity. PHOTO: G. ABEL

The dorsal ridge of the narrow-ridged finless porpoise is narrow, even in large adults. Captivity. PHOTO: J.-P. SYLVESTRE

Narrow-ridged finless porpoises from waters at the northern part of their range, Japan in particular, are very light colored as adults. Captivity. PHOTO: J.-P. SYLVESTRE

In the Yangtze River, there is a unique subspecies of narrow-ridged finless porpoise. These animals are much darker than their more northern counterparts. Yangtze River. PHOTO: WANG DING

porpoises, with most populations having a maximum known longevity of 18–25 years.

Feeding and prey Small fishes, squids, and crustaceans form the diet of narrow-ridged finless porpoises. Some animals have also ingested some plant material, probably incidentally. Common prey species are fishes of the families Apogonidae, Carangidae, Clupeidae, Sparidae, and Engraulidae; cephalopods of the families Loligonidae, Sepiidae, and Octopodidae; and panaeid shrimps.

Threats and status Narrow-ridged finless porpoises are not known to be directly killed in large numbers anywhere, but they are often incidentally killed in fishing gear throughout their range. Gillnets appear to represent a particularly serious threat in nearly all areas of the range. However, they are also taken in trawls, rolling hook longlines, beach seines, set nets, stow nets, and traps in certain areas. Significant numbers have also been live-captured in Japan and China. This coastal species also suffers from serious problems of habitat loss and degradation, vessel strikes, and environmental contamination. The Yangtze River population, a unique subspecies, is apparently declining in number. Estimates of narrow-ridged finless porpoise abundance have only been made for a few areas: China's Yangtze River (<2,000), Korea (29,000, though this may be an overestimate), and Japanese waters (5,000–10,000). Clearly, the species is not in immediate danger of extinction, and global abundance is probably in the high tens of thousands. However, certain populations (e.g., those in the Yangtze River and the Inland Sea of Japan) are indeed seriously threatened by human activities.

IUCN status Vulnerable (overall species), Yangtze River population (Critically Endangered).

Although they may look a bit like little belugas, narrow-ridged finless porpoises are not closely related to those animals. Captivity. PHOTO: H. MINAKUCHI

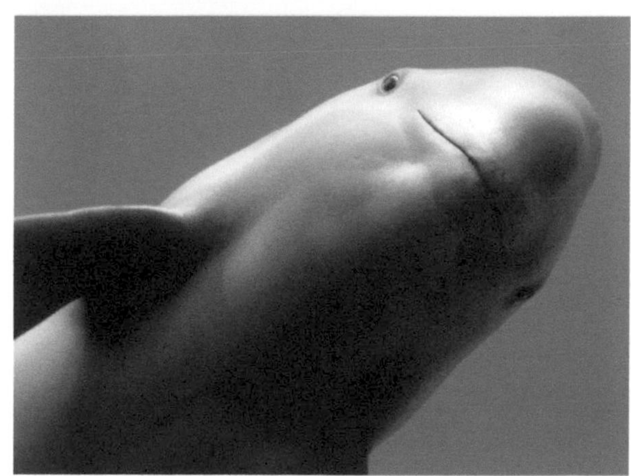

Such a view of the shy narrow-ridged finless porpoise would likely never be seen in the wild. Captivity. PHOTO: J.-P. SYLVESTRE

South Asian River Dolphin—*Platanista gangetica*

(Lebeck, 1801)

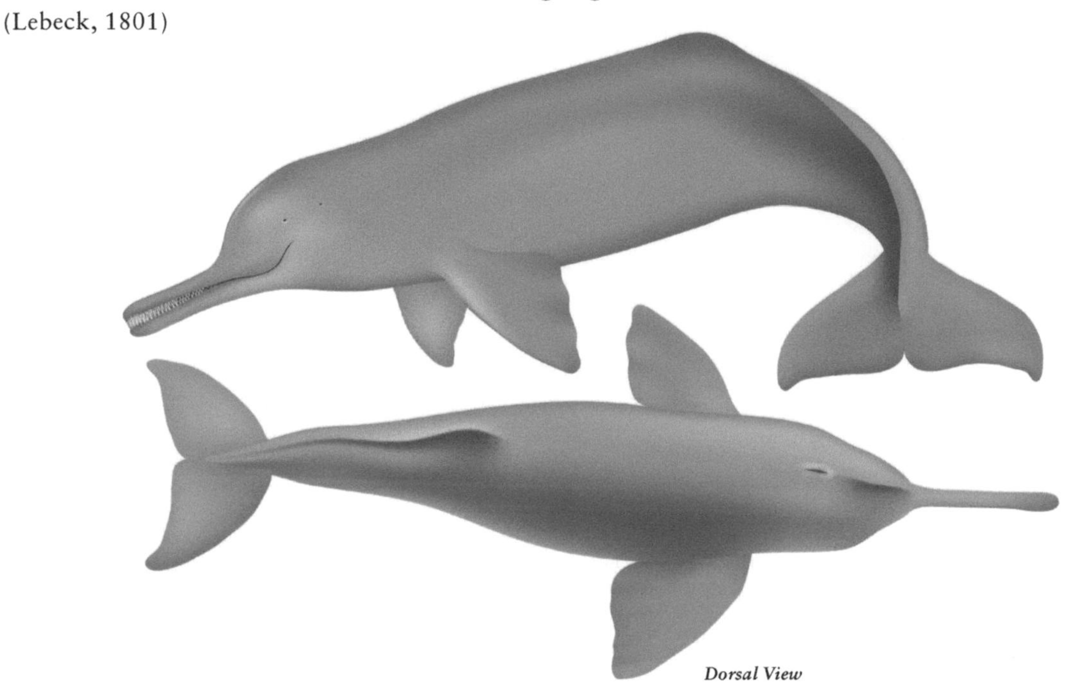

Recently-used synonyms *Platanista indi, Platanista minor.*

Common names Ganges subspecies: En.–susu, shushuk, Ganges River dolphin; Sp.–*platanista del Ganges*; Fr.–*plataniste du Gange*. Indus subspecies: En.–bhulan, Indus River dolphin; Sp.–*platanista del Indus*; Fr.–*plataniste de l'lndus.*

Taxonomic information Cetartiodactyla, Cetacea, Odontoceti, Platanistidae. Until recently, most marine mammal biologists classified the Ganges and Indus River dolphins as separate species *(Platanista gangetica* and *P. minor*, respectively*)*. However, due to the lack of adequate systematic studies, a more taxonomically-conservative view is now favored, with the two forms recognized as subspecies *(P. g. gangetica* in the Ganges/Brahmaputra region, and *P. g. minor* in the Indus). This view is still controversial, and recent studies suggest they are indeed separate species.

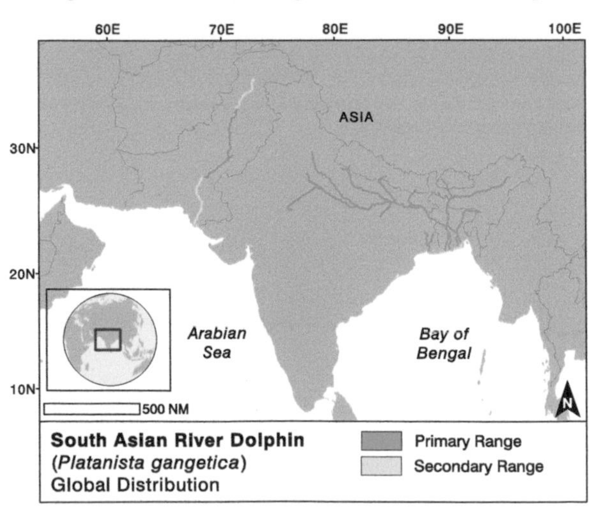

Species characteristics The Ganges and Indus river dolphins are very unusual-looking animals. The body is supple and stocky, with a flexible neck, often characterized by a constriction or crease. The long beak is distinct from the steep forehead, but there is no crease between them. The beak is like a pair of forceps, is laterally compressed, and widens at the tip; it tends to be proportionately longer in females than in males. It curves upward and to one side in long-beaked animals. The blowhole, unlike that of most cetaceans, is a slit that runs along the long axis of the animal's body. There is a distinct, but shallow, longitudinal ridge on the melon, ahead of the blowhole. The melon becomes less rounded as animals approach adulthood. The eyes are extremely small slits or pinholes and are located just above the distinctly-upturned corners of the mouth. The external ear is actually larger than the eye, and is located above it (this is unique for odontocetes). The dorsal fin is a low and wide-based triangle about ⅔ of the way to the flukes, which are concave along the

This photo of two South Asian river dolphins shows well the distinctive features of the species—long beak, and low dorsal fin/ridge. Indus River. PHOTO: M. ARSHAD, COURTESY OF G. BRAULIK

South Asian river dolphins are quite cryptic. If there is any water disturbance in the area, they can be difficult to detect. Nepal. PHOTO: G. ABEL

There is a short dorsal fin in the South Asian river dolphin, almost more of a dorsal ridge than a fin. Nepal. PHOTO: G. ABEL

This spyhopping susu affords a rare view of the extraordinarily-long beak and protruding teeth of this species. Ganges River. PHOTO: R. MANSUR MOWGLI

South Asian river dolphins have long, upturned beaks and teeth that extend outside the closed mouth. Indus River. PHOTO: M. ARSHAD, COURTESY OF G. BRAULIK

The bhulan is a subspecies of the South Asian river dolphin, found only in the Indus River of Pakistan. New research indicates that it may in fact be a separate species. PHOTO: M. ARSHAD, COURTESY OF G. BRAULIK

This South Asian river dolphin is swimming straight towards the camera. There is a vertical ridge on the forehead of this species, something quite unusual for a cetacean. PHOTO: M. ARSHAD, COURTESY OF G. BRAULIK

South Asian river dolphins are sometimes called the "blind dolphins," because the eyes are so reduced that they apparently can do little more than detect general light levels. Indus River. PHOTO: M. ARSHAD, COURTESY OF G. BRAULIK

rear margin. The broad flippers are squared-off at their distal ends, and usually have flat trailing edges, but they are sometimes scalloped. The outlines of the bones of the flipper are visible, which is unusual for cetaceans.

South Asian river dolphins have a relatively simple color pattern (which is consistent with their nearly-blind state). The animals are light brown or brownish-gray, often with a slightly lighter ventral surface. They become increasingly blotchy as they get older. Young animals may have a pinkish cast to the belly.

The 26–39 upper teeth and 26–35 lower teeth are curved. They are sharply pointed in young individuals. The anterior teeth are longer and extend outside of the closed mouth, especially in larger individuals. The teeth may become peg-like from dentine accumulation and wear in older animals. Tooth wear is generally minimal in younger animals. Female South Asian River dolphin adults are up to 2.6 m and males 2.2 m in length. They can reach weights of at least 85 kg. Newborns are apparently between 70 and 90 cm long.

Recognizable geographic forms The relationship between Ganges and Indus River dolphins has been controversial, with many authors suggesting that they should be considered separate species. The river basins they occur in have been largely separate for millions of years, and recent genetic studies found convincing evidence for species-level difference. They are currently considered to be separate subspecies:

Ganges River dolphin (P. g. gangetica)—The Ganges River dolphin occurs in the Ganges and Brahmaputra river systems of India, Bangladesh and Nepal (and possibly Bhutan). Female susu adults are up to 2.6 m and males 2.2 m in length. They can reach weights of at least 85 kg.

Indus River dolphin (P. g. minor)—The Indus River dolphin is found only in the Indus river system of Pakistan. Bhulans are considered to reach slightly smaller sizes than the maximums of 2.6 m (females) and 2.2 m (males) for susus. However, it is unclear if these differences are real, or simply artifacts of limited sample sizes. No other external differences are known between animals from the two river systems.

Can be confused with Susus can be confused with several other small cetaceans that are found in overlapping areas of the Sundarbans mangrove forest and near the mouths of the Hooghly, Karnaphuli and Sangu rivers. They might be confused with Irrawaddy dolphins, finless porpoises, bottlenose dolphins, or Indo-Pacific humpback dolphins. The prominent dorsal fins of bottlenose and humpback dolphins, complete lack of a dorsal fin in finless porpoises, and absence of a beak in Irrawaddy dolphins should make all these species easily distinguishable. Also, adult bottlenose and humpback dolphins are much larger than South Asian river dolphins. Bhulans do not presently overlap in distribution with any other cetacean species.

Distribution The extensive range of the susu includes the Ganges, Brahmaputra-Megna, and Karnaphuli-Sangu river systems and many of their tributaries in India, Bangladesh and Nepal (there is some suggestion that it may possibly extend into Bhutan in the wet season as well, although this has not been confirmed). Susus live not only in the main river channels, but also in seasonal tributaries and appended lakes during the flood season. Though formerly much more widely distributed in the Indus and some of its tributaries, the bhulan's range is now restricted to the middle and lower Indus River. It is centered between Taunsa and Sukkur barrages.

Ecology and behavior As is true for most of the river dolphins, susus and bhulans generally live in small groups of less than 10 individuals, and are most often seen alone or in pairs. Bhulans have occasionally been reported in loose aggregations of up to 30 individuals, defined by common use of clumped resources. Other than the mother/calf bond, affiliations between individuals are thought to be ephemeral. These are active animals, but they do not often engage in leaps. At least in captivity, these dolphins appear to spend much of their time swimming on their sides, and they constantly emit echolocation clicks. This might be related to the fact that they spend time in relatively shallow, turbid waters; however, some of the pools they occur in may be >30 m in depth. South Asian river dolphins are nearly blind, and can probably only detect light levels, and perhaps direction.

Almost nothing is known of the reproductive biology of this species, except that males reach sexual maturity at about 1.70 m and females at slightly longer body lengths. Calving apparently can occur at any time of the year, but for the susu, there may be peaks in December to January and March to May. Newborn bhulans have been observed mainly in April and May. Calves are weaned within about one year of birth.

Feeding and prey Susus and bhulans feed on several species of fish and invertebrates (prawns and possibly clams). They apparently do much of their feeding on or near the bottom.

Threats and status South Asian river dolphins have been subjected to direct killing in both the Indus and Ganges/Brahmaputra systems. In addition to use of their meat as human and livestock feed, the oil of susus is used as a fish attractant and for medicinal purposes. Like almost all small cetaceans, they suffer from entanglement in various types of fishing nets, and they are sometimes also killed by vessel strikes. The damaging effects of environmental contaminants, while not sufficiently understood, are probably important threats to these dolphins. Perhaps the most important problem, and one that affects this species to a much greater degree than other riverine cetaceans, is the placement of dams and barrages across rivers in this species' range. These structures artificially fragment populations and reduce available habitat by altering riverine ecology. The susu has a moderately-large range that spans several countries and is apparently not in immediate danger of extinction, probably numbering at least in the low thousands. There are thought to be from 965 to 1,200 bhulans in Pakistan. The bhulan's range is extremely restricted and declining, and the overall population is fragmented into five subpopulations by barrages, making it one of the most endangered types of cetaceans in the world.

IUCN status Endangered (overall species, Indus and Ganges subspecies).

This young bhulan in the Indus River of Pakistan demonstrates the bizarre head and body shape of the species. PHOTO: G. PILLERI, COURTESY OF S. LEATHERWOOD

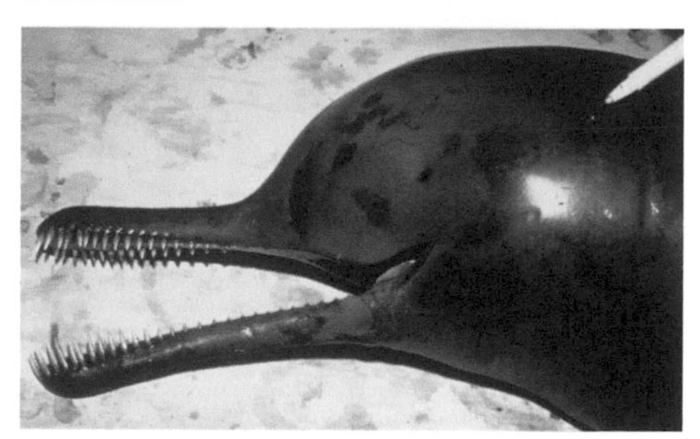

A close-up of the head of a young Indus bhulan. Note the forceps-like beak with long teeth, tiny eye, and external ear (indicated by pen). PHOTO: COURTESY OF S. LEATHERWOOD

Amazon River Dolphin—*Inia geoffrensis*

(Blainville, 1817)

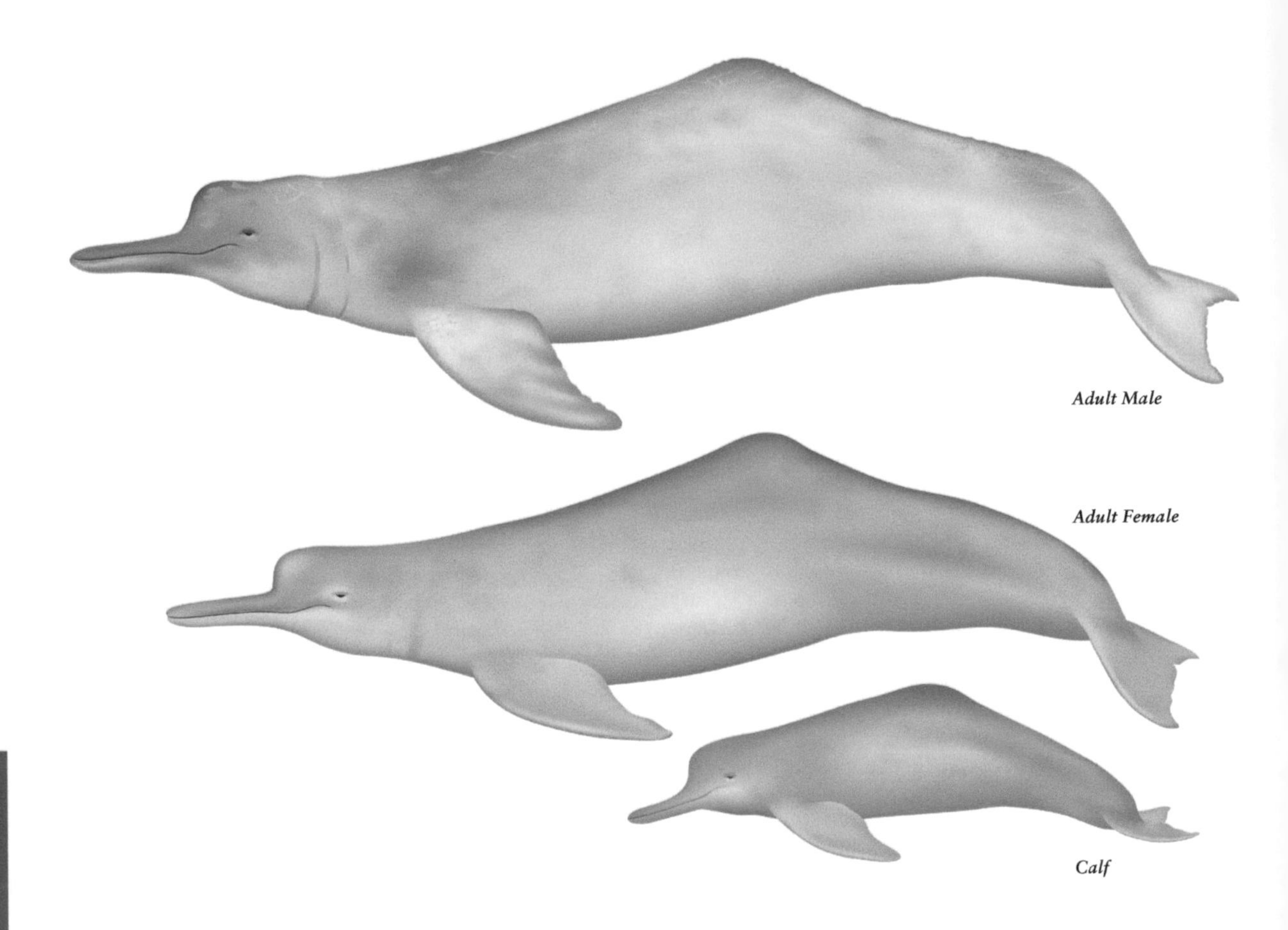

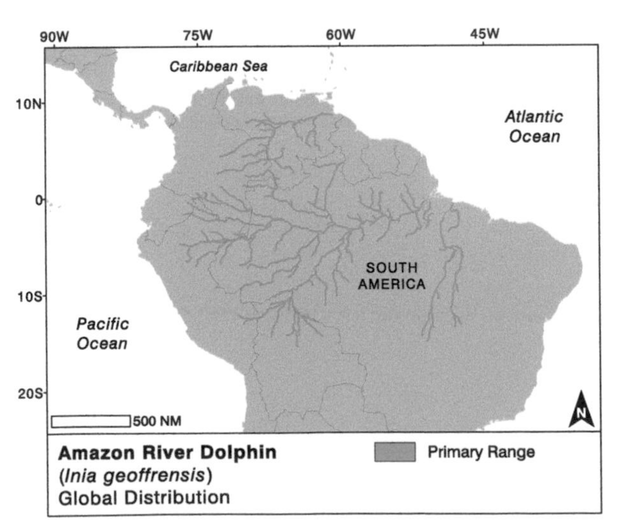

Recently-used synonyms *Inia araguaiaensis, Inia boliviensis*

Common names En.–Amazon River dolphin, boto, bufeo; Sp.–*bufeo, delfín del Amazonas*; Fr.–*inia, boto de l'Amazone*. (Note: *boto* is the Portuguese name used in Brazil, and although it is not used by locals in Spanish-speaking parts of the range, it has been adopted as an English common name.)

Taxonomic information Cetartiodactyla, Cetacea, Odontoceti, Iniidae. Two subspecies are currently recognized: *I. g. geoffrensis* in the Amazon River system of Brazil, Peru, and Ecuador and *I. g. humboldtiana* in the Orinoco basin of Venezuela and Colombia. The validity of these subspecies is questionable, however, and reported differences in morphology are not well established. Recent molecular genetic studies suggest that they may not be valid. The Bolivian animals have recently been suggested to be a distinct species, *I. boliviensis*, and another new species

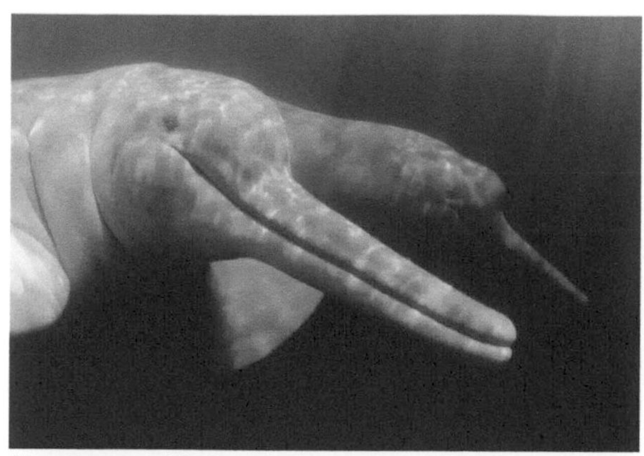

The appearance of the boto is rather primitive, and can be almost reptilian. Brazil. PHOTO: M. CARWARDINE

The lumpy bodies of slow-swimming Amazon River dolphins are quite different from the very streamlined bodies of most oceanic dolphins. Brazil. PHOTO: M. CARWARDINE

The Amazon River dolphin has a long beak and a bulging forehead. Rio Negro, Brazil. PHOTO: H. MINAKUCHI

Botos eat a wide variety of fish species, often chasing them into the flooded forest. Brazil. PHOTO: M. CARWARDINE

Truly a dolphin of the forest, Amazon River dolphins swim into the flooded forest in the high-water season and search for food among the trees and roots. Rio Negro, Brazil. PHOTO: H. MINAKUCHI

Botos have unfused neck vertebrae and therefore can bend their necks more than most cetaceans. Brazil. PHOTO: M. CARWARDINE

Botos can be playful, although they do not often breach out of the water. This is a rare photo of an animal urinating. Brazil. PHOTO: M. CARWARDINE

The river dolphins of Bolivia have been described as a separate species, *I. boliviensis*. However, at this point, the evidence for species status is not certain, and therefore this form is still recognized as a subspecies of *I. geoffrensis*. PHOTO: F. TRUJILLO

A new species, *I. araguaiaensis*, has recently been described from southern Brazilian waters. However, it is not clear yet whether this is a valid species. Tocantins River, Brazil. PHOTO: F. TRUJILLO

The lighter colored parts of the bodies of Amazon River dolphins may take on a pinkish hue when the animals are in very warm water, or when they are particularly active. PHOTO: F. TRUJILLO

Younger botos are largely gray in color, with only small areas of white or pinkish skin visible. PHOTO: F. TRUJILLO

The Amazon River dolphin is quite easy to identify, with its low dorsal ridge, large flippers, bulbous forehead, long slender beak, and often-pinkish color. Captivity. PHOTO: J.-P. SYLVESTRE

As is true of all the river dolphins, Amazon River dolphins tend to have large, paddle-shaped flippers. Captivity. PHOTO: J.-P. SYLVESTRE

(*I. araguaiaensis*) was recently (2014) described based on limited data from the Araguaia River in Brazil. However, there are doubts about the legitimacy of both these putative species, and following the Society for Marine Mammalogy, we do not recognize them here.

Species characteristics The Amazon River dolphin, or boto, is probably the best-studied and most well-known of the river dolphins. These animals are moderately robust, but extremely flexible (e.g., unfused cervical vertebrae allow great movement of the neck). They have long beaks with a series of bristles, and steep bulbous foreheads, which are capable of changing shape by muscular action. More than any other species of cetacean, botos have visible "cheeks." There is no true dorsal fin, but only a dorsal ridge about ⅔ of the way back from the beak tip that is low and wide based. The flippers are large and triangular, with blunt tips, and the flukes are deep and triangular, with a concave trailing edge and prominent notch. The eyes are small, but not as small as those of the susu or bhulan.

Botos are gray to white/pink above and light below; some individuals may appear totally pink. In general, young animals are mostly uniform dark gray; they become progressively more pinkish with age (especially in males). The extreme color is so unusual that the boto is often called the pink dolphin. The pink color is thought to result from flushing of blood to the body surface on animals that have lost most of their pigmentation—and is not the result of red pigments. Albino botos have been documented.

The mouth is lined with 23–35 stout teeth in each row. This is the only genus of cetaceans with differentiated teeth; those at the front of the jaw are typically conical, but those near the rear are flanged on the inside (for crushing of hard-bodied prey). Botos are the largest of the river dolphins. Adult size ranges to 2.3 m (females) or 2.8 m (males). Males can reach maximum weights of 207 kg, about 35% more than females. At birth, botos are about 80 cm long.

Amazon River dolphins have a very unique appearance, with their almost-reptilian jaws. Captivity. PHOTO: J.-P. SYLVESTRE

Calf/juvenile—⅓ to ¾ adult length; body color uniformly gray, scarring light to absent.

Adult female—Adult size (1.6–2.25 m), body grayish pink, with light to moderate scarring.

Adult male—Large size (2.0–2.5 m), body color mostly pink with dark blotches, moderate to heavy scarring (especially on the trailing edges of the flippers and flukes, which are often ragged), and "cobblestoning" of the skin is common.

Recognizable geographic forms Morphological differences have been proposed for the two subspecies, but the validity of these distinctions are questionable. At this point, the subspecies can only be reliably distinguished based on their locality of origin.

Can be confused with The only other dolphin that inhabits the range of the boto is the tucuxi (*Sotalia*). This latter spe-

The Amazon River dolphin has a long beak, and is the only dolphin species with differentiated teeth (heterodont). Rio Negro, Brazil. PHOTO: H. MINAKUCHI

Like most other species of odontocetes, the blowhole of the Amazon River dolphin is offset to the left of centerline. Brazil. PHOTO: M. CARWARDINE

cies is much smaller, has a true dorsal fin, and more spritely, dolphin-like movements, making confusion unlikely.

Distribution Botos are endemic to the Amazon and Orinoco drainage basins of South America. Their distribution extends to the upper reaches (impassible falls or rapids) of these rivers and their tributaries in Colombia, Ecuador, and Peru, as well as the lower reaches in Brazil and Venezuela, a total area of several million km^2. They are found widely, not only in the main river channels, but also in smaller tributaries, lakes, and (seasonally) the flooded forest. Botos are especially common at the confluences of river channels, and entrances to lakes. Bolivian bufeos are endemic to the upper Amazon drainage basin of the Cochabamba, Santa Cruz, Beni, and Pando regions of Bolivia. Their distribution extends to the upper reaches (impassible falls or rapids) of the upper Madeira and Beni/Mamore rivers and their tributaries.

Ecology and behavior Botos are not highly social. Loose aggregations of up to 12–19 botos have been observed, generally at river bends and confluences, for purposes of mating and feeding. However, most botos are seen singly or in small groups of 2–3 (most pairs are mother and calf).

Botos generally move slowly (typically about 1.5–3.2 km/hr), although they are capable of short bursts of up to 22 km/hr. They typically surface at a shallow angle, showing the melon, tip of rostrum, and the dorsal ridge simultaneously. However, they do perform high, arching rolls as well. Their responses to humans can range from shyness to curiosity, although they do not ride vessel bow waves. Aerial behavior is rare, and they generally do not lift the flukes out of the water before a dive. Botos swim into the flooded forest in the high-water season, and can often be heard searching for prey among the roots and trunks of partially-submerged trees. Photo-Identification/tagging studies have shown that some individuals are resident in specific areas year-round. Natural predators probably include caimans, crocodiles, and jaguars.

Reproduction is diffusely seasonal, apparently in each area peaking during the season of maximum flooding (which varies among areas). In Brazil, most births apparently occur in May to July. Females mature sexually at about 5 years of age and 1.6–1.75 m in length, and males do so much later (about 2.0 m in length). Gestation lasts about 11 months and lactation lasts over one year.

Feeding and prey These animals feed on a large variety of fishes (over 43 species!), generally near the bottom. Some of their prey species have hard outer shells, and dolphins have been observed breaking up their larger prey before swallowing them. They sometimes feed in a coordinated manner, occasionally with other species (such as *Sotalia*). *Inia* is the only genus of cetacean with heterodont teeth (the rear teeth have flanges that allow them to crush prey).

Threats and status The boto is unquestionably the species of river dolphin in least danger of extinction. It is still widespread and relatively numerous in many portions of its range, with some of the highest densities known for any species of cetacean. Threats include incidental catches in fishing gear, prey depletion, damming of rivers (although this is, at present, much less of a problem than for the susu and bhulan), and environmental pollution from organochlorines and heavy metals. Botos have received some protection from persecution by local people due to their prominent involvement in the folklore and culture of the Amazon; they are often seen as reincarnated humans and are attributed supernatural, mischievous powers. Unfortunately, these beliefs seem to be disappearing. In Bolivia, bufeos do not figure in local myths and legends, and therefore do not receive protection from persecution by local people. The Bolivian bufeo is considered to be vulnerable to human activities, and certain populations may be declining and therefore threatened. There are thought to be about 13,000 botos in the central Amazon of Brazil (Mamiraua floodplain system), and about 350 each in two small areas of the upper Amazon of Brazil, Peru, and Colombia. Although, abundance has only been estimated in a few areas and accuracy of many of these estimates is questionable, it is likely that there are well over 15,000 Amazon River dolphins throughout their range. Thus, while certain populations may be threatened, the species is in no immediate danger of extinction.

IUCN status Data Deficient.

Franciscana—*Pontoporia blainvillei*

(Gervais and d'Orbigny, 1844)

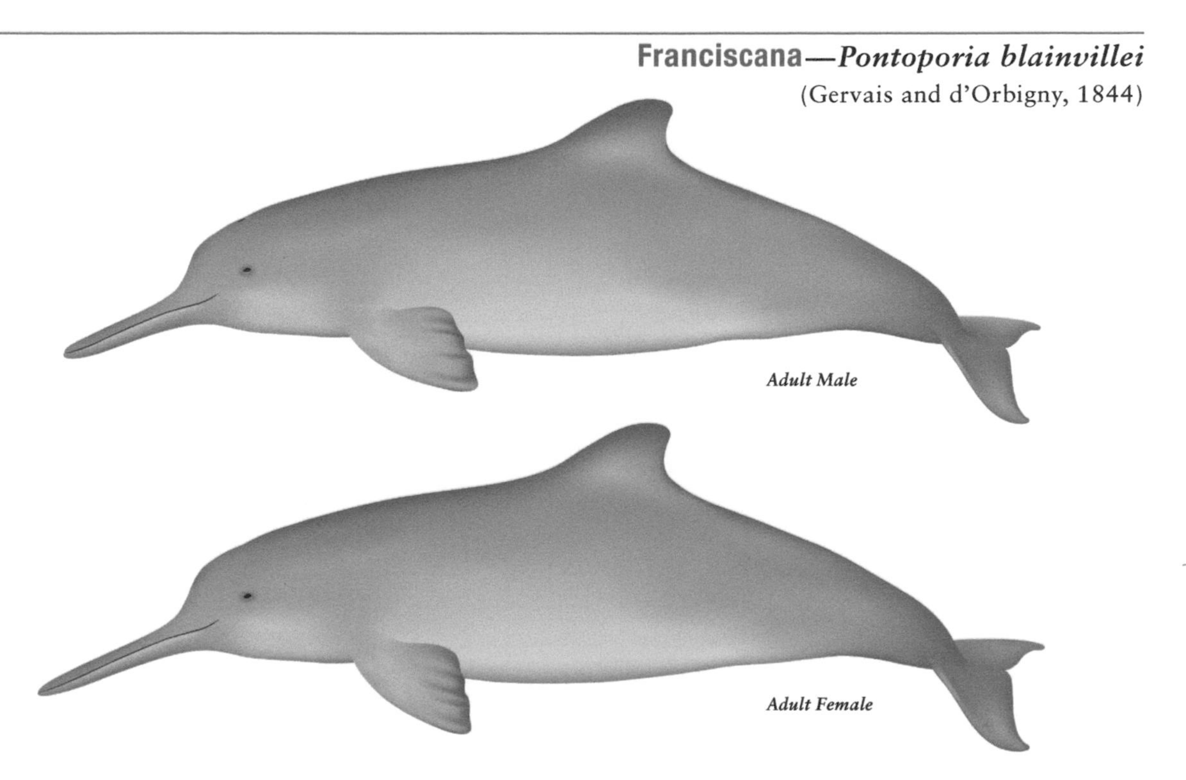

Recently-used synonyms *Stenodelphis blainvillei.*
Common names En.–franciscana, La Plata river dolphin; Sp.–*franciscana*; Fr.–*dauphin de La Plata.*
Taxonomic information Cetartiodactyla, Cetacea, Odontoceti, Pontoporiidae.
Species characteristics Although not a true freshwater dolphin, this primarily-marine species nonetheless shares many features in common with the true "river dolphins." The franciscana's beak is extremely long and narrow, relatively the longest of any living species of cetacean (about 12–15% of total length in adults). The beak grows to be longer in adult females (17–22 cm) than in males (15–19 cm). Although in calves, the beak is much shorter and stouter than it is in older individuals. The forehead is steep and rounded. The dorsal fin is low to moderately tall, with a rounded tip. It is triangular, or sometimes slightly falcate. The flippers are broad and spatulate, sometimes with an undulating trailing edge. In many animals (especially younger ones), there are visible ridges along the surface, corresponding to the finger bones. Newborns have proportionately larger flippers, dorsal fins, and flukes.

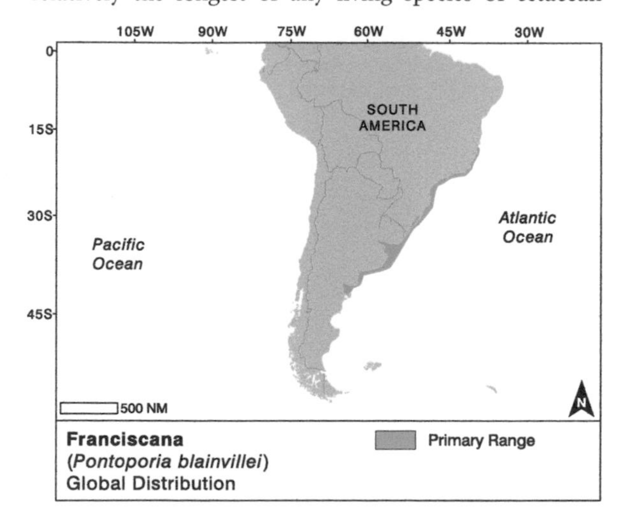

Franciscana
(*Pontoporia blainvillei*)
Global Distribution

Franciscanas have a relatively simple, countershaded color pattern. They are brownish to dark gray above, and lighter brown (or even yellowish) to light gray below and on the lower flanks. The back is darker than the belly and sides, and a faint dorsal cape may sometimes be visible.

The long beak is lined with 50–62 fine, pointed teeth per row, more than in nearly any other species of cetacean. Franciscana females grow larger than males, with males reaching up to 1.63 m, and females 1.77 m in length. Maximum length ranges from 1.30–1.62 m for females and 1.13–1.36 m for males from different populations. Maximum recorded weight is about 53 kg. At birth, franciscanas measure about 71–80 cm, with an average of about 73 cm in length and 6.1 kg in weight.
Recognizable geographic forms Morphological studies have suggested the existence of at least two "forms" of franciscana, a smaller northern and a larger southern one (those

The beak of the franciscana can be extraordinarily long in adults, especially females. PHOTO: PROJECTO TONINHAS UNIVILLE

The combination of a very long beak and short, rounded dorsal fin will help to distinguish the franciscana. PHOTO: PROJETO TONINHAS UNIVILLE

The dorsal fin of the franciscana is bluntly rounded at the tip, and this will help to distinguish it from the sympatric Guiana dolphin, which has a more pointed fin tip. PHOTO: P. BORDINO

Franciscanas are generally cryptic and difficult to see well, but careful attention to dorsal fin shape will help to accurately identify them. PHOTO: PROJETO TONINHAS UNIVILLE

Young franciscanas have somewhat shorter beaks than the adults. PHOTO: PROJECTO TONINHAS UNIVILLE

There are several other small cetaceans that overlap the franciscana's distribution, so attention will be needed to avoid misidentification. PHOTO: PROJETO TONINHAS UNIVILLE

A dead franciscana, showing the species' distinctive body shape and exceedingly long beak. PHOTO: R. BASTIDA

in the far north may be intermediate in size). Other than length, there are no known obvious external morphological differences between them (although this may be due to lack of sufficient data). So, although the franciscana is clearly divided into three genetically-distinct populations (currently managed as four separate stocks or "Franciscana Management Areas"), it is not currently possible to reliably distinguish distinct geographic forms. There is also some evidence of geographic variation in coloration, but this remains unconfirmed.

Can be confused with If not seen well, franciscanas may be confused with marine dolphins of the genus *Sotalia,* but can be identified by their very long beaks and more rounded dorsal fins, broad flippers, and small eyes. Young franciscanas, with relatively short beaks, will cause the most confusion. Burmeister's porpoises overlap in distribution as well, but will be distinguishable by their lack of a prominent beak and oddly-shaped dorsal fin. Bottlenose dolphins should be easy to distinguish, based on their much larger size and tall, falcate dorsal fins.

Distribution Franciscanas are found only along the east coast of South America (Brazil, Uruguay, and Argentina), from Golfo San Matias, central Argentina (42°35′S), to Espirito Santo, southeastern Brazil (18°25′S). They are primarily shallow-water, coastal animals, generally ranging no farther offshore than the 30 m isobath. Some sightings in water beyond 50 m deep and 55 km offshore have been recorded, but the density is very low in such areas. They may also be found in some estuaries, and they sporadically enter the estuary of the La Plata River. Much of the franciscana's habitat is characterized by turbid waters.

Ecology and behavior Although there has been increased research interest in recent years, there is still little known about franciscana behavior. This is due to two factors: the difficulty of observing them in nature (they are somewhat cryptic), and the paucity of research effort. They are found singly or in small groups of up to 15 individuals. In general, they appear to avoid vessels, and do not ride bow waves. They are not considered to be aerially active, and the average swimming speed is about 4.7 km/h. Mean dive duration is about 22 sec. Cooperative feeding behavior has been observed in Argentina.

Although sometimes described as a "river dolphin," the franciscana is not truly a freshwater species. Franciscanas do not migrate, although seasonal inshore/offshore movements have been documented in some areas. Predation by both large sharks and killer whales has been documented.

Peak calving for this species occurs from October to December for most stocks, but in the Rio de Janiero area (northernmost stock) breeding occurs throughout the year. Until recently, there had not been much research on the life history of this species. Reproductive parameters

A live-stranded franciscana swims in its tank, giving a good dorsal view. Note the very long, forceps-like beak. PHOTO: R. BASTIDA

have only been estimated for some of the stocks. For these, sexual maturity in females occurs at ages between 2 and 5 years, and for males between 3 and 4 years. Gestation lasts about 11 months. The relatively-small testes of this species suggest that sperm competition is not a factor in their mating system. Although most individuals appear to live to less than 10 years, longevity is up to at least 15 years for males, and 21 years for females.

Feeding and prey Franciscanas feed on several species of shallow-water fish (e.g., sciaenids, engraulids, gadids, and carangids), cephalopods, and crustaceans. They feed mostly near the bottom, and appear to be opportunistic, with at least 58 fish species, six cephalopod species, and six crustacean species known from the diet. Shrimps are commonly eaten by juveniles, which then switch to a more piscivorous diet after one year.

Threats and status The main problem facing the species is incidental mortality in gillnet fisheries, which has numbered at least 2,500 animals in some recent years, and affects all four management stocks. Other threats include various forms of habitat degradation and pollution. Total abundance of the species is not known, but has been estimated for the two southern stocks. In 1996, there were an estimated 42,000 animals in southern Brazil and Uruguay, and 40,200 in Argentina. While the overall abundance of the species would seem to be high, in most areas the gillnet mortality alone is thought to be unsustainable. Therefore, the franciscana is considered to be at some risk of extinction.

IUCN status Vulnerable (overall species, southern Brazil/Uruguay stock).

Northern Fur Seal
Antarctic Fur Seal
Juan Fernandez Fur Seal
Guadalupe Fur Seal
Galapagos Fur Seal
South American Fur Seal
New Zealand Fur Seal
Subantarctic Fur Seal
Cape & Australian Fur Seals
California Sea Lion
Steller Sea Lion
Galapagos Sea Lion
Australian Sea Lion
South American Sea Lion
New Zealand Sea Lion
Walrus

6. Pinnipeds

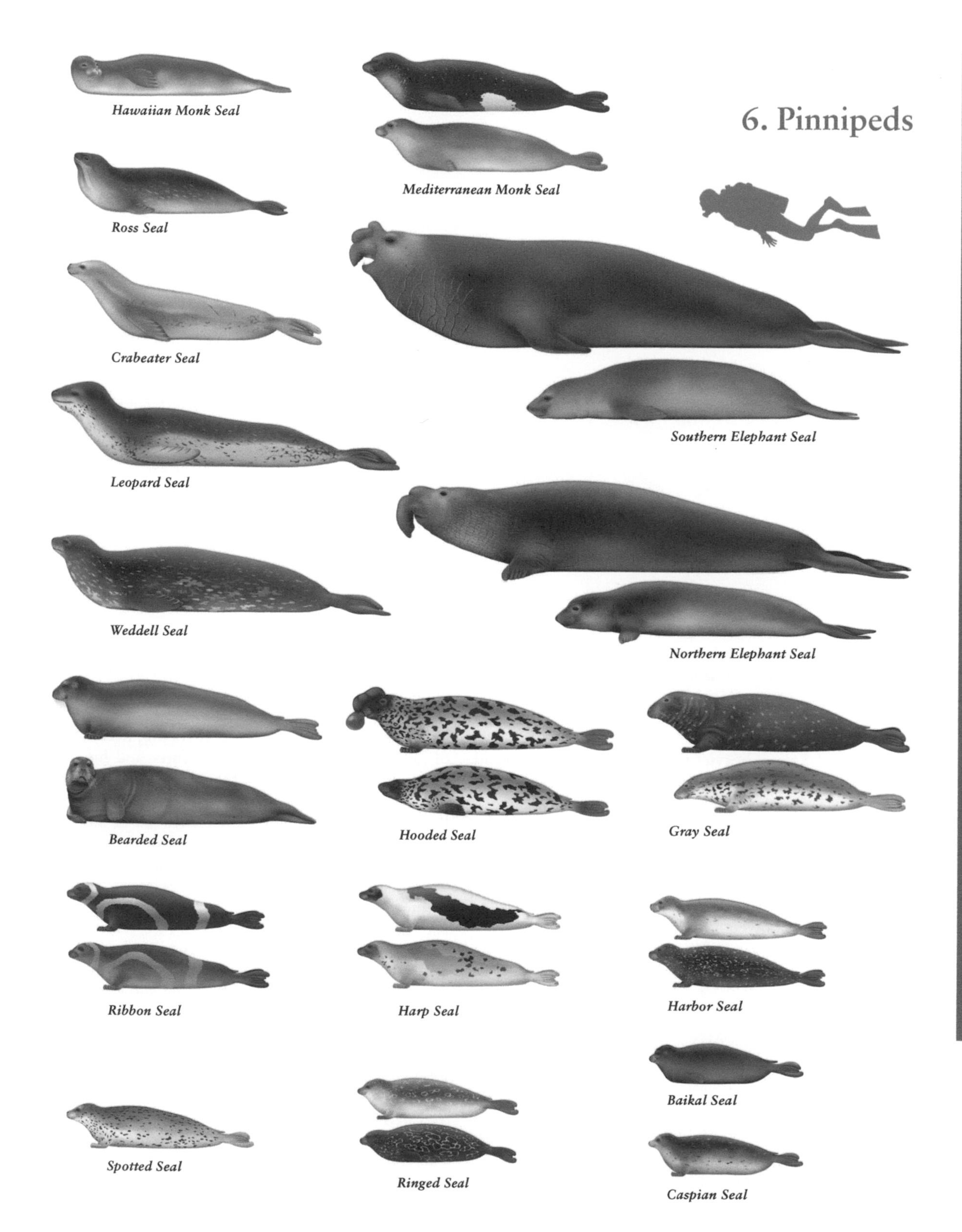

True Seals *Phocidae*

Northern Fur Seal—*Callorhinus ursinus*

(Linnaeus, 1758)

Recently-used synonyms None.
Common names En.–northern fur seal; Sp.–*lobo fino del norte;* Fr.–*otarie des Pribilofs.*
Taxonomic information Carnivora, Otariidae.
Species characteristics Northern fur seals show extreme sexual dimorphism. Adult males are 30–40% longer and more than 4.5 times heavier than adult females. The head appears foreshortened in both sexes because of the very short down-angled muzzle and small nose, which extends slightly beyond the mouth in females and more so in males. The pelage is thick, with a dense, cream-colored underfur. The underfur is usually obscured by the longer guard hairs, although it is partially-visible when the animals are wet and guard hairs part slightly in lines, or when they suffer from alopecia, also called mange, and "blotchy" or "rubbed pelage," involving loss of patches of guard hairs. Features of both fore- and hindflippers are unique and diagnostic of the species. Uniquely in this species amongst all otariids, fur is absent on the top of the foreflippers and there is an abrupt "clean shaven line" across the flipper, at the bend just below the wrist, where the fur ends. Also, the hindflippers are proportionately the longest of any otariid because of extremely long, cartilaginous rod extensions on all of the toes. There are small claws on digits 2–4, well back from the flap-like end of each digit. The toes of the hindflippers are so long that when held up they tend to droop down in a long graceful arc to the tips. The ear pinnae are long and very conspicuous (naked of dark fur at the tips) in older animals. The mystacial vibrissae can be very long, and regularly extend beyond the ears. Adults older than 6–7 years have all pale vibrissae, juveniles and subadults have a mixture of pale and black vibrissae, including some that have dark bases and pale ends, and pups and yearlings have all-black vibrissae. The eyes are proportionately large and conspicuous, especially on females, subadults, and juveniles.

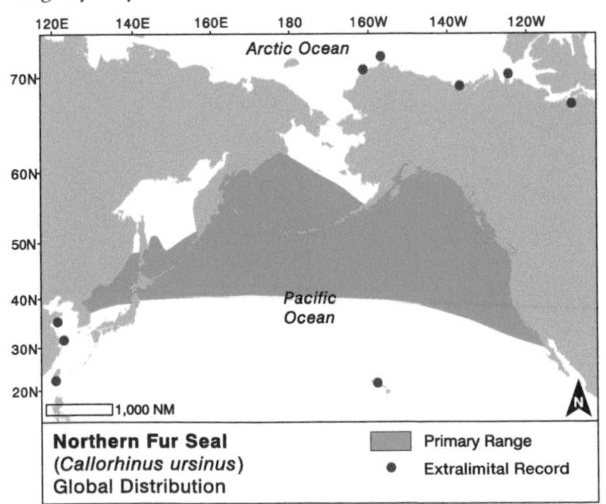

Adult males are stocky in build, and have an enlarged neck that is thick and wide. A mane of coarse longer guard

A bull with a large group of females and pups. Adult female color variation in this photograph is primarily due to staining from time spent on the rookery. Pribilof Islands, Alaska. PHOTO: T. COLLOPY

Adult male or bull northern fur seal, showing the short muzzle, long vibrissae, and mane. Pribilof Islands. PHOTO: J. GIBBENS

Adult female or cow. The fur stops abruptly at the bend in the flipper in this species and the hindflippers are extremely long. Pribilof Islands. PHOTO: J. GIBBENS

This view of copulating northern fur seals puts sexual dimorphism in this species in perspective. Pribilof Islands. PHOTO: J. GIBBENS

Adult male and adult female, nose to nose. Priblilof Islands. PHOTO: J. GIBBENS

Adult female nursing a young pup in the black natal coat. The all white vibrissae indicate the female is more than 7 years old. Pribilof Islands. PHOTO: M. A. WEBBER

A crèche of pups all in their black natal pelage. When females depart for foraging trips, the pups gather in these groups where they groom, sleep, and interact with each other. Pribilof Islands. PHOTO: T. COLLOPY

Adult male with females in front and behind. The orange tint to the pelage is staining from a long tenure on the rookery. Saint Paul Island, Pribilof Islands. PHOTO: M. A. WEBBER

Adult male with females showing sexual dimorphism. The male has a large mane with light tinged guard hairs. Lovushki Island, Kuril Islands, Russia. PHOTO: O. SHPAK

Adult male. The long hindflippers, mane and short muzzle, as well as the large neck and shoulders, of adult males are conspicuous. The all white vibrissae indicate he is over 7 years old. Saint Paul Island, Pribilof Islands. PHOTO: M. A. WEBBER

Adult female. On warm days, northern fur seals wave their hindflippers to increase air flow over dilated blood vessels in their skin to facilitate cooling. Saint Paul Island, Pribilof Islands. PHOTO: M. A. WEBBER

Subadult male beginning to show the enlarged chest and neck of an adult. Saint Paul Island, Pribilof Islands. PHOTO: M. A. WEBBER

Subadult males about the size of adult females, have more robust features, but lack all white vibrissae. Pribilof Islands. PHOTO: J. GIBBENS

This subadult male is larger and longer in the neck than a smaller, but similarly shaped, adult female. Pribilof Islands. PHOTO: T. COLLOPY

hairs extends from the top of the head to the shoulders and covers the nape, neck, chest, and upper back. While the skull of adult males is large and robust for their overall size, the head appears short because of the combination of a short muzzle, and the back of the head behind the ear pinnae being obscured by the enlarged neck. Adult males have an abrupt forehead formed by the elevation of the crown from development of the sagittal crest, and thicker fur of the mane on the top of the head. The canine teeth are much longer and have a diameter greater in adult males than in adult females, and this relationship holds to a lesser extent for all age classes. Adult females, subadults and juveniles, are moderate in build. It is difficult to distinguish the sexes until about 4–5 years old. The body is modest in size and the neck, chest, and shoulders are sized in proportion with the torso.

Daylight hours are spent rafting, and grooming. Flippers are held up to reduce heat loss to the water. One female (upper left), is in the "jug-handle" posture with one foreflipper draped over both hindflippers. The adult male (upper right), has alopecia (guard hair loss, showing underfur on the head). The juvenile (lower), was grooming. California. PHOTOS: (UPPER LEFT) S.N.G. HOWELL, (UPPER RIGHT) J. POKLEN, (LOWER) A. SCHULMAN-JANIGER

Adult females and subadults have more complex and variable coloration than adult males. They are dark silver-gray to charcoal above. The flanks, chest, sides, and underside of the neck (often forming a chevron pattern in this area) are cream to tan with rusty tones. In particular, the pale chest on females and juveniles can impart a look of the animal wearing a vest or "t-shirt" of pale color. There are cream to rust-colored areas on the sides and top of the muzzle, chin, and as a "brush stroke" running backwards under the eye. In contrast, adult males are medium gray to black, or reddish to dark brown all over, and are darker and redder-brown when soiled from being ashore on territories for long periods during breeding. A small percentage are paler almost grayish as adults. The mane can have variable amounts of silver-gray or yellowish to whitish tinting on the guard hairs. Pups are blackish-brown at birth, with oval areas of buff on the sides, in the axillary area, and on the chin and sides of the muzzle. After 3–4 months, pups molt to the color of adult females and juveniles.

Color variants include albino (1 in 30,000–100,000), partially albino (more common than albinos), piebald morphs, and patchy-looking animals that have lost dark guard hairs, exposing pale underfur as a result of alopecia. Variation in the color of pups ranges from creamy, pale gray, silvery, and brown to chocolate. It is not known if these pups molt to normal pelage patterns as juveniles. Northern fur seals have also been reported to have patches of extensive covering of red and brown algae that tint the fur, and infestations of grayish gooseneck barnacles of the genus *Lepas*, all of which can discolor an animal and potentially affect field identification. Cryptorchid males may occur occasionally, in which the testes don't develop normally. They lack a mane entirely, and have unusual proportions, including incomplete development of musculature so their appearance is like large dark adult females.

There have been two confirmed cases of hybridization, and in both the cross was northern fur seal X California sea lion. Both pups had a mixture of features. The most recent individual, seen at a stranding facility, had a sea lion-like muzzle, fur on the dorsum of the foreflippers, and a long, wide hallux, but fur seal-like long ear pinnae, long hindflippers, and dense pelage with underfur. Genetic tests confirmed it was the result of a mating between a northern fur seal bull and California sea lion cow.

Adult males reach lengths of 2.1 m and weights of 270 kg, while females are up to 1.5 m and 50 kg. Pups are 60–65 cm and 5.4–6 kg at birth. Longevity ranges from 10–27 years, with females attaining greater ages. The dental formula is I $^{3}/_{2}$, C $^{1}/_{1}$, PC $^{6}/_{5}$.

Recognizable geographic forms None.

Can be confused with Northern fur seals can be confused with two other otariid species in their range: the Guadalupe fur seal, and California sea lion. Steller sea lions, which occur in many areas with northern fur seals, are so much larger and different in features and coloration as to not be a source of confusion. Coloration of Guadalupe and northern fur seals is similar. Guadalupe fur seals have fur on the dorsum of the foreflipper beyond the wrist, proportionately shorter hindflippers, and a long tapering muzzle with a somewhat bulbous nose that makes the muzzle seem slightly-upturned on the end. At

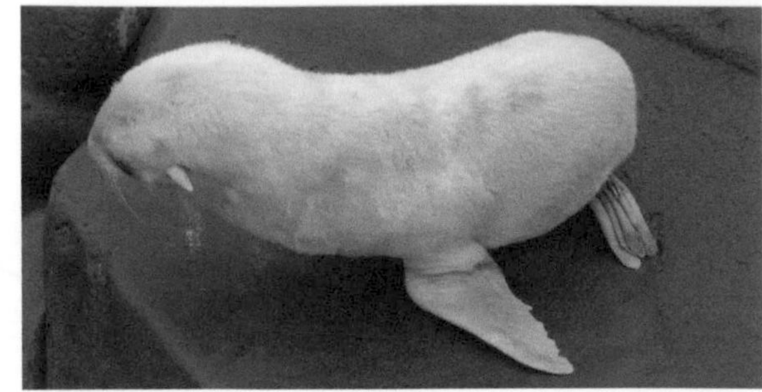

A juvenile hybrid northern fur seal (father) X California sea lion, found beached, shows characteristics of both species. The long pinnae and hindflippers, thick pelage with underfur, and light chest are fur seal features and the wide blunt muzzle and fur past the bend on the top of the foreflippers are sea lion features. PHOTOS: (UPPER LEFT) R. J. WILSON/TMMC (LOWER LEFT & CENTER) THE MARINE MAMMAL CENTER

Variations on albinism. The pup in the upper right seemed sight-impaired but was thriving while still being cared for by his mother. The subadult male in the lower right photo is at least 3 years old. Lack of dark vibrissae in this case may not indicate age. Pribilof Islands. PHOTOS: (UPPER RIGHT) M. A. WEBBER, (LOWER RIGHT) G. THOMSON/USFWS

sea, Guadalupe fur seals will groom and rest in a posture characteristic of fur seals of the genus *Arctocephalus* with the head down and both hindflippers held in the air and apart forming a Y-shape. Northern fur seals are not known to use this position. However, both northern and Guadalupe fur seals species routinely rest in a "jug handle" position with the palm of one foreflipper draped over the soles of the hindflippers, which are rotated forward to meet the foreflipper, so observers must look for other identification features when finding fur seals at rest at the surface in areas of shared range between these two species.

Northern fur seals can be readily distinguished from California sea lions based on differences in thickness, length, and coloration of pelage, the lack of fur on the top of the foreflippers beyond the wrist, much longer hindflippers with longer digits, very short pointed muzzle, and longer prominent ear pinnae. Dark adult male California sea lions have a conspicuous sagittal crest, and lack a mane and the light tipped guard hairs on their neck. The loud repetitive bark of male California sea lions is distinctive and different from all northern fur seal vocalizations.

Distribution Northern fur seals are a widely-distributed pelagic species in the waters of the North Pacific Ocean and the adjacent Bering Sea, Sea of Okhotsk, and Sea of Japan. They range from northern Baja California, Mexico, north and offshore across the North Pacific to northern Honshu, Japan. The southern limit of their pelagic distribution across the North Pacific is approximately 35°N. Vagrants have been reported from China on the coast of the Yellow Sea, Taiwan in the west Pacific, Hawai'i in the central Pacific, and eastern Beaufort Sea in the Arctic. The vast majority of the population breeds on the Pribilof Islands in the Bering Sea, with substantial numbers on the Commander Islands as well. Still other sites are used, including rapidly increasing numbers at Bogoslof Island near the eastern Aleutians, San Miguel Island in California, sites in the Kuril Islands and Tyuleniy Island off Sakhalin Island in Russia. Numerous other sites were formerly occupied and may still be visited.

Ecology and behavior This is a highly polygynous species. Males arrive at the rookeries up to one month before females and vocalize, display, and fight to establish and maintain territories. Breeding on the Pribilof Islands occurs from mid-June through August, with a peak in early July (the median date in southern California is approximately 2 weeks earlier than at the Pribilofs). Northern fur seals become sexually mature at 3–5 years old, at which time females usually produce one pup a year for most of the rest of their lives. Males do not become physically mature, and large enough to compete for a territory that will be used by females, until they are 8–9.

Northern fur seals usually give birth a day after arrival at the rookery. Average time from birth to estrus is 5.3 days, and 8.3 days for departure on the first feeding trip. Females breeding at the Pribilof Islands are relatively far from their foraging areas at the edge of the continental shelf. As a result, they consistently make longer foraging trips than most other female otariids, with a mean trip length of 6.9 days. Once foraging begins, the mean depth of dives is 68 m and duration is 2.2 min with maximum depth recorded of 207 m, and duration of 7.6 min. Pups are visited 8–12 times and attended for a mean of 2.1 days before being abruptly weaned at 4 months old.

An unusually proportioned male with a collar of marine debris embedded in the skin. All white whiskers indicate an older animal, yet the lack of neck development suggest he may be a cryptorchid. Lovushki Island, Russia. PHOTO: J. WAITE

Adult male after fighting with three bulls during an attempt to obtain a territory without females in between them. Pribilof Islands. PHOTO: M. A. WEBBER

Adult male with alopecia involving loss of guard hairs that reveals the creamy underfur that can be tinged pinkish to orangish. Pribilof Islands. PHOTO: M. A. WEBBER

Northern fur seals are one of the most pelagic of pinnipeds. Adults are at sea most of the year, only coming ashore for the breeding season for 35–45 days (on average) for adult females and males, respectively. They do not haul out between breeding seasons, and once weaned, juveniles go to sea and do not haul out until they return, usually to the island of their birth, 2–3 years later. Many animals, especially juveniles, migrate from the Bering Sea south into the North Pacific to spend the winter feeding. At sea, northern fur seals are most likely to be encountered alone or in pairs, with groups of 3 or more being uncommon. They forage relatively far from shore, over the edge of the continental shelf and slope. Diving is very active at dawn and dusk, otherwise northern fur seals spend their time rafting at the surface, sleeping or grooming. They employ a wide variety of resting postures, including raising one or more flippers into the air, and draping one of their foreflippers over both of the hindflippers to form a posture known as the "jug handle" position. Predators include killer whales, sharks, and Steller sea lions.

Feeding and prey The diet varies by location and season, and includes many varieties of epipelagic and vertically-migrating mesopelagic schooling and non-schooling fish and squid. Prey species of importance in the waters off California and Washington include anchovy, hake, saury, several species of squid and rockfish, and salmon. In Alaskan waters, walleye pollock, capelin, sand lance, herring, Atka mackerel, and several species of squid are important prey. Foraging and distribution is often correlated with oceanic eddies and fronts in areas of high cholorphyll in surface waters.

Threats and status Northern fur seals have a long, complex history of commercial harvesting for their pelts, which began with the discovery of the main breeding colonies in the late 18th century, and lasted until 1984. There were many intervals of decline and recovery over this long period. International treaties and agreements were enacted to manage this species, including the International Fur Seal Treaty of 1911 that is regarded as the first international treaty for wildlife preservation. In the 1950s, the population numbered up to 2.5 million animals, and it may have been even more abundant when there were more rookeries before the onset of exploitation by Europeans and Americans. The current population is estimated at 1.1 million (with 650,000 breeding on the Pribilof Islands) and is declining. Entanglements in commercial fisheries gear, including derelict and discarded gear and marine debris, have caused significant annual mortality in the past. This mortality was highest during the period of active high seas driftnet fishing in the North Pacific in the 1980s. Entanglement in debris is ongoing and affects juveniles and subadults more than adults. Northern fur seals compete for walleye pollock with one of the largest commercial fisheries in the world. Mortality from interactions with numerous fisheries and entanglement in debris, large annual harvests of prey species in commercial fisheries, long-term ecosystem change in the North Pacific, and possible changes in the foraging behavior of its predator, the killer whale, may all be factors causing the current decline.

IUCN status Vulnerable.

Note on Hybrid Southern Hemisphere Fur Seals Genus *Arctocephalus*

Hybrids of a number of species of otariids are known, and genetic, behavioral, and morphological characteristics have been published in the scientific literature. Evidence for probable spotted seal X harbor seal and spotted seal X ribbon seal hybrids has been found in a museum collection. Only one wild hybrid phocid has been verified. It resulted from a harp seal and hooded seal mating. In general, most otariid hybrids have features that are a blend of the basic characteristics of the parent species.

In addition to southern fur seals (genus *Arctocephalus*), that are known to hybridize, several eastern Pacific otariid species hybridize, including: northern fur seal X California sea lion (see northern fur seal account) and South American sea lion X California sea lion. However, the most frequently encountered hybrids are from crosses of Southern Hemisphere fur seals, especially where several species occur at the same breeding islands, such as at Macquarie and Marion Islands, and the Juan Fernández Islands. Examples of the known hybrid crosses are shown. The mating patterns, reproductive success, and behavior of hybrids are poorly known. Recent studies have shown that hybrid males may have reduced fertility and that females may depart their territories to breed with other males.

However, vocalizations of southern fur seals are well known, and hybrids are often first recognized or identified based on their unusual vocalizations. Vocalizations of pinnipeds, and particularly hybrid fur seals, are not covered in depth in this guide.

Special thanks are due to the researchers who made these previously-unpublished photographs, notes, and observations available for this brief overview.

Photograph notes and credits

A) Subantarctic X Antarctic fur seal adult male hybrid with a mixture of features. The crest of longer guard hairs on the forehead and pale-colored muzzle and lower jaw, and dark ear pinnae, are characteristic of subantarctic fur seal bulls. The dark neck and chest are characteristics of an Antarctic fur seal bull. The pale-color of the muzzle reaches to the eyes and partially around them, which is an intermediate pattern between subantarctic and Antarctic fur seals. The overall color is also intermediate between the two species, being paler than the subantarctic fur seal, and darker than the Antarctic fur seal. Marion Island, Prince Edward Islands. PHOTO: P. J. N. DE BRUYN

B) Subantarctic X Antarctic fur seal subadult or adult male hybrid. The pelage is not as dark as on a subantarctic fur seal, and lacks the light-colored fur in the chest and face. The ears are pale, as is the case for Antarctic fur seals. The muzzle is short and sharply pointed as in the subantarctic fur seal. The identification of this animal as a hybrid was supported by his vocalizations. Marion Island, Prince Edward Islands. PHOTO: P. J. N. DE BRUYN

C) Subantarctic X Antarctic fur seal adult female hybrid with her pup, with a subantarctic fur seal female in the background. This hybrid is presumed to be a first generation cross because of the small numbers of Antarctic fur seals breeding at Marion Island. She has a fairly even mix of features from both species. Her lighter chest suggests subantarctic fur seal, while her darker upper neck and face, paler back, and large muzzle and head, suggest Antarctic fur seal. Her darker pup looks more like a subantarctic than Antarctic fur seal (the unlikely possibility that this pup was adopted cannot be ruled out). Marion Island, Prince Edward Islands. PHOTO: P. J. N. DE BRUYN

D) Subantarctic X Antarctic fur seal pup hybrid (left) and subantarctic fur seal pup (right). Subantarctic and Antarctic fur seal pups are difficult to separate in the field without experience. Additionally, pups of each species change slightly as they age and approach their first molt. During this time, their features can change to be like those of known hybrids. The identification of this hybrid pup was confirmed based on vocalizations. It has features similar to those found on Antarctic fur seal pups including an extensive area of pale color on the muzzle, and gray in the face. The muzzle is too large and wide, and the overall pelage color is not black enough for this to be a subantarctic fur seal pup. Marion Island, Prince Edward Islands. PHOTO: P. J. N. DE BRUYN

E) Antarctic X New Zealand fur seal adult male hybrid confirmed by mtDNA analysis, as well as by vocalizations. His overall dark gray color is darker than that found in either Antarctic or New Zealand fur seals. His pale muzzle and face are similar to New Zealand fur seals. The muzzle is of medium length, without an upturned end, which is more like an Antarctic fur seal. The ear pinnae are pale and lighter than the surrounding fur, like on Antarctic fur seals. This bull has a slight crest of fur on his forehead, which is a feature occasionally seen on New Zealand fur seals and not found on Antarctic fur seals. Note the yellow tag on the foreflipper. Macquarie Island. PHOTO: B. PAGE

F) Antarctic X New Zealand fur seal adult male hybrid confirmed by mtDNA analysis and vocalizations. This animal has the same muzzle shape and facial color, and a slight crest of fur on his forehead, as on the animal in photo E, but his ear pinnae are dark. Macquarie Island. PHOTO: B. PAGE

G) Antarctic X New Zealand fur seal adult female hybrid. The identity of this female was confirmed by mtDNA analysis and vocalizations. The gray pelage and pale ear pinnae are like that of Antarctic fur seals. The slightly up-turned muzzle is typical of New Zealand fur seals. Macquarie Island. PHOTO: B. PAGE

H) A probable subantarctic X Juan Fernandez fur seal adult male hybrid. This animal has a crest of fur on his forehead and large tear-drop shaped eyes like a subantarctic fur seal. The body color is not as dark as a subantarctic fur seal, and not at all like the dark brown to blackish-brown of a Juan Fernandez bull. The lighter reddish-brown fur on the tops of the flippers is like the color found on Juan Fernandez fur seal flippers. The pale muzzle and face color is similar to the coloration found on other hybrid males. The hind- and possibly foreflippers appear small, and are proportionately more like the flippers found on subantarctic fur seals. Juan Fernández Islands. PHOTO: J. FRANCIS

A
B
C
D
E
F
G
H

Antarctic Fur Seal—*Arctocephalus gazella*

(Peters, 1875)

Recently-used synonyms *Arctocephalus tropicalis gazella*, *Arctophoca gazella*.

Taxonomic information Carnivora, Otariidae. Recently, a proposal was made to reassign this species to the resurrected genus *Arctophoca*.

Common names En.–Antarctic fur seal, Kerguelen fur seal; Sp.–*lobo fino antártico;* Fr.–*otarie antarctique.*

Species characteristics Antarctic fur seals show strong sexual dimorphism, with adult males 4–5 times the mass and 1.4–1.5 times the length of adult females. The muzzle is of medium length and width, straight, tapering to a moderately pointed end. The rhinarium is modest in size and does not extend much past the mouth. The ear pinnae are long, prominent, and pale in color usually with lighter tips. The eyes are almond-shaped. The pale vibrissae are long and conspicuous. On adult males they are some of the longest found on any pinniped, reaching 35–50 cm. Adult males develop a mane of thicker, coarser guard hairs on the chest, neck, and top of the head. The neck and shoulders are also greatly enlarged with fat and the development of muscles in older adult males. Adult females, subadults and juveniles are difficult to tell apart from males until the latter begin to grow larger when they are 4–6 years old. Juveniles and subadults have dark, to mixed light and dark vibrissae, becoming lighter as they age, and become pale when they are adults.

Adult male and adult female Antarctic fur seals showing the extent of sexual dimorphism in this species. Males have a thick, grizzled mane which adds bulk to their larger size. South Georgia. PHOTO: M. A. WEBBER

The fore- and hindflippers are proportionately long at 28–33% and 22–28%, respectively, of the total body length of all age classes. The foreflippers have a dark, sparse, short fur that extends beyond the wrist onto the middle of the dorsal surface of the flipper in a V-pattern that does not reach the rounded tip. The rest of the dorsal surface and the palms of both foreflippers are covered with a hairless black leathery skin. The first digit is the longest, widest and thickest, and angles distally, giving the flipper a swept-back look. Digits 2–5 are successively shorter. There is a small opening in the skin at the end of each digit for a claw that is usually reduced to a vestigial nodule, and rarely emerges above the skin. The claw openings are set back from the free edge of the flippers by cartilaginous rods that extend the length of each digit, and expand the size of the flippers. The hindflippers are long and dark, with short, sparse hair covering part of the proximal end of the flipper and the entire sole is covered in black, leathery skin. Each digit has a cartilaginous rod that adds a flap-like extension to each toe. The bones of the three central toes terminate at the position of the small nails that emerge through the skin on the dorsal surface, set back from the end of the flipper. The claws of digits 1 and 5 are usually vestigial, like the claws on the foreflippers, and may or may not emerge from small openings set back from the end of the flaps. All of the flaps at the end of the flipper are of relatively equal length. The first toe or hallux and the fifth toe are somewhat wider than toes 2–4.

Adult male and female. The cow has light color on the muzzle and below the eye, light chest and lower sides, and dark tops to the base of the foreflippers that are typical. PHOTO: J. BASTIDA

This adult male Antarctic fur seal demonstrates the agility of otariids on land and also illustrates the development of the mane and upper body of bulls. PHOTO: M. NOLAN

Adult females and subadults are gray to brownish-gray above and cream to light gray below with shades of ginger and reddish-brown. There is usually an area of pale color of variable extent on the sides between the flippers. Pale color extends from the chest variably up the neck and on the sides of the neck to as high as the throat, eyes, and muzzle. The dark fur on the top of the foreflippers extends into the area where the foreflipper inserts at the shoulder. Additional lighter areas often surround where the ear pinnae attach and are also on the scrolled pinnae. There is a variable amount of cream to reddish-brown color on the muzzle in the mystacial area. Newly molted juveniles have the same color pattern as adult females, but are silvery-gray in all lighter areas. At birth, pups are blackish, though they may be pale on the face and muzzle, and some animals are paler below. Adult males are dark grayish-brown to charcoal, with off-white to silver grizzling on the guard hairs of the back, mane, and flanks. The long guard hairs of the mane often bunch up and reveal the fawn color underfur.

A cream- to honey-colored morph of the Antarctic fur seal is seen at a rate of 1–2 per thousand at South

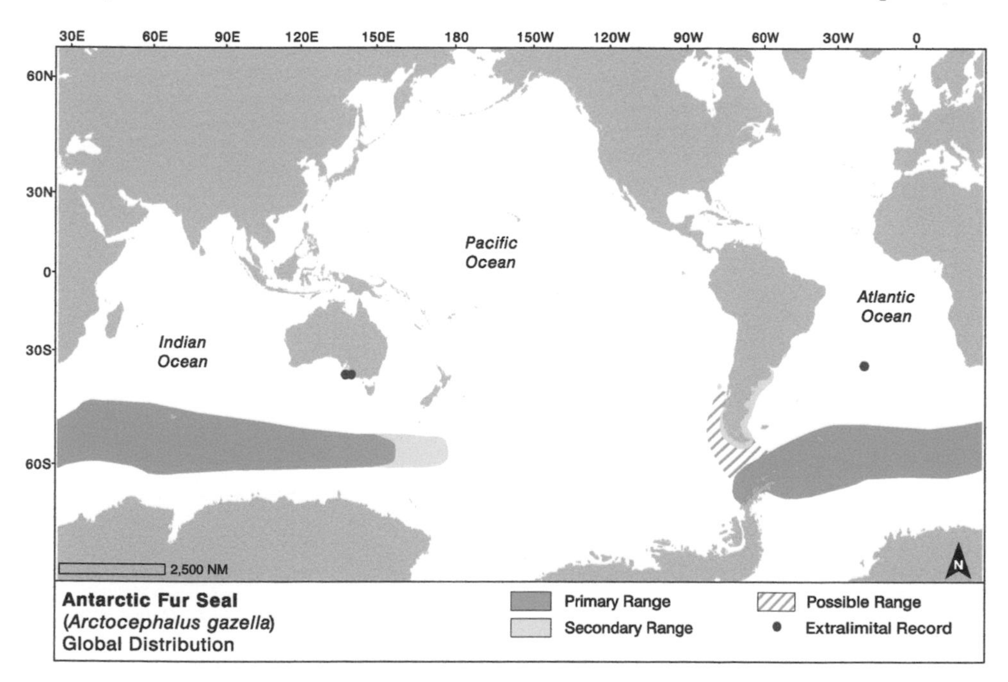

Adult male with thickened neck and mane and long vibrissae. PHOTO: M. NOLAN

Adult female at South Georgia Island; although wet, the pale chest and lower sides are visible. PHOTO: M. A. WEBBER

Adult female with a young pup in natal pelage. Females tend their pups for several days after birth before the first of many foraging trips. South Georgia Island. PHOTO: M. A. WEBBER

Two subadult males sparring. Both lack development of the neck that will come later in life. The animal on the right is pale, but not leucistic. PHOTO: B. RAFTEN

Two males, subadult on the left and adult on the right, sparring on South Georgia. In otariids, size almost always wins out in aggressive encounters. PHOTO: M. NOLAN

Juvenile with wet pelage grooming with the sub-terminal nails on the three central toes. PHOTO: M. NOLAN

Newly molted juvenile Antarctic fur seals are very pale colored. Marion Island, Prince Edward Islands. PHOTO: P. J. N. DE BRUYN

This molted Antarctic fur seal pup has a leucistic pelage, also known as a "golden phase." PHOTO: M. NOLAN

Juvenile or small subadult leucistic Antarctic fur seal. Unlike albinos, this leucistic morph has normally-pigmented skin. PHOTO: J.-P. SYLVESTRE

Adult male leucistic Antarctic fur seal that is unusually whitish. South Georgia Island. PHOTO: J.-P. SYLVESTRE

Three images of unusual color morphs of Antarctic fur seal pups from Cape Shirreff, Livingston Island. (Left to right): "tricolor morph" with orangish markings; "tiger morph" with stripes, with a normal blackish pup to the right; and "brown morph." It is not known if these unusual patterns persist beyond the pup stage. PHOTO: J. ACEVEDO

Georgia. These golden animals are not albinos, but leucistic, with normally-pigmented eyes and paler, but not unpigmented, skin. The guard hairs on these pale animals lack pigment, and the underfur and skin are paler than in normally-pigmented animals. Additionally, four other unusual variations in coloration have been reported from Cape Shirreff on Livingston Island: a light morph found on two pups with dark stripes from the middle of the back towards the ventral surface on the head, neck and torso named a "tiger morph;" a "brown morph" found on one female pup whose body was light brown with pale guard hairs on the head and neck, becoming blond on the face, with pale vibrissae, which is abnormal for a pup; a "tricolor morph" pup that was dark brown to black but had light brown patches on the back and extensive amounts of light brown on the neck, chest and undersides; and a "piebald morph" male pup with light yellowish patches on the chest, chin and nose. These variations appeared in one colony at the rate of about 1 per 3.6 to 5.7 thousand pups born, and provides insight into the frequency of uncommon color patterns.

Adult males on breeding territories can be up to 2 m long, but are usually around 1.8 m. Mean weight is 188 kg, but they can weigh 133–204 kg. Adult females are 1.2–1.4 m long and weigh 22–51 kg, with a mean weight of 26 kg. Pups are 63–67 cm long and 6–7 kg at birth. The dental formula is I $^{3}/_{2}$, C $^{1}/_{1}$, PC $^{6}/_{5}$.

Recognizable geographic forms None.

Can be confused with Antarctic fur seals can be confused with southern otariids that share their range, or where wandering Antarctic fur seal vagrants have been found. Subantarctic, New Zealand, and South American fur seals, and the Juan Fernandez fur seal, as well as the South American and New Zealand sea lions, are the most likely species to consider. Subantarctic fur seals are uniquely colored, with a very pale chest and face, and have short, dark ear pinnae. Males also have a crest of longer pale fur in the dark crown. The flippers are proportionately small on all comparable sizes of subantarctic fur seals. Separating New Zealand and South American fur seals from Antarctic fur seals can be problematic, especially with females and immature animals. Adult male New Zealand and South American fur seals are stockier in build with a longer, more pointed muzzle that appears slightly upturned at the nose. Adult male Antarctic fur seals are generally darker with paler grizzling than on either the grayer brown New Zealand or light to dark brown South American fur seals. New Zealand fur seals have brown ear pinnae while South American fur seals have pale pinnae. Separating adult female, subadult

Subadult male Antarctic fur seal on an ice floe. This species will haul out on sea ice far from shore. PHOTO: M. JØRGENSEN

An Antarctic fur seal, probably a juvenile or young adult female, undertakes a more acrobatic type of leap. South Atlantic Ocean. PHOTO: J.-P. SYLVESTRE

An Antarctic fur seal, probably a subadult male, "porpoises" in the South Atlantic Ocean. Low, long "porpoising" leaps are used when traveling rapidly, and the animal's speed hits a "cross-over" rate where it is more efficient to leap and breathe while briefly airborne. PHOTO: M. NOLAN

A subadult male Antarctic fur seal with a king penguin. Although krill is the species' mainstay, fish, squid and birds are also part of the diet. South Georgia Island. PHOTO: M. NOLAN

and juvenile New Zealand and South American fur seals from Antarctic fur seals of similar size and age is more problematic due to overlap of coloration. Note the muzzle size and rhinarium size and shape. Adult male Juan Fernandez fur seals have a longer muzzle, distinctive bulbous rhinarium, and unique coloration of the crown, nape and upper neck. Separation of adult female, and subadult, and juvenile Juan Fernandez fur seals from Antarctic fur seals is problematic. In general, adult female Juan Fernandez fur seals are somewhat larger than female Antarctic fur seals, and have a longer muzzle, larger rhinarium and downward oriented nares, and a more rounded crown. Antarctic fur seals can be differentiated from all Southern Hemisphere sea lions by their coloration, narrower pointed muzzle, proportionately large eyes, long pale vibrissae, and thick pelage. Sea lions have large, blocky heads with blunt-ended and comparatively wide muzzles, and are paler in color than comparably-sized animals. Adult males of all sea lions are much larger and distinctive.

Hybrid Antarctic X subantarctic, and Antarctic X New Zealand fur seals are known. The few photographs available show that they share characteristics of both species (see Note on Hybrid Southern Hemisphere Fur Seals Genus *Arctocephalus*, pages 366–367).

Distribution Antarctic fur seals are widely distributed in waters south of the Antarctic Convergence, and in some areas slightly to the north. Most of the population breeds on South Georgia and Bird islands, but colonies are widely spread out and can be found in the South Shetland, South Orkney, South Sandwich, Prince Edward, Marion, Crozet, Kerguelan, Heard, McDonald, Macquarie, and Bouvetøya islands. Vagrants have been found at Gough and Tristan da Cunha islands, southern South America, the Juan Fernández Islands, at Kangaroo Island in South Australia and at Mawson Station on the Antarctic continent. Males haul out extensively in the mid- and late summer on islands along the Antarctic Peninsula. Ashore, they prefer rocky habitats, but will readily haul out on sandy beaches and move into vegetation zones such as tussock grass. They disperse widely at sea. Distribution and movements in winter are not well known. Males and subadults occur south to the edge of the consolidated pack ice, and can be found hauled-out on sea ice.

Ecology and behavior Antarctic fur seals are highly polygynous. Females become sexually mature at about 3 years of age, and males at about 7 years. Many males arrive at the colonies in late October 2–3 weeks before the first females arrive and establish themselves on territories. Males continue to arrive and challenge for territories through much of the season. Territories are acquired and held with vocalizations, threat postures and fighting. In prime areas, territories can be as small as 20 m^2 and have up to 19 females. The mean length of tenure for bulls at South Georgia is approximately 34 days. Male vocaliza-

A young subadult male Antarctic fur seal with elements of adult female coloration. Sexing juveniles can be difficult without seeing the pattern of genital openings. PHOTO: M. A. WEBBER

Juvenile with net debris necklace. Discarded and lost pieces of net, rope, monofilament, and packing bands attract curious seals, entangle and eventually kill them. This is a significant source of mortality for many pinniped species. South Georgia Island. PHOTO: M. NOLAN

tions include a bark or whimper, a guttural threat, a low-intensity threat, and a submissive call. Females growl and have a pup-attraction call that is a high-pitched wail that usually includes trilling.

Females begin to arrive in mid-November and most pupping and breeding occurs from late November to late December. They usually give birth 1–2 days after arrival at the colony, attend their pup for 6–7 days, come into estrus and mate, and then depart minutes to hours after mating for their first foraging trip. The length of foraging trips and attendance periods varies by year depending on the availability of the lactating female's chief prey, krill. Generally, 4–5 days at sea are followed by 2–3 days attendance on land. Lactating females routinely dive to 8–30 m and are submerged for less than 2 min, but have been recorded to depths of 181 m, and to undertake dives that have lasted 10 min. Average dive depth and duration increase during the lactation period. Pups are weaned in about 4 months. After they wean their pups, females disperse widely, possibly migrating north and are not seen at the colonies much until the next breeding season. Bulls also depart breeding areas, but subadults and adult males can be seen around the rookeries at South Georgia all year.

Like other southern fur seals, Antarctic fur seals "porpoise" when swimming rapidly. When rafting they often assume many of the typical fur seal resting postures. At other times, they can be found busily engaged in grooming. Predators include killer whales, leopard seals, and at Macquarie Island, New Zealand sea lions. Leopard seals have had a dramatic effect on several re-colonized areas in the South Shetland Islands and have caused a decline at one site due to their extensive predation on pups.

Feeding and prey The diet varies by season and location. Adult females at South Georgia feed heavily on adult krill. At Heard Island, krill is not available and lactating females prey primarily on fish such as myctophids and mackerel icefish. In the winter, males and subadult males at South Georgia take krill and a variety of fish that eat krill, while squid and myctophids are only a small percentage of the diet. At Heard Island in the winter, squid and myctophid fish dominate the diet. Foraging patterns of females in summer indicate nocturnal feeding. Antarctic fur seals will eat penguins. Adult males have been documented chasing, killing and eating king penguins on land on Marion Island. They are also known to take macaroni and gentoo penguins in the water at Heard, Macquarie, and South Georgia islands.

Threats and status As was the case for all other southern fur seals, sealers drove the species to the brink of extinction by the late 19th century. The colony at South Georgia was thought to be as small as 100 animals in the 1930s. Today the total abundance is estimated to be between 4.5–6.2 million animals, and still growing, with 95% of the population using the colony at South Georgia. The Antarctic Treaty and the Convention for the Conservation of Antarctic Seals protect this fur seal below 60°S. Various efforts to launch commercial fisheries for krill near South Georgia have been unsuccessful. Trawling activities developing around Macquarie Island may affect the prey base of the primarily fish-eating Antarctic fur seals that breed on those islands. Antarctic fur seals become entangled in marine debris such as discarded fishing line, nets, packing bands and anything that can form a collar. It was estimated from a 1988–89 study that the numbers entangled might be as high as 1% of the total population, with the majority of the impact on juveniles and subadults, particularly males. The 19th century population bottleneck led to reduced genetic diversity, leaving it vulnerable to disease and stresses of climate change. In particular, the Antarctic fur seal's primary prey base, krill, could be reduced as a result of ocean acidification, or the distribution could be altered by climate change.

IUCN status Least Concern.

Juan Fernandez and Guadalupe Fur Seals—*Arctocephalus philippii*

(Peters, 1866)

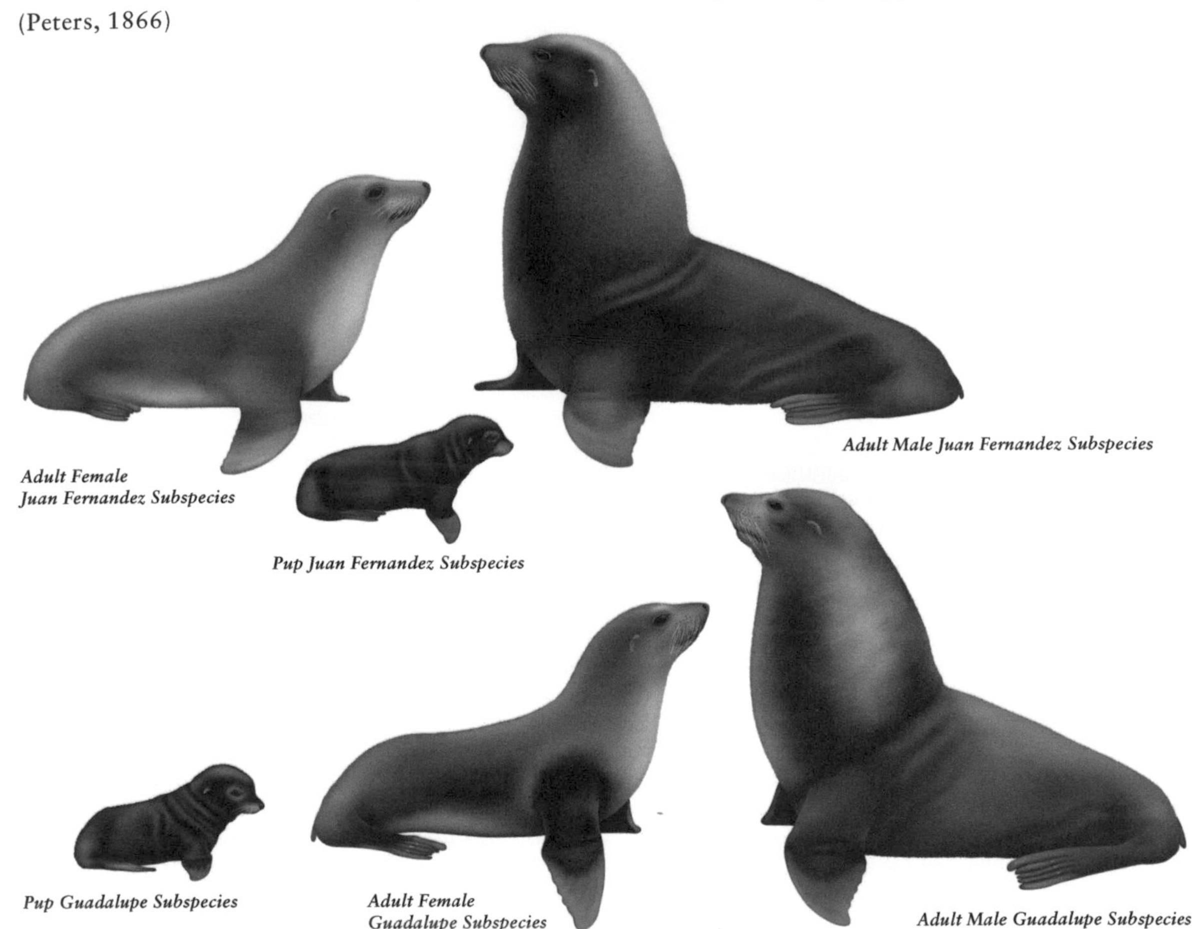

Recently-used synonyms *Arctocephalus philippii* and *Arctophoca philippii* (for the Juan Fernandez subspecies), and *Arctocephalus townsendi,* and *Arctophoca townsendi* (for the Guadalupe subspecies).

Common names En.–Juan Fernandez fur seal; Sp.–*lobo fino de Juan Fernández,* Fr.–*otarie de Juan Fernandez;* En.–Guadalupe fur seal; Sp.–*lobo fino de Guadalupe;* Fr.–*otarie de Guadalupe.*

Taxonomic information Carnivora, Otariidae. Until recently, Guadalupe fur seal and Juan Fernandez fur seal were treated as distinct species, however the Society for Marine Mammalogy now regards them as two subspecies of *Arctocephalus philippii: A. p. philippii* (Juan Fernandez fur seal) in the South Pacific near the Juan Fernández Archipelago and the west coast of South America, and *A. p. townsendi* (Guadalupe fur seal) in the North Pacific near Guadalupe Island and on the west coast of the Baja California Peninsula, and California.

Species characteristics Juan Fernandez and Guadalupe fur seals are sexually dimorphic: male Juan Fernandez fur seals are about 1.4 times longer and approximately 3 times heavier than adult females. Male Guadalupe fur seals are 1.5–2 times longer and approximately 3–4 times heavier than adult females. Both subspecies have thick pelage with dense underfur, which is cream-colored in the Guadalupe fur seal. Adults have moderate length pale vibrissae and long prominent ear pinnae.

Adult male Juan Fernandez and Guadalupe fur seals have a slightly rounded crown with the apex above the ear pinnae. The muzzle is long, straight and somewhat flattened on top, and tapers in width and thickness to the fleshy nose on the end. The black, naked skin of the nose, or rhinarium, is large and somewhat bulbous and the nares orient downward. The mane extends from the shoulders to the crown, and the long erect guard hairs on the head accentuate the crown and forehead. This is especially noticeable where the longer fur of the crown meets the shorter, more swept-back, fur of the muzzle between the eyes. From the crown, the long coarse guard hairs of the mane cover the nape and neck to the shoulders, and on the sides and under-sides of the neck from throat to chest. The neck, chest, and shoulders to the foreflippers are enlarged with muscle and fat. The abdomen is smaller, and tapers to the narrow pelvis. The canine teeth of adult males are larger and thicker than those of females. Adult females, subadults, and juveniles also have a long tapering muzzle, with nares angled slightly downward. However, because the rhinarium is not enlarged, the muzzle is more pointed. The ear pinnae are long and stand out from the head enough to make them conspicuous. The crown of the female Juan Fernandez fur seal is slightly domed, giving the head a rounded look in profile. The crown of the Guadalupe fur seal female is flatter than in males, and the mane absent; as a result they have a flat-topped looking head.

The foreflippers of both subspecies have a dark, sparse, short fur that extends beyond the wrist onto the middle of the dorsal surface of the flipper in a V-pattern that does not reach the rounded tip. The rest of the dorsal surface and the palms of both foreflippers are covered with a hairless black leathery skin. The first digit is the longest, widest and thickest, and angles posteriorly, giving the flipper a swept-back look. Digits 2–5 are successively shorter. There is a small opening in the skin at the end of each digit for a claw that is usually reduced to a vestigial nodule, and rarely emerges above the skin. The claw openings are set back from the free edge of the flippers by cartilaginous rods that extend the length of each digit, and expand the size of the flippers. The hindflippers have dark, short, sparse hair covering part of the dorsal surface of the proximal end of the flipper, and the rest of the dorsal surface, and the entire sole is covered in black leathery hairless skin. The hindflippers are long and each digit has a cartilaginous rod that adds a flap-like extension to each toe. The bones of the three central toes terminate at the position of the small nails that emerge through the skin on the dorsal surface, set back from the end of the flipper. The claws of digits 1 and 5 are vestigial, like the claws on the foreflippers, and may or may not emerge from small openings set back from the end of the flaps. All of the flaps at the end of the flipper are of similar length. The first toe or hallux and the fifth toe are somewhat wider than toes 2–4.

A partly-wet adult male Juan Fernandez fur seal showing the pale crown, nape and back of the neck, on the dark body. PHOTO: J. FRANCIS

Adult males of both subspecies are dark grayish-brown to grayish-black. The longer guard hairs of the mane are tipped off-white to ginger on the crown, nape and, to a variable extent, on the sides and back of the neck, creating a grizzled cape. Adult females, subadults, and juveniles are dark brown to grayish-black above and paler in variable patterns below, especially on the chest and underside of the neck, which can be tan to creamy gray. Both sexes have a variable amount of lighter buff to reddish-brown color on the muzzle, which often extends into the face above the eyes. Pups are born in a black natal coat that lightens to a dark brown, and they probably molt to a juvenile coat several months after birth. A single juvenile with alopecia involving hair loss over large areas of the upper surface was found stranded in Oregon. The extent of this condition in the population is unknown but probably very low.

Adult male Juan Fernandez fur seals are estimated to be 2 m long and weigh 140 kg. A series of lactating females measured over several years averaged 1.42 m in length and 48.1 kg in weight. Pups are approximately 65–68 cm and 6.2–6.9 kg

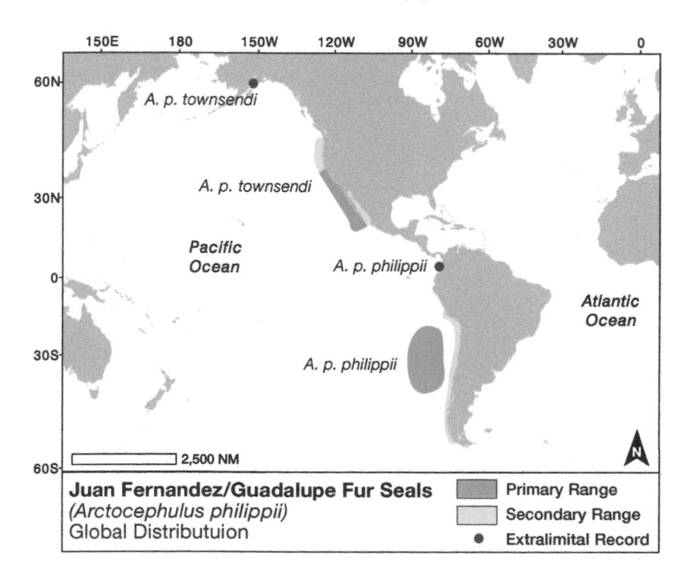

On the way to full adult size, this small adult male Juan Fernandez fur seal is getting thicker in his upper body and neck and his crown is becoming paler. Juan Fernández Islands. PHOTO: M. GOEBEL/NOAA FISHERIES/SWFSC

An adult female Juan Fernandez fur seal shows the long pointed muzzle of this species. PHOTO: M. GOEBEL/NOAA FISHERIES/SWFSC

Adult female Juan Fernandez fur seal. The thin neck and upper body, and proportionately small head, separate this animal from a similar sized subadult male. PHOTO: M. GOEBEL/NOAA FISHERIES/SWFSC

An adult female sniffs a young unmolted pup. The sense of smell in otariids is very important for mother and pup recognition when the adult female returns from foraging trips. PHOTO: M. GOEBEL/NOAA FISHERIES/SWFSC

Head and muzzle detail on a young pup Juan Fernandez fur seal still in natal pelage. PHOTO: M. GOEBEL/NOAA FISHERIES/SWFSC

A Juan Fernandez fur seal pup in black natal pelage is left temporarily while its mother is foraging. PHOTO: M. GOEBEL/NOAA FISHERIES/SWFSC

An adult female Juan Fernandez fur seal. Pale streaks of underfur show through the guard hairs on this wet fur seal. PHOTO: M. GOEBEL/NOAA FISHERIES/SWFSC

at birth. Adult male Guadalupe fur seals may reach 2 m in length. Two adult males measured approximately 1.8 and 1.9 m, the latter estimated to weigh 160–170 kg. Adult females average 1.2 m and reach approximately 1.4 m, and weigh 40–50 kg. Pups are 50–60 cm long at birth. The dental formula is I $^{3}/_{2}$, C $^{1}/_{1}$, PC $^{6}/_{5}$.

Recognizable geographic forms Juan Fernandez and Guadalupe fur seals are similar in appearance, there is little chance for them to overlap in range, therefore the best way to tell them apart is by location.

Can be confused with The range of the Juan Fernandez fur seal is near the distribution of South American, Antarctic, and subantarctic fur seals, and South American sea lions.

South American sea lions have a large, blocky head with a short muzzle that is blunt on the end, and smallish inconspicuous ear pinnae. They are also pale tan to light golden brown, with sleek fur that is not dense or shaggy, except in the case of the mane on the otherwise unmistakable adult males, and are heavy-bodied. Adult male Juan Fernandez fur seals have a longer muzzle, distinctive bulbous rhinarium, and unique coloration of the crown, nape and upper neck that separate them from other southern hemisphere fur seals. Separating adult female, subadult, and juvenile southern fur seal species at a distance is usually difficult and often impossible without extensive experience with the species in question. In general, adult female Juan Fernandez fur seals are longer and heavier than female Antarctic, subantarctic, and South American fur seals, and have a proportionately longer tapering muzzle with downward oriented nares, and a more rounded crown. Adult subantarctic fur seals are uniquely cream-colored in the face and neck. They have a shorter almost attenuated muzzle in comparison, and proportionately smaller flippers. Adult males have a crest of fur on the forehead. Hybrid Juan Fernandez X subantarctic fur seals are known. The few photographs available show that they share characteristics of both species (see Note on Hybrid Southern Fur Seals Genus *Arctocephalus,* pages 366–367).

The Guadalupe fur seal can be confused with two other otariids, the California sea lion and northern fur seal. Coloration of Guadalupe and northern fur seals is similar. Guadalupe fur seals have fur on the dorsum of the foreflipper beyond the wrist, proportionately shorter hindflippers, and a longer tapering muzzle that appears slightly upturned on the end in males and flat in females. At sea, both species actively groom while at the surface. Guadalupe fur seals will rest in a posture characteristic of fur seals of the genus *Arctocephalus* with the head down and both hindflippers held in the air and apart forming a Y-shape. Both species sleep in a "jug-handle" position with the palm of one foreflipper draped over the soles of the hindflippers.

Guadalupe fur seals can be readily separated from California sea lions based on differences in coloration, pelage length and thickness, length and pointedness of the muzzle, relative size and prominence of the ear pinnae, overall size, hindflipper length, and shape of toes. In California sea lions, the muzzle tapers to a blunt end, and the first toe or hallux is larger and longer than all of the other toes. Adult female, subadult, and juvenile California sea lions are tan to pale brown and much lighter in coloration than Guadalupe fur seals. Adult and subadult male California sea lions become dark brown and are similar in color to Guadalupe fur seal bulls. However, dark male sea lions have a sagittal crest in contrast to the head of Guadalupe fur seal bulls that is slightly domed and lacks a conspicuous sagittal crest. California sea lion bulls lack the light hairs on their neck, and they do not have a mane. The loud repetitive bark of male California sea lions is distinctive and different from all Guadalupe fur seal vocalizations. With the ongoing expansion of the Guadalupe fur seal's range, observers should also be aware of the possibility of hybrids (Guadalupe fur seal X California sea lion or Guadalupe fur seal X northern fur seal) in the eastern North Pacific.

Adult male Guadalupe fur seals have an enlarged neck and shoulders, and a thick mane of guard hairs. The long muzzle tapers to a point and the fleshy end or "rhinarium" is large and the nostrils point downward at an angle. PHOTOS: F. R. ELORRIAGA-VERPLANCKEN/CICIMAR-IPN

Adult male and female Guadalupe fur seals demonstrate the degree of sexual dimorphism in length, mass, and features. The males can outweigh the females 3 to 1. PHOTO: F. R. ELORRIAGA-VERPLANCKEN/CICIMAR-IPN

A Guadalupe fur seal with a California sea lion of similar size. Only male California sea lions darken with age. Also the fur seal's tapering muzzle is much more pointed than the blunt, wider muzzle of the sea lion. PHOTO: F. R. ELORRIAGA-VERPLANCKEN/CICIMAR-IPN

Adult Guadalupe fur seal female with her pup in natal pelage. Guadalupe Island, Mexico. PHOTO: P. COLLA/OCEANLIGHT

A nursing older Guadalupe fur seal pup that has molted out of its natal pelage. PHOTO: F. R. ELORRIAGA-VERPLANCKEN/CICIMAR-IPN

A Guadalupe fur seal pup not yet 1 year old but molted out of the natal pelage. At less than a year old, the muzzle is proportionately short, and tapers to a wider blunt end than on 1 and 2 year-old juveniles. PHOTO: F. R. ELORRIAGA-VERPLANCKEN/CICIMAR-IPN

Two subadult male Guadalupe fur seals sparring. The male on the left is just beginning to develop the large fleshy end of his nose. East San Benitos Island, Mexico. PHOTO: I. SZCZEPANIAK

Distribution The range of the two subspecies is disjunct. The Juan Fernandez fur seal is only found ashore regularly in the Juan Fernández Archipelago in the eastern South Pacific, west of Chile. The Archipelago includes the Juan Fernández Island group, and the San Félix Islands approximately 600 km to the north. This is a highly pelagic species at sea, and the full extent of their foraging areas is not well known. Vagrant Juan Fernandez fur seals have been found on the west coast of South America from Colombia to southern Chile. These fur seals prefer rocky and volcanic shorelines with ledges and boulders, and they also use grottos, over-hanging rock ledges and caves.

Guadalupe fur seals commonly float inverted in the water with their hindflippers in the air. PHOTO: P. COLLA/OCEANLIGHT

Adult male Guadalupe fur seal in the water vigorously shaking. PHOTO: M. NOLAN

The majority of the Guadalupe fur seal population is centered on Guadalupe Island, off the west coast of central Baja California, where nearly all pups are born. However, in 1997 a small colony with 9 pups was discovered at the San Benitos Islands, near the Baja California coast, at the site of a former rookery. This colony continues to grow. Also in 1997, a pup was born on San Miguel Island in the Channel Islands. Guadalupe fur seals are regularly seen on other southern California islands, and the Farallon Islands off northern California, and have been found stranded in Oregon and Washington with one extreme northern record for Cook Inlet, Alaska. They have been sighted in the Gulf of California, and as far south as Zihuatanejo, Mexico. Their distribution at sea is poorly known, but like the Juan Fernandez fur seal are considered pelagic and known to forage over large offshore areas.

A Guadalupe fur seal rafting in a "jug-handle" position. Notice the long narrow tapering muzzle and prominent ear pinnae. PHOTO: R. L. PITMAN

An adult male Guadalupe fur seal lying in a mat of *Sargassum*. Gulf of California, Mexico. PHOTO: M. NOLAN

Ecology and behavior Both subspecies are polygynous. For the Juan Fernandez fur seal, the breeding season lasts from mid-November to the end of January, and the colonies are essentially vacated by early September (based on the observations of sealers from the late 18th century), and no later than mid-October. Males defend territories that are typically around 36 m^2 in size and that have an average of 4 females. Most adult females give birth within a few days of arriving at the rookery. Mean time from birth to departure on the first foraging trip is 11.3 days. Juan Fernandez fur seal females also travel long distances to find adequate quantities of prey and, on average, have the longest foraging trips of any otariid. Although females can be gone for as little as 1 day, the mean is 12.3 days per foraging trip and the longest trip recorded lasted 25 days. Mean length of pup attendance is 5.3 days with a range of 0.3–15.8 days. Based on the onset of pupping and the observations of vacant colonies in early September, it has been calculated that pups are weaned in 7–10 months. Juan Fernandez fur seals travel long distances to their foraging areas. The mean distance traveled away from the breeding colony is 653 km, and all tagged females traveled at least 550 km to forage. Most trips were southwest and west of the Juan Fernández Islands, far offshore to deep oceanic areas. Despite this, the mean depth of dive of 12.3 m, and the mean duration of 51 seconds is shallow and short even for an otariid, and indicates surface feeding. The deepest dives are made to 90–100 m and the longest dives are just over 6 min. Nearly all foraging dives occur at night.

Guadalupe fur seal males establish territories that are occupied by an average of 6 females. Pups are born from mid-June to August with a median birth date of 21 June.

Beached juvenile Guadalupe fur seal in Oregon with severe alopecia, or loss of guard hairs, exposing the underfur. Note the orange plastic tag on of the right foreflipper. PHOTO: THE MARINE MAMMAL CENTER

Male tenure on territories lasts at least 31 days. Males defend territories with vocalizations, displays and fighting with neighboring bulls. Once territories are established, fighting between males is rare. Females select male territories that provide cover and shade from the sun for pupping, and territories with females are fronted by water including tidal pools. Many animals breed in small caves, grottos, and cliff and boulder areas on the rugged east coast of volcanic Guadalupe Island. Adult females enter the water daily, presumably for cooling, while otherwise ashore attending their pups. Females returning to the rookery for the first time usually arrive at night or early in the morning. Estrus occurs 5–10 days after a female gives birth, and females can leave for their first foraging trip right after mating, or stay on the colony for another few days before departing. Adult female Guadalupe fur seals range widely and travel great distances while foraging between stints ashore with the pup. Trips with distance of 700–4,000 km with durations of 4–24 days have been reported. Pups are weaned at 9–11 months, and females with pups can be seen on or around the island throughout the winter and into the spring.

At sea, these fur seals can be quite animated as they groom at the surface. They also rest in a number of postures including: head down with hindflippers elevated and swaying in the air, like many southern fur seals; asleep at the surface with foreflipper over both hindflippers in a "jug-handle" position, and with both foreflippers or all four flippers held in the air. When traveling rapidly, the Juan Fernandez fur seal has been observed to "porpoise." Killer whales and sharks are likely predators on both subspecies, and possibly leopard seals that infrequently visit the Juan Fernández Archipelago. A cookie-cutter shark bite has been reported on a male Guadalupe fur seal.

Feeding and prey Juan Fernandez fur seals feed on vertically-migrating prey at night. Their diet is one of the least diverse of any otariid, and along with the long foraging trips made by lactating females, suggests the low productivity of their oceanic feeding areas. Diet varies between years and probably reflects abundance and availability of prey. Myctophids are the most important fishes in the diet and onychoteuthid squid are the most important cephalopods. The prey preferences and foraging activity of the Guadalupe fur seal are poorly known. Stomach contents retrieved from stranded animals included a variety of squid, bony fishes, and crustaceans, including vertically-migrating species.

Threats and status Both subspecies of fur seal have a long association with humans. Juan Fernandez fur seals were hunted to near extinction by sealers trading pelts in China. Intensive sealing began in the late 18th century and ended in the late 19th century. It is likely that several million seals were killed during this period. Small numbers were seen in the early 20th century, but the species was thought to have gone extinct shortly thereafter. They were rediscovered in the middle of the 20th century and have since been making a slow comeback. The total population is 12,000 animals, based on a 1991 census.

The Guadalupe fur seal was also brought to the brink of extinction by the late 19th century, and was not reported again until 1926. Following this "rediscovery," all animals that could be found were collected as specimens, and once again the species was thought to be extinct. Guadalupe fur seals were suspected to have survived based on scattered, unconfirmed reports in the 1930s, and were dramatically rediscovered again with the sighting of a bull on San Nicholas Island in southern California in 1949. An expedition to Guadalupe Island in 1954 confirmed the survival of the species. Since the 1950s, the species has recovered from an estimated population of 200–500 animals to approximately 15,000–17,000, and is probably increasing by about 13% annually.

The limited size of the populations of both Juan Fernandez and Guadalupe fur seals, and the fact that they passed through genetic bottlenecks, may make them vulnerable to catastrophic events and stress from disease outbreaks, oil spills, and environmental regime shifts. Although Guadalupe fur seal numbers are increasing rapidly and their range is expanding, the subspecies is still at risk because the total population is small and nearly all pups are born on one island. Guadalupe Island is relatively close to, and down current from, large human population centers with extensive oil tanker traffic. The subspecies also shares all of its range with California sea lions, which have suffered from viral disease outbreaks in the past, and because of their coastal nature could be a vector for diseases from terrestrial sources. It has few conflicts with commercial fisheries at the present time, although gillnet and set-net fisheries may take small numbers, as could entanglement in marine debris. For the Juan Fernandez fur seal, while no fisheries conflicts have been identified, individuals have been seen with plastic bands around their necks since 1982, though the level of mortality from these entanglements is unknown.

IUCN status Near Threatened (for both subspecies).

Galapagos Fur Seal—*Arctocephalus galapagoensis*

Heller, 1904

Recently-used synonyms *Arctocephalus australis galapagoensis, Arctophoca galapagoensis.*
Common names En.–Galapagos fur seal; Sp.–*lobo fino de Galápagos;* Fr.–*otarie des Galapagos.*
Taxonomic information Carnivora, Otariidae. Formerly considered a subspecies of South American fur seal, *Arctocephalus australis galapagoensis.* Recently, a proposal was made to reassign it to the resurrected genus *Arctophoca,* along with other southern fur seals, and conduct further research to determine if it is a valid species, or a subspecies of the South American fur seal.

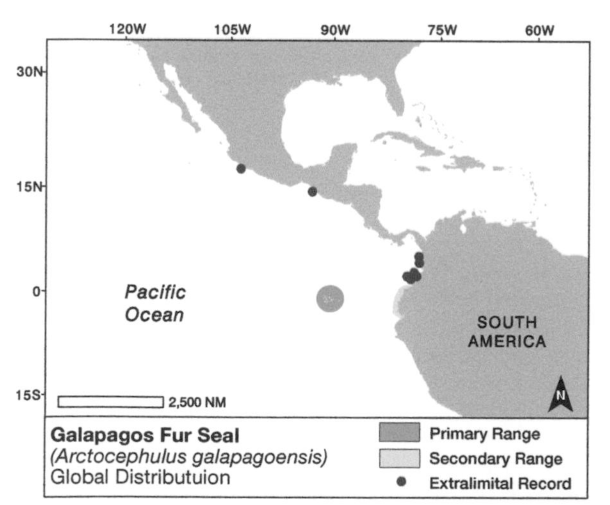

Species characteristics Galapagos fur seals are the smallest otariid, and the least sexually dimorphic. Adult males are larger than adult females, about 1.1–1.3 times longer and twice the weight. Galapagos fur seals are small and compact, and adult males are stocky in build. The muzzle is small: short, straight, and rapidly tapers in width and thickness to the blunt-ended small nose. Mystacial vibrissae are pale in adults. The eyes are proportionately large and the ear pinnae long and prominent.

Adult males are much thicker in the neck and shoulders than females, despite the fact that they lack a mane of longer guard hairs. Adult males do not have a conspicuous sagittal crest, but do have a slightly rounded crown and a short sloping forehead. The canine teeth of adult males are larger and thicker than those of females. Adult females and juveniles lack the thicker neck and shoulders of adult males, and have a flatter crown and barely noticeable forehead. Many adults have scars from shark attacks.

The foreflippers are short and wide and have a dark, sparse, short fur that extends beyond the wrist onto the middle of the dorsal surface of the flipper in a V-pattern that does not reach the rounded tip. The rest of the dorsal surface and the palms of both foreflippers are covered with a hairless black leathery skin. The first digit is the longest, widest and thickest, and angles distally, giving the flipper a swept-back look. Digits 2–5 are successively

Adult male Galapagos fur seal. Note the enlarged chest and neck. PHOTO: G. MEYER

The all-pale vibrissae suggest an adult, and the lack of an enlarged neck, chest and shoulders suggest a female. PHOTO: P. COLLA/OCEANLIGHT

The pale color of the muzzle on adult male Galapagos fur seals can extend into the face. PHOTO: G. MEYER

Adult male (upper) and female (lower) Galapagos fur seals. Note the differences in head, muzzle and neck. PHOTO: ORIONNA

Dependent, but molted, pup resting on its mother (upper) and newborn pup (lower). PHOTO: (UPPER) ORIONNA; (LOWER) F. FELIX

shorter. There is a small opening in the skin at the end of each digit for a claw that is usually reduced to a vestigial nodule, and rarely emerges above the skin. The claw openings are set back from the free edge of the flippers by cartilaginous rods that extend the length of each digit, and expand the size of the flippers. The hindflippers have dark, short, sparse hair covering part of the proximal end of the flipper, and the rest of the dorsal surface, and the entire sole is covered in black leathery hairless skin. The hindflippers are long and each digit has a cartilaginous rod that adds a flap-like extension to each toe. The bones of the three central toes terminate at the position of the small nails that emerge through the skin on the dorsal surface, set back from the end of the flipper. The nails of digits 1 and 5 are vestigial, and may or may not emerge from small openings set back from the end of the flaps. All of the flaps at the end of the flipper are of relatively equal length. The first toe or hallux and the fifth toe are somewhat wider than toes 2–4.

Galapagos fur seals are medium to dark brown above and can have silvery gray to ginger grizzling. In both sexes, most of the muzzle is pale tan and in adult males, this color extends onto the face and forehead over the eyes, giving them a small pale mask. In adult females and subadults, the chest is pale grayish-tan, sometimes continuing to the back of the neck, and the belly is tan to reddish-brown. Bulls

Adult female Galapagos fur seal and a pup that is only several days old, born on the mainland coast of Equador near Guayaquil. PHOTO: F. FELIX

Adult female and older molted pup in the Galápagos Islands. Pups can receive care from their mothers for very long periods of time. PHOTO: ORIONNA

Two male Galapagos fur seals sparring. Young male fur seals gain experience for the demands of gaining access to females later in life by regularly practice fighting with each other. PHOTO: M. NOLAN

The Galapagos fur seal is the smallest otariid. This male has not yet reached full size. The pale areas on the muzzle are conspicuous on this wet animal. PHOTO: R. FLOERKE

are dark above with lighter tones below. Pups are blackish-brown, sometimes with grayish to whitish margins around the mouth and nose. Pups molt this natal coat for one that resembles that of the adult female.

The few adult males measured to date have been 1.5–1.6 m and weighed 60–68 kg. Adult females have shown a range of curvilinear lengths of 1.1–1.3 m and an average weight of about 27.3 kg, with a maximum of 33 kg. Newborns are 3–4 kg. The dental formula is I $^{3}/_{2}$, C $^{1}/_{1}$, PC $^{6}/_{5}$.

Recognizable geographic forms None.

Can be confused with Galapagos fur seals share their restricted range in the Galápagos Archipelago with the Galapagos sea lion. The fur seal can be readily distinguished from the sea lion by its small and compact stocky body, short and more pointed muzzle, long pale vibrissae (on adults), thick long fur and shaggy look when wet, long prominent ear pinnae, proportionately large eyes, and equal length toes on the hindflippers. The South American sea lion has also been recorded in the Galápagos from a single stranding. All of the features that separate Galapagos fur seals from Galapagos sea lions can be used to separate them from this even larger sea lion species. Galapagos fur seals are considerably smaller than South American and Juan Fernandez fur seals, and can be separated from them based on overall size, muzzle length and color.

Distribution Galapagos fur seals are found throughout the Archipelago. Lactating females make trips of relatively short duration, suggesting they do not get far from their colonies. Foraging by males and all animals outside the breeding season is unknown. They are present on nearly all of the islands in the Archipelago, and prefer to haul out near the shoreline on rocky coasts with large boulders and ledges that provide shade and the opportunity to rest in crevices and spaces between the rocks. Most of the colonies are also located in the western and northern parts of the Archipelago, close to productive upwelling areas offshore. Galapagos fur seals are also known to haul out and breed on the coast of mainland Ecuador in small numbers. Two Galapagos fur seals that reached Mexico (Guerrero and Chiapas) are extralimital records.

A number of adult and subadult male Galapagos fur seals occupy a rocky shore with grottos and ledges, which is the typical habitat for this species. PHOTO: B. SIEGEL

Galapagos fur seals may travel the shortest distances from their colonies and haulouts to forage, usually at night, but also routinely enter the water during the day to get the pelage wet for cooling. PHOTO: M. NOLAN

Ecology and behavior The behavior of the Galapagos fur seal has been extensively studied. It has a fairly long pupping and breeding season, lasting from mid- August to mid-November. The peak of pupping shifts from year to year, but usually occurs sometime from the last week of September through the first week of October. Galapagos fur seals mature at 3–5 years old. Males do not become physically mature, and large enough to compete for a territory that will be used by females, until they are much older. Males hold territories that average 200 m^2, which is large compared to the average size of territories held by other otariid males, and especially so when considering the Galapagos fur seal's small size. Colonies are located close to foraging areas and the average length of female trips is the shortest for a fur seal, with a mean trip length of 1.5 days. Most foraging occurs at night and the mean depth of foraging dives is 26 m and duration is less than 2 min, with maximum depths recorded of 115 m, and duration of 5 min.

Pups are visited around 300 times before weaning, with attendance periods of 0.5–1.3 days. Weaning occurs at 18–36 months, with most pups being weaned in their third year. Pups born prior to the weaning of an older sibling rarely survive, with most starving to death and a small percentage being killed by the older pup. Females will allow multiple pups to nurse, but this rarely lasts long enough for the youngest pup to get strong enough to survive. In a few cases, offspring nursed until they were 4–5 years of age.

In the water near haulouts, Galapagos fur seals raft in postures typical of many of the southern fur seal species. There is no evidence for migration, and they do not seem to spend prolonged periods of time at sea. Predators at sea include sharks and killer whales. On land, feral dogs on Isabela Island have attacked and killed this species.

Feeding and prey Food habits are poorly known. Galapagos fur seals consume a variety of small squids including *Onychoteuthis banksi*, and a number of species of ommastrephids. A variety of fish species are also taken including myctophids and bathylagids. They seem to feed mostly at night, possibly exploiting vertically migrating species when they are at the surface.

Threats and status As with all the southern fur seals, there was a severe population decline as a result of 19th century exploitation by sealers and whalers. The species was near extinction early in the 20th century, and has since recovered. A census conducted in 1988–89 estimated 10,000–15,000 animals, which is considered a decline from the 1970s. El Niño events dramatically raise pup mortality, may have an impact on the survival of other age classes, and cause population declines when upwelling and marine productivity dramatically decline around the Archipelago. Tourism in the Galápagos, an Ecuadorian National Park, is heavy but regulated, and fur seals are protected. Episodes of entanglement in local net fisheries have been reported, but are thought to have been largely mitigated by no-fishing zones. Feral dogs on Isabela Island destroyed colonies on the south end of the island by killing seals of all ages. Subsequently, a feral dog control program was put in place, but this problem could erupt again if any remaining animals find their way to colony sites again. Both feral and pet dogs could transmit diseases to pinnipeds. Despite their population size, the Galapagos fur seal will always be vulnerable to a variety of threats because the species has a very restricted range.

IUCN status Endangered.

South American Fur Seal—*Arctocephalus australis*

(Zimmerman, 1783)

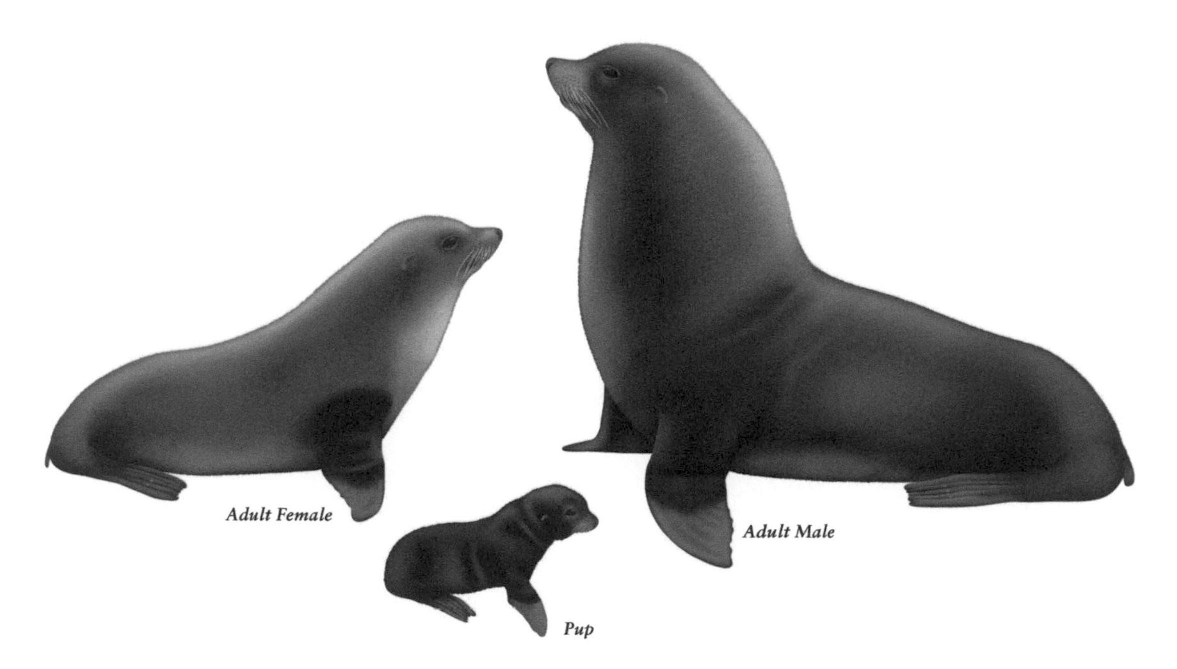

Recently-used synonyms *Arctophoca australis, Arctophoca australis australis, Arctophoca australis gracilis.*

Common names En.–South American fur seal, southern fur seal; Sp.–*lobo fino austral, lobo de dos pelos*; Fr.–*otarie d'Amérique du sud.*

Taxonomic information Carnivora, Otariidae. Two subspecies are recognized: *Arctocephalus australis australis* (South American fur seal) from Brazil, eastern South America, the Falkland Islands and southern Chile; and an unnamed subspecies (Peruvian fur seal) from Peru to northern Chile. Recently, because this species and New Zealand fur seal may be closely related, a proposal was made to assign them to the resurrected genus *Arctophoca*, lumping them into a larger species group comprising four subspecies: *Arctophoca arctophoca forsteri* in New Zealand and Australia; *A. a. australis* in the Falkland Islands; *A. a. gracilis* from Brazil; and an unnamed subspecies in Peru. The group was given the new common name of "southern fur seal," however, this proposed taxonomy is still the subject of debate and has not been adopted by the Society for Marine Mammalogy.

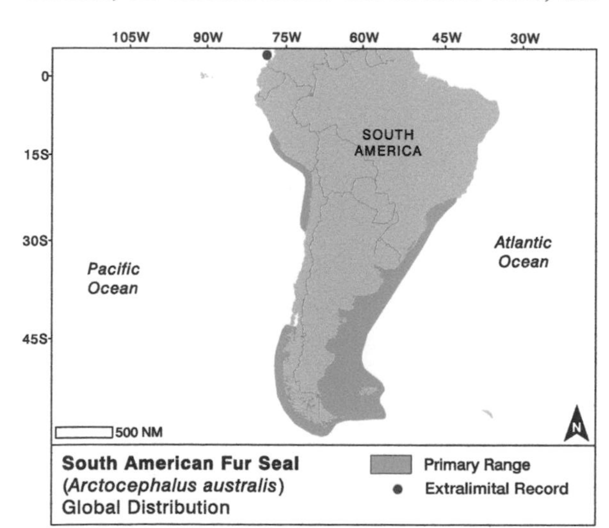

Species characteristics South American fur seals are sexually dimorphic. Adult males are about 1.3 times the length and 3.3 times the weight of females. Both adult males and females have a stocky robust build, large prominent eyes, long conspicuous ear pinnae, and moderate length pale vibrissae. The muzzle is moderately long, straight on top in profile, and tapers in width and thickness to the nose. The rhinarium is somewhat enlarged, contributing to the slightly upturned look to the end of the muzzle in profile on bulls. The crown is rounded and there is a forehead that is steeper in adult males. Adult males have a very thick neck and large shoulders. They have a mane of longer guard hairs from the crown to the shoulders, including the neck and upper chest. The canine teeth of adult males are larger and thicker than those of females.

Adult male and female South American fur seals from the *A. a. australis* subspecies. New Island, Falkland Islands. PHOTO: M. A. WEBBER

Adult male, female, and pup South American fur seals in Peru are from an as yet unnamed subspecies of *A. australis.* PHOTO: S. CARDENAS ALAYZA

Adult female South American fur seal with soiled pelage. Most fur seals of the genus *Arctocephalus* have a straighter leading edge of the foreflipper than any sea lion species. There is an orange plastic tag near the base of the foreflipper. Southern Chile. PHOTO: G. DE RANGO/TMMC

Adult female with an unmolted pup still in the black natal coat. The female has pelage that is stained and darkened from being on the rookery. Southern Chile. PHOTO: G. DE RANGO/TMMC

Juvenile South American fur seal with wet pelage. The guard hairs parting to reveal the underfur. Southern Chile. PHOTO: G. DE RANGO/TMMC

Pup South American fur seal in natal pelage. This individual was probably left to its own devices while the mother went to sea to feed. Southern Chile. PHOTO: G. DE RANGO/TMMC

The foreflippers have a dark, sparse, short fur that extends beyond the wrist onto the middle of the dorsal surface of the flipper in a V-pattern that does not reach the rounded tip. The rest of the dorsal surface and the palms of both foreflippers are covered with a hairless black leathery skin. The first digit is the longest, widest and thickest, and angles distally, giving the flipper a swept-back look. Digits 2–5 are successively shorter. There is a small opening in the skin at the end of each digit for a claw that is usually reduced to a vestigial nodule, and rarely emerges above the skin. The claw openings are set back from the free edge of the flippers by cartilaginous rods that extend the length of each digit, and expand the size of the flippers. The hindflippers have dark, short, sparse hair covering part of the proximal end of the flipper, and the rest of the dorsal surface and the entire sole is covered in black leathery hairless skin. The hindflippers are long and each digit has a cartilaginous rod that adds a flap-like extension to each toe. The bones of the three central toes terminate at the position of the small nails that emerge through the skin on the dorsal surface, set back from the end of the flipper. The claws of digits 1 and 5 are usually vestigial, like the claws on the foreflippers, and may or may not emerge from small openings set back from the end of the flaps. All of the flaps at the end of the flipper are of relatively equal length. The first toe or hallux and the fifth toe are somewhat wider than toes 2–4.

Adult male South American fur seals sparring. Notice the steep forehead, slightly upturned muzzle, and long pale-tipped guard hairs of the wet male on the right. PHOTO: C. A. MORGAN

Subadult and adult male South American fur seals from the Mar del Plata coast, Argentina, demonstrate the additional growth that male otariids go through between sexual and physical maturity. PHOTOS: R. BASTIDA

Adult females and subadults are brown to grayish-black above and paler, often with mixed shades of tan, gray, and rusty brown below. The chest, ventral part of the neck, and sides of the neck are light colored to a variable extent. The head and face are dark, but the sides of the muzzle in the mystacial area are paler. The ear pinnae are pale colored, and in females and juveniles the fur at the base of the pinnae can also be lighter colored. The fur on the top of the flippers is generally quite dark. As they age, males darken and become more uniformly-colored, generally grayish brown to dark brown, with grizzled frosting. Some bulls are paler. At birth, pups have a longer blackish natal coat, but there may be some paler markings on the face and muzzle, and some animals are paler below.

Adult males are 1.9 m in length and 120–160 kg in weight (perhaps up to 200 kg) and females reach 1.4 m and 40–50 kg. Pups are 60–65 cm and 3.5–5.5 kg at birth. The dental formula is I $^3/_2$, C $^1/_1$, PC $^6/_5$.

Recognizable geographic forms Although two subspecies are recognized, there is no known reliable way to tell them apart in the field.

Can be confused with South American fur seals share most of their range with South American sea lions. Five other otariids: Galapagos, Antarctic, subantarctic, and Juan Fernandez fur seals, and Galapagos sea lions can be found as vagrants within the range of the South American fur seal. South American sea lions have large, blocky heads with a short, blunt muzzle, and smallish inconspicuous ear pinnae. Similar-sized South American sea lions are mostly tan to light brown, heavy bodied, with short, sleek fur that is not dense or shaggy except

Juvenile South American fur seals at play and resting. Southern Chile.
PHOTO: G. DE RANGO/TMMC

for the mane on bulls. Female, juvenile, and subadult Galapagos sea lions are also pale tan to light golden brown, with sleek fur that is not dense or shaggy. Adult male Galapagos sea lions are dark brown to black, and thus similar in color to South American fur seals, but have a conspicuous sagittal crest on the crown above the eyes and lack a conspicuous mane. All Galapagos sea lions have a proportionately wider muzzle that is blunt on the end, with less conspicuous vibrissae, and proportionately smaller eyes and shorter ear pinnae.

Adult male Juan Fernandez fur seals have a longer muzzle with a bulbous, enlarged rhinarium and significantly downward-angled nares. The mane of adult males is grizzled principally on the crown, nape and upper neck whereas adult male South American fur seals are grizzled more evenly throughout the entire mane. Galapagos fur seals are very small and stocky. The muzzle is quite short and tapers rapidly to a sharper point. Adult male Galapagos fur seals have a paler "mask" of short lighter colored pelage on the muzzle and face to the eyes. Because of their small size, the head of adult females seems proportionately wide. Distinguishing adult male Antarctic fur seals from adult male South American fur seals is difficult due to their similar size. Adult male Antarctic fur seals are darker brown or gray, have a shorter muzzle without any enlargement to the rhinarium and no slight upward angle at the end due to the size and shape of the rhinarium. Also, Antarctic fur seals have proportionately longer fore- and hindflippers. Subantarctic fur seals are uniquely colored with a pale face and neck. Both sexes have short, dark ear pinnae and adult males have a crest of longer pale fur on the front of the dark crown. The flippers are proportionately small in both sexes. With the exception of adult female subantarctic fur seals, identifying lone adult female, subadult, or juvenile southern fur seals is usually difficult without extensive experience. This is especially true in an area like southern South America where vagrants and hybrids of a number of species are possible.

Distribution South American fur seals are widely-distributed from central Peru to Northern Chile, then after a gap of nearly 20° of latitude, from southern Chile, around the southern tip of the continent, and then from Argentina and the Falkland Islands north to southern Brazil. Distribution at sea is poorly-known. These seals are thought to forage primarily in continental shelf and slope waters. However, there are records from more than 600 km offshore.

Ecology and behavior South American fur seal males and females become sexually mature from 5–6 and 2–4 years old respectively, and most females have given birth to a pup by the time they are four years old. Pupping and breeding takes place from mid-October through mid-January. There is probably some variation in timing associated with local oceanic conditions between and within years that may effect only some colonies, and lead to differences in timing at sites in the Pacific and Atlantic, and along the large latitudinal gradient of the distribution of this species. Colonies are generally on rocky coasts on ledges above the shoreline or boulder strewn areas. Most areas provide some source of shade such as at the base of cliffs, and easy access to the water or tidal pools. Males are polygynous and territorial, and fighting can result in dramatic wounds and scars. At Uruguayan colonies, individual bulls can occupy territories for up to 60 days and have up to 13 females on their territories. Male vocalizations include a bark or whimper, a guttural threat, and a submissive call. Females growl and have a pup-attraction call that is a high-pitched wail. Pups are born shortly after females return to the colonies. Estrus is 7–10 days later, and following mating, a female begins to make foraging trips punctuated by time attending the pup ashore. Pups are weaned at 8 months to 2 years. Females will nurse a yearling and newborn pup. Time spent on trips and attending the pup likely varies with location and changes in marine productivity such as during El Niño years for animals in Peru. Female attendance in Uruguay is affected by weather, with females spending less time ashore during the day when ground temperature exceeds 36° C and conversely, staying ashore longer during storms. Survival rates of pups can be quite low when marine productivity is low such as during El Niño events, and storm surges can sweep large numbers of pups off colonies. Locally, predation by adult male South American sea lions can be significant at some colonies. Data collected on adult females during an El Niño resulted in mean dives to 29 m, with a maximum of 170 m, and mean duration of 2.5 min, and a maximum dive length of 7 min.

No migration is known. Colonies on islands off Uruguay are occupied by portions of the population year-round. Evidence suggests that increases in some colonies are occurring as animals relocate between areas such as Uruguay to Patagonia, and from southern Argentina to

A large subadult male South American fur seal from the Mar del Plata coast, Argentina. The testes become scrotal in males as they sexually mature. PHOTO: R. BASTIDA

Two views of rafting South America fur seals from Peninsula Valdes in Argentina. In the lower picture the animal is in the classic "jug-handle" position used by all fur seals. PHOTOS: P. FAIFERMAN

rookeries in southern Chile. At sea, these fur seals may be seen traveling or rafting at the surface in groups. They will "porpoise," or leap clear of the water when moving rapidly at sea, sometimes traveling like this in large groups. While resting at the surface they spend considerable time grooming and assume many poses typical of southern fur seals, including waving both hindflippers in the air while the head is submerged. Groups often form in the water at the base of a colony. Predators include killer whales, sharks, South American sea lions, and leopard seals. Vampire bats are known to drink blood from the naked skin of the flippers of sleeping South American fur seals.

Feeding and prey Demersal and pelagic fishes make up the majority of the diet in Uruguay and include weakfish, cutlassfish, anchoveta, anchovy, and cephalopods. In southern Chile and the Falkland Islands, South American fur seals take sardines, mackerel, and crustaceans such as lobster krill. Their diet in Peru is strongly influenced by El Niño events. Anchoveta is the dominant prey when marine productivity is high, but other fish consumed include deep sea smelt and sardines. More than 20 species of prey, including cephalopods and lanternfish, have been reported from Peru.

Threats and status Humans have hunted South American fur seals for thousands of years. Exploitation began after discovery by Europeans and the onset of commercial sealing in the 18th century. Harvest levels declined in the 20th century, and hunting ended in many locations. A managed harvest of small numbers of adult males continues in Uruguay. The effect of overfishing by large-scale commercial fisheries, and the ongoing take of numerous small-scale coastal fisheries has an unknown effect on the amount of food available to fur seals. Small numbers of fur seals are taken for food in Chile and Peru, while others are found shot on beaches either from fisheries conflicts, as a source of bait for king crab fisheries, or possibly as failed efforts to harvest them for subsistence. South American fur seals breed at a relatively small number of large rookeries, which could make these sites (and in turn the species or either subspecies) vulnerable to epizootic outbreaks, oil spills, or intensive fisheries harvest of important prey. The total population along the coast and offshore islands of eastern South America is estimated at 250,000–300,000, with the majority of animals in Uruguay. The Falklands population is estimated to be 15,000–20,000, the Chilean at 30,000, with about 12,000 in Peru. The species' abundance in the Pacific varies widely depending on the occurrence of El Niño events, which can cause mass mortalities from which the population usually rebounds after marine conditions return to greater productivity. Numbers in Chile have fallen by over 80,000 in the last decade or more, while the population trend in Patagonia is on the rise.

IUCN status Least Concern.

New Zealand Fur Seal–*Arctocephalus forsteri*

(Lesson, 1828)

Recently-used synonyms *Arctocephalus doriferus* (population in Australia only), *Arctophoca australis forsteri*.
Common names En.–New Zealand fur seal, long-nosed fur seal, Antipodean fur seal; Sp.–*lobo fino de Nueva Zelanda;* Fr.–*otarie de Nouvelle-Zélande.*
Taxonomic information Carnivora, Otariidae. Recently, because this species and South American fur seal appear to be closely related, a proposal was made to assign them to the resurrected genus *Arctophoca*, lumping them into a larger species group comprising four subspecies: *Arctophoca arctophoca forsteri* in New Zealand and Australia; *A. a. australis* in the Falkland Islands; *A. a. gracilis* from Brazil; and an unnamed subspecies in Peru. The group was given the new common name of "southern fur seal," however, this proposed taxonomy is still the subject of debate and has not been adopted by the Society for Marine Mammalogy.

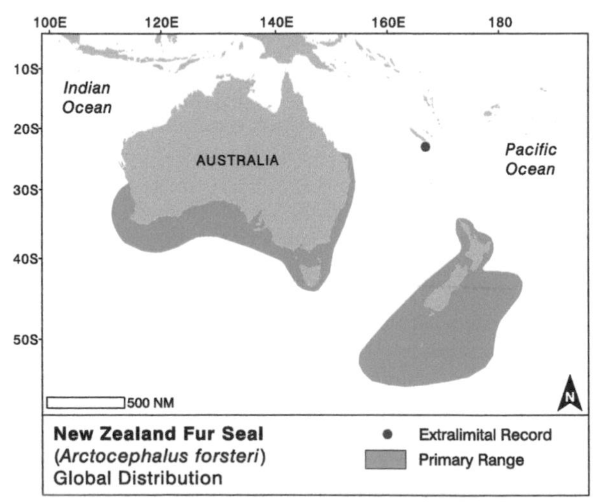

Species characteristics New Zealand fur seals are sexually dimorphic. Adult males can be up to 1.3 times the length and 3–5 or more times the weight of adult females. The muzzle is moderately long, flat, and tapers in width and depth to a flat or slightly upturned and pointed nose. In adults, the vibrissae are pale and medium to long. The ear pinnae are long and prominent, and the eyes are almond-shaped. Adult males have a mane of elongated, coarse guard hairs covering the greatly enlarged neck, chest, and shoulders that extends to the nape and the top of the head. The crown is somewhat rounded due to the modest sagittal crest, and there is a slight forehead. The canine teeth of adult males are larger and thicker than those of females. Adult females, subadults and juveniles have a flatter head with a minimal forehead.

The foreflippers have a dark, sparse, short fur that extends beyond the wrist onto the middle of the dorsal surface of the flipper in a V-pattern that does not reach the rounded tip. The rest of the dorsal surface and the palms of the foreflippers are covered with a hairless black leathery skin. The first digit is the longest and widest, and

A large adult male New Zealand fur seal in excellent condition with a thick chest, neck and mane, and pale short fur on the muzzle and face. PHOTO: D. ORBACH

Adult male New Zealand fur seal showing the long straight-topped muzzle, Australia. PHOTO: J. GIBBENS

Subadult male New Zealand fur seal, South Island, New Zealand. PHOTO: M. JØRGENSEN

Adult female, in typical adult female pelage, and young pup New Zealand fur seal. PHOTO: B. PAGE

Adult female New Zealand fur seal with pup in natal pelage. PHOTO: K. ABERNATHY

Juvenile New Zealand fur seal, Australia. PHOTO: J. GIBBENS

New Zealand fur seals can ascend streams and spend time in freshwater pools in the forest, grooming and apparently enjoying themselves! Near Kaikoura, New Zealand. PHOTO D. ORBACH

A juvenile New Zealand fur seal grooming alongside a freshwater pool. Near Kaikoura, New Zealand. PHOTO D. ORBACH

Two subadult or small adult female New Zealand fur seals just hauled-out and a juvenile or large molted pup (far right), Australia. PHOTOS: J. GIBBENS

has a relatively straight leading edge that angles back to the tip of he flipper, giving the flipper a triangular look. Digits 2–5 are successively shorter. There is a small opening in the skin at the end of each digit for a claw that is usually reduced to a vestigial nodule, and rarely emerges above the skin. The claw openings are set back from the free edge of the flippers by cartilaginous rods that extend the length of each digit. The hindflippers are relatively long, and have dark, short, sparse hair covering part of the proximal end of the flipper. The rest of the dorsal surface, and the entire sole, is covered in black leathery hairless skin. There is a cartilaginous rod that adds a flap-like extension to each toe beyond the bone. The bones of all the toes terminate at about the position of the small nails that emerge through the skin on the dorsal surface, set back from the end of the flipper on toes 2–3. The claws of digits 1 and 5 are vestigial like the claws on the foreflippers, and may or may not emerge from small openings set back from the end of the flaps at about the same distance from the tips as the nails on toes 2–3. All of the flaps at the end of the flipper are of relatively equal length. The first toe or hallux and the fifth toe are slightly wider than toes 2–4 but unlike in all sea lions, not much longer.

Adults are dark olive-gray or rusty brown or metallic gray above and paler below, depending on how long before or after their molt. New Zealand fur seals molt from March to May and appear grayer after the molt, gradually becoming duller and browner in color as time goes on until the next molt a year later. Juveniles up to two years old are an exception being dark brown due to an often-incomplete molt in year two. The muzzle is variably paler, tan to rusty tan or grayish on the end around the mouth and in the mystacial area. The ear pinnae are brown. Adult females are paler than males on the undersides, chest and neck, including the lower sides of the abdomen. Pups are dark brown to blackish, except for a pale muzzle and undersides. They molt and change to a silver-gray juvenile pelage at 4 months.

Some estimates for adult male weights range up to a maximum of 250 kg, while other estimates are only to 200 kg. The maximum weight of a male at the Open Bay Islands, New Zealand was 154 kg. Adult males 8–12 years old were recorded as weighing 124 kg maximum. Adult males reach lengths of 2 m, and females are 1.5 m long and weigh 30–50 kg. Newborns are 3.3–3.9 kg and 40–55 cm. By the time they are weaned at about 9–10 months, the males average 14.1 kg, and the females 12.6 kg. The dental formula is I $^{3}/_{2}$, C $^{1}/_{1}$, PC $^{6}/_{5}$.

Recognizable geographic forms None. Although there is no information on their external differences, recent studies indicate the New Zealand and Australian populations of this species are more genetically distinct than some other fur seal species are from each other.

Can be confused with New Zealand fur seals share their range with five other otariids: Antarctic, subantarctic, and Australian fur seals, and New Zealand and Australian sea lions. Shape of the head and muzzle, presence of a dense underfur, coloration, size and prominence of ear pinnae, straighter leading edge of the foreflipper, and length of the toes on the hindflipper readily distinguish New Zealand fur seals from both sea lions. Australian fur seal bulls are larger and heavier but less stocky, with a long head, and wider and blunter muzzle. The guard hairs of the mane are longer, and the neck and shoulders much more massive and muscular than on New Zealand fur seal males. Generally, both male and female Australian fur seals are paler above than New Zealand fur seals. The eyes of Australian fur seals are set somewhat more to the side while the New Zealand fur seal has eyes positioned more forward. Australian fur seals also have more curvature to the leading edge of the foreflipper giving it a more paddle-like appearance while the New Zealand fur seal and other species of *Arctocephalus* have a straighter edge to the flipper giving it a more oar-like appearance.

Distinguishing New Zealand fur seals from Antarctic fur seals is difficult due to similar size, shape and coloration. Adult male Antarctic fur seals are darker than adult male New Zealand fur seals. They are darker brown or dark gray and have extensive light grizzling on the mane. They also have long, pale-colored ear pinnae, a straight muzzle that is wider and shorter, smaller rhinarium, and steeper forehead. Separating females, subadults and juveniles out of range or when isolated is problematic. Subantarctic fur seals are uniquely colored with a pale face and chest, have ear pinnae that are short and black and have a short, sharply-pointed muzzle. There is also a crest of longer, pale fur in the dark crown. Adult females have similar coloration to the males, but have less well-defined pale facial markings that can be shaded with dull yellow orange to light brown. In both sexes the top of the foreflippers is darker as is the area where the flippers attach to the sides. The fore- and hindflippers are proportionately small.

Hybrid New Zealand X Antarctic fur seals are known. The few photographs available show that they share characteristics of both species (see Note on Hybrid Southern Hemisphere Fur Seals Genus *Arctocephalus*, pages 366–367).

Subadult male New Zealand fur seal just beginning to show enlargement of the neck and chest. PHOTO: B. SABERTON

Juvenile New Zealand fur seals and a Australian fur seal (back, left). Note the straighter leading edge of the foreflipper of the New Zealand fur seal on the right. PHOTO: T. HASSE

Distribution New Zealand fur seals are distributed in two geographically-isolated and genetically-distinct populations. In New Zealand, they occur around both the North and South Islands, with small breeding colonies on the north, and larger colonies on the west and southern coast and islands of the South Island, and on all of New Zealand's subantarctic islands. They are less common off the North Island, with no colonies, but occur as far north as the Three Kings Islands. They are present, but do not breed, on Macquarie Island in April and May. They prefer waters of the continental shelf and slope, and are found widely in waters over the Campbell Plateau, south of New Zealand. A separate population occurs in the coastal waters of southern and western Australia, from just east of Kangaroo Island west to the southwest corner of the continent in Western Australia. Most of the pups are born at five sites in South Australia. Vagrants have been recorded in New Caledonia, and a bone was recovered in a 14th century archaeological site in the Cook Islands. On land, New Zealand fur seals prefer

A large subadult or adult male New Zealand fur seal rafting off of the South Island of New Zealand. This species will also raft in the "jug-handle" position like other fur seals. PHOTO: W. RAYMENT

New Zealand fur seal grooming. Fur seals spend a considerable amount of time attending to their thick pelage. South Island, New Zealand. PHOTO: M. NOLAN

A New Zealand fur seal at the surface breaking up and tearing a fish into pieces that are more readily swallowed. PHOTO: M. NOLAN

rocky habitat in windy locations with shelter from the sun. They will also readily enter vegetation and will even occasionally travel up creeks and streams and spend time in freshwater.

Ecology and behavior New Zealand fur seals are polygynous. Males arrive at colonies in late October before females and acquire and defend territories with vocalizations, ritualized displays, and fighting. Males become sexually mature at 4–5 years of age, but are not able to hold territories until they are physically mature when 9–12 years old. Females become sexually mature at 4–5 years old and begin to reproduce as soon as they reach maturity. Male territories include an average of 5–8 females with numbers varying between colonies. Pupping and breeding occurs from mid-November to January. The number of animals ashore declines rapidly in January. Male vocalizations include a bark or whimper, a guttural threat, a low-intensity threat, a full threat, and a submissive call. Females growl and call their pups with a high-pitched wail. Estrus occurs 7–8 days after a female gives birth, and they usually spend another 1–2 days ashore with their pup before departing and beginning a cycle of foraging trips and periods of pup attendance ashore. The female uses a high-pitched trilling call to attract her pup when she returns to the rookery. Pups are weaned when they are about 10 months old. Foraging trips are shorter when the pup is young and become longer as the pup gets older. The overall mean time ashore attending the pup is a little over 3 days, and is roughly equal to the mean time of all foraging trips combined: of approximately 3.3 days.

Foraging dives by lactating females are almost entirely at night to an average depth of 15 m with a maximum depth of 163 m. Mean dive duration is 50 seconds with a maximum of 6.2 min. Maximum depth of dive is 275 m and length of dive is approximately 11 min.

Although some individuals wander widely, New Zealand fur seals are considered non-migratory. At sea they actively groom and raft in a variety of postures typical of southern fur seals, including the "jug-handle" position. They will also porpoise when traveling rapidly. Predators include killer whales, sharks, male New Zealand sea lions (on pups), and possibly leopard seals at subantarctic islands.

Feeding and prey The diet varies by location and time of year. Nearly all foraging by lactating females occurs at night. Important prey species include both vertically migrating species, and other species that occur throughout the water column and on the bottom. Key prey species include arrow and other squid species, barracouta, anchovy, various lanternfish species, jack mackerel, red cod, swallowtail, hoki, octopus, and penguins. They also feed on shearwaters and possibly other flying marine birds.

Threats and status Humans have likely harvested New Zealand fur seals since first contacts occurred. There is evidence that Polynesian colonization of New Zealand and harvest of seals led to declines and loss of colonies on the coast of North Island. European sealers nearly exterminated the species in the 19th century, but with governmental protection in the late 19th century, beginning in New Zealand and followed by Australia, the species has rebounded to occupy most of its former range. Trawl and other fisheries are a source of entanglement and drowning. Tourism and disturbance at colonies can lead to disruption of breeding behavior and site abandonment, although most colonies are on offshore islands and are relatively inaccessible. Viewing of these fur seals has become a popular attraction. The total population is estimated at approximately 200,000–220,000 animals, with the population split about evenly between New Zealand and Australia.

IUCN status Least Concern.

Subantarctic Fur Seal—*Arctocephalus tropicalis*

(Gray, 1872)

Recently-used synonyms *Arctocephalus tropicalis tropicalis, Arctocephalus gazella, and Arctophoca tropicalis.*
Common names En.–Subantarctic fur seal, Amsterdam fur seal; Sp.–*lobo fino de subantártico;* Fr.–*otarie subantarctique.*
Taxonomic information Carnivora, Otariidae. Recently, a proposal was made to reassign this species to the resurrected genus *Arctophoca.*
Species characteristics Subantarctic fur seals are sexually dimorphic, with adult males being 3 times heavier and 1.2–1.3 times longer than females. In both sexes, the muzzle is straight, short, and narrow, and tapers rapidly in width and thickness to a pointed end. The rhinarium is small. A forehead, more abrupt in males, separates the muzzle from the elevated, but somewhat flattened, crown. The vibrissae are very long and pale, and often reach well past the ears and down the neck. The short, dark, ear pinnae with naked tips lie close to the head and are not particularly prominent. The eyes are rounded and teardrop shaped.

Adult males are heavily-built; their enlarged chest and shoulders make the neck appear short. They develop a prominent tuft, or crest of long guard hairs on top of the head above and behind the eyes and forehead. The guard hairs on the neck and chest are somewhat longer than those on the rest of the body, but there is no conspicuous mane and grizzling is absent. Adult females, young subadult males, and juveniles do not have the crest of hair on the crown.

Overall, both the fore- and hindflippers are proportionately short and broad in all age classes and both sexes. A series of vagrant animals found in South Africa had foreflippers that measured from 19–26% of body length and hindflippers that were 12–16% of body length. The foreflippers have a dark, sparse, short fur that extends beyond the wrist onto the middle of the dorsal surface of the flipper in a V-pattern that does not reach the rounded tip. The rest of the dorsal surface and the palms of both foreflippers are covered with a hairless black leathery

A subadult male (left) and an adult male (right) subantarctic fur seal ashore. They are readily identified by the extensive whitish coloration from the chest through the face. PHOTO: H. LAWSON

skin. The first digit is the longest, widest and thickest, and angles posteriorly, giving the flipper a swept-back look. Digits 2–5 are successively shorter. There is a small opening in the skin at the end of each digit for a claw that is usually reduced to a vestigial nodule, and rarely emerges above the skin. The claw openings are set back from the free edge of the flippers by cartilaginous rods that extend the length of each digit, and expand the size of the flippers. The hindflippers have dark, short, sparse hair covering part of the proximal end of the flipper, and the rest of the dorsal surface, and the entire sole is covered in black leathery hairless skin. Each digit has a cartilaginous rod that adds a flap-like extension to each toe. The bones of the three central toes terminate at the position of the small nails that emerge through the skin on the dorsal surface, set back from the end of the flipper. The claws of digits 1 and 5 are vestigial like the claws on the foreflippers, and may or may not emerge from small openings set back from the end of the flaps. All of the flaps at the end of the flipper are of relatively equal length. The first toe, or hallux, and the fifth toe are somewhat wider than toes 2–4.

Subantarctic fur seals are uniquely and strikingly colored. The back and rump of adult males are dark gray to brownish-black, becoming darker as they change from subadult to adult. From the chest through the muzzle, face, and area around the eyes, the pelage is continuously cream-colored, with yellow to orange shading possible. The ear pinnae are dark, and attach in the dark pelage of the upper parts. The crest of longer fur above the eyes and behind the forehead is pale and is situated in the dark fur of the crown. Adult females have similar coloration to males, but have less well-defined pale facial markings, which can be shaded with a dull yellow-orange to light reddish-brown. In both sexes, the top of the foreflippers is darker, as is the area where the flippers attach to the sides. The underside of the abdomen is dark ginger to reddish-brown.

Adult males reach lengths of 1.8 m and weights of 70–165 kg. Females arc 1.19–1.52 m and 25–67 kg, averaging about 50 kg. Pups are 60 cm and 4.0–4.4 kg at birth. The dental formula is I $^{3}/_{2}$, C $^{1}/_{1}$, PC $^{6}/_{5}$.

Recognizable geographic forms None.

Can be confused with Subantarctic fur seals wander widely and can show-up as vagrants in the range of nearly all Southern Hemisphere pinnipeds. They are differentiated from all sea lions by their distinctive coloration, short and sharply-pointed muzzle, large eyes, long white vibrissae, proportionately short and wide flippers, and thick pelage. Adults can be differentiated from all southern fur seals by their distinctive coloration and proportionately small flippers. Identifying juveniles and subadults from other species can be problematic. Hybrid subantarctic X Antarctic fur seals and subantarctic X New Zealand fur seals are known.

Distribution Subantarctic fur seals are widely-distributed in the Southern Hemisphere, mainly in the South Atlantic and Indian oceans, with about 95% of the population breeding at three locations: Gough Island, the Prince Edward Islands, and Amsterdam Island. They also breed on subantarctic islands north of the Antarctic polar front, including Marion, Crozet, and Macquarie islands,

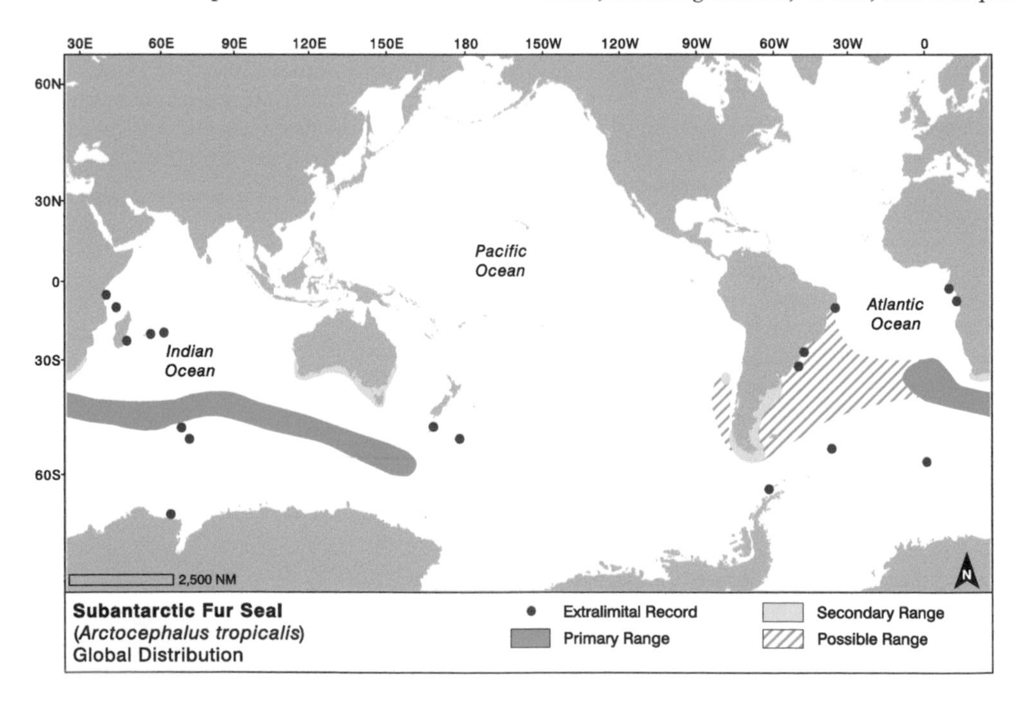

The distinctive coloration, short muzzle, and crest of an adult male subantarctic fur seal are in evidence on this bull. PHOTO: M. JØRGENSEN

Adult female subantarctic fur seals have similar chest coloration to adult males. Pups are dark, glossy black. PHOTO: I. CHARRIER

The short flippers and short pointed muzzle are visible on these plump juvenile subantarctic fur seals. PHOTO: I. CHARRIER

Juvenile subantarctic fur seal. Notice the very short, sharply pointed muzzle, and relatively short flippers. PHOTO: T. JACOBSEN

Three views of the development of the head and coloration of male subantarctic fur seals. A subadult (left), an older subadult with a small crest of fur on his forehead (center), and a fully grown bull with a large crest of hair and richly colored pelage (right). PHOTOS: (LEFT TO RIGHT) G. DE RANGO/TMMC, H. LAWSON, AND G. HOFMEYER

Two views of subantarctic fur seals at sea. The orangish chest and neck is visible on this "porpoising" animal (upper). The juvenile (lower), is rafting in the "jug-handle" position. PHOTOS: M. JØRGENSEN

and north of the subtropical front at Tristan da Cunha, and Saint Paul Island. The northern limit of their range is not well known, but the species is well documented to wander widely and vagrants have appeared in the Juan Fernández Islands, Chile, Argentina, Uruguay, Brazil (as far north as Sergipe State), Gabon, Angola, South Africa, Tanzania, the Comoros, Madagascar, Mauritius, Rodrigues Islands, Australia and Tasmania, and New Zealand's South Island. They have also been recorded south of the Antarctic Convergence at South Georgia, Cape Shirreff on Livingston Island, and on the Antarctic continent at Davis and Mawson bases.

Ecology and behavior Subantarctic fur seals are polygynous, with males defending territories with vocal and postural displays and fighting. Typical territories include 1–5 females and are located on boulder-strewn beaches, at the foot of cliffs, shoreline ledges and terraces, and in shallow shoreline caves and grottos. Most areas have sources of shade or are exposed to prevailing winds. Male vocalizations include a bark or whimper, a guttural threat, a low-intensity threat, possibly a full threat, and a submissive call. Females growl and have a pup-attraction call that is a mournful wail without trilling.

Pups are born from late October to early January, with a peak in mid-December. Females give birth within 6 days of arriving at the colony and estrus and mating occur 8–12 days later. Females spend the time between the birth of their pups and estrus with their newborn before departing for their first foraging trip. Females wean their pups in approximately 11 months. Trip length by lactating females increases over the course of the summer from 5.7 to 10.8 to 23–28 days in the fall and winter. Also during the summer, dives become deeper and slightly longer, starting at a mean of 16.6 m and increasing to a mean of 19 m. Dive duration at this time is generally just over 1 min. and maximum depth is 100 m. In the winter, they dive deeper to a mean depth of 29 m for 1.5 min, with a maximum depth of 208 m and duration of 6.5 min reported. Subantarctic fur seals are ashore for their annual molt between February and April. Otherwise, little is known of their behavior at sea. Except for females with pups, most of the population spends much of the winter and spring (June-September) at sea. Predators include killer whales, sharks, and at Macquarie Island, New Zealand sea lions.

Subantarctic fur seals share breeding islands with Antarctic fur seals at the Prince Edward Islands, Crozet and Macquarie Island. New Zealand fur seals also breed near subantarctic fur seals at Macquarie Island. Hybrid subantarctic X Antarctic, and subantarctic X New Zealand fur seals are known. The few photographs available show that they share characteristics of both species (see Note on Hybrid Southern Hemisphere Fur Seals Genus *Arctocephalus,* pages 366–367). Hybrids have grown to maturity and bred. With the exception of some information on appearance and vocalizations little is known of the foraging and behavior of these hybrids.

Feeding and prey Generally, they are known to feed on varieties of pelagic and vertically migrating species such as myctophids, nototheniids, cephalopods, and krill, and they will also prey on penguins. It has been estimated that their diet is 50% cephalopods, 45% fish, and 5% krill at the Prince Edward Islands. At Amsterdam Island, they feed on cephalopods, fish, and rockhopper penguins.

Threats and status As with all other southern fur seals, subantarctic fur seals were over-exploited by sealers in the 19th century, and were hovering on the brink of extinction at the beginning of the 20th century. Since then they have rebounded to refill much of their former range. The total population is believed to be greater than the 310,000, animals estimated in 1987, since it appears to have been experiencing steady growth since then. Subantarctic fur seals live in some of the most remote oceanic areas, and breed on many of the most isolated islands on Earth. All of the breeding islands are managed as protected areas or parks by the governments that claim them. Human visitation and disruption from scientific research activities is minimal. Fisheries takes and entanglement in marine debris are not well understood, but are not listed as a major threat in status reviews. Their population bottleneck in the 19thcentury leaves them vulnerable to disease or environmental stressors due to climate change. Concern is always needed for species that gather seasonally in large aggregations on relatively few sites, even if they are widely dispersed, due to the potential impact from the loss of even one of these large rookeries on the total population.

IUCN status Least Concern.

Cape and Australian Fur Seals—*Arctocephalus pusillus*

(Schreber, 1775)

Recently-used synonyms *Arctocephalus doriferus, Arctocephalus tasmanicus, Arctocephalus forsteri* (all for the Australian subspecies).

Common names En.–Cape, South African, Australian, Brown, or Afro-Australian fur seal; Sp.–*lobo marino de dos pelos de Sudáfrica*; Fr.–*otarie du Cap*; En.–Australian, Tasmanian, or Afro-Australian fur seal; Sp.–*lobo fino de Australia*; Fr.–*otarie à fourrure austral.*

Taxonomic information Carnivora, Otariidae. There are two widely-separated subspecies: *A. p. pusillus* (Cape fur seal) in southern Africa, and *A. p. doriferus* (Australian fur seal) in southern Australia. The low genetic divergence between the two subspecies indicates they split relatively recently. Recently, a proposal was made to reassign southern fur seals to the resurrected genus *Arctophoca*, with the exception of this species, leaving it as the sole representative of the genus *Arctocephalus*.

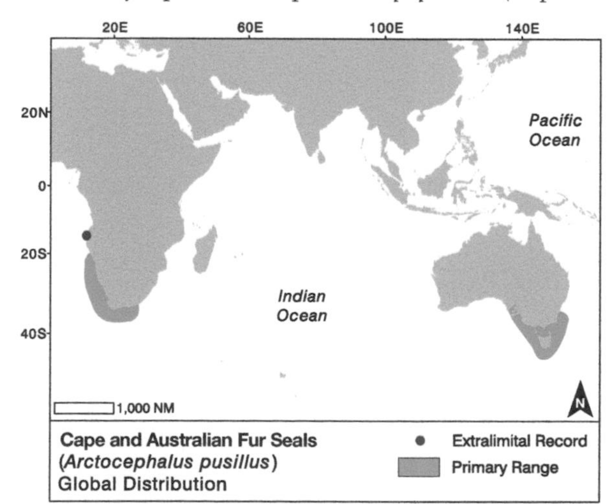

Species characteristics Cape and Australian fur seals are the largest fur seals and have a high degree of sexual dimorphism. Adult males are 3.5 to almost 4.5 times heavier, and approximately 1.3–1.9 times longer, than females. Both sexes of Australian fur seals are on average heavier than their South African counterparts. Adult female Australian fur seals average slightly longer, and males slightly shorter than females and males of the Cape fur seal subspecies, and the length relationship between female and male Australian animals is closer, with males being about 1.1–1.8 times longer than females.

Both subspecies have been described as the most sea lion-like fur seals. Even their barking vocalizations are more sea lion-like. In both, the head is large and wide, and the crown rounded in adult males and flatter in females. The sloping forehead is more prominent in adult males, and less steep, but still present, on females. The muzzle is robust and long, flat and wide on top, tapering only somewhat in width and thickness to the large conspicuous nose. The rhinarium is large, wide and rounded in adult males, less so in adult females, and extends beyond

Adult male, adult female and young pup provide a good measure of the expression of sexual dimorphism in the Cape fur seal. South Africa. PHOTO: J GIBBENS

Adult female Cape fur seal with a young pup in black natal pelage. PHOTO: J. GIBBENS

Adult female and male Cape fur seals differ considerably in head size, muzzle shape and the lack of a mane on the female. PHOTO: J. GIBBENS

Postures, ritualized behaviors and vocalizations reduce the number of violent contacts between males after they establish their territories among their neighbors. PHOTO: J GIBBENS

This subadult male Cape fur seal has darkened in color and is beginning to develop a larger neck and chest. PHOTO: R. L. PITMAN

This adult female Cape fur seal shows the long muzzle and pale color typical for this species. Skeleton Coast, Namibia. PHOTO: M. JØRGENSEN

the end of the mouth. The ear pinnae are long and prominent. The vibrissae are moderately long, pale, and regularly extend to the ear pinnae. Adult males are greatly enlarged in the neck and shoulders, with a mane of guard hairs that are up to 5 cm long from the nape and neck to shoulders and chest. Adult females, subadults, and juveniles are robust, but normally proportioned in the neck and shoulders. The foreflippers have a dark, sparse, short fur that extends beyond the wrist onto the middle of the dorsal surface of the flipper in a V-pattern that does not reach the rounded tip. The rest of the dorsal surface and the palms of both foreflippers are covered with a hairless black leathery skin. The first digit is the longest, widest and thickest, and curves posteriorly, giving the flipper a swept-back look. Digits 2–5 are successively shorter. There is a small opening in the skin at the end of each digit for a claw that is usually reduced to a vestigial nodule, and rarely emerges above the skin. The claw openings are set back from the free edge of the flippers by cartilaginous rods that extend the length of each digit, and expand the size of the flippers. The hindflippers have dark, short, sparse hair covering part of the proximal end of the flipper, and the rest of the dorsal surface, and the entire sole is covered in black leathery hairless skin. The hindflippers are long and each digit has a cartilaginous rod that adds a flap-like extension to each toe. The bones of the three central toes terminate at the position of the small nails that emerge through the skin on the dorsal surface, set back from the end of the flipper. The claws of digits 1 and 5 are vestigial, like the claws on the foreflippers, and may or may not emerge from small openings set back from the end of the flaps. All of the flaps at the end of the flipper are of relatively equal length. The first toe or hallux and the fifth toe are somewhat wider than toes 2–4.

Adult females and juveniles are similar and are tan to light grayish or silvery-brown with yellowish to orange highlights above, a paler chest and reddish-brown to brown shades on the abdomen. Females and juveniles appear to be grayish overall, with a lighter shade on the chest, when wet. Males are usually darker than females and darken as they age. Cape fur seals are generally darker than animals from Australia. The guard hairs have a slight grizzled appearance, especially on bulls. Females, subadults and juveniles can be lighter on the chest. The mystacial area can be paler. The tops of the flippers are very dark. Part of the ear pinnae and the area around the insertions of the pinnae are paler, however the tips of the pinnae are naked and dark in older animals. Pups are born in a blackish natal coat, with variable hints of silver overall, and can be paler below. They molt at 4–5 months to an olive-gray coat. As juveniles they have a silvery-gray coat similar to that of adult females. Alopecia has been reported in Australian fur seals and can occur at a high rate in some colonies amongst juveniles. It also is seen on a small percentage of adult females, but has not been reported on adult males. With alopecia, a lighter colored patch or patches, in a bilaterally symmetrical pattern, appear on the dorsal surface of the back, neck and head when the guard hairs break exposing the paler underfur. In severe cases ulcerated damaged skin is exposed, and animals with this problem can also be in poor overall physical condition.

Part of a large colony of Cape fur seals in South Africa with many females away foraging, as evidenced by the crèches of pups in the center. PHOTO: J. GIBBENS

Adult malc Cape fur seals reach 2–2.3 m in length and weigh an average 247 kg, with a maximum weight of 353 kg. Adult females are 1.20–1.6 m in length and an average 57 kg in weight, with a maximum of 107 kg. Pups are 60–70 cm and about 6 kg at birth. Adult male Australian fur seals are 2–2.3 m in length, attain an average weight of 279 kg, and can weigh up to 360 kg. Adult females are 1.4–1.7 m in length, and attain an average weight of 76 kg, and can weigh up to 110 kg. Pups are 60–80 cm and 5–12 kg at birth. Longevity in Cape fur seals has been reported as up to 25 years. The dental formula is I $^3/_2$, C $^1/_1$, PC $^6/_5$.

Recognizable geographic forms Although the two subspecies differ slightly in average size and weight of adults, and possible slight differences in coloration, there is no other information available on consistent external differences between Cape and Australian fur seals.

Can be confused with Cape and Australian fur seals both share their ranges with vagrant subantarctic fur seals. Subantarctic fur seals are uniquely colored with a pale head and chest. The ear pinnae are dark and short. Adult males have a crest of longer pale fur in the dark crown. Subantarctic fur seals also have proportionately small flippers with a straighter, less curving leading edge.

For the Australian fur seal, New Zealand fur seals and Australian sea lions pose a regular chance for confusion. New Zealand fur seals are grayer and darker, have a thinner and proportionately longer, more pointed, and slightly upturned looking muzzle due to the larger rhinarium (fleshy end of the nose), a more flattened crown, and darker brown ear pinnae. They are stocky, but smaller

Portrait of an adult male Australian fur seal showing the rounded crown. PHOTO: J. GIBBENS

Adult male, female, and older molted pup Australian fur seals. The large size, thick neck and mane of the adult male are obvious differences between the sexes. PHOTO: J. GIBBENS

A huge and magnificent bull Australian fur seal that has just hauled out. PHOTO: J. GIBBENS

Adult male and adult female Australian fur seals. PHOTO: J. GIBBENS

Adult female and a molted pup Australian fur seals that have just hauled out. The female appears to be vocalizing. PHOTO: J. GIBBENS

Adult female Australian fur seal with pup in natal pelage. Mothers identify their pups based on scent, vocalizations, and behavior. PHOTO: J. GIBBENS

overall. The mane of Australian fur seal males is more conspicuous. On land, Australian fur seals move one foreflipper forward at a time, and the head sways from side to side with an alternating gait used by many sea lions. New Zealand fur seals extend both foreflippers ahead at the same time in a bounding gait. Australian sea lions have large, blocky heads with a wide, blunt muzzle, distinctive coloration, lack the longer, thicker pelage of the Australian fur seal, and have shorter ear pinnae. Also, when compared to other Southern Hemisphere fur seals, the leading edge of the foreflipper of both the Cape and Australian fur seals curves more towards the tip whereas the leading edge of the New Zealand fur seal's foreflipper is straighter overall.

Australian fur seals are born in a natal pelage with darker color on the muzzle and face. PHOTO: J. GIBBENS

Australian fur seal in the water. This pup has molted to the juvenile pelage and the pale color in the face is conspicuous. PHOTO: J. GIBBENS

A female Australian fur seal showing the long muzzle and slight forehead. PHOTO: R. KIRKWOOD

A subadult male Australian fur seal showing the long pointed muzzle. PHOTO: R. KIRKWOOD

Distribution Cape fur seals are found along the south and southwest coasts of southern Africa from South Africa to Angola. Australian fur seals are found along the coast and continental shelf and slope waters from western Victoria, east along the coast to the southeast corner of Australia, then north through most of New South Wales. Their range also includes Tasmania, and the islands of Bass Strait. Australian fur seals will travel up to 160 km offshore. On land, they prefer rocky habitat.

Ecology and behavior Both subspecies are highly polygynous. Adult males arrive at the colonies first. Breeding is from late October to the beginning of January. Males become sexually mature at about 5 years, and can establish territories at 8–13 years. They defend territories with vocalizations, ritualized postures and fighting. On average, the territories of bulls are 62 m^2 and hold about 9 females. Male vocalizations include a gentle staccato bark or whimper, and a guttural threat. Females have a threat and a bawling pup attraction call.

Female Australian fur seals come ashore and give birth 1.5–2 days after arrival. The peak is in the first week of December, although there is some variation between colonies. Females attend the pup for 8–9 days before coming into estrus, mating, and departing on their first foraging trip. Foraging trips get longer as the season progresses from summer to winter, changing from a mean of 3.71 to 6.77 days. Periods of attendance stay the same from birth to weaning, and have a mean length of 1.7 days. Pups are usually weaned at 10–12 months even though some pups begin to forage at 7 months, and others are nursed for 2–3 years. Gestation is about 12 months, including a 3-month diapause (delayed implantation). The data are similar for Cape fur seals, except that foraging trip intervals are much shorter, probably reflecting greater availability of food near the colonies in the productive, upwelling waters off southwest South Africa. Australian fur seals are benthic continental shelf foragers, which sets them apart from other fur seal species. They begin to forage upon departure from the rookeries. Most foraging dives by lactating Australian fur seal females are usually to 65–85 m with a maximum of 164 m, and last 2–3.7 min with a maximum recorded duration of 8.9 min. Also unlike many other fur seals, considerable foraging occurs during the day. Two lactating female Cape fur seals dove shallower, averaging 41 and 49 m respectively, but had much deeper maximum dive depths of 191 and 204 m.

On land this species, like sea lions, will lay in contact with each other. At sea, they are found alone or in small groups, often gathering in huge rafts adjacent to rookeries. They adopt a variety of poses while resting in the water, including the "jug-handle." They also purposely entangle themselves in rafts of kelp, possibly using the kelp as an anchor and for camouflage. When traveling rapidly, they sometimes "porpoise." Neither of the subspecies is migratory; they move more locally within their restricted ranges. Predators include killer whales and great white sharks at sea, and black-backed jackals and brown hyenas for South African fur seals at their mainland colonies.

Subadult male Cape fur seal at sea. PHOTO: M. DE BOER

A group of Cape fur seals at sea with a male porpoising. PHOTO: M. DE BOER

Adult female Cape fur seal grooming at sea. PHOTO: M. DE BOER

Adult female Cape fur seal rolling to the right while porpoising. PHOTO: M. DE BOER

Adult female Cape fur seal grooming its muzzle and head with its long foreflippers. PHOTO: M. DE BOER

Cape fur seals often turn to one side and "side-surface" to take a breath when they are traveling. PHOTO: M. DE BOER

A Cape fur seal in the head-down posture with hindflippers elevated used by many southern fur seals. They can stay in this posture for as long as they can hold their breath while grooming. PHOTO: M. DE BOER

Adult male Cape fur seal "side surfacing." PHOTO: M DE BOER

Seal watching is becoming a popular tourist attraction at many locations around the world, including here in South Africa with Cape fur seals. PHOTO BY: L. MURRAY

Disentangling a juvenile Australian fur seal. Discarded and lost net, line, and packing bands are a major threat to pinnipeds worldwide. PHOTO: R. KIRKWOOD

Feeding and prey Both subspecies are opportunistic feeders that take a wide variety of prey. For Cape fur seals, which feed in mid-water, the diet consists of fish (75%), cephalopods (17%), and crustaceans (8%). Important species are Cape hake, horse mackerel, pelagic goby, pilchards, anchovy, squid (genus *Loligo*), rock lobster, shrimp, prawns, and amphipods. Cape fur seals have also been reported to occasionally take jackass penguins and several species of flying seabirds. Australian fur seals tend to feed benthically on the continental shelf, and principally take redbait, barracouta, jack mackerel, gurnard, red cod, leatherjackets, and arrow squid.

Threats and status Of all pinnipeds, the Cape fur seal may have the longest history of interaction with modern humans, with remains in South African middens dated to approximately 130,000 years ago. Cape and Australian fur seals were hunted heavily in the 19th century and both populations were driven to very low levels. With protection, both have recovered, although the South African subspecies to a much greater extent than the Australian, with Cape fur seals numbering approximately 2 million animals in 2004. Australian fur seals are estimated to number 110,000–120,000, and have 15–20 breeding rookeries, with more than half the animals breeding at two of these sites: Lady Julia Percy Island in the Bass Strait, and Seal Rocks, New South Wales. If just two other large sites are included, Kanowna Island and the Skerries, the total increases to 80% of reproduction in this subspecies at four sites.

Fur seal harvests in South Africa were suspended in 1990, but are ongoing in Namibia. The Cape fur seal is considered to be detrimental to commercial fisheries, costing large sums in damaged gear and stolen and damaged catch annually. Some are taken incidentally in fishing operations every year. Also, significant numbers of Cape fur seals are known to become entangled in marine debris such as packing bands, discarded lines and nets, and other material that can become a collar around an animal's neck. Rates of entanglement vary by colony, but have been estimated to be between 0.12–0.66%. There is a potential for human disturbance from tourism at several large colonies. The population of Cape fur seals is located beside one of the busiest shipping lanes for oil tankers, and is at risk from the potential for a catastrophic oil spill. Conflicts exist with commercial fisheries off South Africa, where this fur seal forages around trawlers bringing in their nets.

A juvenile Australian fur seal with alopecia has symmetrical guard hair loss exposing the underfur. The guard hairs become brittle and break off with this condition. PHOTO: M. LYNCH

Australian fur seals are protected from harvest. There are conflicts with local commercial fisheries from fur seals stealing catch, damaging gear, and becoming entangled in nets and traps. They are considered a pest species by some, and are shot under permit to protect fishing gear and catch. Mortality is highest and more significant for younger age classes. They also live close to human population centers and agricultural areas and are exposed to a wide variety of pollutants through the food chain.

IUCN status Least Concern (for both subspecies).

Steller Sea Lion—*Eumetopias jubatus*

(Schreber, 1776)

Recently-used synonyms None.
Common names En.–Steller, Steller's, northern sea lion; Sp.–*lobo marino de Steller;* Fr.–*lion de mer de Steller.*
Taxonomic information Carnivora, Otariidae. There are two recognized subspecies: western Steller sea lion (*E. j. jubatus*), and Loughlin's Steller sea lion (*E. j. monteriensis*), as recently adopted by the Society for Marine Mammalogy. The US National Marine Fisheries Service considers the species to be monotypic, managing it as western and eastern stocks, referred to as Distinct Populations Segments.

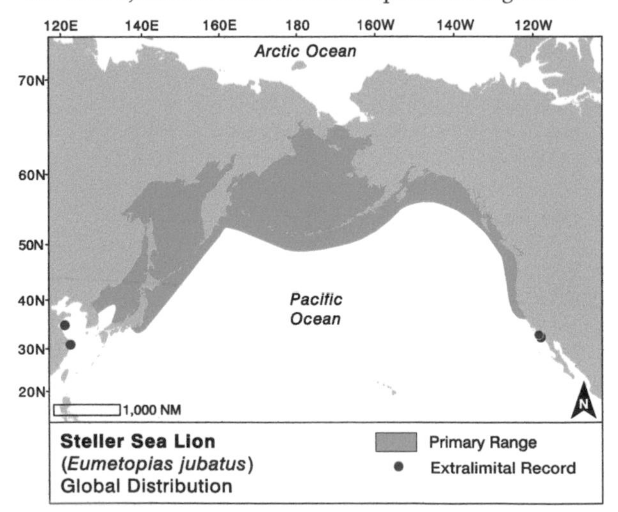

Species characteristics Steller sea lions are the largest otariids and the fourth largest pinniped. Both males and females are robust and powerfully built at all ages. Sexual dimorphism is evident, with adult males that weigh three times as much as females, and grow 20–25% longer. In addition to being larger, males have a mane of longer guard hairs extending from the back of the head to the shoulders and all around the neck. Breeding bulls are also very thick and wide in the neck and shoulder area. There is a sagittal crest on the skull that imparts a small to moderate forehead to adult and subadult males. Adult females and juveniles have no crest and have only a minor forehead dip in front of the eyes and appear to be almost flat from the crown to the tip of the nose. Males also have larger canines that are both longer and thicker than those of females.

Steller sea lions have a massive, wide head. The muzzle is thick, wide, long, and blunt on the end. The eyes are widely-spaced apart, and set well back from the end of the muzzle and, like the ear pinnae, appear small when compared with the size of the head. The vibrissae are pale, conspicuous, and long. Both the fore- and hind-flippers are long and broad. The foreflippers have sparse, short fur that extends beyond the wrist onto the middle of the dorsal surface of the flipper in a V-pattern that does not reach the rounded tip. The rest of the dorsal surface, and the palms of both foreflippers are covered with a hairless black leathery skin. The first digit is the longest, widest and thickest, and curves posteriorly, giving

Adult male western Steller sea lions are huge and make adult females of their species, which are bigger than nearly all other male otariids, look small. Kuril Islands, Russia. PHOTO: O. SHPAK

An adult male Loughlin's Steller sea lion with two juveniles. British Columbia, Canada. PHOTO: V. SHORE

Adult male Loughlin's Steller sea lion vocalizing. British Columbia, Canada. PHOTO: V. SHORE

Adult males gain a huge amount of weight on the run-up to their breeding season. Loughlin's Steller sea lion, British Columbia, Canada. PHOTO: V. SHORE

Several large female western Steller sea lions surrounded by juveniles in the Aleutian Islands. PHOTO: S. EBBERT

Typical for adult female Steller sea lions, this western adult lacks the enlarged chest and neck of bulls. PHOTO: C. KURLE/NOAA FISHERIES/AFSC

Steller sea lions can form large aggregations of both sexes and all ages on haulouts. Loughlin's Steller sea lions. PHOTO: M. NOLAN

Steller sea lions share the coast from California to Alaska with California sea lions. The pale adult female Loughlin's Steller sea lions are as large as the dark adult male California sea lions. British Columbia. PHOTO: V. SHORE

A group of western Steller sea lions with many juveniles, subadults, and small adult males. Cape Olyutorsky, Russia. PHOTO: J. WAITE

Adult male Loughlin's Steller sea lion next to a fairly large adult male California sea lion. British Columbia. PHOTO: V. SHORE

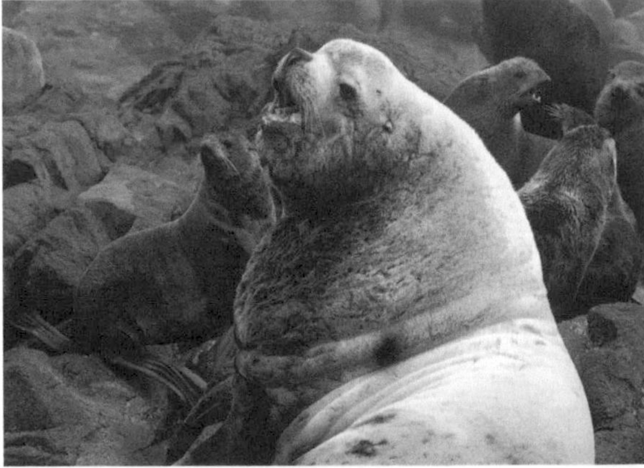

Adult male western Steller sea lion hauled out with northern fur seals. Stellers occasionally prey on juvenile fur seals. Kuril Islands, Russia. PHOTO: O. SHPAK

An adult female with a large unmolted pup and possibly a still cared-for juvenile near her back. Western Steller sea lion, Aleutian Islands. PHOTO: S. EBBERT

A pale adult female vocalizing. Females returning from foraging trips vocalize to find and attract their pup. Kamchatka, Russia. PHOTO: J.-P. SYLVESTRE

A small group of young and older western Steller sea lion pups interacting. The younger pups are still in their dark natal pelage. Kuril Islands, Russia. PHOTO: O. SHPAK

A juvenile that still has a lot of growing-up to do. Females can nurse juveniles in their third year. Western Steller sea lion, Aleutian Islands. PHOTO: S. EBBERT

the flipper a swept back look. Digits 2–4 are successively shorter. There is a small opening in the skin at the end of each digit for a claw that is usually reduced to a vestigial nodule, and rarely emerges above the skin. The claw openings are set back from the free edge of the flippers by cartilaginous rods that extend the length of each digit, and expand the size of the flippers. The hindflippers also have cartilaginous rods that extend the length of each toe. The bones of the three central toes terminate at the position of the small nails that emerge through the skin on the dorsal surface, set back from the end of the flipper. The first and fifth toes are longer than the three middle toes, and the first toe, or hallux, is longer and wider than the fifth toe. The hindflippers have short hair covering part of the proximal end of the flipper, and the rest of the dorsal surface, and the entire sole is covered in black leathery hairless skin.

Coloration in adults is pale yellow to light tan above, darkening around the insertion of the flippers to brown and shading to rust below. Unlike most pinnipeds, when wet, adult Stellers are the same color or paler depending on lighting, appearing light grayish-tan. Juveniles are darker than adults and are dark tan to light brown. Pups are born with a thick blackish-brown natal coat that is molted by about 6 months of age. Scars from bites and healed wounds are darker than the background color.

Adult males reach a length of 3.3 m, and average 1000 kg in weight. Adult females are about 2.5 m and 273 kg. Newborn pups are 1 m and 16–22 kg. Adult males add a considerable amount of weight in late winter and spring, before the summer breeding season, and slim down afterwards.

Longevity is 20 years in males and up to 30 years in females. The dental formula is I $^{3}/_{2}$, C $^{1}/_{1}$, PC $^{5}/_{5}$. There is, unique for otariids, a wide, interdental gap or diastema between the 4th and 5th post-canines.

Recognizable geographic forms None.

Can be confused with Steller sea lions are readily separated from northern and Guadalupe fur seals that occur within their range by their massive size, large blocky head and

Two male western Steller sea lions sparring in the Aleutian Islands. PHOTO: D. WEBER

A group of female and juvenile Steller sea lions at sea; Steller sea lions often travel in groups. Kuril Islands, Russia. PHOTO: J.-P. SYLVESTRE

Steller sea lions hauled-out on sea ice in the Karaginsky Gulf, Russia. PHOTO: M. CAMERON/NOAA FISHERIES/AFSC

An adult male Loughlin's Steller sea lion on with a juvenile California sea lion in southern California. They once bred in the Channel Islands, but are now rarely seen south of Monterey Bay. PHOTO: A. SCHULMAN-JANIGER

An adult male Loughlin's Steller sea lion feeding on a large skate in San Francisco Bay. A flock of gulls is drawn to the activity. PHOTO: B. KEENER

Steller sea lions feed on a wide variety of small schooling fish but also take numerous species of larger prey as shown here, clockwise from upper left: octopus, halibut, shark, and salmon. PHOTOS: (CLOCKWISE FROM UPPER LEFT) M. MALLESON, M. NOLAN, V. SHORE, M. NOLAN

muzzle, and pale color. California sea lions are the most likely species to be confused with Steller sea lions. Careful attention to head and muzzle size and shape, overall coloration, and length and width of fore- and hind-flippers permits separation. Smaller Steller sea lions, in the size range of large California sea lions, appear more muscular and powerfully built than the California sea lions, which look more rounded and streamlined. Also, smaller Steller sea lions have little or no sagittal crest development and a nearly flat-topped head, whereas comparably-sized adult and subadult male California sea lions have a moderate to large sagittal crest and more pronounced forehead. The eyes of Steller sea lions also seem smaller and set farther apart, due to the proportionately larger head and wider muzzle.

Distribution Steller sea lions are found from central California (formerly southern California), north to the Aleutian Islands, and west along the Aleutian chain to Kamchatka, and from there south along the Kuril Islands to northern Japan, the Sea of Japan, and Korea. Stellers also occur in the Sea of Okhotsk. From the Aleutians they range north across the Bering Sea to the Bering Strait. Occasional sightings in southern California can now be viewed as vagrants, as the species rarely is seen south of Monterey Bay in central California. Vagrants have also been reported from the coast of China in the Yellow and Bohai seas. Throughout their range they are usually found from the coast to the outer continental shelf and slope. The division between western and eastern DPSs or subspecies (western and Loughlin's) is given as 144°W based on genetic analysis.

Ecology and behavior Steller sea lions are polygynous, and breed in the late spring and summer. Adult males arrive at the rookeries before females, and those that are nine years or older, fight for and often obtain a territory, which they aggressively and vociferously defend. Stellers have deep voices and produce powerful low-frequency rolling roars that can be heard for long distances over the noise of wind and waves. Roaring males often bob their head up and down while vocalizing. This is in contrast to the side-to-side head wave of California sea lion males when they produce their characteristic repetitive bark.

Pups are born from May through July, and females stay continuously ashore with their newborns for the first week to ten days after giving birth. Following this period of attendance, females make foraging excursions, primarily at night for periods of 18–25 hours, followed by time ashore to nurse their pup. Females come into estrus and mate about two weeks after giving birth. Weaning often takes place before the next breeding season, but it is not unusual to see females nursing yearlings, older juveniles, or multiple offspring. Steller sea lions can leave haulouts in large groups. Sightings at sea are most often of groups of 1–12 animals. They aggregate in areas of prey abundance, including near fishing vessels, where

Female (top), and male (bottom) Loughlin's Steller sea lions seen passing under the Golden Gate Bridge. Note the pale color, even when wet, and proportionately long foreflippers on both, and the large head and wide neck of the male. California. PHOTOS: B. KEENER

This juvenile Loughlin's Steller sea lion has had a fisheries encounter and is hooked with a salmon trolling flasher. PHOTO: M. NOLAN

This subadult male Loughlin's Steller sea lion was branded as a pup for re-identification in long-term studies. PHOTO: M. NOLAN

Loughlin's Steller sea lions have recovered so well that they were removed from the US Endangered Species List, while the western subspecies has not recovered and remains endangered. PHOTO: M. NOLAN

they will feed on netted fish and discarded by-catch. They are not considered migratory, and juveniles and subadults make most long distance trips. Adults usually forage and live near their natal colonies and return to these sites to breed. The area used by adult females for foraging in winter is considerably larger than the area used in the summer. Though they can reach depths of 400 m, diving is generally to 200 m or less and dive duration is usually 2 min or less, with both parameters varying by season and age of the animal. Adult females tend to dive deeper in winter than summer. Diving ability of pups and juveniles increases with age, and they routinely dive to depths of around 140 m for periods of 2 min as yearlings. The diving of adult males has not been studied. Predators include killer whales and sharks.

Feeding and prey Steller sea lions feed on many varieties of fish and invertebrates. Much of the information on diet comes from animals living in Alaska, where they feed on walleye pollock, Pacific cod, Atka mackerel, herring, sand lance, several varieties of flatfish, salmon and rockfish, and invertebrates such as squid, octopus, bivalves and gastropods and numerous other species. Adult females with young pups feed extensively at night, switching to foraging at any time after the breeding season. Adult males are known to occasionally kill and consume young northern fur seals, harbor seals, and in one case, a juvenile California sea lion.

Threats and status Steller sea lions have been important to the native people living near them for millennia. Native Alaskans currently take between 150–300 a year for food and other uses. The worldwide population of Steller sea lions is estimated to be 143,000, with different trends in the eastern and western portions of its range. The western population declined by approximately 57% between 1981 to 2011, most severely (-81%) in the large populations from the Gulf of Alaska, west throughout the Aleutian Islands to Russia. Since 2000, the trend has been a slight increase in numbers overall, but the western Steller sea lion DPS remains Endangered under the US Endangered Species Act. Total abundance in the United States and Russia of this subspecies was estimated to be 78,000 in 2011. In contrast, the Loughlin's Steller sea lion or eastern DPS had an estimated abundance of 65,000 in the United States and Canada in 2011, is increasing in numbers, and was taken off the US Endangered Species Act list in 2013. The reasons for the decline and continuing low numbers of western animals are still unclear, and the subject of ongoing research. Factors hypothesized include the direct and indirect effects of large-scale commercial fisheries on key prey species, long-term ecosystem shifts, changes in distribution of prey and availability of prey types, and changes in killer whale predation.

IUCN status Near Threatened overall. Endangered for the western subspecies and Least Concern for the eastern (Loughlin's) subspecies.

California Sea Lion—*Zalophus californianus*

(Lesson, 1828)

Recently-used synonyms *Zalophus californianus californianus.*
Common names En.–California sea lion; Sp.–*lobo marino de California;* Fr.–*lion de mer de Californie.*
Taxonomic information Carnivora, Otariidae. Currently the California, Galapagos and Japanese sea lions are considered to be three separate species. Until recently, most authors classified them as subspecies of *Z. californianus.* At present, no subspecies of the California sea lion have been proposed, but analysis of mitochondrial DNA delineated five genetically distinct geographic populations: Pacific temperate (waters off the US), Pacific subtropical (waters of the Pacific coast of Baja California, Mexico), and the southern, central and northern portions of the Gulf of California, indicating little reproductive exchange between these area.

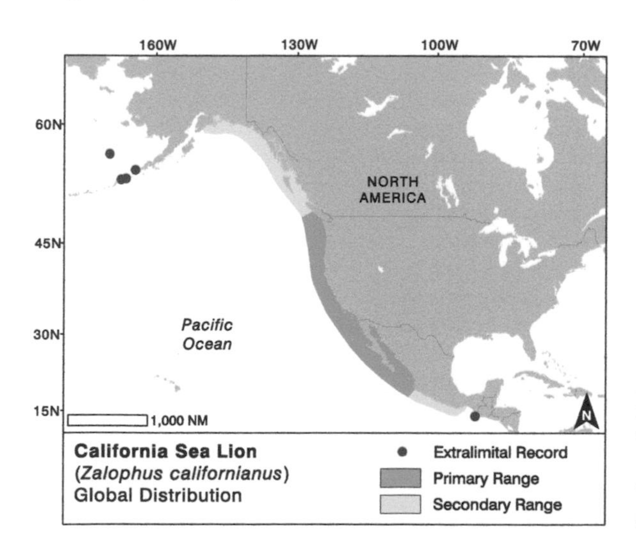

Species characteristics The California sea lion is familiar to people all over the world as the performing seal often seen in oceanaria, zoos, circuses, and commercial advertisements. It is sexually dimorphic, with males achieving 3–4 times the weight of adult females, and 1.2 times their length. Males become very robust in the neck, chest, and shoulders. As males become sexually mature, the sagittal crest enlarges, first appearing as a bump on the crown, then growing to become a prominent ridge, often steep in the front, creating a tall forehead. Adult and subadult male California sea lions bark in often long, repeated sequences. The bark is loud, moderate in pitch, and often delivered while the head is waved from side to side. Females and juveniles do not produce the repetitive bark. Juveniles and subadults of both sexes growl, and when alarmed produce a loud shriek-like bark that is high pitched. The growl of adults is low-frequency and roar-like, and can be explosive, when the animal is angered or startled.

Both sexes have a long and somewhat narrow muzzle that tapers to a blunt nose. In profile, the face of younger animals is dog-like. Adult females and juveniles do not

Adult male, adult female and large pup California sea lions. The male is a pale adult. Mexico. PHOTO: F. R. ELORRIAGA-VERPLANCKEN/CICIMAR-IPN

Two adult male California sea lions sparring in the Gulf of California, with a pale adult male looking on from back left. PHOTO: M. NOLAN

Typical blackish adult male California sea lion at San Miguel Island, California. Note the tall pale colored sagittal crest, and pale muzzle and face. PHOTO: M. A. WEBBER

Pale adult male with a large crest, neck, and pale face mask with a subadult male whose crest is just starting to be conspicuous. Pier 39, San Francisco. PHOTO: B. KEENER

have a sagittal crest, and in profile have a flat head that smoothly transitions into the muzzle with a slight drop. In contrast to adult males, adult females have a long, relatively thin neck and a wider body behind the foreflippers. The leading edge of the foreflippers gradually curves back to the rounded end at the tip. They have a sparse short fur that extends beyond the wrist onto the middle of the dorsal surface of the flipper in a V-pattern that does not reach the rounded tip. The rest of the dorsal surface, and the palms of both foreflippers are covered with a hairless black leathery skin. There is a small opening in the skin at the end of each digit for a claw that is usually reduced to a vestigial nodule, and rarely emerges above the skin. The claw openings are set back from the free edge of the flippers by cartilaginous rods that extend the length of each digit, and expand the size of the flippers. The hindflippers also have cartilaginous rods that extend the length of each toe. The bones of the three central toes terminate at the position of the small nails that emerge through the skin on the dorsal surface, set back from the end of the flipper. The first and fifth toes are longer than the three middle toes, and the first toe, or hallux, is longer and wider than the fifth toe. The hindflippers have short hair covering part of the proximal end of the flipper; the rest of the dorsal surface and the entire sole is covered in black, leathery, hairless skin. The coloration of California sea lions is variable. When dry, the coat of most adult males is dark brown to black, and when wet most appear blackish. Males begin to darken as subadults, and complete this change when they reach physical maturity. Some males do not darken completely or even extensively, and remain various shades of tan to light brown on the sides, belly, and rear quarters with darker areas in the neck and shoulders. Adult females are tan above, and various shades of tan to light brown below. They are usually the same color on the underside of the neck as on the back, and somewhat darker, on the rest of the ventrum and around the base of the flippers. Adult and large subadult males are light-colored on the muzzle, around and above the eyes, on the ear pinnae, and on the sagittal crest. Coloration of juveniles and young subadult males is similar to adult females. Pups are born with a thick

Adult females among a crèche of dark pups, most of which are in their natal pelage. Mexico. PHOTO: F. R. ELORRIAGA-VERPLANCKEN/CICIMAR-IPN

California sea lion pups in transition from their dark natal coats to the lighter tan color of juveniles. PHOTO: F. R. ELORRIAGA-VERPLANCKEN/CICIMAR-IPN

Adult female California sea lion (right) with an older pup that has molted into its juvenile pelage about six months after birth. Mexico. PHOTO: F. R. ELORRIAGA-VERPLANCKEN/CICIMAR-IPN

An adult female California sea lion with a very young pup, possibly before her first foraging trip. San Miguel Island, California PHOTO: M. A. WEBBER

brownish-black natal coat that transitions to a lighter brown, which is shed 4–6 months later and is replaced by a juvenile pelage with adult coloration. California sea lions appear duller and grayer as they get close to the time of the annual molt.

Adult males reach lengths of 2.4 m and weights in excess of 390 kg. Females are 2 m long and 110 kg, on average. Pups are about 80 cm long and 6–9 kg at birth. Males may live to be 19 years old, and females 25 years. The dental formula is I $^{3}/_{2}$, C $^{1}/_{1}$, PC $^{5}/_{5}$ (~79%), PC $^{6}/_{5}$ (~21%). The canine teeth of adult males are larger and thicker than those of females.

Recognizable geographic forms None based on external appearance, however five discrete populations have been identified through genetic analysis.

Can be confused with California sea lions share their range with three other otariids (Steller sea lions, northern fur seal, and the Guadalupe fur seal) and may potentially overlap with vagrants of an additional three species (South American sea lion—a hybrid has been reported—and Galapagos sea lion and fur seal). See the Galapagos fur seal and sea lion accounts for discriminating Galapagos fur seals from the Galapagos sea lion. Galapagos sea lions would be virtually indistinguishable externally from California sea lions, and genetic analysis or measurements of skull features would be needed to separate these two species. Also, see the Galapagos sea lion and South American sea lion accounts for a similar comparison.

Juvenile, subadult, and smaller adult female Steller sea lions fall within the size range of California sea lions. Careful attention to head and muzzle size and shape, overall coloration, and length and width of fore- and hindflippers permits differentiation. Smaller Steller sea lions in the size range of large California sea lions will look like they are more muscular and powerfully built than similar-sized California sea lions, which look more rounded and streamlined with a smaller, narrower and more pointed muzzle. Also, smaller Steller sea lions have no sagittal crest development and a nearly flat-topped head, whereas comparably-sized adult and subadult California sea lion males have a moderate to large sagittal crest and more pronounced forehead. Steller sea lion

An adult female California sea lion. Los Islotes, Gulf of California, Mexico. PHOTO: M. NOLAN

Group of juvenile and subadult California sea lions hauled out at Los Islotes, Gulf of California, Mexico. PHOTO: M. NOLAN

California sea lions will leap from rocks and dive into waves to ride receding waters away from shore. Here an adult female and her older pup head back to the water. Los Islotes, Gulf of California, Mexico. PHOTO: M. NOLAN

eyes appear smaller and set farther apart, due to the proportionately larger head and wider muzzle, and Steller sea lions have much longer off-white vibrissae.

Both northern and Guadalupe fur seals have thick pelage and look shaggier than the more sleek California sea lions. Males of both fur seals are dark in color, often with light tips to the long hairs that impart a grayish cast to the dark color. Adult females, juveniles and subadults are multicolored with dark gray or brown dorsal, tan to buff ventral coloration, and a pale band across the chest. Both fur seals have a more pointed face and proportionately longer ears that stand out farther from the head when they are in the water or otherwise wet. Adults of both fur seals have conspicuous, long, pale vibrissae. Northern fur seals have a very short pointed muzzle, very long hindflippers with long cartilaginous extensions all of equal length and width. The fur on the dorsal surface of the foreflippers stops abruptly at the bend, just below the wrist, and the top of the flippers have a smooth "clean shaven look." This is in contrast to the California sea lion, which has fur on the top of the flipper extending down in a "V" beyond the bend.

Guadalupe fur seals have a long, pointed muzzle. The somewhat bulbous nose contributes to a slight upturned end of the muzzle in silhouette. The hindflippers are longer than those on California sea lions, but lack the extreme length of northern fur seal hindflippers, but have the equal length and width toes of the northern fur seal in contrast to the unequal length and width toes of the California sea lion. The foreflippers have short hair extending in a "V" beyond the bend in the flipper on the dorsal surface.

Distribution The California sea lion occurs in the eastern North Pacific from the Tres Marías Islands off Puerto Vallarta, Mexico, north throughout the Gulf of California, and around the end of the Baja California Peninsula north to the Gulf of Alaska. Vagrants have been reported from the Bering Sea in the north to Acapulco in the south. Most rookeries are south of Point Conception, Southern California. Many islands free of predators and sources of human disturbance throughout their range are used as haulouts. The California sea lion population is currently expanding, and extending its breeding range northward. Females, which were only very rarely found north of Point Conception in the early 1980s, are now routinely found in northern California, where former breeding sites have been reoccupied. Adult and subadult males are now found in small numbers in Alaska, in the northern part of the species' range. To the south, extralimital records of California sea lion have been reported from near Manzanillo, Colima, Mexico, along the coast of Oaxaca (15°39'N), near the Guatemala-Mexico border (14°42'N) and a subadult male (that could have been a Galapagos or California sea lion) was seen at the Cocos Islands (5°N). The discovery of three Galapagos sea lions in Chiapas, Mexico (15°41'N), verified by the analysis of the skull of one individual, suggests that all "California

Part of a large rookery of California sea lions on San Miguel Island, off California, on a hot day during the breeding season. PHOTO: M. A. WEBBER

California sea lions can spend a good deal of time in the water near rookeries and haulouts socializing in the shallows. Los Islotes, Gulf of California, Mexico. PHOTO: M. NOLAN

California sea lions will "porpoise" when traveling rapidly at sea. Los Islotes, Gulf of California, Mexico. PHOTO: M. NOLAN

California sea lions raft and sleep at sea with both their hind- and fore-flippers held up, but they do not use the "jug-handle" position of fur seals. Monterey Bay, California. PHOTO: I. SZCZEPANIAK

An adult female followed by a subadult male California sea lion pass beneath the Golden Gate Bridge in San Francisco. Color and head shape differentiate these similar-sized individuals. PHOTO: B. KEENER

California sea lions can engage in lively play sessions involving high-speed maneuvers and acrobatic leaps. Los Islotes, Gulf of California, Mexico. PHOTO: M. NOLAN

Two dark adult male California sea lions hauled out with two Loughlin's Steller sea lions. British Columbia. PHOTO: V. SHORE

Adult and subadult male California sea lions hauled out with a group of molting elephant seals that tolerate annoyances like this from sea lions. PHOTO: F. R. ELORRIAGA-VERPLANCKEN/CICIMAR-IPN

A young Guadalupe fur seal standing next to a sleeping juvenile California sea lion hauled out on San Pedro Mártir Island, Gulf of California, Mexico. PHOTO: M. NOLAN

A dark phase harbor seal with juvenile California sea lions at San Benitos Island, Mexico. California sea lions haul out with all five other pinnipeds that share their range. PHOTO: F. R. ELORRIAGA-VERPLANCKEN/CICIMAR-IPN

Where they see people regularly, California sea lions can become acclimated and interactive, such as with this snorkler. These interactions can pose risks to both species. Los Islotes, Gulf of California, Mexico. PHOTO: M. NOLAN

Part of a large, densely packed group of several hundred California sea lions feeding in association with humpback whales on anchovy shoals in Monterey Bay, California. The diving by the sea lions was tightly coordinated among themselves and with the whales. PHOTO: I SZCZEPANIAK

sea lions" found south of their normal range in Mexico and Central America should be confirmed to species by genetic testing or examination of skeletal remains.

California sea lions are usually found in waters over the continental shelf and slope, however they occupy several landfalls far offshore in deep oceanic areas, such as Guadalupe Island and Alijos Rocks off Baja California. Large numbers of adult and subadult males and juveniles undertake a post-breeding season migration north from the major rookeries in southern and Baja California and winter from central California to Washington State and British Columbia. Smaller numbers reach from southeast Alaska to the Alaska Peninsula. California sea lions occupy the Gulf of California. Those residing there do not appear to make long migrations and the genetically verified stock structure in the Gulf suggests minimal movement away from core areas. Throughout their range, California sea lions frequent coastal areas including bays, harbors, and river mouths, and regularly haul out on buoys, jetties, boat docks, and even on anchored boats. They traditionally use certain coastal headlands and cliff bases that limit access and approach by terrestrial predators.

Ecology and behavior Pupping and breeding take place from May through July. Males are highly polygynous and hold territories both on land and in shallow water near shore for periods up to 45 days. Females stay ashore with their newborn pups for about 7 days before they depart for the first of many foraging trips that usually last 2–3 days and are followed by attendance of the pup at the rookery for 1–2 days. Most pups are weaned at 10 months, but before this, they can start making foraging trips to sea with their mothers. Some pups continue to receive care as yearlings, and even as 2-year-olds. Estrus occurs around 27 days after giving birth, one of the longer intervals for an otariid. California sea lion females can gather in "milling groups" near bulls, where they roll in the surf and sand, mounting each other and even nearby bulls. Often the females from the milling group disperse after one or more of them copulate with a bull.

This species is not known for deep or long dives. The diving pattern of lactating adult females is consistent with a number of other otariid species. The deepest dive recorded to date was to approximately 274 m and the longest dive lasted just under 10 min. Typical feeding dives are shallower than 80 m, and last less than 3 min. Lactating adult females are active for most of the time they are at sea and feeding bouts occur during the day and at night, with peaks of activity at dawn and dusk. Feeding dives occur in bouts suggesting sea lions are frequently exploiting patches of prey. California sea lions haul out and travel at sea with Steller sea lions, where the two species co-occur. They will also haul out near Guadalupe and northern fur seals and elephant seals, and in a few locations, harbor seals. Predators of California sea lions include killer whales, sharks, coyotes and feral dogs.

A subadult California sea lion with a severe injury from a monofilament gill net that has cut it deeply; in almost all cases these entanglements will kill the animal as its grows. Los Islotes, Gulf of California. PHOTO. M. NOLAN

Feeding and prey California sea lions are opportunistic and feed on a wide variety of prey, often taking what is abundant locally or seasonally in the areas they occupy. Principal prey taken in the Pacific includes: Pacific whiting, market squid, red octopus, jack and Pacific mackerel, blacksmith, juveniles of various species of rockfish, herring, northern anchovy, and salmon. Sea lions in the Gulf of California have northern anchovy, Pacific whiting, and rockfish as prey in common with animals in the Pacific, and also take various species of midshipmen, myctophids, and bass, as well as sardines, cutlassfish, alopus, and cusk eels. Because of their boldness and taste for commercially-important fish species, such as salmon and rockfish that are easily taken from fishing lines, they are considered a nuisance by many sport and commercial fishermen. California sea lions will also ascend rivers long distances following spawning runs of anadromous fish, and take advantage of man-made structures, such as canal locks and fish ladders that concentrate prey.

Recently, unusually large groups of up to several hundred juvenile to subadult, and possibly adult female California sea lions have been observed forming large tightly packed groups foraging on northern anchovy

California sea lions interact with humans through our fisheries, recreation and their use of our structures. Above, sea lons fill an abandoned dock to capacity in Monterey Bay; routinely use buoys; and an adult male has learned to seek handouts from fishermen by climbing onto the swim-steps of returning boats at Cabo San Lucas, Mexico. PHOTOS: (LEFT) I. SZCZEPANIAK, (TOP RIGHT) B. KEENER, (LOWER RIGHT) B. PRATT

(assumed) shoals in association with lunge-feeding humpback whales in Monterey Bay, California. The sea lions dive in a coordinated effort before the feeding whales and surface before them, usually right over the area where the whales surface moments later. This behavior is often repeated over periods of several hours, and in the same areas for many months with the pattern of comings and goings of individual and small groups of sea lions from the large aggregations not being known.

Threats and status California sea lions are abundant and the population is growing. They were historically important to native people living in coastal areas and on islands. Native Americans middens in Southern California and on the Channel Islands contain large numbers of California sea lion and other pinniped bones document the importance of marine mammals in subsistence cultures prior to the arrival of Europeans. In the 19th and early 20th centuries, California sea lions were harvested for a variety of products, and hunted for bounties to such an extent that the population may have been reduced to as few as 1,500 by the end of this period. Protection that began in the mid-20th century, and was solidified with the Marine Mammal Protection Act of 1972 in the US (and under similar measures in Mexico) provided the impetus for recovery of the population, which now numbers an estimated 350,000 animals in the US and Mexico. Of these, the majority are part of the population off the US, with 75,000 to 87,000 off the west coast of Baja California, and approximately 30,000 in the Gulf of California spread between the three recently identified genetically distinct populations. Of note, some of the populations in the Gulf of California have declined 20–35% over the past 15–20 years. California sea lion populations may be at, or approaching, carrying capacity. Episodically, large numbers of starving pup and juvenile California sea lions come ashore on beaches in California in years associated with periods of low marine productivity, mirroring in a less dramatic way the large population swings of the similar Galapagos sea lion during El Niño events. California sea lion mortality occurs in conflicts with fisheries, by poaching, and through entanglement in marine debris. Sea lions also accumulate pollutants through the food chain, and large amounts of DDT and PCBs discharged in the past continue to accumulate in coastal marine food chains, as evidenced by toxins in their tissues and organs. Large amounts of agricultural and urban runoff and waste continue to be discharged into coastal marine habitats annually from numerous sources, and this pollution is having affects on sea lion immune systems and overall health. They also die from periodic outbreaks of planktonic organisms that cause paralytic shellfish poisoning.

IUCN status Least Concern.

Galapagos Sea Lion—*Zalophus wollebaeki*

Sivertsen, 1953

Recently-used synonyms *Zalophus californianus wollebaeki.*
Common names En.–Galapagos sea lion; Sp.–*lobo marino de Galápagos;* Fr.–*lion de mer de Galapagos.*
Taxonomic information Carnivora, Otariidae. Currently the Galapagos, California, and Japanese sea lions are considered to be three separate species. Until recently, most authors classified them as subspecies of *Z. californianus.*
Species characteristics Galapagos sea lions are similar to California sea lions, but show differences in size, behavior, and skull characteristics. Galapagos sea lions are sexually dimorphic. Adult males are larger than females and display several significant secondary sexual characteristics. The degree of sexual dimorphism may be less than in California sea lions.

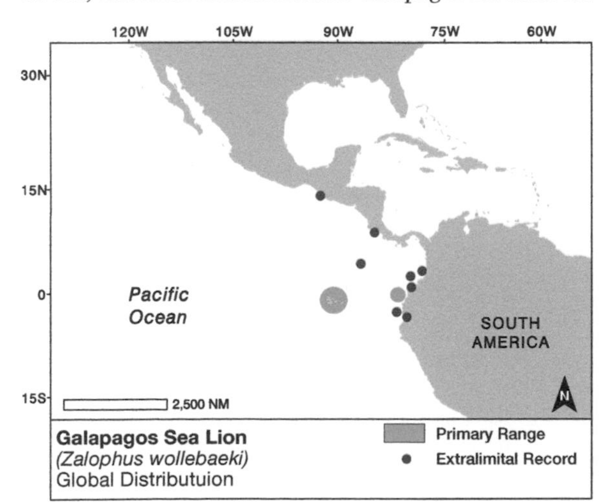

Adult males are robust in the neck, chest, and shoulders and are proportionately smaller in the abdomen. As males mature sexually the enlarging sagittal crest becomes evident as a bump on the crown. The crest grows until the male reaches physical maturity, at which time it forms a ridge above and behind the eyes, and creates a somewhat steep forehead. Galapagos sea lion males are said to lack the pale pelage coloration on top of the sagittal crest common in California sea lions, although this feature appears to be present in at least one published photograph. Also, the skull of adult male Galapagos sea lions has a 20–25% smaller sagittal crest, a shorter muzzle, is about 10% shorter overall and is narrower than the skull of male California sea lions. As in all otariids, the canine teeth of adult males are larger and thicker than those of females.

Adult and subadult male Galapagos sea lions bark in often long repeated sequences. The bark is loud, rapidly repeated, and distinctive. Females and juveniles do not produce the repetitive bark. Juveniles, subadults and adults of both sexes will also growl. Adult females and juveniles do not have a sagittal crest. When viewed in profile, juveniles have a nearly flat head with little or no forehead. Adult females have a slight forehead formed

Small Galapagos sea lion male with a group of females and their young. Note the large juvenile nursing in the foreground. Galápagos. PHOTO: J. GOULD

A typical view down a beach used by Galapagos sea lions, where a female is nursing a juvenile. PHOTO: M. NOLAN

by a gentle slope from the crown to the muzzle. In contrast to adult males, adult females have a long relatively thin neck and a wide torso. Both sexes have a long and somewhat narrow muzzle that tapers to a slightly pointed nose. In profile, the face of younger animals is dog-like. The foreflippers have a sparse short fur that extends beyond the wrist onto the middle of the dorsal surface of the flipper in a V-pattern and does not reach the rounded tip. The rest of the dorsal surface, and the palms of both foreflippers are covered with a hairless black leathery skin. The first digit is the longest, widest and thickest, and curves posteriorly, giving the flipper a swept back look. Digits 2–4 are successively shorter. There is a small opening in the skin at the end of each digit for a claw that is usually reduced to a vestigial nodule, and rarely emerges above the skin. The claw openings are set back from the free edge of the flippers by cartilaginous rods that extend the length of each digit, and expand the size of the flippers. The hindflippers also have cartilaginous rods that extend the length of each toe. The bones of the three central toes terminate at the position of the small nails that emerge through the skin on the dorsal surface, set back from the end of the flipper. The first and fifth toes are longer than the three middle toes, and the first toe, or hallux, is longer and wider than the fifth toe. The hindflippers have short hair covering part of the proximal end of the flipper, and the rest of the dorsal surface, and the entire sole is covered in black leathery hairless skin.

The color of Galapagos sea lions is highly variable. When dry, the coat of adult males ranges from grayish and golden brown to the more common dark brown, and most bulls appear blackish or very dark when wet. Darkening begins when males are subadults and is generally complete when a bull reaches physical maturity. Adult males can have light gray coloration on their backs. Adult females, juveniles and young subadult males are pale colored above, and can be many shades of tan to light brown. There are often light colored areas on the muzzle, and around and above the eyes in both sexes. The sparse short fur covering a portion of the tops of the flippers can be the same color or darker than the color of the body. Pups are born with a longer brownish-black natal coat that fades to pale brown by three to five months. Pups go through their first molt at around five months and emerge with the pelage of adult females and juveniles.

Detailed information is lacking on lengths and weights achieved by Galapagos sea lions, but they are somewhat smaller than California sea lions. Weight estimates for adult males are up to 250 kg. Four adult females still caring for pups weighed from 50–100 kg. Pups are about 6 kg at birth, and weigh about 25 kg by the time they are weaned. The dental formula is I $^{3}/_{2}$, C $^{1}/_{1}$, PC $^{6}/_{5}$, but PC $^{5}/_{5}$ occurs in approximately 25% of animals.

Recognizable geographic forms None.

Can be confused with Galapagos sea lions share the archipelago with Galapagos fur seals. As vagrants, they may show up in the range of South American sea lions and South American fur seals, and there is a record of the former from the Galápagos. Galapagos and South American fur seals have thick pelage and look shaggier, especially when wet, than Galapagos sea lions. Both fur seals are dark gray to brown and darker than the pale

A dry adult male Galapagos sea lion with a moderate sagittal crest, pale muzzle, and paler-colored body. PHOTO: G. MEYER

A wet adult male Galapagos sea lion with a moderate sagittal crest. PHOTO: G. MEYER

Adult female with a pup peeking from behind her. Note the flat-topped head, and slimmer muzzle. PHOTO: J. GOULD

Adult female with a newborn pup in its natal coat, and a juvenile, possibly 1–2 years old, in tow. PHOTO: M. NOLAN

Adult female nursing a pup that is probably 6–12 months old, which has molted out of its natal coat. PHOTO: M. NOLAN

Adult female with a still-dependent 1–2 year old juvenile. Española Island. PHOTO: M. ELLIS

A young pup nursing. Note the contrast between the dark face and paler gray-brown color of the rest of the body. PHOTO: M. NOLAN

A large subadult, or small adult male, whose coat has darkened considerably compared to the juvenile color. PHOTO: J. GOULD

Two subadult males sparring by pushing and shoving each other. Note the low sagittal crest on each, and the dark color of the animal in the background. PHOTO: M. NOLAN

A wet, healthy juvenile in excellent condition has just come ashore. PHOTO: J. GOULD

Three pups in natal pelage playing in an intertidal area, no doubt waiting for their mothers to return from a foraging trip. PHOTO: J. GOULD

tan similar-sized juvenile and female sea lions, with the exception of a fully-grown male Galapagos sea lions. Both fur seals have a more pointed muzzle and proportionately larger eyes and longer ears that stand out farther from the head when they are wet. Adults of both fur seals have long pale conspicuous vibrissae. Galapagos fur seals are the smallest otariids, reaching only 1.5–1.2 m for males and females, respectively, and are stockier with a shorter neck and body than Galapagos sea lions. South American sea lions have large blocky heads with a short, thick, blunt-ended slightly upturned looking muzzle. To add to that, adult male South American sea lions have huge fore quarters with a thick mane and a very large lower jaw.

Distribution Galapagos sea lions are found throughout the Galápagos Archipelago on all the major islands and on many smaller islands and rocks. A colony was established in 1986 at de la Plata island, just offshore of mainland Ecuador, and vagrants can be seen from the Ecuadorian coast north to Gorgona island in Colombia. There is also a record from Cocos Island approximately 500 km southwest of Costa Rica. Identification of this animal as a Galapagos or California sea lion was never confirmed, and it was assumed to be Galapagos sea lion based on location and proximity to the Galápagos. Additionally, fishermen in Costa Rica occasionally report seeing sea lions, which could be either California of Galapagos sea lions.

Recently, a record of Galapagos sea lions was published from Chiapas, Mexico (15°41'N), well within the range of vagrant California sea lions from the southern limit of their range in Nayarit, Mexico and around the Tres Marías Islands. The sighting was of three sea lions, including at least one adult male and a female. Identification of any "California-like" sea lions in southern Mexico and Central America should be considered provisional unless supported by genetic samples or examination and measurement of cranial features because of the overlap of sizes, features and coloration between both sexes of Galapagos and California sea lions.

Ecology and behavior Galapagos sea lions are non-migratory. They are unafraid of humans when ashore and will investigate and climb on backpacks and other things people leave lying around. Haulout sites can be on steep rocky shorelines, ledges and offshore stacks, but rookeries are mostly on gently sloping, sandy and rocky beaches. To avoid overheating, sea lions will use shade from vegetation, rocks, and cliffs, and wade into tidal and drainage pools or move into the ocean, as needed during the heat of the day.

Pupping and breeding take place across an extended period from May through January. Because of the protracted breeding season, non-migratory nature of this species, and the extended care females provide to the pups, there are dependent pups on the rookeries year-round. Females usually wean pups in 11–12 months, but some continue to suckle yearlings along with newborn pups. Pups are attended continuously for 6–7 days, after which the female goes to sea to feed and begins a cycle of daily, diurnal foraging trips that last an average of 12 hours. Pups will enter the water and begin to develop swimming skills 1–2 weeks after birth. Females return at night to nurse their pup, departing again the next morning. Females and pups recognize each other and reunite based on calls and scent. Galapagos sea lion females feed during the day, in contrast to Galapagos fur seals, which primarily feed at night. In addition to foraging niche separation, female sea lions reduce their thermoregulatory challenges by being at sea during the heat of the day. Galapagos sea lions are polygynous; males hold territories both on land and in shallow water near shore that they aggressively and vociferously defend. Males spend extended time on territories because of the protracted breeding period. Shark predation is evident from animals seen with injuries and scars from attacks, and killer whales are presumed to be another predator.

Feeding and prey Little information exists on Galapagos sea lion prey, however their diet may vary throughout the archipelago; they are known to eat sardines, myctophids (lanternfish), and squid. Galapagos sea lions have been seen smashing octopus on the surface of the water, presumably to stun or break them up to facilitate swallowing. Foraging dives by lactating females occur predominantly during the day, and are only to relatively shallow depths. This usually precludes Galapagos sea lions from foraging on vertically-migrating species, such as myctophids, midshipmen, and other deeper living prey routinely taken by California sea lions and Galapagos fur seals. However, during El Niño events prey includes green-eyes and myctophids, suggesting a change in foraging strategy during periods of environmental flux and stress.

A group of juvenile Galapagos sea lions playing in the shallows near shore provides a glimpse of their maneuverability, grace and speed in the water. PHOTO: M. NOLAN

A Galapagos sea lion waits patiently for a hand-out at a fish cleaning station. The lack of fear of humans in this species is remarkable, but can also be of concern because of the potential for negative interactions and disease transmission. PHOTO: M. NOLAN

Threats and status The majority of the Galapagos sea lion population lives in the archipelago, which is an Ecuadorian National Park surrounded by a marine resources reserve. Tourism occurs on a large scale but is strictly controlled to protect wildlife. The population fluctuates between 20,000 and 50,000 animals. Die-offs and cessation of reproduction during El Niño events, when marine productivity collapses, have caused episodes of population decline. Irruptions of a sea lion poxvirus have occurred during El Niño events, adding to the stress on individuals from starvation. Feral and uncontrolled dogs have been reported to kill sea lion pups, and could transmit diseases to the population.

IUCN status Endangered.

South American Sea Lion—*Otaria flavescens*

(Shaw, 1800)

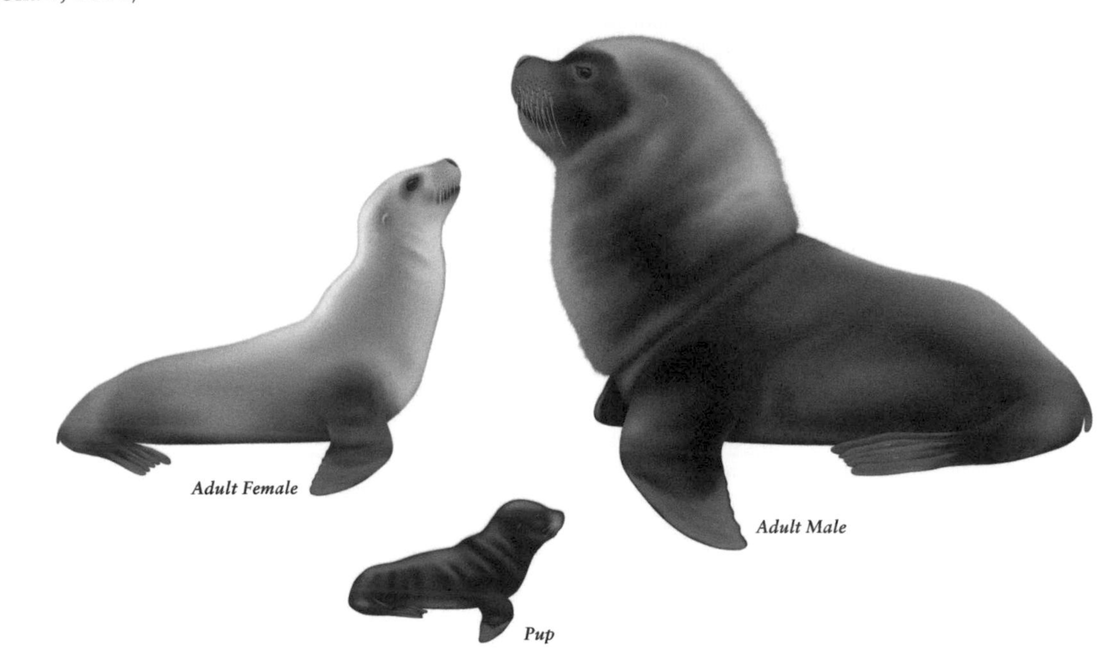

Recently-used synonyms *Otaria byronia.*
Common names En.–South American sea lion, or southern sea lion; Sp.–*lobo marino común;* Fr.–*lion de mer d'Amérique du sud.*
Taxonomic information Carnivora, Otariidae.
Species characteristics South American sea lions are stocky, heavy-bodied otariids that exhibit strong, sexual dimorphism. In both sexes the muzzle is wide, short, and blunt. The nose is large and slightly upturned in females, and even more so in adult males. The lower jaw is wide and deep in both sexes. In adult males it juts forward, and is also accentuated by the longer hair of the mane on the chin. The ear pinnae are small and lie close to the side of the head; they are inconspicuous, particularly in adult males.

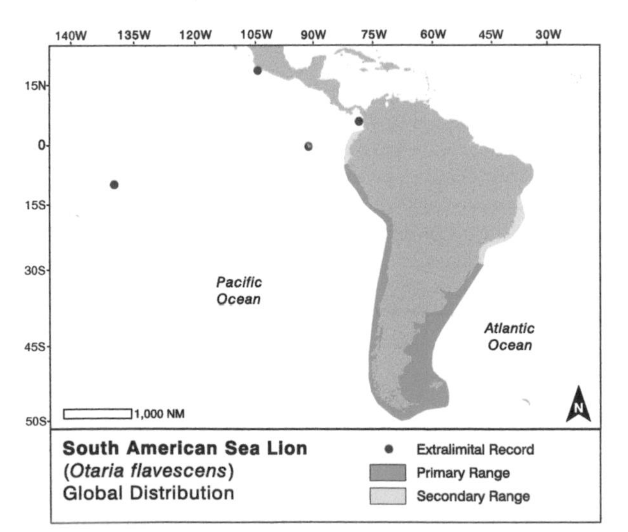

Adult males are unmistakable. They have a mane of long, coarse, erectile guard hairs, extending from forehead and chin to shoulders and mid-chest. The neck, head, jaws, and canine teeth are much larger than those of females. The large head and lower jaw, great bulk of the neck and shoulders, and thick mane make the front of the body appear disproportionately large and wide, despite the fact that the hindquarters are also large. Adult females and subadults of both sexes do not have a mane. Their pelage is predominantly yellowish-gold to tan. They are not all uniformly colored, but can be patterned with areas of slightly different hues. Most males darken with age, becoming brownish-orange to dark brown, although the mane, when it grows-in, and under parts frequently remain lighter. Males can have a darker face, giving them a slightly masked appearance. Some males remain the pale color of females and subadults throughout their lives. Pups are born black above and paler below. Pups undergo their first molt approximately one month after birth, becoming dark brown. This color fades during the rest of the first year to a pale tan to light brown, with paler areas in the face.

A group of South American sea lions with a bull. Note the adult female on the left with a young pup in natal pelage interacting with a juvenile. Beagle Channel. PHOTO: M. JØRGENSEN

A small adult or large subadult male vocalizing as he approaches an adult female. PHOTO: M. NOLAN

Part of a rookery in Peru with a large male at the waterline. PHOTO: S. PRORAK/DREAMSTIME

A bull approaching a female, possibly to investigate her readiness to mate. PHOTO: M. NOLAN

A female nurses a 1–2 year old juvenile in the Beagle Channel. Note their pale coloration. PHOTO: M. NOLAN

Two adult females interacting. Note their overall pale color with some light and dark variation, but no indication of the coat becoming darker. PHOTO: M. NOLAN

A juvenile with a thick head and neck. PHOTO: M.NOLAN

A profile view of a molting subadult male in the Falkland Islands. PHOTO: S. EBBERT

A large subadult male without a mane. PHOTO BY: M. NOLAN

A small adult male with a partial mane that is a rich brown color. Note the upturned nose and robust chin that characterize adult males of this species. PHOTO: D. PYLYPENKO

This leucistic adult male has normally pigmented skin on his flippers and a bleach mark on his back. Researchers have seen him for many years between Mar del Plata and Puerto Quequen, Argentina. PHOTO: J. BASTIDA

The foreflippers have a sparse short fur that extends beyond the wrist onto the middle of the dorsal surface of the flipper in a V-pattern that does not reach the rounded tip. The rest of the dorsal surface and the palms of both foreflippers are covered with a hairless black leathery skin. The first digit is the longest, widest and thickest, and curves posteriorly, giving the flipper a swept back look. Digits 2–4 are successively shorter. There is a small opening in the skin at the end of each digit for a claw that is usually reduced to a vestigial nodule, and rarely emerges above the skin. The claw openings are set back from the free edge of the flippers by cartilaginous rods that extend the length of each digit, and expand the size of the flippers. The hindflippers also have cartilaginous rods that extend the length of each toe. The bones of the three central toes terminate at the position of the small nails that emerge through the skin on the dorsal surface, set back from the end of the flipper. The first and fifth toes are longer than the three middle toes, and the first toe, or hallux, is longer and wider than the fifth toe. The hindflippers have short hair covering part of the proximal end of the flipper, and the rest of the dorsal surface, and the entire sole is covered in black leathery hairless skin.

Adult males are up to 2.6 m long and weigh 300–350 kg. Females are up to 2 m and weigh 144 kg. Pups are 11–15 kg and 75–85 cm at birth. The dental formula is I 3/2, C 1/1, PC 6/5.

Recognizable geographic forms None.

Can be confused with South American sea lions share most of their range with South American fur seals. Wandering South American sea lions have been recorded at the Galapagos, and could also occur within the range of the Juan Fernandez fur seal, or vice versa. Also, Antarctic and subantarctic fur seals occur as vagrants on the coast of South America.

South American sea lions are heavy-bodied animals with proportionately large heads and flippers, and are stockier and more muscular looking than similar size fur seals of any species. All fur seals have longer fur over the entire body and look shaggy when wet. The muzzle of all fur seals is thinner and tapers to a pointed end, and the proportionately large eyes and long ear pinnae are more conspicuous. Fur seals are browns and grays, and darker than most South American sea lions of similar size, and have proportionately smaller flippers. One instance of a probable hybrid *Otaria byronia* X *Zalophus* sp. from an unknown location has been reported from a museum collection. The adult skull has a sagittal crest, as in *Zalophus*, and a wide short upper jaw and nasals, as in *Otaria*.

Galapagos sea lions are smaller overall and slimmer in build, with a much smaller head, proportionately longer and narrower muzzle, smaller lower jaw, and a smaller nose that is not turned up at the end. Coloration in smaller animals is similar, but bulls of each species are unmistakable in color and secondary sexual characteristics.

Distribution South American sea lions are widely-distributed, occurring more or less continuously from northern Peru south to Cape Horn, and north up the east coast of the continent to southern Brazil, including the Falkland Islands. They breed as far north as Isla Lobos de Tierra, Peru (6°S), and range in small numbers to Ecuador (2°S). Primarily a coastal species, they are found in waters over the continental shelf and slope, occurring less frequently in deeper waters. South American sea lions venture into fresh water, around tidewater glaciers, and up rivers. Vagrants have been found near Bahia, Brazil (13°S), the Galápagos Archipelago, and in one extreme case, Colima, Mexico (19°N).

Ecology and behavior South American sea lions are considered non-migratory, although some may wander long distances away from rookeries during the non-breeding season, and southerly locations such as the Falkland Islands are largely abandoned during the winter. However, most rookeries are continuously occupied by at least some animals, and the species has been described as sedentary.

South American sea lions are a highly polygynous species, with various strategies employed by males during the breeding season that are driven by substrate and terrain at the rookery, and thermoregulatory requirements imposed by weather conditions at the site. Adult males tend to establish territories through vocalizing, posturing, and fighting, when rookeries provide shade, have tidal pools that can be used for cooling, or funnel interior areas through narrow beaches between rocks or ledges to the sea. At more homogeneous locations with long shorelines, male strategy changes to focus on identifying, defending, and controlling individual cows in estrus, wherever they are found. Bulls actively and aggressively work to keep these cows close by grabbing, dragging, and throwing them inland, away from the shoreline.

A South American sea lion with a dusky dolphin. Several species of otariids follow and forage with small cetaceans, presumably taking advantage of the cetacean's ability to find food with biosonar. Peninsula Valdes, Argentina. PHOTO: P. FAIFERMAN

A lone South American sea lion off of Peninsula Valdes, Argentina. PHOTO: P. FAIFERMAN

A group of South American sea lions at sea, surfacing synchronously. PHOTO: H. LAWSON

Dramatic images of a male South American sea lion attacking and killing a young southern elephant seal. Sea Lion Island, Falkland Islands. PHOTO: M. JØRGENSEN

The start of the breeding season varies somewhat by location and latitude, with pups being born slightly earlier at more southern rookeries. At most sites, both sexes arrive in mid-December with peak numbers of males ashore just after mid-January and peak numbers of females ashore from mid- to late January. Females give birth 2–3 days after arrival, and pups are born from mid-December to early February with a peak in mid-January, coinciding with the timing of peak numbers of females ashore. Estrus occurs 6 days after parturition, and females make their first foraging trip 2–3 days later. From this point on, a cycle of foraging and pup attendance starts and lasts until pups are weaned at 8–10 months old. As is the case for many sea lions, it is not unusual for females to continue to care for yearlings. Pups gather in large pods on the rookeries while waiting for their mothers to return from 1–4 day long foraging trips. Females usually stay ashore for 2 days between trips.

At sea, South American sea lions are commonly seen rafting alone or in groups. They have been reported in association with feeding cetaceans and seabirds. The mean depth of lactating female foraging dives is about 61 m and the mean duration is just over 3 min. The maximum depth recorded for a dive is 175 m and duration 7.7 min. Two tagged adult males foraged on the continental shelf prior to the onset of breeding, making 5–6 day trips that covered an average of 600 km before they returned to land to haul-out. Lactating females routinely dive between 19–62 m, on dives lasting 2–7 min. The maximum depth of dive for the species is 250 m.

Predators include killer whales, sharks and leopard seals, and possibly the puma. At Peninsula Valdes, Argentina, killer whales are known to surf in on waves, partially beaching themselves while grabbing young sea lions off the shoreline. Vampire bats are known to drink blood from the skin of the flippers of sleeping animals.

Feeding and prey South American sea lions are opportunistic feeders, taking a wide variety of prey that varies by location. Their diet includes many species of benthic and pelagic fishes. Important prey species include South Pacific hake, herring, elephantfish, Peruvian anchovy, grenadier, South American pilchard, cusk-eels, and butterfish. Invertebrates taken include lobster krill, squid, octopus, jellyfish, and marine snails.

A small percentage of adult male South American sea lions regularly prey on South American fur seals. When adult male sea lions hunt for fur seals, they hunt alone and focus their attacks on pups and juveniles. About 17% of attacks are successful, but success varies widely between individual males. Subadult males also attack fur seals, but tend to abduct fur seals to serve as female sea lion substitutes, herding them and attempting to mate with them, usually killing them in the process. Female and juvenile sea lions have not been recorded to hunt fur seals. Sea lions have been observed killing young southern elephant seals in the Falkland Islands. They are also known to take several species of penguins, but the importance of penguins in the diet is unknown.

Threats and status The total population is estimated to be 200,000–300,000. South American sea lions were hunted by native people for thousands of years, and were taken by Europeans as early as the 16th century for food, oil and hides. Large commercial harvests, in several countries, drastically reduced sea lion numbers in the last several hundred years. Most populations are currently recovering from past harvests. Sea lions are taken incidentally or killed intentionally in a number of fisheries and fish farming operations throughout their range. Intensive trawl fishing in the coastal waters of southern South America has been implicated in the severe decline of sea lions in the Falkland Islands, where the population has fallen from 30,000 in the 1960s to approximately 15,000 in the 1980s, and possibly to as low as 3,000 in the 1990s before rebounding to about 6,000. In mainland Argentina there are about 100,000 South American sea lions, 12,000 in Uruguay, 90,000–100,000 in Chile, and 60,000 in Peru.

IUCN status Least Concern.

Australian Sea Lion—*Neophoca cinerea*

(Peron, 1816)

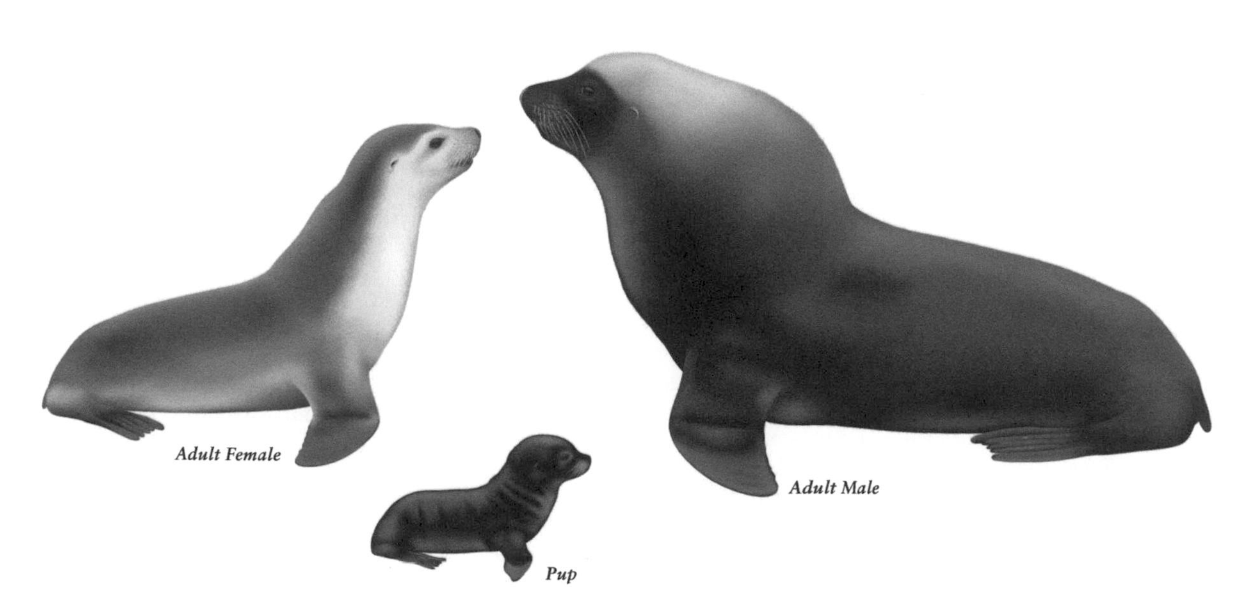

Recently-used synonyms None.
Common names En.–Australian sea lion; Sp.–*otaria de Australia;* Fr.–*lion de mer d'Australie.*
Taxonomic information Carnivora, Otariidae.
Species characteristics Australian sea lions are sexually dimorphic, with adult males reaching 1.25 times the length and 2.5–3.5 times the weight of adult females. The head is proportionately wide and long in both sexes, and is very large in adult males. The muzzle tapers to a blunt, somewhat rounded end. The eyes are widely set apart. The slight forehead is more prominent in adult males due to an enlarged, but posteriorly-positioned, sagittal crest that elevates the crown. The ear pinnae are short and lie close to the head. Mystacial vibrissae are moderate in length, the longest ones reaching 18 cm. Two vibrissae are located above and behind each eye. In adult males, the neck and shoulders are greatly enlarged, and the canine teeth are thicker at the base and up to 2.5 times longer than those of adult females.

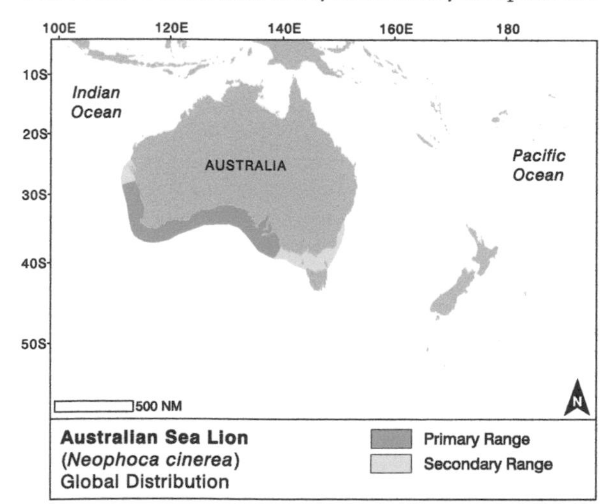

The foreflippers have a sparse short fur that extends beyond the wrist onto the middle of the dorsal surface of the flipper in a V-pattern that does not reach the rounded tip. The rest of the dorsal surface and the palms of both foreflippers are covered with a hairless black leathery skin. The pale pelage on the top of the foreflippers on adult females and juveniles creates a striking contrast with the black naked skin, whereas the darker fur of males is less conspicuous on the flippers. The first digit is the longest, widest and thickest, and curves posteriorly, giving the flipper a swept-back look. Digits 2–4 are successively shorter, and the foreflipper has a more rounded leading edge making them look somewhat paddle-like. There is a small opening in the skin at the end of each digit for a claw that is usually reduced to a vestigial nodule, and rarely emerges above the skin. The claw openings are set back from the free edge of the flippers by cartilaginous rods that extend the length of each digit, and expand the size of the flippers. The hindflippers also have cartilaginous rods that extend

Adult male and female Australian sea lions showing their differences in size and coloration. Note the pale crown of the male. PHOTO: I. CHARRIER

the length of each toe. The bones of the three central toes terminate at the position of the small nails that emerge through the skin on the dorsal surface, set back from the end of the flipper. The first and fifth toes are longer than the three middle toes, and the first toe, or hallux, is longer and wider than the fifth toe. The hindflippers have short hair covering part of the proximal end of the flipper, and the rest of the dorsal surface, and the entire sole is covered in black leathery hairless skin.

Pups are dark gray when born and at around four weeks old transition to a light brown color above and paler color underneath. The face and bridge of the nose is darker while the forehead partially onto the crown is somewhat lighter than the rest of the upper parts. By five months they molt into a juvenile pelage pattern that is like that of the adult females. The crown is paler and there is a dark mask across the face. Adult females and juveniles are countershaded being fawn to silvery gray above and tan to pale yellow on the underside of the neck, chest, and abdomen. The underside of the abdomen can be darker in some animals. The demarcation between light and dark zones is high on the neck and dips down at the insertion of the foreflippers, remaining low on the sides between the fore- and hindflippers. The coloration of the face is variable, with the light coloration of the underside of the neck rising up to include the mystacial area and side of the face, frequently encircling the eyes, and reaching the ears. The foreflippers are often somewhat darker than the rest of the back and sides, but still contrast with the black naked skin parts of the flippers. Some animals are darker overall and have little discernible contrast between coloration above and below. Subadult males are colored like females, but darken as they mature, showing dark spots on the chest as the transition begins. The first evidence of this appears as dark spotting on the chest and darkening of the muzzle. Adult males have a mane of somewhat longer and coarser guard hairs from the crown to the shoulders. Their pelage is dark brown, with a cream-colored crown and nape of variable extent that has the appearance of a light wig. This light color gradually transitions to the darker body pelage on the back and sides of the upper neck. A scattering of light colored hairs sometimes reaches the underside of the neck. This area of pale coloration accentuates the darker "mask" across the face, which extends from ear to ear and covers the sides of the face, muzzle, lower jaw, and throat. Younger bulls are incompletely marked and can have a whitish ring around the eyes.

Adult males reach lengths of 1.85–2.5 m and weights of 180–250 kg, and possibly 300 kg. Females are 1.3–1.8 m in length and 61–105 kg in weight. When born, pups are 62–68 cm long and weigh 6.4–7.9 kg. Longevity can be 26 years for females, and 21–22 years for males. The dental formula is I $^{3}/_{2}$, C $^{1}/_{1}$, PC $^{5}/_{5}$, and occasionally PC $^{6}/_{5}$.

Recognizable geographic forms None

Can be confused with The range of Australian sea lion overlaps that of New Zealand fur seal, is adjacent to that of the Australian fur seal, and wandering subantarctic fur seals periodically show up as vagrants on Tasmania and the southern coast of in mainland Australia. The Australian fur seal probably has the greatest chance of confusion due to their large size and pale coloration for a fur seal. All of the fur seal species have a tapering pointed muzzle and more conspicuous ear pinnae. The head of Australian sea lions is long and wide with a wide muzzle that ends in a blunt nose. Both New Zealand and Australian fur seals have a more tapered muzzle ending in a more pointed nose, and both have thicker pelage, although the Australian fur seal's coat is not as thick as the New Zealand fur seal. Australian sea lions lack the underfur and long guard hairs of both fur seals and their shorter pelage (exclusive of the mane on adult males) looks shorter and less dense than the more luxuriant pelage of dry fur seals, or the shaggy look of fur seals when they are partially wet and dry when just hauling out. There are differences in color between the species. Adult female, juvenile and pup Australian sea lions have a countershaded coat that is silvery gray to dark tan above and light tan below. Adult males darken as they mature to a dark brown, but then lighten in the area of the face, crown and nape to have the appearance of a pale wig. There are also differences in the size and shape of the toes on the hind flippers between Australian sea lions and all fur seals. The foreflippers of Australian sea lions are more rounded on the leading edge, a feature they have in common with Australian fur seals, but which contrast with the more straight edged foreflppers of New Zealand fur seals and all other fur seals of the genus *Arctocephalus*.

Distribution Australian sea lions occur in southern and southwestern Australia from just east of Kangaroo Island west to Houtman Abrolhos in Western Australia. Vagrants have come ashore in eastern Australia as far north as central New South Wales. Australian sea lions breed at 78

Adult male and female Australian sea lions showing the degree of sexual dimorphism in this species. The male has a wound above his right flipper. Kangaroo Island. PHOTO: XVALDES/DREAMSTIME

Adult male Australian sea lion. Note the wide muzzle and dark face, and pale crown, nape, and upper back. PHOTO: I. CHARRIER

Large adult female Australian sea lion with a newborn pup. The rusty orange color on the pale ventral surface of the female is soiling from the afterbirth, feces, and dirt from being hauled-out ashore. PHOTO: T. HASSE

An adult female Australian sea lion with a soiled darkened coat with her molted pup. PHOTO: B. PAGE

Adult female with juvenile-sized fully molted pup on Kangaroo Island. Notice the countershaded color pattern on both animals. PHOTO B. KEENER

A group of juvenile Australian sea lions. PHOTO: B. SABERTON

Compare the coloration and muzzle shape between the Australian sea lion (left) and New Zealand fur seal. PHOTO: B. PAGE

Molting juvenile Australian sea lion. PHOTO: B. SABERTON

sites (50 in South Australia and 28 in Western Australia) primarily on islands, but also at some mainland sites. The majority of pups are born at eight South Australian sites where 100 or more pups are born each year, and one location, Dangerous Reef, produces more pups annually than all of the sites in Western Australia combined. Australian sea lions haul out at an additional 151 sites (90 in South Australia and 61 in Western Australia).

Ecology and behavior Australian sea lions are unusual among pinnipeds in having a supra-annual pupping interval, with females producing pups every 17–18 months. Pups can be born at all times of the year, with females at a given site being loosely synchronous, and pupping over an approximately 5-month period. Neighboring sites are frequently on entirely different breeding schedules. Males are sequentially polygynous, establishing territories around individual females or small groups, herding them in an effort to keep them from departing until the onset of estrus, 7–10 days after they give birth, when they mate. This pattern is repeated until the male is compelled to go to sea and forage, after which he returns and repeats the strategy. Males defend their territories with guttural clicking, growling and barking vocalizations, posturing, and by fighting with rivals. Pups are continuously attended for the first 9–10 days after birth. Over the next 5 months, females make foraging trips that average 2 days, followed by pup attendance periods that average 1¼ days. Females suckle their pups for 15–18 months, usually weaning them a month before giving birth again. Some females care for their offspring for up to two years, and can be seen with a juvenile and a new pup. Adult female Australian sea lions behave aggressively toward pups that are not their own, as do adult males to all pups. Adult males will bite pups and throw them high in the air. Pups can be trampled when males are fighting or moving rapidly to confront rivals and control females, and this can be a significant source of pup mortality. Pups will play at the shoreline and in tide pools while their mothers are away, and they swim on their own after their postnatal molt.

Adult female Australian sea lions are benthic, diurnal foragers. They routinely transit to foraging locations by swimming along the bottom. Mean depth of dives for lactating females ranged from 41.5–83.1 m, with maximum dives to depths of 60–105 m. Mean duration of dives ranged from 2.2–4.1 min, and the longest dive recorded lasted 8.3 min. Australian sea lions are fast, powerful swimmers and will porpoise when moving rapidly at the surface. These sea lions are considered non-migratory, and probably spend most of their lives near their natal colony. The greatest distance traveled by a tagged animal is approximately 250 km. Predators include great white sharks and presumably killer whales. At the Seal Bay breeding colony on Kangaroo Island, South Australia, they have been known to succumb to the bite of the highly venomous tiger snake, which they occasionally encounter when seeking shade in scrubby beachside and upland vegetation.

Feeding and prey The diet of Australian sea lions is poorly known. They are thought to concentrate their efforts on shallow-water benthic prey, but take a wide variety of fishes, such as rays, small sharks, Australian salmon, flathead, swallowtail, leatherjacket and whiting. Other important prey includes octopus, squid, cuttlefish, small crabs, and occasionally penguins, flying seabirds, and small sea turtles. Fishermen complain of sea lions robbing lobster traps and fishing nets.

Threats and status The total population of Australian sea lions was recently estimated at 14,780. A significant sea lion viewing industry has developed, which is regulated at rookeries in parks to minimize disturbance during the breeding season. Extensive disturbance can cause Australian sea lions to abandon colony sites. Traditionally, Australian aborigines and early colonists took sea lions for food and other products. Overharvesting by sealers in the 17th and 18th centuries reduced the population and extirpated them from areas around Bass Strait and Tasmania. Although now protected, the population has not rebounded fully or reoccupied all of its former range. Their unusual breeding strategy includes intense site fidelity which, coupled with a large number of small colonies and other aspects of the species biology, make these small colonies individually vulnerable to extirpation which could threaten the stability of the entire population. The impact of fisheries interactions, including by-catch and entanglements, and the shooting of some sea lions are considered important threats.

IUCN status Endangered.

New Zealand Sea Lion—*Phocarctos hookeri*

(Gray, 1844)

Recently-used synonyms *Neophoca hookeri.*
Common names En.–New Zealand or Hooker's sea lion; Sp.–*lobo marino de Nueva Zelanda;* Fr.–*lion de mer de Nouvelle-Zélande.*
Taxonomic information Carnivora, Otariidae.
Species characteristics New Zealand sea lions are large, heavy-bodied and sexually dimorphic. Adult males are 1.2–1.5 times longer and 3–4 times heavier than adult females. The head is large and blocky with a short, wide muzzle that is blunt to rounded on the end in both sexes. The ear pinnae are small and inconspicuous. The vibrissae are short to moderate in length, reaching as far back as the pinnae on some animals. The head of adult males looks foreshortened as the neck widens rapidly behind the position of the ears, obscuring much of its length. Bulls have a strong sagittal crest well behind the eyes, creating a domed appearance to the crown and a downward slope to the end of the short muzzle with little or no discernible forehead. Depending on posture and viewing angle, the silhouette of the head in many males appears slightly convex from the crown to the end of the muzzle. Bulls have greatly enlarged, wide, thick, rounded necks, increasing in size from the nape and throat to the shoulders. The large neck is covered in a mane of longer coarser guard hairs. The hindquarters are not as enlarged as the neck, but are heavy and powerfully built. Females and subadults have a flat-topped head with a slight downward slope to the end of the short muzzle. Their stocky body lacks the enlarged neck and longer guard hairs of the mane.

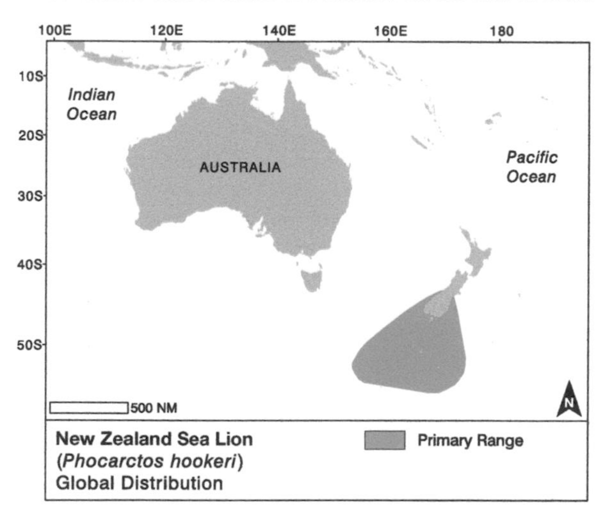

The foreflippers have a sparse short fur that extends beyond the wrist onto the middle of the dorsal surface of the flipper in a V-pattern that does not reach the rounded tip. The rest of the dorsal surface and the palms of both foreflippers are covered with a hairless black leathery skin. The first digit is the longest, widest and thickest, and curves posteriorly, giving the flipper a swept back look. Digits 2–4 are successively shorter. There is a small

Adult male with adult females showing the degree of sexual dimorphism in New Zealand sea lions.
PHOTO: M. JØRGENSEN

opening in the skin at the end of each digit for a claw that is usually reduced to a vestigial nodule, and rarely emerges above the skin. The claw openings are set back from the free edge of the flippers by cartilaginous rods that extend the length of each digit, and expand the size of the flippers. The hindflippers also have cartilaginous rods that extend the length of each toe. The bones of the three central toes terminate at the position of the small nails that emerge through the skin on the dorsal surface, set back from the end of the flipper. The first and fifth toes are longer than the three middle toes, and the first toe, or hallux, is longer and wider than the fifth toe. The hindflippers have short hair covering part of the proximal end of the flipper, and the rest of the dorsal surface, and the entire sole is covered in black leathery hairless skin.

Pups are born in a thick, long, natal coat, and are dark brown with a lighter crown, nape, and mystacial area, and have a pale stripe on the top of the muzzle, originating on the crown. Female pups are lighter than male pups. Pups begin to molt their birth coat at 2 months and at the end of the molt look like juveniles and females. Adult females and juveniles of both sexes are buff to cream or silvery-gray above and light tan to pale yellow below. The diffuse demarcation between light and dark is high on the neck usually rising over the insertion of the foreflippers, and is lower on the sides of the abdomen. There is considerable, but subtle, variation in the coloration of the head and muzzle. The light coloration often extends above the ears, above the eyes, and down the sides of the muzzle. The crown and top of the muzzle are usually darker. The area around the insertion of the foreflippers and the tops of the flippers is often darker grayish to light brown. In some adult females there may be little discernible contrast between coloration above and below, especially when they are faded just before they molt. Males darken as they mature, and pass through a sequence of color phases before becoming fully dark blackish-brown adults, sometimes with rusty tinting. From age 1–2 years they can look like females, although some are light rusty-brown. At 3–4 years they become darker brown. They also begin to show increased bulk in the neck and shoulders and signs of a mane at this age. At 5–6 years their coloration and body shape is that of the adult male and they are blackish-brown over the entire body. Physical maturity is reached in the 7th year. Some older males have white hairs in the mane, giving them a subtle grizzled appearance. From age 2, New Zealand sea lions go through an annual molt that lasts approximately 2 months at some time between February to June. One-year-olds experience only a partial molt.

Adult males are 2.4–3.5 m in length and are 320–450 kg in weight. Adult females reach lengths of 1.8–2 m and weights of 90–165 kg. Pups are 70–100 cm long and weigh 10–11 kg at birth. Their longevity is between 23–26 years. The dental formula is I $^{3}/_{2}$, C $^{1}/_{1}$, PC $^{6}/_{5}$.

Recognizable geographic forms None.

Can be confused with Three fur seals, New Zealand, Antarctic, and subantarctic, are known to occur in or near the present range of the New Zealand sea lion. New Zealand sea lions can be distinguished from all fur seals, based on a combination of features including coloration, fur characteristics, head and muzzle shape, short ear pinnae, more curving leading edge of the foreflipper, and size and shape of the outer toes on the hindflippers.

Distribution The primary habitat of New Zealand sea lions is several subantarctic islands south of New Zealand, and their surrounding waters. They breed primarily in the Auckland Islands and at Campbell Island. New Zealand sea lions also regularly occur in small numbers at Stewart Island and on the southeast coast of the South Island of New Zealand, mostly males aged from 2–11 years. Wandering New Zealand sea lions reach Macquarie Island. Historically, this species had a more extensive range that appears to have included most of New Zealand.

Ecology and behavior The breeding season for New Zealand sea lions begins in November when adult males return and establish themselves on territories through displays, vocalizing, and fighting. Adult females arrive in early December and give birth shortly after returning to

A brown adult male New Zealand sea lion showing the enlarged neck, chest, and shoulders of bulls. PHOTO: A. AUGE

A black adult male New Zealand sea lion. Notice the thick, wide, and blunt-ended muzzle. PHOTO: L. MEYNIER

Pale-colored New Zealand sea lion bull. The huge neck and mane make the large head look small. PHOTO: A. AUGE

Adult male New Zealand sea lion showing the distinctive dome-shaped profile from the muzzle to the back of the head. PHOTO: M. JØRGENSEN

Adult female New Zealand sea lions have a less rounded head and are countershaded light gray to tan above and paler below. PHOTO: M. JØRGENSEN

The new gray fur is replacing the rusty-brown fur on this molting New Zealand sea lion. PHOTO: M. JØRGENSEN

Adult female New Zealand sea lion ashore showing the countershaded pattern and less developed and thickened neck. Otago Peninsula, South Island, New Zealand. PHOTO: W. RAYMENT

A group of New Zealand sea lion pups. The pale bridge of the nose, top of the head, and nape, are markings unique to pups of this species. PHOTO: K. WESTERSKOV/NATURAL IMAGES

Two adult female New Zealand sea lions. Females of this species are only lightly countershaded. PHOTO: L. MEYNIER

An adult female with a pup in the shade of an upland forest haulout site on the Snares Islands, New Zealand. PHOTO: M. RENNER

New Zealand sea lion females with older molted pups (one with a plastic tag on its foreflipper), and one unmolted pup still in its natal coat (foreground). PHOTO: F. RIET

the rookery. Males may have as many as 25 females within their territories. The bulls are frequently challenged by newly arriving males and neighbors, and turnover of males is a regular occurrence. Many territorial bulls depart in mid-January with the end of the pupping period. The onset of estrus occurs 7–10 days after a female gives birth. Prior to this she continuously attends her newborn pup. Following mating, females begin a phase of short foraging trips, each trip followed by a period of attending their pup, typical of many otariids. Lactating females undertake the longest foraging trips of any sea lion, averaging 423 km round-trip lasting an average of 2.8 day, followed by 1.2 days of pup attendance ashore. The maximum distance away from the rookery averages 102 km, with the majority of females foraging over the island shelf and slope. Also typical of many otariids, pups gather into groups known as créches while their mothers are feeding. Females and pups recognize each other through vocalizations and scent, and a small percentage of females will allow additional pups to nurse along with their own pup, which is unusual behavior for a pinniped. Pups are weaned at approximately 10 months. Pups are trampled and killed by adult males challenging other males during territorial disputes. After the primary territorial males abandon the rookery by mid-January, females move their pups away from the rookery, up to 1.5 km inland to vegetated upland habitat, to avoid harassment by the remaining males.

New Zealand sea lions do not appear to be migratory, although they disperse widely over their range during the non-breeding season. Some animals can be found at major rookeries and haulouts year-round. At sea, they are active divers that forage on both pelagic and benthic prey. In fact, female New Zealand sea lions are the champion divers among otariids, executing the deepest and longest dives as they forage far (an average maximum of about 100 km) from their rookeries. They spend more than 50% of their time at sea submerged. The deepest dive documented is 597 m, and the longest lasted 14.5 min. The average dive depth is 129.5 m, and the average

Adult male New Zealand sea lion in the water studies a diver at the Snares Islands, New Zealand. PHOTO: M. RENNER

A bull New Zealand fur seal tearing apart a king penguin that it killed on Macquarie Island for a meal. Bulls are also known to prey on fur seal pups at this site. PHOTO: N. TAPSON

An adult male New Zealand sea lion. Otago Peninsula, South Island, New Zealand. PHOTO: W. RAYMENT

duration is about 4 min. Predators include sharks, leopard seals, and presumably killer whales. Instances of infanticide, and cannibalism of pups, by males were reported for the Dundas Island colony in 1999.

Feeding and prey New Zealand sea lions take a wide variety of vertebrate and invertebrate prey. Frequently-taken species include opalfish, octopus, munida, hoki, oblique-banded rattail fish, salps, squid and crustaceans. Prey is taken in both benthic and pelagic habitats. Antarctic, subantarctic, and New Zealand fur seal pups are taken as prey by adult male sea lions particularly at Macquarie Island. Penguins are also occasionally consumed.

Threats and status New Zealand sea lions once were more abundant, with a more extensive range that included the North and South islands of New Zealand. The Maori people of New Zealand traditionally hunted sea lions, as did Europeans upon their arrival. Commercial sealing in the early 19th century decimated the population in the Auckland Islands, but sealing continued until the mid-20th century, when it was halted. The population has yet to fully recover from this period of overexploitation. At the present time, New Zealand sea lions have a highly restricted distribution, with a small population that numbers under 10,000 adult animals. Two remaining breeding colonies account for 71% (Dundas, Enderby and Figure of Eight islands in the Auckland Islands group) and 29% (Campbell Island) of births. A very limited recolonization has occurred on the Otago Peninsula of New Zealand's South Island, where 4–5 pups are born annually. The overall population has declined recently for reasons not completely understood, but evidence points to by-catch in the commercial arrow squid trawl fishery, indirect competition with fisheries, and disease. A series of bacterial outbreaks at the Auckland Islands led to the mass mortality of pups in 1998 (53%), 2002 (32%) and 2003 (21%). The New Zealand government has responded with a variety of management actions that are aimed at documenting, reducing, and limiting the annual incidental take of the sea lions.

IUCN status Vulnerable.

Walrus—*Odobenus rosmarus*

(Linnaeus, 1758)

Recently-used synonyms *O. r. laptevi.*
Common names En.–walrus, Pacific walrus, Atlantic walrus, Laptev walrus; Sp.–*morsa*; Fr.–*morse*.
Taxonomic information Carnivora, Odobenidae. Two subspecies are recognized: the Atlantic walrus *O. r. rosmarus* from the eastern Canadian Arctic, and Greenland east to Novaya Zemlya; and the Pacific walrus *O. r. divergens* in the Bering, Chuckchi, and East Siberian seas, adjacent Arctic Ocean, and the Laptev Sea north of Siberia. Formerly, the isolated population in the Laptev Sea was described as a separate subspecies, *O. r. laptevi,* but genetics revealed it is has close ties to the Pacific walrus, so it was lumped with that subspecies.

Arctic Ocean
Pacific Ocean
Atlantic Ocean
1,000 NM

Walrus
(*Odobenus rosmarus*)
Global Distribution

● Extralimital Record
Primary Range
Secondary Range

Species characteristics Walruses are very large, heavy-bodied unmistakable looking pinnipeds. Males are longer and heavier than females. Adults have a short coarse pelage that becomes sparse as they age, more so for males than females. The skin is thick, rough and heavily marked with creases and folds. Older males have many lumps, called tubercles, on the neck and chest, giving them a warty appearance, and most males become virtually hairless. The neck, chest, and shoulders are massive, and the body tapers towards the hindflippers. The head and the muzzle are short, but very wide. The "bloodshot" eyes are small, somewhat protruding, and set far apart. The end of the muzzle is very wide, flattened and has large, fleshy, forward-facing mystacial pads sprouting several hundred short, stiff, pale vibrissae. The nostrils are located on top of the muzzle. Walruses have no ear pinnae. The foreflippers are short and squarish, resembling otariid foreflippers, but unlike otariids, each digit has a weakly developed claw. The hindflippers are phocid-like, with longer first and fifth digits, and strong expandable webbing between the digits, each with a small claw. Unlike in phocids, the hindflippers can be rotated forward under the body. The tail is attached to the body by a web of skin.

Walrus coloration varies with age and activity. Most walruses are yellowish to reddish-brown. When walruses

Pacific walruses hauled out on Round Island. Note the thickness of the tusks, pink color of the skin and numerous tubercles on these adult males. The animal at back right has two broken tusks. Bristol Bay, Alaska. PHOTO: T. COLLOPY

Pacific walrus bull with very long tusks and many tubercles. Round Island, Alaska. PHOTO: J. GARLICH-MILLER/USFWS

A large adult male Atlantic walrus with thick rolls of fat on the neck and shoulders. Svalbard. PHOTO: M. JØRGENSEN

This large adult male Pacific walrus shows the wide, deep, face of bulls of this species. Note the robust lower jaw, and thick tusks. Bering Sea. PHOTO: T. FISCHBACH/USGS

Adult female Atlantic walrus at Svalbard. PHOTO: L. MURRAY

Female Pacific walrus with broken tusks and her yearling calf that has budding tusks. Bering Sea. PHOTO: J. GARLICH-MILLER/USFWS

A group of Atlantic walrus females with their young and juveniles hauled-out on sea ice. Walruses are very social in or out of the water and will huddle in close proximity or lie on top of each other. Svalbard. PHOTO: L. MURRAY

Female Pacific walrus and calf. The faces of cows are narrower than those of bulls. Barrow, Alaska. PHOTO: T. FISCHBACH/USGS

An adult female Atlantic walrus with a young, dark calf. PHOTO: I. CHARRIER

Female Pacific walruses will push their calves along, at the surface, until they are able to swim well. PHOTO: J. GARLICH-MILLER/USFWS

Adult female Atlantic walruses with juveniles and a calf riding on the back of its mother. Svalbard. PHOTO: M. NOLAN

remain in cold water for long periods of time, their coloration fades to a pale grayish hue, due to reduced blood flow to the skin (peripheral vasoconstriction). Conversely, when they come out of the water, vasodilation causes blood and color to return to the skin, and under warm conditions, causes the animals to appear pink to reddish, giving them the appearance of having a sunburn. Calves are darker with slate-gray fur.

Adult males achieve lengths of 3.6 m and weights of 880–1560 kg, and females are about 3 m in length, weighing 580–1039 kg. They live to the age of about 40 years. Pups are 1–1.4 m and 33–85 kg at birth. The dental formula is: I $^{1}/_{0}$, C $^{1}/_{1}$, PC $^{3}/_{3}$, with the upper canine teeth developing into tusks that grow throughout life, a trait unique among pinnipeds. The tusks are longer and thicker, with more grooves and fracture lines in similar aged males than in females, and one or both tusks can be partially, or entirely, broken off in adults of both sexes. Individuals can accumulate considerable wear on the leading edge of their tusks and numerous dental anomalies have been recorded, including growth of additional tusks. In Pacific walruses, tusks also tend to be less curved in side view and more divergent at the tips in males, and more curved and convergent at the tips in females, but

A juvenile Atlantic walrus at the edge of a herd. PHOTO: M. NOLAN

A pair of subadult Atlantic walruses. Notice the short tusks. Svalbard. PHOTO: M. JØRGENSEN

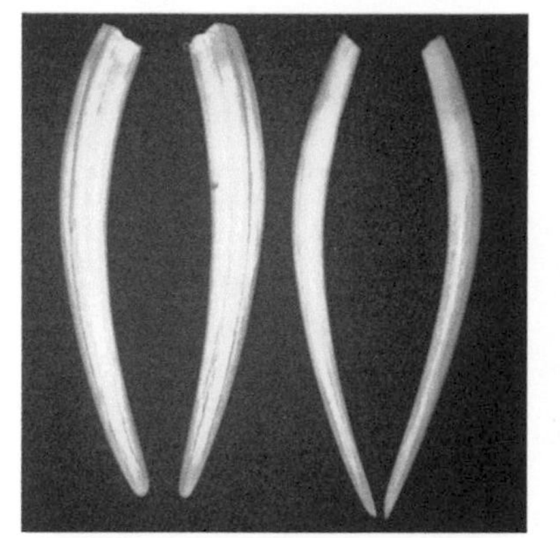

Male Pacific walrus tusks (left) are thicker with more ridges and cracks than those of females (right). PHOTO: US FISH AND WILDLIFE SERVICE

An unusual adult male Atlantic walrus with three tusks. Svalbard. PHOTO: M. JØRGENSEN

An adult male Pacific walrus is gray from vasoconstriction after returning from sea. Cape Seniavin, Alaska. PHOTO M. SNIVELY/USFWS

there is overlap in these features. Tusks about 1 m long can be found on male Pacific walruses. Walrus calves are born without tusks, which usually become visible below the lips around age 2 in the Pacific population.

Recognizable geographic forms It is possible that transits of the Canadian Arctic Archipelago by walruses could result in either Pacific or Atlantic walruses showing up in the other's range, presenting an interesting identification challenge. Features to look for include: Pacific walruses generally have longer, thicker, and more curving tusks than similar sized and aged Atlantic walruses, although there is overlap in tusk length and shape. Pacific walruses have longer and wider skulls than Atlantic walruses, indicating a larger head and wider muzzle, although there is overlap in adult females. Pacific walruses are often described as having a broader more squared off end of the muzzle, compared to the more sloping end of the face of Atlantic walruses. Adult male Pacific walruses regularly have more numerous, and more pronounced, tubercles on the neck, chest, and shoulders than Atlantic walruses. The walrus population in the Laptev Sea is little studied and poorly known, but is thought to be somewhat intermediate between Atlantic and Pacific walruses in overall size and tusk length and shape. At one time, Pacific walruses were thought to be longer and heavier than Atlantic walruses, but this has not been born out by measurements from the field that show that they are quite comparable in both values for males and females, and cannot be told apart based on length and weight.

Can be confused with Walruses are unmistakable on land, ice, and in most sightings at sea. At sea they should not be confused with any other species if more than the back is seen as they surface.

Distribution Walruses have a discontinuous circumpolar distribution in the Arctic and subarctic. The Pacific subspecies is found from the Bering and Chukchi seas to the East Siberian Sea in the west and the western Beaufort Sea in the east, and as an isolated population in the Laptev Sea north of central Russia. The Atlantic subspecies occurs in numerous subpopulations from the eastern Canadian Arctic and Hudson Bay, to the Kara Sea. In the western Atlantic, walruses occurred as far south as the Gulf of Saint Lawrence in historic times, but are now normally found only well north of this area in the Canadian Arctic. Both subspecies of walrus are found in relatively shallow continental shelf areas, and rarely occur in deeper wa-

This large bull Atlantic walrus is resting in shallow water and has impressive, thick tusks and many tubercles on his neck. Svalbard. PHOTO: T. MELLING

An Atlantic walrus flexes its lower back and hindflippers while resting near shore. PHOTO: T. MELLING

Atlantic walruses surfacing, showing the body from above. Notice the "spout" which can be seen sometimes when walruses surface. PHOTO: M. NOLAN

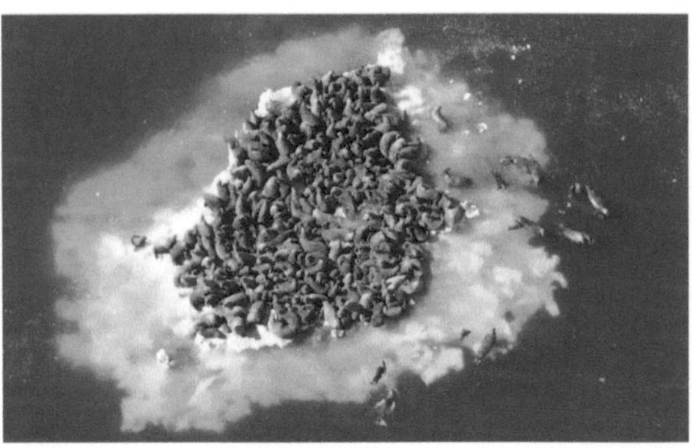
Pacific walruses packed onto an ice floe is not uncommon for this gregarious species. Bering Strait. PHOTO: M. A. WEBBER/USFWS

ters. They regularly haul out on sea ice, sandy beaches, and rocky shores, to rest, molt, give birth and nurse their young. Walruses are not known as extreme wanderers, but have been reported as vagrants from Honshu, Japan, Cook Inlet and Resurrection Bay in south-central Alaska, New England, US, Iceland, the United Kingdom (several records in the Shetlands), and from Norway south to the Bay of Biscay in western Europe.

Ecology and behavior Sexual maturity occurs at 7–10 years for males, and 7–8 years for females, although some cows ovulate at a younger age. Calves are born from mid-April to mid-June on sea ice after a lengthy gestation of 15 months that includes some delay of implantation. The period of calf dependency and maternal care is one of the longest among pinnipeds, routinely lasting 2 years, and often more, before weaning. The calf and female form a tight bond as they spend considerable time in close contact. Courtship and mating has been little studied, because walruses mate in the harsh winter environment of the Arctic. It is believed that walruses are polygynous, and males may form a type of lek with small aquatic territories adjacent to females on ice floes, where they vocalize and display. There is also intense male-male fighting at this time. Bulls probably do not have realistic chances to breed until they are 15 years old or older, when they are more experienced, physically mature and able to outcompete other bulls for access to females. Males produce an unusual bell-like sound when they are in the water.

A pair of elastic pharyngeal pouches can be inflated with air and provide flotation when a walrus is resting in the water. Walruses can walk and climb on land by using all four flippers to move the body, in the manner of an otariid, although they are far less agile and move much more slowly. In the water, they primarily rely on side-to-side sculling of the hindflippers for swimming in the manner of a phocid, but also pull themselves through the water with foreflipper strokes like an otariid.

In the Pacific, walrus migration follows the seasonal advance and retreat of the sea ice. However, some walruses, particularly males, summer far from the sea ice, using land-based haulouts in Bristol Bay, Alaska, some islands in the Bering Sea, and in Russia from the Chukotka Peninsula to Kamchatka. Walruses also haul out on shore, away from ice, in years of reduced pack ice. Walruses are among the most gregarious of pinnipeds. When hauled-out, they are regularly found in tightly huddled masses, often lying on top of each other; at sea they are often seen in large herds, and extremely large

Rare images of walruses from the Laptev population. Recent genetic research revealed they are closely aligned with Pacific walruses, so the distinction as a subspecies was abandoned. Adult male (top left), various age males (top right), and females (bottom left and right). Andreya Island, Peter's Archipelago. PHOTOS: WHITE WHALE PROGRAM/RUSSIAN ACADEMY OF SCIENCES

aggregations have been estimated at 60,000 or more animals. Tusks are used for aggressive displays, fighting, and pulling themselves onto ice floes, not for digging up food as was one believed.

Feeding and prey Walruses feed on a variety of prey, chiefly benthic invertebrates including: clams, worms, snails, shrimp, and slow-moving fish. Some walruses, known as "rogues," prey on ringed, spotted and bearded seals, and possibly even small whales.

Threats and status The current population level of Pacific walruses is uncertain, but roughly estimated to be 200,000, with another possible 4,000–5,000 in its isolated Laptev population. The better-known Atlantic population is estimated to number about 18,000–20,000. The overall trend in the two subspecies is thought to be declining. Walruses were severely depleted by episodic commercial hunting that was heaviest from the 18th through the mid-20th centuries. Native peoples of the Arctic have depended on walruses for food, hides, ivory, and bone since first contact, and subsistence harvests continue today in many parts of the species' range. Conflicts with fisheries are low, however, industrial development, dispersal of pollutants, and human disturbance along with global warming and associated reduction in sea ice extent and duration all have the potential to cause significant impacts on walrus populations. Recently, record-breaking summer sea ice minimums for modern times have occurred, when the ice can retreat north of continental shelf, beyond walrus feeding areas in the Chukchi Sea. Walruses are not capable of spending long periods of time at sea without hauling out, so this sea ice retreat forces walrus herds to use land based haulouts from which they make foraging trips for days to several weeks. Being highly gregarious, these haulouts can become very large. Walruses are known for their tendency to stampede into the water if startled or threatened, and they will fall down steep ridges, and trample and roll each other during group panics that can result in mass injuries and mortalities, especially of calves. There is also concern that these very large herds may have a greater impact on the prey of a smaller area, compared to more dispersed herds traveling with advancing and retreating sea ice, using it as a kind of mobile base throughout the seasons. Because of their more consistent use of land based haulouts and smaller subpopulation size, Atlantic walruses might ultimately benefit from climate change. Alterations of walrus behavior and movements are indicators of the complex ways global warming is affecting the Arctic ecosystem and its inhabitants.

IUCN status Data Deficient.

Hawaiian Monk Seal—*Neomonachus schauinslandi*

Matschie, 1905

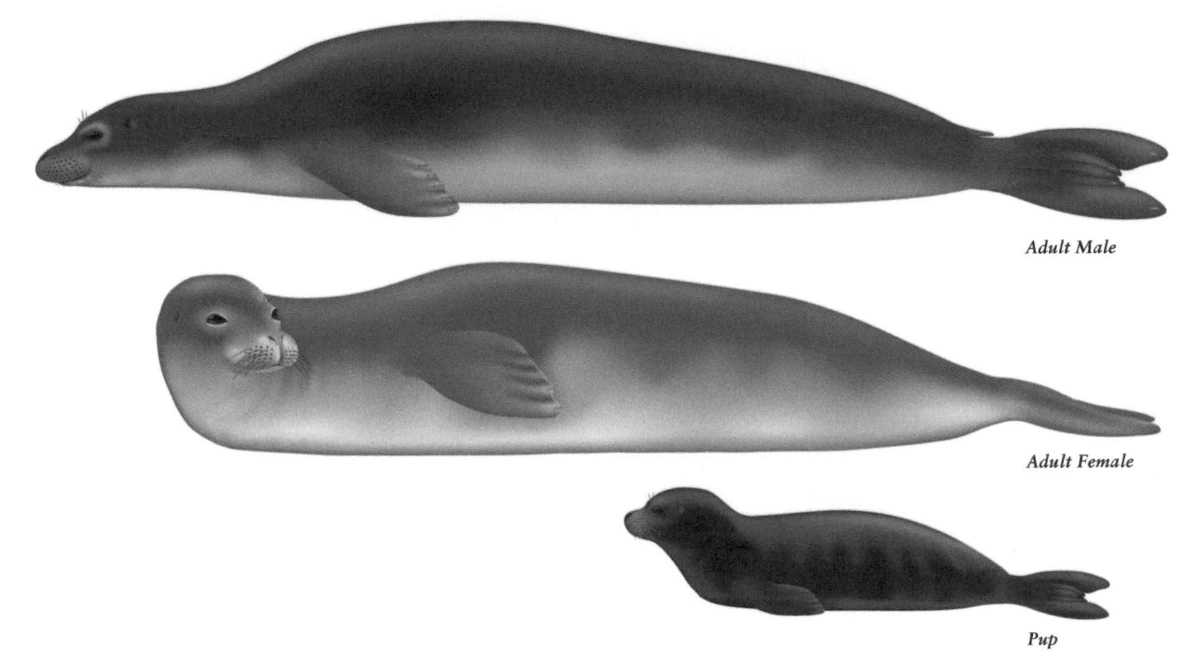

Recently-used synonyms *Monachus schauinslandi.*
Common names En.–Hawaiian monk seal; Sp.–*foca monje del Hawaii;* Fr.–*phoque moine d'Hawaï.*
Taxonomic information Order Carnivora, Family Phocidae. Until recently, the three monk seal species were in the genus *Monachus.* However, the Hawaiian monk seal and extinct Caribbean monk seal are more closely related to each other genetically and morphologically than either is to the Mediterranean monk seal. In 2014, the new genus *Neomonachus* was proposed for the two allied species of the New World, and accepted by the Society for Marine Mammalogy.

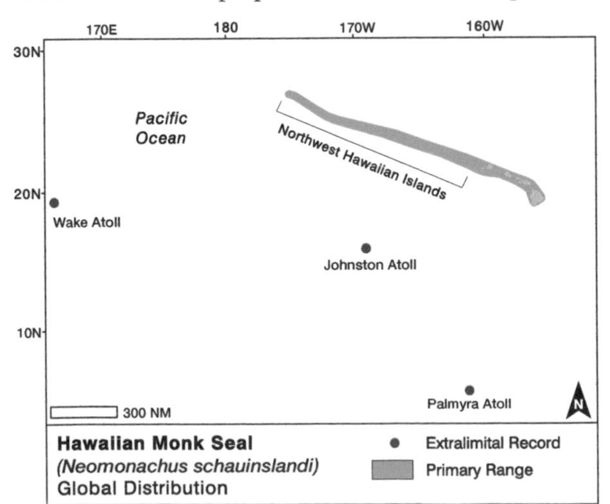

Species characteristics Hawaiian monk seal females grow slightly longer, and are heavier, than males. The long, fusiform body is robust, with relatively short fore- and hindflippers. The relatively small head is wide and somewhat flat, and the eyes are spaced fairly widely apart. The muzzle is also wide, U-shaped, and somewhat flattened. The mystacial pads are large and fleshy, extending beyond the nostrils. The nostrils are in a sub-terminal position, angled up, rather than in a terminal position facing forward, as is typical of most northern hemisphere phocids. The vibrissae of Hawaiian monk seals are smooth and not beaded. Hawaiian monk seal vibrissae vary from short to moderately long, and are black at the base, often having lighter yellowish-white tips. There can be a scattering of all-light vibrissae throughout. There are four functional mammae.

Hawaiian monk seals have an epidermal molt, as is the case in both elephant seals and the Mediterranean monk seal where skin with attached fur peels off in large patches. The seals can look very irregular at this time, showing areas of faded dark fur, clean lighter new fur, and patches of dried skin peeling off. Just after the annual molt, most females and subadults are silvery to slate-gray above, fading to cream or light silver-gray below. During the course of the 11–12 months between molts, the coat fades to dull brownish above and yellowish-tan below.

Adult female (left) and male Hawaiian monk seals having an aggressive interaction at French Frigate Shoals. Female Hawaiian monk seals are longer and heavier than males. Also, note the algae staining on the neck of the male. PHOTO: M. ROMANO

Adult female Hawaiian monk seal with a young pup in the black lanugo coat of this species. PHOTO: K MATSUOKA/NOAA FISHERIES

This young Hawaiian monk seal pup in lanugo has natural pale birthmarks on its ventral surface and left foreflipper. French Frigate Shoals, Hawai'i. PHOTO: B. BECKER/NOAA FISHERIES

Adult female with a healthy and growing pup in lanugo that is chocolate brown. PHOTO: M. SULLIVAN/NOAA FISHERIES

Adult female with large molted pup. Note the weight loss in the female from raising the pup, and the light algae staining on her coat. PHOTO: M. SULLIVAN/NOAA FISHERIES

A very large, molted pup showing how healthy pups look when weaned at the correct weight. PHOTO: A. DIETRICH/NOAA FISHERIES

Many young monk seals are emaciated and juvenile mortality rates are up at some sites. PHOTO: D. LUERS/NOAA FISHERIES

Juvenile Hawaiian monk seal going through the epidermal molt of this species. PHOTO: J. BRACK/NOAA FISHERIES

Adult male with algae staining coming ashore; same individual that is in the image on the top of page 447. PHOTO: M. ROMANO

Adult female Hawaiian monk seal sharing a beach with basking Hawaiian green sea turtles at French Frigate Shoals. PHOTO: M. ROMANO

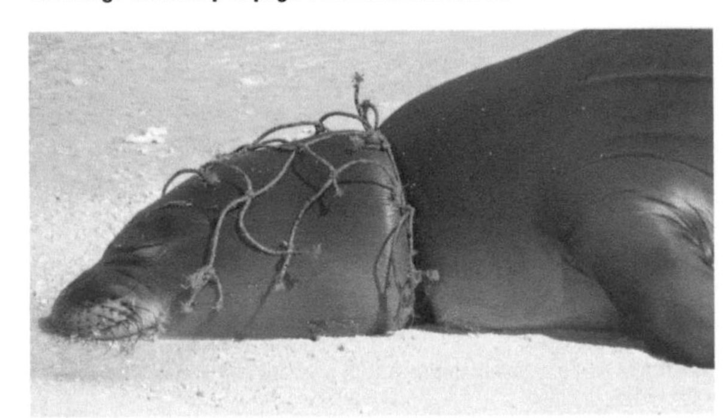

Juvenile monk seal badly entangled in the loop of a net fragment. PHOTO: I. NIMZ/NOAA FISHERIES

Juvenile Hawaiian monk seal still alive after a devastating shark attack, probably by a tiger shark due to the shape of the bite. PHOTO: M. ROMANO

Males and some females become completely dark brown to blackish as they age. There can be a variable amount of highlighting on the mystacial area and on both the upper and lower lips. Adults and juveniles can have a greenish cast from algal growth. Pups are born in a black woolly coat, which is molted completely by about 6 weeks. The first molt from the lanugo coat is a shedding of individual hairs, but each successive annual molt is a more dramatic epidermal molt of juveniles and adults. Many animals have pale birthmarks that take the form of smallish irregular patches, which look like pale bleached areas on the darker pelage, and can occur anywhere on the body. If the pale bleached areas occur on the ends of flippers, the nails there can be whitish.

Hawaiian monk seals appear docile most of the time, but aggressive interactions can be violent. PHOTO: M. SULLIVAN/NOAA FISHERIES

Two large adult Hawaiian monk seals at the shoreline, in the midst of a very dynamic fight. PHOTO: M. SULLIVAN/NOAA FISHERIES

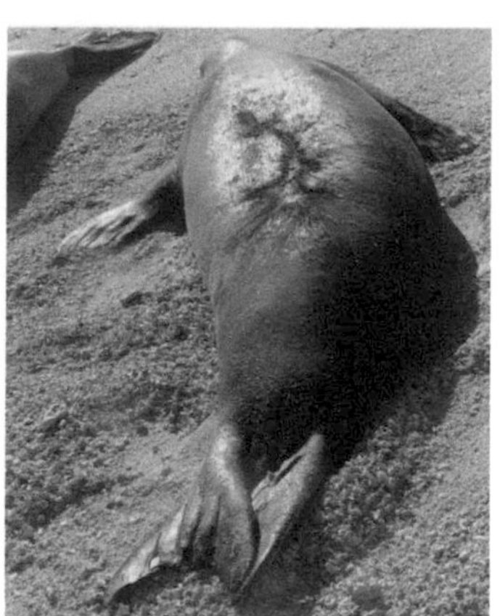

An adult female Hawaiian monk seal with wounds from having been mobbed by males. PHOTO: A. DIETRICH/NOAA FISHERIES

Adults and some subadults have varying amounts of scars, particularly on the back and neck. Males can be heavily scarred on the lower jaw and neck, and some adult females become severely injured when mobbed by groups of males attempting to mate with them. These injuries can include massive gashes, and skin torn away on the back that, when healed, develop into long, jagged, irregular scars. Injuries and scars from shark attacks are also regularly seen. Injuries include oval and semi-circular scars from tearing wounds, evulsions of flesh, and long parallel rakes formed by the penetration and tearing by individual teeth.

Adult males are about 2.1 m in length, and females are about 2.4 m. Males weigh an average of 172 kg, and females weigh up to 272 kg. Newborns are about 1 m in length and 16–18 kg in weight. The dental formula is I $^{2}/_{2}$, C $^{1}/_{1}$, PC $^{5}/_{5}$.

Recognizable geographic forms None.

Can be confused with No other pinnipeds regularly occur within the tropical and subtropical central North Pacific habitat of this seal. However, a northern elephant seal has been recorded at Midway Island. Adult northern elephant seals are much larger than Hawaiian monk seals, and only juveniles and small subadults could be confused with monk seals. The large body size and more rounded shape of the larger head and muzzle of northern elephant seals are diagnostic. Young or female elephant seals close in size to monk seals will show nostrils that are terminal and forward-facing. Female northern elephant seals have only two abdominal mammae.

Distribution Hawaiian monk seals are distributed throughout the northwestern chain of Hawaiian Islands from Nihoa Island to Kure Atoll. Primary breeding sites are Kure Atoll, Midway Atoll, Pearl and Hermes Reef, Lisianski Island, Laysan Island, and French Frigate Shoals. While 90% of the population is found in the northwestern islands, the seals are also now regularly seen on all of the main Hawaiian Islands (particularly Kauai), where small numbers of births have been recorded. Some adult males involved in mobbing incidents with females were translocated to the main Hawaiian Islands and Johnston Atoll, about 1,000 km south of the Hawaiian Archipelago. Hawaiian monk seals are considered non-migratory, and most are typically faithful to their atoll or island of birth. However, small numbers do relocate temporarily or permanently to other sites in the island chain. Long distance wanderers have been recorded at Palmyra, Johnston, and Wake atolls.

Ecology and behavior Hawaiian monk seals haul out and breed on beaches of sand and coral rubble, and on rocky

Juvenile monk seal over coral reefs in the Northwestern Hawaiian Islands. PHOTO: M. SULLIVAN/NOAA FISHERIES

An adult Hawaiian monk seal swallowing a large fish. PHOTO: R. UYEYAMA/US NAVY

terraces. They sometimes leave the beach if vegetation is available for shade. The long breeding season lasts from late December to mid-August, although most pups are born between March and June. Males in this polygynous species patrol the water adjacent to rookeries, or haul out beside females with pups. There are up to three times more breeding-age males than females at some colonies; this contributes to mobbing of estrus females, which are often injured or even killed in these incidents. When approached by another seal or human on land, Hawaiian monk seals often roll to present their underside to the intruder, arch their back, raise a flipper in the air, and open their mouth. They are generally solitary, both on land and at sea. Even when seals gather together on land, they are not generally gregarious and only mothers and pups regularly make physical contact.

Evidence suggests that foraging activities and prey preferences can vary considerably by location, age and sex of the seal, and season. They are known to forage diurnally and nocturnally in the shallow waters surrounding islands and seamounts, and within atolls. Their movements and habitat at sea are not well known, but recent work with satellite tags, dive recorders, and video camera tags on adult and subadult males at French Frigate Shoals has shown that they frequently venture far offshore to deep, outer slopes of islands and reefs, and to nearby seamounts and banks, where they appear to feed on the bottom. They also travel to deep ocean areas, also presumably to forage. In contrast, seals at Pearl and Hermes Reef forage almost entirely within or around this atoll. Most dives that have been recorded have been relatively shallow ones of 100 m or less, with a record of one subadult male diving to at least 500 m, the maximum depth the device could record.

Feeding and prey Hawaiian monk seals consume over 40 species of prey, including many reef fish, octopus, squid and crustaceans, which have been identified from analysis of scats and spews collected from beaches. Some of the most common varieties were various representatives of the fish families *Labridae, Holocentridae, Balistidae*, and *Scaridae*, and many species of eels. Some of the octopus and squid species taken are found over reef environments, while others are from offshore and pelagic mid-water habitats.

Threats and status The vast majority of Hawaiian monk seals live in protected areas, isolated from most direct human contact. Despite this, and as a direct result of exploitation in the 19th century when they were brought to the edge of extinction, the species is critically endangered. The current population numbers about 1,200 animals, and is in decline. Recovery of the species has been affected to an unknown degree by military activities such as development and occupation of bases on several islands, and dumping of waste that started before World War II. Also, development of several fisheries in Northwestern Hawaiian Island waters has led to conflicts from entanglements and animals returning to haulouts with long-line hooks in their mouths. The effect of these fisheries on the ecosystem and prey of monk seals is poorly known and a subject of debate. Marine debris, particularly lost and discarded fishing gear, has been shown to be a significant source of mortality for monk seals. Efforts have been underway to remove this debris from breeding beaches for many years, and more recently from waters surrounding these sites. Large predatory shark species seasonally congregate in large numbers in the Northwestern Hawaiian Islands to prey upon fledgling albatrosses, as well as seals. While the role of shark predation is not completely understood, actions have been taken to reduce shark numbers around certain monk seal breeding beaches in order to reduce mortality on newborn and juvenile seals. Future sea level rise resulting from global climate change poses a threat by reducing the seal's terrestrial habitat on low-lying islands and atolls.

IUCN status Critically Endangered.

Mediterranean Monk Seal—*Monachus monachus*

(Hermann, 1779)

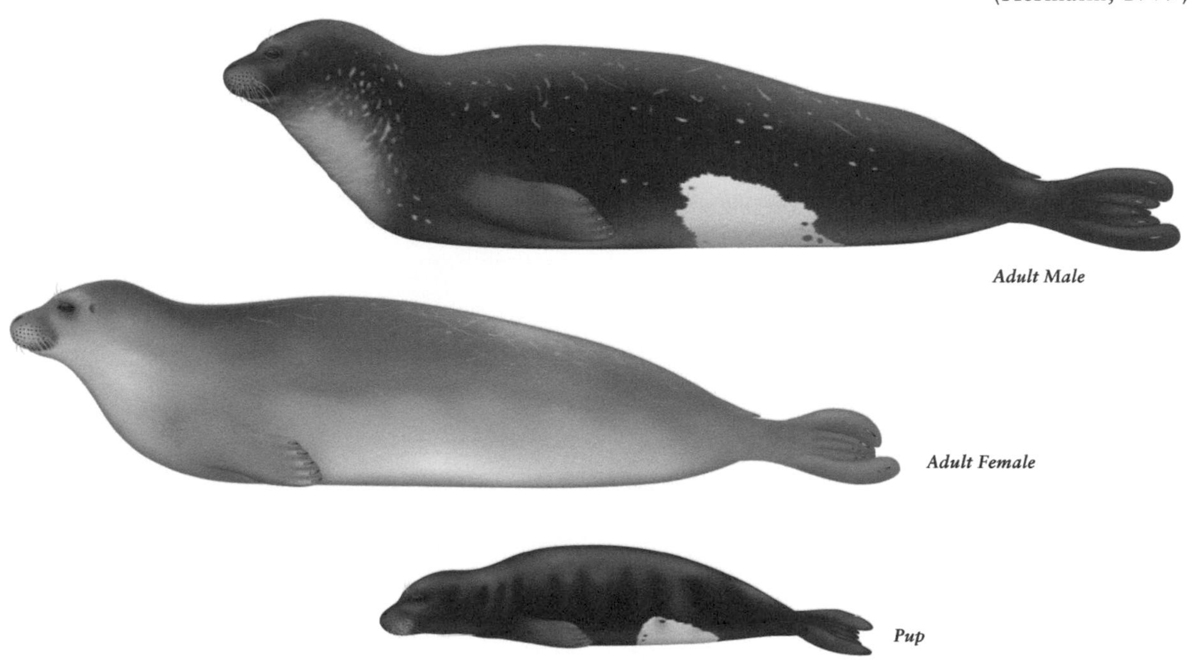

Recently-used synonyms None.
Common names En.–Mediterranean monk seal; Sp.–*foca monje del Mediterráneo;* Fr.–*phoque moine de Méditerranée.*
Taxonomic information Carnivora, Phocidae.
Species characteristics Adults are robust, with short flippers, a long fusiform body, and a moderately-sized head. The head and muzzle are somewhat flat and wide, with the eyes spaced widely apart. The mystacial pads are large, fleshy, and extend beyond the nostrils. The nostrils are in a sub-terminal position, pointed up, rather than in a terminal position facing forward, as is typical of most northern hemisphere phocids. The vibrissae of Mediterranean monk seals are smooth and not beaded. Coloration is variable, with differences existing between adult males and females and between age classes. Variation also occurs within each age class, and has been proposed to be present between the now-isolated sub-populations.

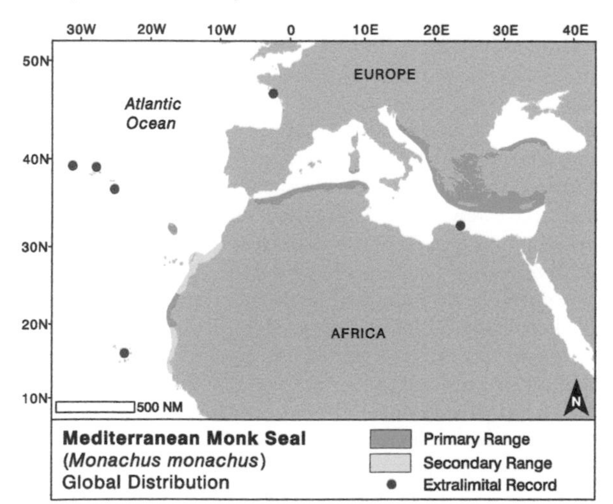

Adult males are dark brown to black over most of the body. They retain a pale area on the throat and upper neck. The back and hindflippers are frequently patterned with pale scratches and speckles, which are scars from healed bite wounds from other seals. They can also have a concentrated, scratched area on the nape. The pale belly patch on the abdomen rises up as a wedge on both sides about halfway to the top of the back. Most adult females are brownish-gray to dark gray above and paler below, the colors separated by either a gradual blending or a sharp demarcation. The dark coloration is made paler by light tips to the hairs. Female coloration is highly variable. They have a hood of darker color on the head. This "frontal hood" stops above the eyes but can have an extension down the muzzle to the nose creating a pale "mask" around the eyes. If the hood extends down the side of the head beyond the ear openings, it is called a "lateral hood." The lumbar area can have a pale, irregular "sash" from a concentration of scratches and scars, and localized depigmentation. They frequently have pale

Adult male Mediterranean monk seal ashore at the Cabo Blanco Peninsula colony. PHOTO: M. A. CEDENILLA/CBD-HABITAT

Adult female Mediterranean monk seal with a 1.5 month old pup at the Cabo Blanco Peninsula. PHOTO: M. A. CEDENILLA/CBD-HABITAT

A group of Mediterranean monk seals at Cabo Blanco Peninsula including two adult males (middle), an adult female and pup (foreground), and an adult female (top). PHOTO: M. HAYA/CBD-HABITAT

Adult male (top) and three adult females, one with a pup, at the Cabo Blanco Peninsula. PHOTO: M. A. CEDENILLA/CBD-HABITAT

scratches on the back, which are scars from bite wounds from other seals. Juveniles and subadults can be a range of gray and brown tones above, and off-white to light gray below. Pups are born in a brownish-black, woolly lanugo coat that is molted at 40–100 days to a juvenile color pattern, comprised of a mixture of weakly developed countershaded coloration features and pup markings. Most pups have a large, irregular, light gray, ventral patch on the abdomen that can rise up on the sides at mid-abdomen, well behind the foreflippers. The pale ventral patch disappears at the second molt from youngster to juvenile pelage at about 7–9 months, when the animals essentially become fully countershaded. Interestingly, this belly patch reappears, later in life, on adult males. Animals from the eastern Mediterranean are generally darker than those found in the Atlantic Ocean. Little is known about the adult and subadult molts other than they are epidermal molts involving loss of a layer of skin with the fur, as is the case for Hawaiian monk seals and both species of elephant seals.

Adults are up to 2.8 m in length (averaging 2.6 m in males and 2.4 m in females) and weigh 240–300 kg. Males may grow to be slightly heavier (maximum recorded is 400 kg) and 5% longer than females. Pups are 80–90 cm and 15–26 kg at birth. The dental formula is I $^{2}/_{2}$, C $^{1}/_{1}$, PC $^{5}/_{5}$.

Recognizable geographic forms None.

Can be confused with Mediterranean monk seals do not regularly share their range with any other pinniped. The nearest occurring species are eastern Atlantic Ocean populations of the more northerly distributed harbor and gray seals. As vagrants, these species, along with wandering hooded, harp, and bearded seals from the Arctic, have occurred in the Mediterranean monk seal's range. Mediterranean monk seals can be readily distinguished from harbor, gray, younger age classes of harp, and adult hooded seals by the lack of spots or blotches covering most or all of their body, distinctively wide and somewhat flattened head and muzzle, subterminal, upward-facing nostrils, smooth vibrissae, and four mammae. Like Mediterranean monk seals, bearded seals have few to no spots, and share smooth vibrissae and four abdominal mammae as features in common with Mediterranean monk seals. However, bearded seals have a proportionately smaller head with much more densely packed, longer, and predominantly downward pointing vibrissae. The short, rounded foreflipper with a short first digit is also unique to the bearded seal and gives the end of the flipper a rounded edge. Adult harp seals are usually prominently or lightly banded and have a dark face. Mediterranean monk seals have typically proportioned foreflippers, with the first digit longer than all other digits.

Adult female Mediterranean monk seal. PHOTO: M. A. CEDENILLA/CBD-HABITAT

Adult male Mediterranean monk seal. PHOTO: M. A. CEDENILLA/CBD-HABITAT

Adult male Mediterranean monk seal. PHOTO: M. A. CEDENILLA/CBD-HABITAT

An adult female with a pup less than two months old. This female has a "hood" of dark coloration and a light "mask" around the eyes. The pup is in its lanugo coat. Cabo Blanco Peninsula. PHOTO: M. A. CEDENILLA/CBD-HABITAT

Mother with a pup older than two months that has molted into the juvenile pelage. Note the scars on the back of the female and the "frontal hood" and pale face "mask" on both animals. Cabo Blanco Peninsula. PHOTO: M. A. CEDENILLA/CBD-HABITAT

Molting adult female Mediterranean monk seal. PHOTO: M. HAYA/CBD-HABITAT

A very light-colored adult female Mediterranean monk seal. PHOTO: M. HAYA/CBD-HABITAT

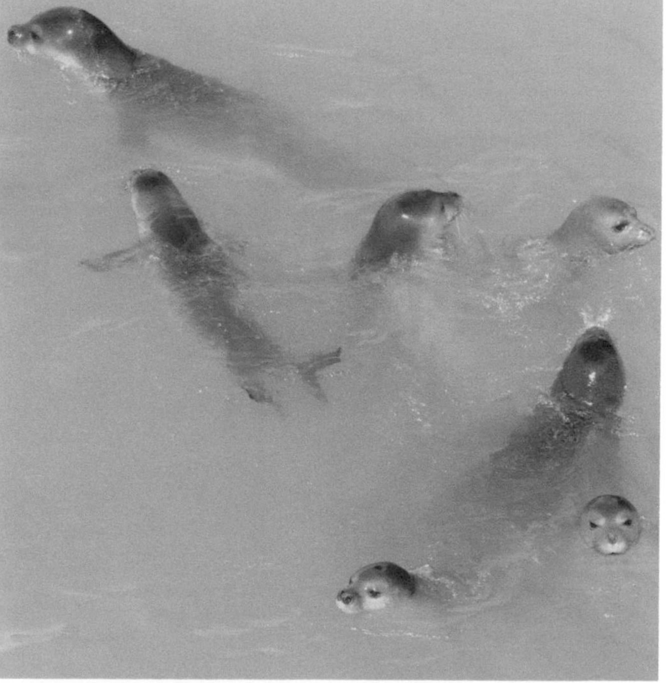
Juvenile and subadult Mediterranean monk seals. Sadly, this small group represents 1.2–1.4% of the remaining world population. PHOTO: M. A. CEDENILLA/CBD-HABITAT

An adult Mediterranean monk seal resting by two younger pups in lanugo. Note the large belly patch on the pup on the right. Desertas Islands, Madeira. PHOTO: R. PIRES

An adult female and a pup about 2 months old finishing its molt out of lanugo, in a shaded grotto at the Cabo Blanco Peninsula. PHOTO: M. A. CEDENILLA/CBD-HABITAT

A 6 month-old Mediterranean monk seal from the Aegean Sea released after rehabilitation. Eastern Mediterranean animals are usually darker overall than Atlantic animals. PHOTO: MONACHUS GUARDIAN

Distribution The remaining Mediterranean monk seals are widely-dispersed in small groups and as individuals in the Mediterranean, Aegean, and Ionian seas, and possibly still in the Black sea, and the Sea of Marmara. Historically, they occurred nearly continuously throughout the region, but now are only found in small remnant aggregations with two main colonies left. They occur in the eastern North Atlantic from the Strait of Gibraltar south and west to Mauritania at 19°N, with wanderers reaching Dakar, Senegal at 15°N. The largest surviving colony is on the border of Mauritania and Western Sahara at Cabo Blanco, a peninsula otherwise known as Cap Blanc, or Ras Nouadhibou. They are also found at Desertas Island in the Madeira Island group. Although Mediterranean monk seals can still be seen on occasion throughout much of their former range, this is becoming increasingly rare. On land, they choose rocky coastlines, with a preference for sea caves and grottos that are generally inaccessible from land approaches, and sometimes have only submarine entrances. In West Africa, they will come ashore on open beaches.

Ecology and behavior This seal is considered non-migratory, with individuals spending most of their time within a very limited home range. They make extensive use of grottos and sea caves for hauling out and breeding. The single pups are born across an extended pupping season that can last from May through at least November, and possibly January, with an October peak, but pupping can occur year-round.

Threats and status The Mediterranean Monk Seal is the most endangered pinniped species in the world, with a total abundance of 500–600 animals. The risk of extinction is high. Its small population is widely-scattered in isolated subpopulations most of which are reproductively isolated. There are about 250–300 in the eastern Mediterranean, of which about 150–200 are in Greece and about 100 in Turkey. Another 220 seals are in the colony at Cabo Blanco, the only large viable colony, and about 30 on Desertas Island in the Madeira group. Pup survival is low, affected by storms and high swells, as well as by low genetic variability in the sub-populations, resulting from the species being so widely-distributed in small breeding groups. Human disturbance of animals at haulout sites and in foraging areas could also contribute to lower reproductive success. The Mediterranean monk seal is legally protected throughout its range, however, in areas of heavy commercial fishing, monk seals may be viewed as a pest species, competing for fish, and in some instances have been shot. They are also susceptible to entanglement in fishing nets, marine debris and discarded and lost nets and line. An outbreak of a phytoplankton-based paralytic toxin was the probable cause of a large-scale die-off in the Cabo Blanco area in 1996–1997; a morbillivirus may also have been a factor in this mortality event.

IUCN status Critically Endangered.

Ross Seal—*Ommatophoca rossii*

Gray, 1844

Recently-used synonyms *Ommatophoca rossi.*
Common names En.–Ross seal; Sp.–*foca de Ross*; *Fr.–phoque de Ross.*
Taxonomic information Carnivora, Phocidae.
Species characteristics Ross seals are typically countershaded, blending along the sides, in colors that range from dark brown to dark gray and black. Most striking are a series of dark streaks, unique to this pinniped, originating from the masked face and dark lower jaw that extend down the neck and sides parallel to each other. There may also be spots, particularly on the sides, where the dark coloration of the back merges with the light color of the undersides. Coloration becomes duller with more brown and tan tones in late summer before the epidermal molt. The molt occurs in January and possibly February, and involves shedding small pieces of skin with the fur. Scars are often seen on the neck, possibly from intraspecific fighting, and a small percentage of animals have scars from wounds believed to be from leopard seal or killer whale attacks. Pups are born in a two-toned lanugo that is dark brown to blackish above and lighter gray, yellowish to silver below.

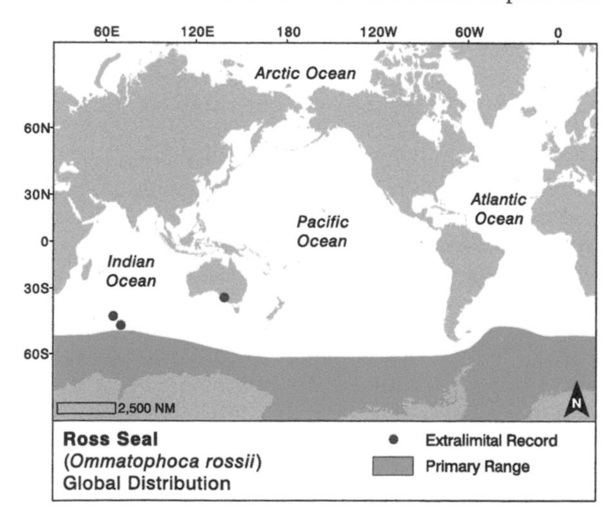

The head and neck are thick and wide, while the rest of the body appears proportionately short and slender. The muzzle and mouth are short and wide, giving the end of the head a blunt appearance. The vibrissae are sparse, short, and inconspicuous. The eyes are proportionately larger than in any other seal and set widely apart. The foreflippers are relatively long, with short claws as is typical of all four Antarctic phocid species. The first digit is very long and robust and the rest of the digits are successively shorter, the length tapering rapidly to create a pointed flipper similar in profile to the foreflipper of a sea lion. The hindflippers are long, to $^{1}/_{5}$ or more of the standard length and are proportionately the longest of any phocid seal. The first and fifth digits of the hindflippers are very long, and much thicker than the other digits.

Ross seals are the smallest of the four Antarctic phocids. Ross seals can reach about 2.5 m or more in length and weigh up to 216 kg, although relatively few have been measured and weighed. Males that have been examined are 1.7–2.1 m and females are slightly larger at about 1.9–2.5 m and 150–201 kg. Pups at birth are 1.0–1.2 m and 17–27 kg. Longevity is to 20 years or more. The dental formula is I $^{2}/_{2}$, C $^{1}/_{1}$, PC $^{5}/_{5}$.
Recognizable geographic forms None.
Can be confused with Of the three other phocids that share the Ross seal's range (Weddell, crabeater, leopard seals), the Weddell is most similar in appearance. Compared to Weddell seals, Ross seals are shorter, and much smaller in the torso, but have a proportionately much wider neck and head, and have irregular dark streaks on the sides of the neck and body, and proportionately longer foreflippers. Differences in how they respond to being disturbed when out of the water may help distinguish the species: Weddell seals regularly roll to one side raising the head

Four views of a molting Ross seal, sex unknown. The old fur is being replaced by richly colored, new pelage. This animal has a dark face with broad streaks that are different on each side. Note the large, unusually-shaped head, bulging eyes, short muzzle, thick neck, and medium length foreflippers. PHOTOS: P. A. OLSON/NOAA FISHERIES /SWFSC

and often a foreflipper "in salute" (crabeaters also do this occasionally), while Ross seals often assume a "singing posture," arching the back with the head held up, mouth open, vocalizing while inflating the throat to show off the thick neck, often with the foreflippers under the chest.

Distribution Ross seals have a circumpolar distribution in the Antarctic. They are usually found in dense consolidated pack ice, but can also be found on smooth ice floes in more open areas. Results of tagging efforts revealed that Ross seals migrate north out of the pack ice zone into open water to forage. Vagrants have been reported from Kerguelen Island, Heard Island, the Falkland Islands, South Orkney, and southern Australia.

Ecology and behavior Sexual maturity for females is probably reached at 3–4 years of age, and at 3–7 years for males. Pups are born in November and December, with a peak from early to mid-November. Weaning takes place at about 1 month, although little is known of the relationship between mother and pup. Nursing-age pups have been seen swimming between ice floes. Mating is thought to occur in the water, but has not been observed. When hauled-out, Ross seals are generally encountered alone. Occasionally, a small number of individuals may be found in the same area, but they are usually spaced widely apart. Ross seals may haul out more often from morning to late afternoon. During the period of the molt, they may be out of the water for longer periods of time. Ross seals are known to reach depths of up to 500 m. One tagged female dived nearly continuously, reaching average depths of 110 m and with average dive durations of 6.4 min, with a maximum dive depth of 212 m and duration of 9.8 min.

Few behaviors have been noted, except for its characteristic habit of raising up the neck and head, pointing the muzzle skyward, and opening the mouth when approached by a human. This has been described as the "singing posture," and may be used for in-air vocalizations, but more often serves as an aggressive or defensive posture, along with the chugging and trill vocalizations that are produced when approached. Vocalizations are also made from other positions, including in the water, and the species' repertoire includes many calls that have been variously described to include pulsed chugs, clucks, loud cries, trills, and tonal siren calls.

Ross seal showing throat coloration and making an open-mouth threat gesture. PHOTO: R. L. PITMAN

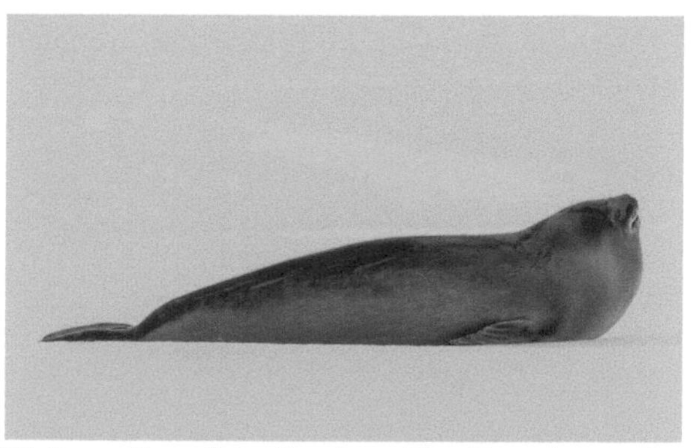

This Ross seal has puffed-up its neck, possibly getting ready to vocalize. PHOTO: J. VAN FRANEKER/IMARES

Ross seal with an open mouth display. Note the length and shape of the foreflipper, and markings on the side of the neck. PHOTO: N. TAPSON

Ross seal in the "singing position" that is a signature of this species. PHOTO: J. VAN FRANEKER/IMARES

Feeding and prey The diet of Ross seals consists primarily of squid, but also includes moderate amounts of fish, such as Antarctic silverfish, and krill.

Threats and status Ross seals typically haul out in dense consolidated pack ice that can usually only be reached by icebreakers. Small numbers have been collected for commercial purposes, scientific studies and museums, but otherwise interactions with humans have been few. Ross seals are protected by the Antarctic Treaty and the Convention for the Conservation of Antarctic Seals. When wandering outside the pack ice zone, they could come in contact with commercial fishing operations, but there are no reports of interactions to date. Loss of Antarctic ice due to climate change could negatively affect this species' habitat and prey base.

The Ross seal is not seen very often due to its solitary and widely dispersed nature, consolidated pack ice habitat, and because it is the least abundant phocid in the Antarctic. No detailed surveys have been conducted recently, but data gathered between 1968–1983 gives a population estimate of around 130,000.

IUCN status Least Concern.

Ross seal on alert with the color pattern of the throat in view. PHOTO: J. DURBAN

Crabeater Seal—*Lobodon carcinophaga*

(Hombron and Jacquinot, 1842)

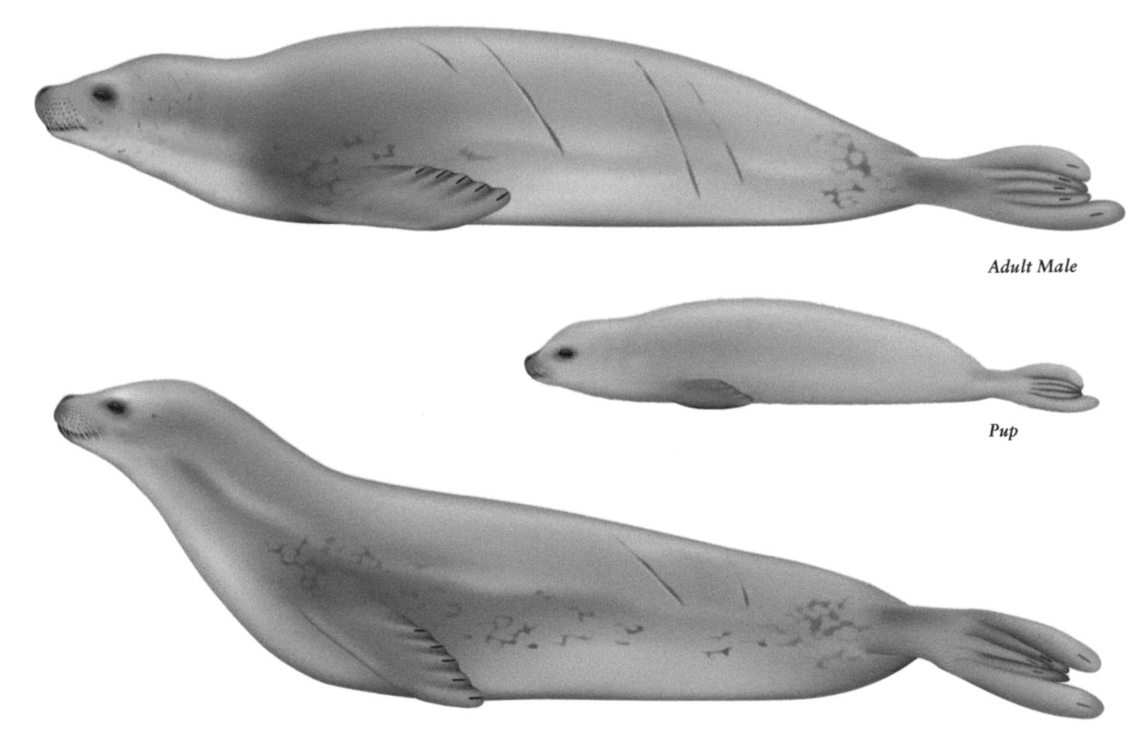

Adult Male

Pup

Adult Female

Recently-used synonyms *Lobodon carcinophagus.*
Common names En.–crabeater seal; Sp.–*foca cangrejera*; *Fr.–phoque crabier.*
Taxonomic information Carnivora, Phocidae.
Species characteristics The head and muzzle are moderately long and slender relative to the animal's length. The eyes are set well apart and the head tapers to the base of the straight-sided muzzle, forming a slight forehead in profile. The nostrils are on top of the muzzle, just back from the end, and in profile look slightly enlarged. This gives the end of the muzzle a slight tipped-up appearance, which is enhanced by the crabeater's tendency to raise the end of the muzzle to the level of the eye or higher when disturbed. The line of the mouth is virtually straight from the gape to the end of the muzzle. The vibrissae are short, pale to clear, and inconspicuous.

The foreflippers are long, wide, and somewhat sickle-shaped. The first digit tapers to a pointed end, and is similar to those of otariids, but the foreflipper is fully furred. Crabeater seals can spread the digits of the fore-flippers while swimming and stretching, greatly enlarging the surface area. Many crabeaters bear long dark scars, either singly or in parallel pairs, that have been attributed to attacks by leopard seals. Many older males have numerous smaller scars and injuries to the face and sides of the mouth and head, presumably from intraspecific fighting. The coat of a freshly molted crabeater has a rich sheen, and can be light to dark shades of colors ranging from silvery gray to tawny brown. Irregular patches of spots and rings can be found on the shoulders, sides, tops of the flippers, and around the insertion of the flippers. These markings produce a reticulated or web-like pattern on the sides of many crabeaters between their fore- and hindflippers. The flippers can be so heavily marked with spots and rings that they appear darker than the rest of the body. As time progresses from the molt, crabeaters fade dramatically, becoming light tan, pale gray, or whitish. As these seals become older they become paler overall, and some look faded all year long. Pups are born with a soft woolly coat that is grayish-brown in color and has been described as light, coffee-with-cream brown, with darker coloring on the flippers. Molt begins in about 2–3 weeks and the pup sheds into a subadult pelage similar to that of adults.

Adults average approximately 2.3 m in length and 200 kg in weight. Pups are about 1.1 m and 36 kg at birth. The dental formula is I $^{2}/_{2}$, C $^{1}/_{1}$, PC $^{5}/_{5}$. All of the post-canine teeth are ornate, with multiple accessory cusps. Upper and lower teeth interlock to form a sieve to strain krill from the seawater. A ridge of bone on each mandible fills in the gap behind the last upper post-canine teeth to prevent the loss of krill from the back of the mouth when the seal is feeding.

A group of crabeater seals on a large ice floe with more than half alert. The multi-colored pattern on several individuals is a matter of wet versus dry pelage. Antarctic Peninsula. PHOTO: M. NOLAN

Recognizable geographic forms None.

Can be confused with Crabeater seals are most likely to be confused with leopard, Weddell, Ross, or young elephant seals. Leopard seals have a much larger, almost reptilian-looking head, with a very broad muzzle, and proportionately longer foreflippers. Weddell seals have a proportionally small round head with close-set eyes, a short, narrow muzzle, and a proportionately long and heavy-set body with short foreflippers and distinctive spots. Ross seals have a wide head and neck, with a short muzzle, giving the snout a blunt appearance. The foreflippers are proportionately about the same length as in crabeater seals. They also have irregular dark streaks originating from the head, lower jaw, and thick neck that extend down the sides of the body. The presence of long scars and sets of parallel scars are much more common and readily visible on the pale, relatively unmarked pelage of crabeaters than on the other three Antarctic phocids. Of all Antarctic phocids, only crabeaters occur routinely in closely associated groups in the water or on ice floes. Young southern elephant seals of similar size are gray to brownish, lack any markings, and have larger wider heads with a large, wide and fleshy muzzle with big and forward pointing terminally placed nostrils, very large eyes and prominent dark vibrissae. Differences in how they respond to being disturbed when out of the water may help distinguish the

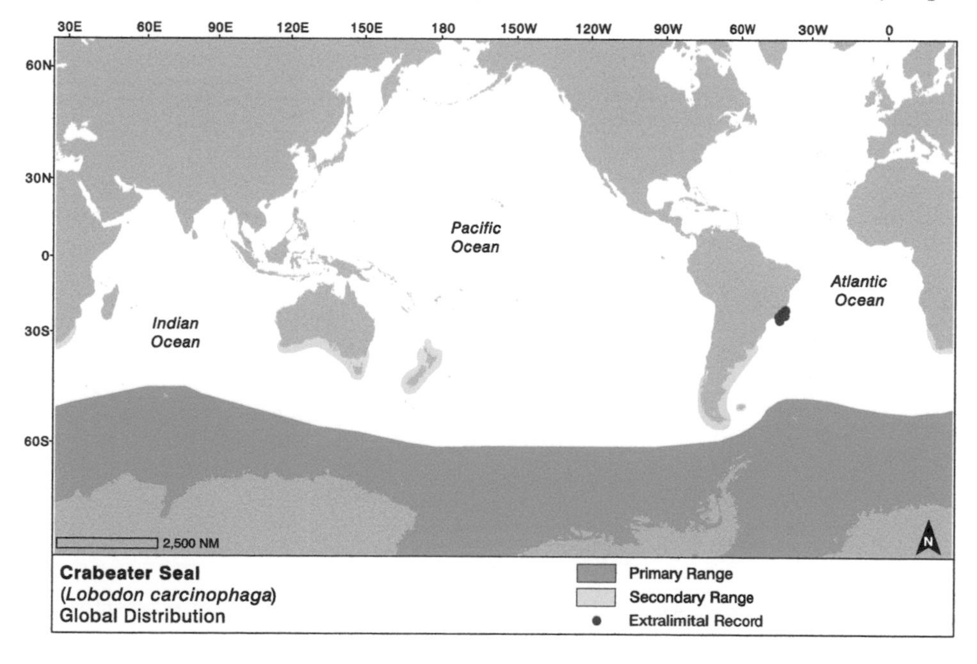

Adult crabeater seal flaring its foreflipper to scratch, showing the shape and webbing between the digits of the flipper. PHOTO: M. NOLAN

A profile view of an adult crabeater seal, probably a male due to the face and neck scarring, flexing a foreflipper. PHOTO: M. NOLAN

A rare photo of an adult female crabeater seal and young pup in lanugo. PHOTO: M. N. NORMAN

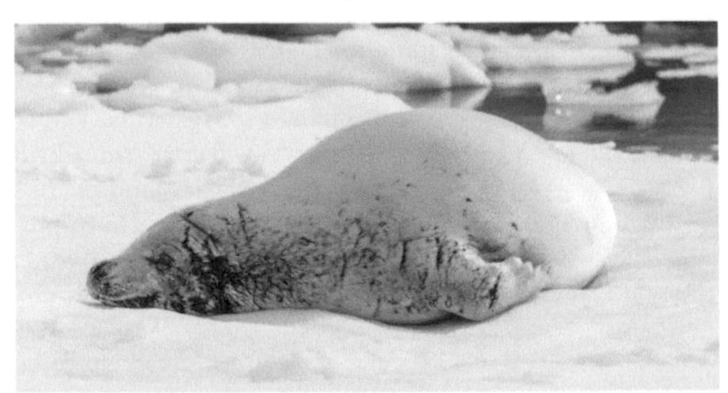

Adult, probably a male, with extensive injuries to the face, head, neck and flippers from breeding interactions with females and fighting with males for access to females. PHOTO: M. NOLAN

An adult crabeater seal lifting unusually high to look around. PHOTO: N. TAPSON

A molting adult male crabeater seal with long scars on the body from leopard seal attacks. PHOTO: M. NOLAN

Crabeater seal in alert posture. PHOTO: M. DE BOER

A beautifully patterned crabeater seal, possibly freshly molted, in Andvord Bay, Antarctica. PHOTO: M. NOLAN

Crabeater seals are one of only a few species of phocids that will travel in large groups at sea. Another is the harp seal of the Arctic. Here such a group glides along on calm seas; they are unmistakable in the Antarctic due to this behavior, and by their coloration and head shape. PHOTO: D. WALKER

Notice the slightly up-turned end of the muzzle on this crabeater seal in a lead. McMurdo Sound. PHOTO: R. L. PITMAN

Two crabeater seals surfacing after a sparring match (see page 460, bottom row). PHOTO: M. NOLAN

species: Ross seals often assume a "singing posture," arching the back with the head held up, mouth open, vocalizing while inflating the throat to show off the thick neck, often with the foreflippers under the chest; Weddell seals regularly roll to one side raising the head and often a foreflipper "in salute" (crabeaters will do this occasionally as well); and crabeater seals rise up with their head level, or angled slightly up, and their belly down.

Distribution The distribution of Crabeater seals is circumpolar in the Antarctic and tied to seasonal fluctuations of the pack ice. They can be found up to the coast of Antarctica, as far south as McMurdo Sound, during late summer ice break-up, and as vagrants as far north as New Zealand and the southern coasts of Africa, Australia, and South America including north to Rio Grande do Sul State in Brazil. Crabeaters have been known to wander over 100 km inland and die in the dry valleys adjacent to McMurdo Sound.

Ecology and behavior Sexual maturity in females is reached between 2.5–4.2 years of age. Pups are born from September to December with a peak in October, and are weaned in approximately 3 weeks. There are no specific rookeries; females haul out on ice singly to give birth. Adult males attach themselves to female-pup pairs and stay with the female until her estrus 3–4 weeks after giving birth. Mating behavior is poorly understood. Males try to prevent females from leaving the ice suggesting they mate out of the water. However, most ice living phocids mate in the water. Females are reported to bite males around the mouth and flippers and this may account for much of the abundant small scars on the faces of older males. Mortality is high in the first year and may reach 80%. Much of this mortality is attributed to leopard seal predation, and up to 78% of crabeaters that survive through their first year have injuries and scars from leopard seal attacks. Leopard seal attacks appear to fall off dramatically after crabeaters reach 1 year of age.

Crabeaters are frequently encountered alone or in small groups on the ice or in the water. However, much larger groups of up to 1,000 have been observed hauled out together, and they are the only species of Antarctic phocid that is gregarious, although Weddell seals will breed and haul out in colonies. They will swim together in herds estimated to be up to 500 animals, breathing and diving almost synchronously. The molt is in January and February. They have a pattern of feeding from dusk until dawn, and hauling out in the middle of the day. Most animals routinely haul out during their annual molt, but

Crabeater seals interacting. Males accumulate considerable scarring on the face, neck, and upper body from females driving them away from their pups, and fighting with males to keep them away from females with a pup. PHOTOS: M. NOLAN

Male and female crabeater seals during courtship. Females often bite and wound males on the head, face and neck (left). Whether crabeater seals mate on ice or in the water is not known, so this interaction could also be part of an aggressive encounter (right). PHOTOS: D. PYLPENKO

A sequence of two adult crabeater seals having an aggressive encounter in the water, described as mock fighting by the photographer. PHOTO: M. NOLAN

even then only about 20% of them will be hauled-out at any one time in an area. Crabeaters frequently use their foreflippers for propulsion, and surface and roll forward to begin a dive. They are known for their ability to move rapidly on ice, with sinuous serpentine motions of the back, aided by the flippers. When agitated, they arch their back and raise their neck and head in an alert posture.

They can dive to 430 m and stay submerged for 11 min, although most feeding dives are to 20–30 m, and shorter. Foraging occurs primarily from dusk to dawn, and tagged seals have been recorded to regularly dive continuously for periods up to 16 hours. Dives at dusk and dawn are deeper than at night, and indicate that crabeater feeding reflects the daily vertical migrations of krill.

Feeding and prey Crabeater seals feed primarily on Antarctic krill (*Euphausia superba*) and 95% of their diet may be made up of this species. Small amounts of fish and squid are also part of the diet.

Threats and status Crabeater seals are undoubtedly the most abundant species of seal, and are one of the most numerous large mammals on earth. Population estimates have ranged up to 75 million animals. There is no recent population estimate available, but a more conservative estimate is 11–12 million animals. A mass die-off was reported from an area near a base on the Antarctic Peninsula in 1955. About 3,000 animals were trapped in areas 5–25 km from the nearest open water and most died over a 2–3 month period. None of the animals examined appeared to be starving, and numerous abortions of fetuses were noted. A disease outbreak was suspected, but not identified; in the 1980s, a study of archived samples revealed antibodies to a distemper-like virus. Several brief episodes of hunting ended when the operations became economically unsuccessful. Crabeater seals are protected by the Antarctic Treaty and the Convention for the Conservation of Antarctic Seals. There are currently no direct threats from human activity throughout most of the species' normal range, but the commercial harvesting of krill may pose a threat to crabeater seals if conducted on a large scale. The effect of global climate change on Antarctic pack ice, and crabeater seal prey, poses an unknown threat to this ice-dependent species.

IUCN status Least Concern.

Leopard Seal—*Hydrurga leptonyx*

(Blainville, 1820)

Recently-used synonyms None.
Common names En.–leopard seal; Sp.–*foca leopardo;* Fr.–*léopard de mer.*
Taxonomic information Carnivora, Phocidae.
Species characteristics Leopard seals have a sinuous body and massive head and jaws. Because of the shape of the head, they appear almost reptilian. Females grow slightly longer and heavier, but not enough to allow the sexes to be distinguished in the field based on size. The long body usually appears slender, but thickest through the shoulders and upper chest. There is no trace of a forehead. The head is widest at the eyes, which appear small and set far apart and well back from the end of the broad U-shaped muzzle. The nostrils are situated on top of and just back from the end of the muzzle. The lower jaw is massive, both long and wide, and the throat and neck are thick. Leopard seals have an enormous gape, and the teeth, especially the canines, are conspicuously large. The vibrissae are clear to pale, generally quite short and inconspicuous. The foreflippers are long, wide and somewhat sickle-shaped, tapering to a rounded end, and similar to those of otariids, but fully furred. The first digit is long and wide, creating a thick strong leading edge to the flipper. Leopard seals can spread the digits of the foreflippers while swimming and stretching, greatly enlarging the surface area. The foreflippers are situated

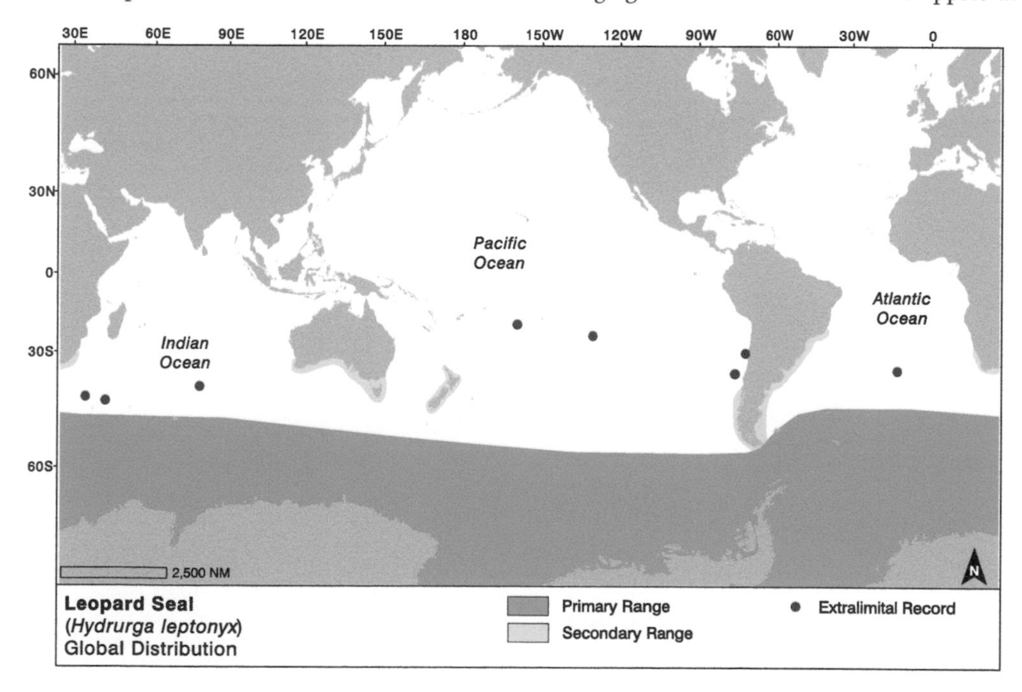

Leopard Seal
(Hydrurga leptonyx)
Global Distribution

Leopard seals are usually long and lean, and can look sinuous. PHOTO: B. RAFTEN

An adult leopard seal with a strongly countershaded pelage pattern sits up in an alert posture. PHOTO: M. NOLAN

Leopard seal showing the large wide head. PHOTO: L. MURRAY

Upper body of a leopard seal with heavy spotting on the chest, neck and throat. PHOTO: M. NOLAN

A large leopard seal stretching and gaping, showing off the very long otariid-like foreflippers of this species. PHOTO: M NOLAN

Leopard seals propel themselves with both phocid-like side-to-side sculling of the hindflippers and otariid-like strokes of the foreflippers. PHOTO: M. NOLAN

A juvenile leopard seal charging ashore in New Zealand. PHOTO: W. RAYMENT

The impressive dentition of an impressive predator. PHOTO: M. NOLAN

The huge unmistakable head of an adult leopard seal. PHOTO: M. NOLAN

An adult leopard seal craning to look at people, a behavior also used to search the ice for prey. Fortunately, these seals know the difference between the two. PHOTO: M. NOLAN

A huge, very successful leopard seal ashore at South Georgia. PHOTO: H. LAWSON

farther back on the body from the muzzle than on other Antarctic phocids, creating the relatively long neck of the leopard seal.

Leopard seals have a countershaded color pattern. They are dark gray above and light silvery-gray below, blending high on the sides of the torso and neck, where the light ventral coloration rises to just below the level of the eye. A swath of lighter color from the neck extends forward on the upper jaw above the mouth to the end of the muzzle, which highlights the line of the mouth. Leopard seals are spotted to varying degrees, most noticeably on the sides, neck, and belly. Pups have essentially the same markings in similar proportions as adults, although their coat is softer, longer and thicker. Dense constellations of spots may occur without any pattern or symmetry. One area where these clusters of spots normally occur is around the insertions of the foreflippers.

Adult males reach lengths of 2.8–3.3 m and weights of 300 kg, while females are 2.9–3.6 m. The largest individuals may be 3.8 m, weighing up to 500 kg. Newborns are 1.0–1.6 m and 30–35 kg. They may attain ages of 26 years or more. The dental formula is I $^{2}/_{2}$, C $^{1}/_{1}$, PC $^{5}/_{5}$. The canine teeth are very long and sharply pointed. The post-canine teeth are ornate with several large lobes, and resemble those of crabeater seals.

Recognizable geographic forms None.

Can be confused with Seen well, the unmistakable features of a leopard seal, including the large head and muzzle, long neck and very long foreflippers, long thin body, irregular spotting pattern, and usually a strong countershaded coloration, distinguish it from all other Antarctic seals. At a distance, however, they may be confused with crabeater seals, which are shorter, with a proportionately smaller head and shorter muzzle demarcated by a small angular forehead, and generally lighter in color, with minimal or no countershading. Also, crabeaters often have long scars, and usually have small numbers of spots concentrated on the shoulders, flanks, tops of the foreflippers, and around the insertions of the flippers. Both Weddell and Ross seals have smaller heads, and very short and blunt muzzles, with proportionately shorter foreflippers that are situated farther forward on the body than on the leopard seal.

Distribution Leopard seals are widely-distributed in cold Antarctic and subantarctic waters of the Southern Hemisphere (50°S to 80°S), from the coast of the continent north throughout the pack ice, and at most subantarctic islands. Vagrants regularly reach warm temperate latitudes. They haul out on ice and land, often preferring ice floes near shore, when available. Vagrants have been reported many times at locations far away from the Antarctic and Subantarctic, including on the coast of New Zealand, Australia, Lord Howe Island, north-central

Two views of a thin leopard seal mother with a dependent female pup, still with some patches of brownish lanugo on its upper back. Leopard seals may be unique in having a countershaded lanugo that is brownish-gray above and creamy-white below. Antarctic Peninsula. PHOTOS: M. NOLAN

A large leopard seal pup along the Antarctic Peninsula with some longer pale lanugo remaining on the chest and belly. PHOTO: S. LAST

A leopard seal pup nursing on a small floe. Note the remnants of the brownish-gray lanugo on the back and shoulders. PHOTO: H. LAWSON

Chile, Rio de Janeiro, Brazil, South Africa, the Falkland Islands, Juan Fernández Islands and Easter Island. Other northern extralimital occurrences are from Pitcairn Island (25°S) and the Cook Islands (21°S).

Ecology and behavior Pups are born on the ice from early November to late December, and the period may be as long as early October to early January. Births at South Georgia occur from late August to the middle of September. Pups are probably weaned at 3–4 weeks, and female estrus occurs at or shortly after weaning. Unlike crabeater seals, male leopard seals do not haul out with female-pup pairs. Mating is believed to occur in the water. Males reach sexual maturity at about 3–6 years of age, and females at 2–7.

At sea and on the ice, where they spend most of their time, leopard seals tend to be solitary. They float at the surface, and crane their neck high in the air to view objects of interest, and will hide around ice floes while underwater, stalking prey. Leopard seals can submerge by sinking straight down, which is typical of most phocids (which use their hindflippers for propulsion), or by rolling forward, the more typical method of otariids (which use their foreflippers for propulsion). Leopard seals are fast, powerful, agile swimmers. Swimming is primarily accomplished with long, powerful, coordinated sweeps of the foreflippers, rather than the side-to-side strokes of the hindflippers typical of most phocids. Leopard seals mostly sleep or are otherwise inactive when out of the water, but will move in a serpentine slithering manner across ice, and toboggan like penguins. They are curious and unafraid of humans and will approach, occasionally brush against, and even mouth, small boats. Explorers and scientists from the heroic age of exploration to the present occasionally report being stalked by leopard seals swimming along the ice edge or leads, where people are working and walking by. Leopard seals have lunged at ankles and, rarely, briefly grabbed people. They will closely approach human divers and appear to be very curious during these encounters. On one occasion, a leopard seal held a snorkeler underwater long enough to drown the person. Despite being a top predator, leopard seals themselves are prey for killer whales.

Feeding and prey Leopard seals are probably best known for their habits of preying upon penguins. Their diet is actually quite varied and changes with seasonal and local abundance of prey. Leopard seals will consume krill, fish, squid, several species of penguins, a variety of other types of seabirds, and young seals, including phocids, such as young crabeater and southern elephant seals and otariids, such as young fur seals. They will also occasionally scavenge from carcasses of whales. Most prey is caught in the water, but penguins are sometimes taken while hauled-out. The seals sling penguins in an arc with a rapid snap

This leopard seal was stalking a group of gentoo penguins on a floe by moving around the perimeter, possibly trying to get them to flush into the water. PHOTO: M. NOLAN

Leopard seal with a fledgling Adélie penguin. PHOTO: M. NOLAN

Leopard seal showing its very long neck while craning to look onto an ice floe. PHOTO: M. NOLAN

Leopard seal with an Adélie penguin. PHOTO: M. NOLAN

Leopard seal slinging a chinstrap penguin to tear it apart. PHOTO: M. NOLAN

A leopard seal with a fish. Leopard seals eat a wide variety of prey species. PHOTO: M. NOLAN

of the head and neck, smashing them at the surface to tear them open. Smaller pieces are then swallowed. The loud crack produced when a leopard seal slaps a penguin on the water's surface has been reported to be audible in air for more than 1 km. Naive, newly fledged penguins are most vulnerable, but adult birds are taken as well. Leopard seals patrol and regularly station themselves just off penguin rookeries, waiting to ambush and chase penguins transiting to and from these colonies. Their predation on crabeater seals is evidenced by the many adults that bare long parallel scars from leopard seal canines.

Threats and status The leopard seal has not been subject to significant historical commercial sealing operations, and is not listed as endangered or threatened. The population is estimated to be 220,000 to 440,000, with highest densities occurring in pack ice with large brash ice blocks and cake ice floes that are 2–20 m in diameter. Leopard seals are protected by the Antarctic Treaty and the Convention for the Conservation of Antarctic Seals. There are currently no threats from human activity throughout most of the species' normal range. The effect of global climate change on the temporal patterns of Antarctic ice formation and the overall extent of pack ice, and leopard seal prey, poses an unknown threat to this ice-dependent pinniped species.

IUCN status Least Concern.

Weddell Seal—*Leptonychotes weddellii*

(Lesson, 1826)

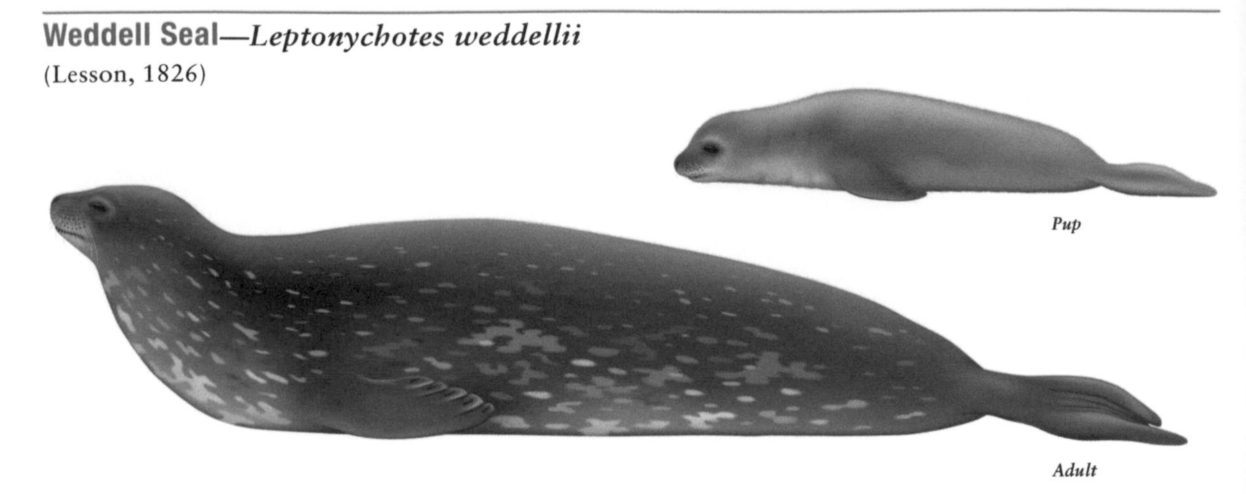

Recently-used synonyms *Leptonychotes weddelli.*
Common names En.–Weddell seal; Sp.–*foca de Weddell;* Fr.–*phoque de Weddell.*
Taxonomic information Carnivora, Phocidae.
Species characteristics Weddell seals are long, heavy-bodied phocids. The chest and abdomen of the fusiform body are long and thick. The neck is short and the head is small and rounded. The eyes are forward-facing and large. The muzzle is short and somewhat narrow with a sparse number of inconspicuous, short vibrissae. Taken together, the head and muzzle shape, large eyes, and upturned mouth line impart a feline appearance to the face. Weddell seal foreflippers are proportionately short and narrow, and owing to the short neck, originate in the forward ⅓ of the body. In freshly molted pelage, adult Weddell seals are generally dark gray above with silver to light gray highlights, and pale gray to off-white below. There is a variable amount of light and dark spotting, streaking, and blotching. These markings are fewer or absent on the top of the back, and become heavier on the sides and ventral surface, where the markings can fuse into irregular shapes. Dorsal color fades from a rich, very dark gray just after the molt, to a dull brownish-gray over the 11–12 months prior to the start of the next molt. The pelage on the end and sides of the muzzle has light gray to whitish highlights, and sometimes there are similar crescent-shaped highlights above and behind the eyes. Pups are born in a woolly silver-gray coat, with a darker swath along the mid-line of the back. This lanugo coat is molted for the adult pelage starting 1–4 weeks after birth, and the molt is usually completed in 2–3 weeks. A fully albino juvenile was reported from Cape Shirreff, Livingston Island.

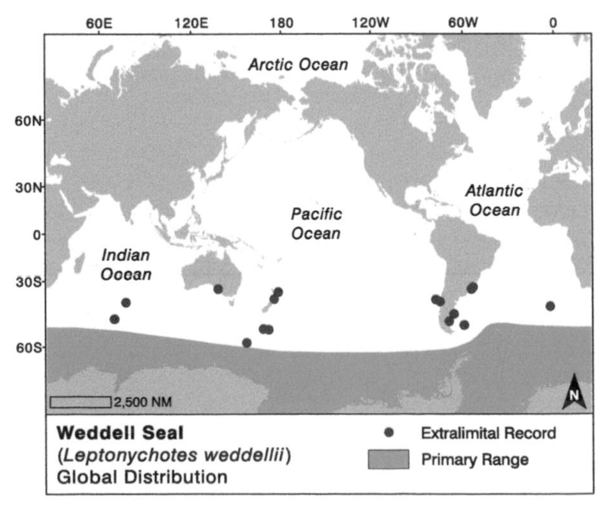

Adult males are up to 2.9 m long, and females attain lengths 3.3 m. Adults weigh 400–600 kg. Females are longer and heavier than males, but not by enough to permit separation of the sexes in the field based on appearance. The weight of adult females fluctuates during the year, as they experience significant weight loss during lactation. Pups are 1.2 m long and 25–30 kg at birth. They live to be about 25 years old. The dental formula is I $^{2}/_{2}$, C $^{1}/_{1}$, PC $^{5}/_{5}$. The second upper incisors and the upper canines are procumbent, angled somewhat forward as a special adaptation for rasping ice to keep breathing holes open.
Recognizable geographic forms None.
Can be confused with Four phocids share the Weddell seal's range. Of these, Ross and crabeater seals are the most similar, and leopard and southern elephant seals are more readily separated from Weddell seals. Ross seals differ in having a proportionately longer and wider head and neck, very small muzzle, and thickness in the throat and under the jaw; taken together, these features impart a blunt profile to the face. They also have dark streaks on the neck, throat and sides, and proportionately longer foreflippers. Crabeater seals have a longer head and muzzle, with an upturned appearance to the end

A large adult Weddell showing the complex coat pattern in this species. The brown and tan tones indicate this animal's coat is faded and worn, and its annual molt is probably approaching. PHOTO: M. NOLAN

This large Weddell seal nicely shows the short head, muzzle and foreflippers that are key identification features of this Antarctic phocid. Near the Antarctic Peninsula. PHOTO: M. NOLAN

A Weddell seal at rest showing the large body mass in this species. PHOTO: J. BASTIDA

Weddell and crabeater seals together. Notice the longer muzzle on the crabeater and proportionately small head of the Weddell. Weddells are always more spotted and marked than crabeaters. PHOTO: K. ABERNATHY

A Weddell seal's foreflipper is much shorter than on any other Antarctic phocid and the first digit is somewhat longer, but not dramatically so, than digits 2-4. PHOTO: M. NOLAN

The Weddell seal's face is short and the head more rounded than on the other Antarctic phocids. PHOTO: L. BLIGHT

An adult Weddell seal with a fresh-looking dark gray coat with complex light gray, white and dark blotches and irregular markings. PHOTO: D. BAETSCHER

A long, thin-looking Weddell seal beginning to molt out of its old faded coat into a fresh darker gray mottled pelage. Molt in many pinnipeds begins in the face and around the insertion of the flippers. PHOTO: N. TAPSON

A newborn Weddell seal pup in lanugo. PHOTO: D. PYLYPENKO

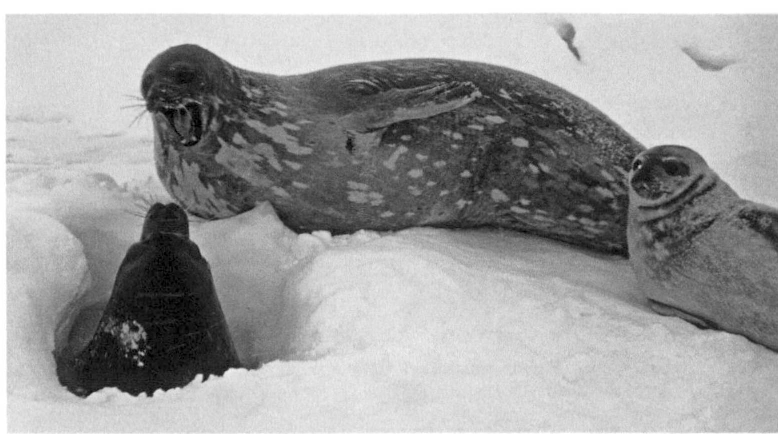

Adult female with her molting pup, and a third seal at a hole in the fast ice. The markings on the female are typical for Weddell seals, and the pup is molting its lanugo. PHOTO: K. WESTERSKOV

The Weddell seal has a small head and short muzzle as seen on this juvenile. Paulet Island, Weddell Sea. PHOTO: M. A. WEBBER

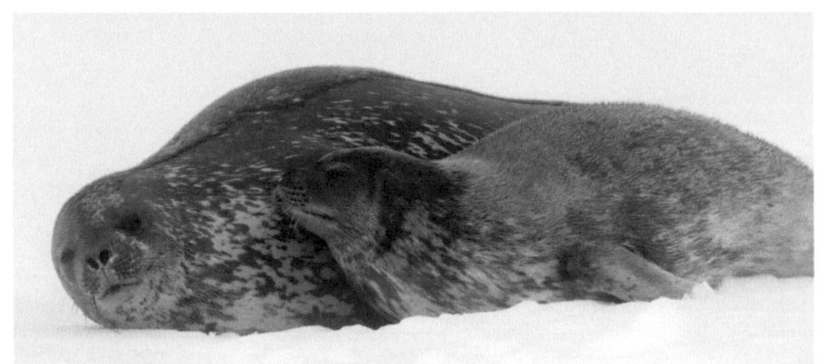

A female Weddell seal with her older molting pup. PHOTO: H. LAWSON

of the muzzle. They are also paler dorsally and on their sides, with few or no spots. Crabeater seals have proportionately longer more pointed foreflippers than than those on Weddell seals. Also, when compared to crabeater and Ross seals, the Weddell seals' foreflippers appear to be positioned more forward on the body owing to their relatively long body and short neck. Differences in how they respond to being disturbed when out of the water may help distinguish the species: crabeater seals rise up with their head level, or angled slightly up, and their belly down; Ross seals often assume a "singing posture," arching the back with the head held up, mouth open, vocalizing while inflating the throat to show off the thick neck, often with the foreflippers under the chest; and Weddell seals regularly roll to one side raising the head and often a foreflipper "in salute" (crabeaters will do this occasionally as well).

Distribution Circumpolar and widespread in the Southern Hemisphere, Weddell seals occur in large numbers on fast ice, right up to the Antarctic continent. They also occur offshore in the pack ice zone north to the seasonally shifting limits of the Antarctic Convergence. Weddell seals are present at many subantarctic islands and islands along the Antarctic Peninsula that are seasonally ice-free. A small population lives on South Georgia. Vagrants have been recorded in many areas further north including: the central coast of Chile, the Juan Fernández Islands, Argentina, the Falkland Islands, Uruguay, New Zealand and southern Australia.

Ecology and behavior The Weddell seal is the world's most southerly breeding mammal. Females become mature at 3–6 years of age and males at 7–8 years. Gestation lasts 11 months, including a delayed implantation (diapause) of 2 months. Pups are born from September through November and nursed for about 4–6 weeks. Animals in lower latitudes pup earlier than animals living at higher latitudes. Males set up territories in the water around access holes in the ice used by females to enter and leave the water. The only copulation that has been observed occurred underwater. After breeding, the adults forage intensively for 1–2 months before spending time hauled-out on the ice as they undergo a molt. After weaning, pups move away from the fast ice to spend 3–4 years in the pack ice, or open water, before returning when sexually mature. Weddell seals are not very social when out of the water, avoiding physical contact most of the

Weddell seal habitat along the Antarctic coast is at the shoreline and foot of glacial ice tongues along the tidal cracks and nearshore fracture lines. PHOTO: J. A. FRANEKER/IMARES

A Weddell seal at sea should be almost unmistakable, based on the small head and short muzzle. PHOTO: D. PYLYPENKO

A Weddell seal with one of its favorite prey species, the giant Antarctic toothfish. PHOTO: K. WESTERSKOV

time. When in fast ice habitat, they tend to congregate in groups along recurrent cracks, leads, and near access holes to the water. There is debate over whether or not this species is migratory. Some individuals remain in residence year-round in the fast ice at latitudes as high as 78°S in McMurdo Sound. Others, particularly newly weaned and subadult animals, move north from the continent and spend the winter in the pack ice.

Weddell seals are superb divers, reaching over 600 m on dives that last up to 82 min. Typically, dives are to 400–800 m for 15–40 min. Deep dives are regularly used for foraging and long dives for searching for new breathing holes, cracks and leads. Seals living in fast ice, or facing freezing over of access holes and leads, abrade and grind the ice to maintain access to and from the water. They bite at the ice and then rapidly swing the head from side to side as they rasp and grind away the ice with their teeth. Predators include killer whales and leopard seals.

Feeding and prey The diving abilities of this species are helpful in finding new breathing holes and obtaining important prey, such as the huge Antarctic toothfish (*Dissostichus mawsoni*), and other notothenioid fishes, as well as the squid and other invertebrates that make up small percentages of its diet.

Threats and status Weddell seals served as an important source of food for men and dogs throughout the historic period of Antarctic exploration, and they continued to be taken for sled dog food into the 1980s. Local populations of seals no doubt suffered declines from these harvests,

A rare albino Weddell seal from Cape Shirreff, Livingston Island. PHOTO: J. ACEVEDO

but in the case of the population in McMurdo Sound, the population has recovered in the years since the harvest ended. The population of Weddell seals has been estimated to be up to 1 million. This is a widespread species and at present it has no immediate threats, though the potential effects of global climate change are unknown. The Weddell seal is protected by the Antarctic Treaty and the Convention for the Conservation of Antarctic Seals.

IUCN status Least Concern.

Southern Elephant Seal—*Mirounga leonina*

(Linnaeus, 1758)

Adult Male

Pup

Adult Female

Recently-used synonyms None.

Common names En.–southern elephant seal; Sp.–*foca elephante del sur*; Fr.–*éléphant de mer austral*.

Taxonomic information Carnivora, Phocidae.

Species characteristics The southern elephant seal is the largest species of pinniped. In both sexes, the body is robust and the neck is very thick. The head, muzzle, and lower jaw are broad, and in females and juveniles in particular the face looks very short and wide due to the short muzzle. The mystacial area and nose are fleshy and blunt, and the bridge of the nose angles down slightly to the nostrils on females and young subadult males. The eyes are large, widely-spaced, and have a forward orientation that is especially noticeable in females and young animals. The vibrissae are beaded, short to medium length, and black. In addition to the mystacial clusters, there are one to two rhinal vibrissae on the bridge of the nose, and several in prominent supra-orbital patches above and slightly behind each eye. Each foreflipper digit bears a large blackish-brown nail. The nails on the hindflippers rarely emerge, but there is a vestigial nail beneath a small opening located above the end of each hindflipper digit.

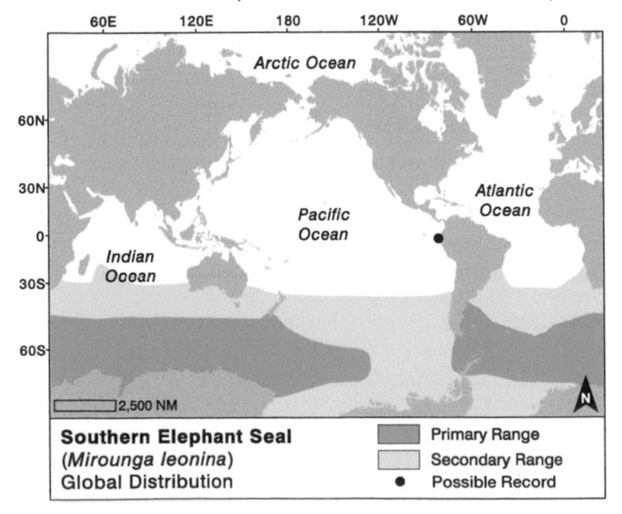

Adult males are unmistakable. The long and enlarged nose, the proboscis, is inflatable. When relaxed, it hangs down to just in front of the mouth. Curiously, the proboscis is much shorter in the southern than in the northern elephant seal, even though the former has a larger body. The proboscis is said to enlarge somewhat during the breeding season. Bulls also develop a chest shield of thickened, creased and heavily scarred skin. The chest shield on southern elephant seals is also not as developed as those on northern elephant seals. There is a varying amount of scarring on the rest of the body, and the proboscis is often heavily torn and scarred. Many bulls become pale in the face, proboscis, and head with increasing age. Bulls are longer and much heavier than adult females, and also have longer and thicker canines.

A long sandy or gravelly beach is typical habitat for southern elephant seals, as here on South Georgia. PHOTO: M. NOLAN

Male southern elephant seals are the largest pinniped, and they even dwarf the females of their own species. PHOTO: M. NOLAN

An adult female bites as a male, interested in copulating, attempts to mount her. PHOTO: M. NOLAN

A pale adult female at risk of being separated from her pup, and having it crushed by an advancing bull. PHOTO: M. NOLAN

An adult male issuing a threat-vocalization at South Georgia. PHOTO: M. NOLAN

A rival bull comes up behind a bull, attempting to copulate with a female. PHOTO: M. NOLAN

Molting subadult males squabbling at South Georgia. PHOTO: M. NOLAN

Southern elephant seals have a very flexible back and can raise more than 2/3 of their body off the ground to fight or threaten rivals. PHOTO: M. NOLAN

Two bulls about to square-off at the shoreline. Both are roaring. PHOTO: M. NOLAN

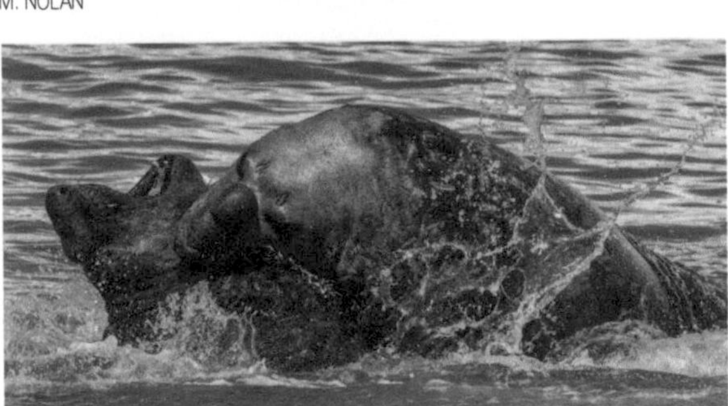

Bulls fight by rearing up and smashing their head and neck into rivals, and by fiercely biting the head, throat and neck. The great mass of large males is used to push smaller animals back, put them off balance and force a retreat. Losers often get bitten on the back and rear as a send-off when they retreat. PHOTOS: M. NOLAN

Adult southern elephant seals have an unspotted pelage of light to dark gray, or tan to dark brown. There is a slight difference between dorsal and ventral coloration, except in newly molted pups, which are somewhat more countershaded. Adult females can be tan, gray, or light brown, and are lightly countershaded. Their coloration fades with time after the last annual molt. Southern elephant seals also become stained rusty orange and brown, from lying in their own excrement on beaches, in muddy wallows, and tussock grass areas.

During the annual epidermal molt, a type of molt they share with northern elephant seals and the monk seals, they look ragged and multicolored, as they slough patches of faded and soiled fur with skin attached to reveal the new, clean, dark silver to gray fur coming in underneath. The molt begins in the axillary region, between the hindflippers, around the tail, and progresses around the body. Pups are born in a long woolly, black lanugo coat that is shed at about 3 weeks of age, to reveal a silver-gray countershaded coat that is yellowish-gray ventrally. Southern elephant seals of all ages will scoop and throw sand on their backs for thermoregulation and as a displacement behavior when stressed.

Adult males are 4.5 m in length (maximum 5.8 m), and weigh 1,500–3,000 kg (maximum 3,700 kg). Older accounts of larger males, up to 6–7 m in length, may include the hindflippers, as opposed to today's standard length measurements from the tip of the nose to the tip of the tail. Adult females are similar in size and weight to northern elephant seal females, weighing about 600 kg up to 800 kg. Pups are 1.2 m and 45 kg at birth. Longevity is rarely more than 10 years for males, and 14 plus years for females. The dental formula is I $^{2}/_{1}$, C $^{1}/_{1}$, PC $^{5}/_{5}$.

Recognizable geographic forms None.

Can be confused with The massive head and the large fleshy proboscis make southern elephant seal bulls virtually un-

Bulls often bite females on the head, neck and back in their efforts to restrain them for copulation. PHOTO: M. NOLAN

During copulation, here in the surf, bulls will drape a flipper over a female to hold her in place. PHOTO: M. NOLAN

An adult female giving birth (left) and having her first interactions with her newborn in its black lanugo. Females recognize their pup based on scent and vocalizations, and early initial contact is essential for imprinting. PHOTOS: M. NOLAN

A young pup in its black lanugo coat nursing. PHOTO: L. BLIGHT

A nearly-spent female with a pup in lanugo at Peninsula Valdes, Argentina. Elephant seals are "capital breeders" and nurse their pups without leaving to forage, losing considerable mass by the time of weaning. PHOTO: P. FAIFERMAN

A weaned pup (or "weaner"), with a probably-weaned large pup in lanugo. PHOTO: M. NOLAN

Once their mothers depart, weaners gather into groups to sleep and socialize. Here a few on South Georgia lie in a flooded tussock grass area. PHOTO: M. NOLAN

Southern elephant seals face many hazards before weaning. Adult males are often oblivious to them and frequently crawl over or stop on them, crushing bones, organs, and causing suffocation (left). Male aggression is also a risk (right). PHOTOS: M. NOLAN

Male aggression is not the only risk of violence to pups. Here an undersized orphan (thin and injured) faces attacks to drive it off by the females it approaches in the colony. PHOTO: M. NOLAN

Adult females and juveniles with deeply stained and worn pelage begin their annual molt. PHOTO: M. NOLAN

mistakable. Three of the four other phocids that occur within the southern elephant seal's range, Weddell, Ross, and leopard seals, can be separated from any age southern elephant seal by their coloration and the presence of spots, blotches of color, or streaks that are lacking on southern elephant seals. With the exception of crabeater seals, all three other species are noticeably countershaded as adults, whereas southern elephant seals have subtle countershading as subadults and adults. Crabeater seals are relatively thin with a narrow head, smaller eyes, a relatively long muzzle, sub-terminal nostrils, and proportionately longer and more pointed foreflippers that lack conspicuous dark nails. The head of female and subadult elephant seals is relatively large and broad and the muzzle is also large, wide and blunt, with more forward-facing nostrils. The vibrissae of elephant seals are black and relatively long, whereas they are inconspicuous and pale all other Antarctic phocids.

Distribution Southern elephant seals have a nearly circumpolar distribution in the Southern Hemisphere. Although they reach the Antarctic continent, and even very high latitude locations such as Ross Island, they are most common north of the seasonally shifting pack ice, especially in subantarctic waters where most rookeries and haulouts are located. Notable exceptions include the northern breeding colonies at Peninsula Valdes in Argentina, and on the Falkland Islands. Also, some pups are born on the Antarctic continent. Southern elephant seals prefer sandy and cobble beaches, but will haul out on ice, snow, and rocky terraces, and regularly rest above the beach in tussock grass and other vegetation, and in mud wallows. At sea, females and males disperse to different feeding grounds. Vagrant southern elephant seals reach the Juan Fernández Islands and the central coast of Chile, Sergipe State in Brazil, southern Africa, southern Australia, New Zealand and Brazil. The northernmost record is at Oman on the Arabian Peninsula. Reports of two sightings of elephant seals (1998 and 2002) came from the Gulf of Guayquil, Ecuador. The authors suspected both animals to be vagrant southern elephant seals, and photos of the 2002 animal suggest this identification, but neither occurrence was confirmed to species.

Ecology and behavior Southern elephant seals spend a large portion of their lives at sea far from land, and only return to land to give birth, breed, and molt. At sea, they travel great distances from their rookeries where they dive nearly continuously, feeding in a belt between the Subantarctic Convergence and the northern edge of the pack ice that lies south of the Antarctic Convergence. Adult males venture further south than females, and are known to forage at the outer edge of the Antarctic continental shelf. They are prodigious divers and routinely reach the same depths

Southern elephant seals will toss or flip sand or gravel on their backs for cooling on warm days. Sand flipping is also a displacement behavior used when an animal is threatened or nervous. PHOTO: H. LAWSON

At sea, southern elephant seals occasionally rest at the surface vertically like a floating bottle with the nose up, and present a bell-shaped profile. PHOTO: R. L. PITMAN

as their northern counterparts. Dive depth and duration vary during the year and between the sexes, but normally range from 400 to 800 m deep and from 15–40 min in duration. A maximum depth of 1430 m was recorded for a female, following her return to sea after the molt. Another post-molt female dove for an astonishing 120 min, coincidentally right at the same duration as the longest dive by a northern elephant seal, and both are the longest dives recorded for pinnipeds. It has been suggested that these very long dives, which are very rare, might be used to avoid more surface-oriented predators like killer whales.

Southern elephant seals are highly polygynous, with males establishing dominance hierarchies on beaches to monopolize access to females. Adult males return first in August, and most of them are ashore by mid-September to establish their status in the breeding hierarchy through displays, vocalizing, and also violent fighting. One of the male's most impressive displays is achieved by rearing up on his hindquarters and lifting almost ⅔ of his bulk straight up to fight with a peer or issue vocal challenges to nearby bulls. Pregnant females arrive from September to October, and usually give birth within a few days of their return. Pups are nursed for an average of 23 days, and abruptly weaned when the female departs to sea. Females come into estrus about 4 days before they wean their pup and mate, starting a new reproductive cycle before completing their current effort. Females first return to mate between 3–6 years of age. Pups remain on the breeding beaches for 8–10 weeks, during which time they complete the molt of their lanugo coat before departing to sea. Vocalizations include a loud booming call made by adult males during the breeding season, variously called a bubbling roar, a harsh rattling sound, and a low-pitched series of pulses with little variation in frequency. Adult females have a high-pitched, yodeling call, which they use when distressed, and to call their pups. They will also utter a low-pitched sputtering growl. Pups call to their mothers with a sharp bark or yap, which is also used when interacting with other seals.

Feeding and prey Prey consists of approximately 75% squid and 25% fish. Antarctic *Notothenia* fishes are thought to be important prey when these seals are near the Antarctic continental shelf. Most feeding by females occurs in deep ocean areas at mid-water depths. Adult males pass through female feeding areas on their way south to Antarctic continental slope and shelf waters, where their deep diving activity suggests they pursue more benthic prey. In two unusual cases, a vagrant bull southern elephant seal pursued and successfully caught and consumed Cape fur seal pups in Plettenberg Bay, South Africa. At Punta Tombo, Argentina, a bull hauled out in a Magellanic penguin colony for three consecutive years, killed and occasionally consumed numerous penguins on land. Observers thought this seal was being more aggressive than predatory in its behavior.

Threats and status The worldwide population of southern elephant seals, estimated in the 1990s, is 650,000. Colonies in the South Atlantic, which include the largest breeding aggregation at South Georgia, are stable or growing, while those in the Southern Indian and Pacific Oceans have decreased by up to 50%. The reasons for these declines, which began in the 1950s and 1960s and are ongoing, are not well understood, but may be related to lower weaning weights due to a reduced food supply. Archaeological digs have revealed that southern elephant seals were hunted by the aboriginal and native peoples of Australia and South America. More recently, they were subjected to intensive commercial harvests starting in the early 19th century and not ending until 1964 at South Georgia. They were prized for their large quantity of blubber that could be rendered to fine, valuable oil. There are few threats and conflicts today, as southern elephant seals live far from human population centers and have minimal interactions with commercial fisheries. However, development of new fisheries at high latitudes in the future, as well as potential effects of global climate change, could have a significant impact on southern elephant seals.

IUCN status Least Concern.

Northern Elephant Seal—*Mirounga angustirostris*
(Gill, 1866)

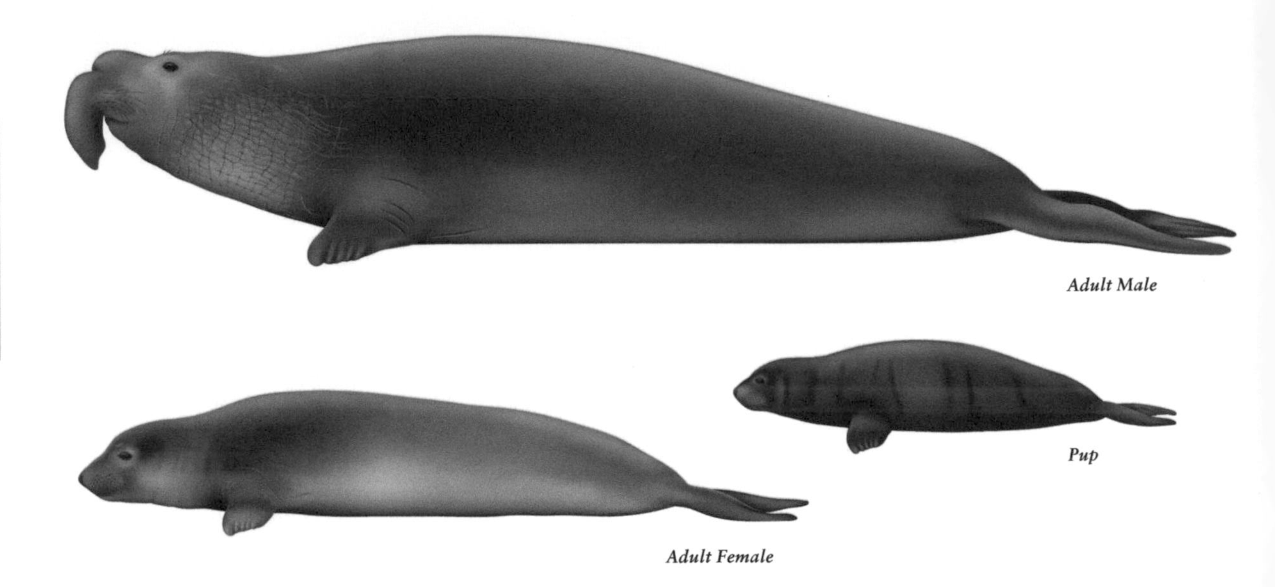

Recently-used synonyms None.
Common names En.–northern elephant seal; Sp.–*foca elephante del norte*; Fr.–*éléphant de mer boréal.*
Taxonomic information Carnivora, Phocidae.
Species characteristics Northern elephant seals are huge and imposing. Significant sexual dimorphism exists in size and secondary sexual characteristics. In both sexes, the body is long and robust, and the neck very thick. Adult males are unmistakable because of their great size, heavily scarred and cornified chest and neck, and large inflatable fleshy nose, called a proboscis. When relaxed, the proboscis hangs down in front of the mouth and resembles a shortened elephant trunk, thus the species' common name. The chest shield is made up of thickened, creased, hardened, scarred, nearly hairless skin. The chest shield and neck become increasingly pink over a large area as the male ages, and in old bulls it can completely ring the neck. Combat, which occurs when bulls are ashore during the breeding season, adds more red coloration from raw open wounds and patches of dried blood crusted on the shield. Wounds and scars accumulate all over the back, sides, and head of males during their lives, and as these heal, the scars become tan to lighter gray on the darker pelage. The proboscis can be bloody and ragged from fighting, and is also often pink and scarred. As subadult males grow, their proboscis lengthens so that it may extend 15–25 cm below the lower lip. A male ashore, asleep on his chest and chin, is readily identified as an older bull when his proboscis is long enough to touch the ground and be bent under or off to the side. In addition to their much longer and larger body size, males also have larger canines that are both considerably longer and thicker than those of adult females. Adult females, juveniles and young subadult males do not have a proboscis, chest shield or extensive scarring, and instead their key feature is a relatively large, wide head and wide muzzle with rounded thick and fleshy mystacial pads. In

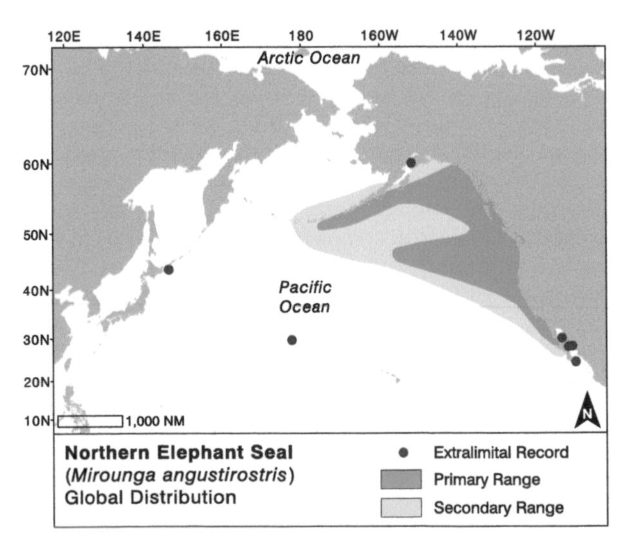

A bull northern elephant seal issues a clap threat call. Notice the long proboscis of the males of this species hangs into the mouth while vocalizing. PHOTO: F. R. ELORRIAGA-VERPLANCKEN/CICIMAR-IPN

Adult male and female, and pup in black lanugo. San Benitos Islands, Mexico. PHOTO: V. SHORE

Head of an adult female northern elephant seal. The face and tip of the nose are longer, and more pointed than in southern elephant seals. PHOTO: V. SHORE

Head and neck of a nearly fully-developed adult male northern elephant seal showing the long proboscis, pale face and head, and pink chest shield. PHOTO: S. EBBERT

profile, the bridge of the nose forms a slight arc down from the bridge of the nose to the terminally placed, and slightly down-angled, nostrils.

The eyes of elephant seals are large, widely-spaced, and have a forward orientation that is especially noticeable in females and younger animals. The vibrissae are beaded, short to medium length, and black. In addition to the mystacial clusters, there are one to two rhinal vibrissae on the bridge of the nose, and several in prominent supra-orbital patches above and slightly behind each eye. Each foreflipper digit bears a large blackish-brown nail. The nails on the hindflippers rarely emerge, but there is a vestigial nail beneath a small opening located above the end of each hindflipper digit.

Northern elephant seal coloration varies from uniformly dark charcoal-gray or dark brown to light gray or tan, with coloration fading over time since the end of the last annual molt. With the exception of adult males, most northern elephant seals are lightly to moderately countershaded, being slightly darker above and paler below. Adult males are typically dark brown to charcoal overall, with the already noted coloration of chest shield, neck, proboscis and body scarring. As males age, the color on the head and face around the eyes and cheeks becomes a paler yellowish-tan. Northern elephant seals are multicolored and ragged looking during their annual molt, when they shed their fur and epidermis together in patches, starting in the axillary region and progressing around the body. During this time they have varying amounts of brighter looking more richly-colored new fur appearing as the faded duller fur is lost. Pups are born in a long woolly black lanugo coat that starts to shed without the epidermis at about 3 weeks, and usually ends after the pup is weaned. After molting, the newly weaned juvenile's coat is made up of short hairs like those found on adults, and it is countershaded dark gray above and silver-gray below. Northern elephant seals of all ages will scoop and throw sand on their backs for thermoregulation and as a displacement behavior when stressed.

Sexual dimorphism is significant in this species, with males being 3–4 times heavier, and nearly 1.5 times the length of adult females. Adult males are 4.2 m long and

Adult male northern elephant seals fighting and confronting each other. Battles can be fierce and bloody. Males sustain numerous injuries, and their thickened chest shields can become raw. PHOTOS: F. R. ELORRIAGA-VERPLANCKEN/CICIMAR-IPN

A young subadult male, probably 4-5 years old, northern elephant seal has virtually no droop or drop to the proboscis. San Benitos Islands, Mexico. PHOTO: M. JØRGENSEN

A subadult male, probably 6-7 years old, with a proboscis hanging in front of the mouth, but without the deep creases and enlarged chambers; it would just touch the ground when lying on its chest. PHOTO: M. A. WEBBER

An older subadult/small adult male, about 8 years old, with a longer nose that reaches the ground, and a moderately-developed chest shield and the start of pale color in the face. PHOTO: M. A. WEBBER

A fully mature bull, 9 years old plus, with a large, long proboscis, a very well-developed chest shield, and extensive pale in the face and on the crown. PHOTO: T. A. JEFFERSON

A bull restrains a cow for mating by biting her neck and head and by draping a foreflipper over her back; this is typical mating behavior for the species. San Benitos Islands, Mexico. PHOTO: I. SZCZEPANIAK

weigh up to 2,500 kg. On average, their length is 3.8 m and weight is 1,844 kg. Bulls are at their heaviest when they arrive for the breeding season after feeding during the summer and fall. The average length of adult females is 2.7 m, with a maximum of 2.8 m. They are 300–600 kg, with an average of 488 kg. A female, weighed after giving birth, reached the maximum known weight of 710 kg. Pups are 1.2 m long and 25–35 kg at birth. Body weight declines dramatically due to the demands of fasting during the breeding season. As "capital breeders," adult females can lose almost half their weight during lactation, when they take their single pup from birth to weaning in approximately 28 days without feeding, during which time the pup can quadruple its birth weight. The situation is similar for breeding males, which also lose about half their mass while ashore for periods that may exceed 90 days. Longevity is about 12–14 years for males, and 18 years for females. The dental formula is: I $^{2}/_{1}$, C $^{1}/_{1}$, PC $^{5}/_{5}$.

Recognizable geographic forms None.

Can be confused with The great size of northern elephant seal bulls makes them unmistakable. The massive head and the large fleshy proboscis are unique features. Only one other phocid, the harbor seal, regularly shares the range of the northern elephant seal, and it is much smaller, being only as large as a small juvenile northern elephant seal and has a spotted coat. Female and subadult male elephant seals can be distinguished from other vagrant seals within their range by a combination of features including: large body size, large size and shape of the head and muzzle, pelage coloration and the lack of spots, blotches or bands. The prominent, all-black vibrissae and large eyes are also distinctive.

Distribution Northern elephant seals are found in the eastern and central North Pacific. Breeding takes place on offshore islands and at mainland sites from central Baja California north through the Channel Islands to northern California, with some pups born at small colonies in

Females can also have aggressive encounters. They vigorously defend their pups and the space around themselves to keep from becoming separated from their pups on crowded rookeries. PHOTO: I. SZCZEPANIAK

A female with a young pup at the San Benitos Islands, Mexico. PHOTO: M. A. WEBBER

A depleted female with a nearly weaned pup at the San Benitos Islands, Mexico. PHOTO: M. A. WEBBER

A group of molted and weaned pups in late March at the San Benitos Islands, Mexico. PHOTO: I. SZCZEPANIAK

Subadult males ashore to molt. The three in front have completed most of the process. PHOTO: F. R. ELORRIAGA-VERPLANCKEN/CICIMAR-IPN

Subadult males at various stages of molting. When ashore to molt, elephant seals seek contact and can form huge tightly-packed aggregations on the beach. PHOTO: M .A. WEBBER

A juvenile ashore to molt. Northern elephant seals molt at different times of the year depending on age and sex, and they usually stay ashore until the molt is complete 3-4 weeks later. PHOTO: I. SZCZEPANIAK

A subadult male flipping or throwing sand on its back probably as a way to cool off, although it could be done as a displacement behavior if stressed. They dig down to the cooler damp soil to throw it up onto their backs. PHOTO: I. SZCZEPANIAK

Oregon and Race Rocks, British Columbia. Northern elephant seals migrate to and from their rookeries twice a year, returning to breed from December to March, and again to molt for several weeks, at different times, depending on sex and age. They also molt at additional coastal sites, as far north as southern Oregon. Their post-breeding and post-molt migrations take most seals north and west to oceanic areas of the North Pacific and Gulf of Alaska twice a year, with some reaching the Aleutian Islands chain (to 178°W). Vagrants have been found as far away as Japan and Midway Island.

Ecology and behavior Northern elephant seals are highly polygynous, but not territorial. Male competition for access to females establishes a hierarchy on the rookery. There is much male-male fighting, vocalizing, and displaying during the breeding season, when bulls may be ashore for months at a time. One of the most impressive displays occurs when a male rears up on his hindquarters, thrusts ½–⅔ of his body upward, rocks his head back so the proboscis falls into his mouth and produces a distinctive very loud clap-threat vocalization as a challenge to other bulls. The vocalization is a loud resonant, metallic series of automobile backfire-like sounds. Females have a throaty sputtering growl, made with the mouth wide open as a threat gesture. Females and pups have a warbling scream that they use to call to each other, and in the case of the pup, when disturbed. Females are sexually mature at 2–4 years, and males at 6–7 years, although they may not be able to breed until they are older and stronger. Gestation is 11 months, including a delay of implantation (diapause) of 2–3 months. Females give birth within 7–8 days of coming ashore, from late December to March, and wean their pups in 27 days on average. Around the time of weaning the female enters estrus and mates, usually before she weans the pup. Shortly after mating, the female weans the pup by abruptly leaving the colony. Mother and pup form a strong bond immediately after birth, and females aggressively bite other pups that approach them, occasionally killing them with bites to the head. Bulls also cause pup mortality by crushing them as they charge through mothers and pups to challenge approaching males. Occasionally, they suffocate pups by stopping on top of them and not moving off soon enough.

Northern elephant seals are one of the deepest-diving pinnipeds. Time-depth recording devices have documented dives to an astounding 1,754 m by an adult male. They also have extreme breath-holding ability and have been recorded to dive for as long as 120 min in an extreme and unusual case. Rest intervals at the surface are usually short, lasting only several minutes. They use a pattern of diving called "drift diving," thought to allow for intervals of rest while they slowly fall through the water column on descent. Each year, adult elephant seals travel to their foraging grounds after breeding, and again after their molt, swimming total distances of 18,000–21,000 km on this double migration. Males feed primarily over the continental margin, and tend to travel farther north in the Pacific than females, which forage in deeper pelagic water, while also making use of seamounts and guyots. Both sexes commonly make foraging dives to a wide range of depths from 300–800 m, but may focus for periods of time at consistent depths where they are presumably finding prey. After leaving the rookeries, these seals routinely spend over 90% of their time underwater, helping explain why they are infrequently seen at sea. They return to land to molt in separate phases. For example, at Año Nuevo Island in central California, the molt cycle is April–May for females and juveniles, May–June for subadult males, and July–August for adult males. Great white sharks and killer whales are predators of northern elephant seals. Research at the Farallon Islands off northern California revealed that great white sharks aggregate around the islands when elephant seals return for their molt. Seals that swim at or near the surface as they approach or depart the islands are particularly vulnerable to ambush attacks by fast-rising sharks that patrol near the bottom.

When at sea, northern elephant seals do not usually spend much time at the surface. However, this adult male, hanging vertically in the water column, has a dry nose and face indicating that he's been in this position for some time. Note the bell-shape of the head and neck which is an identifying feature at a distance. PHOTO: V. SHORE

Adult male resting on the bottom near shore during the breeding season, vocalizing. PHOTO: I. SZCZEPANIAK

Feeding and prey The diet of northern elephant seals includes over 50 species of prey, more than half of which are squid. Other prey includes fishes, such as Pacific whiting, several species of rockfish, ratfish and a variety of small sharks and rays. They have also been reported to feed on pelagic red crabs. The habitat of 70% of their prey is open ocean and includes some species from surface, but mostly from mid-water and benthic zones.

Threats and status The northern elephant seal was hunted to the brink of extinction in a surge of commercial exploitation in the late 1800s. Much speculation exists on the numbers of animals that survived this population bottleneck. Some estimates are as low as 50 animals or fewer at the last breeding colony on Guadalupe Island, off Baja California. Fortunately for this species, the fact that they spend most of their lives at sea, and do not all return to their rookeries at the same time, ensured that enough seals survived the era when sealers undertook wholesale slaughters at rookery sites. Following a slow recovery in the early 1900s, northern elephant seals began to recolonize former sites, a process that continued throughout the 1980s; by 1991 the population was estimated to have reached more than 110,000. The latest estimate of 171,000 suggests that the population is continuing to increase, although there has not been a recent range-wide survey to confirm this figure. Northern elephant seals are fully protected in Mexico and the US. Tourism at mainland locations in the US is popular, but highly regulated and not considered a threat. Incidental take in fisheries is low. Most of the prey of the northern elephant seal is either of low commercial value or minimally harvested in fisheries. As the population expanded, the seals have colonized more mainland beaches, such as Piedras Blancas and Point Reyes in California, and that has posed additional challenges to keep conflicts with humans and domestic animals to a minimum. The risk of transfer of diseases, such as morbillibivirus from other species to northern elephant seals is unknown. Having suffered a population bottleneck, this species has likely lost genetic diversity and may be at higher risk from disease or environmental change brought about by global climate change.

IUCN status Least Concern.

Bearded Seal—*Erignathus barbatus*

(Erxleben, 1777)

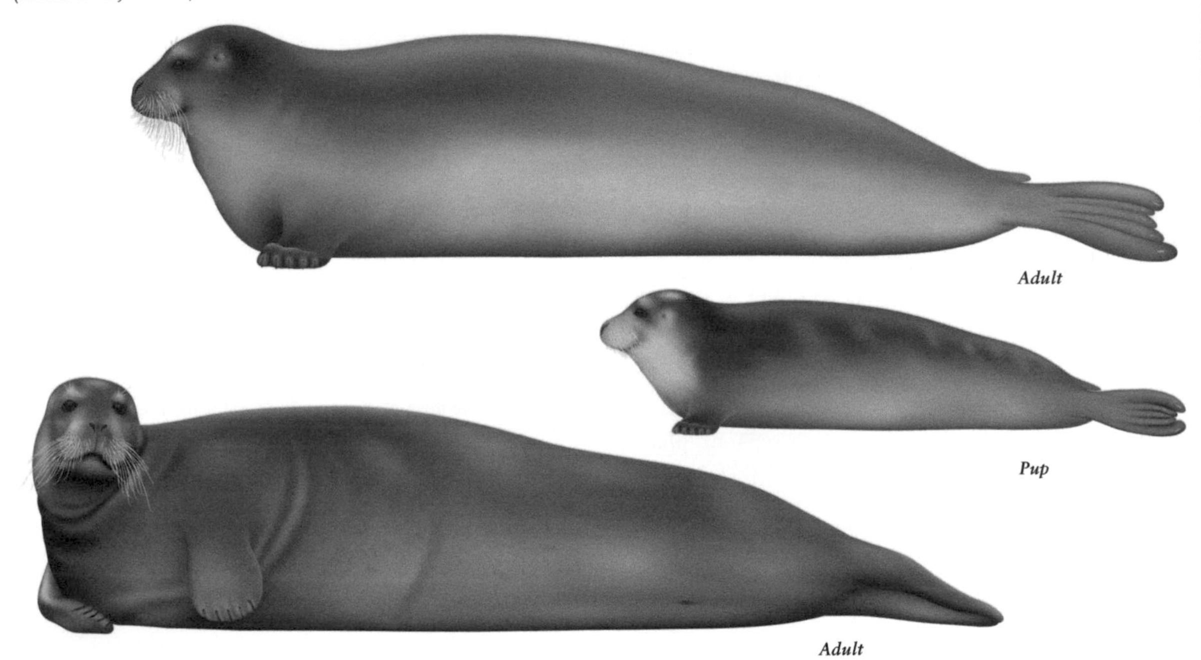

Recently-used synonyms None.

Common names En.–bearded seal, Atlantic or eastern bearded seal, Pacific or western bearded seal; Sp.–*foca barbuda;* Fr.–*phoque barbu.*

Taxonomic information Carnivora, Phocidae. Two subspecies are recognized: *E. b. barbatus (*Atlantic bearded seal) from the central Canadian Arctic east to the central Russian Arctic, including Greenland, Iceland, Jan Mayen and Svalbard, and south to Norway and Newfoundland: and *E. b. nauticus* (Pacific bearded seal) from the central Canadian Arctic west to the central Russian Arctic, south to the Karaginsky Gulf, Russia, and Bristol Bay, Alaska, including the Sea of Okhotsk, and south to northern Japan.

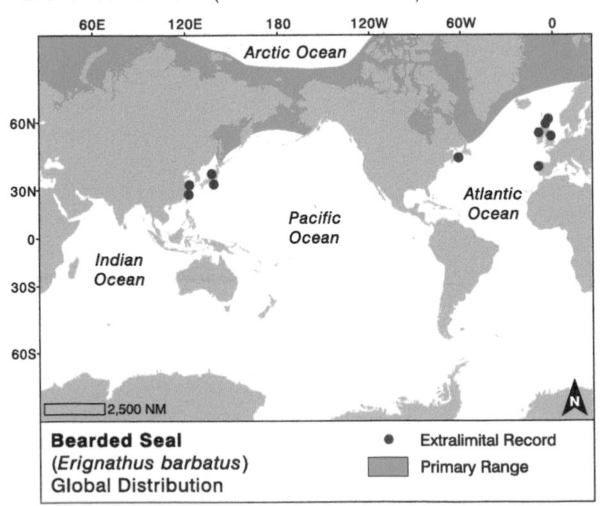

Bearded Seal
(*Erignathus barbatus*)
Global Distribution

Species characteristics Bearded seals are large phocids with relatively small heads and short, rounded to squared-off foreflippers. The head is small and somewhat narrow for the size of the animal, and the eyes are relatively small and close-set. The muzzle is wide and fleshy, with rounded mystacial pads that project slightly beyond the nostrils, which are not quite terminal on the muzzle, but slightly back and dorsal on the muzzle with the nostrils angled slightly upward. This positioning of nostrils is shared to varying degrees with Hawaiian and Mediterranean monk seals and the four Antarctic species: Weddell, crabeater, leopard, and Ross seals. The abundant pale, long, stout, downwardly curved, densely packed vibrissae are quite conspicuous, giving rise to the seal's common name, "bearded." The vibrissae are smooth, not beaded as in other Arctic phocids. When wet, the vibrissae tend to angle down and inwards. When dry, the tips of the vibrissae curl upward sometimes forming loops. The foreflippers are short, relatively broad, and strong, with robust claws. Unlike any other phocid, the bearded seal's foreflippers end in digits of about the same length, or with slightly longer middle digits. The result is a square, or slightly rounded end to the fore-flippers. Also, unlike all other Arctic and nearly all other phocids, bearded seals have four retractable mammae instead of

An Atlantic bearded seal with a dark brown coat. Svalbard. PHOTO: H. LAWSON

Adult Atlantic bearded seal with a generally pale coat and extensive spots and blotches, and an excellent beard of vibrissae. PHOTO: H. LAWSON

Dark gray Atlantic bearded seal color pattern. Svalbard. PHOTO: H. LAWSON

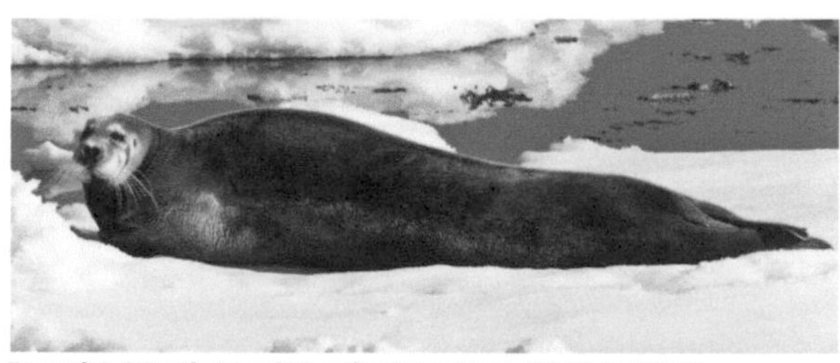

Brownish Atlantic bearded seal color pattern. PHOTO: M. NOLAN

A very pale bearded seal with a rusty tinge to the coat. The short rounded foreflippers are diagnostic for this species. PHOTO: L. MURRAY

Dark brown adult bearded seal with some reddish staining from iron on the face. Svalbard. PHOTO BY: H. LAWSON

Adult male Atlantic bearded seal with a strong rusty tinge to the face, and a good look at the rounded foreflippers. PHOTO: M. NOLAN

Adult bearded seal with yellowish-brown coloration. PHOTO: M. NOLAN

Juvenile Atlantic bearded seal. Svalbard. PHOTO BY: M. NOLAN

Face of a juvenile Atlantic bearded seal in the water. Svalbard. PHOTO: M. NOLAN

Adult Atlantic bearded seal raises its head for a look. Svalbard. PHOTO: M. NOLAN

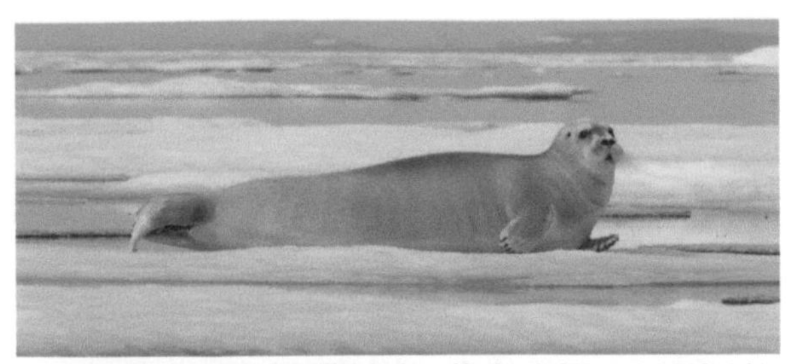

A pale Pacific bearded seal off Kotzebue, Alaska. PHOTO: M. CAMERON/NOAA FISHERIES/ASFC

A dark Pacific bearded seal off Kotzebue, Alaska. PHOTO: J. JANSEN/NOAA FISHERIES/ASFC

A juvenile Pacific bearded seal. Kotzebue, Alaska. PHOTO: J. JANSEN/NOAA FISHERIES/ASFC

A molting Pacific bearded seal with side scars. PHOTO: J. JANSEN/NOAA FISHERIES/ASFC

Bearded seals on a mudflat, including one with a very red-stained face from iron in benthic sediments. Sea of Okhotsk, Russia. PHOTO: O. SHPAK

Adult bearded seal hauled-out in a vegetated area. Sea of Okhotsk, Russia. PHOTO: O. SHPAK

Adult Pacific bearded seal craning its neck to see around ice floes. PHOTO: J. JANSEN/NOAA FISHERIES/NMML

A mixed group of Arctic seal species in the Sea of Okhotsk. The bearded seal is the large seal in the background. At far left is a spotted seal, and the rest of the smaller seals are Okhotsk ringed seals. PHOTO: O. SHPAK

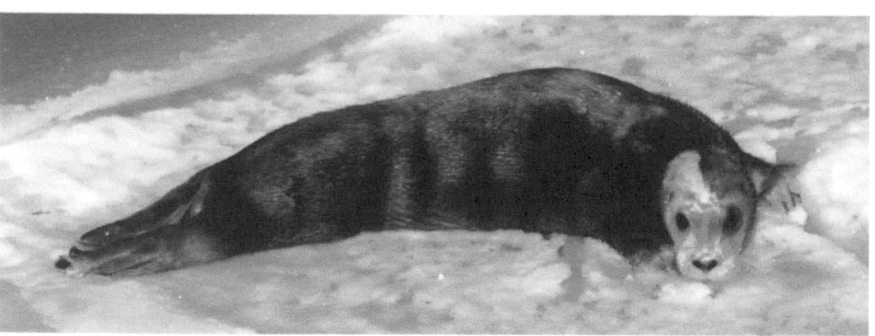

A newborn bearded seal (above), and a bearded seal pup in May, with whitish patches on the muzzle and above the eyes (right). Bering Sea. PHOTO: (ABOVE) G. BRADY/NOAA FISHERIES/NMML; (RIGHT) K. FROST

two, a feature that they also have in common with just the Hawaiian and Mediterranean monk seals.

Adults are slightly darker above than below. Body coloration varies considerably from light to dark gray, or tawny brown to dark brown. There can be rust coloration on the head, back of the neck and top of the foreflippers which is believed to result from the seals feeding by sticking their head, neck and flippers in muds containing iron compounds which adhere to the pelage, and then rust when exposed to the air and stain the hairs. The lanugo is shed in the uterus before birth and pups are born with a somewhat longer, dark brown to dark bluish-gray, wavy coat. There are pale areas on the muzzle, around the eyes, and sometimes on the crown and top of the back. A line of dark color runs from the top of the head between the eyes and along the bridge of the nose.

Adults are 2.1–2.5 m in length, with females slightly longer than males. In general adults reach 250–300 kg, with body weight known to fluctuate between seasons. A maximum weight for the species of over 400 kg has been recorded for very large females. Pups are, on average, about 1.3 m and 33.6 kg at birth. The dental formula is I $^{3}/_{2}$, C $^{1}/_{1}$, PC $^{5}/_{5}$.

Recognizable geographic forms The ranges of the two subspecies of bearded seal meet in the central Canadian Arctic and central Russian Arctic. The Pacific subspecies may be slightly longer and heavier than the Atlantic subspecies, but they are similar, and no distinguishing features are known that would permit their separation in the field.

Can be confused with Bearded seals regularly share their range with seven other Arctic and subarctic phocids, including harbor, spotted, ringed, ribbon, harp, hooded, and gray seals. All of these species are either strongly countershaded, have spots, rings, or many variably sized dark blotches, bands, or a combination of these features. In the North Atlantic, the bearded seal is most likely to be mistaken for the similar-sized hooded seal when hauled-out, except when the blotchy coloration, distinctive face, or diagnostic nose of the (male) hooded seal can be seen, or the dense vibrissae, small head, and square to rounded foreflippers of the bearded seal can be seen.

Distribution Bearded seals have a circumpolar distribution in the Arctic, generally south of 80°N. They are subarctic in some areas, such as the lower Bering Sea, Sea of Okhotsk to northern Japan, and western North Atlantic, where they reach the Gulf of St. Lawrence. Vagrants have been found in western Europe, the British Isles, to Massachusetts on the Atlantic coast of North America, south on both coasts of Japan to central Honshu, and on the coast of central China. Two genetically distinct Distinct Population Segments (DPS) for Pacific bearded seals have been identified: the Beringia DPS and the Sea of Okhotsk DPS.

Ecology and behavior Off Alaska, bearded seal males become sexually mature at 6–7 years of age, and females at 3–6 years. Both sexes reach physical maturity when aged 9–10. Pups are born in the open on the surface of the pack ice on smaller floes, from mid-March to early May. Unlike most other Arctic phocids, females do not stay with the pup continuously or nearly continuously until weaning, but forage extensively during lactation, which lasts for about 2–3 weeks. Another difference with many other phocids is that bearded seal pups can swim shortly after birth, and spend a considerable amount of independent time in the water even before weaning, when they begin to develop their diving skills. Somewhat counter-intuitively, the time females spend away from their dependent pups left alone on the ice is thought to be a strategy for reducing detectability by polar bears, the lone pup providing a much smaller visual cue than both mother and pup together. Females do swim with their pups between feeding bouts, and will attend the pup on ice to nurse 3–4 times per day. Pups are about 2.5 times their birth weight at weaning. Females mate in the water after weaning their pup, and males defend areas around females with elaborate vocalizations and aggressive behaviors while they wait for a female to come into estrus after pup weaning. After the breeding season, many seals migrate northward with the retreating ice, returning south again as the ice advances in fall and winter. The annual molt is in the summer.

Bearded seals are solitary and rarely haul out on the same ice floe together, and even then, maintain healthy

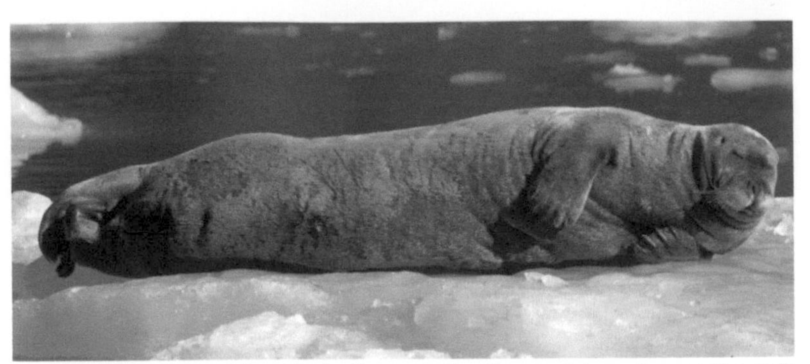

A molting adult male bearded seal. Notice the rounded foreflipper and strong claws, and the penile opening between the umbilical scar and the hindflippers. Svalbard. PHOTO: M. JØRGENSEN

distances from neighbors. Mother and pup pairs are an exception to this rule. At times, bearded seals seem to concentrate in an area, but this may be due to shifting winds and currents driving ice floes together or because of favorable feeding opportunities. Hauled-out bearded seals are exceptionally wary and always position themselves with their head very close to the water at the edge of an ice floe, along a crack, or by an access hole. In the water, bearded seals can be found "bottling," or floating vertically, head-up, asleep. When startled on ice, they bolt into the water and swim with strong strokes of the foreflippers, the head held up in a manner that looks like they are rising and surging through the surface of the water. When in the water and disturbed by a passing ship, bearded seals roll steeply forward to dive, raising the lower back and hindflippers in the air and slapping the flippers on the surface. This is often done repeatedly as the seal swims away. They usually restrict themselves to areas of sea ice and stay in relatively shallow continental shelf areas of continuously moving ice, where leads and polynyas (open water areas) regularly form. In some areas, they are known to haul out on shore, ascend streams, or live a pelagic existence away from ice and land for long periods of time.

Bearded seals primarily feed on or near the bottom, and live in shallow areas overlying the continental shelf. They generally dive to depths of 200 m or less and frequently are found in much shallower waters. The longest dives recorded have been 20–25 min, with most dives lasting less than 10 min. They are quite vocal in the water and are known for their oscillating frequency-modulated songs that can last more than a minute and be heard in the air, and for long distances underwater. Singing has largely been attributed to males, but it is likely that females sing as well. Besides humans, predators include polar bears, killer whales and Greenland sharks.

Feeding and prey Bearded seals feed on a large diversity of demersal fish species and invertebrates that live near, on and in the bottom. Different combinations of prey dominate feeding at different times of year and in different locations, and juveniles and adults have different prey preferences. In the Bering and Chukchi seas, fishes taken include capelin, Arctic and saffron cod, long-snouted pricklebacks, sculpins, flatfishes, several snailfish species, and eelpouts. Invertebrate prey includes several species each of crabs, clams, snails, amphipods, shrimps, marine worms, and octopuses.

Threats and status Native peoples of the Arctic have hunted bearded seals for subsistence for thousands of years. This practice continues to the present, with several thousand being taken annually throughout their circumpolar Arctic range. Bearded seals provide numerous products, including food for humans and sled dogs, oil for lamps, skins for clothing, boats and tent coverings, and leather for sinews, to name a few. Commercial exploitation by Soviet whalers in the Sea of Okhotsk and Bering and Chukchi seas during the mid-20th century resulted in harvests of up to 10,000 animals per year for several decades. In 2012, the US National Marine Fisheries Service proposed listing as Threatened under the Endangered Species Act, two Distinct Population Segments of the Pacific subspecies of bearded seal, the Beringia DPS and the Okhotsk DPS. As of this time, only the Okhotsk DPS is listed and the status of the Beringia DPS is not final. This conservation step was taken in large part because of dramatic decreases in the extent of Arctic sea ice over the last few decades, and predictions for additional losses in the future. Additional threats from sea ice reduction are increases in development of oil and gas exploration and extraction, and the risk of disease transmission to once-isolated populations of bearded seals from other pinniped species traveling from distant areas that have been exposed to diseases such as phocine distemper virus.

The worldwide population of bearded seals is variously estimated at 450,000–500,000, and even up to possibly 700,000, but their numbers are not well known because of the difficulties of conducting counts of animals over vast and remote polar areas. More than half of the world total are thought to live in the Bering and Chukchi seas, however others suggest an almost equally large portion of the population is in the Sea of Okhotsk, both areas being part of the Pacific bearded seal subspecies' range. Large numbers also occur in the Canadian Arctic, but almost nothing is known about numbers east of Canada, in most parts of the Atlantic Ocean, and in the Barents Sea east to where the Atlantic subspecies' range contacts the range of Pacific walruses, north of Chukotka (170°E).

IUCN status Least Concern.

Hooded Seal—*Cystophora cristata*

(Erxleben, 1777)

Adult Male

Pup

Adult Female

Recently-used synonyms None.
Common names En.–hooded seal; Sp.–*foca capuchina;* Fr.–*phoque à capuchon.*
Taxonomic information Carnivora, Phocidae.
Species characteristics Hooded seals are large, sexually dimorphic phocids. The head is wide at the bridge of the nose reflecting the widely separated orbits on the skull. The muzzle is also wide and overhangs the mouth in adult females and subadults. In adult males, there is an unmistakable inflatable nasal cavity in the form of a black bladder. When flaccid, it hangs down in front of the mouth; when inflated, it forms a taut, bi-lobed, crescent-shaped hood that almost doubles the size of the head and substantially elevates its profile. Another secondary sexual characteristic of male hooded seals is that after partially inflating the nose hood, they extrude the elastic nasal septum from a nostril and inflate it like a bright red balloon. The flippers of this species are relatively short, slightly pointed and angular, with a longer first digit. The vibrissae are beaded, relatively short, and inconspicuous; they are dark in pups and light in adults. Hooded seals have two mammae, typical of most northern phocids.

Adults are silvery-white, with numerous, small to large, irregularly shaped, dark blotches and clusters of blotches. The head to behind the eyes and jaws, and the tops of the flippers, are blackish. Pups, called "blue-backs," are born in a handsome countershaded coat of dark blue-gray above and creamy-white below. The dark color continues onto the hindflippers and also extends downward to include the foreflippers. The face and muzzle are very dark, with almost black coloration extending behind the eyes. The pale color rises high on the flanks and neck, and encompasses the lower jaw. "Blue-backs" retain their coat until the following summer, when they molt at 14 months and start to develop adult markings as juveniles.

Adult males reach lengths of 2.6 m and weights of 192–352 kg; females average about 2 m in length and weigh 145–300 kg. Pups are 87–115 cm and 20–30 kg at brith. Hooded seals live to between 25 and 30 years of age. The dental formula is I $^{2}/_{1}$, C $^{1}/_{1}$, PC $^{5}/_{5}$.
Recognizable geographic forms None.
Can be confused with Hooded seals share their range with five other phocids: harp, harbor, ringed, bearded, and gray seals, which can be distinguished by pelage color, and head shape and size. Harp seals are smaller and uniquely marked. Bearded seals have no dark blotches, are larger, have a small head, densely-packed, downward-pointing, smooth vibrissae, and four mammae. Gray seals have similar blotches, but have a very different size and shape to their head and muzzle, which is large and proportionately long.
Distribution Hooded seals are found in the Arctic Ocean, and in high latitudes of the North Atlantic. They breed on pack ice and are associated with it for most of their lives, shifting their distribution with seasonal fluctuations. The major whelping or pupping areas are in the western North Atlantic (the Gulf of St. Lawrence, north of Newfoundland

and east of Labrador, in the Davis Strait), and in the eastern North Atlantic (Greenland, Iceland and Jan Mayen Island). Hooded seals can wander widely, and vagrants have occurred as far south as the Azores, Portugal, southwest Spain, and in the Mediterranean in southern Spain and northwest Africa, and from New England to Florida in North America, and south to the Turks and Caicos Islands in the Caribbean. One exceptional dispersal event ended with an animal ashore in southern California. Presumably, it crossed the Canadian Arctic and Beaufort Sea, and then traveled down the Bering Sea and North Pacific to San Diego, where it beached.

Ecology and behavior Apart from breeding and molting periods, when hooded seals form loose aggregations, little is known of their activities, and it is presumed they are relatively solitary. Sexual maturity is reached at 4–6 and 3–5 years in males and females, respectively. Males are usually not able to breed until they are larger and able to compete effectively with other males. They birth their pups away from floe edges on pack ice in March and early April, and the breeding season is quite brief, usually only 2.5 weeks. Females are usually widely-spaced apart and aggressively defend their pups. Remarkably, pups are weaned in an average of only 4 days, the shortest time for any mammal. Males are territorial and patrol the ice edge, often hauling out near females and forming trios. Bulls actively fight among themselves, and can inflict bloody wounds; they more routinely display by inflating their black nasal bladders and extruding their nasal septum, shaking this bright red "balloon" violently in efforts to ward off competing males. They also vocalize at the same time, producing a loud "pinging" noise. Mating follows weaning of the pup.

The pack ice edge is home to hooded seals throughout the year, and they migrate with it as it retreats north in summer and advances south in the fall. Aggregations form in the Denmark Strait east of Greenland and near Jan Mayen during the late spring molt, which begins after the pupping season. Following the molt, they disperse widely for the summer and winter, primarily living along the ice edge. Hooded seals are deep divers and are capable of long dives. The maximum-recorded depth reached is 1,600 m, and the longest dive has been nearly 1 hour. Typical dive depth while foraging is 260 m and last around 13–15 min. Longevity in this species often reaches to 25 years with some living to 35 years. Polar bears, killer whales, and Greenland sharks are known hooded seal predators.

Feeding and prey Hooded seals typically fast during breeding and molting, but actively feed during much of the rest of the year. Their diet consists primarily of squids and fishes such as Greenland halibut, Atlantic and Arctic cod, several redfish species, herring, and capelin. Newly weaned pups feed on pelagic crustaceans prior to taking on the adult diet.

Threats and status Hooded seals were subjected to episodes of intense commercial hunting from the late 1800s to the late 1900s. Harvests were often conducted in association with harp seal harvests and commercial fisheries for Greenland sharks. Norway, the former Soviet Union, Canada, and Greenland were the main players involved in the commercial harvests, which were focused on newborns because of their highly-prized "blue-back"

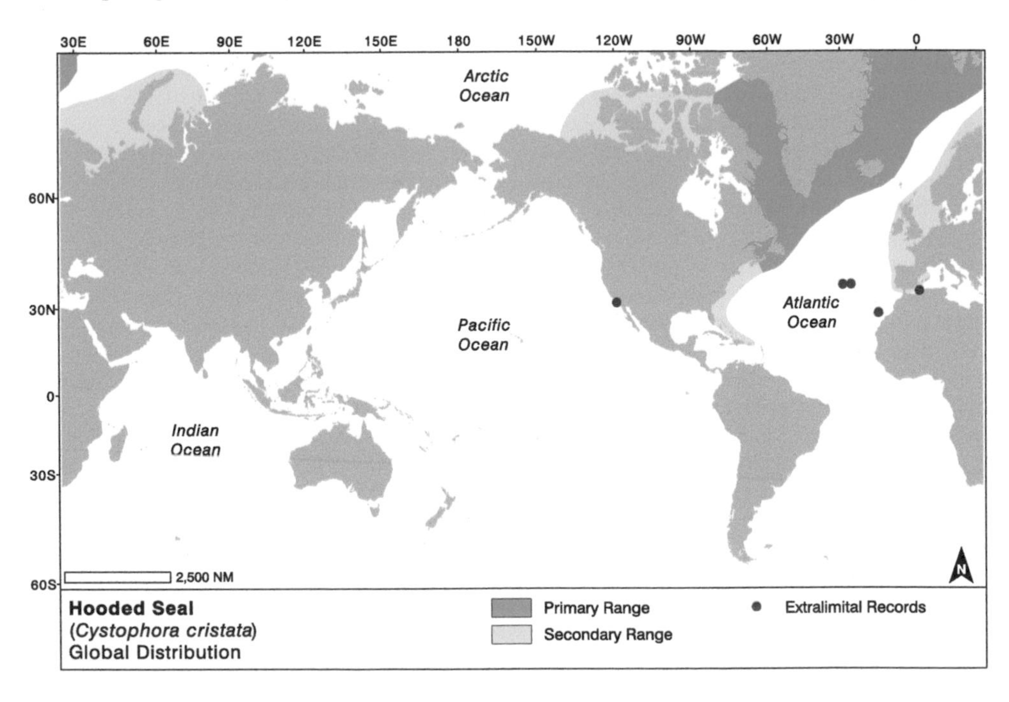

Hooded Seal
(*Cystophora cristata*)
Global Distribution

Adult male hooded seal inflating his nasal septum "balloon" to begin displaying and vocalizing. PHOTO: IFAW/ D. WHITE

A male hooded seal from Sable Island, Canada. The oversized, wide nose overhangs the mouth. PHOTO: R. BAIRD

Adult female hooded seal defending her "blue-back" pup. Females have a dark muzzle and face and many dark blotches of various sizes on the body. PHOTO: IFAW/ F. BRUEMMER

A hooded seal female and her "blue-back" pup in the foreground. In the background, notice the adult male with black hood inflated for display. PHOTO: IFAW/ D. WHITE

Adult female hooded seal with "blue-back" pup. She is blood stained from recently giving birth. Note the band of dark color reaching the foreflipper on the pup. PHOTO: IFAW/ D. WHITE

A juvenile hooded seal has a short wide face, is countershaded and lacks the large irregular dark blotches of the adults. PHOTO: M. NOLAN

pelt. A permanent import ban on hooded seal pelts by the European Economic Community ended most of the commercial harvesting of this species, although Canada took approximately 26,000 as recently as 1996. The number of "takes" is presumed to have been higher than the reported number harvested due to the killing of females that aggressively defend their pups, and the loss of animals that are injured and escape to the water only to die later. The species is subject to by-catch in coastal net fisheries. Hooded seals are important in native subsistence harvests on both coasts of central and southern Greenland, where they are used for food for humans and sled dogs. The western North Atlantic population of hooded seals is estimated to be about 590,000, based on aerial surveys conducted in 2005, and may be stable. In contrast, the eastern North Atlantic population has declined over the past 60 years from about 500,000 to 70,000, and continues to suffer losses. The exact cause of the decline is uncertain, but indirect effects of commercial fishing are suspected. The hooded seal requires pack ice to haul out for pupping and molting, and is therefore vulnerable to the potential effect of sea surface temperature rise, and loss of extensive sea ice, due to global climate change.

IUCN status Vulnerable.

Gray Seal—*Halichoerus grypus*

(Fabricius, 1791)

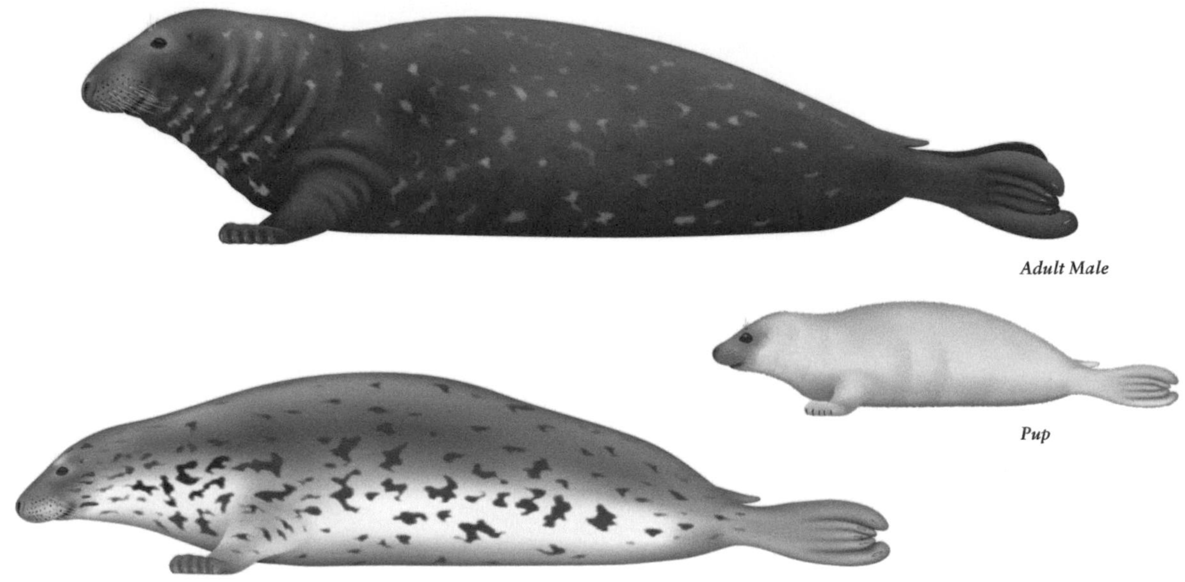

Recently-used synonyms *H. g. atlantica* and *H. g. baltica* were formerly used as subspecies.

Common names En.-gray seal, grey seal; Sp.-*foca gris; Fr.-phoque gris.*

Taxonomic information Carnivora, Phocidae. Two subspecies are recognized: *H. g. grypus* (Atlantic gray seal) from the North Atlantic, including North America, Iceland, Norway, western Russian, the British Isles, and France to Denmark; and *H. g. macrorhynchus* (Baltic) from the Baltic Sea.

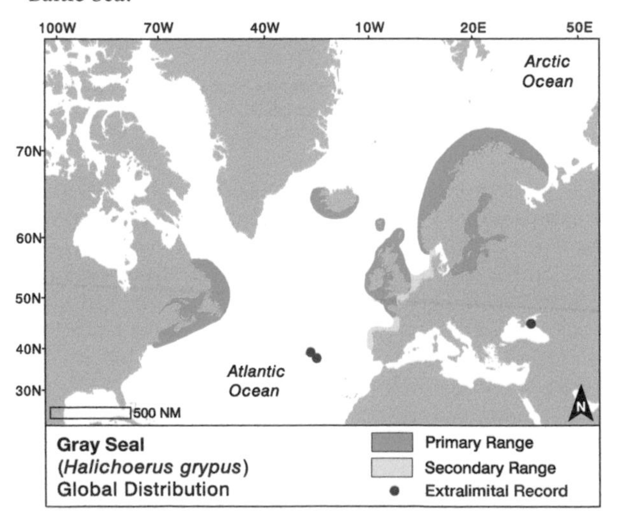

Species characteristics Gray seals are robust and sexually dimorphic; males grow substantially larger than females. In both sexes the muzzle is long and wide as has a fleshy mystascial area. When compared to females, males have a larger and broader head, longer and wider muzzle, and thicker neck. Also, in adult males, the top of the muzzle is convex. In adult females and subadults, the top of the muzzle is variable and can be flat, slightly convex, or slightly concave (producing a barely noticeable forehead). The shape of the head has led to the locally-used common name "horsehead." The nostrils are widely-separated and almost parallel to each other, the folds forming a "W" pattern as opposed to the "V" of seals of the genus *Phoca.* The eyes are small in proportion to the size of the head, widely-separated, and due to the length of the muzzle, farther back from the nose than on other phocids that share the gray seal's range. The foreflippers are proportionately short and wide.

Pelage color and pattern is quiet variable. Most gray seals are shades of gray, tan, or brown, and slightly darker above than below. There are usually numerous irregular dark blotches and spots on the back and sometimes a few below, although some females with very few spots appear to be solid grayish-cream color. Males darken with age, becoming dark brown to blackish, with a variable number of lighter blotches and spots. Orange to reddish coloration can be seen on the neck, undersides, and flippers of some animals. Gray seals appear paler and duller in coloration

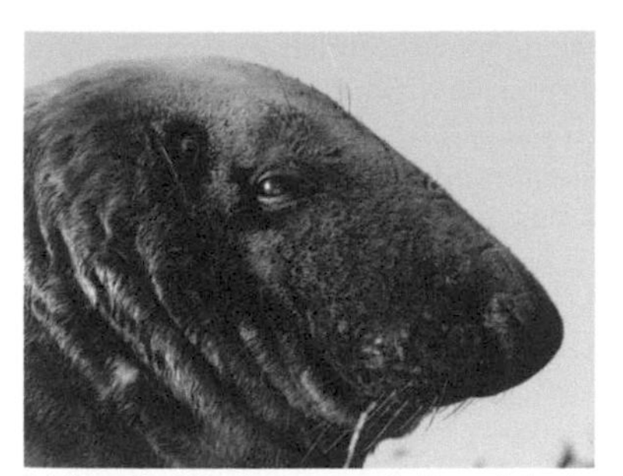
Adult male Atlantic gray seal. Farne Islands, England. PHOTO: D. STILL

Adult male and female gray seals. The difference in overall size and shape of the muzzle is dramatic. PHOTO: G. CRESSWELL

Adult female Atlantic gray seal with a light coat and dark spots. PHOTO: D. STILL

Gray seal bulls generally darken with age. Notice the wide and long muzzle. Sable Island, Canada. PHOTO: R. BAIRD

Mating Atlantic gray seals. Farne Islands, England. PHOTO: D. STILL

Adult female Atlantic gray seal with a young pup in lanugo. PHOTO: T. MELLING

An anomalously colored gray seal pup in lanugo with adult female. Gray seal pups are typically light silver gray. Sable Island, Canada. PHOTO: R. BAIRD

just prior to the annual molt. Newborns have a silky, creamy-white lanugo, sometimes with a grayish tinge. The lanugo is molted in 2–4 weeks, and is replaced by a pelage like that of the female, but with more subtle markings.

Western North Atlantic gray seals males are 2.25 m in length and weigh 300–350 kg, and can reach in excess of 400 kg. Females are 2 m in length and weigh 150–200 kg, and may reach 250 kg. In the eastern North Atlantic (British Isles), males are 2 m on average, with a weight of 233 kg, and a maximum of 310 kg. Females there are 1.8 m in length and weigh 155 kg. Pups are about 1 m in length and weigh 14.8–15.8 kg at birth. The dental formula of adults is I $^{3}/_{2}$, C $^{1}/_{1}$, PC $^{5-6}/_{5}$.

Recognizable geographic forms The two subspecies of gray seal cannot be separated in the field, their distinctiveness limited to features of the skull and genetic makeup. Within the western Atlantic subspecies, gray seals in Canada have been shown to be larger than European animals. Other differences also exist between and within the subspecies such as timing of birthing, and breeding on fast ice versus land.

Can be confused with Harbor, ringed, bearded, hooded, and harp seals share the gray seal's range. Adult gray seals are much larger than harbor, ringed, and harp seals. Between similar sized animals, their larger head, longer muzzle, widely spaced apart eyes, and the large blotchy markings

Adult Atlantic gray seal at sea. Note the long straight and wide muzzle. Off Massachusetts. PHOTO: L. LILLY

Part of a large group of Baltic gray seals hauled out on a flat-topped rocky island. PHOTO: M. KUNNASRANTA

An adult female Baltic grey seal nursing a pup in a captive breeding program in Sweden. PHOTO: A. ROOS

Three juvenile Baltic gray seals at a stranding facility in Russia, showing some of the range in pelage color in this subspecies. PHOTO: V. ALEXEEV

are distinctive. Although they are similar in size, the same features separate gray seals from the proportionately small-headed bearded seal, along with the short squared-off to rounded foreflippers of the bearded seal. Overall coloration and markings, and head and muzzle size and shape, also facilitate separation of gray seals from similar-sized hooded seals. Additionally, hooded seals have a very dark face, and very dark flippers. Newly weaned hooded seals are countershaded, have a dark face, and lack spots or blotches, and are smaller than all but gray seal pups. Gray seal nostrils are widely separated at the bottom of the nose and do not converge to a V-shape typical of the other species, and the male hooded seal's nose is unique and unmistakable.

Distribution Gray seals have a cold temperate to subarctic distribution in the North Atlantic. The North Atlantic subspecies has two somewhat isolated stocks: a western Atlantic stock, centered off northeastern North America, with Sable Island off Nova Scotia as an important breeding site, and an eastern Atlantic stock, at Iceland, the Faroe Islands, Norway, the United Kingdom and Ireland. The Baltic subspecies is isolated in the Baltic Sea, barely reaching eastern Denmark. The western North Atlantic population has shifted it limits south in recent decades to include breeding colonies on islands off Maine and Cape Cod, US. Vagrants reach the mid-Atlantic seaboard of the US, the Azores, and Portugal in Europe, and the southern tip of Greenland. A gray seal found around the Kerch Peninsula, Crimea, in the Black Sea and seen regularly for at least a decade was considered most likely a zoo escape animal, but this was never confirmed.

Ecology and behavior Gray seals are polygynous, but males do not defend territories or herd females. They actively compete for access to females using vocalizations, threat gestures, and occasional fighting. Pupping and breeding occur between late September and early March, depending on location. Gray seals breed earliest in the British Isles, followed by those in Norway and Iceland, and finally by those off Canada, and in the Baltic Sea. They will give birth on land, shore-fast and pack ice depending on location. The single pup is usually attended continuously by its mother and weaned in 15–18 days, at which time many have quadrupled their birth weight of 11–20 kg. After weaning, the pups remain ashore fasting for 2–4 weeks before dispersing to sea and wandering widely.

Many, but not all gray seals disperse from their rookeries during the non-breeding season, but gather again at traditional sites to haul out for the annual molt. They are usually quite gregarious at haulouts, with groups of 100 or more being common, and they will share haulouts with harbor seals. When ashore, gray seals do not generally lie in contact with each other. Gray seals spend most of their time in coastal waters. When not in the water they haul out on isolated beaches and rocky ledges of islands, or on

An adult male Atlantic gray seal killing and eating a juvenile Atlantic harbor seal. Gray seals are dynamic predators, and have also recently been documented to prey on harbor porpoises. Heligoland, Germany. PHOTO: S. FUHRMANN

Adult male Atlantic gray seals fighting in the surf zone at the Farne Islands, England. PHOTO: D. STILL

Two adult male Atlantic gray seals fighting in what has become a wrestling match on the mudflats. England. PHOTO: T. MELLING

ice. At sea, gray seals are usually solitary, or can be seen in small dispersed groups. They will rest at the surface in a vertical "bottle" position, treading water with only the head and upper neck exposed. Generally, gray seals are shallow divers, with most adults foraging to 120 m or less, with durations of under 8 min. The maximum dive depth documented is 412 m, and some dives can last up to 30 min. Predators on pups in the Baltic Sea include white-tailed eagles and greater black-backed gulls.

Feeding and prey Gray seals feed on a wide variety of benthic and demersal prey in coastal areas. They also feed on schooling fish in the water column, and occasionally take seabirds and marine mammals. Prey species taken include: sandeels (*Ammodytes sp.*), whiting, saury, smelt, various kinds of skates, capelin, lumpfish, pollock, cod, haddock, plaice, flounder, salmon, and a variety of cephalods and mollusks. Cannibalism by adult males on pups has been reported, and recent evidence from the North Sea indicates gray seals will occasionally attack and consume harbor seals, and even harbor porpoises.

Threats and status At present, the North Atlantic gray seal population is healthy and growing, with a worldwide population estimated at nearly 400,000–500,000 (including about 300,000 in Canada). The Baltic Sea population, which once numbered 100,000, is now comprised of approximately 22,000 animals, having never fully recovered from sealing and poaching in the early 20th century. North Atlantic gray seals experienced similar hunting pressures during this period, largely because of government-sponsored bounties to hunters. Bounties were established to control gray seal populations that were deemed to damage important commercial fisheries either directly (through feeding) or indirectly (as a vector for the parasitic nematode known as cod, or seal, worm). Prior to bounty and commercial hunting, gray seals were locally important in subsistence harvests throughout the history of their contact with humans. Some gray seal deaths have been attributed to distemper virus outbreaks that caused extensive mortality in harbor seals. As a coastal species, gray seals are exposed to, and ingest, industrial and agricultural pollutants through the food chain. Contaminant loads, including PCBs and DDT, may have an effect on their immune system, and has been linked to reproductive declines in Baltic animals. Entanglement in fishing nets is another source of mortality. Interestingly, human over-exploitation of North Atlantic sharks may have had the effect of helping gray seal populations grow and recover by increasing survival, particularly of newly weaned pups and juveniles.

IUCN status Least Concern.

Ribbon Seal—*Histriophoca fasciata*

(Zimmerman, 1783)

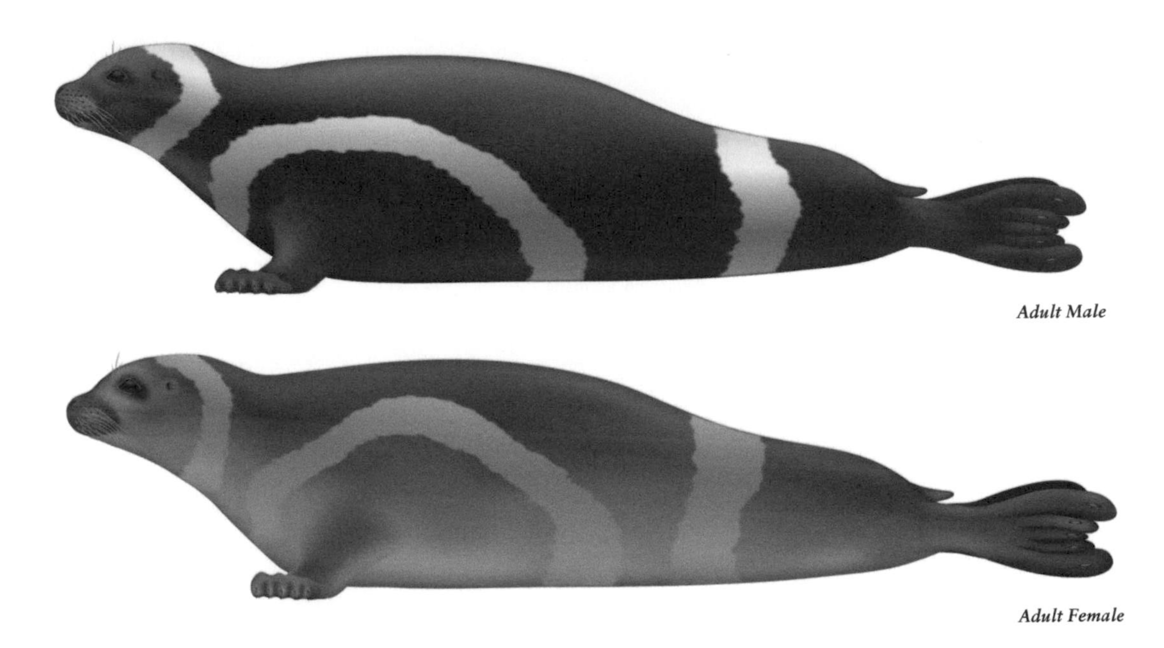

Adult Male

Adult Female

Recently-used synonyms *Phoca fasciata.*
Common names En.–ribbon seal; Sp.–*foca fajada*; Fr.–*phoque à rubans, phoque rubané.*
Taxonomic information Carnivora, Phocidae. Originally described in 1783, this species was placed in the genus *Phoca,* but soon thereafter was assigned to the monotypic genus *Histriophoca.* A 1970 review proposed to move it back to *Phoca.* Numerous sources since then, including the Society for Marine Mammalogy, have retained it in *Histriophoca,* leading to some nomenclatural differences in the literature.

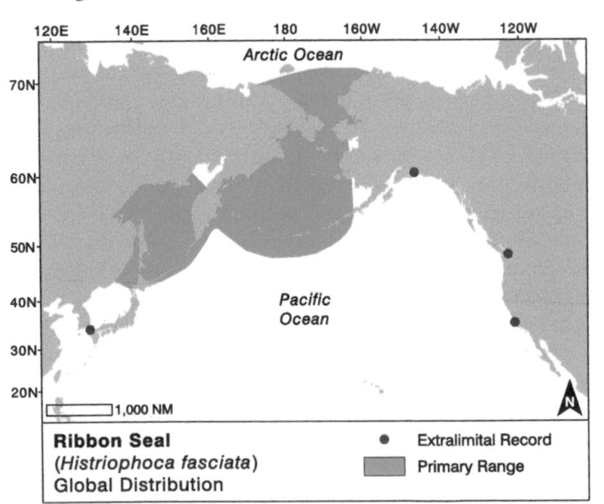

Species characteristics Adult male ribbon seals have the most striking color pattern found on any seal. Adults of both sexes have pale bands of variable width that encircle each foreflipper, the neck and nape, and the pelvic area. The bands have defined edges and vary greatly in width. On some animals, the bands are so wide that they merge. Band color ranges from a shade just paler than the surrounding dark pelage to white. Adult males are black to brownish-black, while females are light to dark brown. The bands are less distinct on females and subadults. Juveniles are plain-looking until they are about 2 years old, lacking the dramatic markings that will characterize them as adults. They are gray to brownish after molting the woolly whitish lanugo at about 5 weeks of age, and have a light colored area of pelage over the ear opening. Ribbon seals are more slender than other Bering Sea ice seals. The head is small, relatively wide, and flat-topped, and the forehead is small in profile. The close-set eyes appear large. The muzzle is short, blunt, and slightly tapering. The vibrissae are light-colored, beaded, and fairly prominent. There are long, hooked claws on all digits of the foreflippers. The ends of the foreflippers are weakly pointed with a somewhat longer first, or outer, digit and successively shorter digits 2–5.

Adult ribbon seals reach a maximum length of about 1.8 m and weights of 90–148 kg. Pups are approximately 86 cm long and 10.5 kg at birth. The dental formula is I $^{3}/_{2}$, C $^{1}/_{1}$, PC $^{5}/_{5}$.

Adult male ribbon seal have a more complex pattern of bands on the ventrum. Bering Sea. PHOTO: G. BRADY/NOAA FISHERIES/AFSC

The coloration of adult male ribbon seals makes them both one of the most handsome and unusual looking mammals. PHOTO: D. WITHROW/NOAA FISHERIES/AFSC

Adult male ribbon seal with tags in the hindflippers. PHOTO: J. LONDON/NOAA FISHERIES/AFSC

An adult male ribbon seal showing the wide bands on the lower back, and the shoulder bands on the back. Note the wide, short, blunt muzzle. Bering Sea. PHOTO: D. WITHROW/NOAA FISHERIES/AFSC

The same adult male, partially wet, and rolled onto its belly in an alert posture. PHOTO: D. WITHROW/NOAA FISHERIES/AFSC

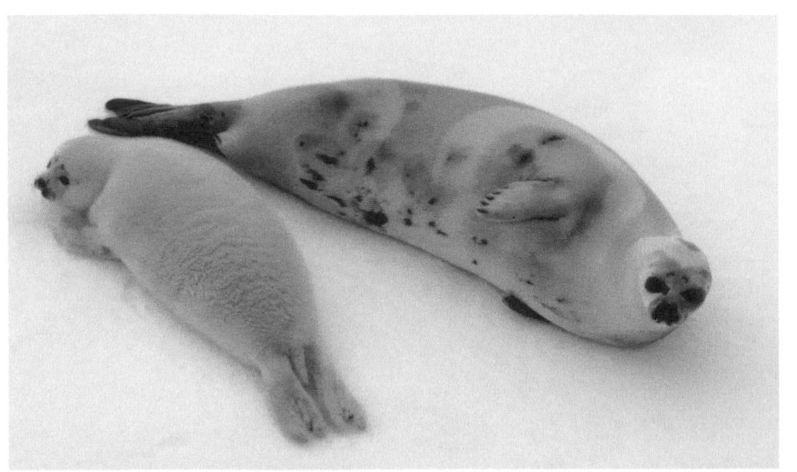

An adult female ribbon seal with an older pup in full lanugo. Both animals show the short muzzle of the species. The dark blotches on the female appear to be where molting has caused temporary fur loss. Bering Sea. PHOTO: G. BRADY/NOAA FISHERIES/AFSC

Adult female ribbon seal with a pup in lanugo. PHOTO: J. JANSEN/NOAA FISHERIES/AFSC

An adult female ribbon seal with her molting pup. Bering Sea. PHOTO: S. DAHLE/NOAA FISHERIES/AFSC

A weaned juvenile ribbon seal. Note the irregular boundary on this countershaded animal, and the light colored mark near the ear. Ozernoy Gulf, Russia. PHOTO: M. CAMERON/NOAA FISHERIES/AFSC

Recognizable geographic forms None. Although three distinct breeding populations are recognized, no morphological or genetic differences have been reported, and no separation into distinct population segments or subspecies has been proposed.

Can be confused with Four other phocids, ringed, harbor, spotted, and bearded seals, share the range of the ribbon seal. Adult ribbon seals are unmistakable with distinctive bands, and have no spots or rings. Juvenile ribbon seals lack bands, are countershaded, and have no rings or spots. The bearded seal has a much greater density of downward-curving vibrissae than the ribbon seal, has foreflippers that are rounded on the end, and except for pups and juveniles, is considerably longer and heavier-bodied.

Distribution Ribbon seal distribution most closely matches that of the spotted seal, as it occurs in the Sea of Okhotsk and Japan Sea, the western North Pacific, and from the Bering Sea northward through the Chukchi Sea, east to about 160°W and rarely to the western Beaufort Sea. Three separate breeding populations of ribbon seals are known: the southern Sea of Okhotsk, the northern Sea of Okhotsk, and the Bering Sea. Ribbon seals inhabit the southern edge of the pack ice from winter to early summer. Most are thought to be pelagic in the Bering Sea during the summer. Records from the North Pacific south of the Aleutian Islands indicate a wider range during the summer when ribbon seals are not associated with sea ice. Records of vagrants that have wandered widely come from Cordova, Alaska, the Puget Sound, Washington, and as far south as 35°N in the Tsushima Strait between the Korean Peninsula and Japan, and in Morro Bay, California.

Ecology and behavior Ribbon seals are solitary for much of their lives. In addition to being pelagic in the summer, they prefer to use sea ice from the continental slope seaward out over deeper oceanic areas. When on ice, they prefer areas of moderate to pack ice coverage (60–80%), and do not like highly concentrated pack or areas of sheet ice coverage, as ribbon seals can only open and maintain access holes in ice up to approximately 15 cm thick. Curiously, tend to haul out and position themselves away from the edge of floes, and will extend their necks to peer at sources of disturbance, but are fairly approachable by boat. Ribbon seals are able to move rapidly on ice using slashing side-to-side motions. Sexual maturity in males occurs at 3–5 years of age, and in females at 2–4 years. Adult females have a high annual reproductive rate, but there is also a fairly high first year mortality of 44%. Pups are born on ice floes from early April to early May. Males are generally nowhere to be seen during the nursing period. Ribbon seals are rarely encountered because of the remote and inhospitable nature of their polar habitat. Predators include polar bear, walrus, killer whale, and Greenland shark.

Feeding and prey Diet varies by area and age of the seal. Ribbon seals in the Okhotsk and Bering seas are known to take 35 different species of fish and invertebrates. After weaning, ribbon seals feed on zooplankton, such as euphausiids, for the first year, then begin to feed mainly on larger shrimp species for their second year. As 2-year-olds, they take up the adult diet, which includes a variety of fishes, squids, and octopuses. Ribbon seals in the Sea of Okhotsk have a diet that is 65% pollock, while those in the Bering Sea consume about the same percentage of squids and octopuses.

Threats and status The total population of ribbon seals is not known with precision. In 2007, it was estimated to be 240,000, with 90,000–100,000 in the Bering Sea and 140,000 as a combined estimate for the two separate breeding populations in the Sea of Okhotsk. Other estimates for the Sea of Okhotsk, based on 1968–1990 counts, give an average of 370,000 per year (320,000 and 50,000 in the northern and southern breeding populations, respectively).

Large scale commercial harvests were carried out in the Sea of Okhotsk and Bering Sea from the 1930s to 1994, and have since been halted, with only small more-localized harvests continuing today. Subsistence hunting by Alaska Natives occurs at low levels in the US. Global warming, accumulation of contaminants, oil and gas exploration and development, entanglements in commercial fishing gear, and depletion of prey species (such as pollock) in commercial fisheries are all ongoing threats and concerns. In particular, since ribbon seals occupy much of the southern extent of pack ice in the Pacific Arctic, changes in the extent of the ice due to global warming could be problematic for this species. Of potential concern is a large increase in vessel traffic in the Arctic due to decreases in sea ice, specifically in the Bering and Chukchi seas. Such vessel traffic could lead to more disturbance and higher risk of accidental sinking and grounding of ships and resultant fuel or oil spills. A 2013 status review by the US National Marine Fisheries Service determined that the ribbon seal was not threatened or endangered.

IUCN status Data Deficient.

Adult male ribbon seal craning to look at something that caught its attention. PHOTO: J. LONDON/NOAA FISHERIES/AFSC

Adult female ribbon seal on alert. Note the paler pelage than on the adult male. PHOTO: J. LONDON/NOAA FISHERIES/AFSC

Adult male ribbon seal showing the coloration of the front of the body. PHOTO: J. JANSEN/NOAA FISHERIES/AFSC

A weaned and molted ribbon seal pup showing the small muzzle and countershaded pattern of the juvenile. PHOTO: J. JANSEN/NOAA FISHERIES/AFSC

Harp Seal—*Pagophilus groenlandicus*

(Erxleben, 1777)

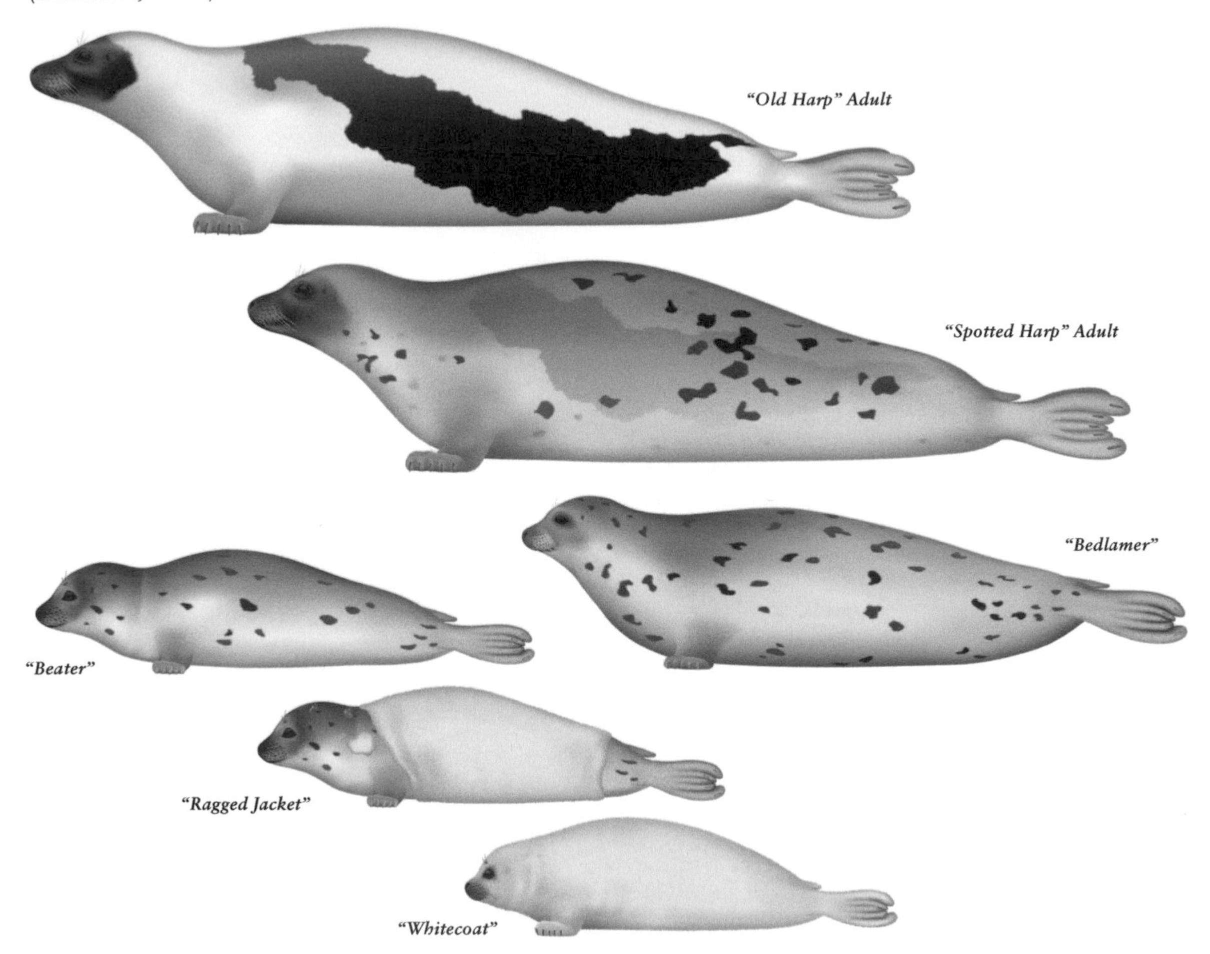

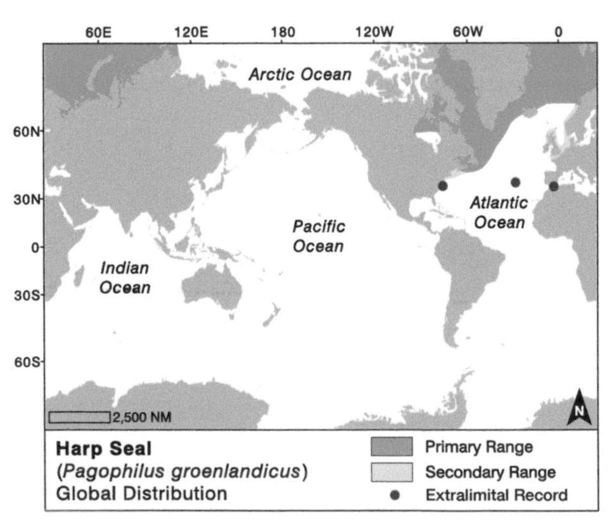

Recently-used synonyms *Phoca groenlandica.*

Common names En.–harp seal; Sp.–*foca pía, foca de Groenlandia*; Fr.–*phoque du Groenland.*

Taxonomic information Carnivora, Phocidae. The species is now considered monotypic, but two subspecies were recognized in the past: *P. g. groenlandicus* from Newfoundland to Norway, and *P. g. oceanicus* from the White Sea of Russia.

Species characteristics The harp seal's head appears somewhat long, wide and flattened. The long muzzle tapers slightly, and in adults, can appear slightly upturned. The eyes are close-set and there is a slight dip to the forehead. The flippers are relatively small. The foreflippers are slightly pointed and angular, with a short row of digit endings; the claws are strong and dark.

The change of pelage patterns over the lifespan of an animal is the most complicated of any pinniped, and in most stages it is the species' most distinctive feature.

A group of "old harps" with black masks swimming in a lead with many more hauled-out behind. PHOTO: IFAW/D. WHITE

The black-masked faces of "old harps." Gulf of Saint Lawrence, Canada. PHOTO: IFAW/D. WHITE

The newborn's pure white coat, which can be stained yellowish for the first few days by amniotic fluid, persists for about 12 days, during which time they are known as "whitecoats." At this stage they are difficult to distinguish from other northern phocids born in a similar lanugo pelage. The "greycoat" and "ragged jacket" stages follow when the underlying juvenile pelage and spotted pattern shows through the lanugo, and appears as the lanugo is molted. From approximately 3 to 4 weeks-old, when they have completely molted out of their lanugo coat, through about age 13 to 14 months, harp seals are known as "beaters," at which time they can be countershaded slightly darker gray above and silver-gray below, with numerous spots and irregular dark blotches all over the body. At their next molt they initially look about the same, but are then known as "bedlamers." They remain at this stage until a dark "harp" pattern begins to form on the back and sides and the spots start to disappear. Harp seals with both spots and a harp are known as "spotted harps." Finally, when all spots disappear and the harp and face become black, they are known as "old harps." The adult's base color is silvery-white to light gray. Initially, the harp pattern of the adult appears as a faint shadow of what it will become, as the seals transition to the "old harp" pattern. It consists of wide and irregular shaped bands with ragged edges starting at the pelvis and dipping down on each side before rising up and connecting over the shoulders. Seen from above, the pattern resembles a large irregular "V" with curving arms. The head of the adult is masked: the face, chin, upper neck and top of the head are black. This mask also has ragged margins. Black marks may also occur at the insertions of the hindflippers.

Harp seals exhibit little sexual dimorphism. Adult males are up to 1.9 m in length and average 120–135 kg in weight, and females are up to 1.8 m and 120 kg. Pups are born at about 85 cm and almost 4 kg. Longevity is about 30 years. The dental formula is I $^3/_2$, C $^1/_1$, PC $^5/_5$.

Recognizable geographic forms The populations in the North Atlantic and the White Sea show differences in skull morphology, and the mean length of adults from the White Sea is slightly more (about 8–9 cm) than their counterparts from the Atlantic, but there is overlap. Although there may be minor differences in the pelage between the two populations, there is no reliable way to separate the two populations in the field, except by going to whelping grounds for each during the breeding season.

Can be confused with Adult harp seals are unlikely to be confused with any other seal. The silvery-white body, with a black harp pattern and mask, is unique. However, the "bedlamer" and "spotted harp" patterns are more generic and pose some confusion with the harbor, ringed, gray, and hooded seals that share their range. The irregular spots of "bedlamer" and "spotted harps" are usually less dense, and randomly scattered over the entire body. Harbor seals tend to have more spots and markings dorsally than ventrally, and are usually countershaded. Harp seals lack the rings that characterize ringed seals. Hooded and gray seals are much larger as adults and have distinctive heads with other unique features. Hooded seals have a larger head with a wider muzzle at all ages. Young hooded seals are strongly countershaded, bluish-gray above and lighter gray below, and have no spots. Adult hooded seals have medium to large irregular spots over the entire body, and dark, sooty faces and foreflippers. Gray seals are heavy-bodied animals at all ages, with a proportionately large head and muzzle that is long and wide, often with a convex bridge. Young gray seals are similar in size to harp seals but even when young have a distinctly different head and irregular dark blotches that are usually larger. Bearded seals are robust heavy bodied animals that lack spots, have very dense mystacial vibrissae, and have rounded-off ends of their foreflippers.

Distribution Harp seals are widespread in the Arctic and North Atlantic oceans and adjacent areas from Hudson Bay and Baffin Island east to Cape Chelyuskin and the White Sea, Russia. Well-known breeding aggregations are in the western North Atlantic: the "Front" (waters off northeastern Newfoundland and southern Labrador),

The back of an "old harp" showing the gap in the harp pattern and the black hood. PHOTO: B. McTAVISH

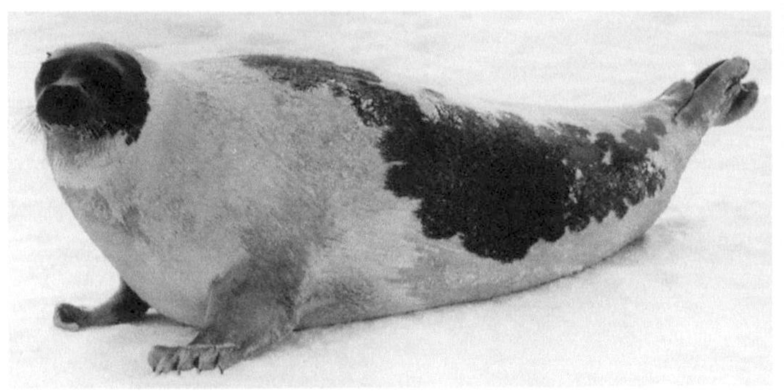

A large, adult harp seal in the "old harp" pelage. Note the irregular edge to the black harp-shaped bands. White Sea, Russia. PHOTO: S. V. ZYRIANOV/PINRO

"Old harp" female and a newborn with amniotic fluid-stained lanugo. Gulf of Saint Lawrence, Canada. PHOTO: IFAW/D. WHITE

A "spotted harp" female nurses a "whitecoat" pup. Gulf of Saint Lawrence, Canada. PHOTO: IFAW/D. WHITE

A "ragged jacket" or molting pup. Gulf of Saint Lawrence, Canada. PHOTO: IFAW/S. COOK

Molted pups and juveniles are called "beaters" because of the way they beat the water when they swim. PHOTO: IFAW/S. COOK

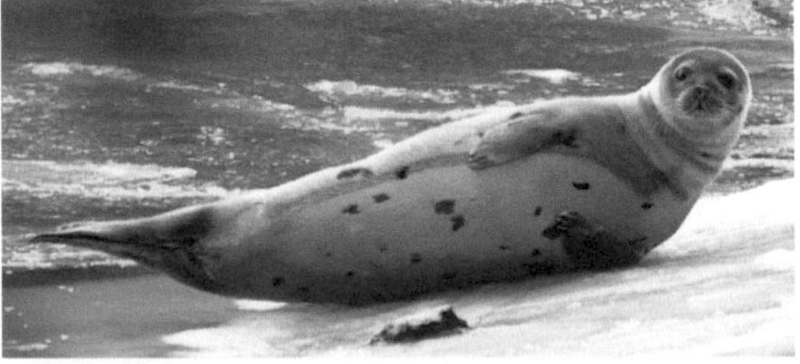

A "bedlamer" harp seal in a river. Periodically, many harp seals come ashore in New England, outside their typical range. PHOTO: G. McCULLOUGH

A "bedlamer" harp seal with very fine spotting, like the spotting on a harbor seal. Cape Cod, United States. PHOTO: CAPE COD STRANDING NETWORK

A group of mixed-age harp seals, several of which are in an alert posture. Harp seals will travel and haul out together outside the breeding season. PHOTO: M. JØRGENSEN

Harp seals travel in groups at sea and churn up the water like a large school of dolphins. Notice that many animals are swimming upside-down. Gulf of Saint Lawrence, Canada. PHOTO: IFAW/S. COOK

and the Gulf of St. Lawrence near the Magdalen Islands. Other breeding areas are in the eastern North Atlantic, off Jan Mayen Island, and in the White Sea. Harp seals live chiefly in pack ice, but can be found away from it in summer. Their range limits off North America have shifted south in the past few decades, and harp seals now routinely reach the Gulf of Maine and Sable Island in much larger numbers than a few decades ago, with a high percentage of these being juveniles both healthy and unhealthy. This shift has been linked to changes in prey availability and distribution. Vagrants reach the US central east coast, and south to North Carolina, as well as the northern to central coast of Europe including the United Kingdom. There are regular appearances of large numbers of animals in the coastal waters of central and northern Norway for feeding. Recently, a vagrant harp seal was found in the Mediterranean Sea at Motril, Province of Granada, Spain.

Ecology and behavior Harp seals congregate to whelp (pup) on pack ice, where they form huge groups, sometimes numbering into the thousands. Males are sexually mature at 7–8 years of age, and females at 4–7 years. Pups are born from late February to mid-March, and are nursed for 12 days. Promiscuous mating occurs in the water and sometimes on the ice, from mid- to late March. Adult animals follow the ice north and haul out for periods of time to undergo an annual molt after the breeding season. Harp seals are migratory and following the breeding season and molt, they follow the ice north in summer to feed in the Arctic. The seals in Canadian waters undertake an annual round-trip migration of over 5,000 km. They are very active in the water and sometimes travel in groups that are quite large and can churn the water like fast moving dolphin schools. They are generally shallow divers, foraging to 100 m or less in dives that last about 16 min, but tagging has shown that these seals can reach 400 m in the Barents Sea. They have been described as coastal, but tagging indicates summer foraging in oceanic areas as well. The maximum life span of a harp seal is approximately 30 years; most animals that reach sexual maturity live to over 20 years. Killer whales and Greenland sharks are known predators, and in Svalbard, harp seals comprise up to 13% of the diet of polar bears.

Feeding and prey Harp seals feed on a wide variety of crustaceans and fishes, with more than 130 species in their diet, including capelin, Arctic cod, and polar cod. Atlantic cod, a mainstay of North Atlantic fisheries and now severely overfished, makes up a small percentage of the diet.

Threats and status As probably the most abundant pinniped in the Northern Hemisphere, the harp seal has been at the center of controversies between environmentalists, sealers and governments for decades. Commercial hunting has been ongoing since the 1600s, with harp seals being particularly sought after when easily reached populations of walrus, gray and harbor seals had been dramatically reduced. Harp seals by the millions were harvested for oil, pelts and meat. In Canada, as recently as 2008, over 200,000 animals were taken for their pelts by licensed sealers; in recent years, annual catches have varied between about 40,000–75,000 animals (out of a total allowable catch limit of 400,000). Globally, harp seal numbers are believed to be more than 9 million, and their numbers may currently exceed the carrying capacity for the environment. The western North Atlantic stock is estimated to have increased to over 7.1 million animals after 2010. The breeding group in the West Ice near Jan Mayen increased to approximately 635,000 by 2005. The White Sea breeding group was estimated (in 2005) to be 2 million animals. Attempts have been made to link harp seals with the demise of the once vast stocks of Atlantic cod, but this species is not an important component of the seal's diet. Despite this fact, efforts are continuously being made to justify reducing numbers of harp seals in response to pressures stemming from this complex fisheries management issue. Overfishing and alteration of marine ecosystems pose an ongoing threat to the health of harp seal populations, as does global warming and changes in sea ice patterns, and accumulation of toxic contaminants in the marine environment.

IUCN status Least Concern.

Harbor Seal—*Phoca vitulina*

Linnaeus, 1758

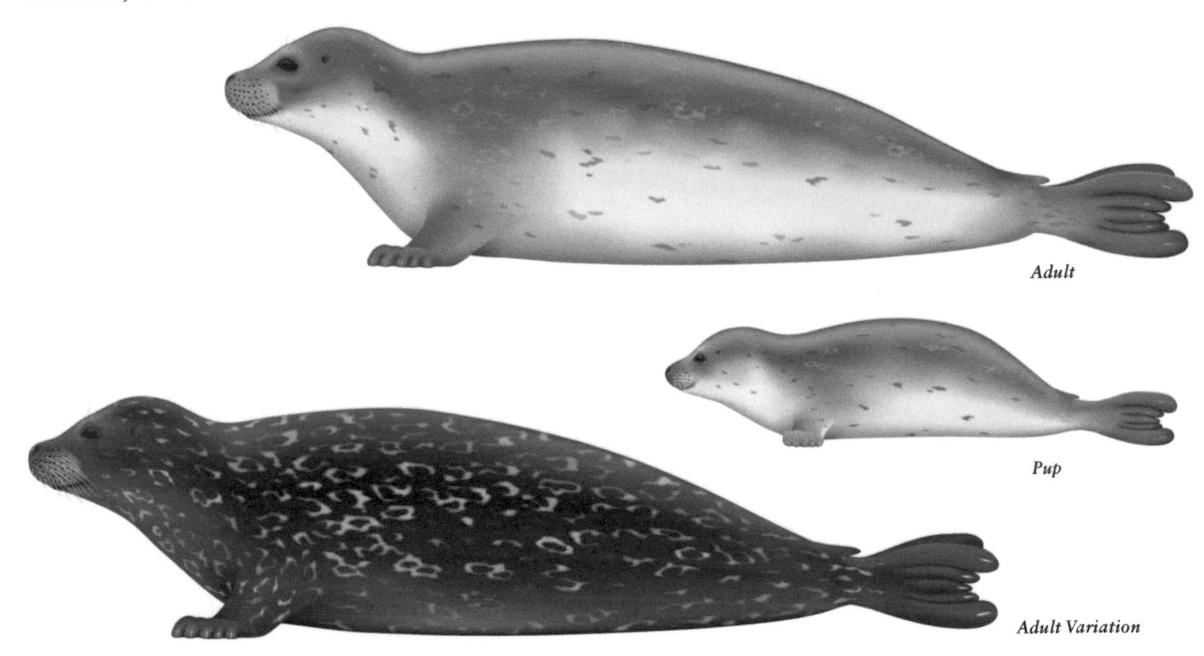

Recently-used synonyms *P. v. concolor, P. v. stejnegeri, P. v. richardsi* was formerly used for the subspecies *P. v. richardii.*

Common names En.–harbor seal, common seal; Sp.–*foca común;* Fr.–*phoque veau marin.*

Taxonomic information Carnivora, Phocidae. Three subspecies are recognized: *P. v. vitulina* (Atlantic harbor seal); *P. v. richardii* (Pacific harbor seal); and *P. v. mellonae* (Ungava harbor seal) from the Ungava Peninsula, Canada. Recently, the Society for Marine Mammalogy lumped *P. v. concolor* (western North Atlantic) with *P. v. vitulina*, and *P. v. stejnegeri* (western North Pacific) with *P. v. richardii*, based on genetic analyses.

A harbor seal in the foreground and a gray seal in back. Coloration is not always useful for telling harbor and gray seals apart, as both species have light and dark phases and can be many shades of gray and brown. Note however, muzzle length and shape differences. PHOTO: G. CRESSWELL

Species characteristics Harbor seals are small phocids with a torpedo-like body shape. The head is medium-sized, and the eyes are relatively close-set. There are prominent light-colored, beaded vibrissae above the eyes and on the relatively straight sided and short muzzle that tapers when viewed in profile. The nostrils are small and terminal, forming a "V" that converges at the bottom. The external ear openings are positioned slightly behind and below the eyes. Harbor seals are not obviously sexually dimorphic, and observation of the pattern of genital openings and the presence or absence of mammary openings, or the association of a pup with its mother are required to separate the sexes. The flippers are relatively short, only about 1/5 to 1/6 of standard length, with long, thin, dark hooked claws on all digits. The ends of the foreflippers are somewhat pointed, with a longer first digit and successively shorter digits two through five.

The variably colored coat has fine- to medium-sized spots, often with smaller numbers of ring-like markings and small irregular blotches. The markings are usually scattered liberally over the body, with higher densities occurring dorsally than ventrally. Fusion of markings imparts a confused speckled appearance to many animals.

Harbor (foreground) and northern elephant seals (background) differ in many ways including adult size, markings and coloration, body proportions and features of the head and flippers. Todos Santos Island, Mexico. PHOTO: I. SZCZEPANIAK

A trio of Pacific harbor seals showing the three common color morphs: dark (front), intermediate (middle), and light (back). Aleutian Islands, Alaska. PHOTO: S. EBBERT

The most common base color pattern is a light to dark gray, or brownish-gray, back, forming a mantle with lighter sides to underneath creating a countershaded pattern. This pattern is the light morph. Dark morph, animals are uniformly dark brown-black above and below, and appear to have only pale rings, irregularly shaped marks, and ovals, although dark spotting that is very difficult to see is usually present on the pelage. There is a tendency for harbor seals in the northern part of their range to have lighter-colored pelage than those in the more southerly part. Intermediate morph animals are darker than light morphs, with browns and grays farther down their sides, and their light-colored areas are often darker than on light morph harbor seals. In some localities, a small percentage of animals have a rust- to red-colored tinge to the head due to iron oxide deposits, which in extreme cases can form a complete hood that reaches the flippers and back. Most pups shed their silvery-gray lanugo coat in the uterus before birth. Exceptions to this include pups born early in the breeding season or those born prematurely.

Adult males reach lengths of 1.9 m and weights of 70–150 kg, while females are 1.7 m and 60–110 kg. Newborn pups are 61–102 cm long and weigh 8–12 kg. Longevity for males is 20–25 years, and 30–35 for females. The dental formula is I 3/2, C 1/1, PC 5/5.

Recognizable geographic forms Although three subspecies of harbor seal are recognized, there are no known external characteristics that are useful to distinguish these types. Field identifications are made by location, as all three subspecies are geographically isolated from each other.

Can be confused with Eight other phocids share the range with one or more subspecies of harbor seal. Features for distinguishing harbor seals from northern elephant, gray, hooded, ringed, spotted, harp and ribbon seals are given in those respective species accounts. In the North Pacific, spotted seals pose the greatest identification challenge, and cannot reliably be separated from harbor seals where they occur in the same habitats. Spotted seals routinely inhabit and pup on sea ice, whereas harbor seals do not, although they will haul out and give birth on glacial ice

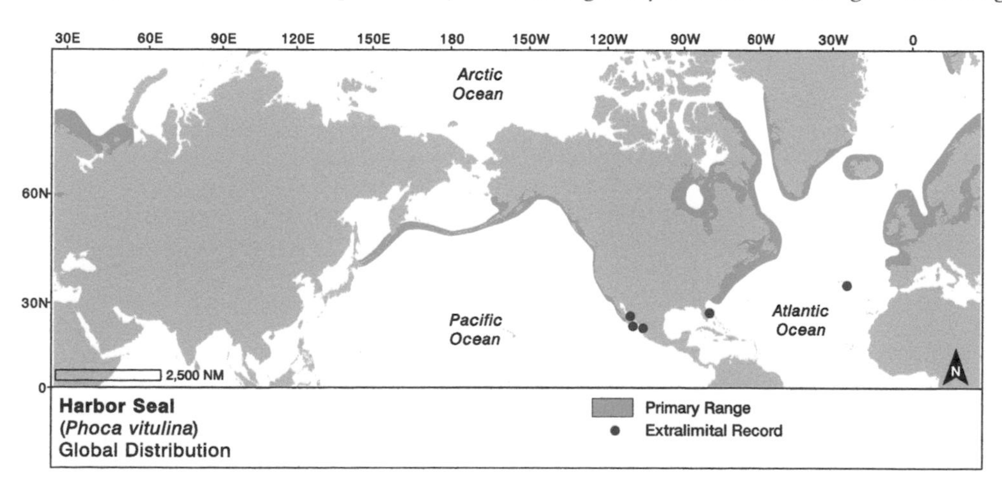

A molting Atlantic harbor seal. The new coat is light gray on this animal. Scotland. PHOTO: L. CUNNINGHAM

The northernmost occurrence of Atlantic harbor seals is Svalbard. A female is in the foreground with a male behind her. PHOTO: H. LAWSON

A lighter phase (and wet) Atlantic harbor seal with her dark phase pup. Scotland. PHOTO: L. CUNNINGHAM

Adult light phase Atlantic harbor seal with a dense covering of small black spots. PHOTO: L. MURRAY

Living in freshwater lakes and rivers, the Ungava subspecies of harbor seal is one of the rarest pinnipeds in the world, and possibly the least photographed. PHOTO: M. KUNNASRANTA

in coastal fjords. Generally, healthy harbor seal pups are born in adult-patterned pelage, whereas spotted seals are always born in a grayish-white lanugo coat. Skull differences may be reflected in head and facial differences in living animals, but this has yet to be studied in detail. There may also be differences in pelage color and markings, and aspects of behavior, that separate these species.

Distribution Harbor seals are widespread in coastal areas of the Northern Hemisphere, from temperate to polar regions. The ranges of the three subspecies are: *P. v. vitulina* in the North Atlantic from northern Portugal to the Barents Sea in northwestern Russia, including the Baltic Sea, north to Svalbard, and west to Iceland, Greenland, the Canadian Arctic and mid-Atlantic United States; *P. v. richardii* in the North Pacific from central Baja California, Mexico, north along the west coast of the United States and Canada, then around the coastal zone of Alaska (including Prince William Sound, Cook Inlet, and Lake Iliamna) north to Nunivak Island, the Pribilof islands in the Bering Sea, and throughout Bristol Bay, west through the Aleutian Islands to the Commander Islands in Russia, from there down Kamchatka, and south through the Kuril Islands to Hokkaido, Japan; and *P. v. mellonae* in rivers and land-locked lakes of the Ungava Peninsula in northern Québec, Canada, about 160 km from Hudson Bay. In the eastern North Pacific, vagrant harbor seals have been reported south to Cabo San Lucas, Mexico, inside the Gulf of California at Los Islotes, off La Paz, and Isla San Pedro Mártir in the north-central Gulf. There is also a record from Sinaloa on the Mexican mainland.

Ecology and behavior Harbor seals are mainly found from the coast to the continental slope. They routinely enter bays, rivers, estuaries, and intertidal areas, and are essentially non-migratory. On land, harbor seals are extremely wary and shy, and readily stampede into the water when disturbed. In contrast, when in the water, they can be curious, often craning their necks to peer at people on shore or in boats. Most harbor seal haulout sites are used daily, based on tidal cycles, although foraging trips can last for several days. Harbor seals are gregarious at haulouts, however, they usually do not lie in contact with each other. A hissing rolling growl is one of the few vocalizations made by this seal, and it frequently accompanies foreflipper slapping, batting and scratching of neighbors when a seal is agitated. This is common at haulouts where mov-

A dark phase Pacific harbor seal that has just given birth, evidenced by the afterbirth that has not been fully discharged. La Jolla, California. PHOTO: V. SHORE

An exceptionally pale light phase Pacific harbor seal. British Columbia. PHOTO: V. SHORE

Adult dark phase Pacific harbor seal. Baja California. PHOTO: F. R. ELORRIAGA-VERPLANCKEN

A "red" Pacific harbor seal with iron oxide staining. San Francisco Bay. PHOTO: B. KEENER

A light phase Pacific harbor seal pup/juvenile. M. NOLAN

Harbor seals in Lake Illiamna, Alaska; it is not known if they are genetically distinct from Pacific harbor seals, but studies are underway. These seals can leave the lake via a river to reach the ocean at Bristol Bay. PHOTO: M. KUNNASRANTA

A dark phase Pacific harbor seal demonstrating how little is visible of this species when at the surface. British Columbia. PHOTO: T. MELLING

A dark phase harbor seal from Kamchatka, Russia, from a population once classified as a separate subspecies, *P. v. stejnegeri*. PHOTO: M. JØRGENSEN

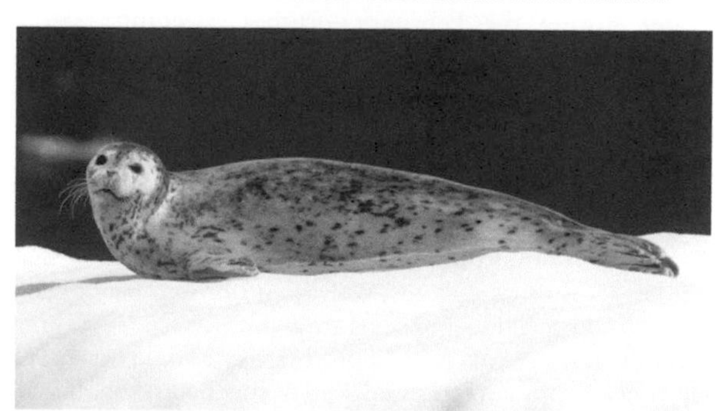

Adult intermediate phase Pacific harbor seal on glacial ice. Southeast Alaska. PHOTO: M. NOLAN

An upside down Pacific harbor seal chasing a fish, with a gull attempting to grab the fish first. San Francisco Bay. PHOTO: B. KEENER

A Pacific harbor seal feasting on salmon, tearing off hunks at the surface while an interested seal looks on. Alaska. PHOTO: M. MALLESON

A tiger-striped harbor seal? Not really. A temporary pattern probably created by a combination of molt and wet creases when the seal straightened from a U-shaped resting position. Southeast Alaska. PHOTO: M. NOLAN

A large adult Pacific harbor seal overpowers and bites a smaller juvenile, and then lets it go. Extreme aggression like this is uncommon in harbor seals. Aleutian Islands, Alaska. PHOTO: S. EBBERT

ing animals bump into or move too close to one another. At sea, they are most often seen alone, but occasionally occur in small groups. Localized aggregations can form in response to feeding opportunities. Although primarily coastal foragers, with average dives to <35 m for 3–5 min, a maximum depth of 800 m has been recorded.

Male harbor seals reach sexual maturity at 4–5 years of age, and females at 3–5 years. Because male-male competition was not observed on land, it was assumed the mating system was promiscuous or weakly polygynous. Underwater recordings revealed the males may gather in "leks" (sites where males attract females), and may also establish marine territories near where females give birth. These "maritories" are defended by underwater calling, chasing, and fighting. Mating usually takes place in the water during the February–October breeding season. Pupping peaks between April and July, and the mother weans the pup within 1 month, at the time she enters estrus. In some regions, pupping occurs earlier in more southerly areas of the range. They molt following the breeding season, and will spend more time hauled-out until the molt is finished. Predators include killer whales, sharks, and other pinnipeds, such as Steller sea lions, and eagles, feral dogs, coyotes, wolves and brown bears.

Feeding and prey Harbor seals are generalist feeders taking a wide variety of fish, cephalopods, and crustaceans obtained from surface, mid-water, and benthic habitats. Freshwater lake and brook trout feature prominently in the diet of the Ungava seals.

Threats and status Harbor seals are exposed to high levels of pollutants, and diseases from terrestrial animals. Oil spills and discharges cause direct mortality, and have long-term impacts on harbor seal health and their environment. Harbor seals live in some of the most heavily fished waters on Earth, and as a result there are entanglement issues as well as effects on the food chains they depend on for their prey. There are also conflicts with smaller, localized fisheries, and historically there have been organized population reduction programs and bounties for taking seals. Mass die-offs from viral outbreaks have claimed thousands of harbor seals. In 2002, an estimated 30,000 seals were killed by a phocine distemper virus outbreak (morbillivirus) in Europe. Proximity to terrestrial carnivores, including human pets, creates an increased risk of exposure to communicable diseases. Immunosuppression from chronic exposure to organochlorine contaminants has been linked to harbor seal susceptibility to diseases.

Despite the fact than many harbor seals live near humans, generally their population levels are not precisely known. Totaling recent estimates yields a global population of 400,000–500,000 (with roughly 300,000 in North America). The subspecies *P. v. mellonae* (with perhaps as few as 80–100 animals) in the rivers and lakes of the Ungava Peninsula, Canada is the population most at risk. Isolated populations numbering in the hundreds at Svalbard, in the Baltic Sea, and freshwater Iliamna Lake, Alaska, are also dangerously low.

IUCN status Least Concern.

Spotted Seal—*Phoca largha*

(Pallas, 1811)

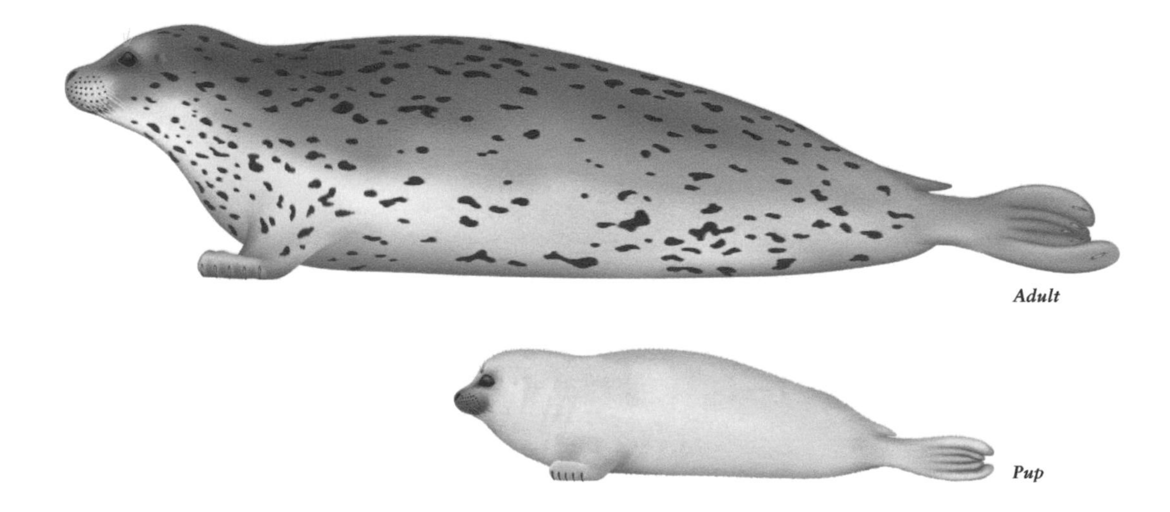

Recently-used synonyms *Phoca vitulina largha, phoca petersi.*
Common names En.–spotted seal, largha seal; Sp.–*foca largha;* Fr.–*veau marin du Pacifique.*
Taxonomic information Carnivora, Phocidae. The spotted seal was once a harbor seal subspecies, but genetic studies confirm it as a distinct species. Spotted X harbor seal hybrids are known from captivity and possibly the wild.
Species characteristics Spotted seals and harbor seals are nearly identical in external appearance, and behavior is the best means of separating the two species. Most spotted seals look like light-phase harbor seals. They have a gray to dark mantle on the back that extends down the sides, and is often darkest along the dorsal midline. The lighter sides and undersides are pale whitish-gray to silvery-gray. They are usually profusely spotted, including on the ventral surface. Most spots are dark and uniform in size, but variation occurs in density and size. They can also have larger spots look as if several spots have fused together. Light rings and partial rings can also found on the dark mantle. Intermediate phase spotted seals are darker overall, or have a more extensive mantle that covers more of the sides. Other spotted seals are paler with only a minimal mantle or hint of a gray on the back. Dark phase spotted seals have not been described in the literature, but some individuals from the Yellow Sea, appear mostly dark down their sides.

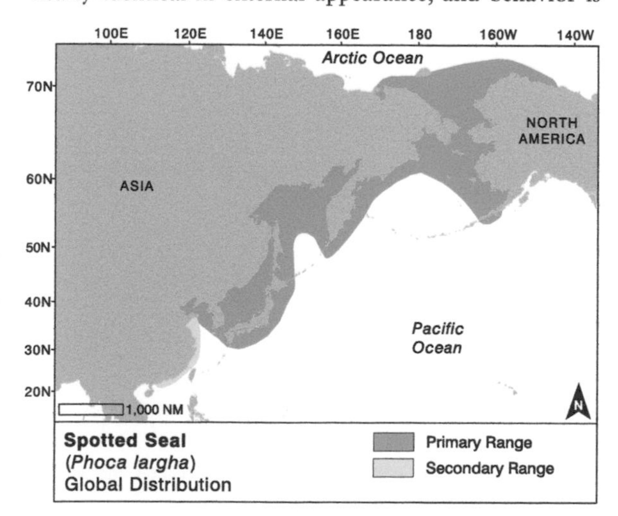

Adult males are 1.61–1.7 m long, and females are 151–1.6 m. Adult males weigh 85–110 kg and females 65–115 kg. At birth, spotted seals are 77–92 cm long and weigh 7–12 kg. Longevity is 35 years or more. The dental formula is I $^{3}/_{2}$, C $^{1}/_{1}$, PC $^{5}/_{5}$.
Recognizable geographic forms None.
Can be confused with Spotted seals share their range with ringed, ribbon, bearded, and harbor seals. Coloration and body size of spotted seals overlap with the harbor seal. Besides genetics and skull characteristics, the only reliable means of separating the species in the field are behavioral.

A group of spotted seals hauled out on an ice floe off Kotzebue in the Chukchi Sea. PHOTO: J. JANSEN/NOAA FISHERIES/AFSC

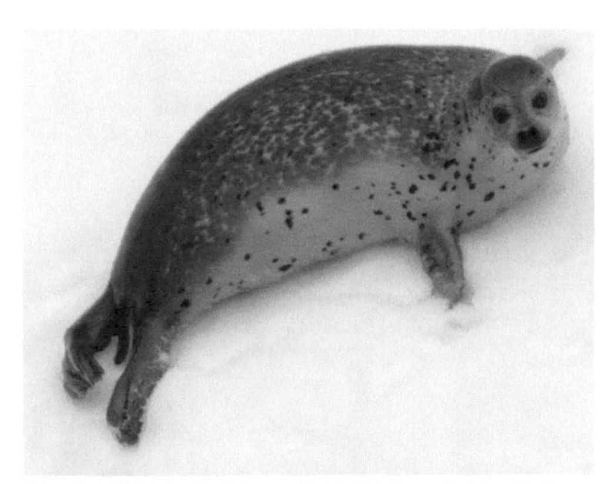

A wet spotted seal with excellent detail on the head shape, countershaded coloration and mix of spots and rings. Southern Bering Sea. PHOTO: G. BRADY/NOAA FISHERIES/AFSC

Adult spotted seal with a much paler mantle and more uniform spotting than the animal in the image to the left. Kotzebue, Alaska. PHOTO: P. BOVENG/NOAA FISHERIES/AFSC

A spotted seal pup that has fully molted. PHOTO: J. JANSEN/NOAA FISHERIES/AFSC

A molting spotted seal pup still has patches of lanugo, and was probably recently weaned. PHOTO: J. JANSEN/NOAA FISHERIES/AFSC

Spotted seal profile and head-on views. Sea of Okhotsk. PHOTOS: O. SHPAK

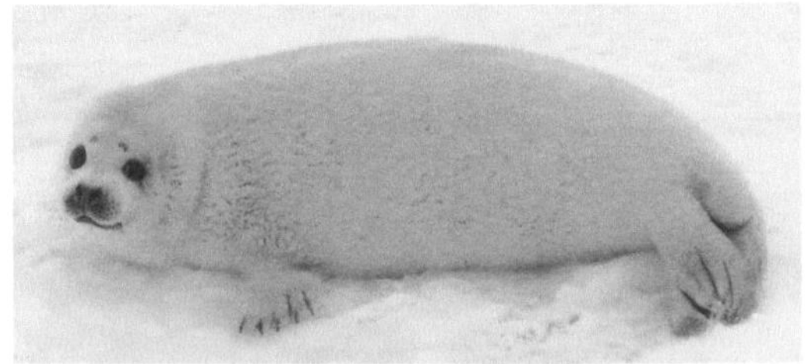

An older spotted seal pup about to begin to molt the light-colored lanugo. Southern Bering Sea. PHOTO: S. DAHLE /NOAA FISHERIES/AFSC

Spotted seal with its belly probably resting on a mudbank as the tide changes. Sea of Okhotsk. PHOTO: O. SHPAK

Spotted seals are variable in coloration like harbor seals, with pale to dark animals. South Korea. PHOTO: H. W. KIM

A very dark and a very light spotted seal, hauled-out in South Korea. Note the head, muzzle and lower jaw shape. PHOTO: H. W. KIM

Part of a group of spotted seals hauled-out at a river mouth in Kamchatka, Russia. PHOTO: T. MELLING

Spotted seals give birth on sea ice and always give birth to pups in a whitish-gray lanugo coat that is shed in 2–4 weeks. Females pup alone and are usually attended by a single adult male. Most harbor seals pup on land with a few in Alaska that give birth on glacial ice in coastal fjords. Their pups are born in the same coat as adults and only rarely (prematurely) born in a lanugo coat, which is shed in utero. Harbor seals are able to swim within hours of birth and regularly do so, moving with their mother between haulout sites and foraging areas. Where they occur together, spotted seals give birth up to 2 months earlier than harbor seals. In Bristol Bay, Alaska, spotted seals can be found hauled out on land with harbor seals.

Spotted seals and ringed seals can be confused, but spotted seals are longer and proportionately leaner, with a longer neck, head, and muzzle. Spotted seals usually have no rings, but if some are present, they are usually located on the sides and the dark mantle of the back, whereas ringed seals have an abundance of rings that are usually larger, and found from top to bottom. Pups of both species, born in a long whitish lanugo coat, would be difficult to tell apart if away from adults. However, spotted seal pups are born on top of ice floes, whereas ringed seal pups are born in lairs under snow and ice. Ribbon seals lack spots and have broad, pale to white bands on a black or brown body, or are gray above and lighter below as juveniles; in all age classes, ribbon seals lack the abundant spots of the spotted seal. Bearded seals have densely packed, long conspicuous vibrissae, are much larger and heavier bodied, with a proportionately small head, lack spots, blotches or rings, and have square to rounded ends on their flippers. Spotted seals prefer to haul out on pack ice with small floes, in contrast to ringed seals that select large floes, and bearded seals that choose mixed ice areas of large and small floes.

Distribution Spotted seals occur in three proposed Distinct Population Segments: the Yellow Sea off of China and Korea, including the Bohai Sea and an isolated group in the Peter the Great Gulf in the Sea of Japan (Southern DPS), the entire Sea of Okhotsk from Hokkaido and the northwest coast of Japan and the Strait of Tartary north (Sea of Okhotsk DPS), and throughout the Bering and Chukchi seas (Bering Sea DPS). Animals from the Bering Sea range north into the Arctic Ocean to about the edge of the continental shelf, west to about 170°E and east to the Mackenzie River Delta, where they reach Canada.

Ecology and behavior Spotted seals lead a mostly solitary existence in seasonal pack ice from the time it forms in the

A mating triad of spotted seals in spring. Southern Bering Sea. PHOTO: G. BRADY/NOAA FISHERIES/AFSC

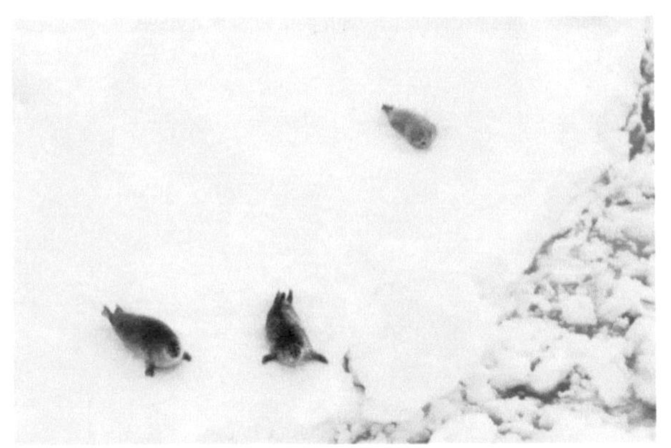

Aerial view of spotted seals in a mating triad. Western Bering Sea. PHOTO: B. SOLOVYEV

fall until it breaks up in late spring. In late summer and fall, spotted seals may become pelagic and range widely, or move into coastal areas, including river mouths, bays and estuaries where they feed on spawning fish. In coastal areas, they haul out on remnant patches of ice, rocks, gravel and mud bars. They will haul out in small aggregations, and in a few areas, among one or more of harbor, ringed, and bearded seals. They breed almost exclusively on sea ice, though some breeding on land has been documented in the Kuril Islands, off Kamchatka, Russia, and in the Peter the Great Gulf. When occupying pack ice, spotted seals use small floes less than 20 m across. Sexual maturity is reached at 4–5 years of age in males, and 3–4 years in females. Breeding takes place from January to mid-April, with pups being weaned in about 3–4 weeks. Peak numbers of pups are born mid- to late March. Adult male spotted seals are thought to be annually monogamous and territorial. They haul out with females and their pups, groups referred to as triads. The male presumably prevents access to the female by other males, and follows her into the water to mate when she weans her pup. It is not known whether they associate and mate with more than one female in a breeding season. As is typical for many pinnipeds, molt is believed to follow the breeding season. Predators include killer whales, Pacific sleeper sharks, walruses, polar bears, brown bears, wolves, Arctic foxes, and Siberian tigers in the Sikhote-Alin Reserve of Russia's Far East.

Feeding and prey Spotted seals feed on a wide variety of organisms. Their diet varies with the age of the seal, location, and seasonal variation in abundance of preferred prey species. Newly weaned pups feed on small crustaceans, advance to schooling fishes, larger crustaceans, and octopuses, and finally "graduate" to higher percentages of bottom dwelling fish species. Common prey species for spotted seals are walleye pollock, Arctic cod, sand lance, capelin, and saffron cod.

Threats and status Subsistence hunting of spotted seals has no doubt occurred since humans first made contact with the species. Population estimates are 200,000 in the Bering Sea; 100,000–130,000 in the Sea of Okhotsk; over 8,000 in the Sea of Japan; and 4,500 in the Bohai Sea off China in 1990, although this number may be drastically lower today. Threats are associated with climate change, fisheries, contaminants, and vessel traffic. Intensive harvesting of commercial fish species in the North Pacific and southern Bering Sea poses an unquantified risk. Entanglement in commercial fishing gear occurs in Japan and the Sea of Okhotsk, and small population culls regularly occur in Japan. Global warming and reduction of the extent of sea ice coverage poses a potentially serious threat to this species that breeds on ice. Increased amounts of vessel traffic associated with decreasing sea ice coverage in the Pacific Arctic is of concern due to the increased risk of accidents, oil spills, and disturbance. In 2010, the US National Marine Fisheries Service determined the Southern DPS, occupying the area from the Bohai Sea, China to Peter the Great Gulf, Russia, warranted listing as Threatened under the US Endangered Species Act, while the Bering Sea and Sea of Okhotsk DPS's were found to be sufficiently large and stable so as to not require listing under this law.

IUCN status Data Deficient.

Ringed Seal—*Pusa hispida*

(Schreber, 1775)

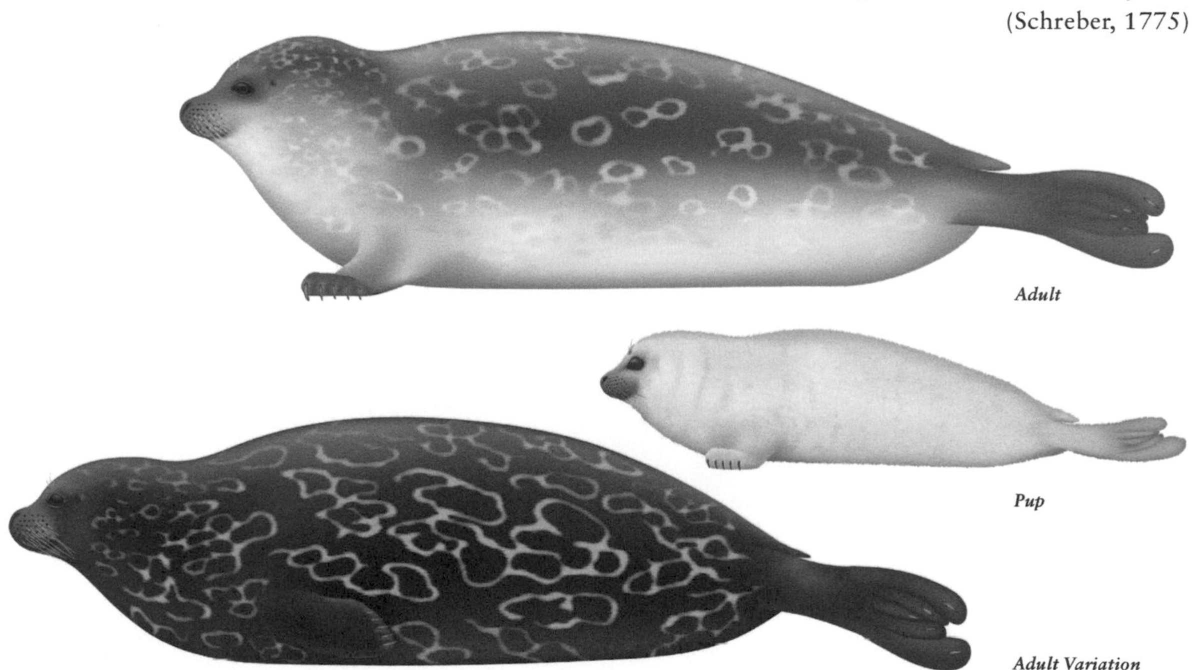

Recently-used synonyms *Phoca hispida*.

Common names En.–ringed seal, Arctic ringed seal, Baltic ringed seal, Okhotsk ringed seal, Lake Ladoga seal, Saimaa seal, jar seal, fjord seal; Sp.–*foca ocelada, foca marbreada*; Fr.–*phoque annelé, phoque marbré*.

Taxonomic information Carnivora, Phocidae. There are five recognized subspecies: *P. h. hispida* (Arctic ringed seal) in the Arctic basin; *P. h. ochotensis* (Okhotsk ringed seal) in the Sea of Okhotsk and Japan Sea; *P. h. botnica* (Baltic ringed seal) in the Baltic Sea; *P. h. ladogensis* (Lake Ladoga seal) in Russia; and *P. h. saimensis* (Saimaa seal) in Finland.

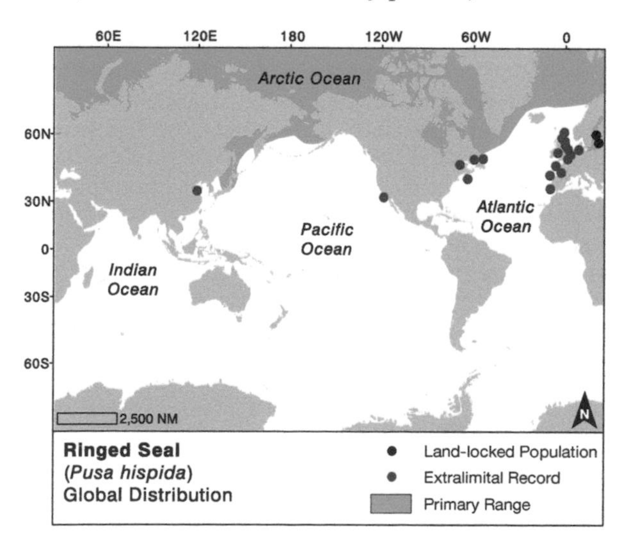

Species characteristics Ringed seals resemble harbor and spotted seals, but are decidedly plumper (axillary girth may reach 80% of length). They also have a smaller, more rounded head on a conspicuously short thick neck. The muzzle is quite short, slightly broader than thick, and blunt. The vibrissae are light-colored and beaded. The eyes are relatively large and conspicuous. More so than in other northern phocids, the size of the head and muzzle, and the close-set forward-facing eyes, impart a cat-like appearance. The foreflippers are relatively small and slightly pointed, as in the harbor seal. The background coloration is variable, but normally it is medium to dark gray above and light gray to silver below, but they can also be dark tan to brown. Ringed seals are conspicuously marked with light gray to off-white rings that encircle "spots" of the darker dorsal and lateral background pelage color. Some ringed seals can be so heavily marked that many rings and spots fuse, creating a confused paint-splattered appearance. The lighter-colored lower sides and undersides have a variable number of dark spots that do not appear to be encircled by rings because of the paler lateral and ventral pelage. The seals in Lake Saimaa are dark brown with rings, and in Lake Ladoga they are charcoal, or blackish when wet. Pups are born with a wooly, thick, whitish lanugo. Fur of the succeeding coat

An Atlantic ringed seal rises up in an open lead. Canadian Arctic. PHOTO: L. MURRAY

A Pacific ringed seal showing the classic pelage pattern of rings and irregular shapes that characterize this species. Kotzebue, Alaska. PHOTO: J. JANSEN/NOAA FISHERIES/AFSC

A very young ringed seal pup in lanugo outside its lair. Barrow, Alaska. PHOTO: L. LOWRY

An adult Saimaa seal, one of the rarest types of pinnipeds. Lake Saimaa, Finland. PHOTO: M. KUNNASRANTA

is finer and slightly longer than that of adults, and is dark gray above, merging to silver below. There may be a few scattered dark spots on the undersides of these juveniles, and few, if any, rings on the back. At this stage, they are known as "silver jars."

Adults of both sexes are 1.10–1.65 m in length and 50–90 kg in weight, with a maximum of 110 kg. Pups average about 60–65 cm and 4–5 kg at birth. Longevity may be up to 50 years. The dental formula is: I $^{3}/_{2}$, C $^{1}/_{1}$, PC $^{5}/_{5}$.

Recognizable geographic forms There are five geographic forms of ringed seal. The primary subspecies that occupy most of the Arctic, *P. h. hispida*, and the Sea of Okhotsk subspecies, *P. h. ochotensis*, are very similar, and can be separated in the field only on the basis of location. *P. h. botnica*, cut off from the main ringed seal population by its isolation in the Baltic Sea, is dark overall with a scattering of light rings and may be the largest subspecies. Both the Saimaa (*P. h. saimensis)* and Ladoga *(P. h. ladogensis)* populations, living in isolated freshwater lakes, are dark when compared to most Arctic ringed seals, *P. h. hispida.*

Can be confused with Ringed seals share their extensive range with seven other phocids. They are not likely to be confused with bearded seals, or most ages of harp, hooded, gray, or ribbon seals because of differences in size, coloration, and body shape. However, care is required to distinguish them from other seals with rings, spots or spot-like markings, such as harbor seals, spotted seals and juvenile harp seals. Separating these species requires careful attention to coloration and markings, the relative size and shape of the head and muzzle, and overall body size and shape. Spotted and harbor seals are very similar in build and length, and adults of both are longer than ringed seals, and the neck, head and muzzle are also proportionately longer and generally larger. Also, most spotted seals have few or no rings. Spotted seals have their pups on the surface of the ice, whereas ringed seals have their pups in lairs under snow layers on sea ice, or inside pressure ridges. Harbor seals do not regularly use ice except where glaciers discharge ice into bays and fjords as semi-permanent feature. Except in rare cases, or when pups are born prematurely, the coloration of harbor seal pups is essentially the same as in adults and subadults. Harbor seals generally show some rings, but when present, there is always a mixture of rings and spots. They are far less uniformly marked with rings on the back and upper to mid-sides than most ringed seals.

Distribution Ringed seals have a continuous circumpolar distribution throughout the Arctic basin, Hudson Bay

Atlantic ringed seal with another seal using the same hole in the ice. Note the dark face of the adult on the ice. Canadian Arctic. PHOTO: L. MURRAY

A Pacific ringed seal sits up in an alert position. Kotzebue, Alaska. PHOTO: M. CAMERON/AFSC

Even with very little of this Atlantic ringed seal showing, rings can still be seen behind the smallish round head. East Greenland. PHOTO: H. LAWSON

An adult Pacific ringed seal with very conspicuous rings on its dark back. Off Kotzebue, Alaska. PHOTO: J. JANSEN/NOAA FISHERIES/AFSC

A juvenile Atlantic ringed seal. The small round head is quite evident on this plump, fit-looking animal. East Greenland. PHOTO: H. LAWSON

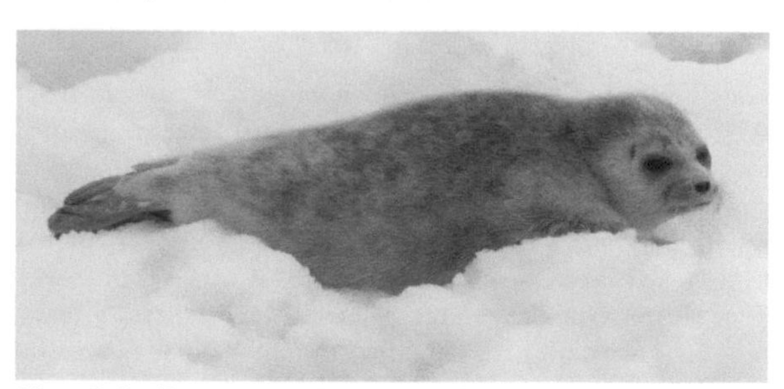

A juvenile Pacific ringed seal. Despite the faint pelage markings, this animal is unmistakable, based on the small round head, forward- facing eyes and very small short muzzle. PHOTO: M. CAMERON/AFSC

and Straits, and the Bering Sea. The other populations do not overlap the large Arctic population and are all isolated: in the Sea of Okhotsk, the Baltic Sea, and in land-locked freshwater Lake Ladoga, Russia and Lake Saimaa, Finland. The distribution of ringed seals is strongly correlated with pack and shore-fast ice, and areas covered at least seasonally by ice. Adults use fast ice for breeding, molting, and over-wintering habitat. Vagrants have been reported from the Yellow Sea of China, Sable Island, Canada, New Jersey, the United Kingdom, the Azores, and Europe south to Portugal.

Ecology and behavior Nearly all ringed seals breed on dense pack ice and shore-fast ice. The onset of sexual maturity is around 7 years in males and 3.5–6.0 years in females. Males are unlikely to be able to breed until they can defend their access to females, which is usually not until they are about 8–10 years old. Using their strong foreflipper claws, females excavate and maintain lairs in snow, in pressure ridges, and beneath other snow-covered features. These lairs have access to the ocean from below, and provide camouflage and some protection from predators such as the polar bear. Most animals maintain 4–6 lairs so that they can escape and find refuge if one is broken into by a polar bear. Pupping generally occurs from March through April, and earlier in the Baltic Sea, with pups being weaned around 20 kg in 4–6 weeks. Males are territorial, and possibly annually monogamous, defending access holes to female lairs with a combination of vocalizations and aggressive interactions with approaching males. Many adults remain in the same localized areas year-round and are solitary for most of their lives. Molting is in June and July and follows a period of post-breeding season replenishment, when ringed seals spend longer periods hauled out, in contrast to the steady, near continuous pattern of foraging they engage in when not breeding or molting. Some ringed

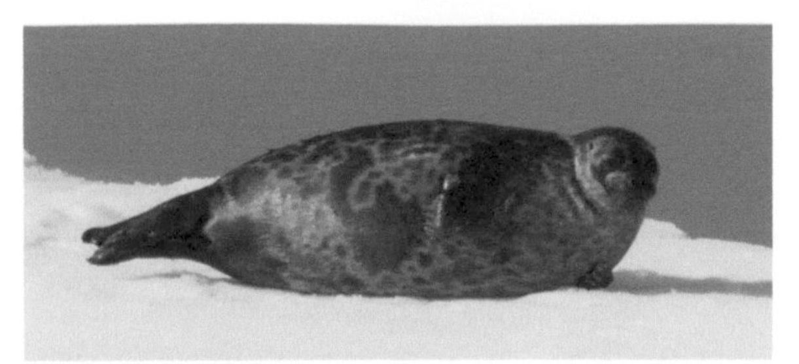

A Baltic ringed seal showing the round, plump shape, small head and muzzle, and rings that together confirm the identification of this animal as a ringed seal of the Arctic/subarctic. PHOTO: M. KUNNASRANTA

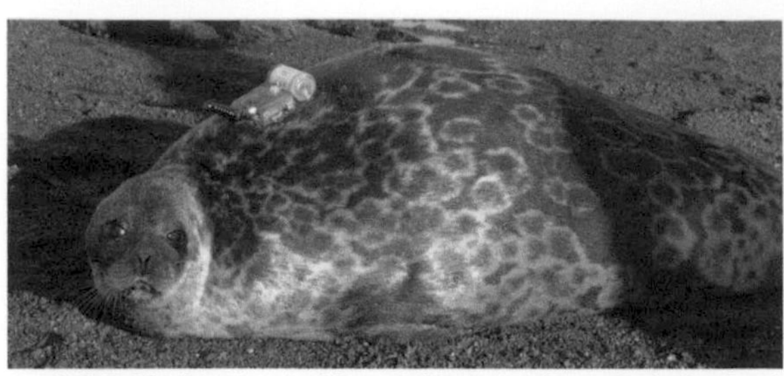

An adult Baltic ringed seal with a satellite transmitter attached to track its movements and diving patterns. Note the very large rings. PHOTO: V. ALEXEEV

A beached molting Baltic pup, possibly separated from its mother too soon. PHOTO: M. JÜSSI

A small Baltic ringed seal pup on the ice near an access hole. PHOTO: M. JÜSSI

Detail of the head of a Baltic ringed seal in a stranding facility in Russia. PHOTO: V. ALEXEEV

seals haul out on land as needed when ice is absent from their preferred feeding grounds, or when they are obliged to do so, such as when their entire habitat is ice free as in summers in Lakes Ladoga and Saimaa, the Baltic Sea and most of the Sea of Okhotsk. Lake Ladoga in particular is known for large summertime haulouts of seals on rocks near shore. Out of water, ringed seals are generally wary, regularly scanning for predators, such as polar bears and humans. Their dives are usually to 100 m or less, with the deepest reported dives being to around 300 m. Longevity in the wild has been reported to be almost 50 years.

The list of ringed seal predators is long, due to their wide distribution around the Arctic. Polar bears are a well-known predator, and the link between the two species has been extensively studied. Brown bears are a suspected predator on Saimaa seals in Finland, and may play an increasing role as the bears' habitat shifts northward with climate change, and ice habitat for seals diminishes or disappears entirely seasonally. Canids, such as Arctic and red foxes, gray wolves, and feral dogs, are all known predators. The impact of Arctic fox predation can be significant, but is less well known than that by polar bears. Arctic foxes have a high latitude distribution, while red foxes are likely to be predators where ringed seals live near shore or in freshwater south of the Arctic Circle. Feral dogs have been known to be a problem predator for Ladoga seals. Lynx and European mink are somewhat unexpected predators, probably most relevant to Ladoga and Saimaa seals. Walruses, killer whales, and Greenland sharks are known marine predators, and common ravens, glaucous gulls, snowy owls and several species of eagles are potential avian predators on newly weaned pups, or pups exposed when snow covering their lair melts prematurely.

Feeding and prey Ringed seals consume a wide variety of small prey, including many species of fishes, cephalopods, and planktonic and benthic crustaceans. In a study off Alaska, these opportunistic feeders took over 155 different species, of which 99 were eaten frequently. Usually the diversity of prey species taken in an area is lower, and reflects availability. In deep water, they forage in the water column and along the undersides of ice floes, while in shallow water they often forage along the bottom. Some common prey species are: Arctic and saffron cod, walleye pollock, sand lance, capelin, prickleback, and Pacific herring. Juveniles take large numbers of crustaceans such as krill, mysid shrimp, and amphipods when available. The freshwater Saimaa and Ladoga seals take many species of fish, including: burbot, perch, roach, smelt, vendace, and whitefish. Ringed seals primarily forage singly. Small groups are sometimes seen, and are likely unassociated individuals drawn to a feeding opportunity.

Threats and status Indigenous peoples of the Arctic hunt ringed seals for food and skins, and have done so for thousands of years. Existing world-wide annual subsistence harvest levels are not well known, but suspected to be 150,000 plus, and while large, is not currently of conservation concern. Commercial harvests for skins was very high in the 1960s, with unknown effects. However the over-exploitation of the relatively small populations in Lake Ladoga and Lake Saimaa did have detrimental

Lake Ladoga ringed seals resting on the bottom near shore. Lake Ladoga, Russia. PHOTO: M. KUNNASRANTA

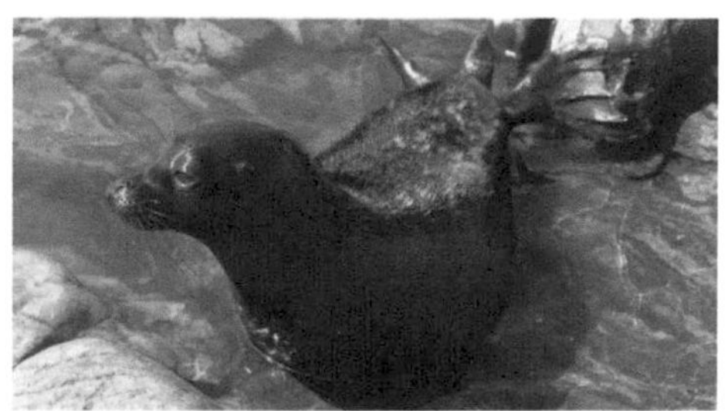

A juvenile Lake Ladoga seal being released form a stranding facility. RUSSIA. PHOTO: V. ALEXEEV

Adult female and nursing pup Saimaa seal ashore under the shade of trees. Finland. PHOTO: M. KUNNASRANTA

A Saimaa seal on the surface of the lake could be vulnerable to predators if it is not near an access hole. Finland. PHOTO: J. HEIKKILA

The dark area in the right foreground is the broken cover of a Saimaa seal lair on the lakeshore. PHOTO: M. KUNNASRANTA

An Okhotsk ringed seal on a mudflat. PHOTO O. SHPAK

A grayish Okhotsk ringed seal. See page 484 for ringed seals hauled out with other species. PHOTO: O. SHPAK

effects. Commercial use of skins from some harvests, such as the ones in Greenland, do not appear to be negatively affecting any ringed seal populations. Heavy burdens of organochlorines can effect reproduction and cause immunosuppression, a risk for ringed seals in some areas, especially the Baltic Sea, which is surrounded by large human population centers.

Estimates of the world-wide population, ranging from 2.5–7.0 million ringed seals (with the majority of those being from the Arctic subspecies), should be taken with caution, given their vast geographic range, coupled with their solitary nature, and the challenges of conducting population assessments in remote polar areas. Abundances for other subspecies are: 800,000–1,000,000 in the Sea of Okhotsk, based on estimates from 1968–2002; 5,000–8,000 for the Baltic; 3,000–5,000 in Lake Ladoga; and 310 in Lake Saimaa. Only about 249,000 of the Arctic subspecies live in US waters. All the subspecies might be in some form of decline. In 2013, the US listed the Lake Ladoga subspecies as Endangered, and the Arctic, Sea of Okhotsk and Baltic subspecies as Threatened, indicating the level of widespread global concern about the future of ringed seals due to the impact of the steadily decreasing extent of sea ice in its Arctic and subarctic habitat. Successful reproduction by ringed seals is tied to the presence of sea ice, essential for building lairs. Without lairs, pups would be dangerously exposed to predators, as they do not immediately take to the water like harbor seals. Additionally, Arctic and other nations are already planning greater use and development of Arctic resources and new vessel routes. More vessel traffic brings more noise, physical disturbance and an increased risk of oil and fuel spills that could have a very detrimental effect on ringed seals. Further decreases in sea ice coverage could also permit more movements by pinnipeds between Atlantic and Pacific Arctic areas, possibly leading to the exchange of diseases between the two areas.

IUCN status Least Concern.

Baikal Seal—*Pusa sibirica*

(Gmelin, 1788)

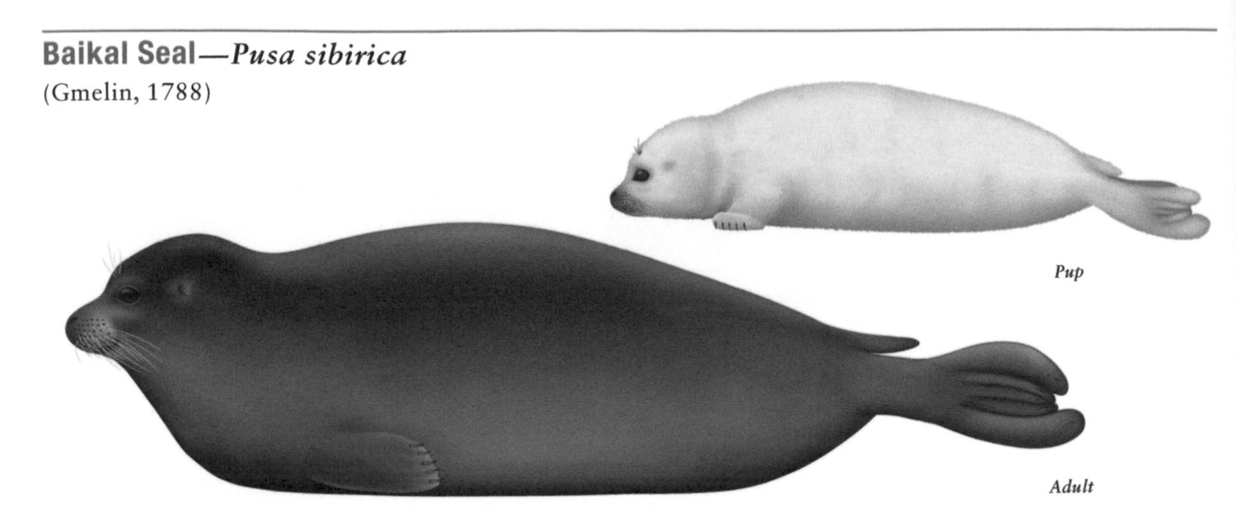

Recently-used synonyms *Phoca sibirica.*

Common names En.–Baikal seal; Sp.–*foca de Baikal*; Fr.–*phoque du lac Baikal*. In Russia, this seal is called *nerpa*.

Taxonomic information Carnivora, Phocidae. The ringed seals is considered the ancestor of the Baikal seal, and are thought to have reached the lake by traveling up the river system and drainage that runs from the lake to the Arctic Ocean some 400,000 years ago during the Pleistocene.

Species characteristics Baikal seals are dark silver-gray above and lighter gray on their undersides. They appear charcoal to black above when wet. Pups are born in a whitish lanugo shed at 4–6 weeks, when they transition to a juvenile pelage that is lighter gray with a darker face. Baikal seals are unmarked, with no rings or spots. These seals are plump, with a proportionately wide body and a small delicate-looking head. The eyes are relatively large, forward set, and have a slight bulging appearance. The muzzle is small, wide, and delicate-looking in juveniles. The vibrissae are beaded, long and prominent, including those over the eyes. The foreflippers are short, broad, and have large strong dark claws. Adult male Baikal seals are approximately 1.3–1.45 m, and adult females are slightly smaller at 1.2–1.3 cm. Weights for both sexes combined and are 50–90 kg. Pups are 60–65 cm in length and 4–4.2 kg at birth. The dental formula is $I\ ^{3}/_{2}$, $C\ ^{1}/_{1}$, $PC\ ^{5}/_{5}$.

90E
100E
110E
120E
60N
50N
RUSSIA
Lake Baikal
MONGOLIA
CHINA
250 NM
N
Primary Range

Baikal Seal
(*Pusa sibirica*)
Global Distribution

Recognizable geographic forms None.

Can be confused with No other pinnipeds occur in Lake Baikal, so there should be no source of confusion.

Distribution Baikal seals are confined to Lake Baikal and only short distances up its rivers and streams. Most seals are found in the northern and central regions of the lake, and a portion of the population moves south in front of advancing ice that forms in the late fall and winter. Individual seals have been known to travel several hundred kilometers up the Angara River, a part of the Yenisei River watershed that reaches the Arctic Ocean. Lake Baikal is the world's largest lake by volume of freshwater.

Ecology and behavior Baikal seals use islands and the rugged shorelines for hauling out during ice-free periods, and in autumn they move into eastern coastal areas where ice forms earliest, becoming ice-living throughout the lake, particularly over deep central areas from January to May. Adults are mostly in the northern parts of the lake and immatures tend to stay in more southern areas. The onset of sexual maturity occurs at 7–10 and 3–7 years in males and females, respectively. Females can bear their first pup at 4 years, but most begin pupping at 5–6 years, while some do not reproduce until they are 9 years old.

Females give birth in snow-covered lairs excavated in the lake ice where they also maintain breathing and access holes in the ice. This affords protection from predators, primarily eagles and canids, since newborn pups remain out of the water for the first 2–3 weeks of life. A female's lair and array of breathing holes do not overlap with those of other females. Pupping occurs from mid-February to the end of March, and pups are weaned in 2–3 months,

A view of Baikal seals on a traditional haulout site. Note the plump body that lacks rings, spots or other markings. PHOTO: S. GABDURAKHMANOV

An adult Baikal seal, possibly a male. Notice the more developed, wider face and larger head. PHOTO: S. GABDURAKHMANOV

Baikal seals are often found alone on the winter ice. However, during the summer they will haul out in groups. PHOTO: S. GABDURAKHMANOV

A very round Baikal seal provides a diagnostic view of this species. PHOTO: S. GABDURAKHMANOV

Baikal seal pup, in lanugo, at the opening of its lair. PHOTO: A. ZAHRADNIKOVA

Three Baikal seals share an aggressive interaction, with an open mouth gesture on the left and flipper wave on the right. PHOTO: S. GABDURAKHMANOV

A large seal waving and scratching with its right foreflipper as it attempts to haul out into a crowded space on a rock. PHOTO: S. GABDURAKHMANOV

Details of the head include the slim short muzzle, forward-facing eyes, and tall superciliary vibrissae. PHOTO: S. GABDURAKHMANOV

The small Baikal seal in the front right of this image is a juvenile. PHOTO: M. KUNNASRANTA

although some pups may receive care for slightly longer periods. Consistent with the distribution of adults, more than half the pups are born in the northern 1/3 of the lake and over 80% of births occur in the upper 2/3. Newly-weaned juveniles emerge from the lairs in April. Twins are born approximately 4% of the time, probably the highest rate for any pinniped. Mating follows about a month after the birth of the pup. Molting follows mating and weaning of pups, and coincides with ice break-up in late spring and early summer, a period of time when Baikal seals spend considerable time hauled out on the remaining ice, sometimes in large aggregations of several thousand animals. Baikal seals are otherwise generally solitary, but will also haul out at sites on land in summer in groups, where animals lie in fairly tight aggregations with individual animals close to each other. Baikal seals routinely dive for 2–4 min, sometimes for 6–10 min, to depths of 10–50 m at night and 100–200 m during the day. The deepest and longest dives are to 300 m and for 40 min.

Feeding and prey Baikal seals forage extensively through the winter. Their diet consists primarily of several species of sculpins and golomyanka or oilfish (genus *Comephorus)*, but includes many varieties of fishes, varies by season, and includes commercially important species such as omul.

Threats and status Modern humans have lived in and traveled through the area of Lake Baikal for tens of thousands of years. The archaeological record indicates that early cultures of the area harvested this species. A commercial harvest of up to 9,000 animals per year ended by 1980. Estimates of the population of Baikal seals varied between 60,000–70,000 during the 1970s–1980s. The current population is estimated at 80,000–100,000.

Baikal seals have high organochorline contaminant burdens due to industrial wastes discharged into rivers feeding the lake and bio-concentrating of these contaminants up the food chain. An estimated 6,500 seals died during a 1987–88 morbillivirus outbreak related to canine distemper that probably originated from domestic dogs. High burdens of organochlorine pollution can negatively impact the immune system of pinnipeds, and it is possible that the pollutants entering the Lake Baikal system and accumulating in the animals may have contributed to the susceptibility of the seals to infection when exposed to the morbillivirus. Subsistence and commercial hunting and poaching continue, as do entanglement bycatch deaths, which lead to a combined annual mortality of 3,000–4,000.

IUCN status Least Concern.

Caspian Seal—*Pusa caspica*

(Gmelin, 1788)

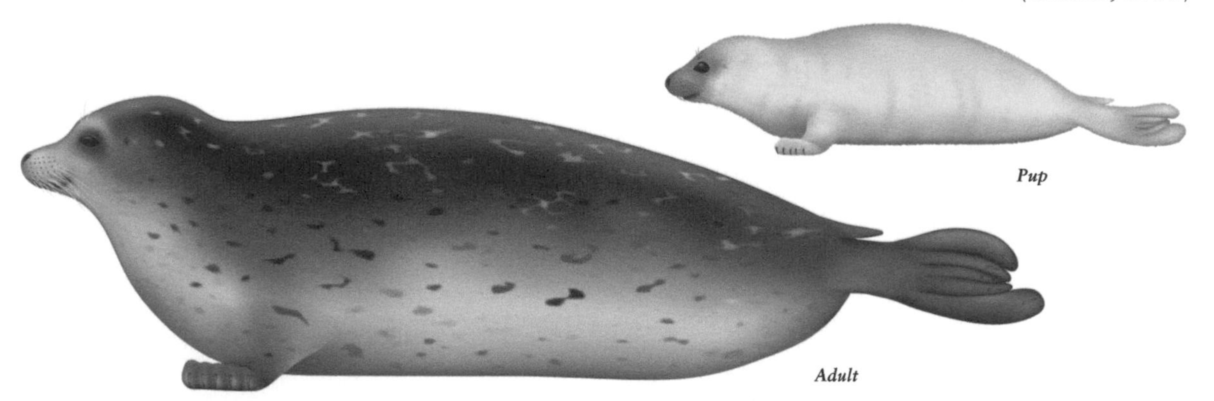

Recently-used synonyms *Phoca caspica.*

Common names En.-Caspian seal; Sp.-*foca del Caspio;* Fr.-*phoque de la Caspienne.*

Taxonomic information Carnivora, Phocidae. It is believed the ancestors of the Caspian seal entered Eurasia 2–3 MYA. Once thought to be closely related to ringed and Baikal seals, recent genetic evidence suggests a closer connection to gray seals and seals of the genus *Phoca.*

Species characteristics Caspian seals are countershaded medium gray to grayish-brown above and paler below. There are often yellowish or tan undertones to the coat color. Caspian seals have a variable covering of light and dark spots and mottling all over their body. Males are typically more heavily spotted than females. Pups are born in a long white lanugo coat that is molted around weaning, at about 3–4 weeks. The juvenile coat is lightly spotted and more silvery and paler than the adult pelage. Caspian seals are relatively short and plump. The head is small in proportion to the body with large conspicuous eyes. The muzzle is relatively long, wide, and somewhat flattened, with whitish vibrissae. Adult males and females reach maximum lengths of 1.5 and 1.4 m, respectively, and weigh 75–86 kg. Pups are 65–80 cm and about 5 kg at birth. The dental formula is I $^3/_2$, C $^1/_1$, PC $^6/_5$.

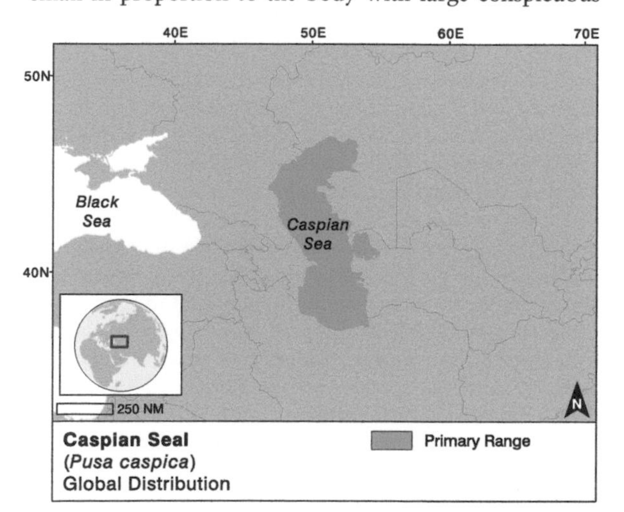

Recognizable geographic forms None.

Can be confused with No other pinnipeds occur in the Caspian sea, so there should be no source of confusion.

Distribution This species is confined to the Caspian Sea and its feeder rivers. Their seasonal movements are prompted by ice formation and foraging. In the late winter, adults migrate to the northern part of the sea, where they haul out on islands awaiting ice formation. In late January, they assemble on the ice for pupping. After pupping, and the ice melt, adults again haul out on islands of the north Caspian, where they undergo their annual molt. After the molt, from late spring, they disperse to all parts of the Caspian, particularly the deep central and southern basins. Non-breeding adults and juveniles may spend the winter in the central and southern Caspian. A small number of females pup on Ogurchinsky Island in the south.

Ecology and behavior Caspian seals are monogamous, an unusual trait among pinnipeds. Females become sexually mature at 5 years old, but usually don't produce a pup until they are 7. The onset of sexual maturity in males is at about 6–7 years. Most pups are born in winter in the northern Caspian on the surface of the ice, and are not concealed in snow lairs. Pups are often born along edges and cracks in the ice, but females also can give birth away from the ice edge where they create a network of holes for access to the sea. Pups are weaned in 4–5 weeks, and females mate around this time after which they molt. Females molt from late February to early March and males molt from late March to early April. Predators include wolves and eagles that prey primarily on pups. Caspian seals are long-lived, with females reported to reach 43.5 years and males, 33.5 years. Most dives for

The face of a captive juvenile Caspian seal in Iran. PHOTO: A. S. SHIRAZI

A countershaded juvenile Caspian seal with a low density of spots. PHOTO: N. ZAKHAROVA AND V. KUZNETZOV

Caspian seals near an access hole in the sea ice. The relatively long muzzle of the adult is noticeable. PHOTO: S. WILSON

This adult female provides a good view of the long and somewhat flattened muzzle characteristic of this species. PHOTO: S. WILSON

This aerial view of a group of Caspian seals shows most of the animals are associated with the ice edge. PHOTO: B. SOLOVYEV

two adult males were less than 50 seconds, with some dives to over 3 min. Most dives were less than 50 m, with a few dives to over 200 m (the Caspian Sea's average depth in the central basin is 190 m). The sea is freshwater in its northern portions, and more saline in the south, with an average salinity 1/3 that of typical ocean water.

Feeding and prey Caspian seals eat many species of fish and crustaceans; the diet varies by season and location. Several species of Caspian kilka (sprat), silversides and gobies, zander and bream, make up much of the diet. Other species taken include sand smelts and crustaceans, and when in rivers, carp, Caspian roach and pike-perch.

Threats and Status Commercial and subsistence hunting have been ongoing for over 200 years. From the late 18th to the mid-19th century, it is estimated that sealers took about 115,000 animals per year, but at least one harvest reached 300,000. The population probably was 1–1.5 million early in the 20th century, and declined to 360,000–400,000 by the late 1980s. Large commercial harvests ended in 1996, however about 3,000–4,000 pups per year are still taken. Their numbers remained fairly high into the early 2000s, but in 2005 they were found to have declined to a very low level of 100,000 seals, or about 10% of their former greatest abundance.

Three large mortality events (1997, 2000, and 2001), attributed to canine distemper virus (morbillivirus), claimed mainly young seals. A further threat to seal survival is by-catch in fishing nets and direct kills by fishermen. There has been considerable industrial development around the shores of the Caspian Sea, and discharge of high levels of many kinds of pollutants. Accumulation of organochlorine contaminants, including DDT, in Caspian seals has been reported at levels high enough to play a role in weakening their immune system, contributing to disease outbreaks, and reducing the reproductive rates in females. Overfishing in this large enclosed ecosystem is an ongoing concern. An invasive comb jellyfish, a predator on zooplankton, arrived in the Caspian Sea in 1999. It is believed to be causing a reduction in fish, including stocks of kilka, and poses an undetermined threat to the Caspian seal. Predation by sea eagles and wolves has an impact that is not well understood. In the past, wolves were implicated in high mortality of pups, but a recent study found eagles took 10% of pup production, or about 2,000 pups, in 2005–2006. Finally, diversion of water, particularly from the Volga River, has lead to fear of falling water levels and harm to the entire Caspian Sea ecosystem.

IUCN status Endangered.

6. Sirenians

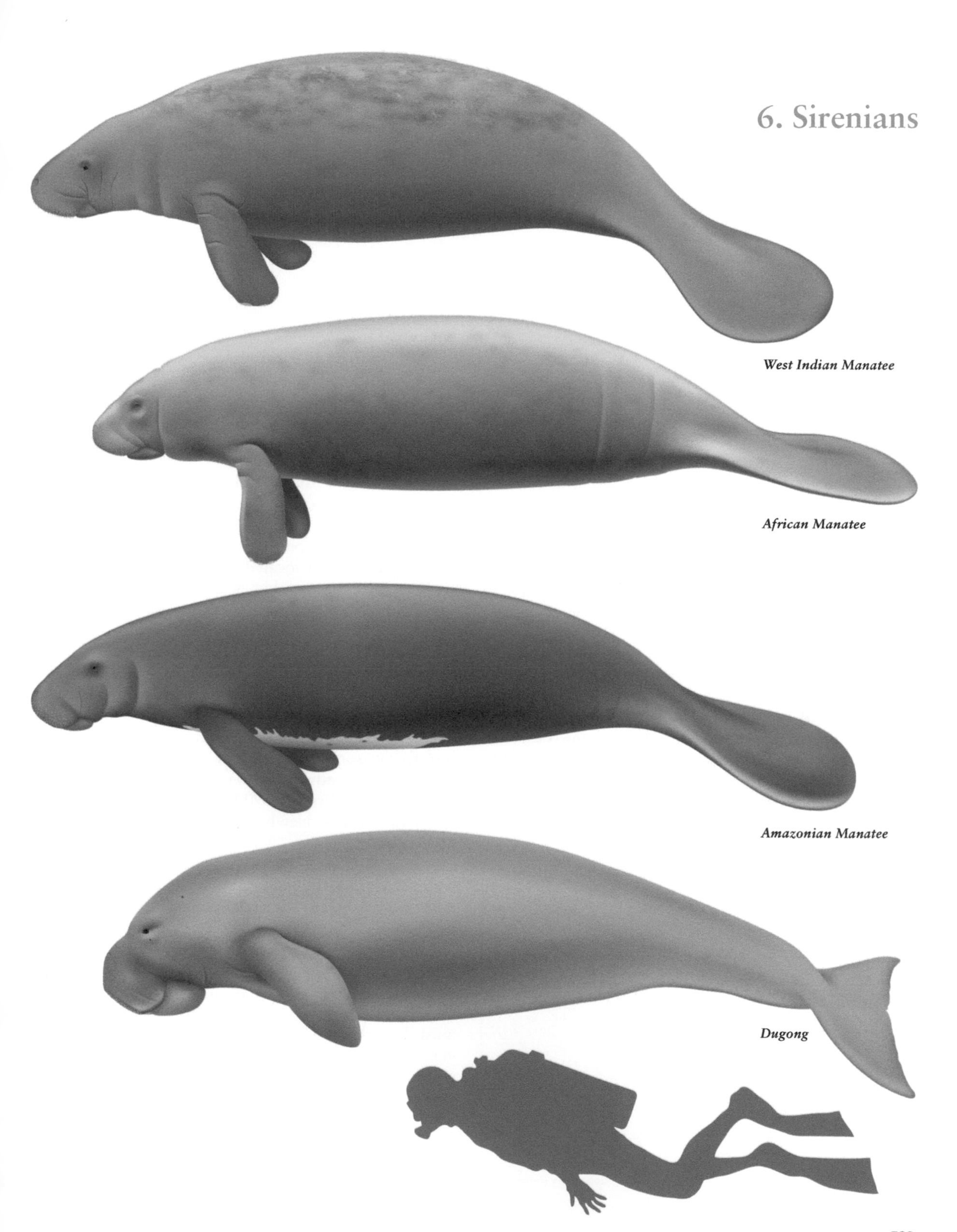
West Indian Manatee
African Manatee
Amazonian Manatee
Dugong

West Indian Manatee—*Trichechus manatus*

Linnaeus, 1758

Recently-used synonyms None.

Common names En.–West Indian manatee; Sp.–*manatí del Caribe, vaca marina*; Fr.–*lamantin des Caraïbes*.

Taxonomic information Sirenia, Trichechidae. There are two subspecies recognized: the Florida manatee (*T. m. latirostris*) in the southeastern US and the Antillean manatee (*T. m. manatus*) from northern Mexico to northern South America, including islands of the Caribbean. Hybrids between West Indian and Amazonian manatees have been documented near the mouth of the Amazon River (and possibly also near the mouth of the Orinoco River).

Species characteristics West Indian manatees are rotund, with broad backs, and have long flexible forelimbs and rounded, paddle-like tails. The head is small, with no discernible neck, and the body exhibits numerous folds and fine wrinkles. The squarish, thickened snout has fleshy, mobile lips (with stout bristles on the upper lip) and two semi-circular nostrils at the front. The rostrum is strongly deflected downwards (29–52°—more so than in the other manatee species), which reflects the species' preference for feeding on submerged vegetation. The skin is very rough and thick, and has fine hairs sparsely distributed over its surface. Each flipper has 3–4 fingernails at the tip. The tail stock is not laterally compressed into a peduncle. The mammary glands of manatee females are located in the axillary region.

The color of the skin is generally gray to brown, sometimes with a green, red, white, or black tinge caused by algal and/or barnacle growth. The short hairs are colorless. Calves appear to be a darker shade of gray, almost black.

There are 5–7 pairs of bicuspid post-canines/molars in each jaw. When forward teeth are worn or lost, they are replaced from behind. At birth, each jaw also has two vestigial incisors, which are lost as the animal ages. West Indian manatee adults are up to 3.5 m (4.0 m maximum) long and weigh up to 1,590 kg. Females tend to be a bit larger than males. Newborns measure about 1.2 m and weigh about 30 kg.

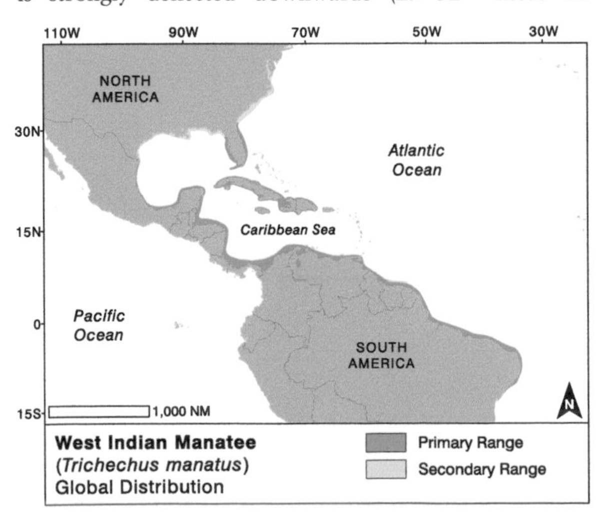

Recognizable geographic forms The two subspecies are separated largely by their distribution and differences in skeletal morphology. Reliable methods of distinguishing them from external appearance are not known.

Can be confused with The West Indian manatee is the only sirenian throughout most of its range, with the exception of the area around the mouth of the Amazon River (and maybe the Orinoco as well). In this region, special care must be taken to distinguish it from the Amazonian manatee. Pay special attention to head shape (shorter and wider rostrum in the West Indian), body color (lighter, generally without white belly patches in the West Indian), skin texture (rougher in the West Indian), and presence/absence of nails on the flippers (present in the West Indian).

Distribution West Indian manatees are found in coastal marine, brackish, and freshwater areas of the south-eastern US coast, tropical/subtropical Gulf of Mexico,

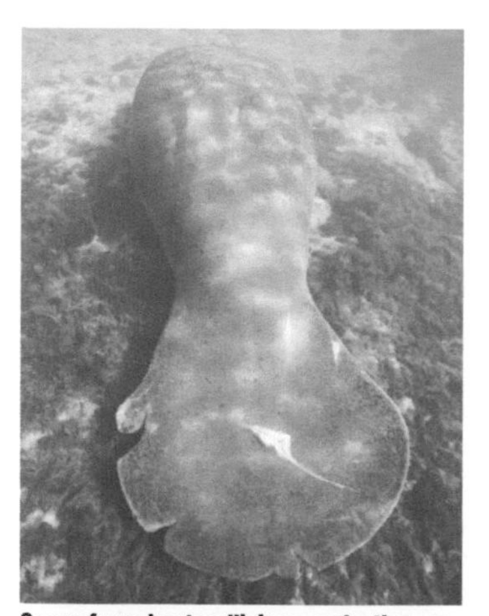

Scars from boat collisions and other anthropogenic sources are common on the bodies of many West Indian manatees—these are used to assist in identification of individuals. Florida. PHOTO: M. NOLAN

The muzzle is deflected downwards in West Indian manatees, an adaptation to assist in feeding on seagrasses and other vegetation along the bottom. Florida. PHOTO: M. NOLAN

The West Indian manatee will likely be easy to identify, as it is the only sirenian in much of its range. However, near the mouth of the Amazon and Orinoco rivers, care must be taken to distinguish it from the Amazonian manatee. Florida. PHOTO: M. NOLAN

A West Indian manatee calf suckles from its mother in Florida waters. The teats are near the axillae in sirenians. PHOTO: T. PUSSER

West Indian manatees (in this case a mother and calf) are found in very clear water in some areas of Florida. PHOTO: L. KEITH

In some freshwater spring areas of Florida, West Indian manatees gather in large numbers. PHOTO: L. KEITH

West Indian manatees in Florida will aggregate in great numbers around the outfalls of power plants and other artificial sources of warm water in the winter months. PHOTO: L. KEITH

A West Indian manatee swims on its back, lifting its flippers and snout above the surface. PHOTO: L. KEITH

Caribbean Sea, and Atlantic coast of northeastern South America. There are two subspecies: the Florida manatee (*T. m. latirostris*) from Texas to Virginia (occasionally to Massachusetts) in the northern Gulf of Mexico and southeast US, and the Antillean manatee (*T. m. manatus*) from northern Mexico to central Brazil and the islands of the Caribbean. Increasing numbers of records of manatees in the US Gulf of Mexico, west of Florida, have been documented in recent years, probably due to both increases in public awareness and manatee dispersal. Extralimital records of Florida manatees have been recorded from Virginia to Massachusetts, the Bahamas, and the western Florida keys, and Antillean manatees at Turneffe Atoll, Belize. A few West Indian manatees transplanted into the Panama Canal may have passed through the locks and made it to the Pacific side, but this is not part of their normal range.

Ecology and Behavior West Indian manatees are slow-moving and lethargic, and can be very difficult to observe, since they generally only expose the tip of the muzzle and possibly the upper back when they surface. They are seen mostly alone or in groups of up to six. However, some feeding groups regularly number up to about 20. For instance, during cold weather, large aggregations assemble near sources of warm water (such as power plant outfalls) in Florida, and these aggregations occasionally contain hundreds of animals. Mating groups of manatees clasp and touch each other, with much rolling and cavorting. Individual manatees are identified by scars and marks on the body (many of these inflicted as a result of boat collisions).

Most individuals in the southeastern US migrate between a summer range and a winter range further south in warmer waters. These movements, which are determined by water temperature changes, may span hundreds or even thousands of kilometers. However, the presence of warm water sources (such as power plant outfalls) has encouraged some animals to delay their migration or even not to migrate. They tend to show strong fidelity among years to specific ranges. Within the species, there is strong population structure, with the Lesser Antilles apparently representing a barrier to gene flow. In the Florida manatee, four populations or management stocks are recognized (Atlantic Coast, Southwest, Upper St. John's River, and Northwest).

West Indian manatee females reach sexual maturity at about 3–4 years. They breed throughout the year, with a peak, at least in Florida, in the spring and summer months (May-September). Generally, a single calf is born (twins occur sometimes—about 1–2% of births) after a gestation period of about 11–13 months. The calf is weaned typically at the age of about 1–2 years, but some females nurse their calves for up to 4 years. Maximum lifespan is about 60–70 years.

Feeding and prey These animals are herbivores, feeding on a wide variety of aquatic plants, such as water hyacinths and marine seagrasses. At times in some areas, they also eat algae, parts of mangrove trees, floating and shoreline vegetation, and even fish that they remove from fishing nets.

Threats and status Although hunting was prevalent in the past, it only occasionally occurs (illegally) today in US and central American waters. Human-caused threats include vessel collisions, incidental kills in fishing nets and lines, disturbance from boat traffic and other human activities, entrapment in flood control structures, pollution (especially from pesticides and herbicides in central America), ingestion of plastic debris, loss and degradation of habitat, and harassment by people. In addition, several natural mortality factors are important, and these include biological toxins (red tides), cold-related mortality, and apparently high levels of perinatal mortality. The dependence of some manatees in Florida on warm water from power plant outfalls concerns many scientists, as the loss of these warm-water sources, even temporarily, could cause mass die-offs. Boat collisions are the number one human-related cause of death (causing about 25% of manatee deaths) in Florida waters, where high-speed watercraft are extremely common and widespread. Abundance of the Florida manatee was estimated to be about 3,300 individuals. The Antillean manatee's numbers are unknown, but at least 400 have been counted on recent surveys in Belize and southern Mexico and about 500 are thought to exist in Brazilian waters. Some populations are growing, while many others are declining. Despite the species still facing significant threats, it is being actively managed in at least portions of its range. It therefore may be facing somewhat less danger of extinction than the other two manatee species.

IUCN status Vulnerable (overall species), Endangered (both subspecies).

African Manatee—*Trichechus senegalensis*

Link, 1795

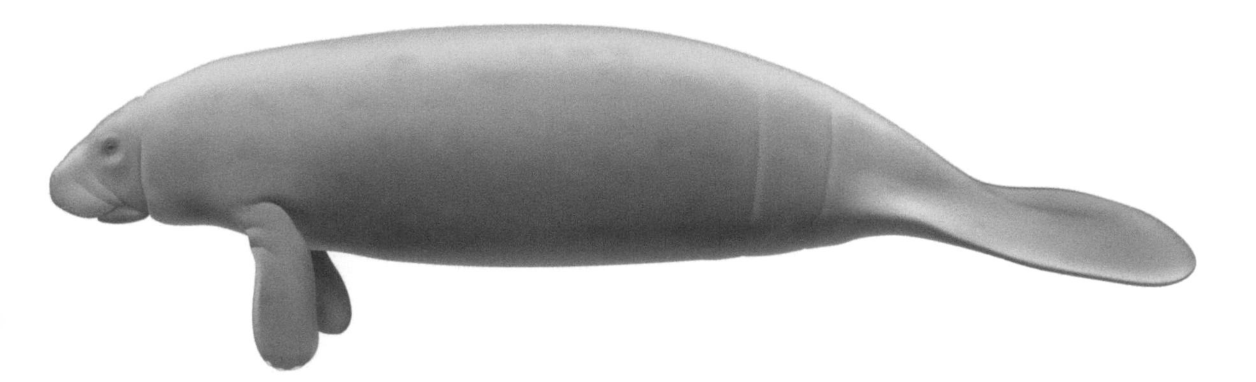

Recently-used synonyms None.
Common names En.–African manatee; Sp.–*vaca marina, manatí de África*; Fr.– *lamantin d'Afrique*.
Taxonomic information Sirenia, Trichechidae.
Species characteristics African manatees are very similar in appearance to West Indian manatees, although they tend to be a bit more slender. They have rounded, paddle-like tails. The head shape is similar to that of the West Indian manatee, but the snout is blunter, and the small eyes protrude from their sockets more. Additionally, the rostrum is less deflected downwards than in the West Indian manatee. There are stiff bristles on the lips. As in other manatees, the flippers are paddle-like; there are 3–4 nails on the upper surface (although there are reports of animals with no nails). The skin is wrinkled, with a sparse covering of short hairs.

The bodies of African manatees are dark gray; the sparse hairs found on the body are white. Individuals in coastal marine waters may have some barnacles growing on the surface.

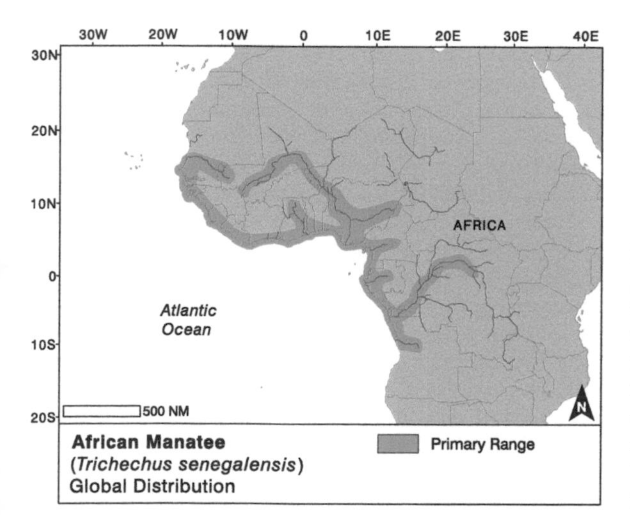

African Manatee
(*Trichechus senegalensis*)
Global Distribution

There are 5–7 functional teeth (all molars) in each tooth row. These are replaced from the rear by newly-erupting teeth. Newborn animals have two vestigial incisors, which are later lost.

Adult African manatees average about 2.5 m in length (although there is a highly-questionable report of an animal that was 4 m long) and possibly over 1,000 kg in weight. Newborns are presumably about 1.0–1.2 m long, based upon similarity to West Indian manatees.
Recognizable geographic forms None.
Can be confused with African manatees should be easy to identify, as they are the only sirenians in their range, and the only marine mammals in most of the riverine waters that they occur in.
Distribution African manatees are found in coastal marine waters, enclosed lagoons, rivers, and estuaries from the Senegal/Mauritania border to central Angola. The species may also occur in the upper reaches of some rivers (e.g., the Niger, Benue, Gambia, and Senegal rivers) up to 2,000 km from the sea. Additionally, manatees are found in several lakes in the land-locked country of Chad. They appear to prefer areas with abundant growth of mangroves and aquatic plants.
Ecology and behavior Manatees are mostly solitary, but individuals are found together in mother/calf pairs and aggregations of up to six to ten for feeding and reproductive purposes. The mating herds may exhibit much rolling and splashing, and tend to have several male animals apparently pursuing a single female. Manatees sometimes rest in the middle of lagoons, along riverbanks, and underneath overhanging vegetation or roots, generally during the day. They tend to feed at night in some areas. They may be able to hold their breath for up to 18 min, although timed dive data are rare. They migrate seasonally in rivers with large seasonal fluctuations in freshwater

African manatees have two rounded nostrils at the front of the snout, as in other species of manatees. Senegal. PHOTO: L. KEITH

Mating groups of African manatees may involve several males jockeying for position to mate with a female. Such groups tend to be boisterous and active. Senegal. PHOTO: L. KEITH

African manatees tend to be very cryptic and often little will be seen above the surface other than the top of the head and a small part of the back. Senegal. PHOTO: L. KEITH

The habitat of the African manatee includes not only coastal waters, but also rivers and lakes. They are capable of living in fresh waters for extended periods of time. Gabon. PHOTO: L. KEITH.

African manatees are very similar in appearance to their West Indian cousins. Although they have a tendency to be slightly more streamlined, captivity has fattened these ones. PHOTO: TOBA AQUARIUM

flow. Radiotagging studies show that some African manatees tend to be rather sedentary, but they do make some larger-scale movements (up to about 140 km in 2–3 days). These animals may occasionally be eaten by crocodiles. There is little else known of their behavior and ecology.

There is some breeding year-round, but there also appear to be seasonal peaks between June and September in some areas, related to the rainfall and water levels. A single calf is born, generally in shallow lagoons. The limited evidence suggests the existence of strong population structure, with several different stocks in different parts of the species' range. The lifespan is not well known, but based on its close relatives, may be greater than 60 years.

Feeding and prey Aquatic vascular plants comprise much of the diet of African manatees. They may also feed on mangrove leaves or plants on the banks of rivers or channels. Fish are sometimes taken from gillnets. Clams and freshwater mollusks have been found in the stomachs of some animals.

Threats and status The folklore of the West African peoples contains many stories of manatees. Although never subjected to large-scale commercial hunting, like their congeners, African manatees have been killed on a smaller scale for many decades throughout most, if not all, of their range. This is done mainly to obtain meat for food, oil and body parts for use in traditional medicines, and also in some cases to reduce perceived interference with fisheries and agriculture. They are harpooned, captured in drop traps, and entangled in specialized nets. Manatees damage fishing gear and invade flooded rice fields and eat rice, damaging the crops. Other serious threats include incidental mortality in fisheries, contamination due to pollution, death due to both dam structures and entrapment behind dams, and damming of rivers. As coastal development in Africa increases, potential threats include boat collisions and ingestion of plastics. Protected areas for manatees exist in at least 11 African countries, including the Ivory Coast, Cameroon, and Nigeria, although their effectiveness is often questionable. There are no estimates of abundance for any portion of the range, however it is likely that the African manatee is the most highly-threatened of all the sirenians. It is likely that there are no more than 10,000 individuals. Despite national protection in virtually all countries of origin, the species appears to be declining in many areas.

IUCN status Vulnerable.

Amazonian Manatee—*Trichechus inunguis*

(Natterer, 1883)

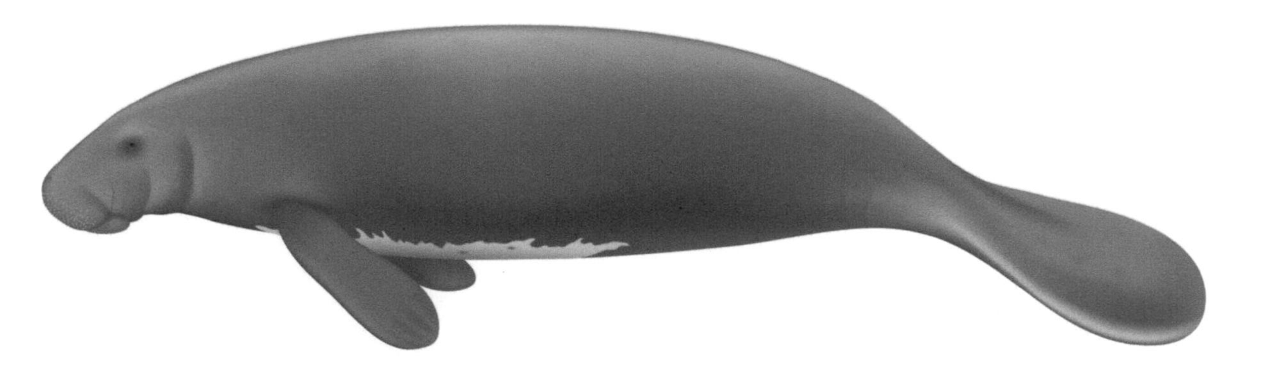

Recently-used synonyms None.

Common names En.–Amazonian manatee; Sp.–*manatí del Amazonas*; Fr.–*lamantin de l'Amazone*. In Brazil, they are called *peixe-boi*.

Taxonomic information Sirenia, Trichechidae. A "dwarf manatee" (*Trichechus pygmaeus*, or *Trichechus bernhardi*) has been described from the Amazon, but expert opinion is that these are simply young of the Amazonian manatee. Hybrids between West Indian and Amazonian manatees have been documented near the mouth of the Amazon River (and possibly also near the mouth of the Orinoco River).

Species characteristics Amazonian manatees are the smallest, most slender of the three species of manatees. They have the rounded, paddle-like tails characteristic of all manatees. The skin of adults and juveniles is smooth, rather than wrinkled as in their relatives (it has been described as "rubbery"). The flippers are longer than in the other species and they lack nails (unlike the other two manatee species, which have nails). There are thick bristles on the lip pads of both jaws, and the rostrum is longer, narrower, and less deflected downwards than in the West Indian manatee. The body has a sparse covering of fine hairs.

Amazonian manatees are black or dark gray; most have white or pink belly and chest patches with irregular shapes and very distinct borders (these only rarely occur in Florida manatees). However, not all individuals have the belly patches, and some may just have a slightly-lighter ventrum.

Five to seven functional postcanines/molars, and two vestigial incisors (the latter are resorbed after birth) are found in each jaw. Typical of manatees, teeth are replaced from the rear throughout life. The teeth are smaller than those of other manatee species.

Amazonian manatees reach lengths of about 2.8–3.0 m, and weights of up to 480 kg. There is no known sexual dimorphism in body size. Length at birth is not well-known, but is probably around 0.8–1.1 m.

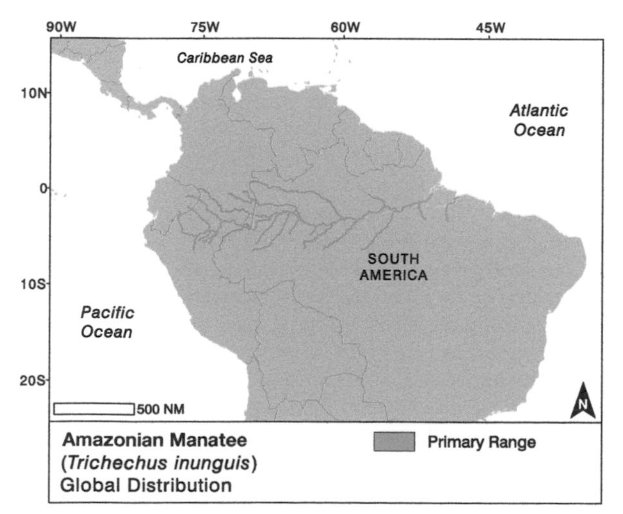

Recognizable geographic forms None.

Can be confused with Amazonian and West Indian manatees may co-occur in or near the mouth of the Amazon River (and possibly also near the mouth of the Orinoco River). Although they may be very difficult to distinguish when seen from a distance, in closer views, the size, shape, and coloration differences listed above should help in identification. Pay special attention to head shape (longer and narrower rostrum in the Amazonian), body color (darker, with frequent white belly patches in the Amazonian), skin texture (smoother in the Amazonian), and presence/absence of nails on the flippers (absent in the Amazonian).

Distribution The Amazonian manatee is an exclusively-freshwater species. It is found in calm waters of the Amazon River and its tributaries and appended lakes in Brazil, Guyana, Colombia, Peru, and Ecuador. It is uncertain if it occurs in the Orinoco drainage. It occurs

The Amazonian manatee is similar in appearance to other manatees, but tends to be darker in color and has no nails on the flippers. Captivity. PHOTO: J.-P. SYLVESTRE

Amazonian manatees are the most distinctive of the three manatee species. They are darker in color, with more distinct white belly patches, and they are generally more slender and smoother than the others. Brazil. PHOTO: D. PERRINE/SEAPICS.COM

A glimpse of an Amazonian manatee in clear water like this would likely never happen in the wild, where they live mostly in turbid waters. Captivity. PHOTO: J.-P. SYLVESTRE

In nature, Amazonian manatees are generally very difficult to detect. They may hide away in weeds along the river banks, and remain quiescent during much of the day. These may be, at least partially, adaptations to avoid detection by hunters. Brazil. PHOTO: D. PERRINE/SEAPICS.COM

from near the headwaters of the rivers (although not in the fast-flowing and turbulent waters) to the Amazon estuary. It can be found in all three water types in the Amazon (i.e., black, white, and clear waters).

Ecology and behavior The Amazonian manatee remained poorly-known until more-intensive studies of its biology began in the mid-1970s. It occurs singly or in feeding groups of up to 8 individuals, although animals in feeding aggregations generally show little social interaction. The large herds often seen in the past are a rarity today, although aggregations do form for mating. They are active both day and night (however, hunting disturbance may have caused some animals to become more nocturnal in their habits), and their activities are strongly influ-

This mother and calf Amazonian manatee show the white belly patches that characterize many adults in this species. Brazil. PHOTO: D. PERRINE/SEAPICS.COM

enced by the seasonal floods. As water levels rise, they move into the flooded forest and "varzea" areas to feed, and they return to the main river channels and perennial lakes during the dry season. Amazonian manatees may survive for long periods in pools separated from the main river during the dry season, and have developed impressive fasting abilities (documented up to 200 days) to survive these events. They are able to hold their breath for up to 20 minutes. Individuals may emit low-frequency sounds in and out of water.

Breeding may occur throughout the year, but there is a calving peak in February to May, when the water level in the river rises and food is plentiful and nutritious. A single calf is born after a gestation of about 12 months. Lactation lasts up to two years. The age at sexual maturity is not well known, but is thought to be between 6 and 10 years. Potential predators are sharks, caimans (crocodiles), and jaguars. The lifespan is thought to be around 60–70 years.

Feeding and prey Amazonian manatees feed on at least 24 species of vascular aquatic plants, but they have also been observed to eat floating palm fruits. Emergent species are preferred, followed by floating, and finally submerged types. Some manatees may fast or eat dead plant material during the dry season. They consume the equivalent of about 8% of their body weight per day.

Threats and status Amazonian manatees have been hunted heavily throughout the Amazon basin since at least 1542. Between 1935 and 1954, a commercial fishery killed 80,000 to 140,000 Amazonian manatees for their meat and hides (which were used to make leather). Manatees are still hunted (often illegally) in at least Peru, Colombia, Brazil, and Ecuador. Additional potential threats to their survival include incidental catch, pollution associated with mining and other industrial and agricultural practices, damming of rivers, and large-scale deforestation. Amazonian manatees apparently do not often get hit by boats (this is a major problem for the West Indian species). About 500–1,000 manatees were thought to occupy Amana Lake in Brazil in the early 1980s. There are no statistically-defensible estimates of the species' overall abundance or population size. However, the population was undoubtedly seriously reduced by the commercial exploitation of the past, and is being further reduced by the current smaller-scale subsistence hunts. Legal protection occurs in most range countries, but enforcement is generally lax to non-existent. The Amazonian manatee is therefore in real danger of extinction in the next couple of decades.

IUCN status Vulnerable.

Dugong—*Dugong dugon*

(Müller, 1776)

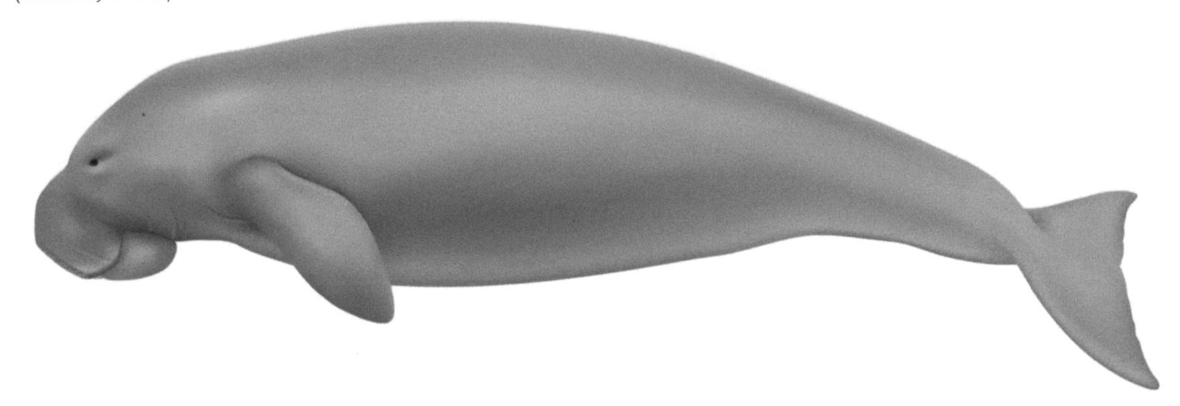

Recently-used synonyms None.
Common names En.–dugong; Sp.–*dugón*; Fr.–*dugong*.
Taxonomic information Order Sirenia, Family Dugongidae. Some biologists have recognized two subspecies: *D. d. dugon* in the Indian and western Pacific oceans, and *D. d. hemrichii* in the Red Sea. These subspecies designations are not well-accepted, and are now almost never used.
Species characteristics The dugong is unique among living sirenians in having whale-like flukes with a slight median notch, instead of the rounded tails possessed by manatees. They have been described as looking something like a cross between a dolphin and a walrus. In general, dugongs are more streamlined and cetacean-like than manatees, with a smooth, fusiform body. The tail stock in front of the flukes is laterally compressed into a peduncle. The paddle-shaped flippers have no nails. There is a strong downward deflection to the muzzle, which ends in a "rostral disk" with short, dense bristles. The nostrils are valve-like and are situated on the top and front of the animal's snout. The eyes are small, and there are no external earflaps. The skin is generally smooth (not wrinkled, although there are folds) and is sprinkled with short hairs. The mammary glands of females are located in the axillae (armpits), unlike those of cetaceans.

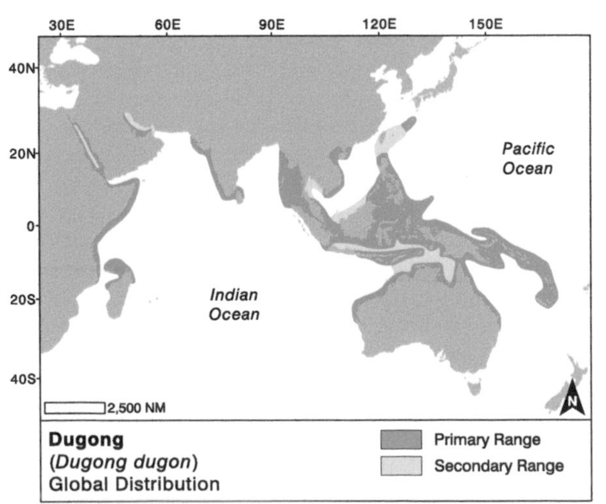

Adult dugongs are brownish to slate-gray on the back (generally appearing more brown when seen from above the surface), slightly lighter on the belly (which sometimes can have a pinkish hue). Older individuals may have a large unpigmented area on the back (these are called "scarbacks"). Calves are a slightly paler gray color than adults.

The dental formula is I $^{12}/_{3}$, C $^{0}/_{1}$, PM $^{3}/_{3}$, M $^{3}/_{3}$. The lower incisors and canines, and the inner pair of upper incisors, are vestigial. The outer pair of upper incisors of males and some females erupt from the gums, and are referred to as "tusks." However, they do not extend outside the closed mouth.

Maximum known size for dugongs is about 3.3 m and at least 570 kg (a specimen reported to be 4.06 m and 1,016 kg is considered to be an error). There is not much difference in size between males and females. At birth, dugongs are between 1 and 1.3 m long and weigh about 25–35 kg.
Recognizable geographic forms None.
Can be confused with This is the only sirenian in the Indo-Pacific. There is some possibility of confusion with the finless porpoise or Irrawaddy/snubfin dolphins, but the single blowhole of the cetaceans and the double nostrils of the dugong will generally allow them to be easily distinguished. Also, pay attention to head and flipper shape (and dolphins of the genus *Orcaella* have a small dorsal fin). Dugongs also tend to move more slowly than most dolphins.
Distribution Dugongs are widely distributed in the Indo-Pacific region in coastal tropical and subtropical waters, covering some 40 countries. They are still present at the

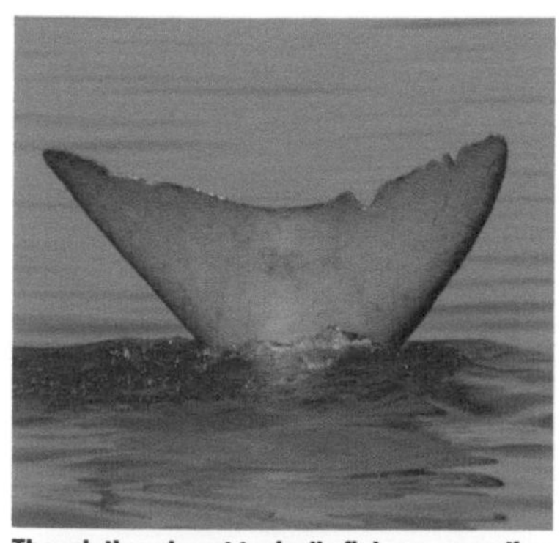

Though they do not typically fluke up on a dive, sometimes dugongs may do so, showing the distinctive-looking tail flukes. New Caledonia. PHOTO: C. GARRIGUE

The dugong has a body shape that is a bit more like that of cetaceans than other sirenians. However, the head shape, with the small eyes and downward-deflected muzzle are more similar to those of manatees. Captivity. PHOTO: J.-P. SYLVESTRE

When seen surfacing at sea, dugongs may appear quite similar to some finless cetaceans; however, the double nostrils are a giveaway that this is a sirenian, not a toothed cetacean. New Caledonia. PHOTO: C. GARRIGUE

Dugongs have a unique head shape; males and even some females have erupted tusks; however, these do not extend outside the closed mouth. Captivity. PHOTO: J.-P. SYLVESTRE

limits of their original range; however, there have been many reductions in their historical range. Dugongs appear to prefer shallow protected bays and channels, and the lee sides of large inshore islands, often in muddy waters. Although primarily coastal, dugongs do occur farther offshore in areas where the continental shelf is wide and protected, in waters up to about 40 m deep. The range is discontinuous: it extends from east Africa north to the Red Sea; in the Persian (Arabian) Gulf; along western India to Sri Lanka; and throughout the Indo-Malay archipelago and the Pacific islands, to the Ryukyu Islands in the north, the central coasts of Australia in the south and Vanuatu in the east.

Ecology and behavior Dugongs occur mostly in small groups. Herds as large as several hundred animals are seen at some locations. Dives up to 11 min have been recorded, although most dives are much shorter.

The specific function of the tusks is uncertain, although many animals bear tusk scars that are believed to be associated with social interactions involving mature males. Although they are not known to migrate *per se,* and most movements are small-scale, dugongs do undertake

Dugongs are bottom feeders, pulling seagrasses and other plants from the sea bottom, and leaving sinuous feeding trails or 'tracks' in the seabed. Red Sea. PHOTO: L. DINRATHS

Unlike most cetaceans, the dugong has no notch in the center of the tail flukes. Red Sea. PHOTO: L. DINRATHS

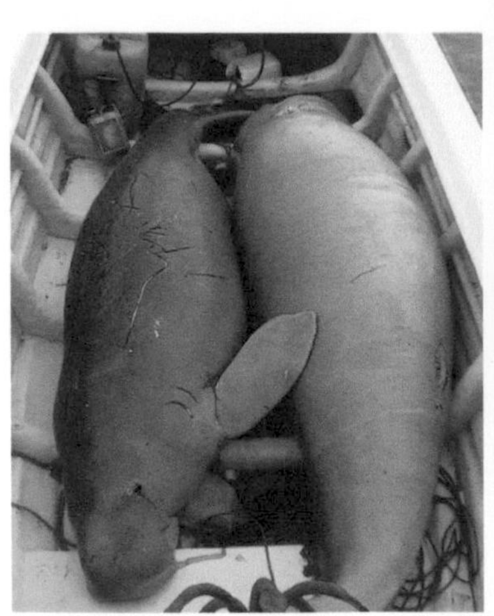

Although it is usually illegal, dugongs are often hunted throughout their range. Sri Lanka. PHOTO: R. NANAYAKKARA

large-scale movements of hundreds of kilometers. They can move up to several hundred kilometers in just a few days. Average cruising speed is about 10 km/hr. Dugongs may roll like a cetacean when traveling or deep diving. These animals are preyed upon by large sharks, killer whales, and crocodiles although the relative importance of predation rather than scavenging is not known. The seagrass communities that dugongs rely on are vulnerable to human destruction and disturbance. The dugong is of cultural significance to many of the peoples of the tropical Indo-Pacific, such as the native peoples of northern Australia/Papua New Guinea, and Okinawa.

The social behavior of this species appears to be more complex than for the related manatees. Although reproductive activity can occur throughout the year in most areas, there are spring seasonal calving peaks in at least some parts of the range. Not much is known about reproductive behavior in the dugong, but it is thought to be polygamous. Groups of males seem to compete to mate with a single estrus female. Females do not produce their first calf until at least 7 years of age (sometimes as late as 17). The gestation period is about 13–14 months, and a single calf is born, with high maternal investment in the calf. Calves are weaned at about 18 months, but may remain with the mother for several more months. Dugongs are very long-lived, to at least 73 years.

Feeding and prey Dugongs have a specialized diet. They are seagrass community specialists, eating mainly various types of bottom vegetation, primarily seagrasses but sometimes including invertebrates. Feeding trails in seagrass beds can be seen in dugong feeding areas exposed by the tides.

Threats and status Historically, dugongs were probably hunted nearly everywhere in their range for meat and oil, but now in most countries this practice is illegal (although it still commonly occurs). The hunting dates back at least 6,000 years in the Arabian region and 4,000 years in Torres Strait. Body parts are also used in traditional medicine and for amulets in some parts of Southeast Asia. Legal hunting still occurs in some parts of the Pacific Islands and northern Australia. Dugongs are highly vulnerable to human impacts, due to their life history and dependence on easily-damaged coastal seagrass beds. Additional threats include loss of and damage to seagrass beds, incidental catches in fishing nets, and captures in shark nets set to protect bathing beaches. There are several natural threats as well: cyclones, storm surges, parasites, and predation. Dugongs are afforded legal protection in much of their range, although these laws are seldom enforced. In Australia, there is a rigorous system of management, which incorporates not only extensive research and the establishment of reserves, but also community involvement by indigenous peoples in the overall management scheme. There have been documented or presumed population declines in many areas of the range, and the species is considered to be at risk of extinction in East Africa, India/Sri Lanka, Japan, and Palau. Except in Australia, dugongs are probably represented mostly by relict populations, separated by areas of complete or near-total local extinction. Thus, Australia is seen as the main hope for the long-term survival of this species (although the Arabian region also appears to contain a reasonably healthy population). There were estimated to be about 100,000 dugongs worldwide in the 1990s; 85,000 are considered to occur in Australia alone. These numbers are almost certainly underestimates. Along the Andaman Sea coast of Thailand, there are estimated to be at least 120 dugongs. About 7,300 are estimated to occur in the Persian (Arabian) Gulf. Few other regional abundance estimates are available.

IUCN Status Vulnerable.

7. Otters & Polar Bear

Marine Otter

Polar Bear

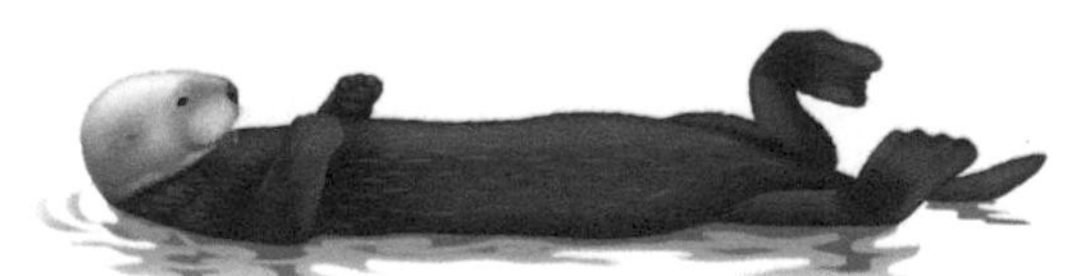

Sea Otter

Sea Otter—*Enhydra lutris*

(Linneaus, 1758)

Recently-used synonyms None.
Common names En.–sea otter; Sp.–*nutria marina*; Fr.–*loutre de mer d'Amérique du nord.*
Taxonomic information Carnivora, Mustelidae. Three subspecies are currently recognized: *E. l. lutris* in the western North Pacific, *E. l. kenyoni* from the Aleutian Islands to Washington State, and *E. l. nereis* from central California to northern Mexico.
Species characteristics The sea otter is the most derived of the otters. The muzzle has a set of thick vibrissae. The large head has a blunt snout, and is connected to the body by a short, stocky neck. The forelimbs are short and similar to those of other otters, with a loose flap of skin under each that serves as a pocket to store food. The forepaws are very dexterous and sensitive. The hind limbs are large and flattened like flippers; they are oriented backwards. Although the tail is relatively short (among otters), it is not noticeably tapered until the tip. It is flattened top to bottom into a paddle-like structure that otters can swim with. The pelage of sea otters is the densest of any mammal (more than 150,000 hairs/cm^2). A layer of sparse guard hairs overlays the dense underfur. Sea otters are completely covered with fur, except for the nose pad, inside the ear flaps, and the pads on the bottom of the paws and flippers. The color of the fur is dark reddish-brown to black. Some individuals, especially older ones, may become grizzled, with the fur around the head, neck and shoulders becoming almost white.

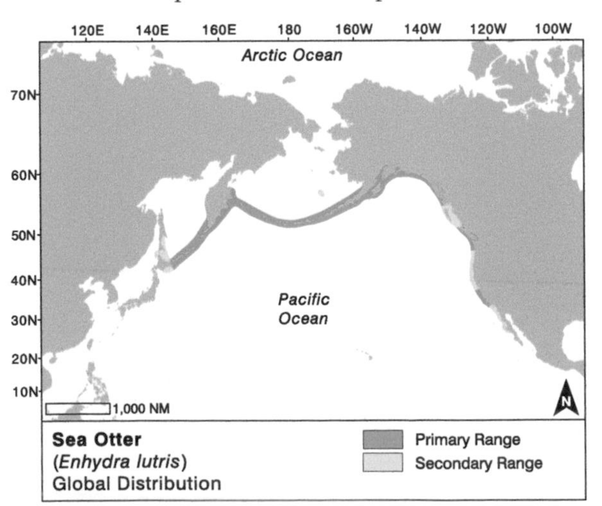

The dental formula is I 3/2, C 1/1, PM 3/3, M 1/2. Male sea otters reach lengths of 1.48 m and weights of 45 kg. Females can be up to 1.4 m and 32.5 kg. Newborns are about 0.6 m long and weigh about 1.8–1.9 kg.
Pup—Size 0.6–1.1 m; fur is light buff in color and very fluffy; younger animals closely associated with mother.
Juvenile—Size 1.1–1.3 m; pelage more sleek, and dark brown to black in color; independent of mother at about 0.5 years.
Adult—Adult size (>1.3 m); pelage dark brown to black. Older individuals become grizzled around the head and shoulders (males more so than females). Females may have bloodied noses or pinkish nose scars.
Recognizable geographic forms The different subspecies are generally not recognizable in the field from external differences.
Can be confused with The sea otter is the only truly marine otter in its range, although North American river otters (*Lontra canadensis*) and Eurasian otters (*Lutra lutra*) are often found in marine waters along the northwest coast of North America and in the western North Pacific, respectively. River otters are smaller and more slender than sea otters, with longer tails. Also, river otters generally swim belly down even at the surface, while sea otters usually (but not always) swim along the surface on their backs.
Distribution Sea otters are found in shallow, nearshore waters of the North Pacific Rim, from the southern Kuril Islands, north along the Kamchatka Peninsula, then east

The heavily-whiskered face and posture of floating belly up at the surface allow sea otters to be distinguished from pinnipeds, even from a great distance. PHOTO: J. TOWERS

Sea otters lie wrapped in kelp off the Monterey Peninsula. Otters rest in the kelp beds to avoid large waves and rough water. PHOTO: T. A. JEFFERSON

A mother and pup sea otter in the Monterey Bay area. The pups will often ride along on the mother's belly like this. PHOTO: M. COTTER

This is undoubtedly a female. Sea otter males will bite the nose of the female to position themselves during mating, often leaving her with nasty wounds. Monterey Bay. PHOTO: M. COTTER

along the Aleutian Islands to the Alaska Peninsula and Prince William Sound, and thence south to southern California (and occasionally as a vagrant to central Baja California, Mexico). Originally, their distribution was nearly continuous from Hokkaido, Japan, to central Baja California, Mexico. However, there are now four disjunct remnants: Kuril Islands to southeast Kamchatka Peninsula (classified as *E. l. lutris*); Commander Islands; Aleutian Islands to Prince William Sound, Alaska (these two groups are classified as *E. l. kenyoni*); and central/southern California (classified as *E. l. nereis*). In addition, there have been several reintroduction attempts (some successful, others not) along the west coast of North America. Southeast Alaska, British Columbia and Washington State are three areas where reintroductions have been successful. One area of recent reoccupation is Glacier Bay in Southeast Alaska, where sea otters were absent until as recently as 1994, but currently number >4000 individuals.

Ecology and behavior Sea otters can be seen singly or in groups (most often resting groups, called rafts). Rafts in California tend to be small, between 2 and 12 animals, and rarely exceed 50–100 individuals. Those in Alaska are also generally small, but can contain over 2,000 otters. Sea otters are thought to be capable of having a major influence on their prey populations, with cascading effects on coastal marine communities, especially where sea otter densities are high. Foraging generally occurs singly, although animals may aggregate in good feeding areas. Females with small pups may leave their pups on the surface while diving, and spend more time feeding at night to avoid eagle predation.

Sea otters are most often seen floating on their backs in and around kelp beds or in protected areas while resting. They bring their prey to the surface to feed on it. Rocks and other hard objects are used as tools to crack open shells of their prey, which they eat using their bellies as "dinner tables." They sometimes store tools in the skin flaps under the front limbs. They use their sense of smell in many aspects of their life, and can be difficult to approach from upwind (presumably because they can smell the intruder). When they enter a group, otters use a ritualized greeting that apparently involves scent recognition. Sea otters spend 10% or more of their time grooming their fur, which maintains its insulation properties. They are not deep divers, as marine mammals go, and most dives are to less than 60 m. However, they are capable of dives of over 100 m and up to about 7 min. Sea otters can haul out onshore, and retain limited ability to walk around (clumsily) on shore and over rocks.

Sea otters are polygynous; males tend to defend large (average about 0.4 km^2) territories that encompass the home ranges of several females. Females aged three and

This sea otter is resting inside Moss Landing Harbor in central California. The extensive white head of this animal indicates that it is probably an adult male. PHOTO: M. COTTER

Sea otters will "stand" vertically in the water before actively diving. Monterey Bay. PHOTO: M. COTTER

The head shape of the sea otter is different from that of the sympatric pinniped species—sea otters have little or no snout, a very flat nose, and shaggy fur even on the head. Monterey Bay. PHOTO: M. COTTER

Otters are not often seen on land, but they do haul out and can walk, although somewhat clumsily, on all fours. Elkhorn Slough, California. PHOTO: T. R. KIECKHEFER

above give birth to a single pup annually (twin births are rare). Pupping occurs throughout the year, but peaks in April to June in Alaska, and December to February in California. During mating, the male bites the nose of the female to position himself; thus, females often have nose scars (these are useful to researchers in identification of individuals).

Feeding and prey Sea otters consume the equivalent of up to 25–30% of their body weight per day to maintain their high metabolic rate. They feed on or near the bottom in shallow waters (often in kelp beds). They have a diverse diet of over 150 prey items. Major prey are benthic invertebrates, such as abalones, sea urchins, and rock crabs. However, sea otters also eat other shellfishes, cephalopods, and some near-bottom fishes in Asia and the Aleutian Islands. Diet is often highly variable among individuals, but most individuals specialize on just one or two of the most common diet items, including *Cancer* crabs, abalone, sea urchins, marine snails, clams, sea stars and kelp crabs.

Threats and status Historically, sea otters have been hunted by aboriginal peoples of the North Pacific for many centuries. Although the primitive methods sometimes were sustainable, there is evidence that some maritime peoples hunted otters to local extinction. Sea otters were commercially hunted heavily between 1750 and 1900 throughout their range for the North Pacific fur trade, and most likely less than 1,000 sea otters survived this carnage. In the 20th century, the survivors thrived and increased under protection and intensive management efforts. However, recently there have been several disturbing declines in abundance. Populations in the Aleutian Islands of Alaska declined by about 80–90% in the 1990s, and this may have been at least partially related to killer whale predation. The decline is continuing, and it appears possible that otters may disappear from most or all of the Aleutians in the next 10 years or so. The California population declined by about 12% to just under 2,000 individuals in the mid- to late 1990s, and since then the numbers have increased to about 2,700. The translocated population at San Nicholas Island is now at about 60 individuals. The southern sea otter population is reported annually as a three-year-running average, and increased in 2013. The reasons for the declines are unknown (although pollution, disease, fisheries, and food limitations have all been suggested). In Asia, sea otters have been much less intensively studied than in the rest of their range. There, they occupy most of their historic range, but abundance is thought to be reduced, and there is apparently frequent poaching in Russian waters. Globally, there were estimated to be about 100,000 sea otters in the 1990s, mostly in Alaskan waters. Current worldwide estimates of abundance are between 82,000 and 95,000. Overall, sea otters are not in danger of global extinction, as they were in the past, but problems persist and overall the species appears to be declining. The main threats today include pollution (including oil spills), disease, by-catch, hunting by native peoples, poaching, habitat loss and predation by killer whales and sharks.

IUCN status Endangered.

Marine Otter—*Lontra felina*

(Molina, 1782)

Recently-used synonyms *Lutra felina.*

Common names En.–marine otter, chungungo; Sp.–*nutria de mar, chungungo*; Fr.–*loutre de mer, loutre marine d'Amérique du Sud.*

Taxonomic information Carnivora, Mustelidae. As a result of recent molecular studies, most mammalogists now use the genus name *Lontra,* instead of *Lutra,* for the New World otters. Two subspecies have been designated, but their validity is doubtful.

Species characteristics Marine otters are very similar in general appearance to freshwater otters. The snout is blunt at the tip and the rhinarium (nose pad) is naked and relatively flat. The head slopes smoothly back from the nose. The ears are small and set far back on the head. The tail tapers to a point, typical of fresh-water otters, and is relatively short—about 30–36 cm long (about 1/3 of the total length). The feet are of moderate size, with well-webbed digits, and strong claws. The ventral surface of the webs are partially furred. The coarse pelage is rough in appearance, and has a dense underfur and a set of long guard hairs (up to 20 mm long). The vibrissae of the upper lip and corner of the mouth are large.

Marine otters are dark brown above and on the sides, with a lighter fawn color below, especially on the throat. The muzzle, throat, and lips are not spotted, as in some other otter species. The nose pad is black, and its structure may be geographically variable (nose pad variations are used to distinguish different species within the otter subfamily). Juveniles are slightly lighter than adults.

The dental formula is I $^{3}/_{3}$, C $^{1}/_{1}$, PM $^{3-4}/_{3}$, M $^{1}/_{2}$, for a total of 36 teeth. Marine otters are the smallest species of the genus *Lontra.* They attain total lengths (including the tail) of 0.87–1.15 m.

Recognizable geographic forms None.

Can be confused with This is the only truly marine otter along the west and southwest coasts of South America, although there are southern river otters (*Lontra provocax*) along some parts of the coast, and marine otters do enter rivers on occasion. However, marine otters tend to occur in exposed outer coast habitats, while river otters are often found in more protected inner waters (most live entirely in freshwater habitats). Also, river otters can be distinguished from marine otters by their larger size, darker color, finer fur, and more peaked nose pads.

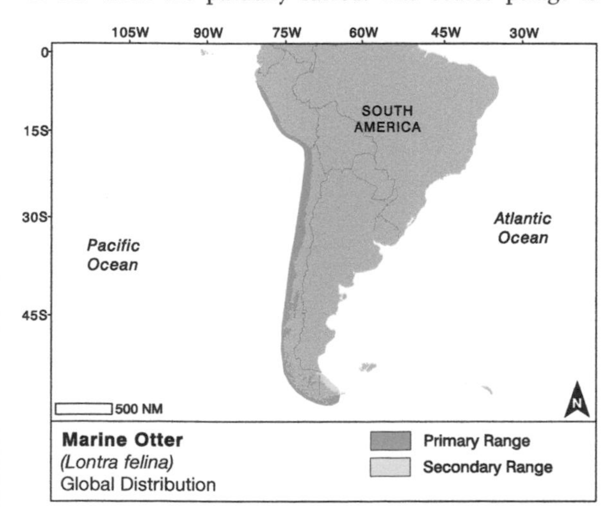

Distribution These coastal otters are found in a narrow strip on open, wave-exposed rocky shores (to about 30 m inland) and adjacent waters (to about 150 m offshore) from the southern tip of Chile (Cape Horn, 56°S) to north of Huanchaco, central Peru (8°S). They were thought to have been extirpated from Argentina, but there still may be groups extant at Isla de los Estrados and in the Strait of Magellan. They are largely restricted to marine waters, but sometimes enter rivers.

Ecology and behavior Very little is known of the biology of the marine otter. They are found mostly singly or in pairs, but groups of three or more are sometimes seen.

Marine otters tend to feed along exposed coastlines, where wave action is high. Chile. PHOTO: P. CACERES

Although most marine mammals are quite large, the marine otter manages to get by at about the same size as terrestrial otters. PHOTO: C. OLAVARRIA

A chungungo eating a prey item on a rocky Chilean shoreline. PHOTO: C. OLAVARRIA

A marine otter hauled-out on rocks. Note the similarity in appearance to terrestrial otter species. PHOTO: J. MANGEL

A marine otter hauled-out near the entrance to a sea cave. Notice the coarse guard hairs around the neck. PHOTO: J. MANGEL

Marine otters are among the least adapted of the marine mammals to an oceanic existence. They spend much time ashore, and when in the water generally don't range far from land. PHOTO: J. MANGEL

Marine otters are not known to be territorial, and they have overlapping home ranges. They are known to be diurnally active, and tracking of marine otters indicated similar activity patterns throughout day and night. There is some evidence that marine otters at Chiloe Island, Chile, may have a narrower diet based on larger prey than at other nearby sites. Marine otters are more agile in the water than on land, although they are capable of climbing rocks. They sometimes float on their back, like sea otters. Hunting dives last an average of about 30 sec, although dives of over 1 min have been recorded. They often occur around large rock outcroppings with caves, which are used for pupping, resting, feeding, and defecation. Predators of the marine otter are probably killer whales and sharks.

Although there are no other truly-marine otter species in the range of the chungungo, southern river otters can overlap them in distribution, and so one must make identifications carefully. Chile.
PHOTO: P. CACERES

Marine otters are thought to be monogamous, sometimes tending toward polygamy. The reproductive season is not well-known, but much of the pupping may occur from January to March. The usual litter of two pups (ranging from 2–4) is born after a gestation period of 60–65 days. Pups remain with the mother for about 10 months.

Feeding and prey Marine otters are opportunistic feeders, taking crabs, shrimps, mollusks, and fish. The diet in southern Chile is made-up of 52% crustaceans, 40% fish, and 8% mollusks. In Peru, there is a shift toward more fish in the diet relative to crustaceans. Prey size averages 95–185 g. They sometimes enter rivers to feed on freshwater prawns.

Threats and status The marine otter is rare and is considered to be threatened with extinction. Historically they were heavily involved in fur trading, but this does not happen much anymore. The main threats to their survival are remaining illegal hunting for fur, habitat destruction by urban and industrial development, mining, pollution, competition with fisheries, bycatch, and explosives use by fisheries. Invasive species and tourism may also be factors in some areas. There are no reliable estimates of global abundance for the species. But by some estimates the total population may be <1,000 individuals and declining. About 750 are thought to occur in Peruvian waters, and the largest populations apparently occur in Chilean waters. The species may already be extinct in Argentina.

IUCN status Endangered.

Polar Bear—*Ursus maritimus*

Phipps, 1774

Recently-used synonyms None.
Common names En.–polar bear; Sp.–*oso polar, oso blanco*; Fr.–*ours blanc.*
Taxonomic information Carnivora, Ursidae. Two subspecies have been recognized by some biologists: *U. m. maritimus* in the Atlantic Arctic, and *U. m. marinus* in the Pacific Arctic. However, there is doubt about their validity, and the Society for Marine Mammalogy no longer recognizes them. Polar bears evolved from brown bears (*Ursus arctos*) relatively recently (in the last 200,000 years).

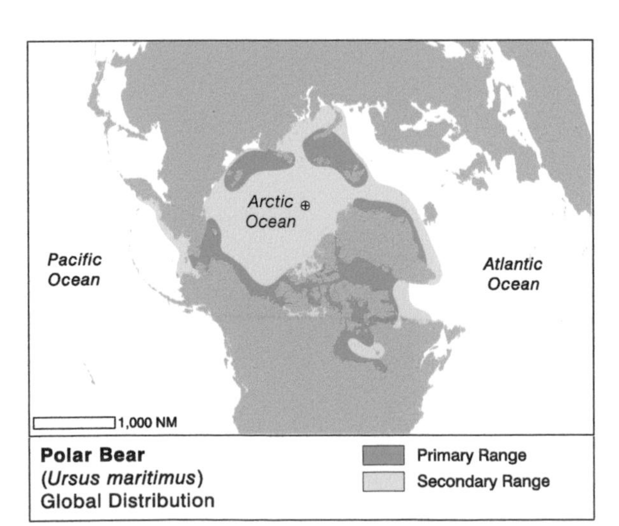

Species characteristics The polar bear is not radically different from other bears in body form. It is similar in size to brown and grizzly bears, but is more slender, and has a longer neck and more elongated head. It also lacks an enlarged shoulder hump. The ears are small, an adaptation to the bitterly-cold environment it lives in. Large, partially-webbed paws on the front limbs are used for swimming (hindlimbs are not used). There are five digits on each foot, each with a heavy, non-retractable claw. Polar bears are covered with fur on all but the nose and the pads on the bottoms of the feet. The guard hairs overlaying the underfur are up to 15 cm long, and they reach greater lengths on the forelegs in males. Females typically have four functional mammae.

Generally, the pelage of polar bears is white, but (depending on lighting and condition) it can appear yellow, light brown, or even light gray. The area around the mouth may be stained with blood after a recent meal. The nose, lips, eyes, and footpads are all black, as is the skin underneath the fur.

The dental formula is I $^{3}/_{3}$, C $^{1}/_{1}$, PM $^{2\text{-}4}/_{2\text{-}4}$, M $^{2}/_{3}$. The back cheek teeth are much smaller than those of other bears. Males may be up to 2.5 m long and may weigh up to 800 kg, which is somewhat larger than females. Females reach lengths and weights of 2.0 m and 350 kg,

While polar bears spend the majority of their time on the ice, they are capable swimmers, and can move distances of 100 km more across water. Melting ice from global climate change may force them to spend more time in the water in future years. Baffin Island. PHOTO: M. NOLAN

Polar bears are largely creatures of the ice and water, spending little of their time on solid land. Spitsbergen. PHOTO: M. NOLAN

The great white bear of the north is generally very easy to identify—the few grizzly and black bears that overlap it in distribution are generally much darker in color. Manitoba. PHOTO: M. NOLAN

Possibly more than any other marine mammal, rising global temperatures are a conservation concern for polar bears, which do almost all of their hunting from the ice. Spitsbergen. PHOTO: M. NOLAN

Polar bears can appear comical at times, striking human-like poses, but these are formidable predators that can be very dangerous to people. Manitoba. PHOTO: M. NOLAN

The primary prey species of the polar bear are pinnipeds (especially the ringed seal)—this bear has consumed nearly every edible part of this harp seal. PHOTO: F. TODD

Polar bears in some parts of their range have become opportunistic and have learned that garbage dumps can supply a ready source of easy meals. Manitoba. PHOTO: M. NOLAN

respectively. Pregnant females sometimes attain weights of up to 500 kg. Adult males are generally about twice as heavy as females. At birth, the tiny cubs weigh only about 0.6 kg.

Recognizable geographic forms There is some slight geographic variation, but the subspecies are generally not recognizable in the field.

Can be confused with There should be no problem recognizing polar bears in most cases. In the few areas where grizzly/brown (*Ursus arctos*) and American black (*Ursus americanus*) bears are found within the polar bear's range, the much lighter color of the polar bear's fur will make it virtually unmistakable. Hybrids between polar and brown/grizzly bears may cause confusion.

A polar bear mother and cub scavenge from a whale carcass, which has nearly been picked clean. PHOTO: M. NOLAN

Distribution Polar bears have a circumpolar distribution in the Northern Hemisphere. Their southern limits fluctuate with the ice cover (they have been recorded as far south as the Pribilof Islands in the Pacific and Newfoundland/Labrador in the Atlantic). The northernmost record is from around 88°N. Polar bears are generally associated with sea ice, but they have been seen swimming at sea many kilometers away from the nearest land or ice.

Ecology and behavior Polar bears tend to be solitary, but breeding pairs and females with up to three cubs may be seen together. They also aggregate in areas of high ringed seal density (their primary prey), and around refuse dumps. These bears can swim well, using their large webbed forepaws. They spend most of their time on ice, but sometimes spend significant periods of time on land. Polar bears in Hudson Bay and southeastern Baffin Island (where snow melts completely in summer), spend several months fasting on land before ice forms in the autumn. They have impressive fasting abilities, due to their fat storage capabilities. Those that have access to sea ice throughout the year tend not to fast. Polar bears typically hunt by waiting near a hole in the ice used by seals for breathing. They then pull the seal out onto the ice when it surfaces, and devour it. These bears have an excellent sense of smell, and do much of their hunting with the assistance of olfaction. They are the top marine predator in the Arctic ecosystem, and they get virtually all of their sustenance from the sea, although they rarely hunt in the water directly.

Polar bears make long migrations of 2,000–4,000 km across the ice, and may move over 100 km in the water. They are capable of moving quite fast on ice, up to 40–50 km/hr for short sprints. This large predator is potentially

A mother polar bear and her two cubs. The normal litter size is two cubs, but sometimes three or even just one cub are born. Manitoba. PHOTO: M. NOLAN

A mother polar bear and her large cub roam the arctic snow in search of food. Manitoba. PHOTO: M. NOLAN

Two polar bears sparring—such fights may erupt over food, territory, or mating access. Polar bears are dangerous to humans as well. Manitoba. PHOTOS: M. NOLAN

quite dangerous to people in the Arctic and does occasionally kill and eat humans.

Mating in polar bears occurs from April to June. Each male may mate with one or several females. Females breed for the first time at 4–5 years of age. In November to December, the pregnant female excavates a den, where typically two cubs (sometimes three and occasionally just one) are born in December and January. Females show some fidelity to denning areas, but not necessarily to individual dens. The cubs are generally nursed for about 2.5 years. Polar bears can live to be in their early 30s.

Feeding and prey The primary prey of polar bears consists of ringed seals, but they also take bearded, harp, and hooded seals, and rarely walruses and beluga whales. These bears sometimes also eat Arctic cod and other forms of animal and vegetable matter. They even scavenge on refuse at dumps in some areas.

Threats and status Polar bears have been hunted by aboriginal peoples of the north for a long time, and are still killed by natives in at least Alaska, Canada, Greenland, and Siberia. About 800–900 per year are taken, and many of these hunts are carefully regulated to make sure they are sustainable. Thus, hunting is not the major threat to these animals. Other potential threats to the animals include pollution, and disturbance from oil and other mineral extraction operations. Issues related to environmental contamination have become increasingly worrisome in recent years, although exactly how contaminants affect the bears is still not well known (some bears with high PCB levels have been found with both male and female reproductive organs). In addition, disturbance from ecotourism could be problematic in some areas, although such tourism has increased the public appreciation for these animals to be kept alive. For instance, in Churchill, Manitoba, Canada, there is a thriving bear-watching industry. Recently, a new threat has been identified—the changes in sea ice and associated impacts caused by global climate change. In fact, there is some speculation that the global warming issue alone could threaten the future survival of the species. There are 19 stocks (IUCN subpopulations) recognized, which together added up to a global population of about 20,000 bears in 2009.

IUCN status Vulnerable (overall species), some subpopulations would qualify as Data Deficient.

8. Extinct Species

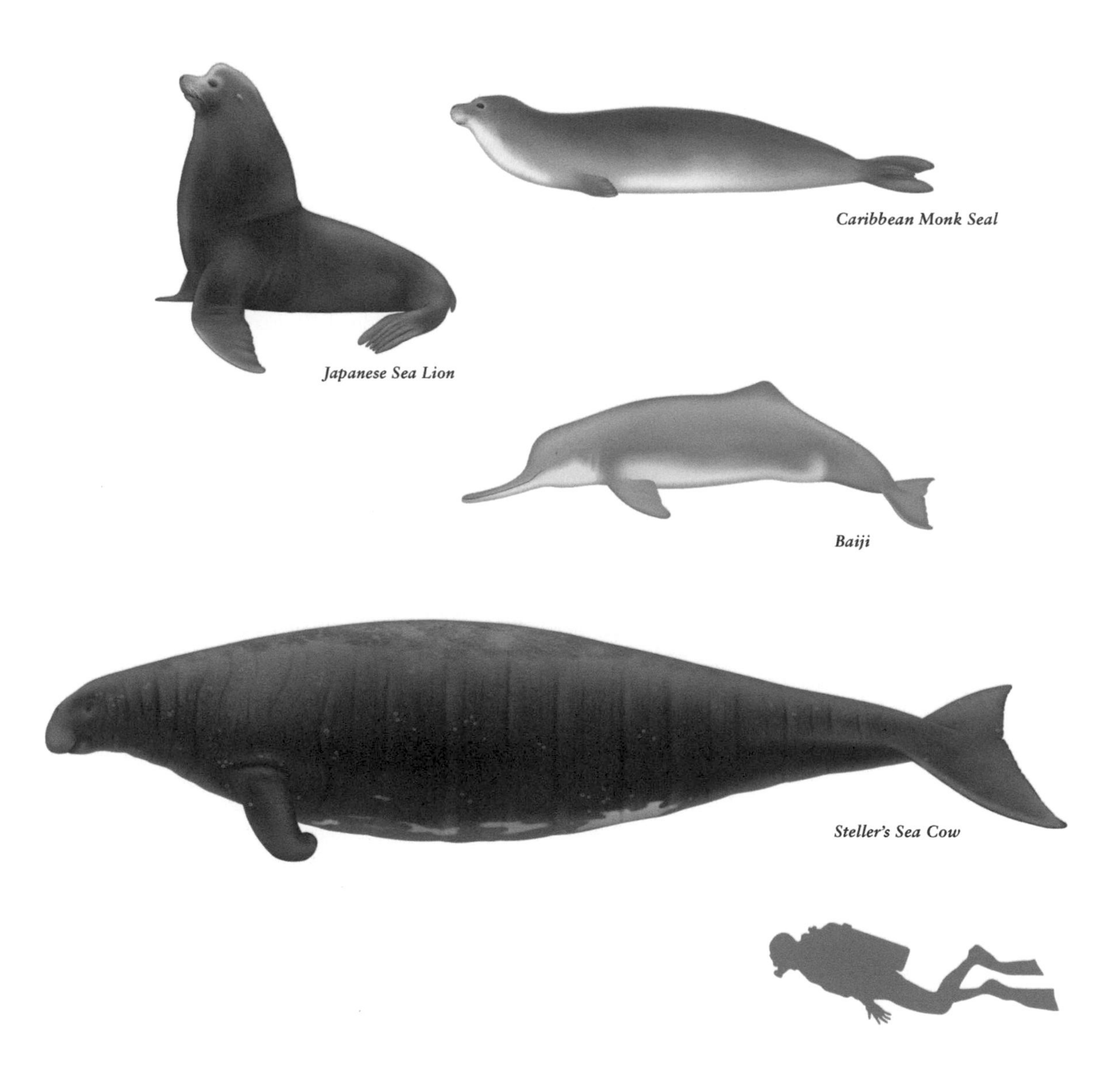

Baiji—*Lipotes vexillifer*

Miller, 1918

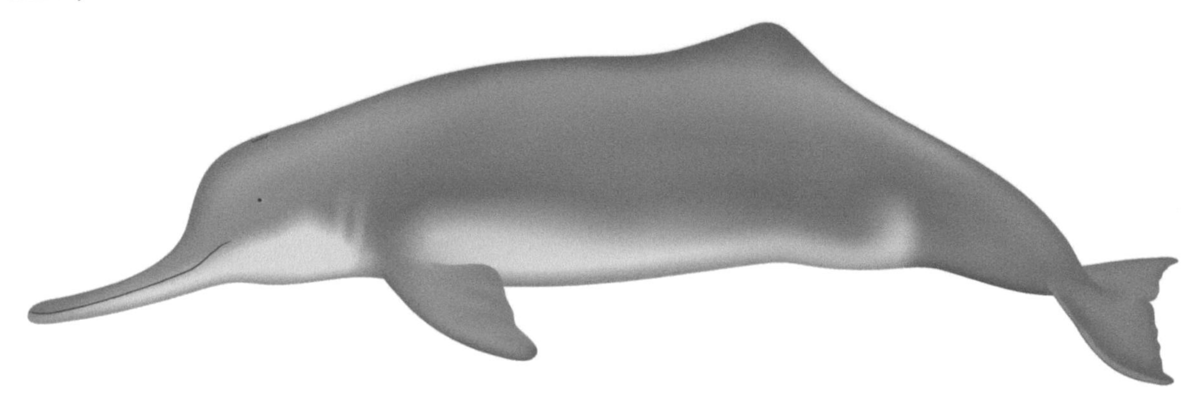

Recently-used synonyms None.
Common names En.–baiji, Yangtze River dolphin; Sp.–*delfín del Yangtze*; Fr.–*dauphin fluviatile de Chine, baiji*.
Taxonomic information Cetartiodactyla, Cetacea, Odontoceti, Lipotidae. Molecular studies suggest the baiji may have had a common ancestor with the boto and franciscana, but not with the South Asian river dolphin. It was a relict species.
Species characteristics Outside of China, very little was known of the baiji's biology until the 1970s and 1980s. Baijis had a fairly typical river dolphin appearance. These animals were moderately robust, with long, slightly-upturned beaks, rounded melons, and broad rounded flippers. There were bulging "cheeks" on the face. The dorsal fin was fairly prominent, but low and triangular, with a wide base—it was set about ⅔ of the way back from the beak tip. The eyes were small and set higher on the face, compared to those of oceanic dolphins. Although their vision was apparently poor, the eyes were functional. The blowhole was a longitudinal oval.

Baijis had rather simple, countershaded color patterns. They were predominantly gray to bluish-gray above and white to ashy-white on the ventral surface. There were light patches and brushings on the side of the face and the side of the tail stock.

Each tooth row of the baiji's mouth contained 31–36 conical teeth. Male baijis reached sizes of 2.3 m and 157 kg, and females reached 2.6 m and over 167 kg. In addition to females being somewhat larger than males, there were some other, minor differences in external morphology between the sexes. Apparently, newborn Yangtze River dolphins were about 92 cm in length.

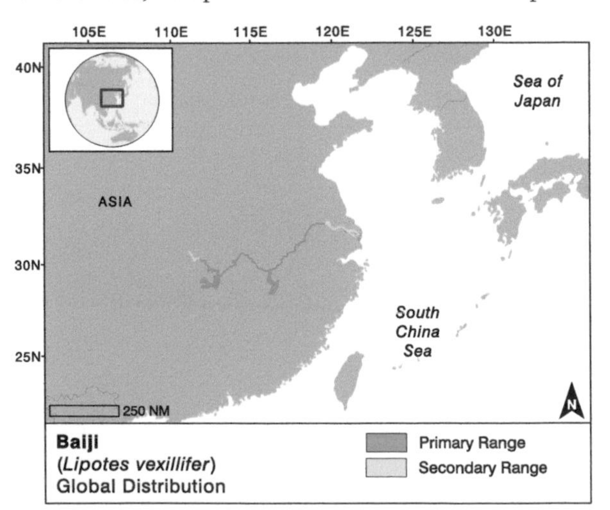

Recognizable geographic forms None.
Distribution In recent times, the baiji was found only in the middle and lower reaches of the mainstream Yangtze River in China (a length of about 1,400 km). The historical range was much broader, including waters of the Yangtze estuary, and several large lakes that dolphins entered during the flood season. They tended to congregate near confluences and sand bars with large eddies.
Ecology and behavior Groups of two to six baijis were most commonly seen, but aggregations of up to 16 animals sometimes formed. Mixed groups with finless porpoises were quite common (in the most recent surveys, about 63% of sightings). The porpoises appeared to behave aggressively towards baijis on some occasions, although it is not known if this was typical. Baiji groups in recent surveys consisted of about 57% adults, 26% juveniles, and 17% calves.

These dolphins were generally shy of boats, and their surfacings were shallow, often exposing only the top of the head, dorsal fin, and a small part of the back. They generally breathed with little surface disturbance, and most submersions were 10–30 sec long, with occasional

A dead baiji, which was killed by rolling hook fishing gear, lies along the banks of the Yangtze River. The low, extremely wide-based dorsal ridge and "cheeks" are distinctive to this species. This animal appears to be emaciated. PHOTO: ZHOU KAIYA, COURTESY OF S. LEATHERWOOD

A captive baiji swims in the clear waters of its tank in China. The waters of its natural habitat are in fact, very murky. Wuhan Institute of Hydrobiology. PHOTO: WANG DING

Qi Qi lived in captivity for 22 years and died in 2002. Now apparently extinct, the baiji lived only in the Yangtze River of China. Wuhan Institute of Hydrobiology. PHOTO: WANG DING

Much of what we know about the biology of the baiji came from studies of just a couple of captive individuals. Most famous of these was Qi Qi, shown here. Wuhan Institute of Hydrobiology. PHOTO: WANG DING

longer dives (the longest ones recorded were slightly under 3.5 min). Baiji movements included both short- and long-distance (200+ km) meanderings.

Breeding occurred mainly in the first half of the year; the peak calving season was February to April. Males reached sexual maturity at ages of about 4 years, and females did so at about 6 years. The oldest known individual was 30 years old. It is assumed that there was only a single population, although this was never tested.

Feeding and prey Baijis were apparently opportunistic feeders. A large variety of freshwater fish species made up the diet of the baiji, the only limitation probably being size.

Threats and status For most of the past several decades, the baiji was widely acknowledged to be the most critically-endangered cetacean in the world. In fact, sightings in the new millennium had become exceedingly rare. Although protected by Chinese law (it was a "Protected Species of the First Order"), and direct killing was considered rare, other threats were rampant. The main threat was probably mortality from non-selective fishing gear, such as rolling hooks (a type of snagging "longline"), and dynamite and electric fishing (the latter became a major threat, accounting for about 40% of baiji mortalities). Habitat deterioration/destruction was another major factor, as the Yangtze River has been dramatically modified to meet the needs of the huge surrounding human population (by some assessments, nearly 10% of the world's total). Other conservation issues included pollution, vessel collisions, and prey depletion. The genetic and demographic consequences of extremely low population size, although often overlooked, were another class of problems. In the latter half of the 1900s and early 2000s, as the threats got more serious, Chinese officials had been attempting to remove animals from the river, and placing them into a "semi-natural reserve." However, even this became infeasible, as living specimens were hard to find. A 3,500-km survey by an international team of experts, which covered nearly all of the species' known range over a 6-week period in late 2006, failed to observe or acoustically detect a single dolphin. Only sporadic, undocumented sightings have been reported since. It is therefore virtually certain that the species is now extinct. Sadly, the baiji became the first species of cetacean to be exterminated by human activities.

IUCN status Critically Endangered (IUCN has not yet officially declared it "Extinct").

Japanese Sea Lion—*Zalophus japonicus*

(Peters, 1866)

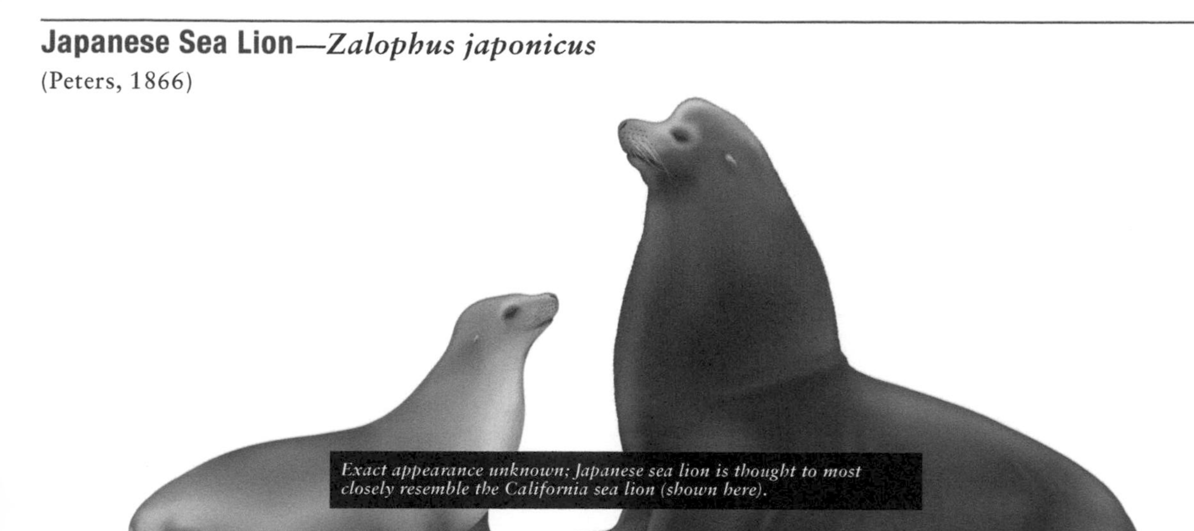

Exact appearance unknown; Japanese sea lion is thought to most closely resemble the California sea lion (shown here).

Recently-used synonyms *Zalophus californianus japonicus.*
Common names En.–Japanese sea lion; Sp.–*lobo marino de Japón;* Fr.–*lion de mer de Japon.*
Taxonomic information Carnivora, Otariidae. Currently, the Japanese sea lion is considered a separate species. Until recently, most authors classified it, along with Galapagos and California sea lion, as subspecies of *Z. californianus.*
Species characteristics Very little information exists on the appearance of this animal, but it was once considered a subspecies of the California sea lion. In an account from otter and seal hunters working in the western Pacific in the early 20th century, the "black sea lion" was said to have been present, in addition to Steller sea lions. This common name may point out that some animals, presumably adult males, as is the case for most adult male California sea lions, were very dark brown to black. A mid-19th century work gives a description of females as "straw colored with a darker throat and chest," which is also consistent with the coloration of female and juvenile California sea lions.

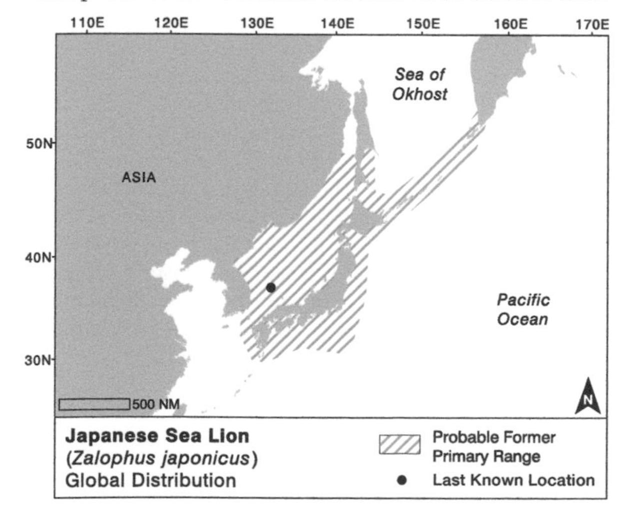

Adult males were 2.5 m, females 1.4 m, and a 4-month-old pup was 65 cm long and 9 kg, according to a Japanese zoologist in the 1950s. A study in 2004 examined 12 skulls and found them to be significantly longer than the skulls of adult male California sea lions. Recent efforts to retrieve data on length and weight and reconstruct parameters suggest the range of standard length (nose to end of the tail) of a small number of adult males to be 2.28–2.49 m and for one sub-adult female to be 1.63 m. A single adult male of total length of 2.88 m (nose to the end of the hindflippers) was estimated to weigh 493.7 kg. If this is representative of the species, the Japanese sea lion may have been 10% longer and 30% heavier than the California sea lion. The difference between Japanese sea lions and California sea lions was large enough that only the largest adult male California sea lion skulls reached the length of the shortest adult male Japanese sea lion skulls. Japanese sea lion skulls are also longer and wider in a number of measurements,

There are very few photos of Japanese sea lions, such as these at an undated, unknown location on a rocky reef. The group includes several large, dark adult males with thick necks and pale faces and muzzles. At least one animal appears to have a sagittal crest, as seen on adult male California sea lions. There are also a number of smaller animals, several of which are partly or wholly pale.
PHOTO: SHIMANE PREFECTURAL GOVERNMENT (NAKAWATASE ALBUM), COURTESY OF T. SUYAMA AND T. YAMADA

and also had a significantly taller sagittal crest. Taken together, these findings also suggest that adult male Japanese sea lions were larger than adult male California sea lions. Only one adult female Japanese sea lion skull was examined in this study, so it was not possible to draw conclusions about females from skulls.

Only a single photograph was found for preparation of this account, and on careful examination, the large animals closely resemble adult male California sea lions, with a prominent sagittal crest and the appearance of a pale colored mask in the face that contrasts with the very dark body. Observers in the former range of this species should be vigilant, and detailed notes should be made of any otariids that cannot be readily identified as Steller sea lions or northern fur seals. For field identification purposes, Japanese sea lion features should be presumed to be similar to those of California sea lion.

Recognizable geographic forms None.

Can be confused with Japanese sea lions shared their range with Steller sea lions and northern fur seals. Assuming that Japanese sea lions were similar to California sea lions, refer to the "Can be confused with" section for California sea lions.

Distribution Japanese sea lions were found at least from the end of the Korean Peninsula and along the northwestern coast and the southern end of Japan, along both coasts, throughout the Sea of Japan to southern Sakhalin Island. They were also found throughout the Kuril Island Archipelago to southernmost Kamchatka, Russia. A recent review lists 28 locations around the islands of Japan and the south end of Sakhalin Island that were used by Japanese sea lions, with four sites known to have served as locations with breeding: Liancourt Rocks (also known as Takeshima, or Dokdo, Islands), Kyuroku-jima Island, Aomori Prefecture, and Shikine-jima and Onbase-jima Islands near Tokyo.

Ecology and behavior Very little information is available on these animals, although they were assumed to be similar to the California sea lion. They were said to be good divers, although studies on diving were never conducted. Breeding was said to occur from April to July.

Feeding and prey Japanese sea lions apparently fed on fishes. Historical information from a review stated that they were caught in "fixed nets," that they ate yellowtail (*Seriola lalandi*) and fed on sardines and squid, but no other information is available.

Threats and status The Japanese sea lion is probably extinct. A comprehensive survey has not been made to determine if the species might still exist, and there has been only a minimal effort to search for specimens, accounts, data, and photographs in Japan, South and North Korea, and Russia. There is a long history of hunting and harvesting for meat, pelts, and fat for rendering of lamp oil that goes back to the Jōmon period (approximately 5,000 years ago, or more). From the mid-19th century until World War II, sea lions were harvested with firearms and nets. During a period of apparent intensive harvesting from 1904–1911, an estimated 14,000 sea lions of both sexes and all ages were harvested at Liancourt Rocks alone. Estimates are that 30,000–50,000 animals may have been present in the mid-19th century. The last population estimates available were of 100 animals on Liancourt Rocks and a total population of 300 by the late 1950s. The species probably became extinct sometime in the late 1950s. Reports of sightings from as recently as December 1975 from Liancourt Rocks were published recently, but no evidence was provided. Hope of the remote possibility of a remnant colony surviving in an isolated area of the former range still exists, however the prevailing opinion is that the species is extinct.

IUCN status Extinct.

Caribbean Monk Seal—*Neomonachus tropicalis* (Gray, 1850)

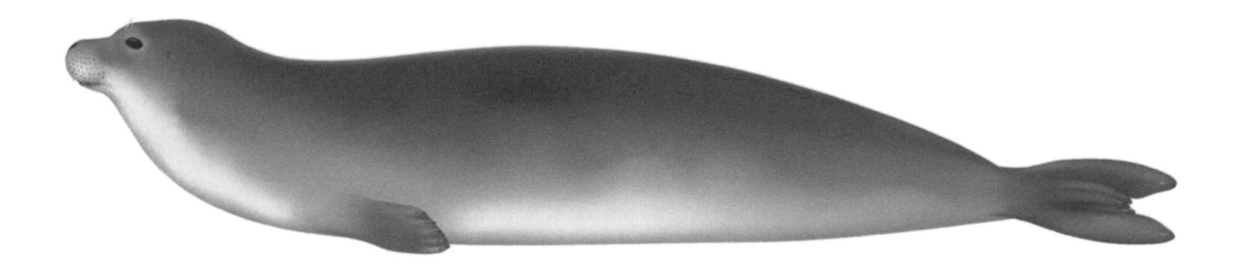

Recently-used synonyms *Monachus tropicalis.*
Common names En.–Caribbean monk seal, West Indian monk seal; Sp.–*foca monja del Caribe, foca fraile del Caribe*; Fr.–*phoque moine des Caraïbes.*
Taxonomic Information Carnivora, Phocidae. Until recently, the three monk seal species were in the genus *Monachus*. However, the extinct Caribbean monk seal and the Hawaiian monk seal are more closely related to each other genetically and morphologically than either is to the Mediterranean monk seal. In 2014, the new genus *Neomonachus* was proposed for the two allied species of the New World, and accepted by the Society for Marine Mammalogy.
Species characteristics Adult Caribbean monk seals were brown with a grayish tinge above, due to the tips of the hairs being lighter. The color became lighter on the sides, and transitioned to yellowish-white below. The end of the muzzle and upper and lower lip areas on the sides of the muzzle were yellowish-white. Adult vibrissae were predominantly whitish and smooth, with a few having darker bases, and still other short vibrissae were entirely dark. No difference in coloration was noted between males and females. Younger animals tended to be paler than adults, appearing more yellowish dorsally, with ochre tones ventrally, and a dusky central area at the end of the muzzle. Like Hawaiian monk seals, Caribbean monk seals were said to occasionally have green algae growing on the pelage. Females had four abdominal mammae. Newborn pups had a long, soft, glossy black lanugo coat that persisted for an unknown period of time. The vibrissae of pups were uniformly dark. Generally, these seals were described as having very few scars from fighting, although one large adult male was observed with gashes and scars that resembled seams.

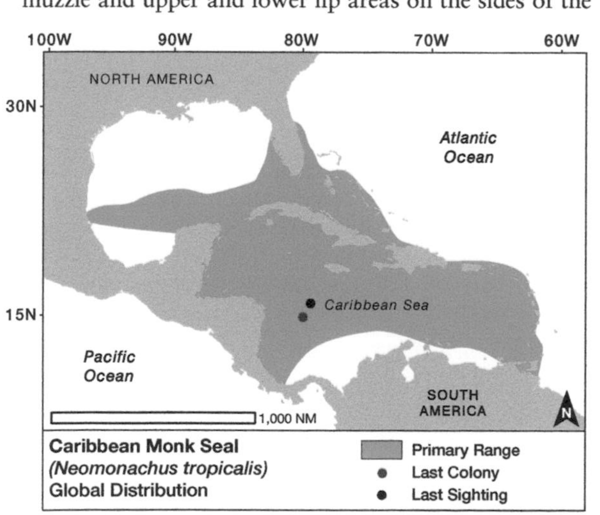

Adult Caribbean monk seals reached at least 2.4 m in length, (females may have been slightly larger than males). Hawaiian monk seals of comparable length to the largest reported Caribbean monk seals weigh 170–270 kg. Pups were probably about 1 m and 16–18 kg at birth. The dental formula was: I $^{2}/_{2}$, C $^{1}/_{1}$, PC $^{5}/_{5}$.
Can be confused with No other pinniped species regularly inhabits the former range of the Caribbean monk seal. Hooded, harbor, and less frequently, harp seals are known to stray occasionally as far south as the central east coast of Florida, near the edges of the Caribbean monk seal's former range. A monk seal could easily be distinguished from all of the above by the following collection of features: their long unspotted and unbanded body, brownish dorsal and pale ventral coloration, broad flat head and muzzle, and smooth unbeaded vibrissae. California sea lions that have escaped from captivity have also been reported from the Gulf of Mexico, but otariid seals differ from monk seals by the presence of external ear pinnae, long oar-like foreflippers, and a long narrow head and dog-like

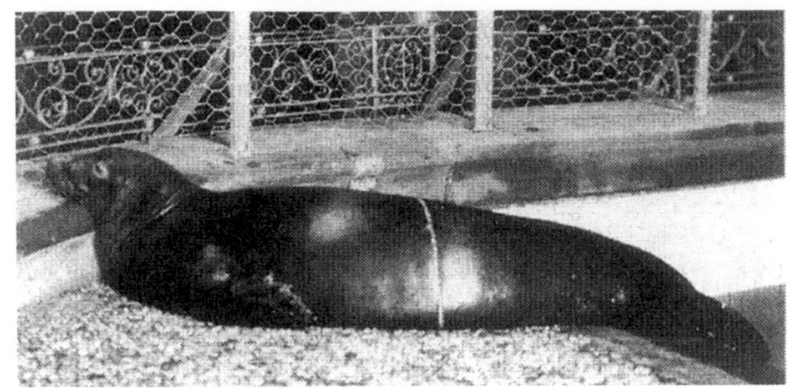

A rare photograph, from around 1910, of a Caribbean monk seal. This was one of three animals collected in 1909 and exhibited at the New York Aquarium. Its age and sex are unknown. Captured at Arrecifés Triángulos or Arrecife Alacrán, Mexico. PHOTO: COURTESY OF THE AMERICAN MUSEUM OF NATURAL HISTORY, DEPARTMENT OF MAMMALOGY ARCHIVES

muzzle. West Indian manatees shared part of the range of Caribbean monk seals and were similar in color. Key differences are the larger size, greater mass and width of adult manatees, and the muzzle and head shape, with the seal having a much smaller head and larger eyes. Also, the large paddle on the manatee is strikingly different than the hindflippers on any seal.

Distribution This monk seal once inhabited most of the Caribbean Sea, southern Florida, the waters of the Bahamas, and the Gulf of Mexico, except for the northern and western Gulf, where sightings are unconfirmed. The last stronghold of the species was the Serrana Bank, a remote reef lying between Jamaica and Nicaragua.

Ecology and behavior Observations made in the field, and from animals collected in the 19th century, provide evidence that pups were born from at least late fall to early winter. A long pupping season is known for both the Hawaiian and Mediterranean monk seals, which also live in subtropical habitats, and it is reasonable to assume this was the case for the Caribbean monk seal. Animals collected in December 1886 included newborn pups, and several females with term fetuses. Also, an animal described as recently weaned was encountered in the spring, and a female with a large fetus was taken in July.

It can be inferred that the Caribbean monk seal was a social species, possibly similar to the Hawaiian monk seal. The 49 specimens collected in December 1886 were all taken from three small cays in three days. The collectors describe finding females (with term fetuses) hauled out near one another, and in another case a pregnant female was hauled out in the vicinity of a female suckling a pup. Hauled-out groups of 20–40 were observed and reference was made to groups of 100 or more in earlier times. An otherwise undescribed group of five animals hauled out together included a large, scarred adult male. On another occasion, collectors encountered a group whose composition and numbers were not given, but the seals were "huddled together." Young Caribbean monk seals were also said to rest in pools of water, presumably for thermoregulation. Several descriptions exist of the vocalizations of the Caribbean monk seal. A young animal briefly held in captivity was said to grunt like a pig, and bark, growl and snarl like a dog. In another account, seals approached by hunters were said to "bark in a hoarse, gurgling, death-rattle tone."

Feeding and prey There is no information on the seals' food and feeding habits. The animals collected in 1886 all had stomachs that had fluid only, or were empty. The assumption is that their diet consisted of fish and crustaceans.

Threats and status The Caribbean monk seal is now extinct, and extensive searches have been unsuccessful. A recent study calculated that prior to European contact, 233,000–338,000 monk seals ranged across the Caribbean in 13–14 breeding colonies. The first references to the species are from the voyages of Christopher Columbus (1494), and exploitation of the New World, particularly after the development of the sugar industry, led to the seals being hunted heavily for oil. Records show that in 1688, monk seals were numerous enough at colonies to enable sealers to take "hundreds" in a single day. Colonies at the edge of the species' range disappeared in the 18th century. By 1850, the commercial hunt was no longer viable, and as the species became rare, large sums were paid for zoo and museum specimens, such as the 49 monk seals collected in December 1886. By 1922, the seals had been extirpated from the northern Caribbean. Yet another factor in the seals' demise may have been intensive commercial fishing that reduced the availability of their prey. The last breeding colony to survive was on the Serrana Bank, and the last confirmed sighting of a monk seal was at Serranilla Bank in 1952.

IUCN Status Extinct.

Steller's Sea Cow—*Hydrodamalis gigas*

(Zimmerman, 1780)

Recently-used synonyms None.
Common names En.–*Steller's sea cow*; Sp.–*vaca marina de Steller*; Fr.–*rhytine de Steller*.
Taxonomic information Sirenia, Dugongidae. Although classified with the dugong in the family Dugongidae, the Steller's sea cow was very distinctive and is generally placed in a separate subfamily.
Species characteristics The external morphology of the Steller's sea cow is not well-known, since the species was exterminated in 1768, before it could be studied in any detail. One thing is clear—this was a very large sirenian. It was rotund, especially in the summer, when feeding was good. It had a relatively-small head (only about 10% of body length) and horizontal flukes with a notch. The short, stubby forelimbs were about 60–70 cm long, with no nails or claws (and in fact, they lacked finger bones as well). The skin was rough on most of the body (so much so that people referred to it as like "bark"), although somewhat smoother on the back. It was very thick, up to at least 3 cm thick in places. There was a sparse covering of white hairs on the body, as well as stiff white bristles on the lips and foreflippers. The nostrils were rounded, and the eyes small and black. There were no external ear pinnae, only minute auditory openings. Steller's sea cow was brownish-black on the body, with a few white patches and streaks, especially on the undersides.

Functional teeth were completely absent, and grinding of food was done by keratinized masticatory plates inside the mouth. These were apparently white in color and "boat-shaped." These massive animals grew to lengths of at least 7.52 m. The weights attained by these animals are not known, but have been estimated to be somewhere between 4,000–10,000 kg. They were, by far, the largest of the sirenians.

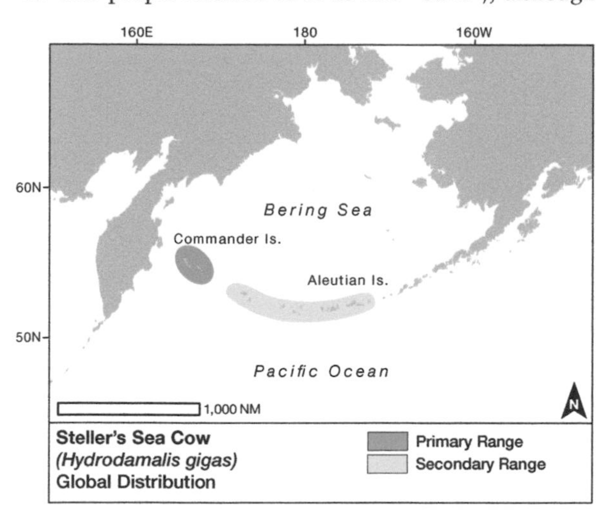

Distribution These animals lived only in the shallow, nearshore waters of the subarctic Bering Sea. They were found along the Commander Islands, and probably also occurred in the western Aleutian Islands. They were said to come very close to shore during high tide. A skull fragment from Monterey Bay, California, and other evidence suggests that historically the species may have ranged farther south.
Ecology and behavior Steller's sea cows often congregated in small herds in shallow waters. Young were said to have been kept at the front of the group, surrounded by adults on the sides and from behind. While largely defenseless, Steller's sea cow's huge size may have kept it safe from most enemies, except killer whales (and of course, humans). Seabirds often picked parasites from the animals' backs.

Not much is known of the behavior of this species. They spent most of their time feeding in kelp beds, near the islands. However, they apparently had difficulty submerging, due to their great buoyancy. They floated with their backs exposed and lifted their heads to breathe

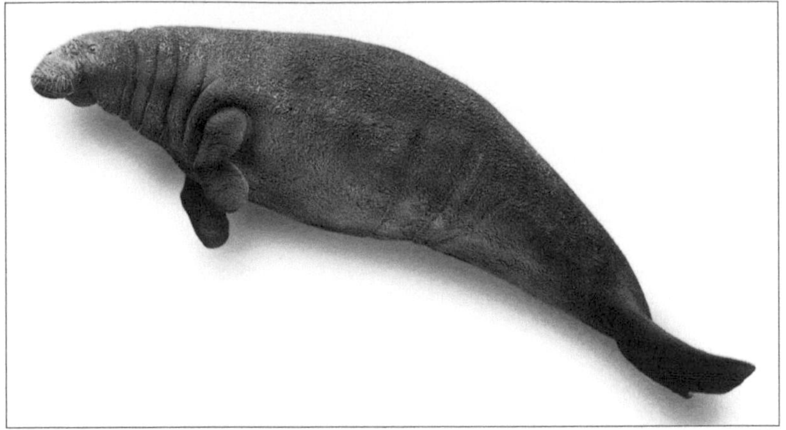

Since the species went extinct in the 1700s, there are no photos of Steller's sea cow. However, this model from the Natural History Museum of Milan provides a good indication of what this unusual marine mammal would have looked like. PHOTO: V. FOGATO

every 4 to 5 min. They pulled themselves through the shallows with their short forelimbs, and used their flukes both vertically and horizontally to move through the water. Apparently, they sometimes slept on their backs. Animals were observed to stand by injured conspecifics. They were thought to have been largely mute, with few sounds heard from them (although of course, no underwater recordings were made).

Virtually nothing is known about reproduction in this species, except that a single young was born, apparently mostly in the autumn. Copulation is said to have taken place in the early spring months, and gestation lasted more than one year.

Feeding and prey Steller's sea cows fed mostly on algae, especially kelp. They apparently only ate the leafy parts of the plants.

Threats and status Steller's sea cow was discovered in 1741, and after just 27 years of relentless exploitation by (mostly Russian) fur traders, it became extinct in 1768. It was hunted with harpoons from boats, and with spears and hooks from shore. It was highly prized for its meat, skin and blubber by explorers and hunters in the North Pacific. The meat was used for food, the skin for shoes and boats, and the fat for food and lamp oil. It was thought by those who observed them that the population was larger than it actually was (pre-exploitation numbers were estimated at about 1,500 to 2,000 animals), and there was no restraint on the wasteful hunting. Later reports of sightings of this species have been convincingly refuted, and there is no doubt that it is long extinct, the first modern marine mammal species to be wiped out through human ignorance and greed.

IUCN Status Extinct.

9. Identification Keys

General Notes on the Use of the Keys

The following general rules on the use of the keys below should be kept in mind:

• The keys (skull keys in particular) assume some previous knowledge/experience with these animals, and at least some knowledge about basic marine mammal anatomy.

• Feel free to jump to any place in the key to begin. For instance, if you are sure that the specimen you are working with is a toothed whale, jump to that part of they key (don't feel the need to go through each step of the key related to baleen whales).

• Use multiple features to make an identification and don't rely on a single feature.

• Use the information in the key with more "latitude" when dealing with young or damaged specimens, or those from a very poorly-studied geographic region. Be aware that some values may fall outside given ranges.

• If possible, try to have calipers, a ruler, or tape measure available to take measurements.

• Use geographical information as little as possible. Be aware that these are wide-ranging animals that sometimes wander far outside their normal ranges.

• Always confirm your final identification by examining photos/illustrations, and comparing to appropriate descriptions in the species accounts.

A. Key to Identification of Cetaceans of the World, Based on External Appearance

1a. Double blowhole; no teeth present; baleen plates suspended from upper jaw

Mysticete, Go to 2

b. Single blowhole; teeth present (though sometimes not protruding from gums); no baleen plates

Odontocete, Go to 14

2a. Long ventral pleats absent (though 2–7 short creases or furrows may be found on throat)

Other baleen whale, Go to 3

b. Long ventral pleats present; dorsal fin present; upper jaw relatively flat viewed from the side and broad viewed from the top

Balaenoperidae, Go to 5

3a. 2–7 short creases or furrows on throat; upper jaw and mouthline flat to slightly arched; dorsal fin or hump present

***Caperea* or *Eschrichtius*, Go to 4**

b. No creases on chin or throat; no dorsal fin or hump; upper jaw and mouthline strongly arched viewed from the side and very narrow viewed from the top; long, narrow black baleen plates with fine black fringes

***Balaena* or *Eubalaena*, Go to 12**

4a. 2 indistinct creases on throat; prominent falcate dorsal fin set about ⅔ of the way back from tip of jaw; upper jaw arched when viewed from side; 213–230 yellowish white baleen plates in each side; maximum body length 7 m; Southern Hemisphere distribution only

Pygmy right whale *Caperea marginata*

b. 2–7 short furrows on throat; mouthline slightly arched; no dorsal fin, but small dorsal hump followed by 8–14 crenulations present; 130–180 white to yellowish baleen plates with coarse bristles per side; body mottled gray and usually covered with patches of reddish to yellowish whale lice and gray to white barnacles; maximum body length 15 m; North Pacific distribution only

Gray whale *Eschrichtius robustus*

5a. Ventral pleats end before navel

Minke or sei whale, Go to 6

b. Ventral pleats extend to or beyond navel

Humpback, Bryde's, Omura's, blue, or fin whale, Go to 8

6a. 22–70 ventral pleats, longest ending before navel (often ending between flippers); 200–300 baleen plates with coarse bristles per side, less than 21 cm long, mostly white or yellowish white (sometimes with dark margin along outer edge); often conspicuous white bands on upper surface of flippers; from above, head sharply pointed; maximum body length 11 m

Minke whale (*Balaenoptera* spp.), Go to 7

b. 32–65 ventral pleats, longest ending past flippers, but well short of navel; 219–402 pairs of black baleen plates with many fine whitish bristles, less than 80 cm long; flippers all dark; from side, rostrum slightly downturned at tip; maximum body length 18 m

Sei whale *Balaenoptera borealis*

7a. Usually, there is a brilliant white band running across each flipper (although occasionally the band may be very indistinct)

Common minke whale *Balaenoptera acutorostrata*

b. Usually, there is no white band across the flippers; distribution limited to the Southern Hemisphere

Antarctic minke whale *Balaenoptera bonaerensis*

8a. Flippers 1/4 to 1/3 of body length, with knobs on leading edge; flukes with irregular trailing edge; 14–35 broad, conspicuous ventral pleats, longest extending at least to navel; top of head covered with knobs, one prominent cluster of knobs at tip of lower jaw; 270–400 black to olive brown baleen plates with gray bristles per side, less than 80 cm long; dorsal fin usually atop a hump; maximum body length 17 m

Humpback whale *Megaptera novaeangliae*

b. Flippers less than 1/5 of body length, lacking knobs; 40–100 slender ventral pleats; head lacking knobs

Bryde's, Omura's, blue or fin whale, Go to 9

9a. Three conspicuous ridges on rostrum; 40–70 ventral pleats extending to umbilicus; 250–370 slate gray baleen plates per side, with white to light gray fringes; head coloration symmetrical; maximum body length 16.5 m; tropical/subtropical distribution only

Bryde's whale *Balaenoptera edeni*

b. Only one prominent ridge on rostrum; 55–100 ventral pleats

Omura's, blue, or fin whale, Go to 10

10a. Head broad and almost U-shaped from above; dorsal fin very small (about 1% of body length) and set far back on body; 260–400 black baleen plates with black bristles per side (all three sides of each plate roughly equal in length); head coloration symmetrical; body mottled gray, with white under flippers; maximum body length 33 m

Blue whale *Balaenoptera musculus*

b. Color pattern asymmetrical, with a dark right lower jaw and whitish left lower jaw

Omura's or fin whale, Go to 11

11a. From above, head V-shaped and pointed at tip; dorsal fin about 2.5% of body length and rising as shallow angle from back; 260–480 gray baleen plates with white streaks per side (front 1/3 of baleen on right side all white); head coloration asymmetrical (left side gray, much of right side white); back dark, with light streaks; belly white; maximum body length 27 m

Fin whale *Balaenoptera physalus*

b. From above, head not strongly pointed at tip; dorsal fin prominent and rising at steep angle from back; 180–210 baleen plates, which are yellowish-white in front and black in the rear; head coloration asymmetrical (left side dark gray, much of right side white); back dark, with light streaks; belly lighter; maximum body length 12 m; distribution limited to Indian and Pacific oceans

Omura's whale *Balaenoptera omurai*

12a. Callosities (roughened areas of skin to which whale lice attach) present on head; 200–270 long (up to 2.8 m) baleen plates per side; body black, often with white ventral blotches; maximum body length over 18 m; temperate to subarctic distribution

Right whale (*Eubalaena* spp.), Go to 13

b. No callosities; white chin patch and often white band just before flukes; 230–360 long (some longer than 4 m) baleen plates per side; maximum body length 20 m; Arctic distribution only

Bowhead whale *Balaena mysticetus*

13a. Northern Hemisphere distribution
Northern right whale (*Eubalaena* spp.), Go to 14

b. Southern Hemisphere distribution
Southern right whale *Eubalaena australis*

14a. North Pacific distribution
North Pacific right whale *Eubalaena japonica*

b. North Atlantic distribution
North Atlantic right whale *Eubalaena glacialis*

15a. Upper jaw extending well past lower jaw; lower jaw very narrow
Sperm whale, Go to 16

b. Upper jaw not extending much or at all past lower jaw; lower and upper jaws about same width
Other odontocete, Go to 18

16a. Body black to charcoal gray, with white lips and inside of mouth; head squarish and large, 1/4 to 1/3 of body length; short creases on throat; S-shaped blowhole at left side of front of head; low, rounded dorsal "hump" followed by a series of crenulations along the midline; 18–26 heavy, peg-like teeth in each side of lower jaw, fitting into sockets in upper jaw; body 4–18+ m
Sperm whale *Physeter macrocephalus*

b. Body less than 4 m; head not more than 15% of body length; blowhole set back from front of head; prominent dorsal fin; 8–16 long, thin, sharply-pointed teeth in each side of lower jaw, fitting into upper jaw sockets
***Kogia* sp., Go to 17**

17a. Throat creases generally absent; dorsal fin short (< 5% of body length); distance from tip of snout to blowhole greater than 10.3% of total length; 12–16 (rarely 10–11) sharp teeth in each half of lower jaw
Pygmy sperm whale *Kogia breviceps*

b. Inconspicuous throat creases; dorsal fin tall (>5% of body length); distance from tip of snout to blowhole less than 10.2% of total length; 7–12 (rarely up to 13) teeth in each side of lower jaw, sometimes 1–3 in each half of upper jaw
Dwarf sperm whale *Kogia simus*

18a. Two conspicuous creases on throat, forming a forward-pointing V; notch between flukes usually absent or indistinct; dorsal fin relatively short and set far back
Ziphiidae, Go to 19

b. No conspicuous creases on throat; prominent median notch in flukes
Delphinoidea or Platanistoidea, Go to 28

19a. Exposed teeth in both upper and lower jaws (may be inconspicuous)
***Tasmacetus* or *Mesoplodon grayi*, Go to 20**

b. 1–2 pairs of teeth in lower jaw only (even these not erupted in many individuals)
Other species, Go to 21

20a. Many teeth in both jaws (17–28 per tooth row); pair of tusks at tip of lower jaw that erupt only in males; maximum body length about 8 m; white ventral field extending onto sides in three areas of body; Southern Hemisphere distribution only
Shepherd's beaked whale *Tasmacetus shepherdi*

b. Small head; extremely long, narrow beak; white rostrum in adults; 2 small triangular teeth well behind tip of lower jaw in males; 17–22 pairs of vestigial teeth in upper jaw of both sexes
Gray's beaked whale *Mesoplodon grayi*

21a. One or two pairs of teeth at or near tip of lower jaw, erupted only in some adults; head either with indistinct beak, or with distinct beak and steep forehead
Other species, Go to 22

b. Usually one pair of teeth well behind tip of lower jaw, erupted only in adult males; small head; prominent beak with forehead rising at shallow angle; sometimes flippers fit into depressions on the body; scratches and scars common on body; maximum body length 6.2 m
***Mesoplodon* sp.,[2] Go to 27**

22a. Two pairs of teeth in lower jaw, one pair at tip exposed outside closed mouth, second smaller pair behind first; long tube-like snout; rounded forehead rises from snout at a shallow angle
***Berardius* sp., Go to 23**

b. One pair of teeth at tip of lower jaw (exposed only in adult males)
Other species, Go to 24

23a. Maximum body length 13 m; Northern Hemisphere distribution only
Baird's beaked whale *Berardius bairdii*

[2] The species of the genus *Mesoplodon* are generally poorly-known. External morphology and pigmentation patterns have not been properly described for most of them, and it is generally not possible for non-experts to identify whales of this genus to species. Even for experts, examination of skulls or genetic evidence may be required to identify anything but mature males.

b. Maximum body length 10–11 m; Southern Hemisphere distribution only

Arnoux's beaked whale *Berardius arnuxii*

24a. Beak indistinct; head small relative to body size; forehead slightly concave in front of blowhole; single pair of teeth directed forward and upward at tip of lower jaw (exposed only in adult males); mouthline upturned at gape; head light colored; maximum body length 7.8 m

Cuvier's beaked whale *Ziphius cavirostris*

b. Tube-like beak distinct; pronounced bulge to steep forehead; tall, pointed dorsal fin; maximum length about 11 m

***Hyperoodon* or *Indopacetus*, Go to 25**

25a. Northern Hemisphere distribution

***H. ampullatus* or *Indopacetus*, Go to 26**

b. Southern Hemisphere distribution only

Southern bottlenose whale *Hyperoodon planifrons*

26a. North Atlantic distribution only

Northern bottlenose whale *Hyperoodon ampullatus*

b. Warm temperate and tropical North Pacific and Indian Ocean distribution

Longman's beaked whale *Indopacetus pacificus*

27a. Moderate beak, not sharply demarcated from forehead; males with white "cap" or "beanie" in front of blowhole; adult males with large flattened tusk in the middle of each side of lower jaw, protruding above upper jaw when mouth is closed; known from North Pacific only (females and subadults require museum preparation for identification)

Hubbs' beaked whale *Mesoplodon carlhubbsi*

b. Mostly dark beak, possibly with white jaws; in adult males, tusks near middle of lower jaw barely breaking gumline; known from the Pacific Ocean only, but may extend into the Indian Ocean (females and subadults require museum preparation for identification)

Ginkgo-toothed beaked whale *Mesoplodon ginkgodens*

c. Mostly dark beak, with white jaws; in adult males, tusks near middle of lower jaw not extending high above; known from the Pacific and Indian oceans only (females and subadults require museum preparation for identification)

Deraniyagala's beaked whale *Mesoplodon hotaula*

d. White markings on beak and forehead absent; lower jaw usually light in color; tusks of males very large, located on bony prominences near corners of mouth, and oriented slightly forward; lower jaw massive (particularly in adult males), with high arching contour; forehead has concavity in front of blowhole (females and subadults require museum preparation for identification)

Blainville's beaked whale *Mesoplodon densirostris*

e. Flattened tusks of adult males near tip of lower jaw (females and subadults require museum preparation for identification)

Hector's beaked whale *Mesoplodon hectori*

f. Small (maximum length about 4 m); dorsal fin small, triangular, and rounded at tip; color dark gray above fading to lighter below; small egg-shaped teeth located on prominences near the middle of the lower jaw in adult males; known only from eastern Pacific (females and subadults require museum preparation for identification)

Pygmy beaked whale *Mesoplodon peruvianus*

g. Pair of small oval teeth at tip of lower jaw of adult males; body gray with dark areas around eyes (females and subadults require museum preparation for identification)

True's beaked whale *Mesoplodon mirus*

h. Two small flattened teeth near front of lower jaw of males; body dark gray above, light gray below; known only from Atlantic Ocean (females and subadults require museum preparation for identification)

Gervais' beaked whale *Mesoplodon europaeus*

i. White areas in head and neck area; 2 large flattened tusks at top of arch in lower jaw of adult males (protruding above upper jaw when mouth closed); known only from North Pacific Ocean, apparently most common in subarctic waters of Alaska (females and subadults require museum preparation for identification)

Stejneger's beaked whale *Mesoplodon stejnegeri*

j. Adult males with white jaws and tusks on slightly raised prominences in middle of jaw; known mostly from South Pacific and Indian oceans (females and subadults require museum preparation for identification)

Andrews' beaked whale *Mesoplodon bowdoini*

k. Complex pattern of black, white, and gray; adult males with pair of tusks that grow outside of mouth from lower jaw, and wrap around upper jaw, preventing it from opening more than a few centimeters; known only from Southern Hemisphere (females and subadults require museum preparation for identification)

Strap-toothed beaked whale *Mesoplodon layardii*

l. Gray with lighter sides and belly; teeth of adult males protrude outside mouth in middle of lower jaw; vestigial teeth sometimes present in both jaws; known only from the temperate and subarctic North Atlantic (females and subadults require museum preparation for identification)

Sowerby's beaked whale *Mesoplodon bidens*

m. Color pattern apparently non-descript; tusks of males flattened and triangular, located just behind tip of lower jaw; known only from the eastern North Pacific (females and subadults require museum preparation for identification)

Perrin's beaked whale *Mesoplodon perrini*

n. Nothing is known of the external appearance of this species; tusks of males are massive and lean backwards at a 45° angle, they are spade-shaped with a prominent denticle at the tip (females and subadults require museum preparation for identification)

Spade-toothed beaked whale *Mesoplodon traversii*

28a. Teeth blunt with expanded crowns, laterally compressed, and relatively small; beak extremely short or nonexistent

Phocoenidae, Go to 29

b. Teeth conical and sharply pointed, unless heavily worn (in cross section, circular or oval)

Platanistoidea, Mondontidae, or Delphinidae, Go to 35

29a. Body light gray to nearly black, with lighter belly; no dorsal fin; narrow dorsal ridge; 15–22 teeth in each tooth row; maximum size to 2.3 m; distribution limited to the Indo-Pacific area

Finless porpoise (*Neophocaena* spp.), Go to 30

b. Dorsal fin present

Other species, Go to 31

30a. Body color dark gray to black; dorsal "ridge" wide (4.8–12.0 cm with 10–25 rows of tubercles); 15–22 teeth in each tooth row; distribution limited to tropical/subtropical areas of the Indian and western Pacific Oceans, from the Taiwan Strait to the south and west

Indo-Pacific finless porpoise *Neophocaena phocaenoides*

b. Body color light gray to dark gray; dorsal ridge narrow (0.2–1.2 cm with 1–10 rows of tubercles); 15–21 teeth in each tooth row; distribution limited to the Pacific Ocean—Chinese, Korean, and Japanese waters from the Taiwan Strait to the north

Narrow-ridged finless porpoise *Neophocaena asiaeorientalis*

31a. Body dark charcoal gray to black; dorsal fin set far back on body, rising at a shallow angle from back, with long leading edge and convex trailing edge; 10–23 pairs of teeth in upper jaw, 14–23 in lower; maximum size to 2 m; distribution limited to coastal South America

Burmeister's porpoise *Phocoena spinipinnis*

b. Dorsal fin upright and set near middle of back

Other species, Go to 32

32a. Body gray to brownish gray, with light belly, and dark lip patches and eye rings; flippers large; dorsal fin tall and slightly falcate; 16–22 teeth per side of each jaw; maximum size about 1.6 m; distribution limited to the Gulf of California, Mexico

Vaquita *Phocoena sinus*

b. Triangular dorsal fin; found outside the Gulf of California

Other species, Go to 33

33a. Body dark gray on back to white below; dark gape to flipper stripe; short, triangular, wide-based dorsal fin; 19–28 pairs of teeth in each jaw; maximum size to about 2 m; Northern Hemisphere distribution only

Harbor porpoise *Phocoena phocoena*

b. Color pattern sharply demarcated black and white

Other species, Go to 34

34a. Black body with striking large white patch on sides and belly; extremely robust, with small head and appendages; deepened caudal peduncle; dorsal fin triangular, with recurved tip; white or light gray trim on dorsal fin and flukes; 23–28 pairs of extremely small teeth per jaw; maximum size to 2.4 m; North Pacific distribution only

Dall's porpoise *Phocoenoides dalli*

b. Body bicolored, black on dorsal half and white on ventral half; black lips; white "spectacle" surrounding eye; dorsal fin triangular; 17–23 pairs of teeth in upper jaw, 17–20 in lower; maximum size to about 2.3 m; distributed only in cold temperate waters of the Southern Hemisphere

Spectacled porpoise *Phocoena dioptrica*

35a. No dorsal fin or prominent dorsal ridge (there may be a slight dorsal ridge)

Monodontiae or *Lissodelphis*, Go to 36

b. Dorsal fin or prominent dorsal ridge present

Other species, Go to 39

36a. Slight dorsal ridge present, sometimes marked with nicks or cuts; jaws short and wide; forehead high and globose; flippers short, broad, and rounded; distribution limited to arctic and subarctic areas

Mondontidae, Go to 37

b. No dorsal ridge present; body extremely slender; small flippers and flukes; beak short but distinct
Lissodelphis, Go to 38

37a. Body gray to brownish gray, mottled; short flippers often upturned at tips; flukes with more or less convex trailing edge; only two teeth in upper jaw, unerupted except in adult males, in which the left tooth develops into a left-spiraled tusk up to 2.7 m long; maximum size up to 5 m (excluding tusk); distribution limited to high Arctic
Narwhal *Monodon monoceros*

b. Body white to dark gray; extremely stocky; melon bulbous; beak short; head and appendages small; "neck" often visible; 9 pairs of teeth in upper jaw, 8 in lower; maximum size 5.5 m
Beluga whale *Delphinapterus leucas*

38a. Body black with white lanceolate pattern on belly; 37–54 fine pointed teeth per side of each jaw; maximum body length 3.1 m; North Pacific distribution only
Northern right whale dolphin *Lissodelphis borealis*

b. Body black above and white below; flippers, beak, and forehead mostly white; 44–49 teeth in each tooth row; maximum size to at least 3 m; Southern Hemisphere distribution only
Southern right whale dolphin *Lissodelphis peronii*

39a. Jaws extremely long; flippers broad and more or less triangular; eyes small; low, broad- based dorsal fin or dorsal ridge; distributed in rivers and lakes, only rarely in estuaries
Platanistoidea, Go to 40

b. Prominent dorsal fin; distribution estuarine or marine
Delphinidae or *Pontoporia*, Go to 43

40a. Blowhole transverse and crescentic; body gray, often with pinkish cast; dorsal hump low and set two-thirds of the way from the snout tip; forehead steep; 23–35 teeth per tooth row; maximum size to 2.8 m; distribution limited to Amazon and Orinoco drainage basins of Brazil, Bolivia, Peru, Colombia, Ecuador, and Venezuela
Amazon river dolphin *Inia geoffrensis*

b. Blowhole longitudinal
Other species, Go to 41

41a. Body bluish gray above and white below; blowhole oval; beak upturned at tip; dorsal fin triangular with blunt tip; 31–36 teeth per tooth row; maximum size to 2.6 m; distribution limited to the Yangtze River of China
Baiji *Lipotes vexillifer*—extinct

b. Body gray with lighter or pinkish belly; blowhole slit-like; eyes extremely small; beak long, narrow when viewed from above, with interlocking teeth protruding outside closed mouth at front half; low dorsal ridge; 26–39 teeth in each row; maximum size to 2.5 m
South Asian river dolphin *Platanista gangetica*, Go to 42

42a. Distribution limited to the Ganges and Brahmaputra River systems of India, Nepal, Bangladesh, and (possibly) Bhutan
Ganges River dolphin *Platanista gangetica gangetica*

b. Distribution limited to the Indus River system of Pakistan
Indus River dolphin *Platanista gangetica minor*

43a. Body dark gray with lighter belly; prominent triangular dorsal fin with rounded tip; flippers broad with curved leading edge and serrated trailing edge; eyes small; 50–62 teeth per tooth row; maximum size to 1.8 m; distribution limited to coastal and estuarine waters of Argentina, Uruguay, and Brazil
Franciscana *Pontoporia blainvillei*

b. Flippers without serrated trailing edge; eyes not particularly small
Delphinidae, Go to 44

44a. Head blunt with no prominent beak
Blackfish or other species, Go to 45

b. Head with prominent beak
Long-beaked delphinids, Go to 56

45a. Only 2–7 pairs of teeth at front of lower jaw (rarely 1–2 pairs in upper jaw), but teeth may be absent or extensively worn; forehead blunt with vertical crease; dorsal fin tall and dark; body gray to white, covered with scratches and splotches in adults; flippers long and sickle-shaped; maximum body length 4 m
Risso's dolphin *Grampus griseus*

b. Teeth (7 or more pairs) in both upper and lower jaws; forehead without vertical median crease
Other species, Go to 46

46a. Flippers broad and paddle-shaped with rounded tips
Blackfish and other species, Go to 46

b. Flippers long and slender with pointed or blunt tips
Other species, Go to 52

47a. Flippers large and paddle-shaped; dorsal fin tall and erect (up to 0.9 m in females and 1.8 m in males); striking black and white coloration, with white postocular patches, white lower jaw, white ventrolateral field, and light gray saddle patch behind dorsal fin; 10–14 large to 2.5 cm in diameter) oval teeth in each tooth row; maximum body length 10 m

Killer whale *Orcinus orca*

b. Dorsal fin low and rounded or triangular; adults less than 3 m; greater than 12 teeth per tooth row

Other species, Go to 48

48a. Body gray with lighter belly; dorsal fin small and slightly falcate; forehead bluff; neck crease often present; 8–22 teeth per row; maximum size 2.8 m; distribution limited to coastal areas and rivers of southeast Asia and northern Australia

***Orcaella* spp., Go to 77**

b. Color pattern with distinct lobes of light and dark; dorsal fin relatively large; no neck crease; 27–35 teeth per row; distribution limited to southern South America, southern Africa, New Zealand and the Kerguelen Islands

***Cephalorhynchus* sp., Go to 49**

49a. Sides light gray; dark gray cape (very narrow just behind blowhole area); belly white, with arms that surround the urogenital area and extend up both sides; white throat patch; white axillary patches; dorsal fin moderately tall and triangular; 22–28 teeth in each row; maximum size 1.8 m; distribution limited to southwest coast of Africa

Heaviside's dolphin *Cephalorhynchus heavisidii*

b. Dorsal fin rounded

Other species, Go to 50

49a. Distinct black and white color pattern, with black head and flippers, and black from the dorsal fin to the flukes; white chin patch; black genital patch; 28–35 teeth per row; maximum size 1.8 m; distribution limited to coastal and inshore waters of southeastern South America and the Kerguelen Islands

Commerson's dolphin *Cephalorhynchus commersonii*

b. Coloration largely various shades of gray

Other species, Go to 50

51a. Body mostly gray, with white belly and "arms" that extend up the sides on the tail stock (clearly demarcated by a dark gray line), and black dorsal fin, flippers, flukes, face, beak tip, and blowhole area; 24–31 teeth per row; maximum size to 1.7 m; distribution limited to coast of New Zealand

Hector's dolphin *Cephalorhynchus hectori*

b. Body gray with clearly demarcated white belly and chin; dark band between the flippers; white spots in axillae; 29–34 teeth in each row; maximum size to at least 1.7 m; distribution limited to west and south coast of South America, especially southern Chile

Chilean dolphin *Cephalorhynchus eutropia*

52a. Dorsal fin low and broad-based, located on forward third of back; head bulbous; body black to dark gray with light anchor-shaped patch on belly and often light gray saddle behind dorsal fin; often a light streak above and behind each eye; deepened tail stock; long sickle-shaped flippers; 7–13 pairs of teeth in front half only of each jaw

***Globicephala* sp., Go to 53**

b. Dorsal fin near middle of back

Other species, Go to 54

53a. Flipper length 18–27% of body length, with prominent "elbow;" 8–13 teeth in each tooth row; maximum size to 6.7 m; distribution limited mostly to cold temperate regions of North Atlantic and Southern Hemisphere

Long-finned pilot whale *Globicephala melas*

b. Flipper length 14–19% of body length; 7–9 pairs of teeth in each tooth row; maximum body length 7.2 m; distribution limited to tropical and warm temperate waters

Short-finned pilot whale *Globicephala macrorhynchus*

54a. Flipper with distinct hump on leading edge; body predominantly black; no beak; 7–12 large teeth in each half of both jaws, circular in cross-section; maximum body length 6 m

False killer whale *Pseudorca crassidens*

b. Body black or dark gray with white lips; white to light gray patch on belly; flipper lacks hump on leading edge; 8–25 teeth in each tooth row

Other species, Go to 55

55a. Less than 15 (8–13) teeth in each half of both jaws; flippers slightly rounded at tip; distinct dorsal cape; head rounded from above and side; maximum body length 2.6 m

Pygmy killer whale *Feresa attenuata*

b. More than 15 (20–25) teeth per side of each jaw; flippers sharply pointed at tip; face often has triangular dark mask; faint cape that dips low below dorsal fin; head triangular from above; extremely short, indistinct beak may be present in younger animals; maximum body length 2.8 m

Melon-headed whale *Peponocephala electra*

56a. Head long and conical; beak runs smoothly into forehead, with no crease; body dark gray to black above and white below, with many scratches and splotches;

narrow dorsal cape; flippers very large; 19–28 slightly wrinkled teeth in each tooth row; maximum body length 2.8 m

Rough-toothed dolphin *Steno bredanensis*

b. Beak distinct from forehead (however, there may not be a prominent crease between beak and melon)

Other species, Go to 57

57a. Beak very short and well-defined (less than 2.5% of body length); body stocky

***Lagenodelphis* or *Lagenorhynchus*, Go to 58**

b. Beak moderate to long (greater than 3% of body length)

Other species, Go to 64

58a. Flippers, flukes, and dorsal fin small; broad dark stripe from eye to anus area (muted in some animals); dorsal fin slightly recurved and uniformly dark; extremely short, but well- defined beak; grooves on palate; 38–44 teeth in each side of each jaw; maximum length at least 2.7 m

Fraser's dolphin *Lagenodelphis hosei*

b. Dorsal fin large; no palatal grooves

***Lagenorhynchus* sp., Go to 59**

59a. Body sharply demarcated black and white, with distinct white hourglass pattern on side, and white belly; dorsal fin strongly falcate; 26–35 teeth in each row; maximum size to about 2 m; distribution limited to colder waters of circumpolar Antarctic currents

Hourglass dolphin *Lagenorhynchus cruciger*

b. Color pattern complex with light gray patches on sides

Other species, Go to 60

60a. Body mostly black to dark gray, with white to light gray patches on the sides, and white belly and beak; dorsal fin large and falcate; 22–28 teeth in each row; maximum size to 3.1 m; distribution limited to cold waters of North Atlantic

White-beaked dolphin *Lagenorhynchus albirostris*

b. Prominent light gray flank

Other species, Go to 61

61a. Back dark gray, belly white, and sides light gray with white (below the dorsal fin) and yellowish-brown (on the tail stock) patches; black eye ring; deepened tail stock; 30–40 teeth in each row; maximum length 2.8 m; distribution limited to cold waters of the North Atlantic

Atlantic white-sided dolphin *Lagenorhynchus acutus*

b. Large light gray flank patch and gray stripes on tail stock with extensions running forward to thoracic area

Other species, Go to 62

62a. Black to dark gray above, white below, with light gray patches on sides; face, beak, melon, and most of the chin grayish-black; body relatively robust; up to at least 37 teeth in each row; maximum size to about 2.5 m; known distribution limited to southern South America and around Palmerston Atoll (although the latter is probably extralimital)

Peale's dolphin *Lagenorhynchus australis*

b. Much of face and lower jaw white to light gray; dorsal fin bicolored

Other species, Go to 63

63a. Appendages relatively large; dorsal fin bicolored and falcate (sometimes extremely hooked); back dark gray with light "suspender stripes" from forehead to tail stock, white belly, light gray flank patches (black lines separate belly from sides); 23–36 pairs of teeth in each jaw; maximum body length 2.5 m; distribution limited to North Pacific

Pacific white-sided dolphin *Lagenorhynchus obliquidens*

b. Belly white, back dark, flank patch light gray (no black line separates flank patch and belly); dorsal fin and flippers bicolored; 27–36 teeth in each tooth row; maximum length to at least 2.1 m; distribution limited to Southern Hemisphere (known mostly from South America, southern Africa, and New Zealand)

Dusky dolphin *Lagenorhynchus obscurus*

64a. Less than 39 teeth per tooth row

***Tursiops*, *Sotalia*, or *Sousa*, Go to 65**

b. Greater than 39 teeth per row

***Delphinus* or *Stenella*, Go to 71**

65a. Moderately robust; 18–29 teeth in each tooth row (teeth may be extensively worn or missing); body to 3.8 m; moderately long robust snout set off by distinct crease; color dark to light gray dorsally, fading to white or even pink on belly

Bottlenose dolphins (*Tursiops* spp.), Go to 66

b. 26 or more teeth in each tooth row; indistinct crease between melon and beak

***Sotalia* or *Sousa*, Go to 67**

66a. Beak relatively short and stubby; generally no spotting on belly; no spinal blaze; 18–27 pairs of teeth; maximum size up to 3.8 m; distributed worldwide

Common bottlenose dolphin *Tursiops truncatus*

b. Beak relatively long and slender; generally with black spotting on the belly; often a spinal blaze below the dorsal fin; 21–29 pairs of teeth; maximum size 2.7 m; distributed only in the Indo-Pacific

Indo-Pacific bottlenose dolphin *Tursiops aduncus*

67a. Back dark gray and belly light; beak long without distinct crease; low triangular to slightly falcate dorsal fin; 26–36 teeth in each tooth row; maximum size to 2.2 m; distribution limited to coasts, rivers, and lakes of the east coast of South America from Panama to southern Brazil, including the Amazon and Orinoco drainage basins

Tucuxi or Guiana dolphin *Sotalia fluviatilis* or *S. guianensis*

b. Body gray to white, sometimes with pink tinge, and light belly; base of dorsal fin of adults often expanded to form longitudinal ridge, especially west of Bay of Bengal; beak long, crease indistinct; 26–39 teeth in each tooth row; maximum size to 2.8 m

Humpback dolphins (*Sousa* spp.), Go to 68

68a. Wide-based dorsal fin with no dorsal hump; distribution limited to Pacific and eastern Indian Oceans, from Australia to Southeast Asia

***Sousa chinensis* or *S. sahulensis*, Go to 69**

b. Small, narrrow-based dorsal fin sits atop an elongated hump on back; distribution limited to Atlantic or Indian Oceans, from West Africa to the Bay of Bengal

***Sousa teuszii* or *S. plumbea*, Go to 70**

69a. Adult color pattern basically montone, with white ground color (there may be extensive dark patches and/or spotting); 32–39 teeth in each upper row and 29–38 in each lower row; distribution limited to Southeast Asia (from the Bay of Bengal to central China and Indonesia)

Indo-Pacific humpback dolphin *Sousa chinensis*

b. Adult color pattern two-tone, with a curved, diagonal dark gray dorsal cape (there may be extensive white patches and/or spotting, especially in older animals); 31–35 teeth in each upper row and 31–34 in each lower row; distribution limited to Australia and New Guinea

Australian humpback dolphin *Sousa sahulensis*

70a. Adult coloration dark gray (there may be some scarring or dark spotting); 27–32 teeth in each upper row and 26–31 in each lower row; distribution limited to the Eastern Atlantic Ocean (from Western Sahara to Angola)

Atlantic humpback dolphin *Sousa teuszii*

b. Adult coloration dark brownish gray (there may be some scarring or spotting); 33–39 teeth in each upper row and 31–37 in each lower row; distribution limited to the Indian Ocean from South Africa to the Bay of Bengal)

Indian Ocean humpback dolphin *Sousa plumbea*

71a. Dorsal fin erect to slightly falcate; back dark and belly white; tan to buff thoracic patch and light gray streaked tail stock form an hourglass pattern that crosses below dorsal fin; cape forms a distinctive V below dorsal fin; chin to flipper stripe; maximum size 2.7 m; 40–67 teeth in each row; palate with two deep longitudinal grooves

Common dolphins (*Delphinus* spp.), Go to 72

b. No hourglass pattern on side; palatal grooves, if present, shallow

***Stenella* sp., Go to 73**

72a. Body relatively stocky; beak shorter; slope of forehead relatively steep; flipper stripe narrow and not approaching gape; often light patches on flippers and dorsal fin; anus stripe faint or absent

Short-beaked common dolphin *Delphinus delphis*

b. Body relatively slender; beak longer; slope of forehead relatively shallow; flipper stripe wide and often contacting gape; light patches on flippers and dorsal fin generally absent; anus stripe may be distinct

Long-beaked common dolphin *Delphinus capensis*

73a. Color pattern black to dark gray on back, white on belly, prominent black stripes from eye to anus and eye to flipper; light gray spinal blaze extending to below dorsal fin (not always visible); shallow palatal grooves often present; 40–55 teeth in each row; maximum size 2.6 m

Striped dolphin *Stenella coeruleoalba*

b. Generally, no eye to anus stripe; distribution limited to tropical and warm temperate waters

Other species, Go to 74

74a. Light to heavy spotting present on dorsum of adults (on some individuals, spots may be absent); no palatal grooves

Spotted dolphin, Go to 75

b. No spotting on dorsum of adults; cape dips to lowest point at level of dorsal fin; eye-to-flipper stripe; shallow palatal grooves often present

***S. clymene* or *S. longirostris*, Go to 76**

75a. Body moderately robust, dark gray above, with white belly; light spinal blaze; slight to heavy spotting on adults (occasionally spotting nearly absent); maximum size 2.3 m; 30–42 teeth per row; distribution limited to warm waters of the Atlantic Ocean

Atlantic spotted dolphin *Stenella frontalis*

b. Dorsal fin narrow and falcate; dark cape that sweeps to lowest point on side in front of dorsal fin; dark gape to flipper stripe; beak tip and lips white; adults with light to extensive spotting and gray bellies (spotting sometimes absent); 34–48 teeth in each half of each jaw; maximum size 2.6 m

Pantropical spotted dolphin *Stenella attenuata*

1) Body and beak relatively robust; heavy spotting that nearly obliterates cape; known distribution limited to within 185 km of the coast in the eastern tropical Pacific

Coastal spotted dolphin (*S. a. graffmani*)

2) Body and beak slender; spotting slight to moderate; maximum body length 2.4 m; distributed more than 30 km from mainland in eastern tropical Pacific and found in oceanic waters worldwide

Offshore spotted dolphin (*S. a. attenuata*)

76a. Body color three-part (dark gray cape, light gray flanks, white belly); cape dips in two places (above eye, and below dorsal fin); snout light gray with dark tip, dark lips, and dark line from tip to apex of melon; often, dark "moustache" on top of beak; more robust than spinner dolphins; 39–52 teeth in each tooth row; maximum size about 2.0 m; distribution limited to tropical Atlantic Ocean

Clymene dolphin *Stenella clymene*

b. Dorsal fin slightly falcate to canted forward; beak exceedingly long and slender; 40–62 very fine sharply pointed teeth per tooth row; maximum size 2.4 m

Spinner dolphin *Stenella longirostris*

1) Color pattern three-part (white belly, light gray sides, dark gray cape); dorsal fin falcate to erect; body more robust than in other forms; post-anal hump of adult males nearly absent; distribution worldwide, except eastern tropical Pacific

Gray's spinner dolphin *S. l. longirostris*

2) Pigmentation monotone gray, with light patches around genital area and axillae; dorsal fin triangular to canted forward (extremely canted in adult males); adult males with deepened tail stock and enlarged post-anal hump; maximum size 2 m; known distribution limited to the eastern tropical Pacific east of 145°W

Eastern spinner dolphin *S. l. orientalis*

3) Pigmentation monotone gray; apparently, no light patches around genital area and axillae; dorsal fin triangular to canted forward (extremely canted in adult males); adult males with deepened tail stock and enlarged post-anal hump; to 2.2 m long; known distribution limited to 80 km offshore from southern Mexico to Panama in the eastern tropical Pacific

Central American spinner dolphin *S. l. centroamericana*

4) Body slightly more robust than above two forms; color pattern largely bipartite, with dark dorsal cape, and white belly and lower sides; dorsal fin slightly falcate to slightly canted (tending towards canted in adult males); post-anal hump of adult males small to moderate; distribution limited to offshore eastern tropical Pacific

Whitebelly spinner dolphin (hybrid *S. l. longirostris* X *S. l. orientalis*)

5) Body very small (<1.6 m), with a relatively large head and flippers; dorsal fin not canted; postanal hump small or absent; tripartite color pattern; distribution limited to the Indo-Pacific, from northern Australia to southeast Asia

Dwarf spinner dolphin *S. l. roseiventris*

77a. No dorsal groove; color pattern tripartite; no extension of white ventral field onto underside of flippers; distribution limited to Australia and Papua New Guinea

Australian snubfin dolphin *Orcaella heinsohni*

b. Dorsal groove present; color pattern bipartite; white ventral field extends onto underside of flippers; distribution limited to Asia (eastern India to southern Vietnam and central Indonesia)

Irrawaddy dolphin *Orcaella brevirostris*

B. Key to Identification of Cetaceans of the World, Based on Skull Morphology

Notes on the Use of This Key

This key is intended to allow the user to identify a skull of any cetacean, even if the area of origin is not known (therefore we do not use geographic information as primary features in the key). Although it is intended primarily for adult skulls, it may be useful for identifying skulls of subadults in some cases as well.

For some parts, this key is based on incomplete information, and thus may not always be reliable. It is presented here because of the lack of other suitable materials for identifying cetacean skulls. Skulls from subadult mammals often do not show the species' diagnostic characters, and geographic variation may yield it unreliable when examining skulls from certain areas. Clearly, a damaged skull may not be possible to identify to species, no matter what aids are used. So, while it will not be possible to identify every cetacean skull with this key, we hope that it will aid readers in identifying a large majority of them. We urge users to contact us to let us know how it can be improved.

1a. Teeth and alveoli absent; skull bilaterally symmetrical; lower jaw lacking bony symphysis; posterior portion of mandible not hollowed; maxillae extend posteriorly underneath frontal; size always large (adults >1 m)

Mysticeti, Go to 2

b. Teeth or alveoli present (although they may not emerge from jaw bones in some beaked whales); skull generally asymmetrical; lower jaw possessing bony symphysis; posterior non-tooth-bearing portion of mandible hollowed-out to form thin-walled "pan bone"; maxillae and premaxillae extend posteriorly over the frontal; skull generally much smaller (<1.5 m, except in *Physeter*)

Odontoceti, Go to 11

2a. Rostrum strongly arched from lateral view (>20° between basicranium and base of rostrum); base of rostrum narrow ($<\frac{1}{3}$ cranial width); mandibles strongly bowed out

Balaenidae, Go to 3

b. Rostrum slightly arched or flat from lateral view (<18° between basicranium and base of rostrum); base of rostrum wide ($>\frac{1}{2}$ cranial width); mandibles only slightly bowed

Neobalaenidae, Eschrichtiidae, or Balaenopteridae, Go to 4

3a. In profile, skull has a continuous curve from occipital condyles to tip of rostrum; occipital shield does not overhang temporal fossae; nasals long (about 3 times their width); high latitude Northern Hemisphere distribution only

Bowhead whale *Balaena mysticetus*

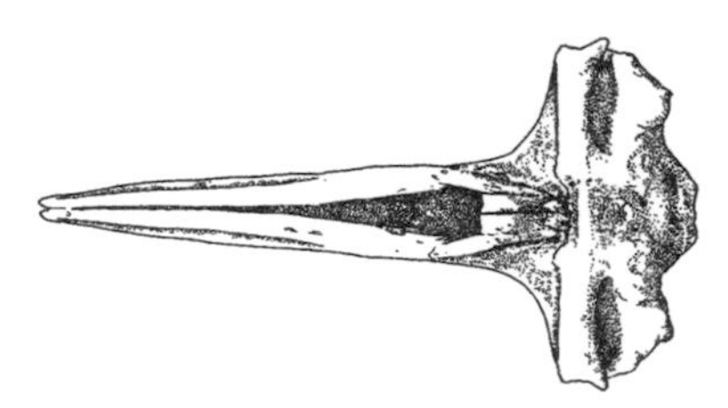

Dorsal View

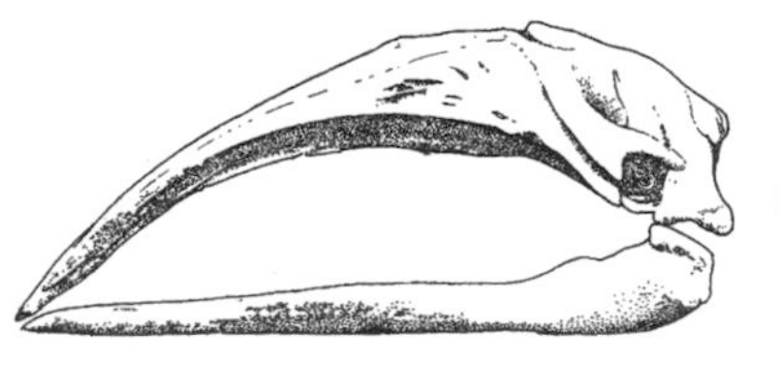

Lateral View

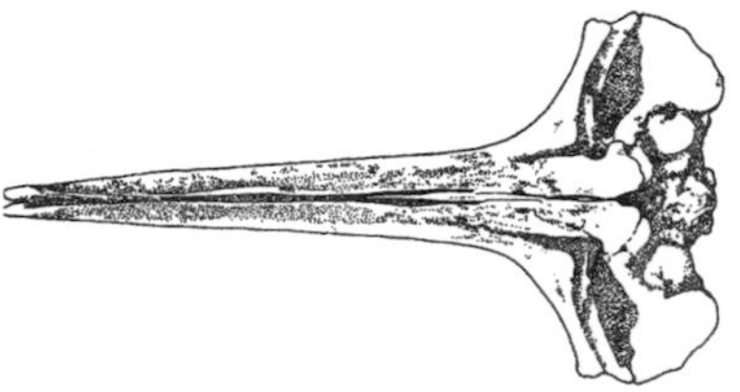

Ventral View

b. In profile, skull has a distinct angled apex between rostral and cranial bones; occipital shield overhangs temporal fossae; nasals relatively short (about twice their width)

Right whales *Eubalaena* spp.

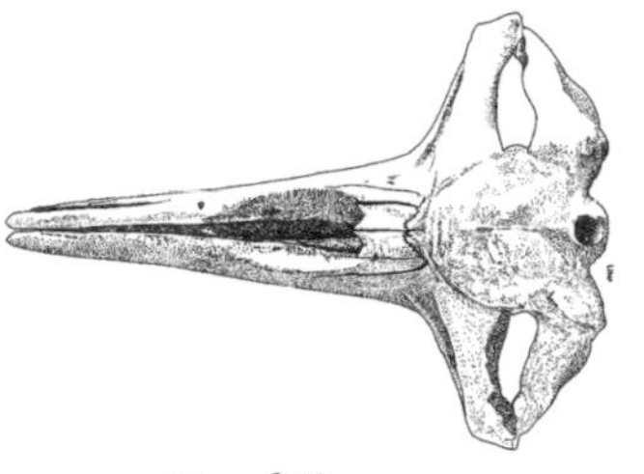

Dorsal View

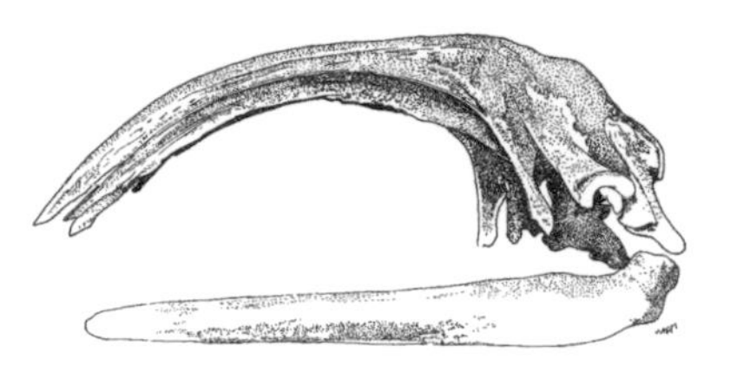

Lateral View

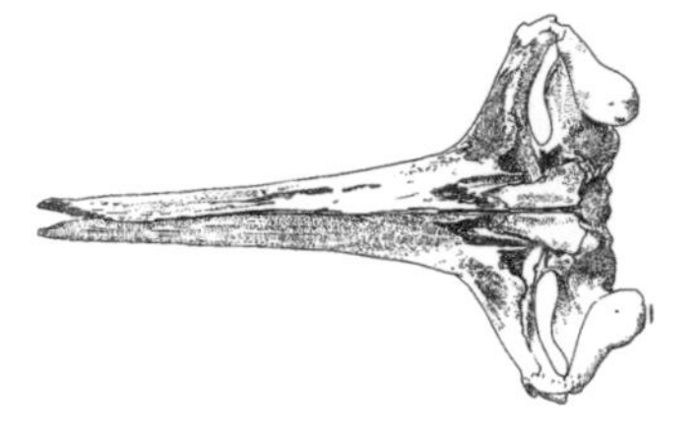

Ventral View

4a. Rostrum moderately arched in lateral view (>14° between basicranium and base of rostrum); large anteriorly-thrust occipital shield (extending forward to base of rostrum); overall posterior margin of cranium strongly concave when viewed from above; high latitude Southern Hemisphere distribution only

Pygmy right whale *Caparea marginata*

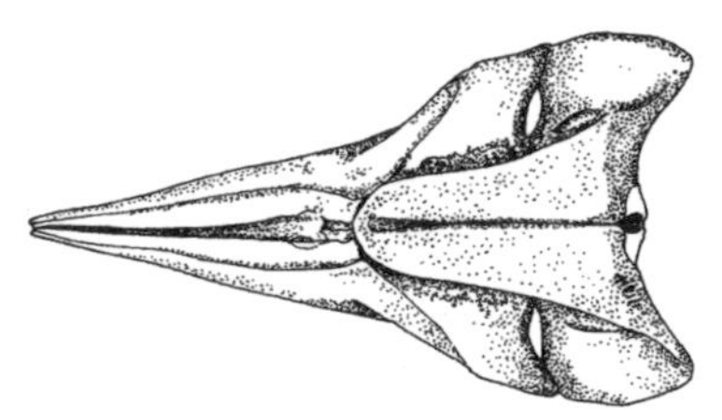

Dorsal View

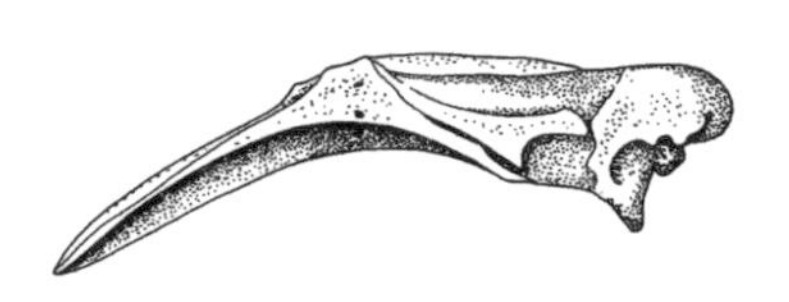

Lateral View

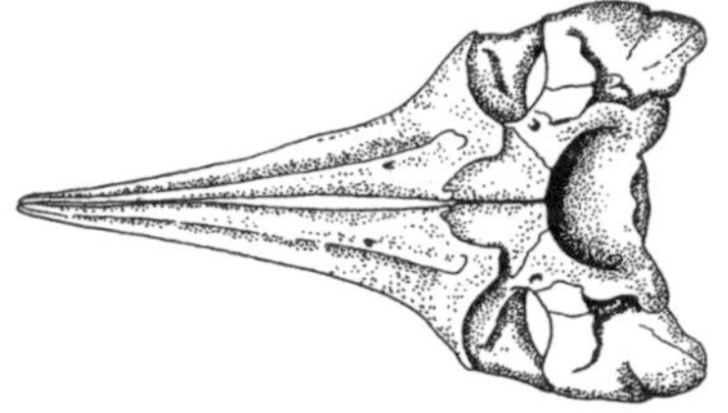

Ventral View

b. Rostrum only slightly arched in lateral view (<13° between basicranium and base of rostrum); no prominent occipital shield (occipital only extends forward up to 75% of length of brain case); overall posterior margin of cranium convex or relatively straight

Eschrichtiidae or Balaenopteridae, Go to 5

5a. Nasals large (nearly 1/2 length of post-rostral cranium); paired occipital tuberosities on posterior portion of cranium; rostrum relatively arched (>8° between basicranium and base of rostrum); North Pacific distribution only

Gray whale *Eschrichtius robustus*

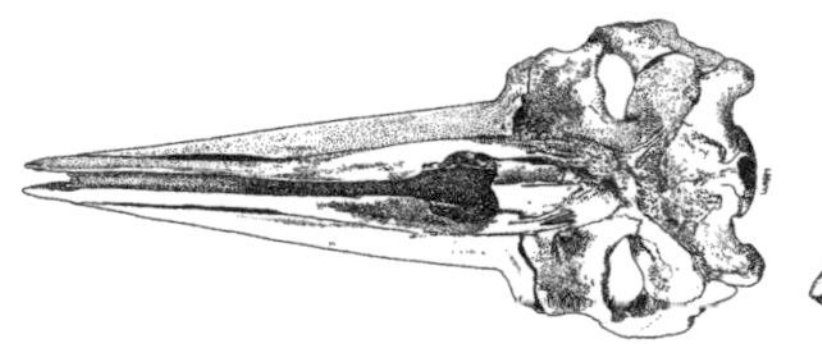

Dorsal View

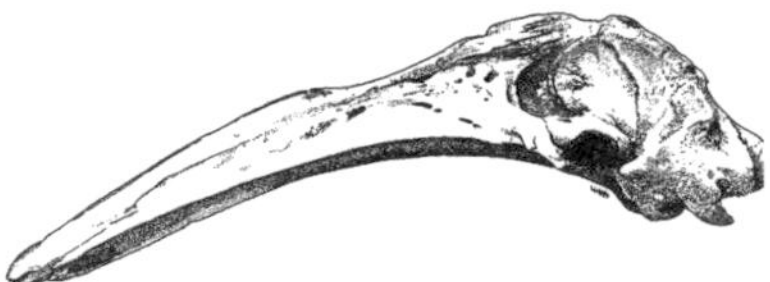

Lateral View

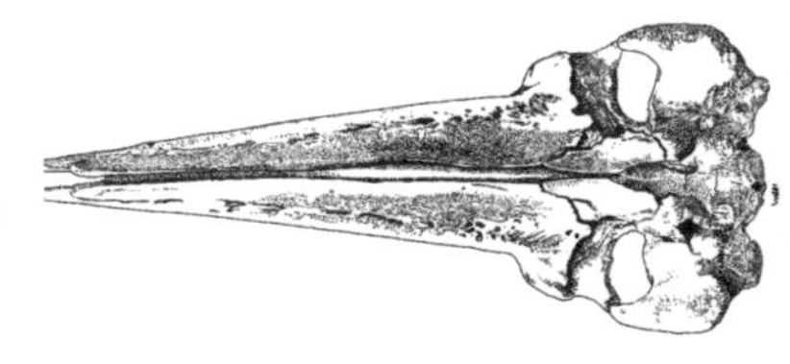

Ventral View

b. Nasals smaller (<1/4 length of post-rostral cranium; occipital tuberosities absent; rostrum relatively flat (<7° between basicranium and base of rostrum)

Balaenopteridae, Go to 6

6a. Base of rostrum about 1/2 cranial width; anterior margin of squamosal rounded or U-shaped
Humpback whale *Megaptera novaeangliae*

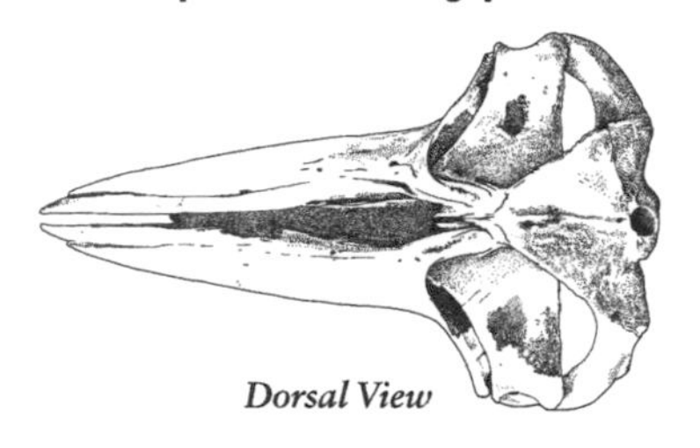
Dorsal View

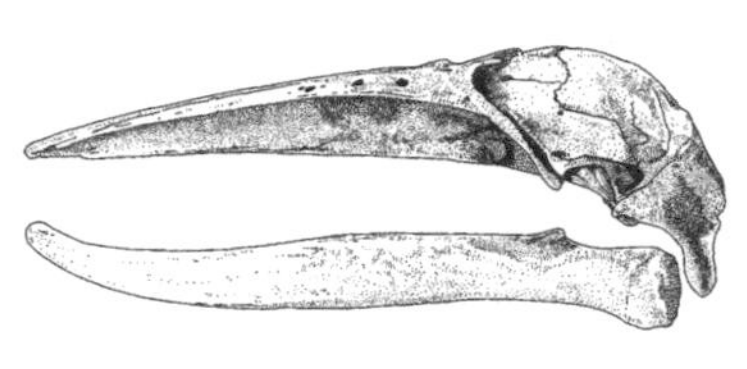
Lateral View

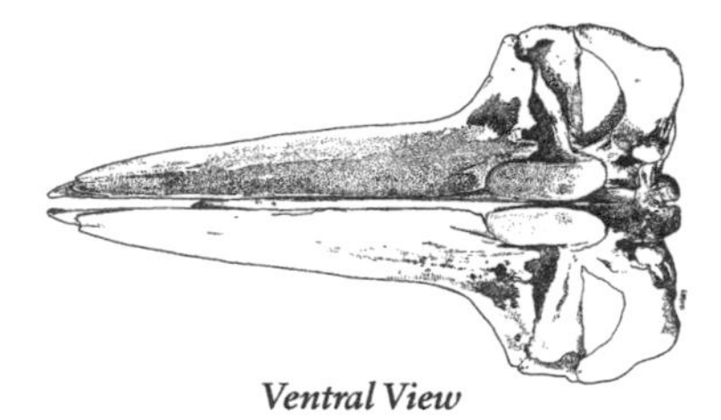
Ventral View

b. Base of rostrum at least 2/3 cranial width; anterior margin of squamosal pointed or V-shaped
***Balaenoptera* sp, Go to 7**

7a. Rostrum U-shaped and somewhat rounded at tip; rostral borders parallel along proximal half; from above, occipital length <1/5 condylobasal length (CBL)
Blue whale *Balaenoptera musculus*

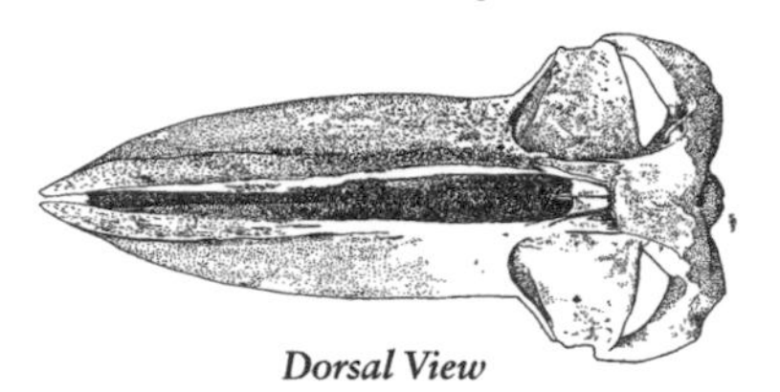
Dorsal View

Lateral View

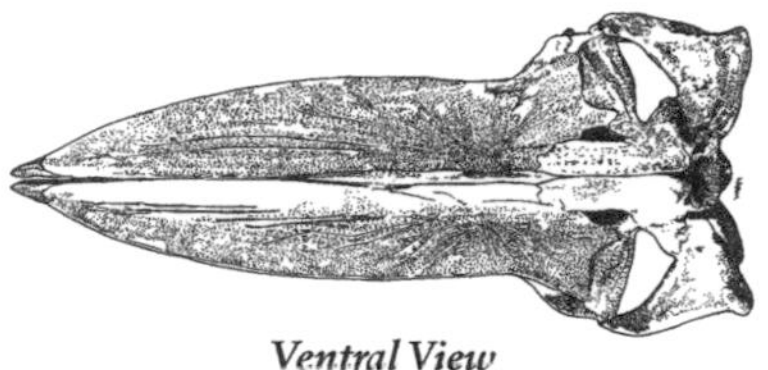
Ventral View

b. Rostrum V-shaped and pointed; rostral borders divergent throughout their length; from above, occipital length >1/5 condylobasal length (CBL)
Minke, Bryde's, sei, or fin whale, Go to 8

8a. CBL of adults <2.0 m
Minke or Omura's whale *Balaenoptera acutorostrata/bonaerensis/omurai*

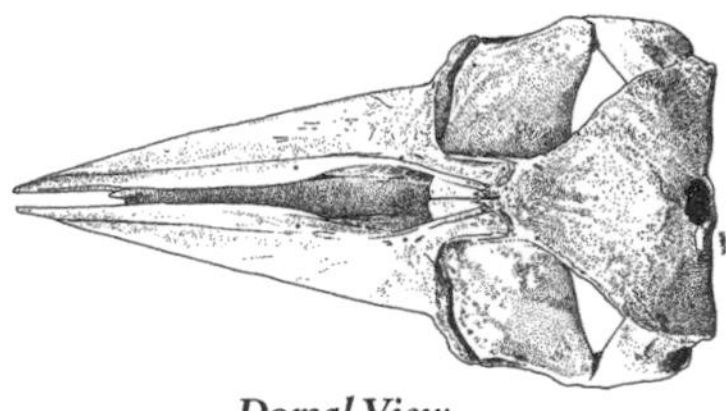
Dorsal View

Lateral View

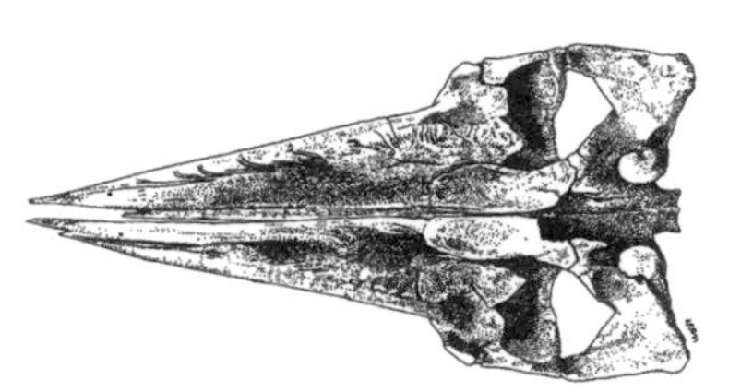
Ventral View

b. CBL of adults >2.0 m
Fin whale, sei whale or Bryde's whale, Go to 9

9a. Nasals small—length (measured along their suture) less than 1/2 length of nasofrontal process; vomer widely expanded at its posterior end
Fin whale *Balaenoptera physalus*

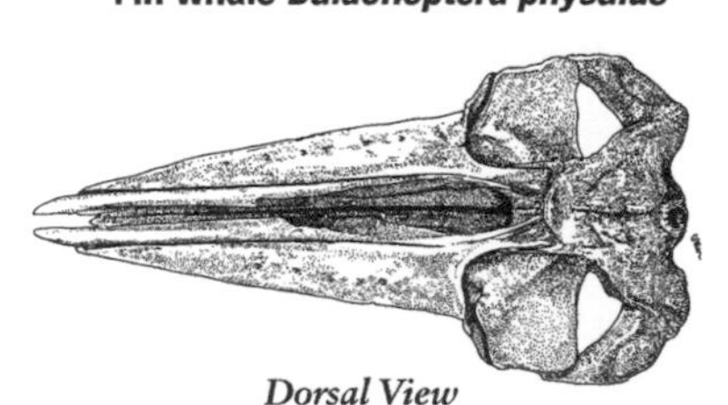
Dorsal View

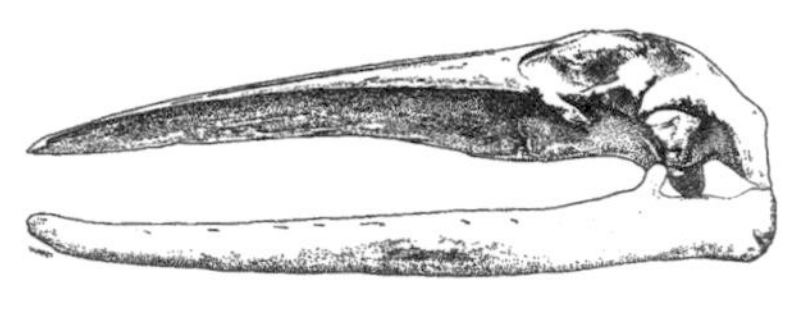
Lateral View

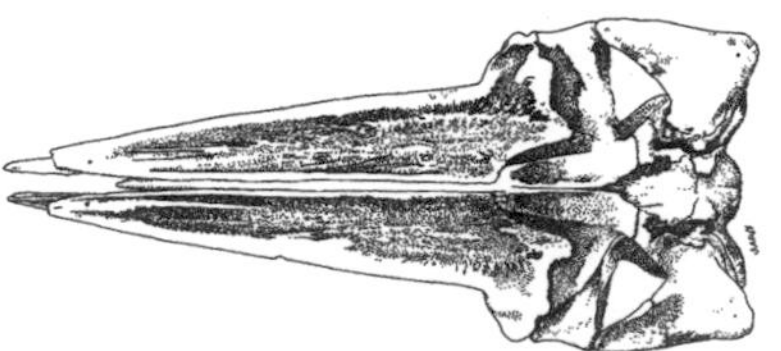
Ventral View

b. Nasals relatively large—length more than 1/2 length of nasofrontal process; vomer not expanded at its posterior end

Sei or Bryde's whale, Go to 10

10a. Anterior edge of nasals concave or straight; rostrum relatively flat in lateral view; from below, palatines do not extend far back and basicranial part of skull exposed behind palatines much longer than broad

Bryde's whale *Balaenoptera edeni*

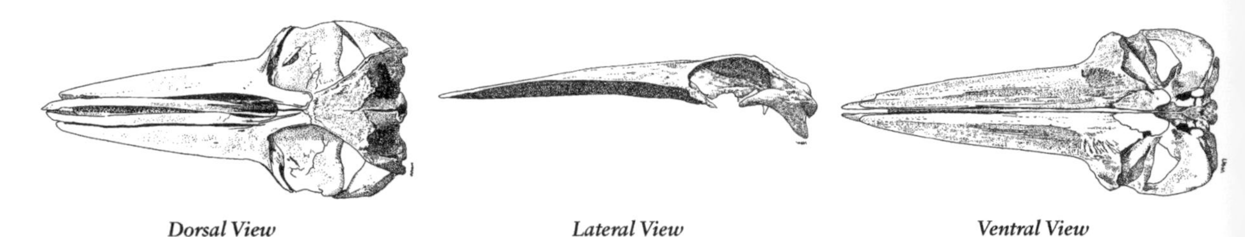

Dorsal View *Lateral View* *Ventral View*

b. Anterior edge of nasals convex; rostrum tip may be downturned in lateral view; from below, basicranial part of skull exposed behind palatines is squarish to broader than long

Sei whale *Balaenoptera borealis*

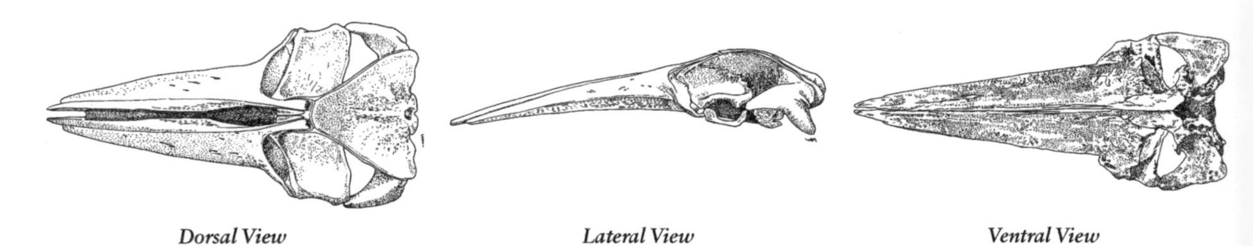

Dorsal View *Lateral View* *Ventral View*

11a. Anterior cranial region basin-like, or with elevated maxillary ridges on vertex; functional teeth generally restricted to lower jaw; lack of distinct ventral cranial hiatus (opening for cranial nerves VII to XI) around the periotics

Physeteroidea or Ziphioidea, Go to 12

b. Anterior cranial region not basin-like, nor with elevated maxillary ridges (except in some river dolphins); teeth in both upper and lower jaws (except in *Monodon* and *Grampus*); cranial hiatus present

Delphinoidea or Platanistoidea, Go to 20

12a. Nares extremely asymmetrical (left naris at least twice as large as the right); rostrum much wider than deep; one or both nasal bones lacking; no strongly elevated vertex; lateral furrow of tympanic bullae absent

Physeteroidea, Go to 13

b. Nares similar in size; rostrum nearly as deep as wide; two nasal bones present; vertex (including nasals, as well as portions of maxillae and premaxillae) strongly elevated; lateral furrow of tympanic bullae present

Ziphiidae, Go to 15

13a. Rostrum long (>60% of CBL); apex of rostrum composed of only premaxillae; zygomatic arches complete; >17 pairs of teeth; mandibular symphysis long (>30% of mandibular length); jugal present; one nasal bone present

Sperm whale *Physeter macrocephalus*

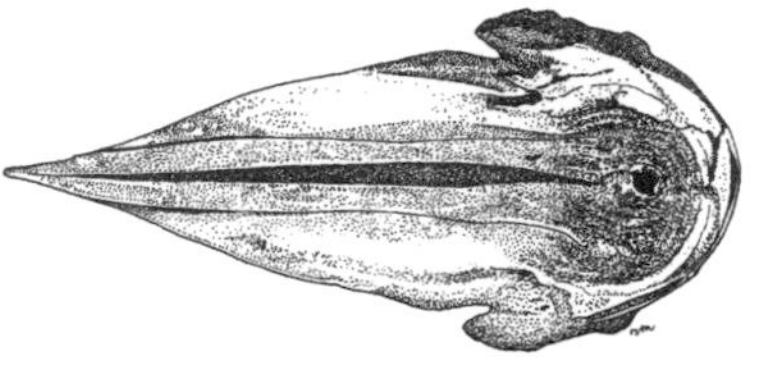

Dorsal View

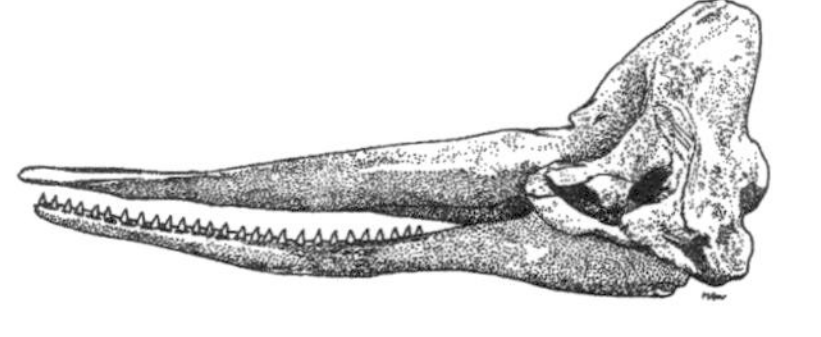

Lateral View

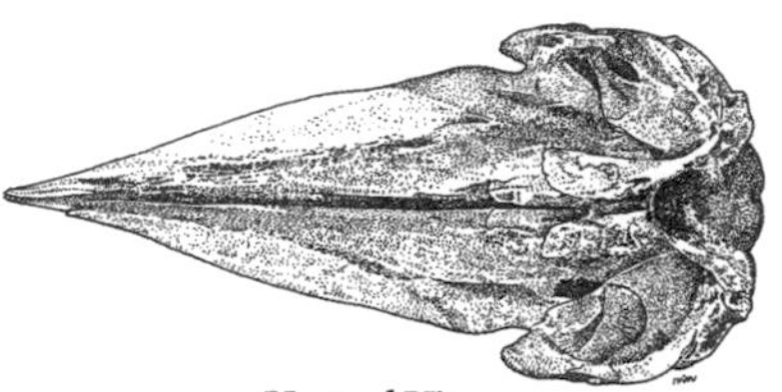

Ventral View

b. Rostrum short (<60% of CBL); apex of rostrum composed of maxillae, premaxillae, and vomer; zygomatic arches incomplete; <17 pairs of teeth; mandibular symphysis short (<30% of mandibular length); jugal absent; nasal bones absent

Kogiidae, Go to 14

14a. Adult skull relatively large (CBL >35 cm); rostrum relatively long (generally >35% CBL); typically 12–16 pairs of teeth (sometimes 10 or 11) only in lower jaw; teeth curved (but not strongly hooked)

Pygmy sperm whale *Kogia breviceps*

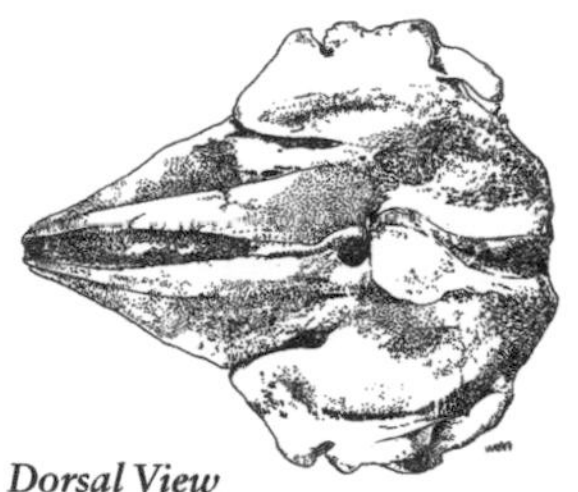

Dorsal View

Lateral View

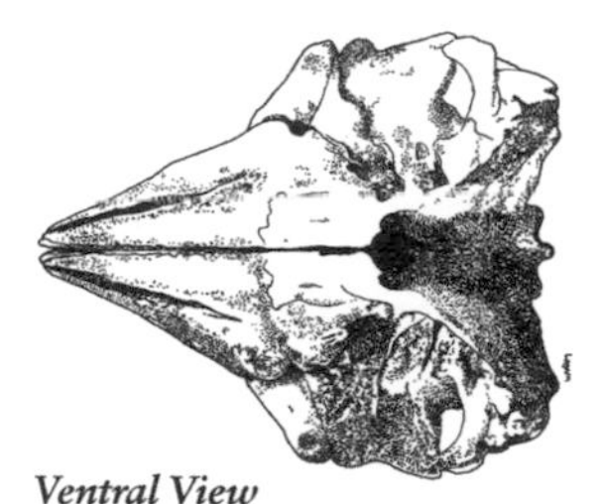

Ventral View

b. Adult skull relatively small (CBL <35 cm); rostrum relatively short (generally <35% CBL); typically 8–11 pairs of teeth (sometimes 12 or 13) in lower jaw, and occasionally up to 3 vestigial pairs in upper jaw; teeth strongly hooked

Dwarf sperm whale *Kogia sima*

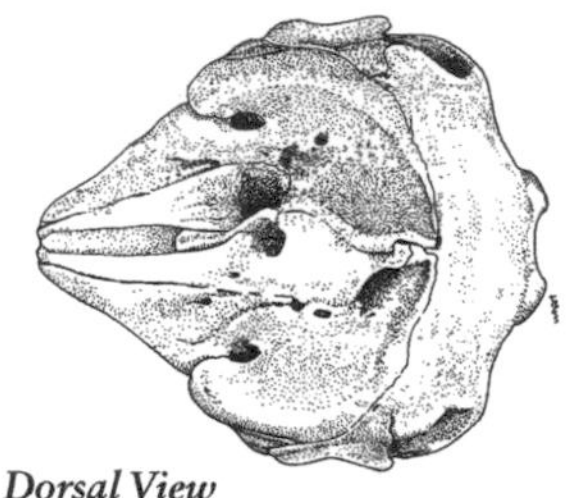

Dorsal View

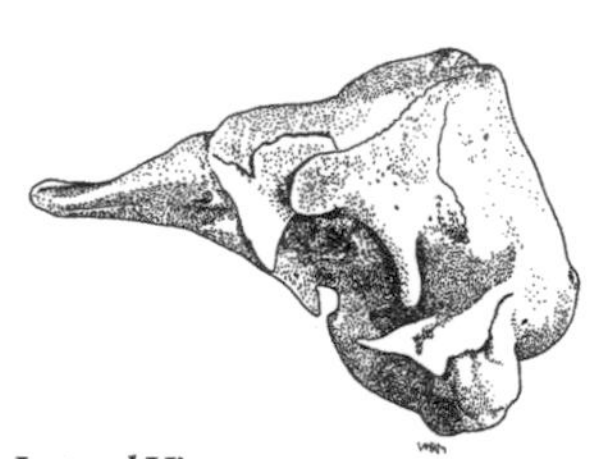

Lateral View

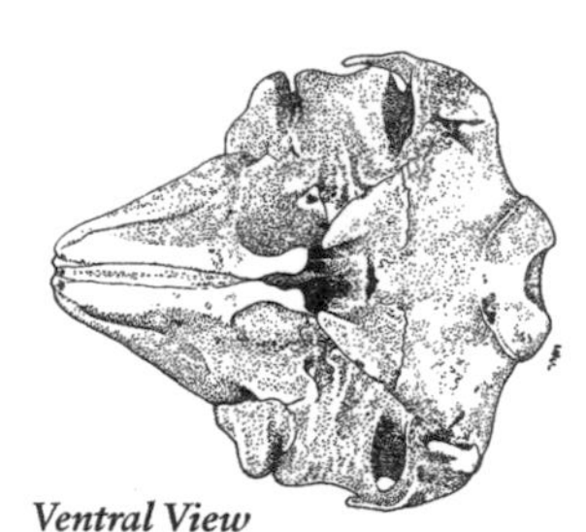

Ventral View

15a. Numerous teeth (17–29 pairs) in both upper and lower jaws; Southern Hemisphere distribution only

Shepherd's beaked whale *Tasmacetus shepherdi*

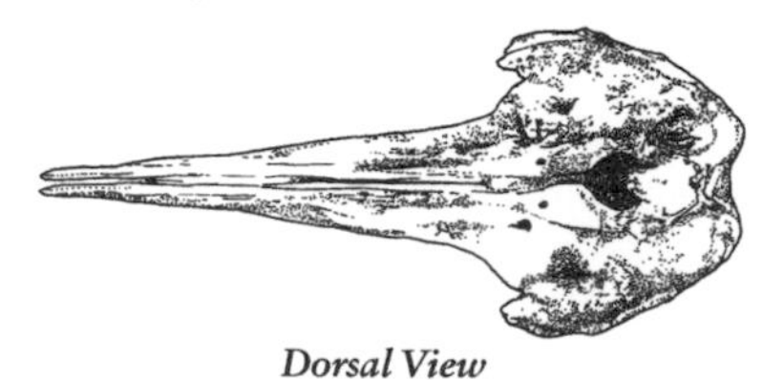

Dorsal View

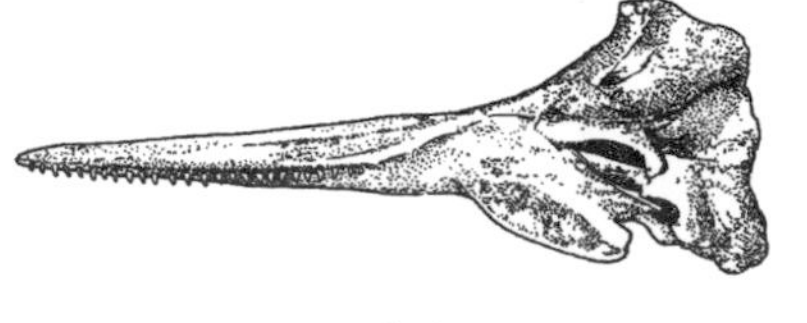

Lateral View

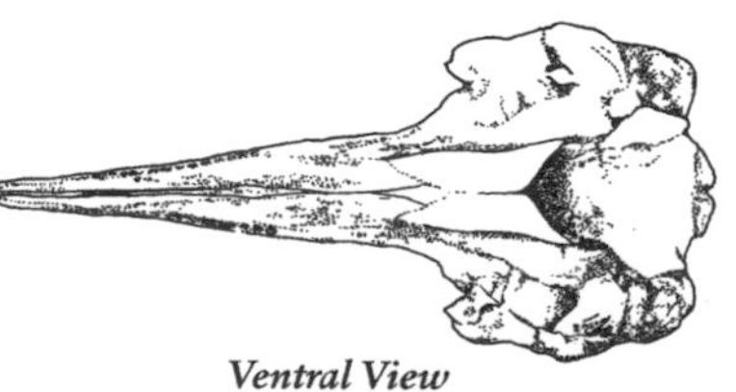

Ventral View

b. No teeth in upper jaw
Berardius, Hyperoodon, Ziphius, Mesoplodon or _Indopacetus_, Go to 16

16a. Two pairs of mandibular teeth
Baird's/Arnoux's beaked whales _Berardius_ spp.

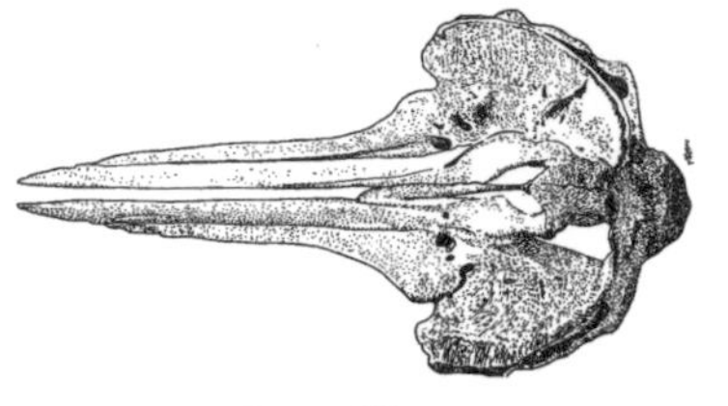
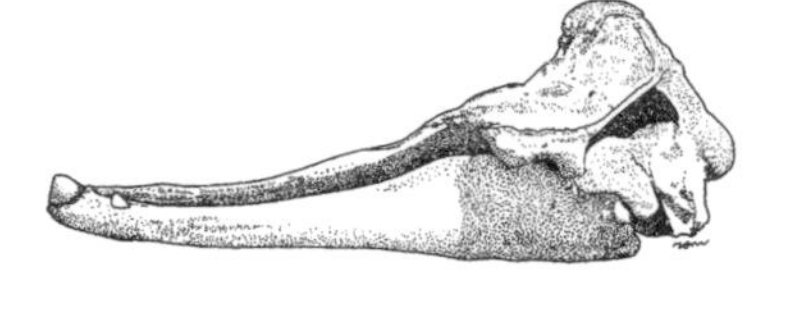
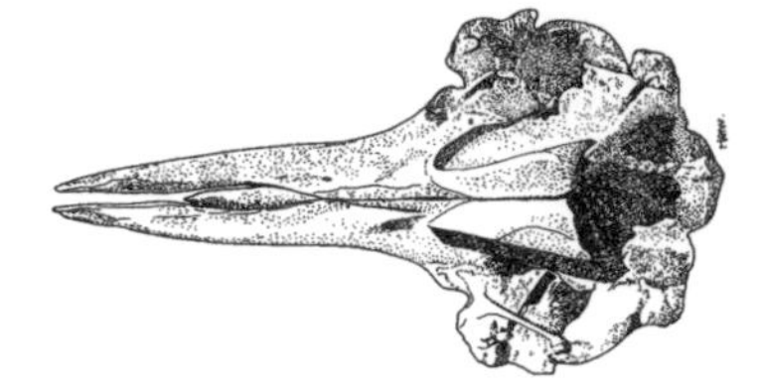

Dorsal View *Lateral View* *Ventral View*

b. No more than one pair of mandibular teeth (occasionally extra rudimentary teeth present)
Mesoplodon, Indopacetus, Ziphius or _Hyperoodon_, Go to 17

17a. Enlarged maxillary crests present (may be relatively small in females)
Bottlenose whales _Hyperoodon_ spp.

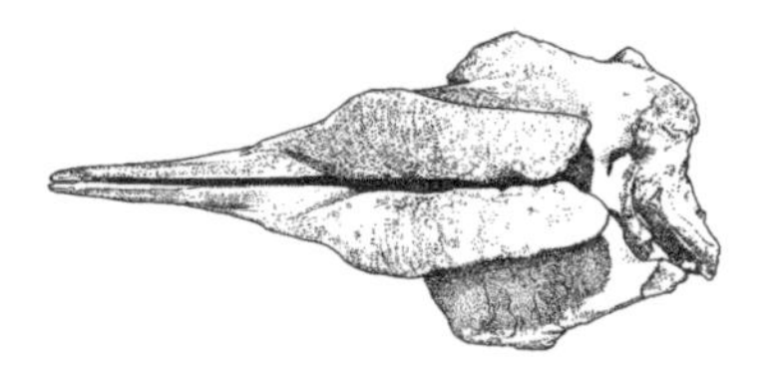
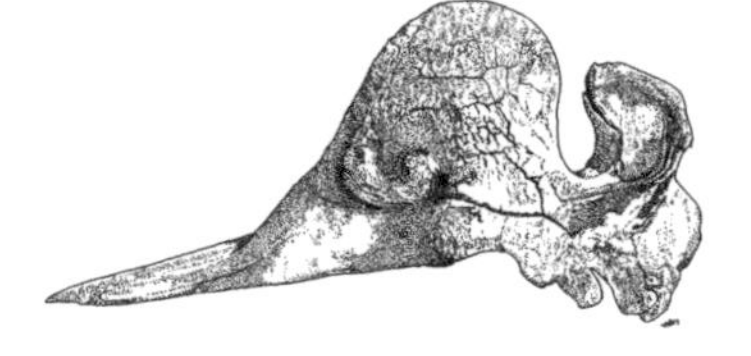
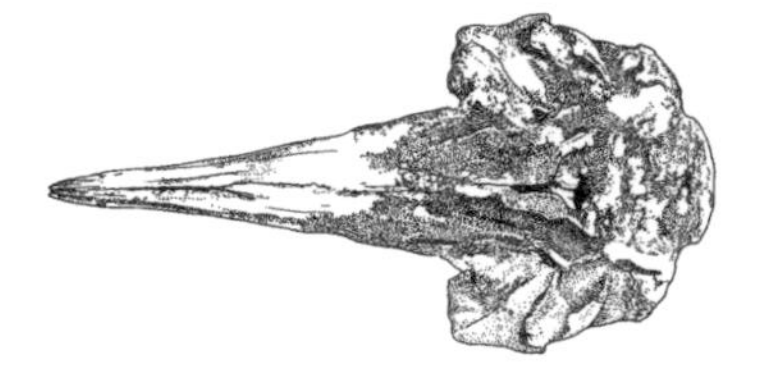

Dorsal View *Lateral View* *Ventral View*

b. No enlarged maxillary crests
Mesoplodon, Indopacetus or _Ziphius_, Go to 18

18a. On vertex, enlarged nasals extend forward past premaxillaries and overhang external bony nares; teeth conical and located at tip of mandibles
Cuvier's beaked whale _Ziphius cavirostris_

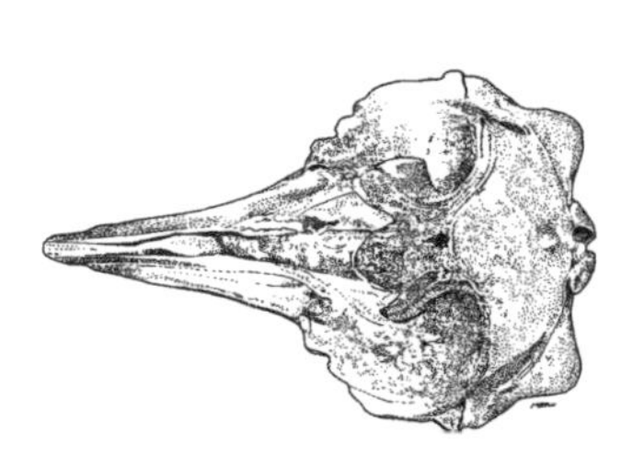
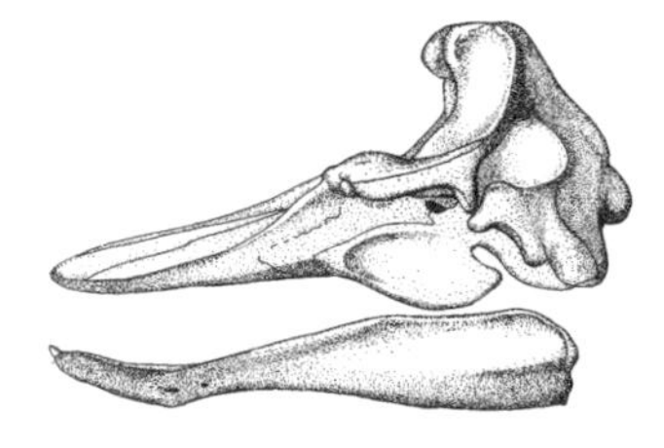
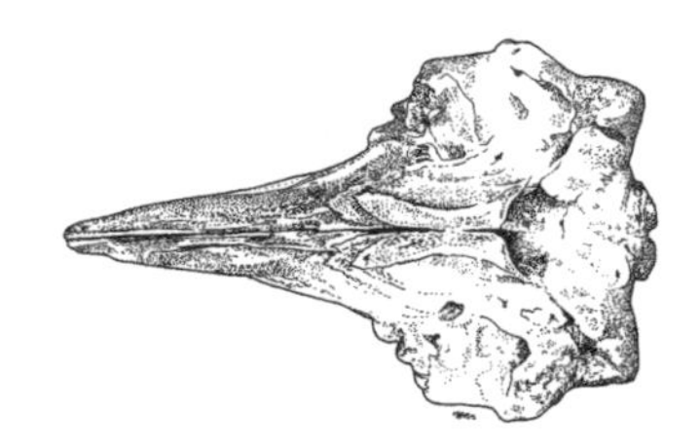

Dorsal View *Lateral View* *Ventral View*

b. On vertex, premaxillaries extend forward past nasals; teeth (if present) flattened (except in *Indopacetus*, *Mesoplodon peruvianus*, and *M. mirus*) and often (but not always) located well back from tip of mandibles
Indopacetus/Mesoplodon spp., Go to 19

19a. In the lateral extension of the maxillary over the orbit, there is a deep groove, about half as long as the orbit; at midlength of the rostrum, there is a lateral swelling caused by the maxillae not coverging for a short distance; teeth are more nearly conical

Longman's beaked whale *Indopacetus pacificus*

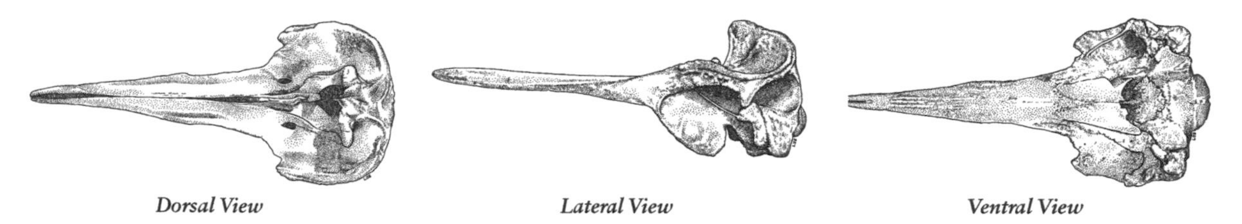

Dorsal View *Lateral View* *Ventral View*

b. No groove in the maxillary over the orbit; no lateral swelling of the rostrum; teeth generally either absent or highly flattened into a single pair of tusks in the mandible

Mesoplodont beaked whales *Mesoplodon* spp. [3]

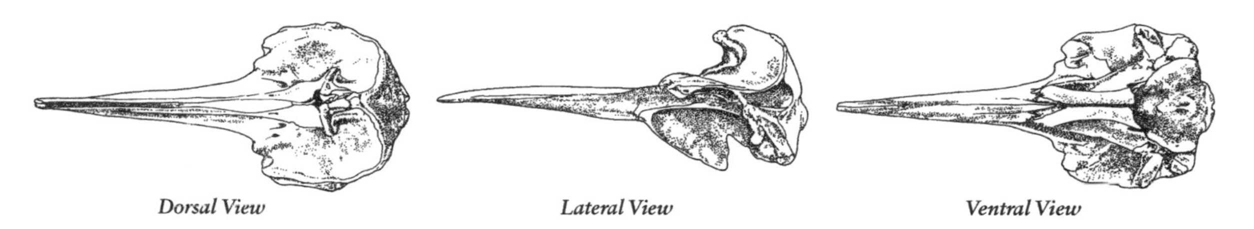

Dorsal View *Lateral View* *Ventral View*

20a. Mandibular symphysis very long (>40% mandibular length); tympanic bullae convex in outer lateral view; distributed only in rivers and coastal waters of Asia and South America

Platanistoidea, Go to 21

b. Mandibular symphysis relatively short (<40% mandibular length); tympanic bullae concave in outer lateral view

Delphinoidea, Go to 24

21a. Large, pneumatized maxillary crests present and overhanging anterior face of cranium; distributed only in the Indian subcontinent

South Asian River dolphin *Platanista gangetica*

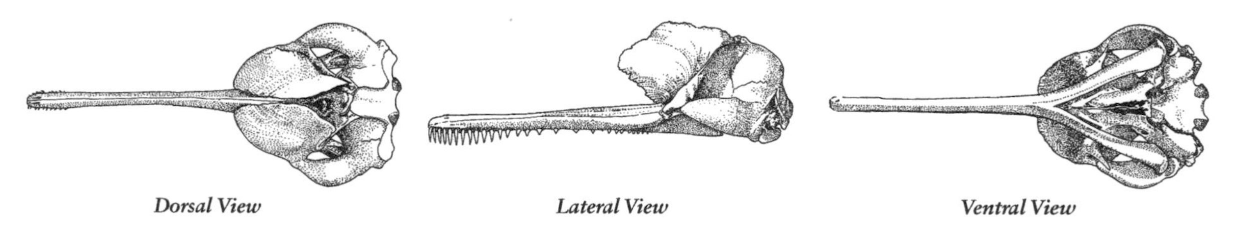

Dorsal View *Lateral View* *Ventral View*

b. Maxillary crests absent or relatively small; distributed only in China or South America

***Inia, Lipotes,* or *Pontoporia,* Go to 22**

[3] The species of the genus *Mesoplodon* are generally poorly-known. Intraspecific variation in skull morphology has not been adequately described for all of them, and it is generally not possible for non-experts to identify whales of this genus to species reliably. Even for experts, detailed measurements of skulls or genetic evidence may be required.

22a. Heterodont dentition; 23–35 teeth in each row, posterior teeth with lateral flanges; zygomatic arches incomplete; distributed only in the Amazon/Orinoco River systems

Amazon river dolphin *Inia geoffrensis*

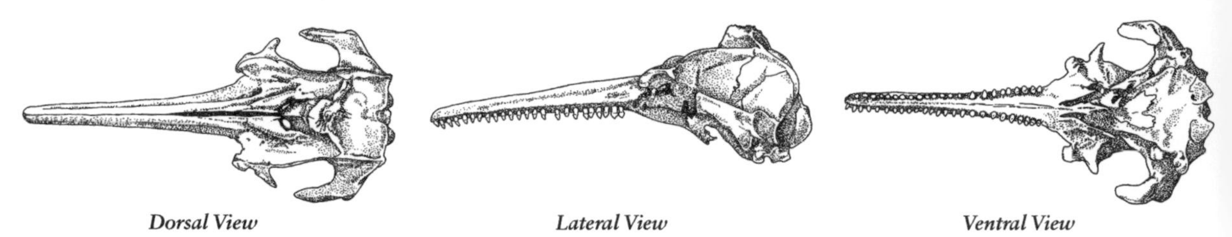

Dorsal View *Lateral View* *Ventral View*

b. Homodont dentition; zygomatic arches complete or with only small gap (<5 mm); distributed only in China or coastal waters of eastern South America

***Lipotes* or *Pontoporia*, Go to 23**

23a. Mandibular symphysis 50% or less of length of mandible; area of bony nares moderately assymetrical; tooth counts <45 in each row; distinct "pinch" at base of rostrum; distributed only in China

Baiji *Lipotes vexillifer*—extinct

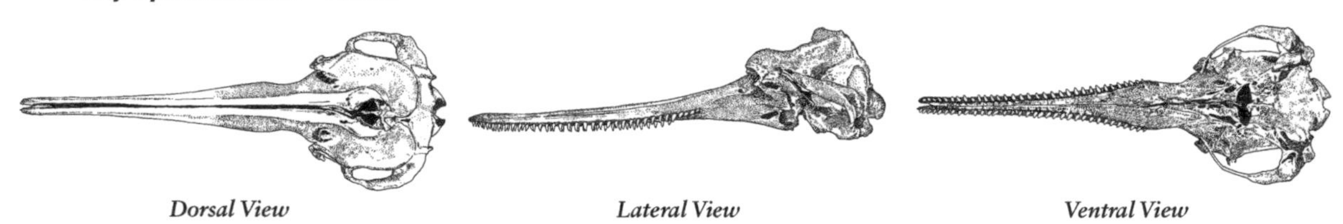

Dorsal View *Lateral View* *Ventral View*

b. Mandibular symphysis >50% length of mandible; area of bony nares relatively symmetrical; tooth counts >45; no "pinch" at base of rostrum; distributed only in eastern South America

Franciscana *Pontoporia blainvillei*

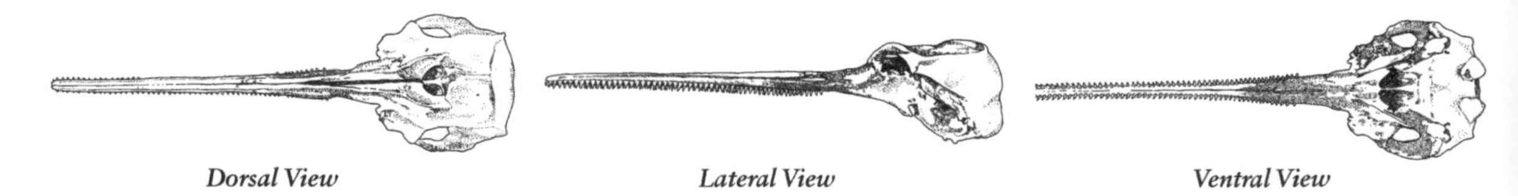

Dorsal View *Lateral View* *Ventral View*

24a. Teeth spade-shaped or peglike; rounded bony bosses on premaxillae anterior to nares; premaxillae do not contact nasals

Phocoenidae, Go to 25

b. Teeth generally conical (except in *Orcaella*); no bosses anterior to nares (except in *Cephalorhynchus*, which have sharpened ridges); right prexaxilla in contact with right nasal

Other Delphinoidea, Go to 31

25a. Rostrum short and wide (length/width ratio <1.28) and rounded at the tip; antorbital notches present and relatively deep (>2 mm); premaxillae level with maxillae in distal quarter of rostrum; distributed only in the Indo-Pacific

Finless porpoise *Neophocaena sp.*, Go to 26

b. Rostrum relatively long and narrow (length/width ratio >1.25) and relatively pointed at the tip; antorbital notches absent or very shallow (<2 mm); premaxillae elevated above maxillae (visible in lateral view)

***Phocoenoides* or *Phocoena*, Go to 26**

26a. Skull relatively small (adult CBL 181–245 mm); rostrum relatively wide and short (length ranges from 62–92 mm); tooth counts 15–22 in upper rows and 16–22 in lower rows; distribution limited to Taiwan Strait and south

Indo-Pacific finless porpoise *Neophocaena phocaenoides*

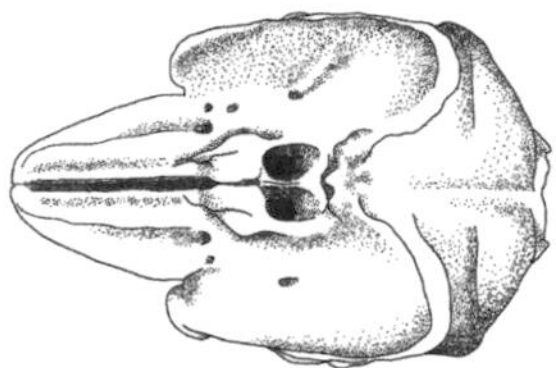

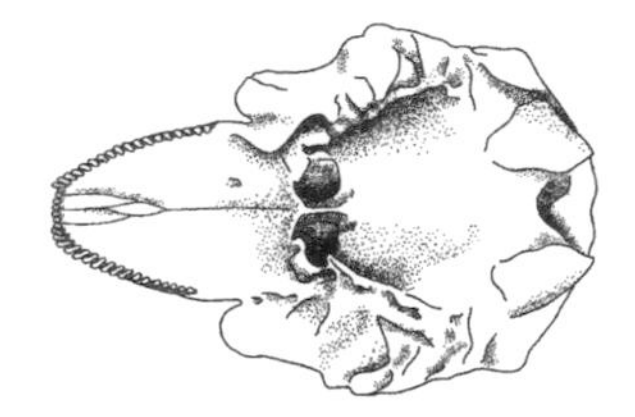

Dorsal View *Lateral View* *Ventral View*

b. Skull may be relatively large (adult CBL 210–295 mm); rostrum relatively narrow and long (length ranges from 77–97 mm); tooth counts 15–21 in upper rows and 15–20 in lower rows; distribution limited to Taiwan Strait and north

Narrow-ridged finless porpoise *Neophocaena asiaeorientalis*

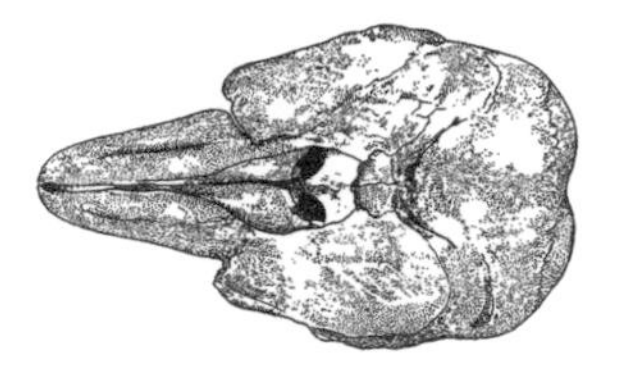
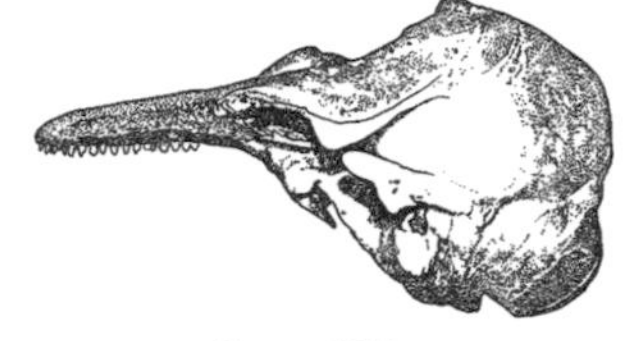
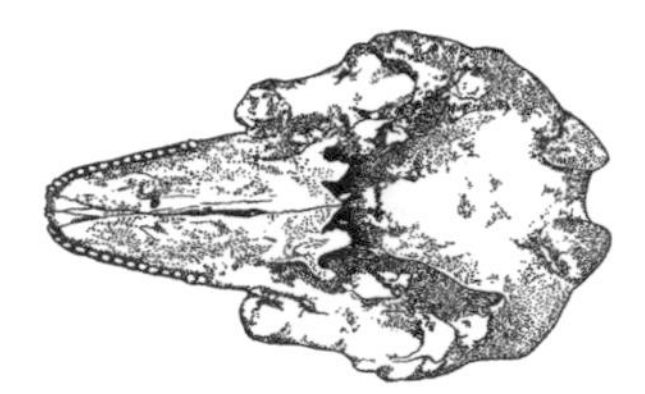

Dorsal View *Lateral View* *Ventral View*

27a. Face of cranium high and nearly vertical, with strong development of supraoccipital crest; frontal not visible where it meets supraoccipital (in dorsal view)

***Phocoenoides dalli* or *Phocoena dioptrica*, Go to 28**

b. Face of cranium low and strongly diagonal, with weak development of supraoccipital crest; frontal on at least one side visible where it meets supraoccipital (in dorsal view)

***Phocoena* spp., Go to 29**

28a. Teeth very small and peglike; tooth counts >23 per row; premaxillary bosses with vertical or near-vertical lateral margins; distributed only in the North Pacirfic

Dall's porpoise *Phocoenoides dalli*

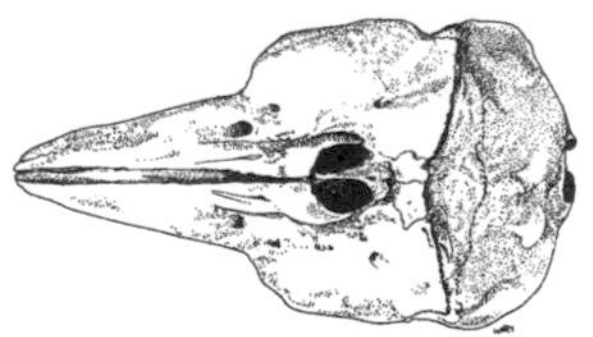

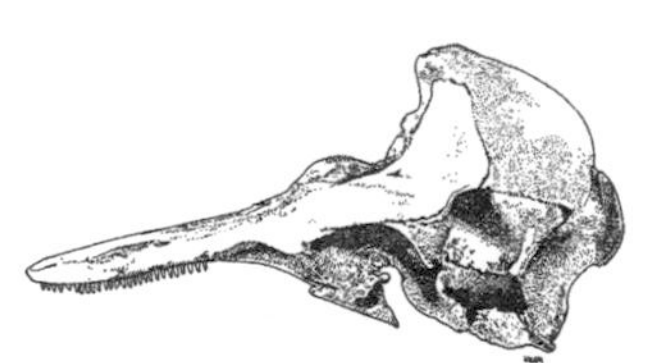
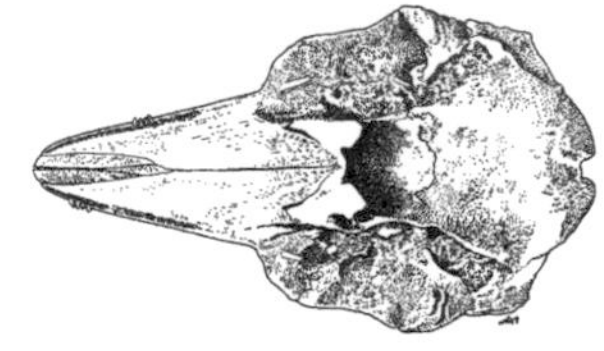

Dorsal View *Lateral View* *Ventral View*

b. Teeth relatively large; tooth counts generally <23 per row (sometimes up to 26); premaxillary bosses with strong vertical overhang; distributed only in the Southern Hemisphere

Spectacled porpoise *Phocoena dioptrica*

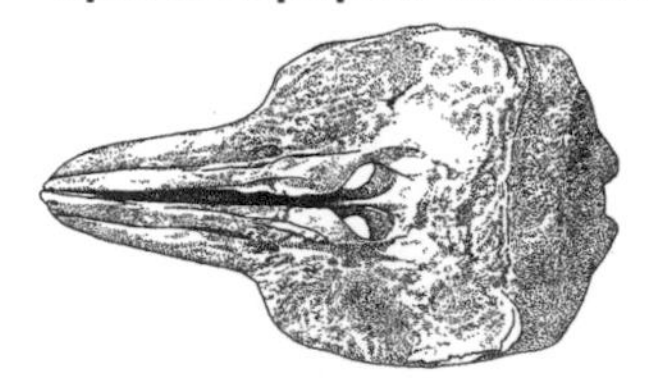

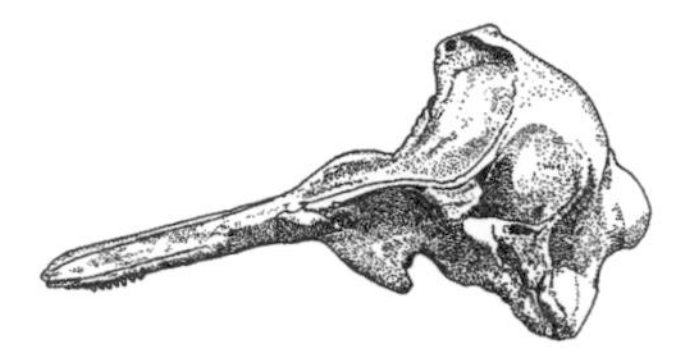
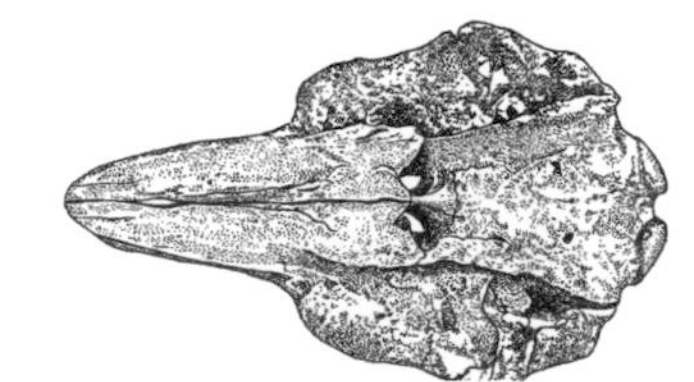

Dorsal View *Lateral View* *Ventral View*

29a. Adult skull small (CBL <245 mm); rostrum short (length/width ratio <1.4); posterior margin of palate usually U-shaped; distributed only in the northern Gulf of California

Vaquita *Phocoena sinus*

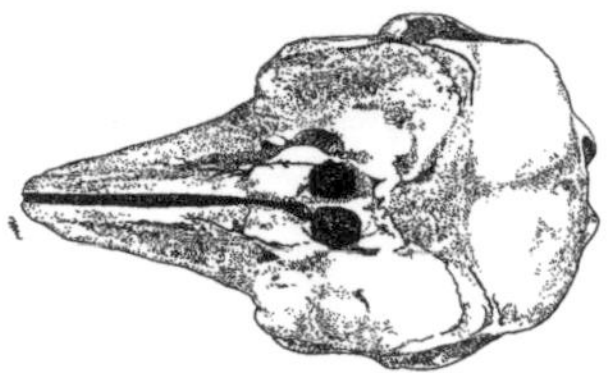

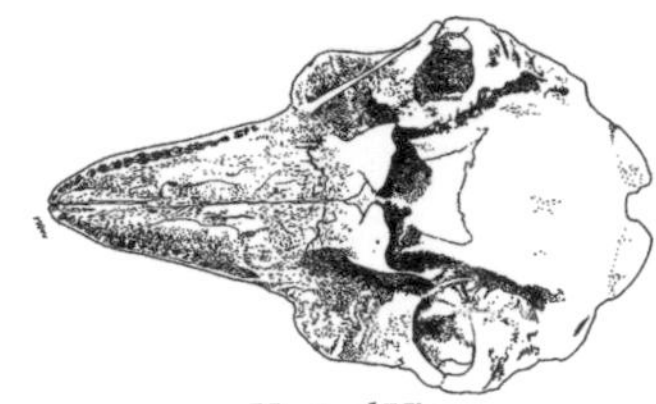

Dorsal View *Lateral View* *Ventral View*

b. Adult skull relatively large (CBL >245 mm); rostrum long (length/width ratio >1.4); posterior margin of palate usually W-shaped

Other species, Go to 30

30a. Rostrum relatively short (length/width ratio <1.8); distributed only in the coastal waters of South America

Burmeister's porpoise *Phocoena spinipinnis*

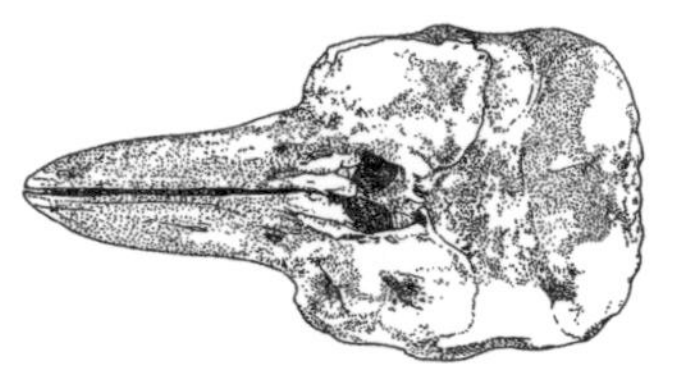
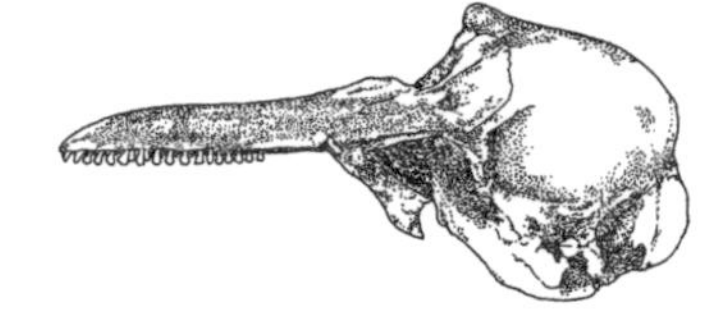
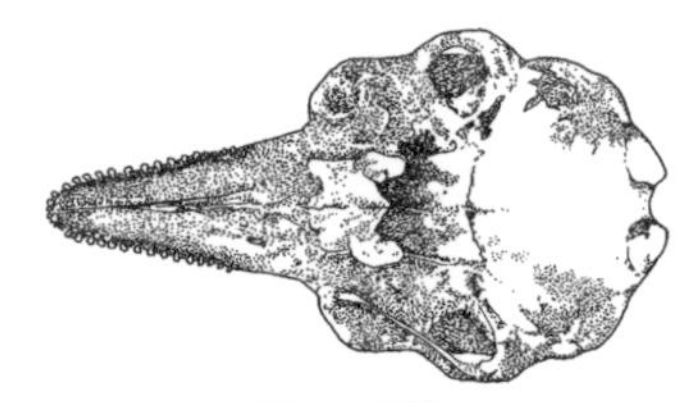

Dorsal View *Lateral View* *Ventral View*

b. Rostrum relatively long (length/width ratio >1.8); distributed only in the Northern Hemisphere

Harbor porpoise *Phocoena phocoena*

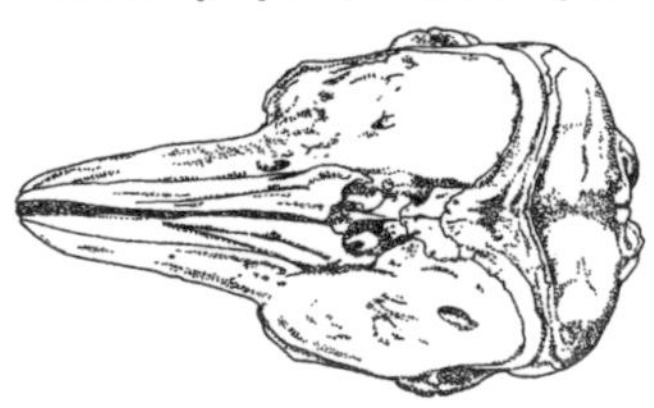
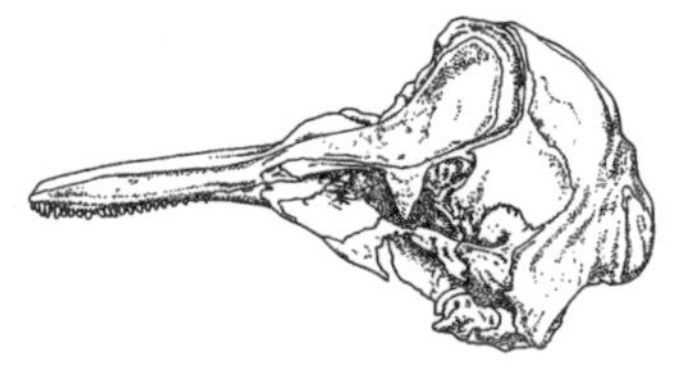
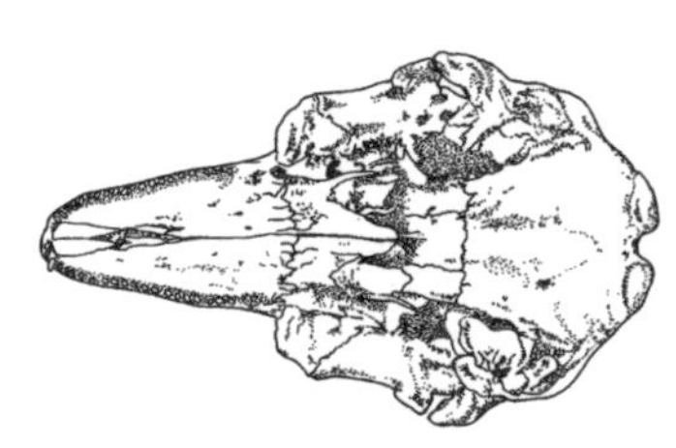

Dorsal View *Lateral View* *Ventral View*

31a. In profile, facial plane very flat or convex, with little or no rise in area of nares; spiracular plate slightly rugose; distributed only in high latitudes of the Northern Hemisphere

Monodontidae, Go to 32

b. In profile, facial plane concave and cranium rising dramatically in area of nares (except in *Grampus*); spiracular plate relatively smooth

Delphinidae, Go to 33

32a. Facial profile flat; no teeth in lower jaw; upper jaw teeth number no more than 2, including a long forward-directed and spiraled tusk in some animals

Narwhal *Monodon monoceros*

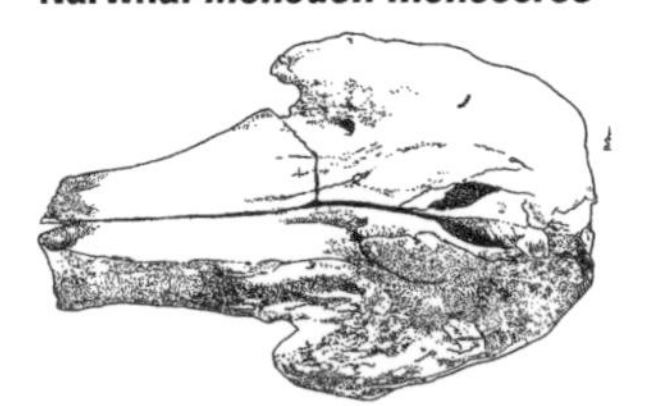
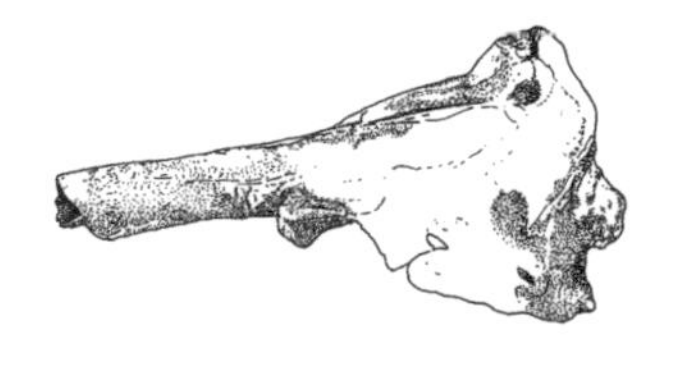
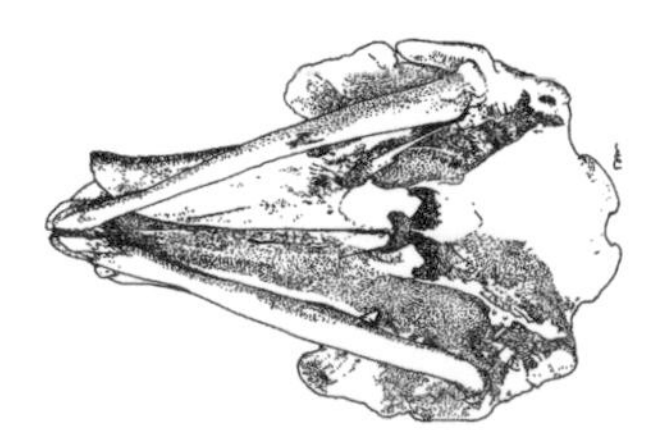

Dorsal View *Lateral View* *Ventral View*

b. Facial profile slightly convex, with downturned tip of rostum; 8–9 teeth present in each row of both upper and lower jaws

Beluga whale *Delphinapterus leucas*

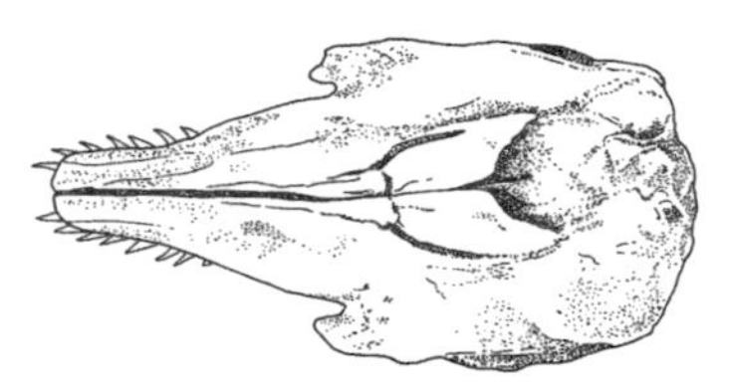
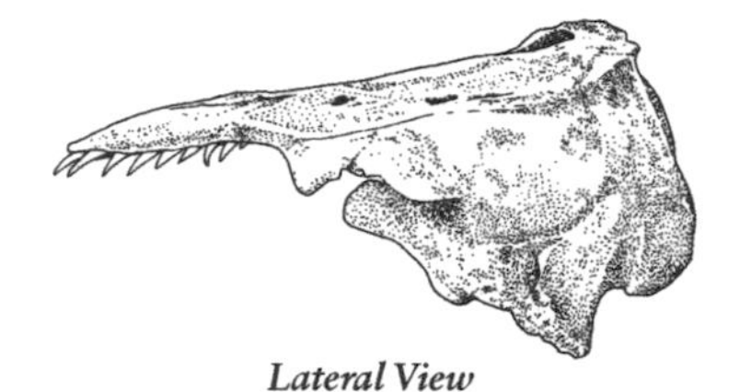

Dorsal View *Lateral View* *Ventral View*

33a. Rostrum very short (<45% of CBL); teeth peg-like, with slightly expanded crowns; tympano-periotic bones attached to cranium by a triangular ventral pad on zygomatic arch; distributed only in the coastal waters of the Indo-Pacific

Irrawaddy and snubfin dolphins *Orcaella* spp.

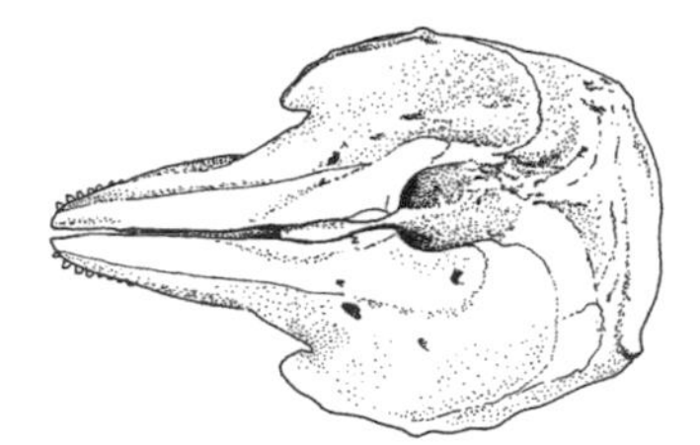
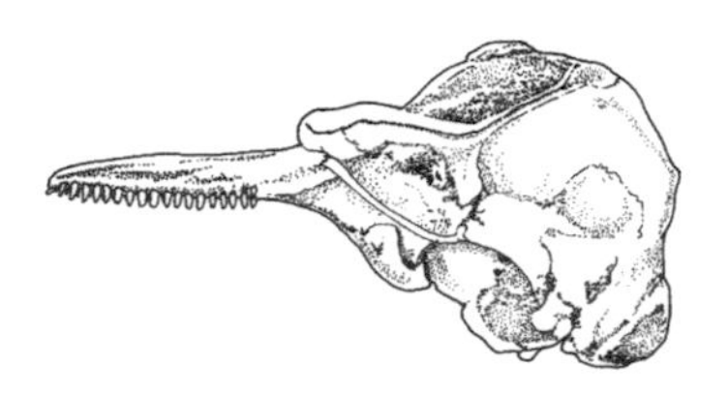
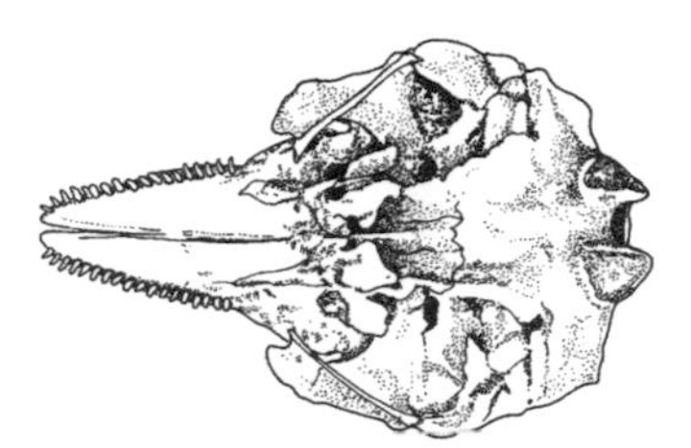

Dorsal View *Lateral View* *Ventral View*

b. Rostrum may be relatively long (>45% of CBL); teeth conical (generally pointed), without expanded crowns; tympanoperiotic bones attached to cranium by a cavity formed by squamosal, exoccipital and basioccipital bones

Other Delphinidae, Go to 34

34a. Mandibular symphysis long (generally >25% mandibular length); margins of rostrum somewhat concave throughout most of their length, when viewed from above

Other species, Go to 35

b. Mandibular symphysis short (<25% mandibular length); margins of rostrum convex, or concave only along middle of their length

Globicephalinae or Delphininae, Go to 39

35a. Teeth with shallow vertical wrinkles; 19–28 teeth in each row; constriction at base of rostrum; orbits very large; prominent cylindrical ridge at 45° angle on ventral aspect of frontal; pterygoids only slightly separated

Rough-toothed dolphin *Steno bredanensis*

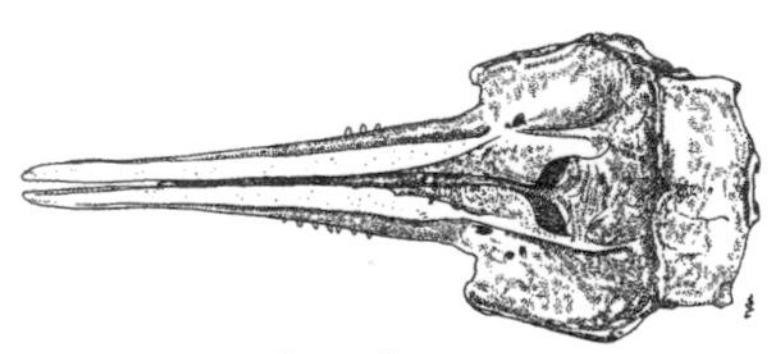
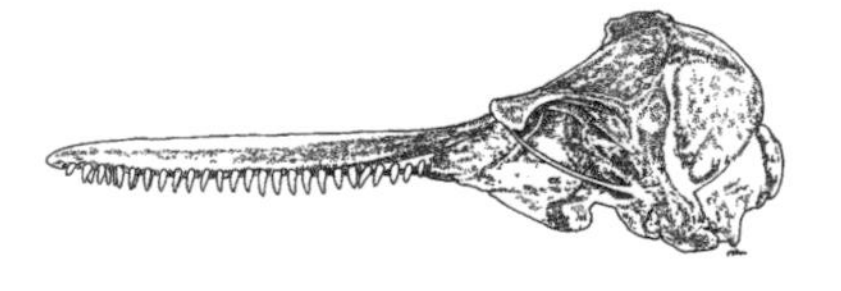
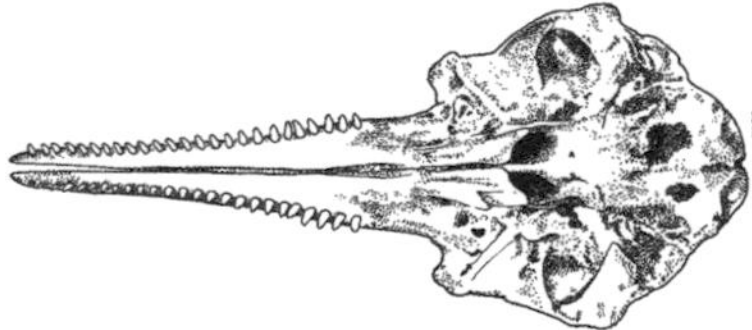

Dorsal View *Lateral View* *Ventral View*

b. Teeth smooth (no wrinkles); 26–39 teeth in each row; orbits relatively small; no prominent ridge on ventral aspect of frontal; pterygoids widely separated; distributed only in the Indo-Pacific, West African, and South American waters

***Sousa/Sotalia*, Go to 36**

36a. Skull relatively small (adult CBL <400 mm); rostral margins relatively straight; distributed only in South America

Tucuxi or Guiana dolphin *Sotalia fluviatilis* or *S. guianensis*

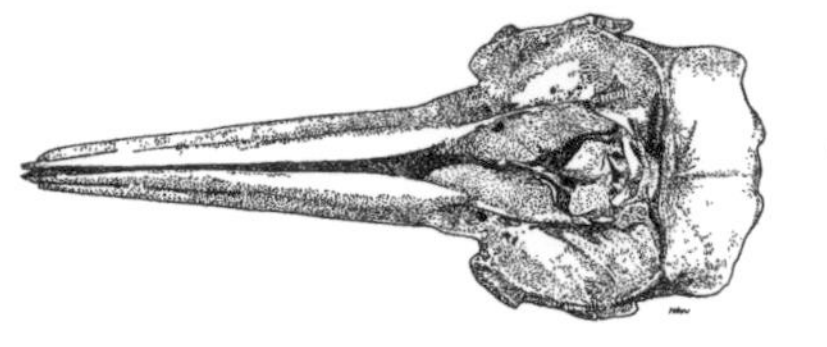

Dorsal View

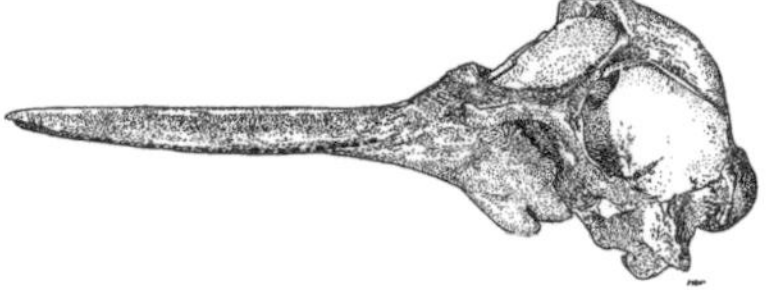

Lateral View

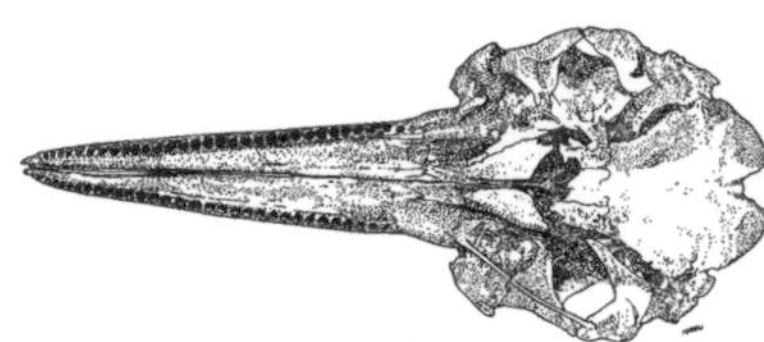

Ventral View

b. Skull relatively large (adult CBL >405 mm); rostral margins strongly concave distributed only in the Indo-Pacific and West Africa

***Sousa*, Go to 37**

37a. Low tooth counts (27–32 teeth per row); rostrum relatively short (<309 mm) and wide; distributed only in Atlantic Ocean

Atlantic humpback dolphin *Sousa teuszii*

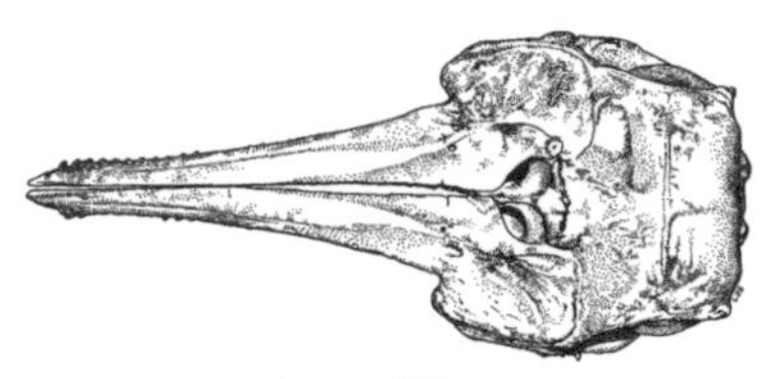

Dorsal View

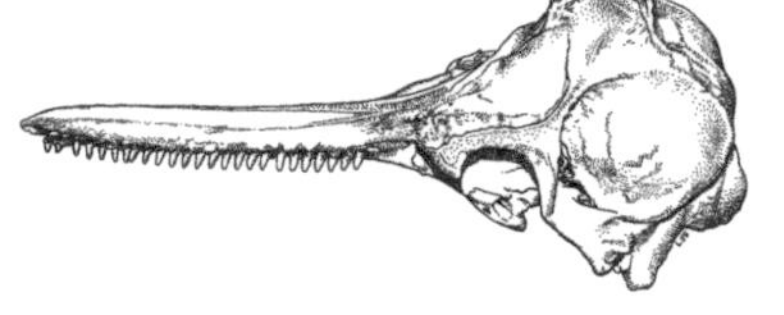

Lateral View

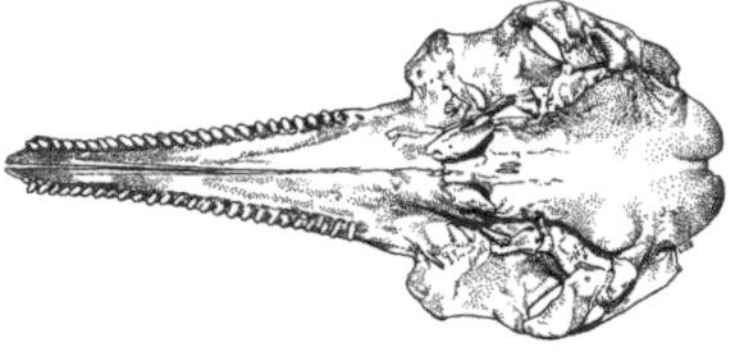

Ventral View

b. Relatively high tooth counts (31–39 teeth per row); rostrum relatively long (up to 377 mm) and slender; distributed only in the Indo-Pacific region

Other humpback dolphins (*Sousa* spp.), Go to 38

38a. Moderately low tooth counts (31–35 teeth per row); distributed only in Australia and New Guinea

Australian humpback dolphin *Sousa sahulensis*

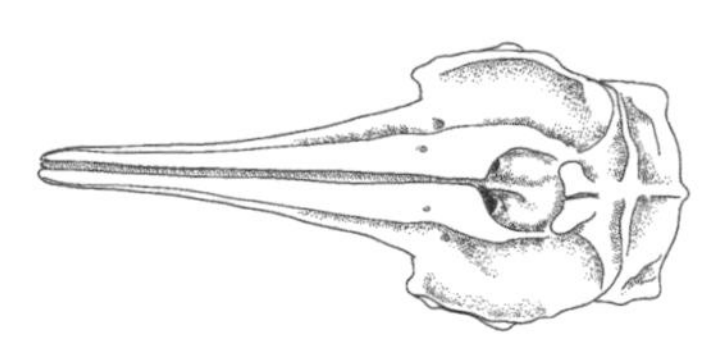

Dorsal View

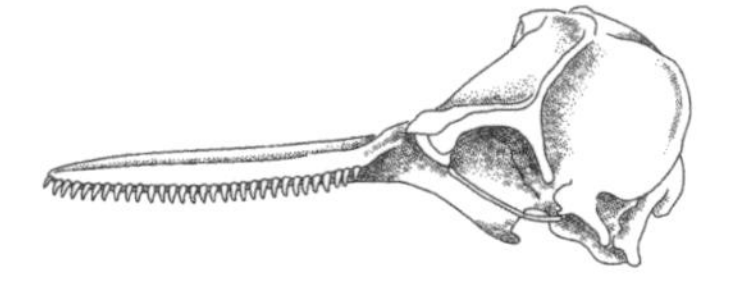

Lateral View

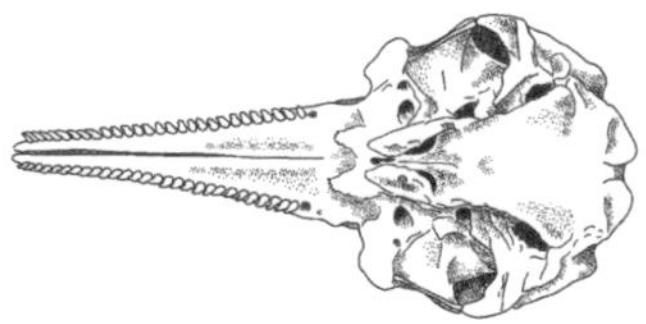

Ventral View

b. Relatively high tooth counts (32–39 teeth per row; distributed in the Indian and western Pacific Oceans (from South Africa to Southeast Asia)

Indian or Indo-Pacific humpback dolphin *Sousa plumbea* or *S. chinensis*

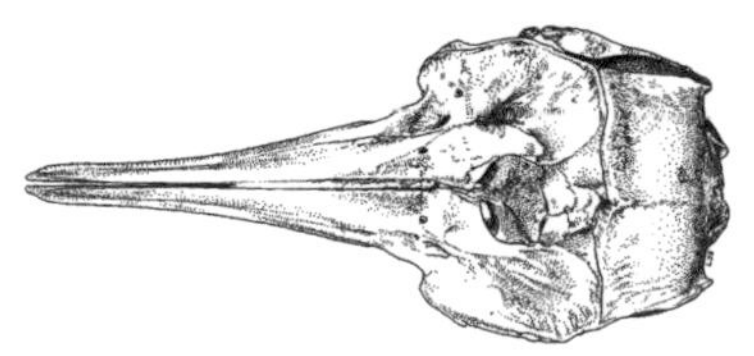

Dorsal View

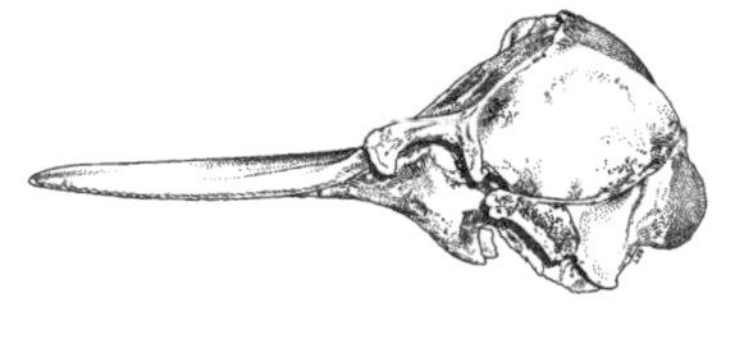

Lateral View

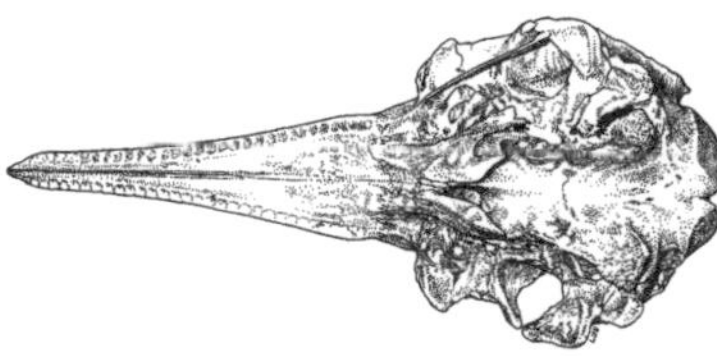

Ventral View

39a. Less than 27 teeth per toothrow

***Globicephalinae*, *Tursiops*, or *Grampus*, Go to 40**

b. Greater than 27 teeth per toothrow
***Lagenodelphis*, *Delphinus*, *Lagenorhynchus*, or *Stenella*, Go to 48**

40a. 2–7 pairs of teeth present near tip of lower jaw only (uncommonly 1–2 pairs in upper jaw); lateral margins of rostrum concave along middle part of their length
Risso's dolphin *Grampus griseus*

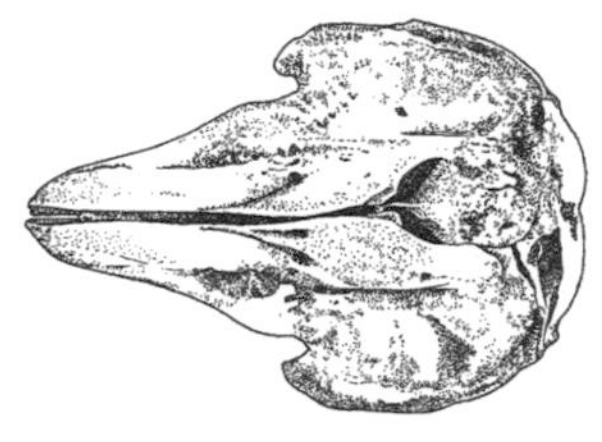
Dorsal View

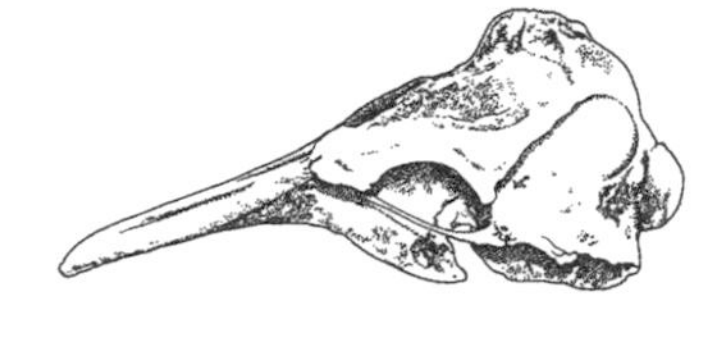
Lateral View

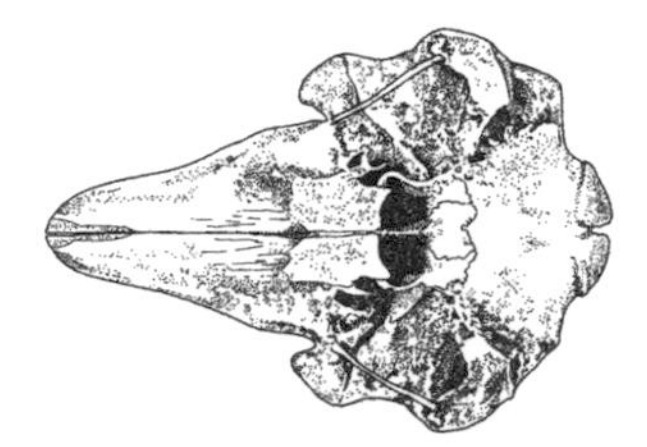
Ventral View

b. At least 7 teeth present in each toothrow of both upper and lower jaws; lateral margins of rostrum generally convex
Globicephalinae or *Tursiops*, Go to 41

41a. Greater than 15 teeth per toothrow
***Tursiops* or *Peponocephala*, Go to 42**

b. Less than 15 teeth per toothrow
***Globicephala*, *Feresa*, *Pseudorca*, or *Orcinus*, Go to 44**

42a. Rostrum relatively wide (length/breadth ratio <2); antorbital notches very deep; teeth absent in posterior 25% of upper jaw; 20–26 teeth per row
Melon-headed whale *Peponocephala electra*

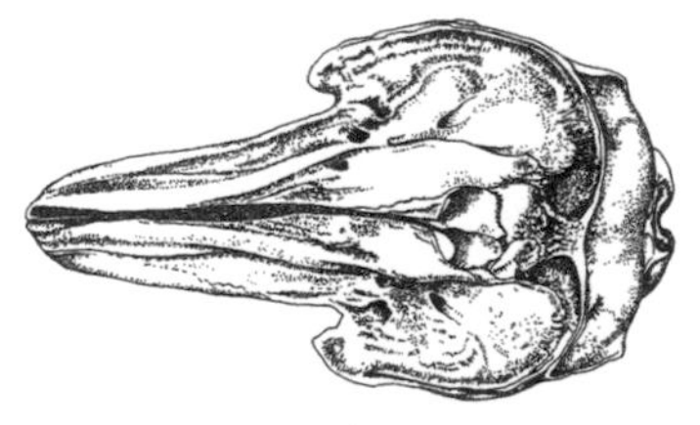
Dorsal View

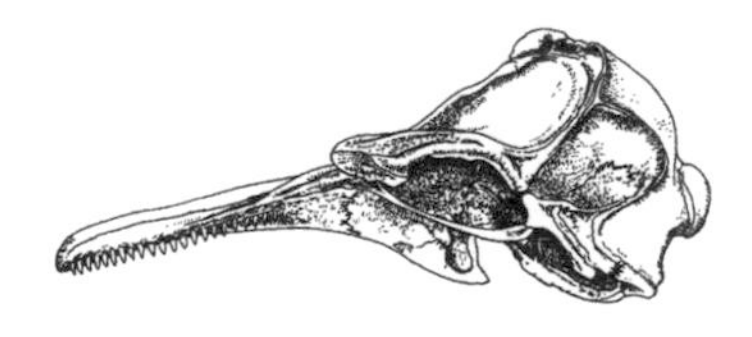
Lateral View

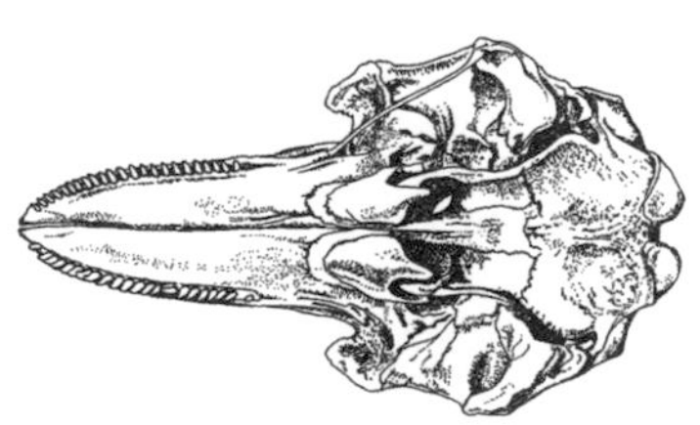
Ventral View

b. Rostrum relatively narrow (length/breadth ratio >2); antorbital notches relatively shallow; teeth present in posterior 25% of upper jaw; 18–26 teeth per row
Bottlenose dolphin *Tursiops sp.*, Go to 43

43a. From lateral view or dorsal view, no obvious premaxillary convexity; premaxillary "pinch"; tip of rostrum to apex of premaxilla convexity divided by length of rostrum <0.606; tooth counts 18–27
Common bottlenose dolphin *Tursiops truncatus*

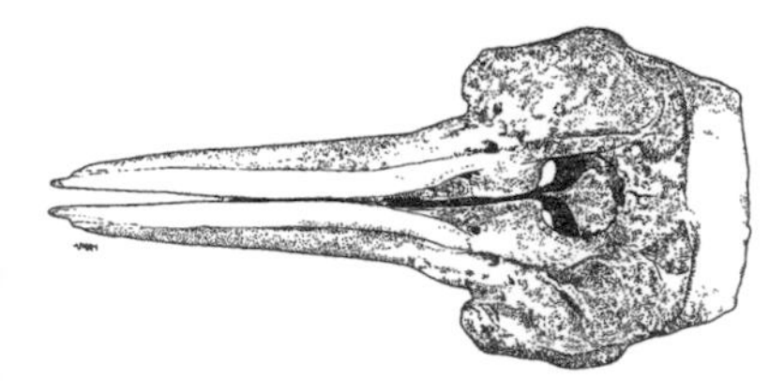
Dorsal View

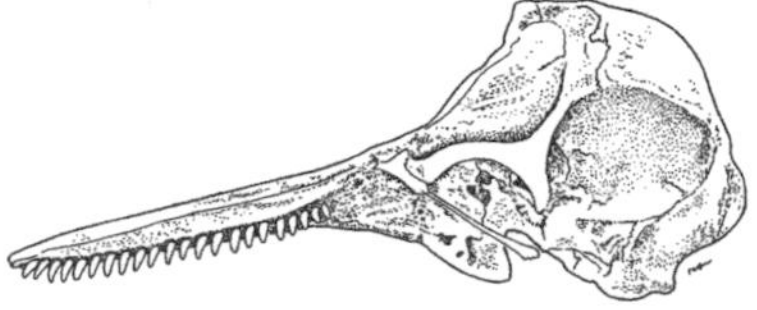
Lateral View

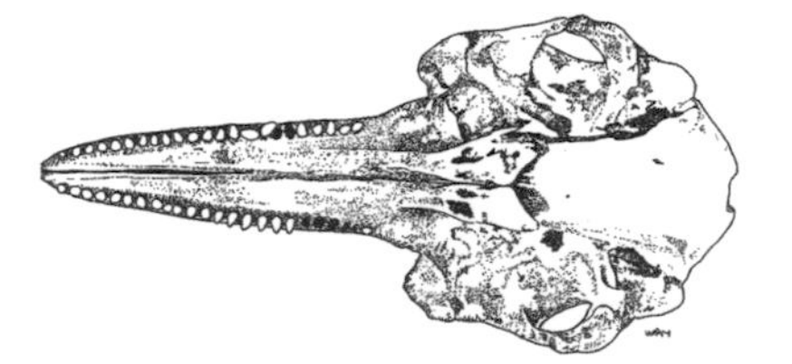
Ventral View

b. From lateral view, premaxillary convexity and from dorsal view, premaxillary "pinch" obvious; tip of rostrum to apex of premaxilla convexity divided by length of rostrum >0.607; tooth counts 21–29; distributed only in the Indo-Pacific

Indo-Pacific bottlenose dolphin *Tursiops aduncus*

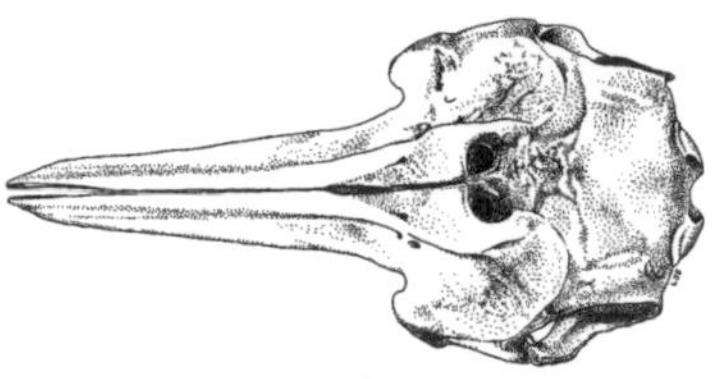

Dorsal View

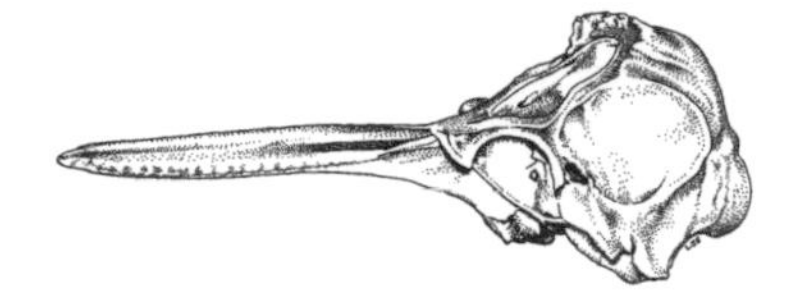

Lateral View

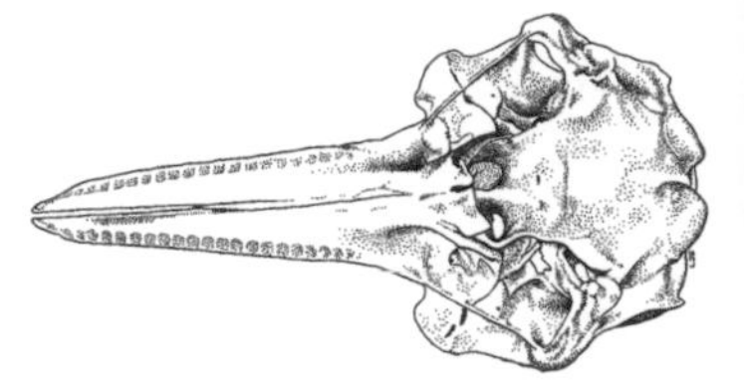

Ventral View

44a. 7–9 teeth present only in anterior half of rostrum; rostrum wide (length/breadth ratio generally <1.3)

***Globicephala*, Go to 45**

b. Teeth present in both anterior and posterior halves of rostrum; rostrum relatively narrow (length/breadth ratio generally >1.3)

***Feresa*, *Pseudorca*, or *Orcinus*, Go to 46**

45a. Rostrum very wide; premaxilla expanded and completely covering maxilla along anterior half of rostrum (or leaving only a very small portion visible in dorsal view)

Short-finned pilot whale *Globicephala macrorhynchus*

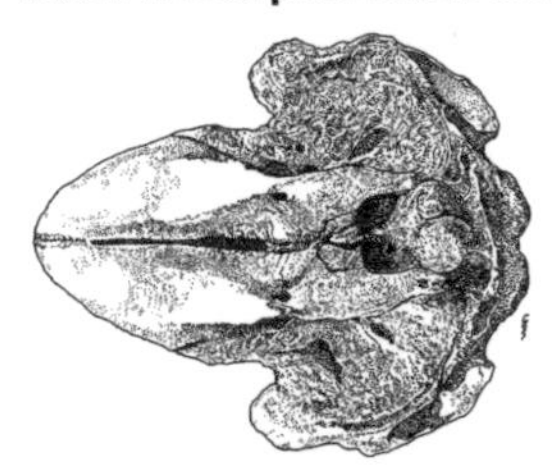

Dorsal View

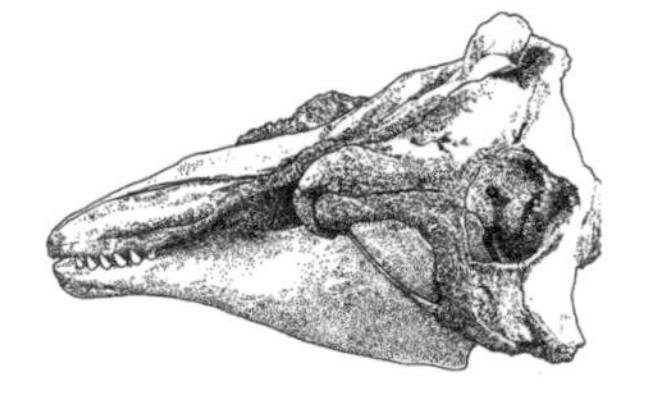

Lateral View

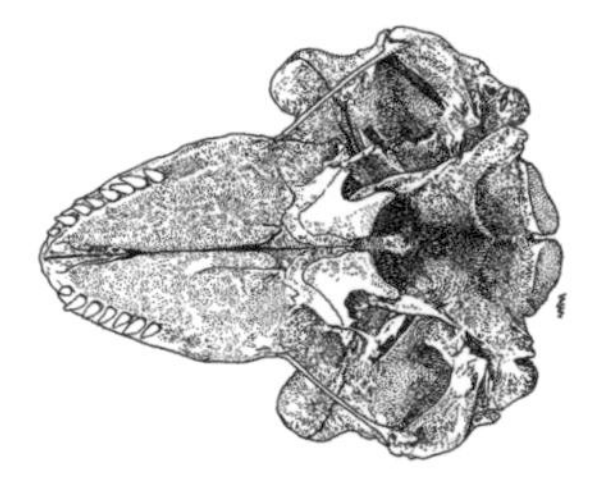

Ventral View

b. Rostrum relatively narrow; premaxilla not completely covering maxilla along anterior half of rostrum (at least a 1 cm portion of maxillae visible in dorsal view)

Long-finned pilot whale *Globicephala melas*

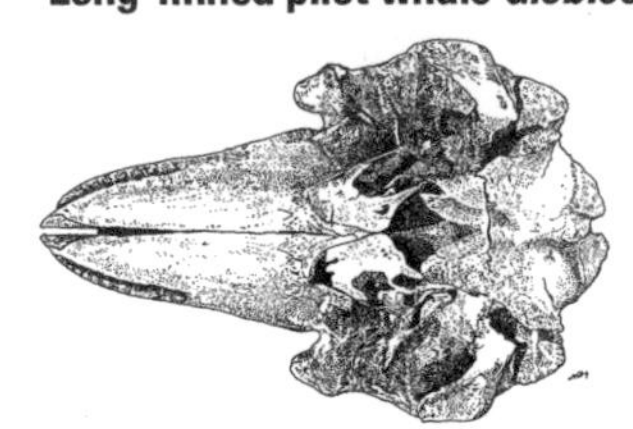

Dorsal View

Lateral View

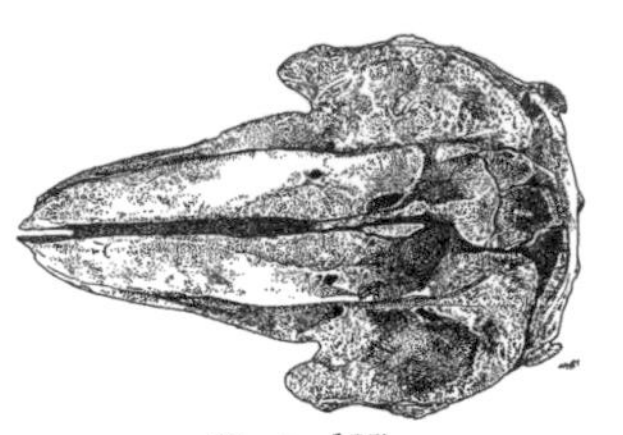

Ventral View

46a. Teeth relatively slender (generally <10 mm in diameter); adult CBL <50 cm; 8–13 teeth in anterior $^{2}/_{3}$ of rostral tooth rows only

Pygmy killer whale *Feresa attenuata*

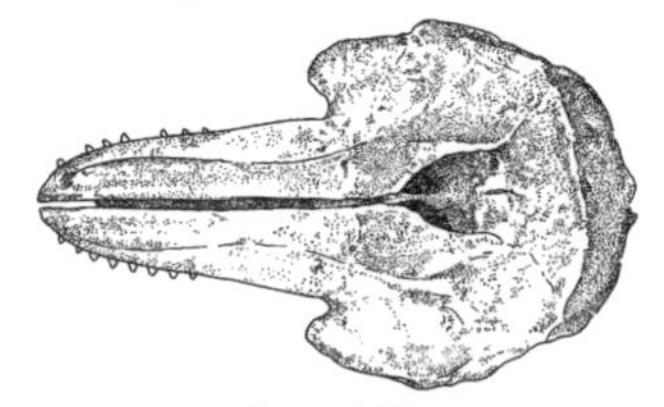

Dorsal View

Lateral View

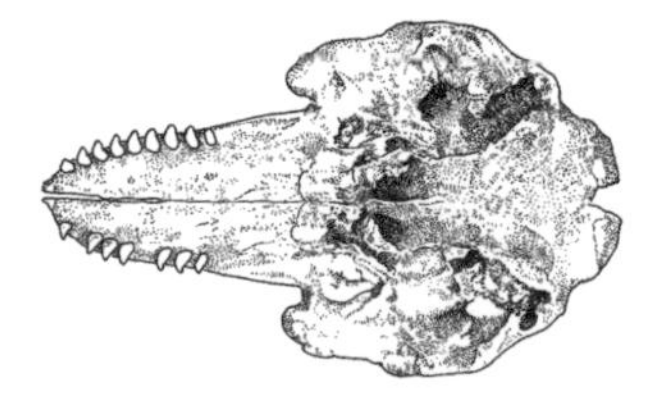

Ventral View

b. Teeth relatively robust (generally >10 mm in diameter); adult CBL >50 cm; teeth present in posterior 1/3 of rostral tooth rows

***Pseudorca* or *Orcinus*, Go to 47**

47a. Teeth round in cross-section (greatest diameter of largest teeth generally <23 mm); adult CBL <78 cm; 7–12 teeth in each row; width across premaxillae >50% of rostral basal width

False killer whale *Pseudorca crassidens*

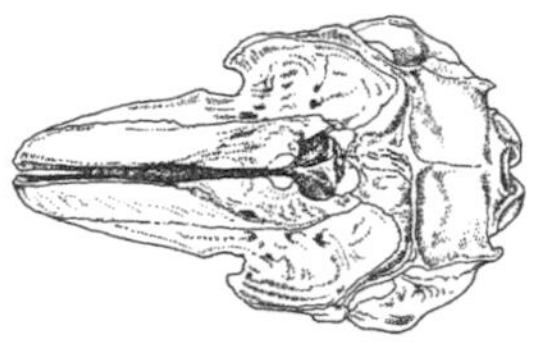

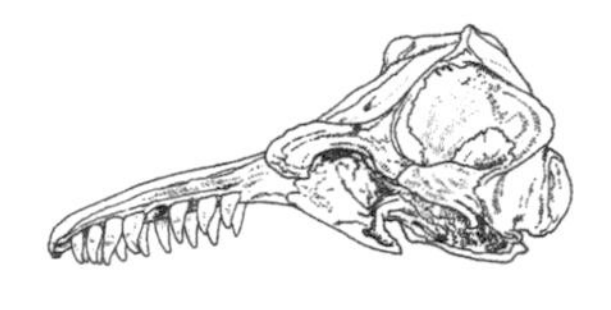

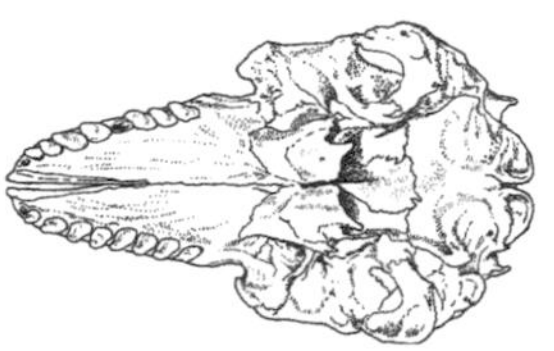

Dorsal View *Lateral View* *Ventral View*

b. Teeth oval in cross-section (greatest diameter of largest teeth generally >23 mm); adult CBL >78 cm; 10–14 teeth in each row; width across premaxillae <50% of rostral basal width

Killer whale *Orcinus orca*

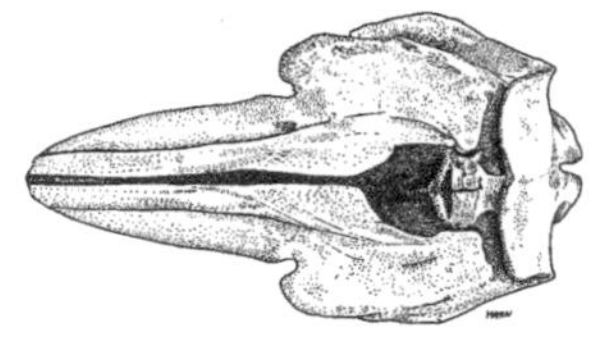

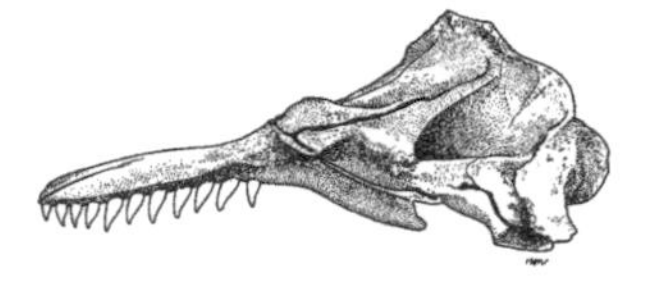

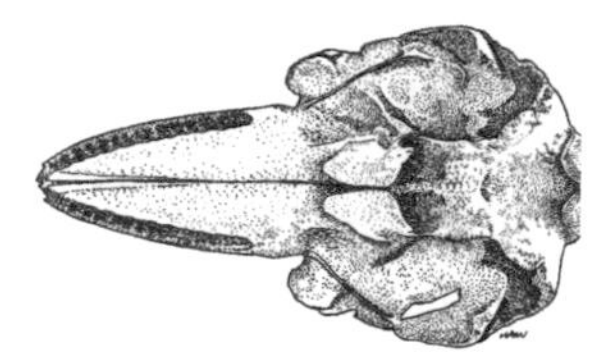

Dorsal View *Lateral View* *Ventral View*

48a. Deep palatal grooves present (>3 mm at midlength of rostrum)

***Lagenodelphis* or *Delphinus*, Go to 49**

b. Palatal grooves shallow (<3 mm at midlength of rostrum) or non-existent

***Cephalorhynchus, Lissodelphis, Lagenorhynchus* or *Stenella*, Go to 51**

49a. Rostrum relatively wide (length/breadth ratio <2.4); <45 teeth per row

Fraser's dolphin *Lagenodelphis hosei*

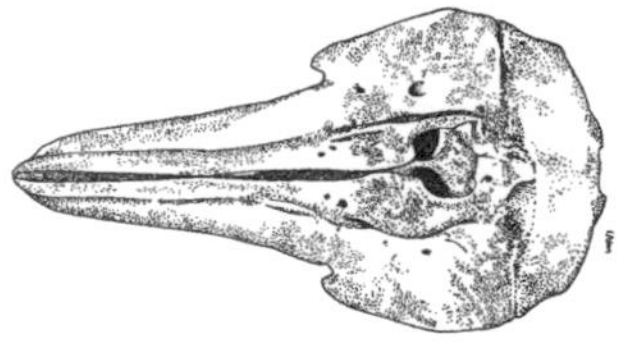

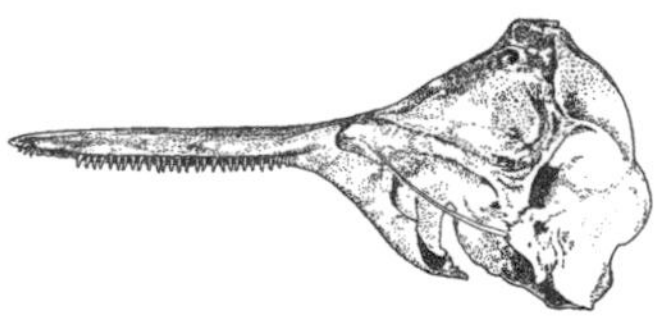

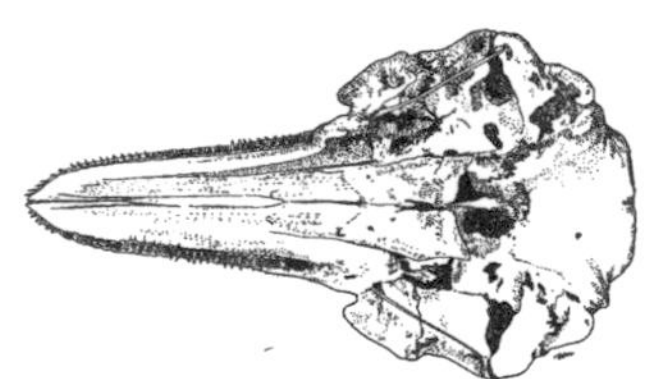

Dorsal View *Lateral View* *Ventral View*

b. Rostrum relatively narrow (length/breadth ratio >2.5); >40 teeth/row

***Delphinus*, Go to 50**

50a. Rostrum relatively short and wide (<275 mm long; length/breadth ratio <3.2; rostrum lenth/zygomatic width ratio 1.25–1.62); 41–54 teeth in each row; trapezoid-shaped palatal ridge with no "pinch"

Short-beaked common dolphin *Delphinus delphis*

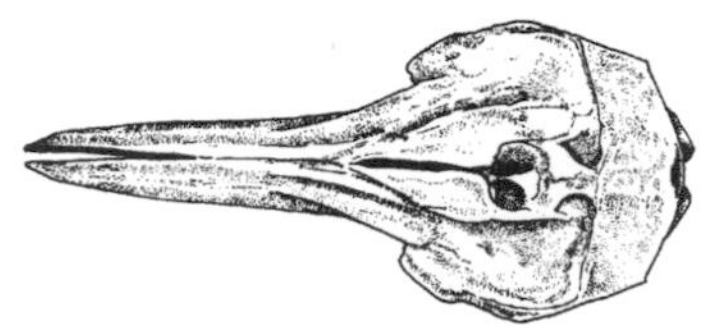

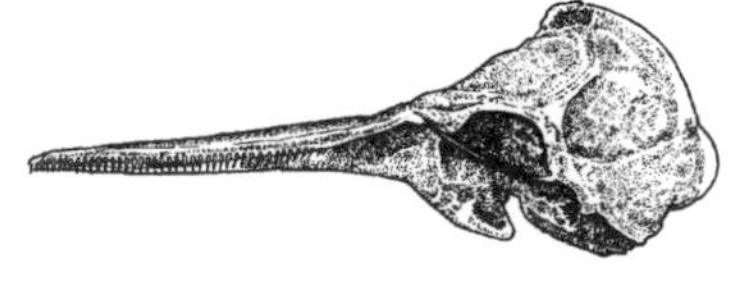

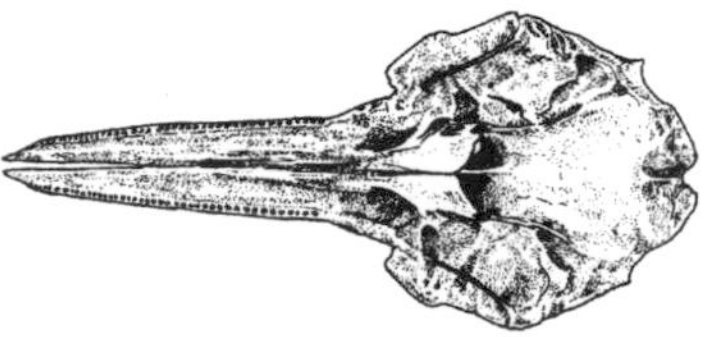

Dorsal View *Lateral View* *Ventral View*

b. Rostrum relatively long and slender (>275 mm; length/breadth ratio >3.2; rostrum length/zygomatic width ratio 1.46–2.06); 47–67 teeth in each row; lanceolate-shaped palatal ridge with distinct "pinch"

Long-beaked common dolphin *Delphinus capensis*

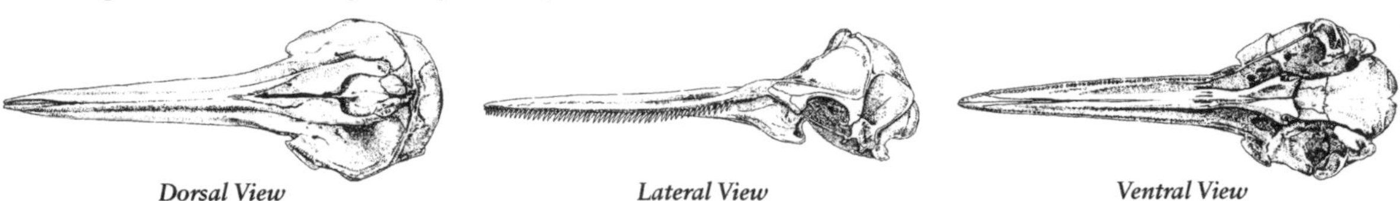

Dorsal View *Lateral View* *Ventral View*

51a. Rostrum relatively narrow (width at base generally <25% CBL, length/breadth ratio >2.1); >30 teeth per tooth row

***Stenella* or *Lissodelphis*, Go to 52**

b. Rostrum relatively wide (width at base generally >25% CBL, length/breadth ratio <2.2); <40 teeth per tooth row

***Cephalorhynchus* or *Lagenorhynchus*, Go to 57**

52a. Premaxillaries converge and meet or nearly meet along dorsal aspect of rostrum; orbits relatively deep; distal portions of mandibles more robust; rostrum relatively long and narrow (length generally >2.3 times its width at base)

***Stenella*, Go to 53**

b. Premaxillaries remain widely separated along dorsal aspect of rostrum (starting at base and moving toward tip); orbits relatively shallow; distal portions of mandibles very narrow; rostrum relatively short and wide (length about 2.2 times its width at base); distributed only in the North Pacific and Southern Hemisphere

Right whale dolphins *Lissodelphis* spp.

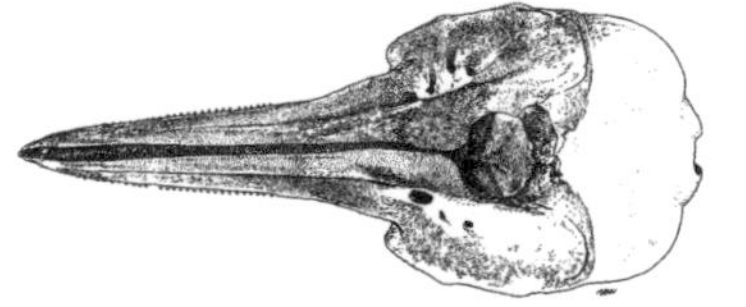

Dorsal View

Lateral View

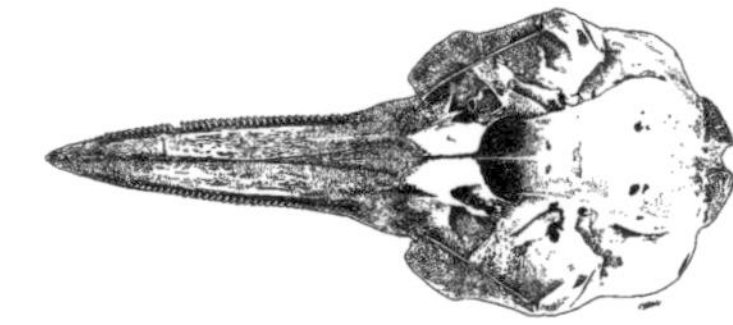

Ventral View

53a. Mandibular symphysis relatively long (usually >17% of mandible length); mandible arcuate; <49 teeth/row; temporal fossa relatively large (length >14% CBL); distal half of rostrum rounded on dorsal surface; no palatal grooves; toothrows converge throughout their length

***S. attenuata* or *S. frontalis*, Go to 54**

b. Mandibular symphysis relatively short (usually <17% of mandible length); mandible sigmoid; >38 teeth/row (usually >40); temporal fossa relatively small (length <17% CBL); distal half of rostrum flattened on dorsal surface; shallow palatal grooves sometimes present; central portions of toothrows parallel or nearly so

***S. longirostris*, *S. coeruleoalba*, or *S. clymene*, Go to 55**

54a. >42 teeth/row; teeth relatively small (<3.2 mm in diameter); rostrum narrow distally (width at 3/4 length <13% length)

Pantropical spotted dolphin *Stenella attenuata*

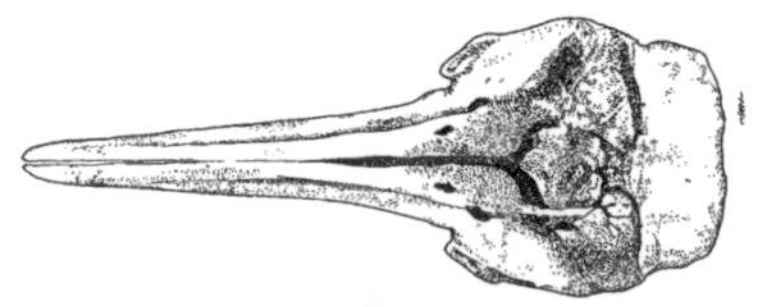

Dorsal View

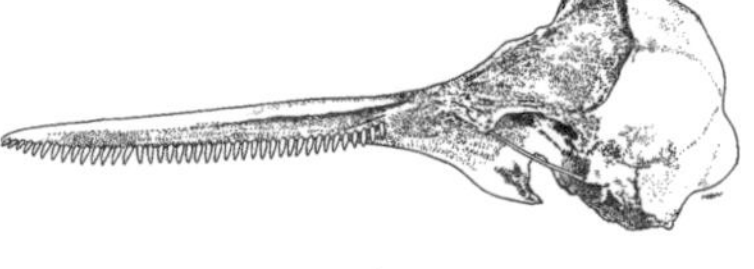

Lateral View

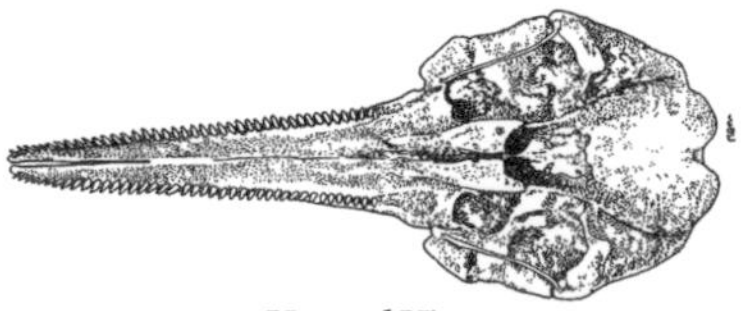

Ventral View

b. <34 teeth/row; teeth relatively large (>4.1 mm in diameter); rostrum broad distally (width at 3/4 length >16% length); distributed only in the Atlantic Ocean

Atlantic spotted dolphin *Stenella frontalis*

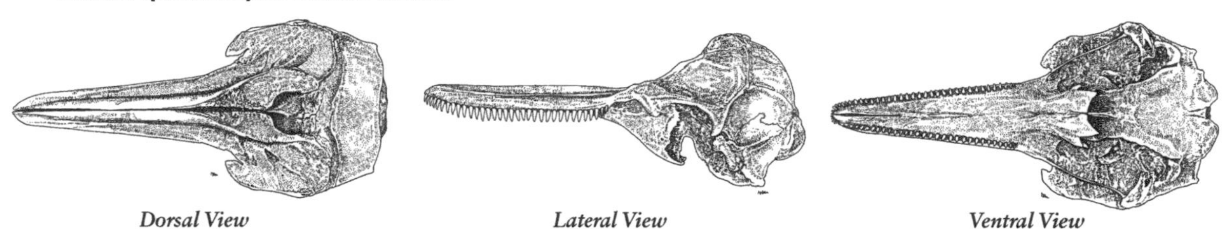

c. 34–42 teeth/row; teeth 3.2–4.1 mm in diameter; rostrum width at 3/4 length 13–16% length

Use discriminant function in Perrin et al. (1987). Marine Mammal Science 3:99-170.

55a. Rostrum relatively long and slender (>61% CBL; length/breadth ratio >3); 40–62 teeth/row; preorbital width <158 mm

Spinner dolphin *Stenella longirostris*

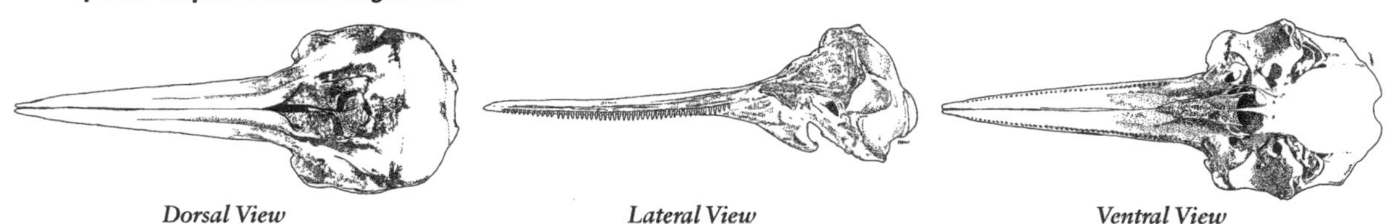

b. Rostrum relatively short and wide (<62% CBL; length/breadth ratio <3); <56 teeth/row; preorbital width >149 mm

***S. coeruleoalba* or *S. clymene*, Go to 56**

56a. Adult skull relatively large (CBL >420 mm); 40–55 teeth/row; palatal grooves usually shallow (0.5 mm or less at 1/2 length of rostrum); upper toothrow >212 mm; preorbital width >177 mm; often a raised area on premaxillae near base of rostrum (visible in lateral view)

Striped dolphin *Stenella coeruleoalba*

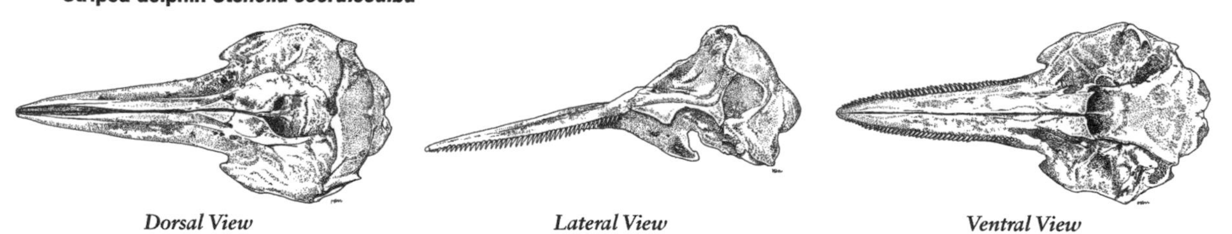

b. Adult skull relatively small (CBL <415 mm); 39–52 teeth/row; palatal grooves usually distinct and relatively deep (0.5 mm or more at 1/2 length of rostrum); upper toothrow <212 mm; preorbital width <175 mm; rostrum relatively flat when viewed from lateral aspect; distributed only in the Atlantic Ocean

Clymene dolphin *Stenella clymene*

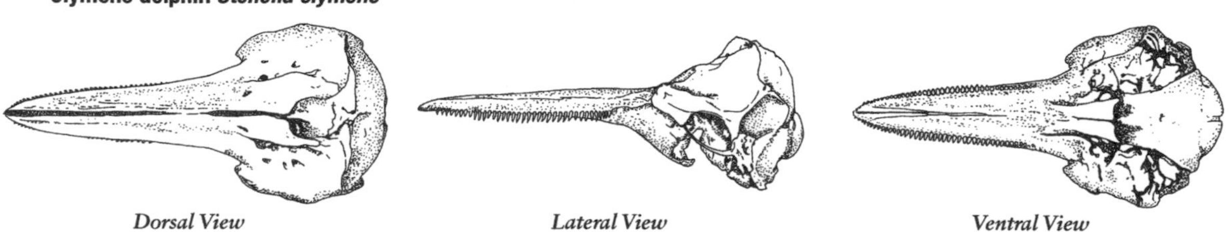

57a. Skull relatively large (adult CBL generally >350 mm); pterygoids in contact along at least a portion of posterior half; zygomatic processes robust

***Lagenorhynchus*, Go to 58**

b. Skull relatively small (adult CBL generally <350 mm); pterygoids separated along entire posterior half; zygomatic processes slender (longer than high); distributed only in coastal waters of the Southern Hemisphere

***Cephalorhynchus*, Go to 63**

58a. Rostrum very wide at base (>30% of CBL); upper tooth counts <29; distributed only in the North Atlantic Ocean

White-beaked dolphin *Lagenorhynchus albirostris*

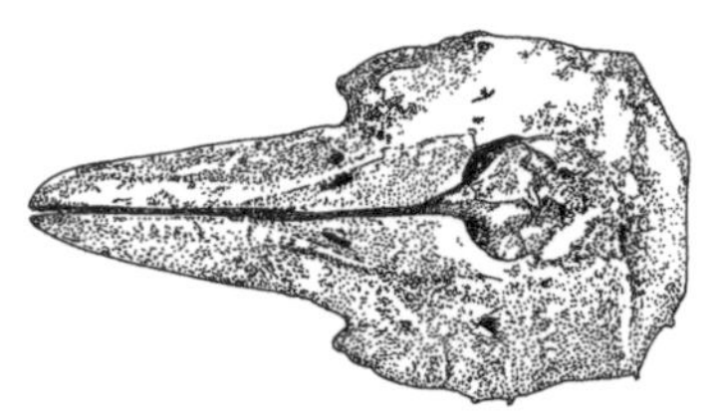

Dorsal View

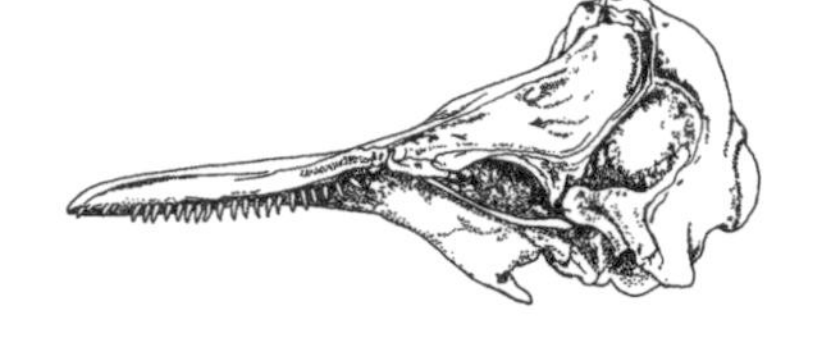

Lateral View

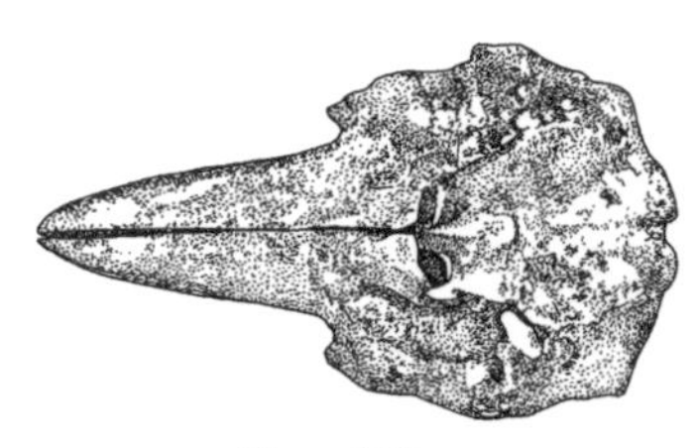

Ventral View

b. Rostrum relatively narrow at base (<31% of CBL); upper tooth counts generally >28

Other species of *Lagenorhynchus*, Go to 59

59a. Length of lacrimal >12% of CBL; upper tooth count generally >35; distributed only in the North Atlantic Ocean

Atlantic white-sided dolphin *Lagenorhynchus acutus*

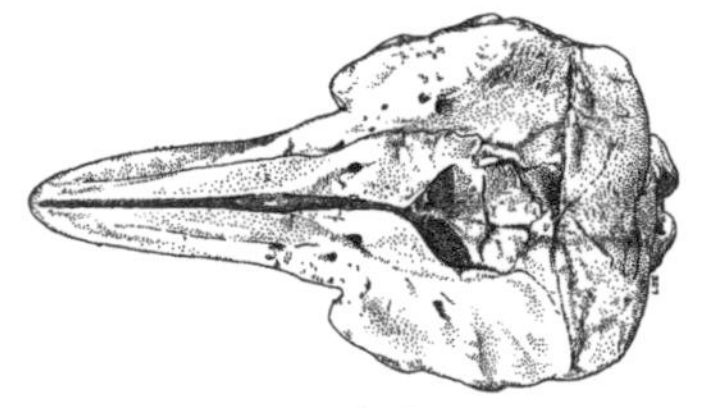

Dorsal View

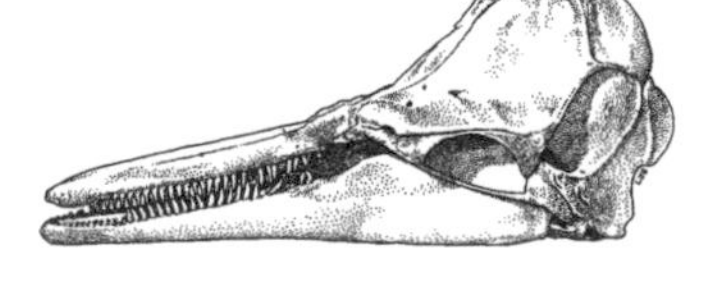

Lateral View

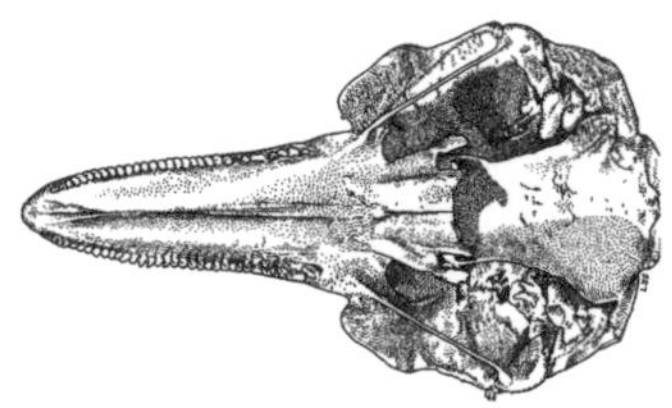

Ventral View

b. Length of lacrimal <12% of CBL; upper tooth count <37

Other species of *Lagenorhynchus*, Go to 60

60a. Height of mandible >18% CBL; distributed only in the high latitudes of the Southern Hemisphere

***L. cruciger* or *L. australis*, Go to 61**

b. Height of mandible <18% CBL

***L. obscurus* or *L. obliquidens*, Go to 62**

61a. Width of external nares <15% CBL; width of rostrum at base generally <26% CBL

Peale's dolphin *Lagenorhynchus australis*

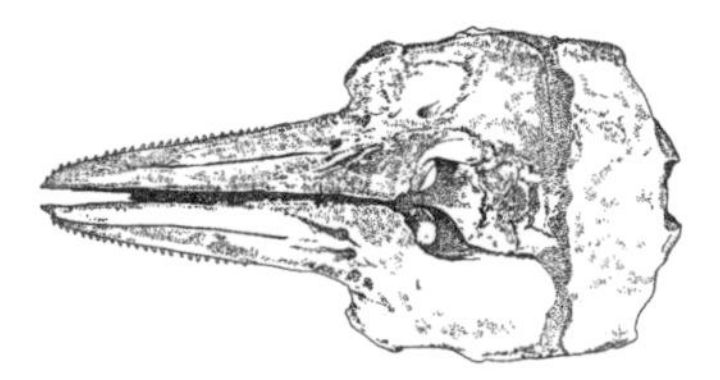

Dorsal View

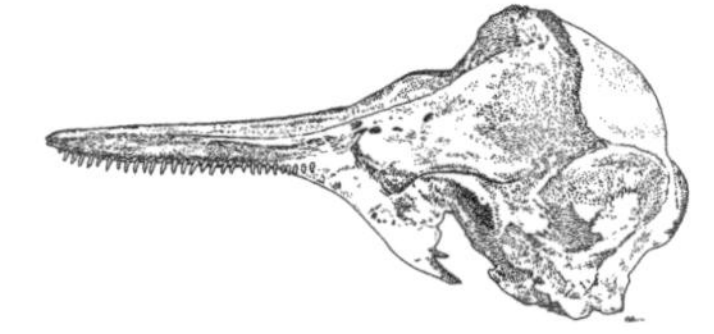

Lateral View

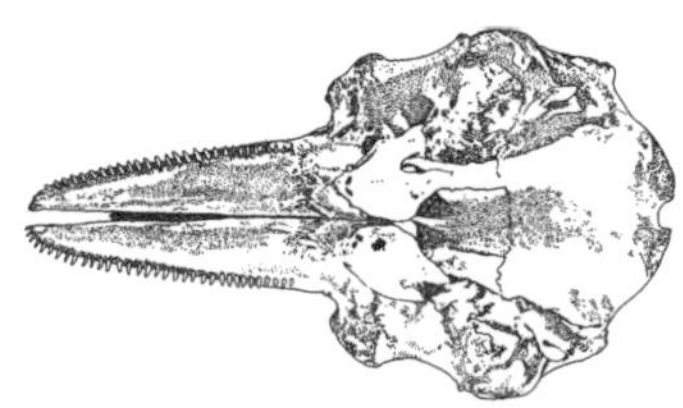

Ventral View

b. Width of external nares >15% CBL; width of rostrum at base >26% CBL

Hourglass dolphin *Lagenorhynchus cruciger*

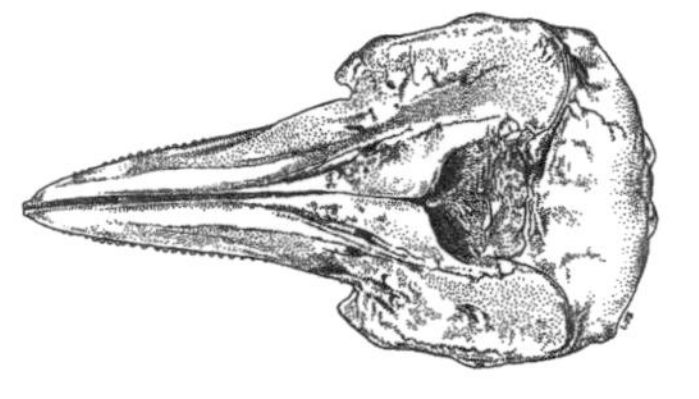

Dorsal View

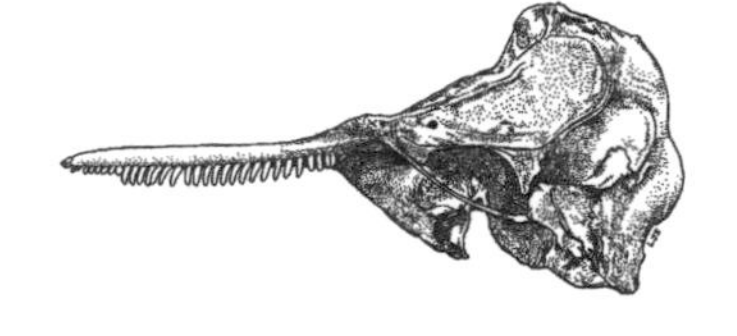

Lateral View

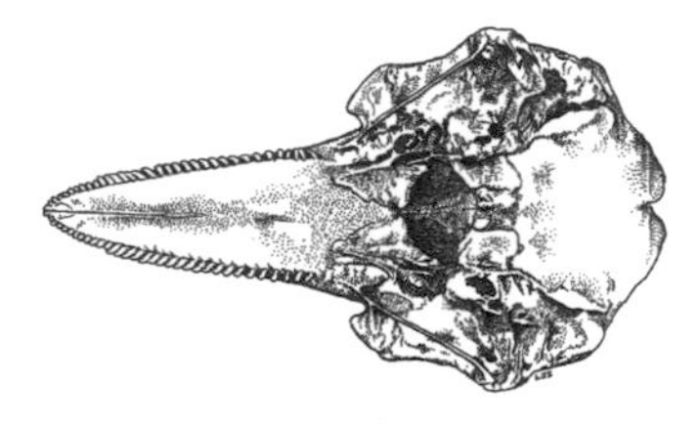

Ventral View

62a. Preorbital width >165 mm; distributed only in the North Pacific Ocean

Pacific white-sided dolphin *Lagenorhynchus obliquidens*

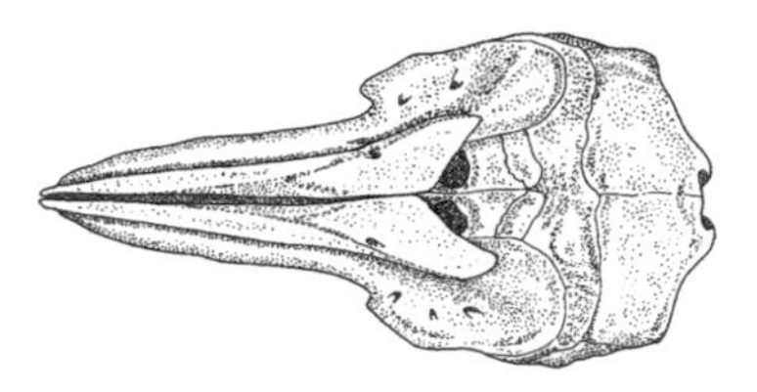

Dorsal View

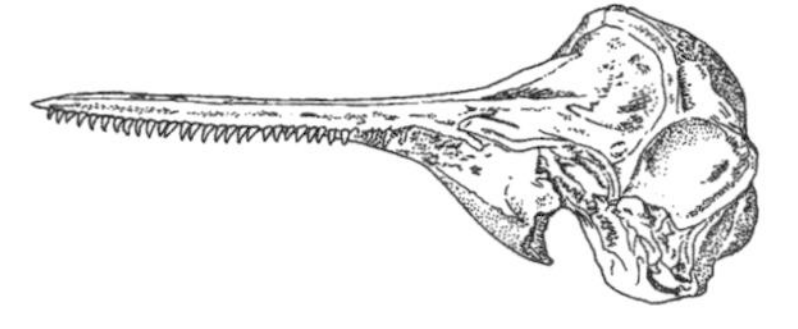

Lateral View

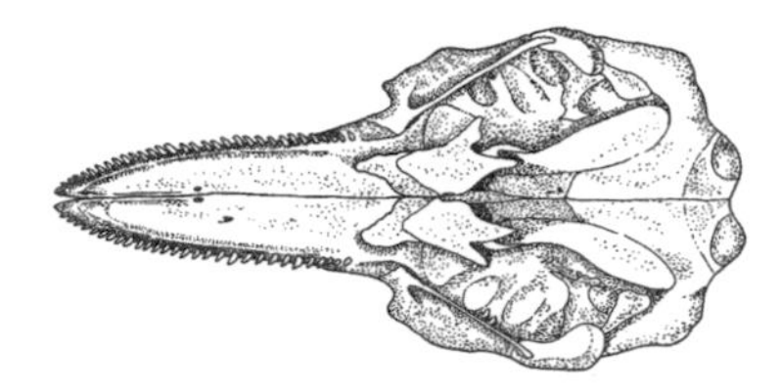

Ventral View

b. Preorbital width <165 mm; distributed only in the Southern Hemisphere

Dusky dolphin *Lagenorhynchus obscurus*

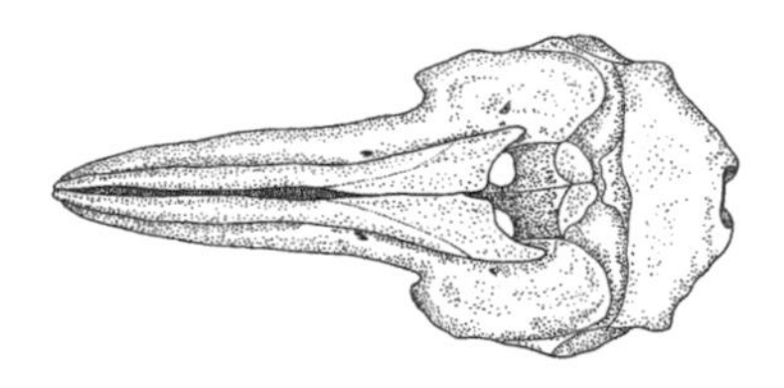

Dorsal View

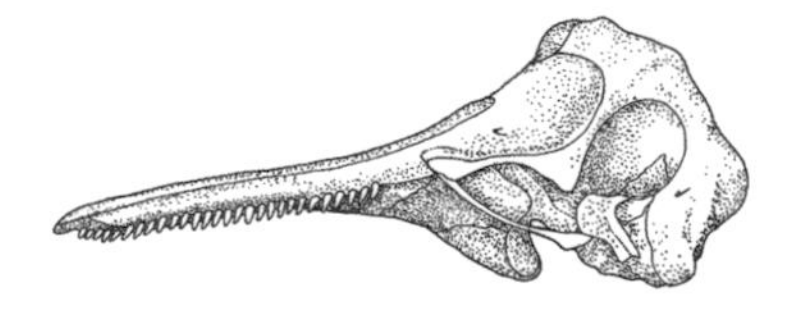

Lateral View

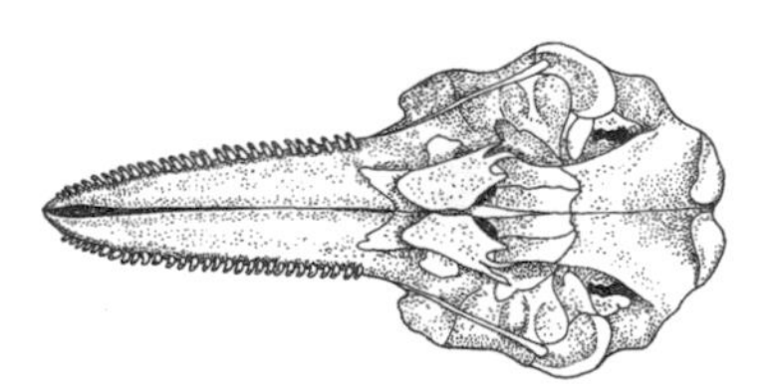

Ventral View

63a. Pterygoids separated by a finger-like projection of the palatines; <116 total teeth; distribution limited to southwestern Africa

Heaviside's dolphin *Cephalorhynchus heavisidii*

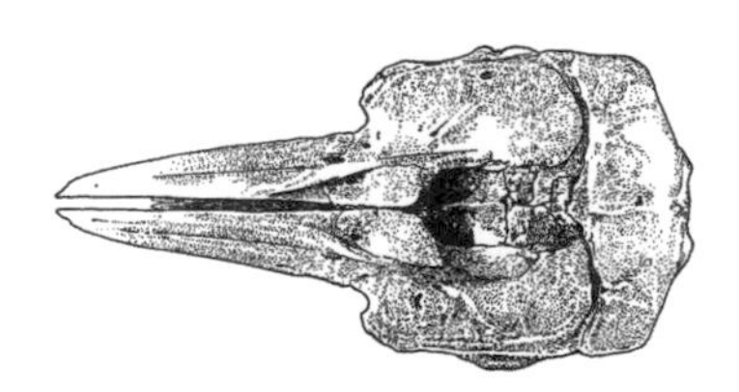

Dorsal View

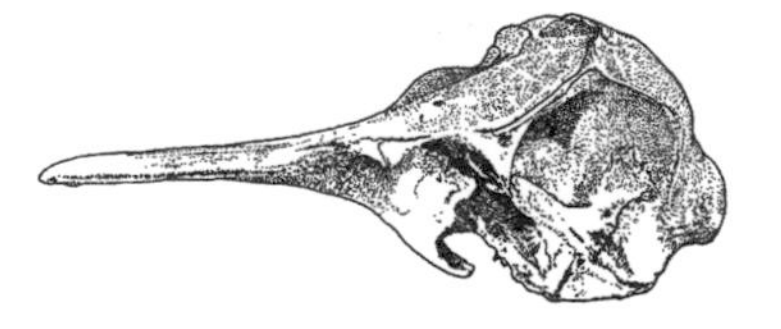

Lateral View

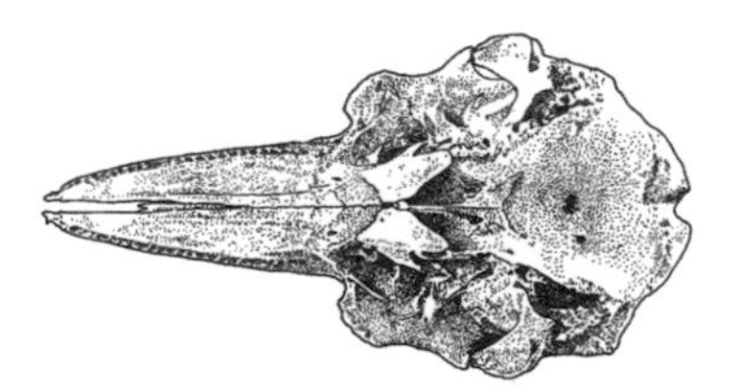

Ventral View

b. No finger-like projection of the palatines; >116 total teeth; distributed only in New Zealand, South America and the Kerguelen Islands

Other species of *Cephalorhynchus*, Go to 64

64a. CBL generally >300 mm; length of rostrum >50% CBL; left premaxilla extends posterior to the midpoint of the nares; distributed only on the west coast of South America

Chilean dolphin *Cephalorhynchus eutropia*

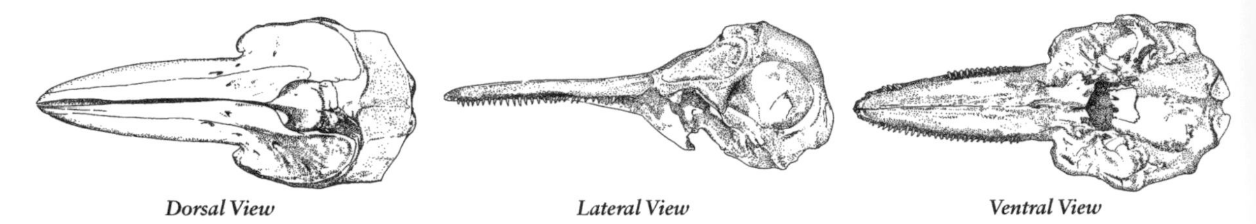

Dorsal View *Lateral View* *Ventral View*

b. CBL generally <300 mm; length of rostrum <50% CBL; left premaxilla does not extend posterior to the midpoint of the nares

***Cephalorhynchus hectori* or *C. commersonii*, Go to 65**

65a. Area of nasals combined <650 mm^2; distributed only in New Zealand

Hector's dolphin *Cephalorhynchus hectori*

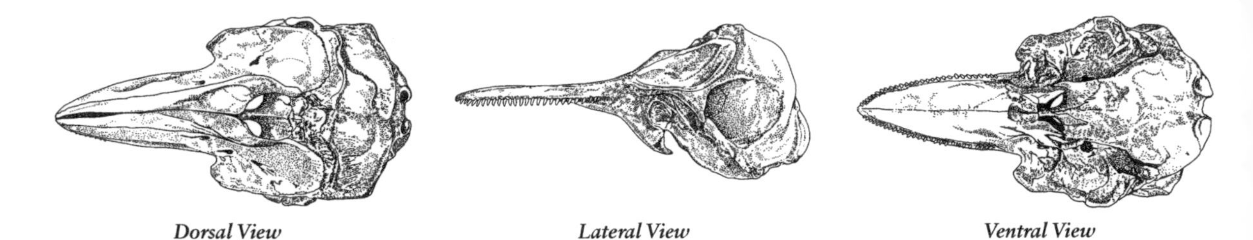

Dorsal View *Lateral View* *Ventral View*

b. Area of nasals combined >650 mm^2; distributed only in southern South America and the Kerguelen Islands

Commerson's dolphin *Cephalorhynchus commersonii*

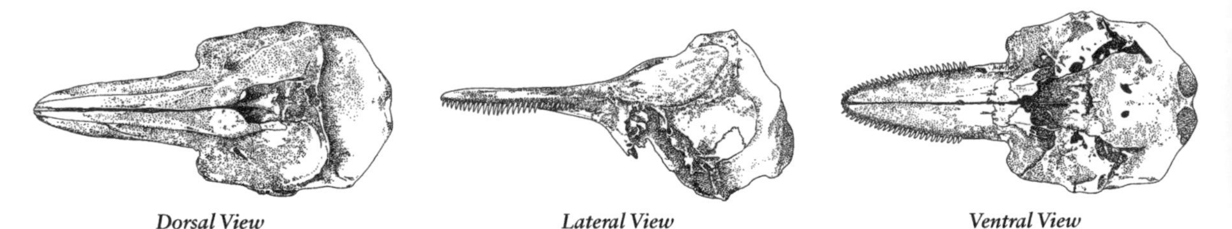

Dorsal View *Lateral View* *Ventral View*

C. Key to Identification of Baleen Whales of the World, Based on Baleen Plates

1a. Baleen plates long (>1.5 m) and dark gray to black in color; very fine, long fringes
Right whale *Eubalaena* spp. or bowhead whale *Balaena mysticetus*

b. Plates much shorter (<1.5 m) with variable color; fringes range from fine to coarse
Other species, Go to 2

2a. Plates yellowish-white, with a dark marginal band of color
Pygmy right whale *Caparea marginata*

b. Plates generally dark in color (if light, not possessing a marginal band)
Other species, Go to 3

3a. Fringes very coarse and white in color, plates number 130–180
Gray whale *Eschrichtius robustus*

b. Fringes finer and with variable color; plates number >200 and may be dark
Other species, Go to 4

4a. Baleen color highly assymetrical (anterior plates on right side much lighter than the rest); plates number up to 480
Fin or Omura's whale *Balaenoptera physalus/omurai*

b. Baleen color symmetrical or only very slightly assymetrical; plates number <400
Other species, Go to 5

5a. Fringes relatively coarse and short; plates with olive brown tinge
Humpback whale *Megaptera novaeangliae*

b. Fringes finer and longer; plates black or white, with shades of gray
Other species, Go to 6

6a. Fringes very fine
Sei whale *Balaenoptera borealis*

b. Fringes much coarser
Other species, Go to 7

7a. Plates very broad-based (base approximately equal to length) and black in color
Blue whale *Balaenoptera musculus*

b. Plates significantly longer than broad and variably colored (may be gray to cream)
Minke or Bryde's whale *Balaenoptera acutorostrata/bonaerensis/edeni*

D. Key to Identification of Pinnipeds of the World, Based on External Appearance

1a. No external ear pinnae

Phocidae or Odobenidae, Go to 2

1b. External ear pinnae present; hindflippers can rotate under the body (permitting walking and sittingupright); all flippers incompletely furred, with only a sparse growth of short hair on top; claws on foreflippers vestigial or absent; claws on hindflipper on the 3 central digits; long terminal flaps beyond claws on the digits of the hindflippers; skin light in color; first two upper incisors with transverse grooves. Features shared with odobenids are hind flippers that can rotate under the body; features shared with phocids (2 b.) are the short free tail, muzzle generally tapering and somewhat pointed, nearly all vibrissae on the sides of the muzzle, and 5–6 post-canines in each row.

Otariidae, Go to 3

2a. Hindflippers can rotate under the body (permitting walking and sitting upright); tail attached to body by web of skin; muzzle very short and broad with a flat end; nearly all vibrissae on the forward end (as opposed to the sides) of the muzzle; 2 massive upper canine teeth enlarged to form tusks that project up to 1 m out of the mouth (except in infants, or when broken off or worn in adults); only 3 post-canine teeth in each tooth row; distribution limited to high latitudes of the Northern Hemisphere

Walrus *Odobenus rosmarus*

b. Hindflippers cannot be rotated under the body (thus cannot walk on land); all flippers completely furred on top and bottom; 5 usually prominent claws (reduced in Antarctic seals, particularly so in leopard and Ross seals, in which they are very small); 5 or no claws visible near the end of each digit on hindflippers; no long flaps of skin beyond claws on hindflipper digits; skin dark in color; upper incisors not grooved transversely. Features shared with otariids (1 b.) are the short free tail, muzzle generally tapering and somewhat pointed, nearly all vibrissae on the sides of the muzzle, and 5–6 post-canines in each row

Phocidae, Go to 16

3a. Dense underfur present; guard hairs (outer visible fur) long, giving a thick plush appearance; terminal flaps on hindflipper digits all approximately equal in length and shape; relatively long prominent ear pinnae; tapering muzzle ends in pointed muzzle and nose in adults

Fur seals, Go to 4

b. No underfur; guard hairs short and stiff; hindflipper digits unequal in length, with the hallux and the 5th digit is wider and usually longer than digits 2–4 (the hallux is widest and longest); ear pinnae relatively short and lying close alongside head; muzzle tapers only slightly, ends in wide blunt muzzle and nose in adults

Sea lions, Go to 11

4a. Fur on the foreflippers stops abruptly at the "wrist" (bend point in the flipper when the animal is sitting upright), with the top of the foreflippers entirely naked; hindflippers long, about 1/4 standard length; very long terminal flaps beyond the claws on the hindflippers; muzzle very short; distribution limited to North Pacific and adjacent seas

Northern fur seal *Callorhinus ursinus*

b. Fur on top of hindflippers beyond the "wrist"; hindflippers about 1/5 of standard length; terminal flaps beyond the claws on the hindflippers moderate in length; muzzle relatively long; distribution limited to Southern Hemisphere and warm temperate North Pacific

***Arctocephalus* spp.[4], Go to 5**

5a. Muzzle short, with somewhat flattened end

Subantarctic, Antarctic, and Galapagos fur seal, Go to 6

b. Muzzle moderate to long with more pointed end (may not be possible to distinguish from 5a. for females and subadults)

Other fur seals, Go to 8

6a. Adults with yellowish to orangish upper chest, neck, and face (to over the eyes); prominent crest of longer guard hairs on crown just behind the eyes

Subantarctic fur seal *Arctocephalus tropicalis*

b. Adults with moderate to no contrast in coloration on upper chest, neck, and face

Antarctic and Galapagos fur seal, Go to 7

[4] The "southern fur seals" (genus *Arctocephalus*) are all very similar in appearance and have overlapping distributions, several species wander widely and vagrants have shown up far from core distribution areas, and it may be difficult to identify them without experience. Also, some species (i.e., *A. gazella*, *A. tropicalis*, *A. forsteri*, and *A. p. philippii*) are known to hybridize in different combinations, providing further challenges to field identification. Skulls may be required to positively identify some species and separate them from related forms.

7a. Adults medium-sized; silver gray with frosted guard hair tips; distribution antarctic and subantarctic only

Antarctic fur seal *Arctocephalus gazella*

b. Adults very small; generally minimal frosting on tips of guard hairs (if present usually not silver gray; distribution largely confined to Galapagos Archipelago

Galapagos fur seal *Arctocephalus galapagoensis*

8a. Large fleshy rhinarium (part of nose without fur); downward-facing nostrils (adult males); distribution confined to area around Juan Fernández Archipelago, off the coast of Chile (Juan Fernandez subspecies), or to eastern North Pacific, from about southern Baja California, north to northern California (Guadalupe subspecies)

Juan Fernandez/Guadalupe fur seal *Arctocephalus philippii*

b. Small to moderate nose; nostrils facing ahead

Other fur seals, Go to 9

9a. Distribution confined to coastal South America, from Peru, south to Cape Horn, and north to Brazil and the Falkland Islands

South American fur seal *Arctocephalus australis*

b. Distribution limited to either New Zealand, southern Australia and adjacent subantarctic islands and waters, or South Africa

Cape/Australian and New Zealand fur seal, Go to 10

10a. Very large, robust build; head massive; curving leading edge of foreflipiper; distribution limited to southwestern and southern Africa (Cape subspecies) and southeastern Australia, including Tasmania (Australian subspecies)

Cape/Australian fur seal *Arctocephalus pusillus*

b. Medium size; moderate build and head size; muzzle straight and flat on top to the end of the nose; leading edge of foreflipper straighter, only slightly curved; distribution limited to New Zealand and adjacent subantarctic islands, and southwestern Australia

New Zealand fur seal *Arctocephalus forsteri*

11a. No mane of longer guard hairs on adult male

***Zalophus* spp., Go to 12**

b. Mane of longer guard hairs on adult male

All other sea lions, Go to 13

12a. Adult males with large bulging sagittal crest above eyes; distribution from Alaska to Mexico, including the Gulf of California

California sea lion *Zalophus californianus*

b. Adult males with moderate to large bulging sagittal crest above eyes; distribution Galapagos Archipelago and adjacent coast of Ecuador

Galapagos sea lion *Zalophus wollebaeki*

c. Adult males dark brown to black, with large bulging sagittal crest above eyes; distribution North Pacific Ocean, from Japan, the Korean Peninsula, Russian Primorye Coast, and the Sea of Japan

Japanese sea lion *Zalophus japonicus*—extinct

13a. Adult males massive; large blocky head; mane of longer guard hairs; adult females large; wide diastema (interdental gap) between 4th and 5th post-canine teeth in both sexes; distribution limited to temperate and subpolar rim of North Pacific

Steller sea lion *Eumetopias jubatus*

b. Adult males large and adult females medium sized; no wide interdental gap (diastema) between 4th and 5th post-canines; Southern Hemisphere (either southwest South Pacific Ocean or both coasts of South America)

Southern Hemisphere sea lions, Go to 14

14a. Extremely heavy (thick) mane of very long guard hairs; very short broad muzzle; massive (deep and wide) lower jaw; distribution coastal South America, from Peru on the west coast, south to Cape Horn, and north to Brazil on the east coast, including the Falkland Islands

South American sea lion *Otaria byronia*

b. Moderate mane of medium-length guard hairs; muzzle blunt, but moderate in length; distribution limited to New Zealand and adjacent subantarctic islands, or southern to southwestern Australia

Australian and New Zealand sea lion, Go to 15

15a. Adult males with mane extending up onto the top of the head and relatively flat topped muzzle; generally brownish, with yellowish back of neck and crown; females often strikingly bicolored, dark above, pale below, with pale color on the face and over the eyes; distribution limited to coastal waters of southern and southwestern Australia

Australian sea lion *Neophoca cinerea*

b. Adult males with mane that stops at nape (head seems disproportionally small); muzzle usually slightly convex in silhouette; color blackish brown; distribution limited to southern New Zealand and adjacent subantarctic islands

New Zealand sea lion *Phocarctos hookeri*

16a. Vibrissae smooth in outline

Bearded and monk seals, Go to 17

b. Vibrissae beaded (sometimes only weakly) in outline

Other phocid seals, Go to 19

17a. Foreflippers square to rounded, with equal length digits, or digits 2–4 slightly longer; vibrissae pale very densely packed, so as to obscure mouthline; circumpolar distribution in the Arctic

Bearded seal *Erignathus barbatus*

b. Foreflippers pointed, with first digit longer and digits 2–5 becoming shorter; black vibrissae with sparse to moderate density; adults medium-sized; muzzle and head moderate in size and somewhat flattened; nostrils pointing slightly upwards; males without enlarged nose; females with 4 mammary teats

Monk seals, Go to 18

18a. Distribution limited to the eastern North Alantic Ocean near northwest Africa including Desertas Islands in Madeira, in the Mediterranean Sea, and possibly the Black Sea

Mediterranean monk seal *Monachus monachus*

b. Distribution limited to the Hawaiian Island chain, North Pacific Ocean

Hawaiian monk seal *Neomonachus schauinslandi*

c. Distribution limited to the Caribbean Sea, islands and reefs, and adjacent coastal areas of Central and South America

Caribbean monk seal *Neomonachus tropicalis*—extinct

19a. Adults very large (males to 4.5 m); muzzle and head very broad and deep; nostrils point ahead or down; adult males with large inflatable proboscis; females with 2 mammary teats; no spots, blotches or bands on pelt

Elephant seals, Go to 20

b. All other phocid seals

Go to 21

20a. Proboscis of adult males relatively large; distribution limited to temperate eastern and central North Pacific

Northern elephant seal *Mirounga angustirostris*

b. Proboscis of males relatively small; distribution circumpolar in polar to temperate waters of the Southern Hemisphere

Southern elephant seal *Mirounga leonina*

21a. Distribution limited to Southern Hemisphere

Antarctic phocid seals, Go to 22

b. Distribution limited to Northern Hemisphere

Other phocids seals, Go to 25

22a. Head and muzzle short and wide; foreflippers about 1/5 or less of standard length; post-canine teeth relatively simple

Weddell or Ross seal, Go to 23

b. Head and muzzle long and narrow; foreflippers long, at least 1/4 standard length; post-canines ornate and multi-cusped

Leopard or crabeater seal, Go to 24

23a. Adults very long (2.5–3.3 m) and massive, with a relatively small head; numerous blotches of light and dark, particularly on sides and belly

Weddell seal *Leptonychotes weddellii*

b. Adults generally <2.5 m; long streaks of color on face, neck, chest, and extending onto the sides; head more normal in size, neck appears thick and enlarged

Ross seal *Ommatophoca rossii*

24a. Head and jaws massive and reptilean in appearance; body long (to 3.3 m), and serpent-like, thickest at shoulders; foreflippers very long, almost 1/3 standard length; foreflipper claws very small

Leopard seal *Hydrurga leptonyx*

b. Head and jaws long, but tapering, with a somewhat flattened muzzle; body moderately robust, more filled out; foreflippers long, but only to about 1/4 standard length; foreflipper claws more normal in size

Crabeater seal *Lobodon carcinophagus*

25a. Distribution limited to either Lake Baikal or the Caspian Sea, far from oceanic areas

Baikal or Caspian seal, Go to 26

b. Distribution oceanic or in lakes or rivers near oceanic areas

Other seals, Go to 27

26a. Distribution limited to Lake Baikal and connecting rivers

Baikal seal *Pusa sibirica*

b. Distribution limited to the Caspian Sea and connecting rivers

Caspian seal *Pusa caspica*

27a. Pelage pattern consists of bands or broad swaths of light or dark color

Ribbon or harp seal, Go to 28

b. Pelage pattern consists of spots, rings, or blotches

Other seals, Go to 29

28a. Light color bands encircling foreflippers, neck, and abdomen; Body orange-brown to black; distribution limited to Bering Sea, Sea of Okhotsk, and adjacent Arctic Ocean

Ribbon Seal *Histriophoca fasciata*

b. Broad swath of black on each side, meeting (generally) over the shoulders to roughly form a "V" pattern[5]; body generally silvery-white, with some animals sooty gray and others with scattered blotches

Harp seal *Pagophilus groenlandicus*

29a. Pelage pattern consists of irregular, small to large, dark brown to black or sometimes tan blotches; distribution limited to North Atlantic and adjacent Arctic areas

Hooded or gray seal, Go to 30

b. Pelage pattern consists primarily of round to oval smaller spots or rings around spots, or a combination of the above

Other seals, Go to 31

30a. Head broad and short with short muzzle on females, and large fleshy nasal bladder (with overhanging nostrils) on males; head dark in both sexes

Hooded seal *Cystophora cristata*

b. Head and muzzle very long and somewhat narrow; in silhouette, nose is rounded down and outwards (convex) in males and straight to slightly rounded in females; adult males dark brown to gray black with lighter (tan) blotches

Gray seal *Halichoerus grypus*

31a. Adults short and plump; head small and round with a short muzzle; pelage pattern with few or no spots not encircled by a lighter ring

Ringed seal *Pusa hispida*

b. Pelage pattern consists mostly of small round to oval spots with some rings and irregular incomplete rings to no rings; adults more elongate and less rounded; head more conspicuous with a longer muzzle

Spotted or harbor seal, Go to 32

32a. Pelage pattern consists of few to no rings; spotting more even from top to bottom; pups born on sea ice in lanugo coat; distribution limited to North Pacific and adjacent Arctic areas

Spotted seal *Phoca largha*

b. Pelage pattern consists of a moderate number of light rings around spots; more heavily spotted above than below; pups born on land or glacial ice in adult pelage; distribution limited to North Pacific and North Atlantic in temperate to subpolar waters including bays, river estuaries, and some lakes

Harbor seal *Phoca vitulina*

[5] Some harp seals never develop the harp pattern and remain blotched as adults. These blotched animals can be separated from gray seals, based on their smaller size, clearly demarcated and shorter muzzle, and closer-set nostrils; and from hooded seals, based on their longer, but thinner, head and muzzle and lack of hood pattern on the head

E. Key to Identification of Pinnipeds of the World, Based on Skull Morphology

1a. Tympanic bullae flat, small, and angular; supraorbital processes present on the frontal bones in most species (except Odobenidae); frontals penetrating anteriorly slightly to moderately between nasals on the midline; antorbital processes present

Otariidae and Odobenidae, Go to 2

b. Tympanic bullae inflated and rounded; supraorbital processes absent; nasals narrow and penetrating deeply back between frontals on midline; antorbital processes absent

Phocidae, Go to 9

2a. Upper canines massive, enlarged to form tusks; no supraorbital processes; frontals and nasals make a relatively straight point of contact; no grooves on upper incisors; only 3 post-canines on each side; mandibular symphysis exhibiting bony fusion; distributed only in the Northern Hemisphere

Walrus *Odobenus rosmarus*

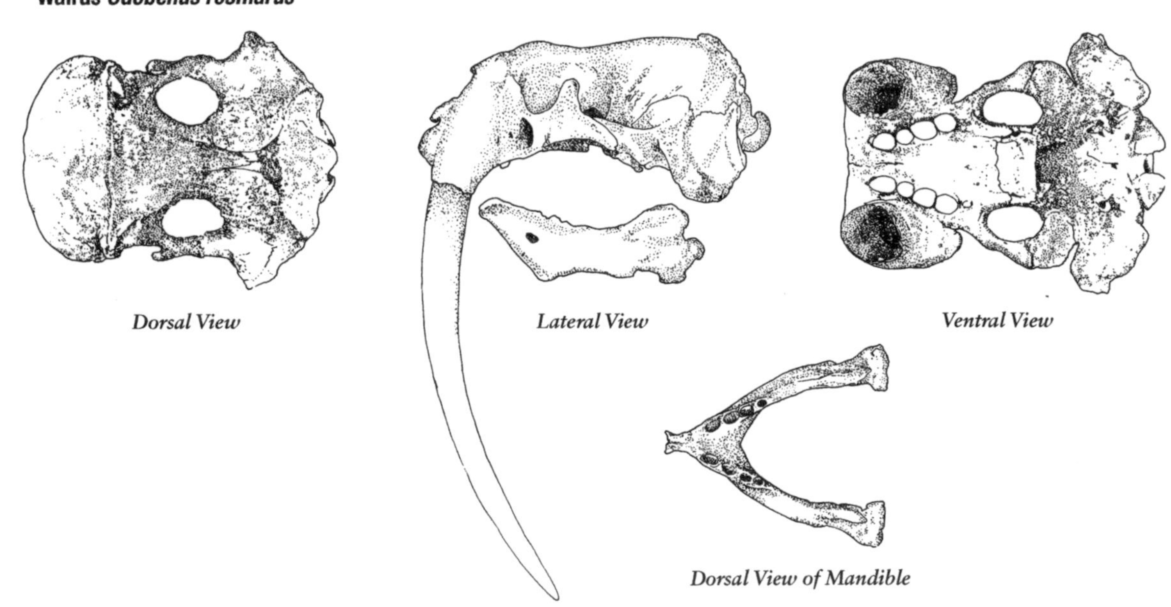

Dorsal View *Lateral View* *Ventral View*

Dorsal View of Mandible

b. Upper canines not enlarged into tusks; supraorbital processes present; frontals and nasals make a W-shape at their point of contact; 2 lower incisors on each side; transverse groove on first 2 upper incisors; 5–6 post-canines on each side; mandibular symphysis not fused

Otariidae, Go to 3

3a. Facial angle <125°; distributed only in the Northern Hemisphere

Northern fur seal *Callorhinus ursinus*

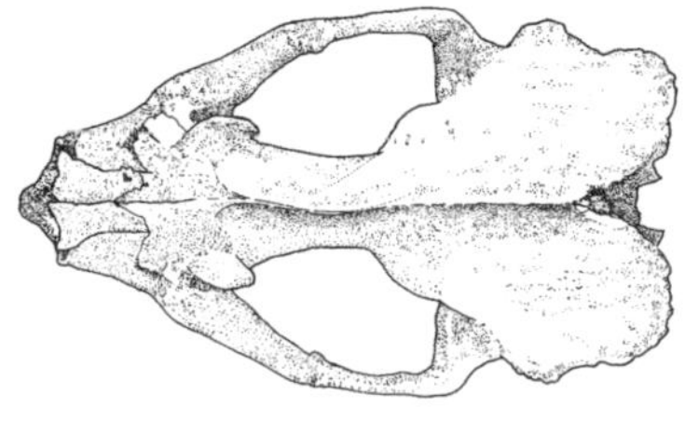

Dorsal View

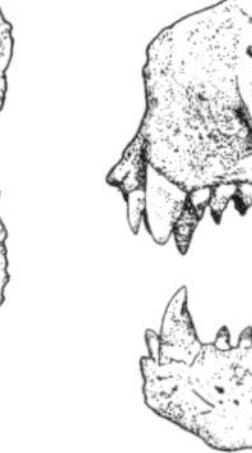

Lateral View

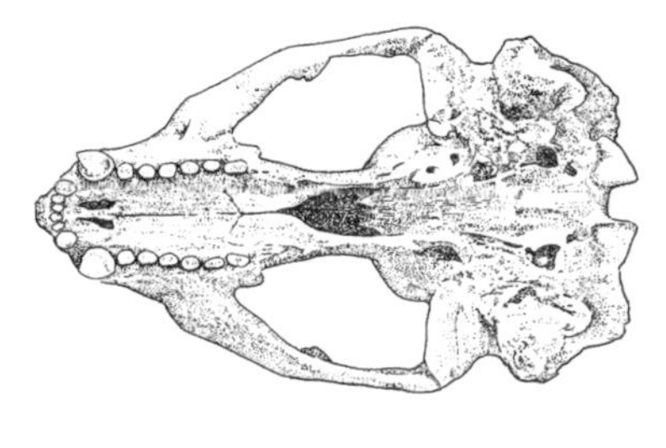

Ventral View

b. Facial angle >125°

Other species, Go to 4

4a. Very long palate, extending posterior to orbitemporal; distributed only in South America

South American sea lion *Otaria flavescens*

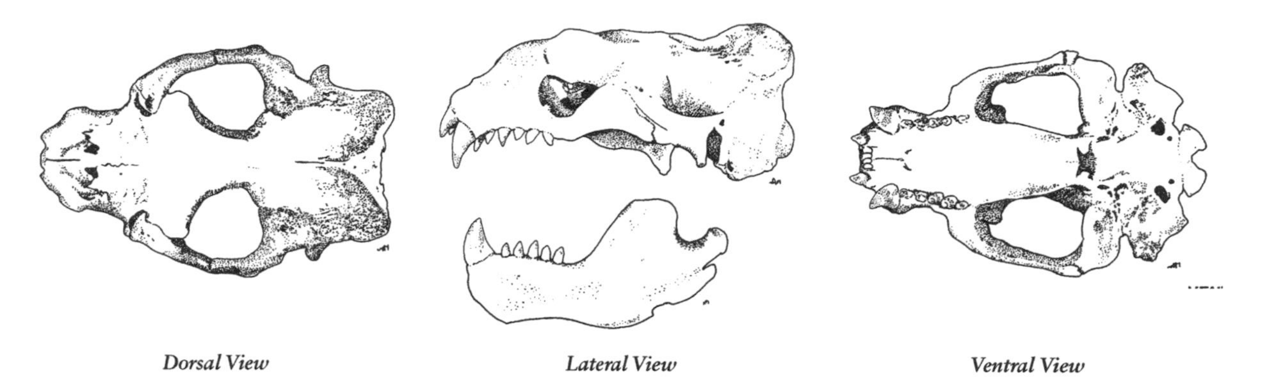

Dorsal View *Lateral View* *Ventral View*

b. Palate much shorter, not extending posterior to orbitemporal

Other species, Go to 5

5a. Large diastema (interdental gap) between 4th and 5th post-canines (width of about 2 teeth); distributed only in the North Pacific

Steller sea lion *Eumetopias jubatus*

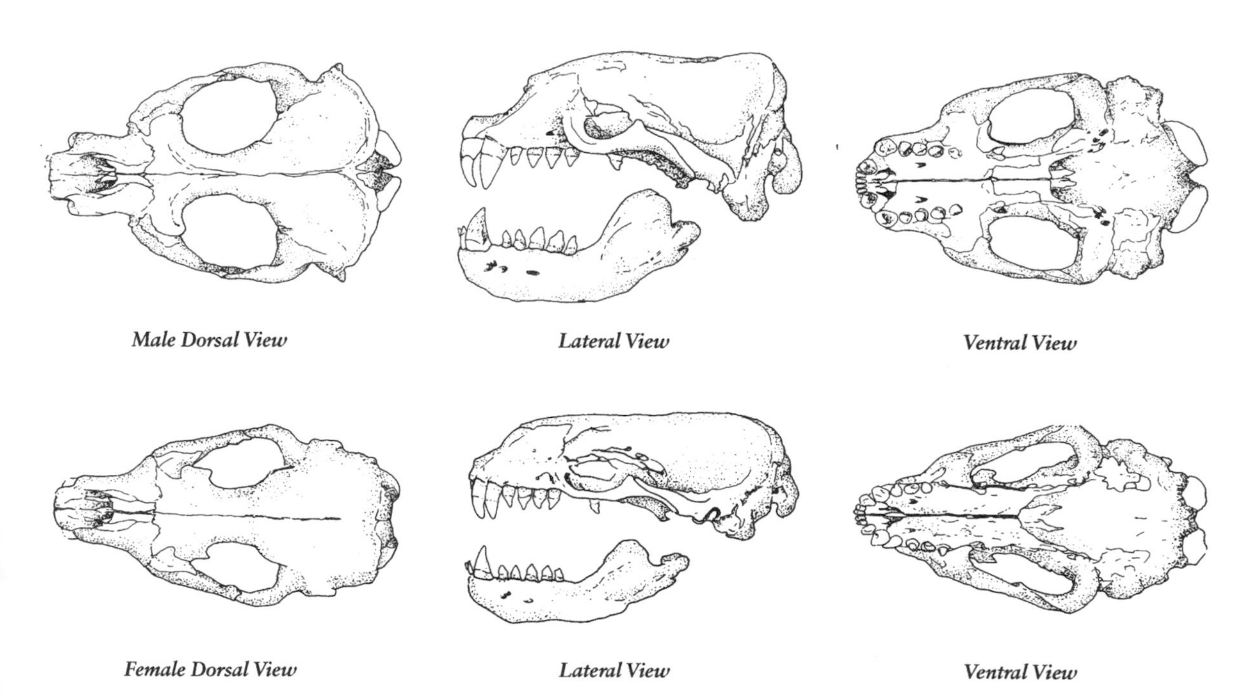

Male Dorsal View *Lateral View* *Ventral View*

Female Dorsal View *Lateral View* *Ventral View*

b. No diastema (or very reduced gap) between 4th and 5th post-canines

Other species, Go to 6

6a. Tympanic bullae with caudal extensions (approximately cylindrical in old individuals); distributed only in the region around New Zealand

New Zealand sea lion *Phocarctos hookeri*

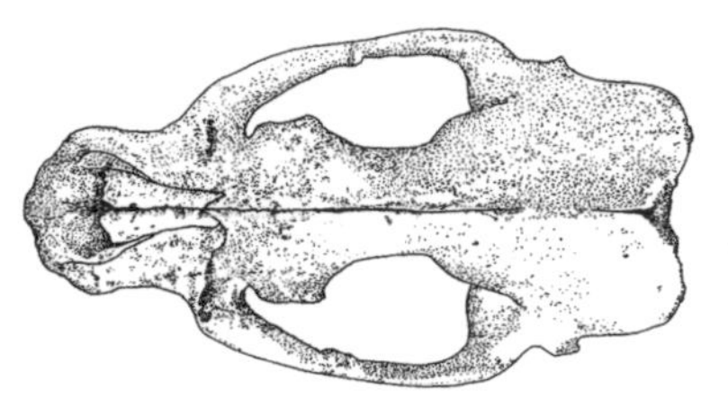
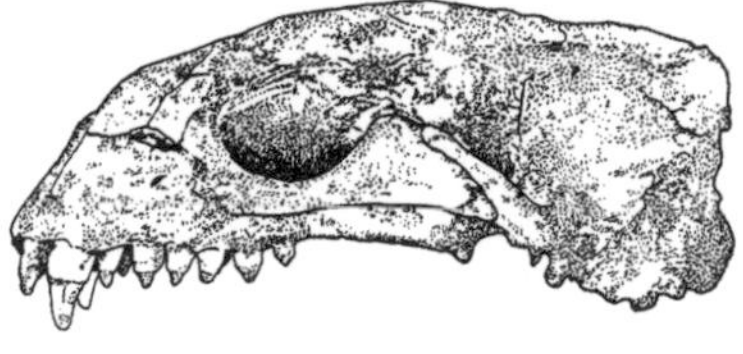
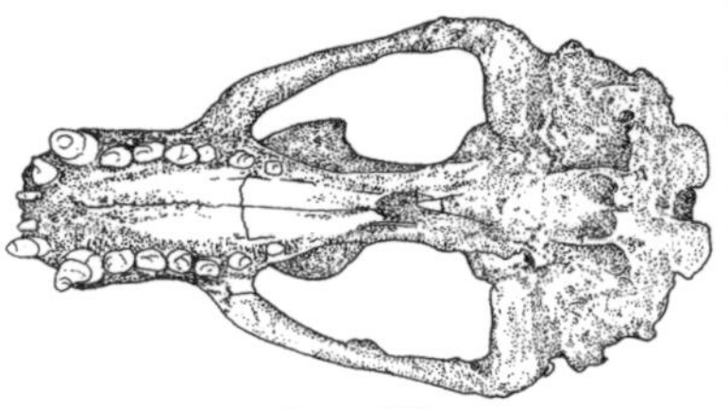

Dorsal View *Lateral View* *Ventral View*

b. Tympanic bullae of irregular form, without caudal extensions

Other species, Go to 7

7a. Anterior process of orbitotemporal very broad; distributed only in the area of southern Australia

Australian sea lion *Neophoca cinerea*

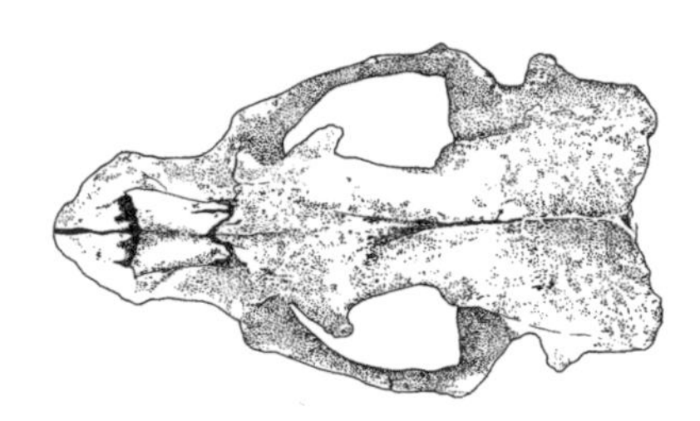
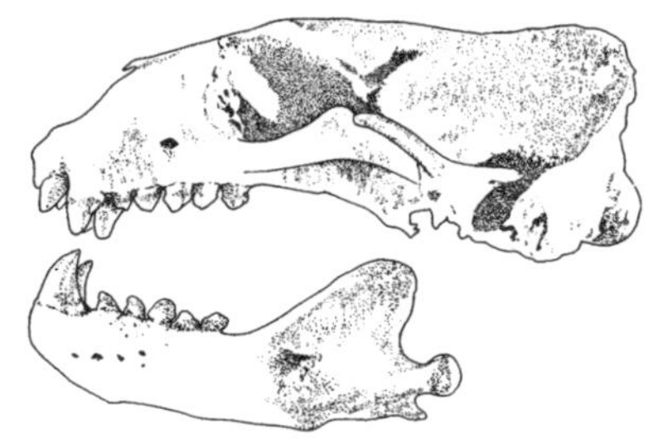
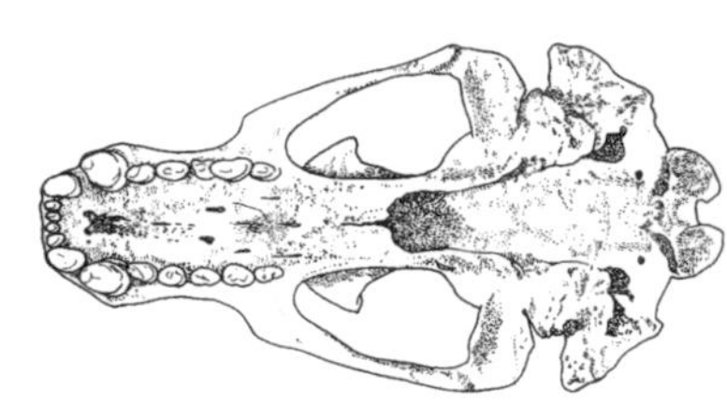

Dorsal View *Lateral View* *Ventral View*

b. Anterior process of orbitotemporal relatively narrow

Other species, Go to 8

8a. Facial part of cranium lengthened; nasals relatively long and narrow; distributed only in North Pacific and equatorial waters of the Pacific

California, Galapagos, and Japanese sea lions *Zalophus* spp.

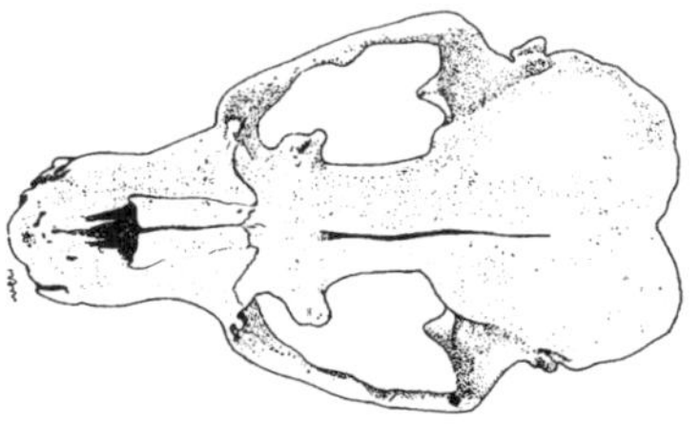

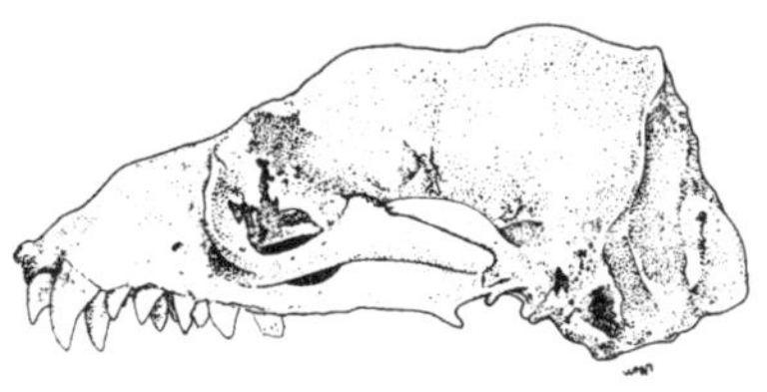
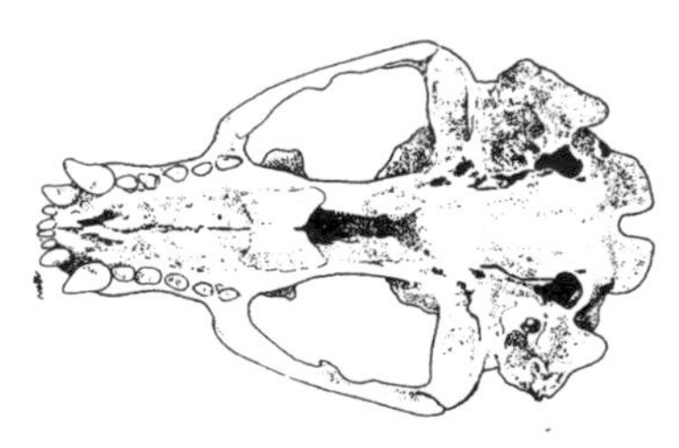

Male Dorsal View *Lateral View* *Ventral View*

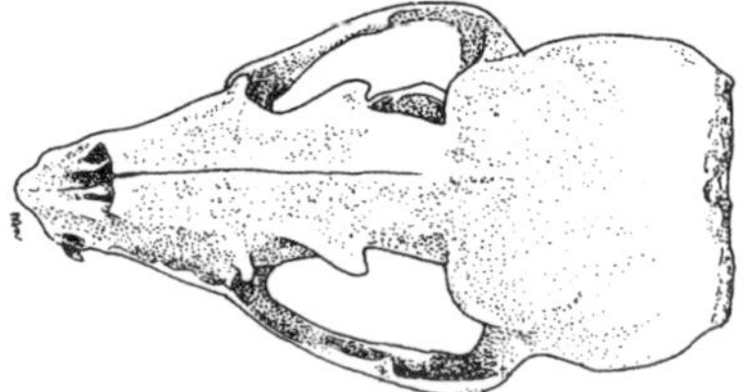

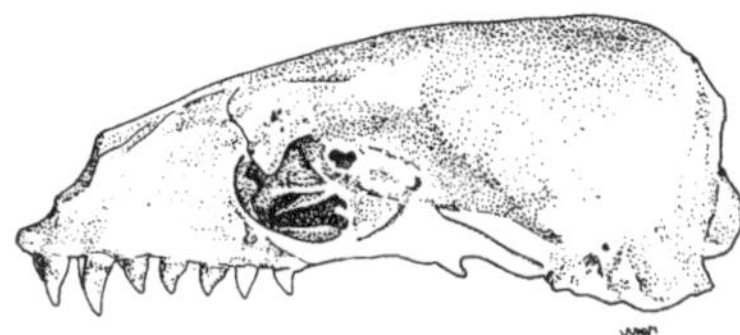
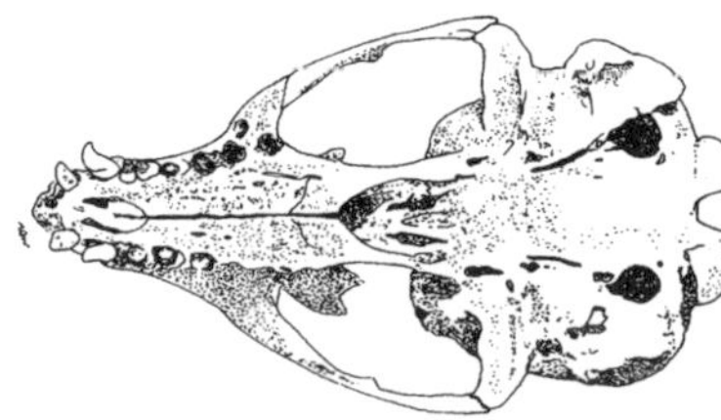

Female Dorsal View *Lateral View* *Ventral View*

b. Short, broad cranium; short facial part; nasals widening distinctly anteriorly
Southern fur seals *Arctocephalus* spp.

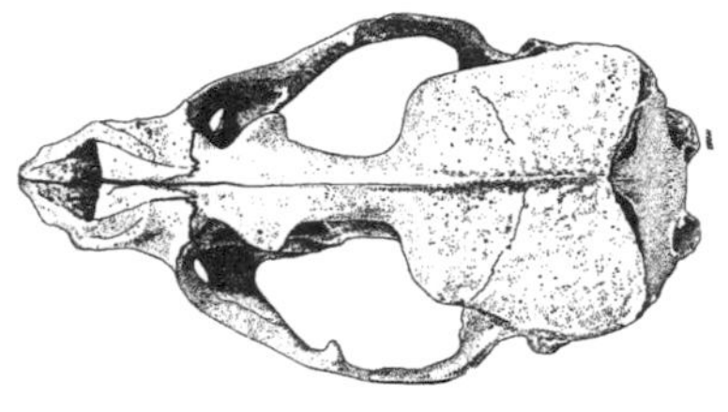

Dorsal View

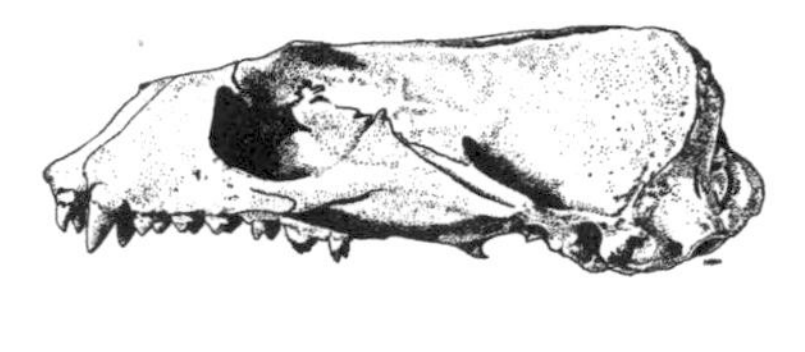

Lateral View

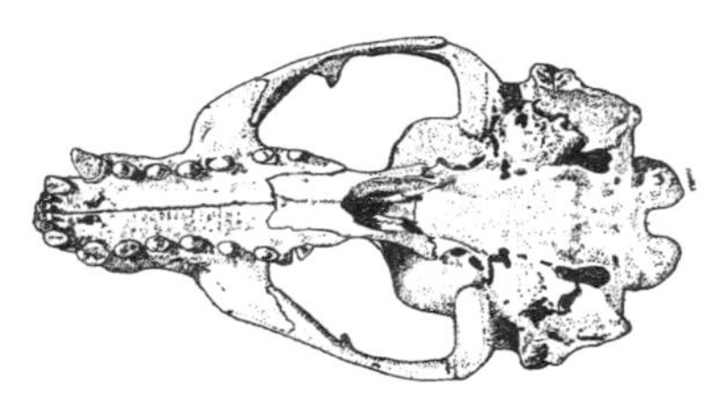

Ventral View

9a. Zygomatic arches heavy and extremely stout; nasals covered over 1/3 of their length; distributed only in high latitudes of the North Atlantic
Hooded seal *Cystophora cristata*

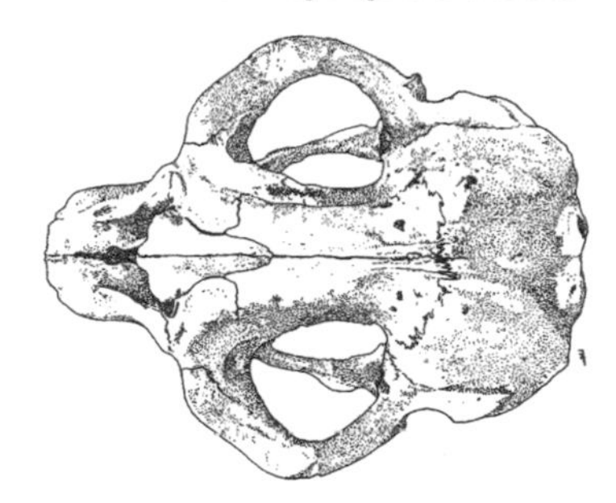

Dorsal View

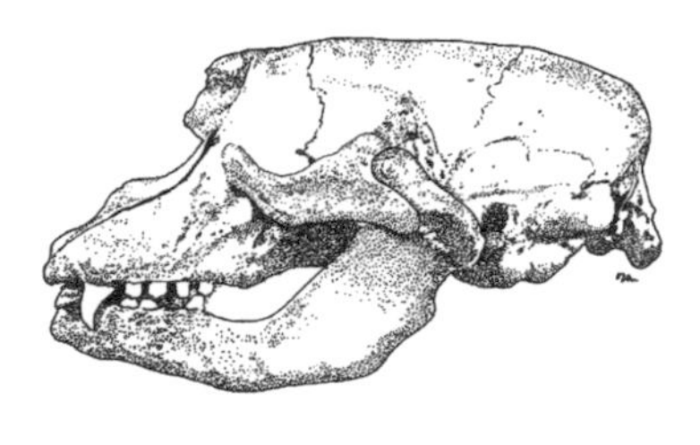

Lateral View

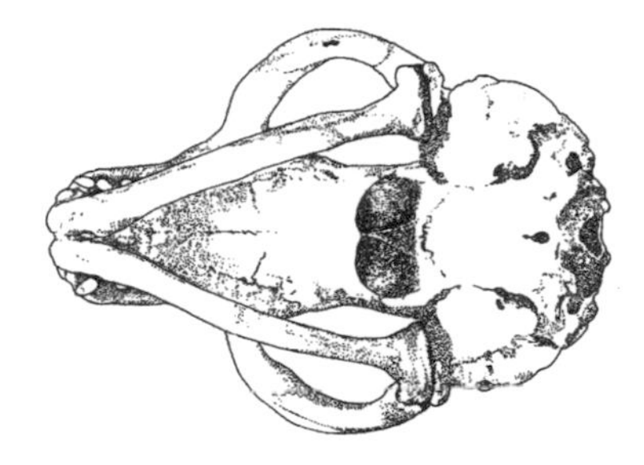

Ventral View

b. Zygomatic arches relatively slender; nasals not or only slightly projecting
Other species, Go to 10

10a. Dorsal profile of cranium highly arched; jugals short and broad; distributed only in high latitudes of the Northern Hemisphere
Bearded seal *Erignathus barbatus*

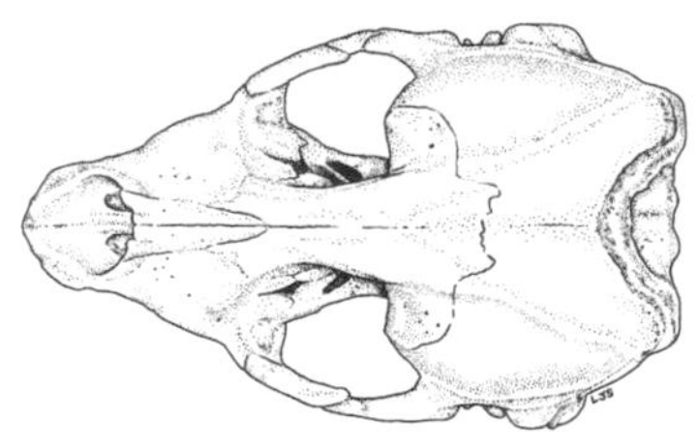

Dorsal View

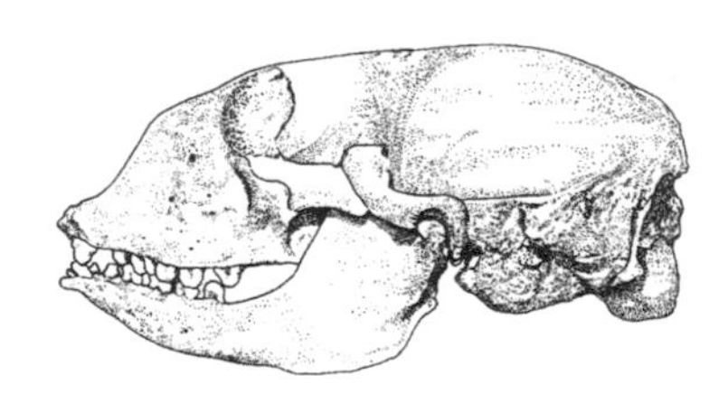

Lateral View

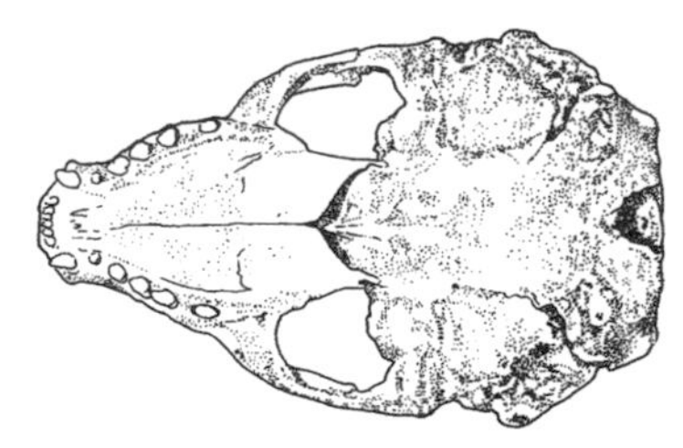

Ventral View

b. Dorsal profile of cranium not at all or only slightly arched; jugals relatively long and narrow
Other species, Go to 11

11a. Three upper incisors in each half jaw; distributed only in high latitudes of the Northern Hemisphere
***Halichoerus*, *Phoca*, *Pusa*, *Pagophilus*, and *Histriophoca*, Go to 12**

b. Two upper incisors in each half jaw
Other species, Go to 16

12a. Anterior surface of muzzle flat or convex
***Halichoerus* and *Phoca*, Go to 13**

b. Anterior surface of muzzle concave
***Pusa, Histriophoca,* and *Pagophilus*, Go to 14**

13a. Muzzle long, high, and wide; naso-frontal area elevated; premaxillae extend back to reach nasals; distributed only in high latitudes of the North Atlantic
Gray seal *Halichoerus grypus*

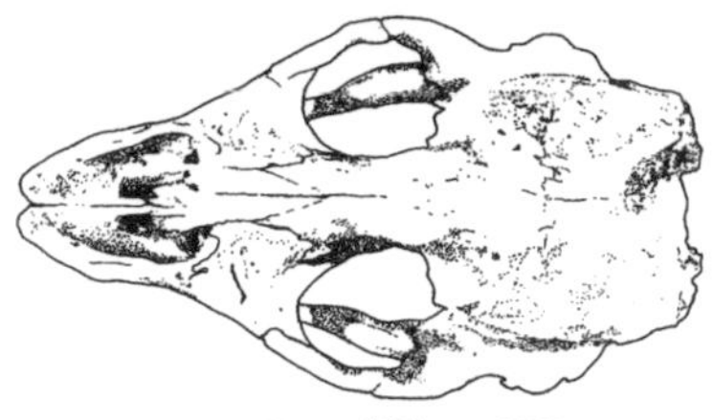

Dorsal View

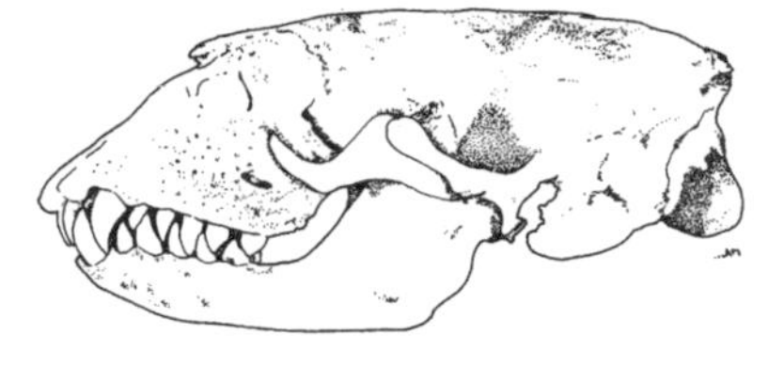

Lateral View

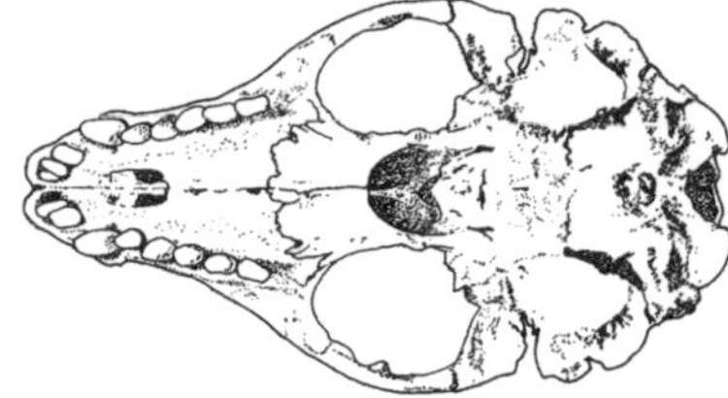

Ventral View

b. Muzzle relatively short and low; naso-frontal area less elevated; premaxillae do not extend back as far
Harbor and spotted seals *Phoca* spp.

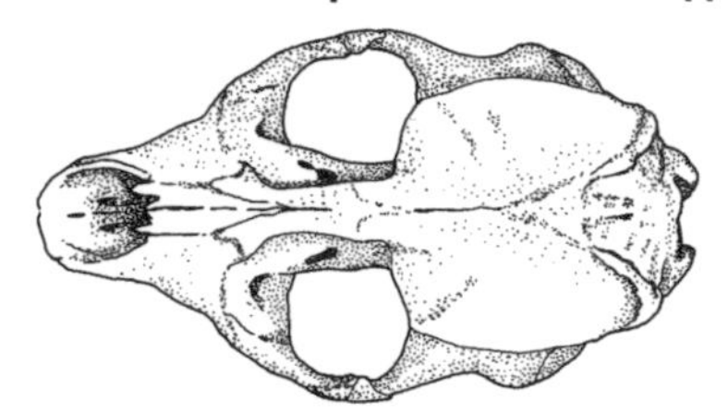

Dorsal View

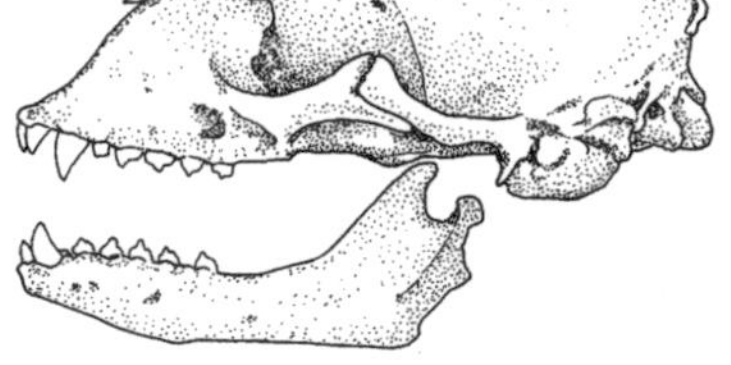

Lateral View

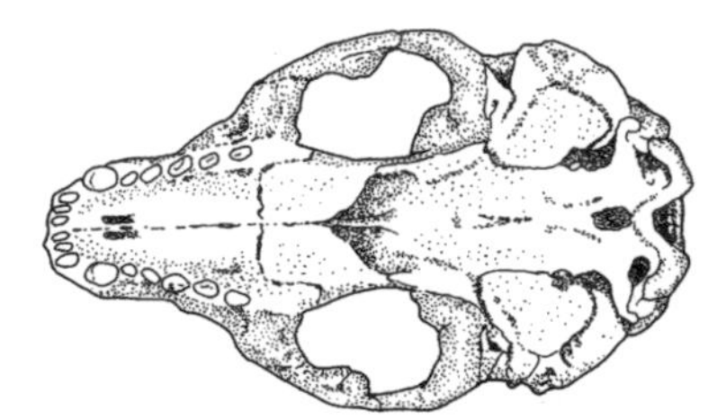

Ventral View

14a. Interorbital area extremely reduced and narrow (interorbital width <7 mm)
Ringed, Caspian, and Baikal seals *Pusa* spp.

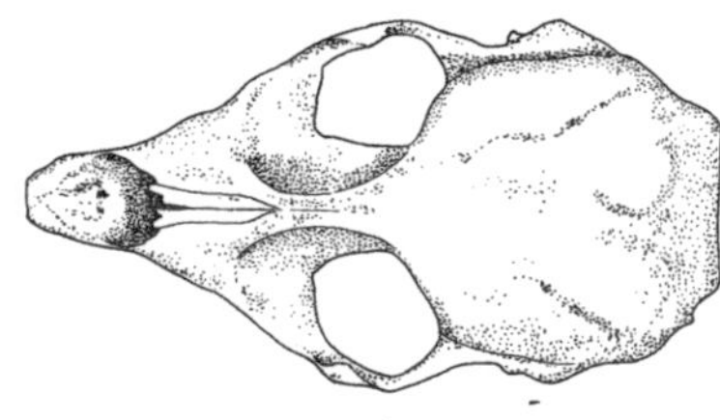

Dorsal View

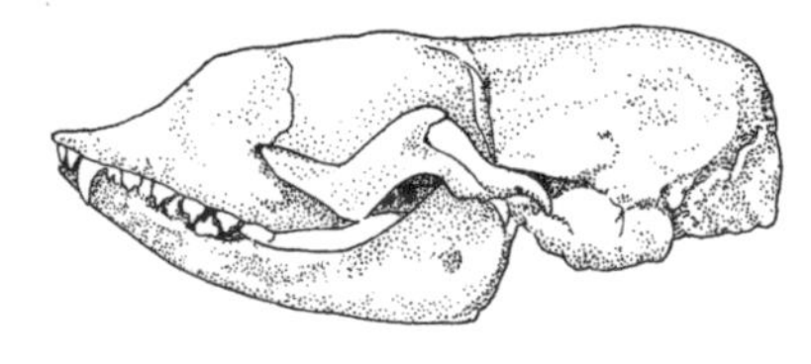

Lateral View

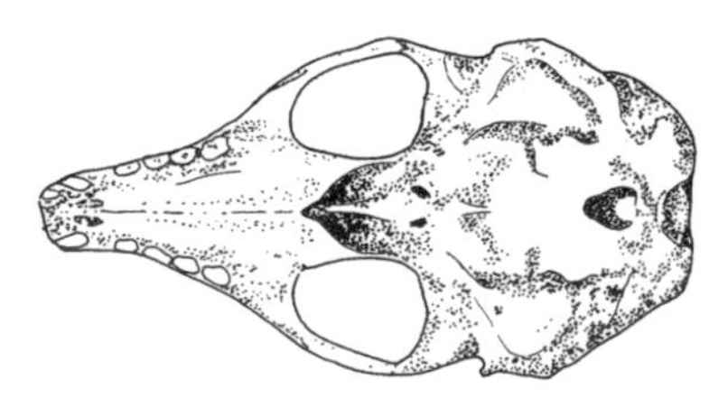

Ventral View

b. Interorbital area relatively broad (interorbital width generally >7 mm)
***Histriophoca* and *Pagophilus*, Go to 15**

15a. Alveolar edge of premaxillae arched in horizontal and vertical planes; posterior palate short and very broad; distributed only in the North Pacific and adjacent Arctic waters
Ribbon seal *Histriophoca fasciata*

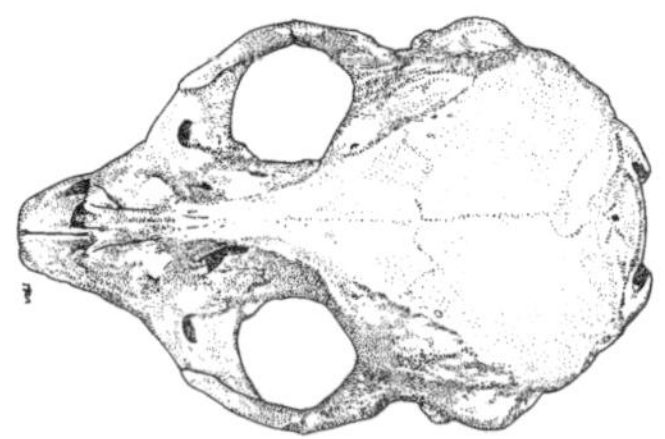

Dorsal View

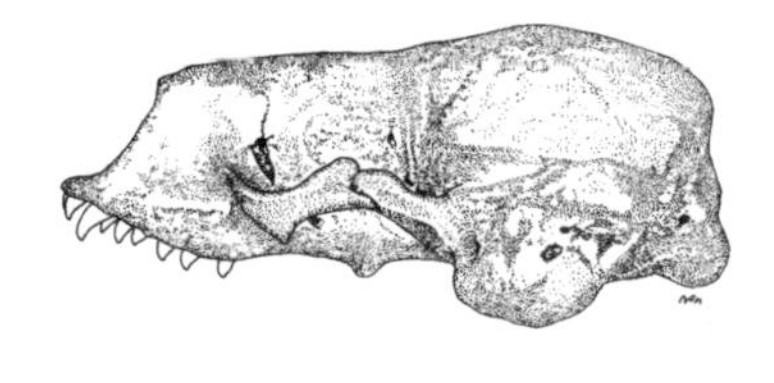

Lateral View

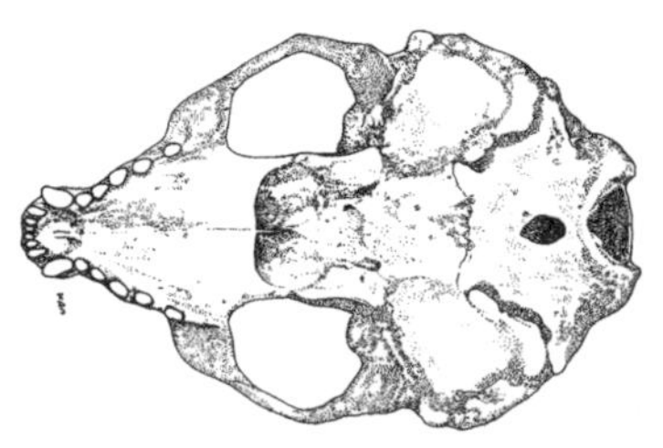

Ventral View

b. Alveolar edge of premaxillae straight or only slightly arched; posterior palate relatively long and narrow; distributed only in the North Atlantic and adjacent Arctic waters

Harp seal *Pagophilus groenlandicus*

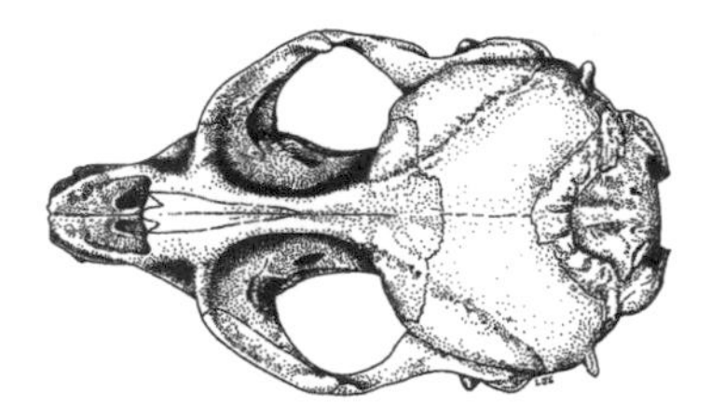

Dorsal View

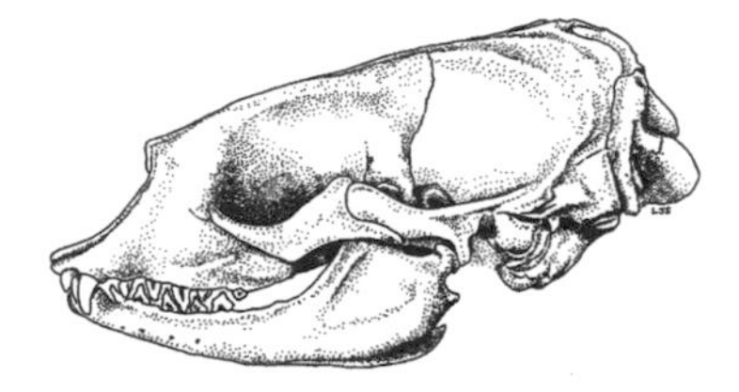

Lateral View

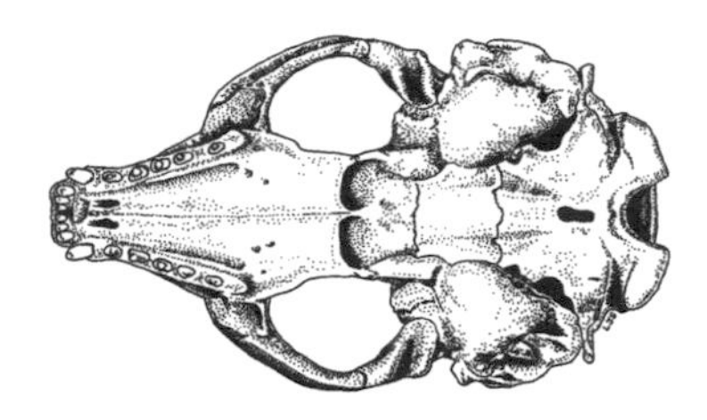

Ventral View

16a. Muzzle low and long, generally with concave upper margin; distributed only in the mid- to high latitudes of the North Pacific and Southern Hemisphere

Elephant seals *Mirounga* spp.

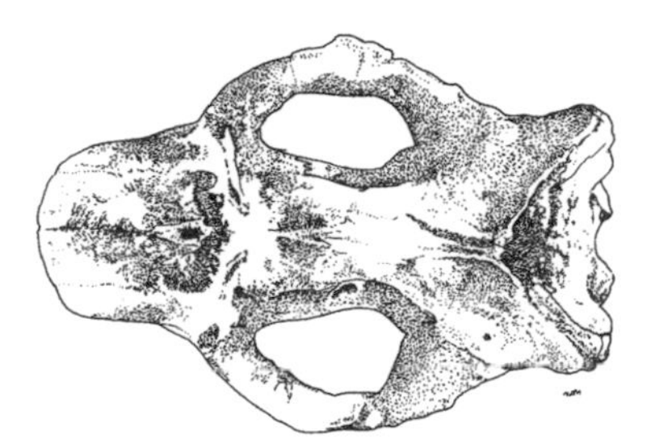

Dorsal View

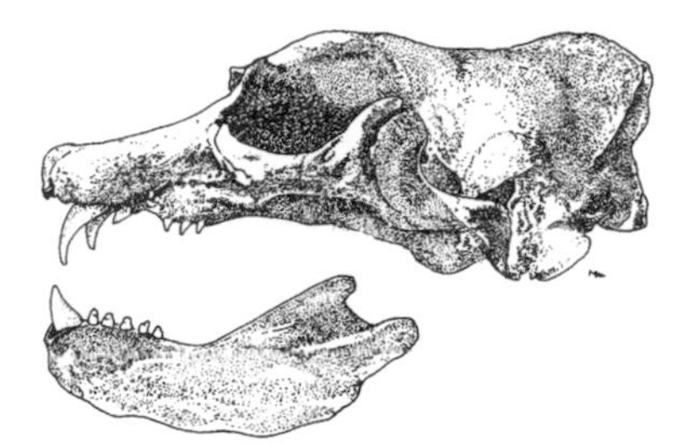

Lateral View

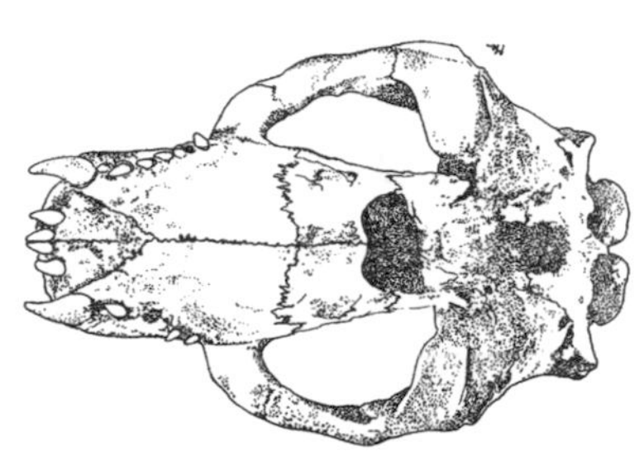

Ventral View

b. Muzzle relatively high and with straight or convex upper margin

Other species, Go to 17

17a. Nasal process of premaxillae broadly in contact with nasals; post-canines wide and heavy (crushing type); distributed only tropical and warm temperate regions

Monk seals *Neomonachus* and *Monachus* spp.

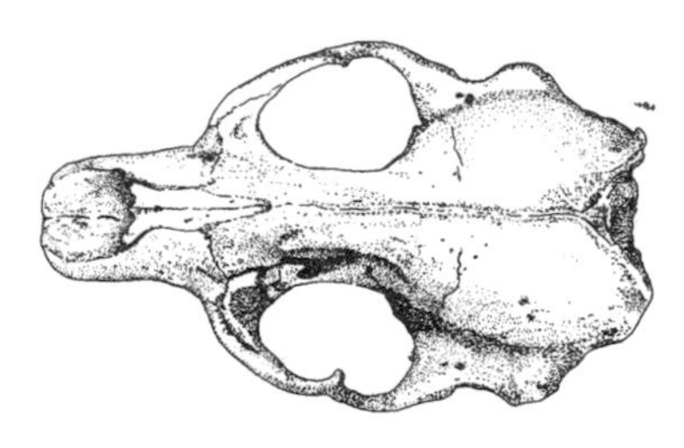

Dorsal View

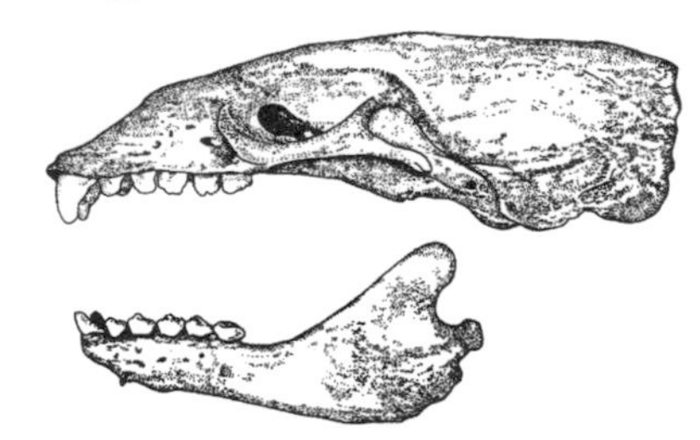

Lateral View

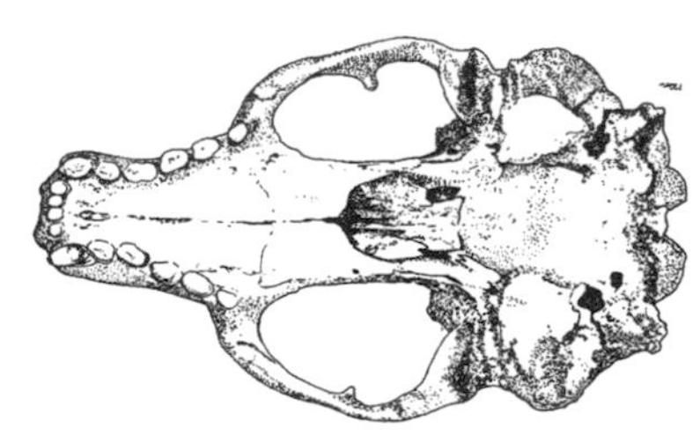

Ventral View

b. Nasal process of premaxillae barely touching or not touching nasals; post-canines not crushing type; distributed only in high latitude areas of the Southern Hemisphere

***Lobodon and Hydrurga*, Go to 18**

18a. Highly specialized cheek teeth, with well-developed accessory cusps; muzzle long; occiput high

***Lobodon* and *Hydrurga*, Go to 19**

b. Cheek teeth reduced, with small or absent accessory cusps; muzzle relatively short; occiput relatively low

***Leptonychotes* and *Ommatophoca*, Go to 20**

19a. Post-canines with 4–5 cusps

Crabeater seal *Lobodon carcinophagus*

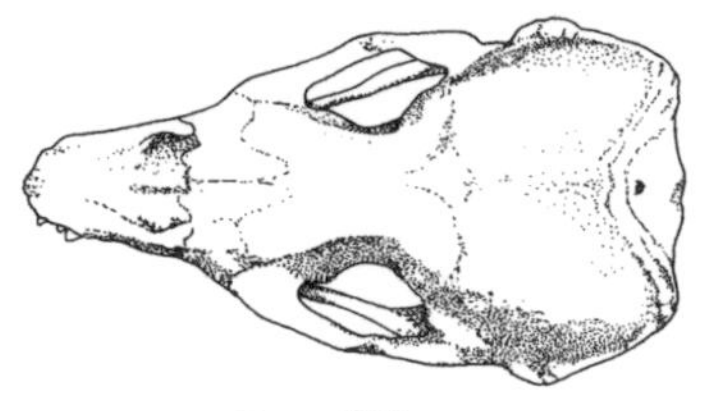

Dorsal View

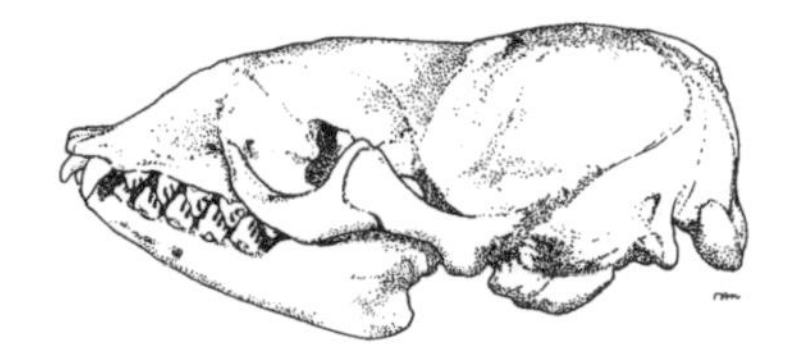

Lateral View

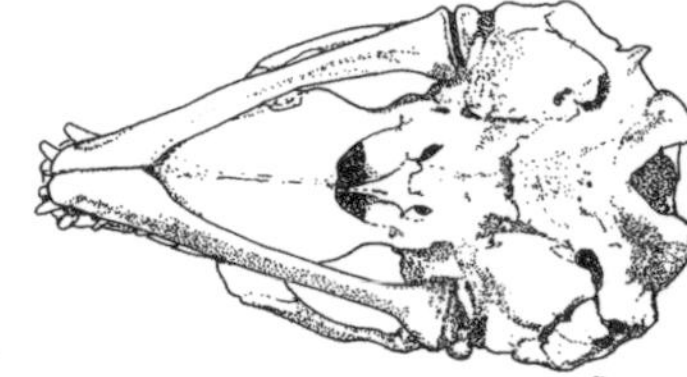

Ventral View

b. Post-canines tricuspid

Leopard seal *Hydrurga leptonyx*

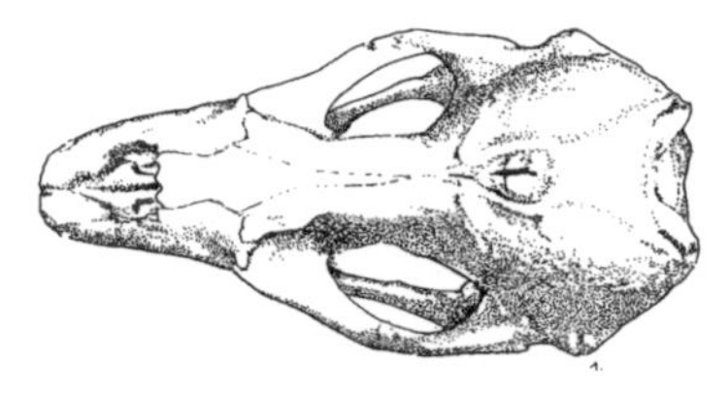

Dorsal View

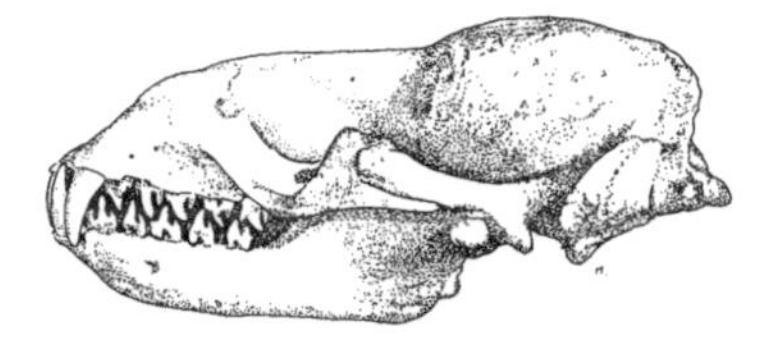

Lateral View

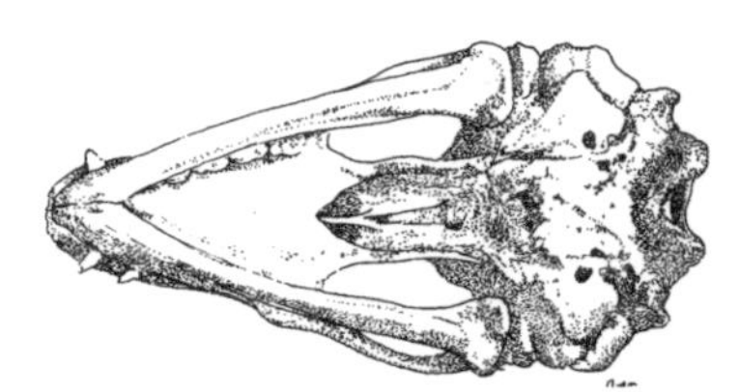

Ventral View

20a. Upper incisors of unequal size (inner ones much smaller than outer ones); upper incisors and canines with oblique roots

Weddell seal *Leptonychotes weddellii*

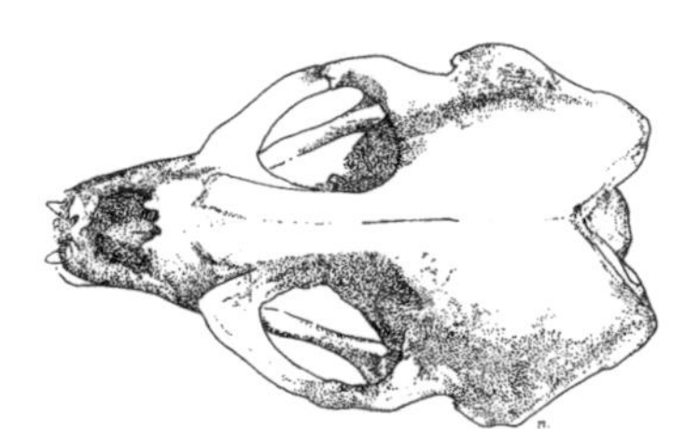

Dorsal View

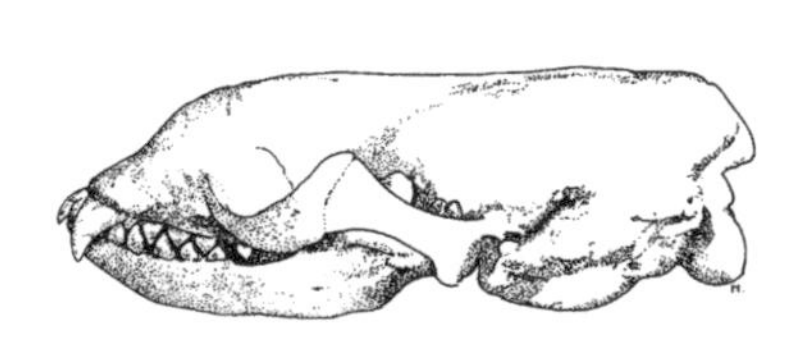

Lateral View

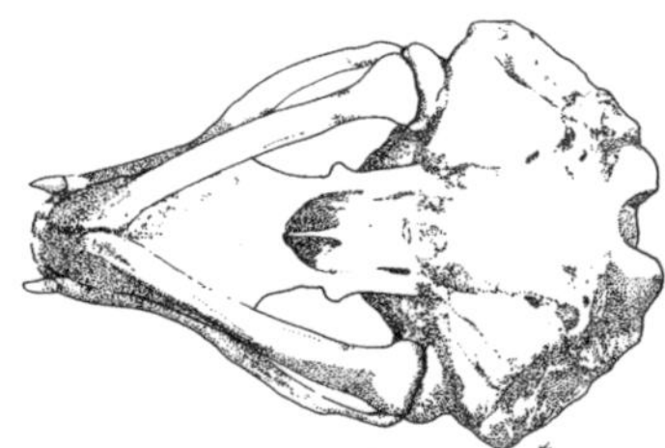

Ventral View

b. Upper incisors all of similar size, with vertical orientation

Ross seal *Ommatophoca rossii*

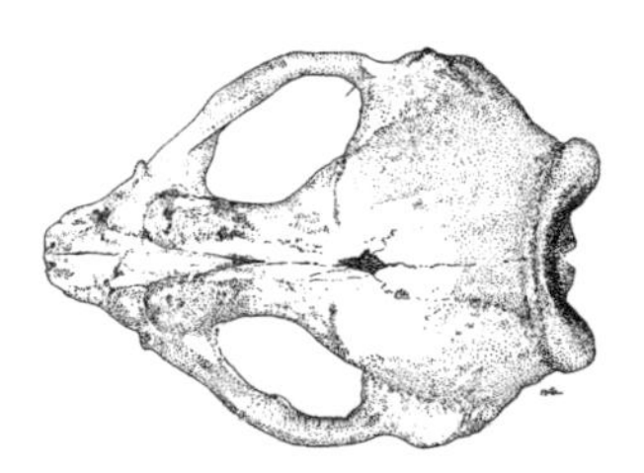

Dorsal View

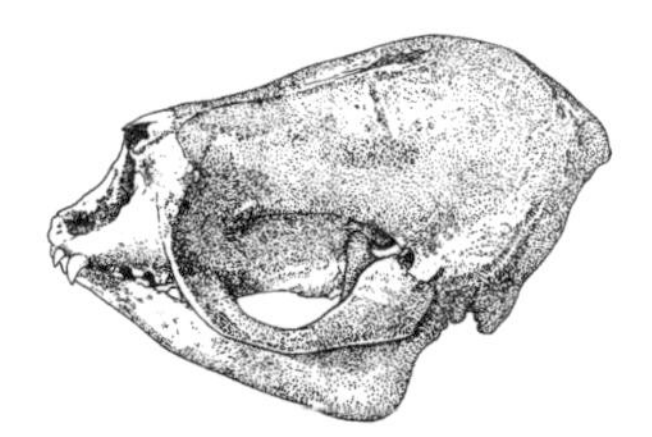

Lateral View

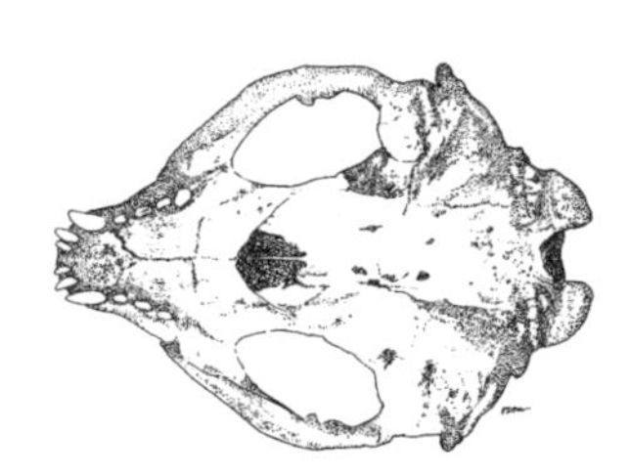

Ventral View

F. Key to Identification of Sirenians of the World, Based on External Appearance and Distribution

1a. Tail split into flukes, with a median notch; tail stock laterally compressed; nostrils on top of snout; incisors (tusks) present; distribution limited to the Indo-Pacific region

Dugong *Dugong dugon*

b. Tail rounded and paddle-like; tail stock not laterally compressed; nostrils at front of snout; incisors not present in adults; distribution limited to Atlantic Ocean and surrounding seas and rivers

Manatee *Trichechus* spp., Go to 2

2a. No nails on flippers; skin of non-calves unwrinkled; light patches on belly and chest; maximum length 3 m; distribution limited to Amazon River and its tributaries

Amazonian manatee *Trichechus inunguis*

b. Nails present on flippers; skin wrinkled; generally, no light ventral patches; occurrence near the Amazon River limited to vicinity of river mouth

West Indian or African manatee, Go to 3

3a. Distribution limited to coastal and inland waters of West Africa

African manatee *Trichechus senegalensis*

b. Distribution limited to waters of the southeastern United States, Gulf of Mexico, Caribbean Sea, and northeastern coast of South America

West Indian manatee *Trichechus manatus*

G. Key to Identification of Sirenians of the World, Based on Skull Morphology

1a. Rostrum long and strongly deflected (generally >60°); outer pair of upper incisors enlarged into tusks

Dugong *Dugong dugon*

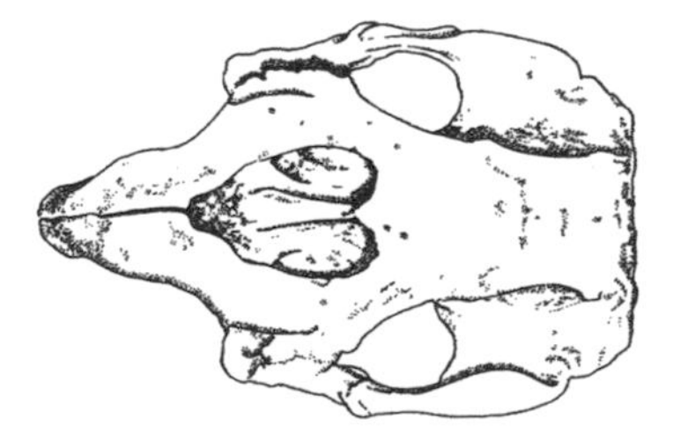

Dorsal View

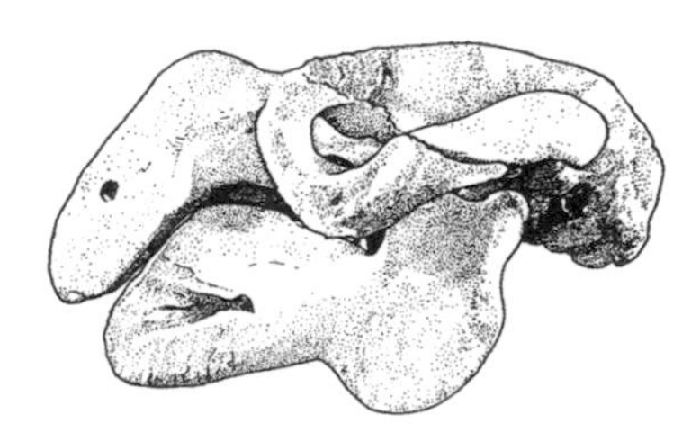

Lateral View

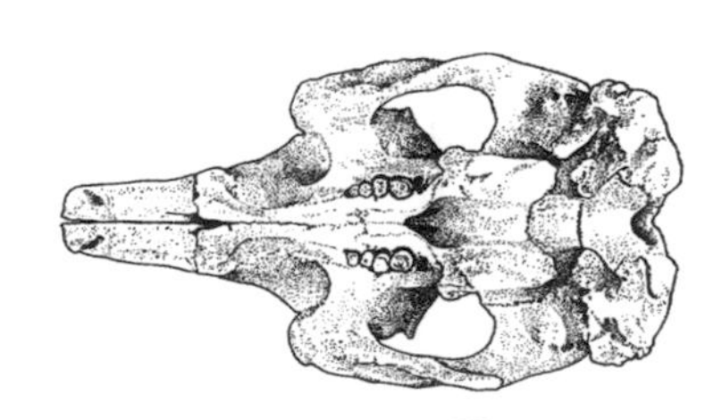

Ventral View

b. Rostrum relatively short and less strongly deflected (15–52°); incisors vestigial and generally lost in adults (no tusks)

Manatee (*Trichechus* spp.), Go to 2

2a. Skull relatively narrow and elongate; rostrum long and narrow; temporal crests laterally overhanging; zygomatic arches at angle of 25° laterally from long axis of skull

Amazonian manatee *Trichechus inunguis*

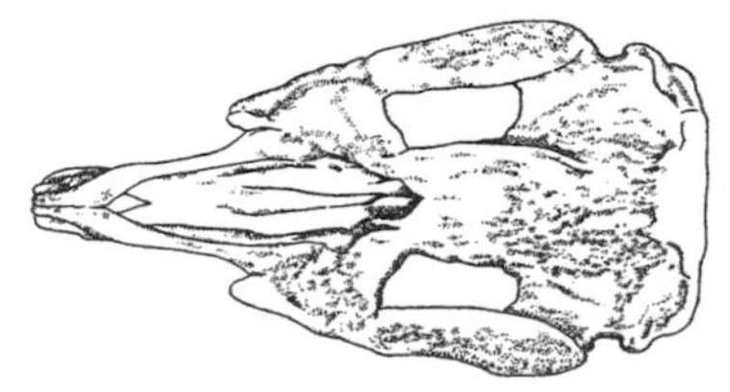

Dorsal View

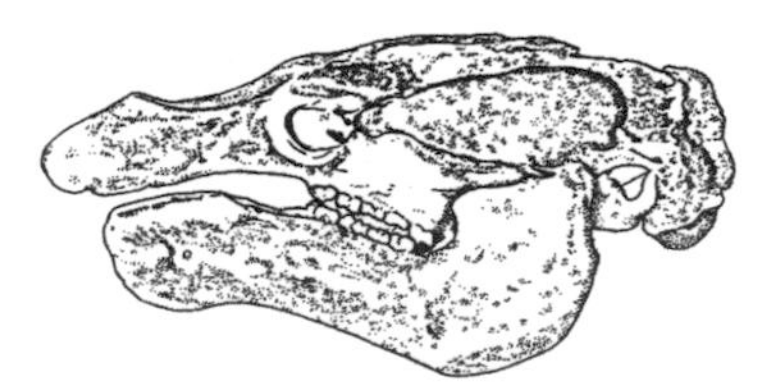

Lateral View

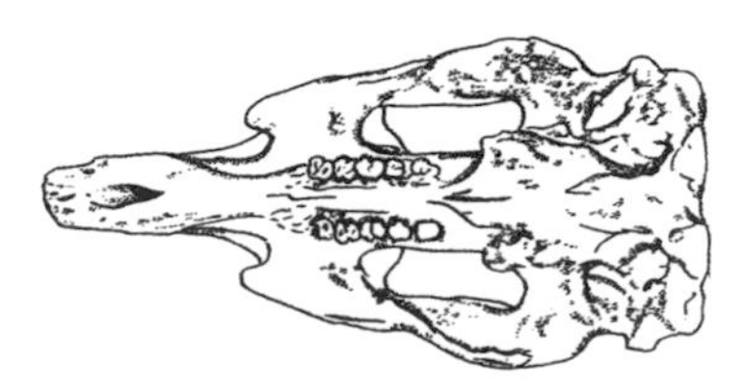

Ventral View

b. Skull relatively broad; rostrum shorter and wider; temporal crests usually rise above level of skull roof; zygomatic arches at angle of 35–40° laterally from long axis of skull

West Indian or African manatee, Go to 3

3a. Rostral deflection 15–40°; nasal bones usually absent; anterior border of frontals usually smooth; distribution limited to Western Africa

African manatee *Trichechus senegalensis*

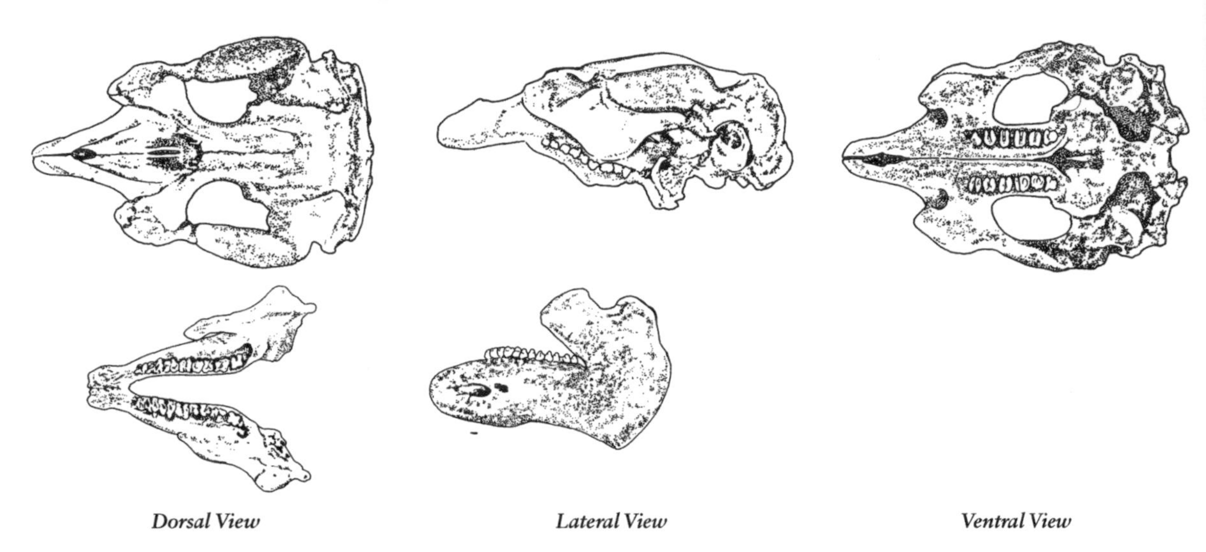

Dorsal View *Lateral View* *Ventral View*

b. Rostral deflection 29–52°; nasal bones usually present; anterior border of frontals usually jagged; distribution limited to western Atlantic Ocean

West Indian manatee *Trichechus manatus*

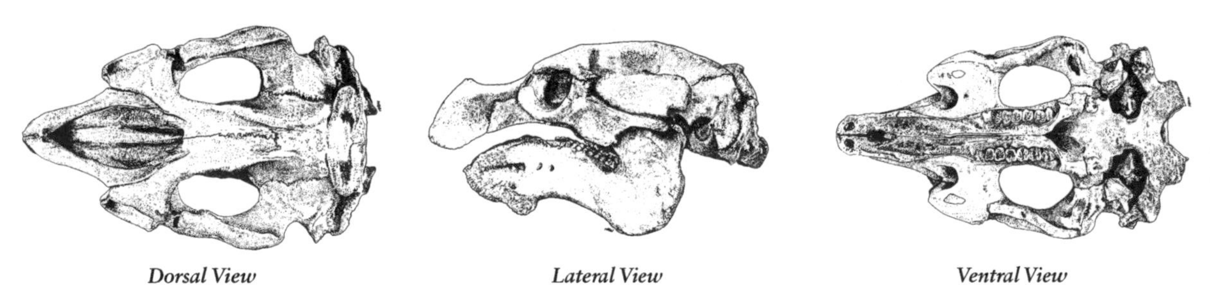

Dorsal View *Lateral View* *Ventral View*

10. Sources for More Information

Marine Mammals—General Biology

Berta, A., J. L. Sumich, and K. M. Kovacs. 2006. *Marine Mammals: Evolutionary Biology* (Second Edition). Academic Press. 547 pp.

Boyd, I. L., W. D. Bowen, and S. J. Iverson (Editors). 2010. *Marine Mammal Ecology and Conservation: A Handbook of Techniques*. Oxford University Press. 450 pp.

Hoelzel, A. R. (Editor). 2002. *Marine Mammal Biology: An Evolutionary Approach*. Blackwell Science Publishing. 432 pp.

Parsons, E. C. M. 2013. *An Introduction to Marine Mammal Biology and Conservation*. Jones and Bartlett Learning. 345 pp.

Perrin, W. F., B. Würsig, and J. G. M. Thewissen (Editors). 2009. *Encyclopedia of Marine Mammals* (Second Edition). Academic Press. 1,316 pp.

Reynolds, J. E., and S. A. Rommel (Editors). 1999. *Biology of Marine Mammals*. Smithsonian Institution Press. 578 pp.

Society for Marine Mammalogy. 1985-present. *Marine Mammal Science* (Journal). Wiley Online Library, misc. pagination.

Wilson, D. E. and R. A. Mittermeier (Editors). 2014. *Handbook of the Mammals of the World. Vol. 4. Sea Mammals*. Lynx Edicions. 614 pp.

Marine Mammals—Evolutionary Biology

Berta, A. 2012. *Return to the Sea: The Life and Evolutionary Times of Marine Mammals*. University of California Press. 205 pp.

Thewissen, J. G. M. (Editor) 1998. *The Emergence of Whales*. Plenum Press. 477 pp.

Marine Mammals—Conservation Biology

Reynolds, J. E., W. F. Perrin, R. R. Reeves, S. Montgomery, and T. J. Ragen. 2005. *Marine Mammal Research: Conservation Beyond Crisis*. Johns Hopkins University Press. 223 pp.

Twiss, J. R., and R. R. Reeves (Editors). 1999. *Conservation and Management of Marine Mammals*. Smithsonian Institution Press. 471 pp.

Cetaceans

Heyning. J. E. 1995. *Whales, Dolphins, Porpoises: Masters of the Ocean Realm*. University of Washington Press. 112 pp.

Mead, J. G., and J. P. Gold. 2002. *Whales and Dolphins in Question*. Smithsonian Institution Press. 200 pp.

Ridgway, S. H., and R. Harrison (Editors). 1985. *Handbook of Marine Mammals, Vol. 3: The Sirenians and Baleen Whales*. Academic Press Inc., London. 362 pp.

Ridgway, S. H., and R. Harrison (Editors). 1989. *Handbook of Marine Mammals, Vol. 4: River Dolphins and the Larger Toothed Whales*. Academic Press. 442 pp.

Ridgway, S. H., and R. Harrison (Editors). 1994. *Handbook of Marine Mammals, Vol. 5: The First Book of Dolphins*. Academic Press. 416 pp.

Ridgway, S. H., and R. Harrison (Editors). 1999. *Handbook of Marine Mammals, Vol. 6: The Second Book of Dolphins and the Porpoises*. Academic Press. 486 pp.

Pinnipeds

Ridgway, S. H., and R. J. Harrison (Editors). 1981. *Handbook of Marine Mammals, Vol. 1: The Walrus, Sea Lions, Fur Seals, and Sea Otter*. Academic Press. 235 pp.

Ridgway, S. H., and R. J. Harrison (Editors). 1981. *Handbook of Marine Mammals, Vol. 2: Seals*. Academic Press. 359 pp.

Riedman, M. 1990. *The Pinnipeds: Seals, Sea Lions, and Walruses*. University of California Press. 439 pp.

Sirenians

Marsh, H., T. J. O'Shea, and J. E. Reynolds III. 2011. *Ecology and Conservation of the Sirenia: Dugongs and Manatees*. Cambridge University Press. 521 pp.

Reynolds, J. E., and D. K. Odell. 1991. *Manatees and Dugongs*. Facts on File. 192 pp.

Ridgway, S. H., and R. Harrison (Editors). 1985. *Handbook of Marine Mammals, Vol. 3: The Sirenians and Baleen Whales*. Academic Press Inc., London. 362 pp.

Otters and Bears

Riedman, M. 1990. *Sea Otters*. Monterey Bay Aquarium. 80 pp.

Stirling, I. 2011. *Polar Bears: The Natural History of A Threatened Species*. Fitzhenry and Whiteside. 334 pp.

VanBlaricom, G. 2001. *Sea Otters*. Voyageur Press. 72 pp.

11. Index—Common Names

Note: Locators in **bold** refer to the opening page of species accounts.

African manatee, **527**, 599, 600
Amazon River dolphin, **350**, 562, 574
Amazonian manatee, 524, **529**, 599, 600
Andrews' beaked whale, 130, **162**, 560
Antarctic blue whale, **49**
Antarctic fur seal, 4, 366, **368**, 589
Antarctic minke whale, 70, 71, **75**, 188, 560
Antillean manatee, **524**
Arnoux's beaked whale, **106**, 560, 572
Atlantic bearded seal, **484**
Atlantic humpback dolphin, 16, **224**, 567, 580
Atlantic spotted dolphin, 241, **251**, 567, 583
Atlantic walrus, **440**
Atlantic white-sided dolphin, **285**, 564, 584
Australian fur seal, 393, **399**, 432, 589
Australian sea lion, **431**, 589, 595
Australian snubfin dolphin, 13, **180**, 566
Australian humpback dolphin, 20, 180, **234**, 565, 578

Baiji, 2, 15, 20, 337, **548**, 562, 574
Baikal seal, **518**, 590
Baird's beaked whale, **102**, 113, 559
Beaked whales, 2, 5, 13, 17, 19, 20, 46, 73, 75, 99, **102**, 573
Bearded seal, 452, **484**, 498, 590, 595
Beluga whale, 174, **176**, 577
Bhulan, 20, **346**, 353, 355
Bigg's killer whale, **186**
Black Sea common dolphin, **268**, 270
Blainville's beaked whale, **125**, 130, 144, 165, 560
Blue whale, 18, **49**, 58, 67, 558, 569, 587
Boto, 20, 217, **350**, 548
Bottlenose dolphin, 4, 155, 212, 214, 215, 217, 223, 226, 228, 232, 234, **237**, 252, 282, 329, 348, 357, 564, 579
Bowhead whale, **42**, 558, 567, 587
Bryde's whale, **63**, 69, 558, 587
Burmeister's porpoise, 331, **333**, 561, 576

California sea lion, 363, 366, 377, 408, 410, **413**, 424, 551, 589, 594
Caribbean monk seal, 446, **553**, 590
Cape fur seal, **399**, 477, 589
Caspian seal, **521**, 590
Central American spinner dolphin, **255**, 257, 566
Chilean dolphin, **319**, 563, 586
Chungungo, see Marine otter
Clymene dolphin, 3, **261**, 270, 566, 583
Coastal pantropical spotted dolphin, **246**
Commerson's dolphin, **309**, 319, 331, 563, 586
Common bottlenose dolphin, 3, 4, 224, **237**, 564, 579
Common dolphin, 228, 234, 262, **268**, 291, 293, 343, 565
Common minke whale, **70**, 75, 558
Common seal, see Harbor seal
Costero, see Guiana dolphin
Crabeater seal, **458**, 590, 598
Cuvier's beaked whale, 103, **109**, 129, 171, 560, 572

dalli-type Dall's porpoise, **322**
Dall's porpoise, 3, 13, 192, **322**, 327, 561, 575
Deraniyagala's beaked whale, 130, **138**, 560
Dugong, 13, 16, 21, **532**, 555, 599
Dusky dolphin, 289, **293**, 301, 306, 429, 564, 585
Dwarf minke whale, 3, **70**
Dwarf sperm whale, 96, 97, **99**, 559, 571
Dwarf spinner dolphin, **255**, 258, 566

Eastern spinner dolphin, **255**, 566
Elephant seal, **472**, 597

False killer whale, **200**, 204, 205, 563, 581
Fin whale, **54**, 61, 65, 67, 68, 69, 298, 558, 569
Finless porpoise, 180, 183, **339**
Florida manatee, **524**
Franciscana, 20, 217, 257, **355**, 548, 562, 574
Fraser's dolphin, **278**, 562, 581
Galapagos fur seal, **381**, 415, 589
Galapagos sea lion, 363, 415, 420, **421**, 589, 594
Ganges River dolphin, **346**, 348, 562
Gervais' beaked whale, 130, **152**, 348, 560
Ginkgo-toothed beaked whale, 130, **136**, 560
Goosebeaked whale, see Cuvier's beaked whale
Gray seal, **492**, 504, 591, 596
Gray whale, 10, 19, **84**, 558, 568, 587
Gray's beaked whale, 130, **132**, 140, 141, 559
Gray's spinner dolphin, **255**, 257, 566
Greenland right whale, see Bowhead whale
Guadalupe fur seal, 363, **374**, 415, 418, 589
Guiana dolphin, 214, **217**, 356
Gulf of California (harbor) porpoise, see Vaquita

Harbor porpoise, 4, 11, 323, **326**, 561, 576
Harbor seal, 418, 481, 495, 502, **504**, 509, 513, 514, 591
Harp seal, 366, 461, 490, **500**, 591, 597
Hawaiian monk seal, **446**, 553, 554, 590
Heaviside's dolphin, **313**, 563, 585
Hector's beaked whale, 130, **140**, 168, 560
Hector's dolphin, **316**, 563, 586
Hooded seal, 366, 488, **489**, 591, 595
Hooker's sea lion, see New Zealand sea lion
Hourglass dolphin, **297**, 564, 585
Hubbs' beaked whale, 130, **144**, 162, 163, 560
Humpback whale, 11, 18, 35, **79**, 25, 558, 569, 587

Indian Ocean humpback dolphin, 224, 226, **230**, 565
Indo–Pacific beaked whale, see Longman's beaked whale
Indo–Pacific common dolphin, **273**, 275
Indo–Pacific bottlenose dolphin, 237, **243**, 251, 253, 565, 580
Indo–Pacific finless porpoise, **339**, 561, 575

Indo–Pacific humpback dolphin, 226, **227**, 565, 578
Indus River dolphin, **346**, 562
Irrawaddy dolphin, 11, 180, 181, **183**, 566

Japanese sea lion, 16, **551**, 589
Juan Fernandez fur seal, 13, 366, 371, **374**, 429, 589

Kerguelen Islands Commerson's dolphin, **309**
Killer whale, 9, 10, 12, 20, 45, 50, **186**, 314, 563, 581

La Plata river dolphin, see Franciscana
Laptev walrus, see Walrus
Largha seal, see Spotted seal
Leopard seal, **463**, 590, 598
Long-beaked common dolphin, **273**, 565, 582
Long-finned pilot whale, **197**, 563, 580
Longman's beaked whale, 119, 120, 156, **169**, 560, 573
Loughlin's Steller sea lion, **406**, 418

Manatees, 1, 8, 9, 12, 13, 16, 21, **324**, 532, 554
Marine otter, 7, 21, **539**
Maui's dolphin, 16, **316**
Mediterranean monk seal, 446, **451**, 553, 590
Melon-headed whale, 204, 205, **207**, 563, 579

Narrow-ridged finless porpoise, **342**, 561, 575
Narwhal, 11, 19, **172**, 176, 562, 576
New Zealand fur seal, 366, 385, **390**, 589
New Zealand sea lion, **435**, 589, 594
North Atlantic right whale, 15, **30**, 539
North Island Hector's dolphin, see Maui's dolphin
North Pacific right whale, **34**, 559
North Sea beaked whale, see Sowerby's beaked whale
Northern bottlenose whale, **114**, 119, 560
Northern elephant seal, 2, 16, 449, 472, **478**, 590
Northern fur seal, **360**, 366, 377, 415, 416, 588, 592
Northern right whale dolphin, **303**, 305, 562
Northern sea lion, see Steller sea lion

Offshore killer whale, **186**
Offshore pantropical spotted dolphin, **246**
Omura's whale, 60, 65, **67**, 560, 587

Pacific bearded seal, **484**
Pacific walrus, **440**
Pacific white-sided dolphin, **289**, 293, 562, 585
Pantropical spotted dolphin, **246**, 251, 252, 258, 260, 264, 566, 582
Peale's dolphin, 294, **300**, 321, 564, 584
Perrin's beaked whale, 130, 140, 141, **142**, 561
Peruvian fur seal, **385**
Polar bear, 1, 2, 7, 8, 10, 12, 16, 21, 22, **542**
Pygmy beaked whale, 130, 142, **146**, 560
Pygmy killer whale, **204**, 207, 563, 580
Pygmy right whale, **46**
Pygmy sperm whale, **95**, 100, 559, 571

Resident killer whale, **186**
Ribbon seal, **496**, 505, 511, 513, 591, 596
Ringed seal, 486, 501, 51, **513**, 518, 545, 591
Risso's dolphin, 3, 4, 191, 196, **210**, 237, 292, 304, 562, 579
Ross seal, **455**, 459, 461, 465, 468, 470, 484, 588, 590, 598
Rough-toothed dolphin, 11, 132, **220**, 238, 564, 577

Scamperdown whale, see Gray's beaked whale
Sea otter, 1, 7, 8, 9, 12, 16, 21, **536**, 541
Sei whale, 49, 50, 52, **59**, 63, 65, 66, 67, 75, 558, 570, 587
Shepherd's beaked whale, **122**, 559, 571
Short-beaked common dolphin, 3, 262, **268**, 273, 274, 565, 581
Short-finned pilot whale, **193**, 197, 237, 563, 580
South African fur seal, see Cape fur seal
South American Commerson's dolphin, **309**, 310, 319
South American fur seal, 333, 371, 372, 377, 381, **385**, 390, 422, 430, 589
South American sea lion, 333, 366, 377, 383, 387, 388, 389, 415, 422, 424, **426**, 589, 593
South Asian river dolphin, 20, 340, **346**, 548, 573
South Island Hector's dolphin, **316**
Southern bottlenose whale, 107, 108, 118, **119**, 170, 560
Southern elephant seal, 430, 459, 466, 468, **472**, 590
Southern right whale, 30, 31, 34, **37**, 559
Southern right whale dolphin, 3, 293, 303, **306**, 313, 562
Sowerby's beaked whale, 117, 130, **149**, 561
Spade-toothed beaked whale, 130, 159, **167**, 561
Spectacled porpoise, 20, 310, **330**, 334, 561, 575
Sperm whale, 9, 15, 19, 32, 82, **88**, 121, 196, 213, 213, 559, 571
Spinner dolphin, 4, 171, 232, 246, 249, **255**, 261, 263, 268, 270, 274, 275, 280, 566, 583
Spotted seal, 486, 498, 505, 506, **509**, 591, 596
Stejneger's beaked whale, 130, 137, 143, **164**, 560
Steller's sea cow, 12, 16, 21, **555**
Steller sea lion, 12, 16, 363, **406**, 413, 416, 418, 419, 508, 551, 552, 589, 593
Strap-toothed beaked whale, 130, 133, **158**, 167, 560
Striped dolphin, 3, 261, 262, 263, **264**, 281, 565, 565, 583
Subantarctic fur seal, 3, 366, 371, 377, **395**
Susu, 16, 20, **346**, 353, 355

truei-type Dall's porpoise, **322**
True's beaked whale, 130, 152, **155**, 170, 560
Tucuxi, **214**, 217, 219, 353, 565, 578

Vaquita, 2, 15, 16, **336**, 561, 576

Walrus, 7, 8, 22, 23, **440**, 488, 516, 546, 588, 592
Weddell seal, 192, 455, 459, 461, **468**, 590, 598
West Indian manatee, **524**, 599, 600
Western Steller sea lion, **406**
White-beaked dolphin, **282**, 285, 287, 564, 584
White whale, see Beluga whale
Whitebelly spinner dolphin, 4, **255**, 566

Yangtze finless porpoise, **342**
Yangtze River dolphin, see Baiji

12. Index—Scientific Names

Note: Locators in **bold** refer to the opening page of species account.

Arctocephalus australis, 381, **385**, 589
Arctocephalus australis galapagoensis, see *Arctocephalus galapagoensis*
Arctocephalus doriferus, see *Arctocephalus forsteri; Arctocephalus pusillus*
Arctocephalus forsteri, **390**, 399, 589
Arctocephalus galapagoensis, **381**, 589
Arctocephalus gazella, **368**, 395, 589
Arctocephalus philippii, **374**, 589
Arctocephalus philippii philippii, see *Arctocephalus philippii*
Arctocephalus philippii townsendi, see *Arctocephalus philippii*
Arctocephalus pusillus, **399**, 589
Arctocephalus tasmanicus, see *Arctocephalus pusillus*
Arctocephalus townsendi, 374
Arctocephalus tropicalis, 366, **395**, 588
Arctocephalus tropicalis gazella, see *Arctocephalus gazella*
Arctocephalus tropicalis tropicalis, see *Arctocephalus tropicalis*
Arctophoca australis, see *Arctocephalus australis*
Arctophoca australis australis, see *Arctocephalus australis*
Arctophoca australis forsteri, see *Arctocephalus forsteri*
Arctophoca galapagoensis, see *Arctocephalus galapagoensis*
Arctophoca gazella, see *Arctocephalus gazella*
Arctophoca philippii, see *Arctocephalus philippii*
Arctophoca townsendi, see *Arctocephalus philippii*
Arctophoca tropicalis, see *Arctocephalus tropicalis*
Australophocaena dioptrica, see *Phocoena dioptrica*

Balaena mysticetus, **42**, 556, 565, 587
Balaenoptera acutorostrata, **70**, 558, 569, 587
Balaenoptera acutorostrata acutorostrata, **70**, 75, 558, 569, 587
Balaenoptera acutorostrata scammoni, **70**
Balaenoptera bonaerensis, **75**, 556, 558, 569, 587
Balaenoptera borealis, **59**, 63, 558, 570, 587
Balaenoptera borealis borealis, **59**
Balaenoptera borealis schlegelli, **59**
Balaenoptera brydei, **63**, 558, 570, 587
Balaenoptera davidsoni, see *Balaenoptera acutorostrata*
Balaenoptera edeni, **63**, 65, 558, 570, 587
Balaenoptera musculus, **49**, 558, 569, 587
Balaenoptera musculus brevicauda, **49**
Balaenoptera musculus indica, **49**
Balaenoptera musculus intermedia, **49**
Balaenoptera musculus musculus, **49**
Balaenoptera omurai, 54, 63, **67**, 558, 569, 587
Balaenoptera physalus patachonica, **54**
Balaenoptera physalus, **54**, 558, 569, 587
Balaenoptera physalus quoyi, **54**
Berardius arnuxii, **106**, 560, 572
Berardius bairdii, **102**, 559, 572

Callorhinus ursinus, **360**, 588, 592
Caperea marginata, **46**, 557, 568, 587
Cephalorhynchus albifrons, see *Cephalorhynchus hectori*
Cephalorhynchus albiventris, see *Cephalorhynchus eutropia*
Cephalorhynchus commersonii, **309**, 310, 563, 586
Cephalorhynchus commersonii commersonii, **309**
Cephalorhynchus commersonii kerguelenensis, **309**
Cephalorhynchus eutropia, **319**, 563, 586
Cephalorhynchus heavisidii, **319**, 563, 586
Cephalorhynchus hectori, **316**, 563, 586
Cephalorhynchus hectori hectori, **316**
Cephalorhynchus hectori maui, **316**
Cystophora cristata, **489**, 591, 595

Delphinapterus dorofeevi, see *Delphinapterus leucas*
Delphinapterus freimani, see *Delphinapterus leucas*
Delphinapterus leucas, **176**, 562, 577
Delphinus bairdii, see *Delphinus capensis*
Delphinus capensis, 258, 268, **273**, 565, 582
Delphinus capensis capensis, 258, 268, **273**, 565, 582
Delphinus capensis tropicalis, **273**
Delphinus delphis, **268**, 273, 276, 565, 581
Delphinus delphis ponticus, **268**
Delphinus tropicalis, see *Delphinus capensis*
Dugong dugon, **532**, 599
Dugong dugon dugon, **532**
Dugong dugon hemrichii, **532**

Electra electra, see *Peponocephala electra*
Enhydra lutris, **536**, 537
Enhydra lutris kenyoni, **536**
Enhydra lutris lutris, **536**
Enhydra lutris nereis, **536**
Erignathus barbatus, **484**, 590, 595
Erignathus barbatus barbatus, **484**
Erignathus barbatus nauticus, **484**
Eschrichtius gibbosus, see *Eschrichtius robustus*
Eschrichtius glaucus, see *Eschrichtius robustus*
Eschrichtius robustus, **84**, 558, 568, 587
Eubalaena australis, 30, **37**, 559, 568, 587
Eubalaena glacialis, **30**, 34, 36, 559, 568, 587
Eubalaena japonica, 30, **34**, 559, 568, 587
Eumetopias jubatus, **406**, 589, 593
Eumetopias jubatus jubatus, **406**
Eumetopias jubatus monteriensis, **406**

Feresa attenuata, **204**, 209, 563, 580
Feresa intermedia, see *Feresa attenuata*

Globicephala brachyptera, see *Globicephala macrorhynchus*
Globicephala edwardii, see *Globicephala melas*
Globicephala macrorhynchus, **193**, 197, 563, 580
Globicephala melaena, see *Globicephala melas*
Globicephala melas, 195, **197**, 563, 580
Globicephala scammoni, see
Globicephala macrorhynchus
Globicephala seiboldii, see *Globicephala macrorhynchus*
Grampus griseus, **210**, 562, 579
Grampidelphis griseus, see *Grampus griseus*

Halichoerus grypus, **492**, 591, 596
Halichoerus grypus atlantica, see *Halichoerus grypus*
Halichoerus grypus baltica, see *Halichoerus grypus*
Halichoerus grypus grypus, **492**
Halichoerus grypus macrorhynchus, **492**
Histriophoca fasciata, **496**, 591, 596
Hydrodamalis gigas, **555**
Hydrurga leptonyx, **463**, 590, 598
Hyperoodon ampullatus, **114**, 560, 572
Hyperoodon planifrons, **119**, 560, 572
Hyperoodon rostratus, see *Hyperoodon ampullatus*

Indopacetus pacificus, 119, **169**, 560, 573
Inia araguaiaensis, see *Inia geoffrensis*
Inia boliviensis, see *Inia geoffrensis*
Inia geoffrensis boliviensis, **350**
Inia geoffrensis geoffrensis, **350**
Inia geoffrensis humboldtiana, **350**
Inia geoffrensis, **350**, 562, 574

Kogia breviceps, **95**, 101, 559, 571
Kogia sima, 96, **99**, 559, 571
Kogia simus, see *Kogia sima*

Lagenodelphis hosei, **278**, 569, 581
Lagenorhynchus acutus, **285**, 564, 584
Lagenorhynchus albirostris, **282**, 564, 584
Lagenorhynchus asia, see *Peponocephala electra*
Lagenorhynchus australis, **300**, 564, 584
Lagenorhynchus cruciger, **297**, 564, 585
Lagenorhynchus electra, see *Peponocephala electra*
Lagenorhynchus fitzroyi, see *Lagenorhynchus obscurus*
Lagenorhynchus obliquidens, **289**, 564, 585
Lagenorhynchus obscurus, **293**, 564, 585
Lagenorhynchus ognevi, see *Lagenorhynchus obliquidens*
Lagenorhynchus superciliosus, see *Lagenorhynchus obscurus*
Lagenorhynchus wilsoni, see *Lagenorhynchus cruciger*
Leptonychotes weddelli, see *Leptonychotes weddellii*
Leptonychotes weddellii, **468**, 590, 598
Leucopleurus acutus, see *Lagenorhynchus acutus*
Lipotes vexillifer, **548**, 562, 574
Lissodelphis borealis, **303**, 562, 582
Lissodelphis peronii, **306**, 562, 582
Lobodon carcinophaga, **458**, 590, 598
Lobodon carcinophagus, see *Lobodon carcinophaga*
Lontra felina, **539**
Lutra feline, see *Lontra felina*

Megaptera nodosa, see *Megaptera novaeangliae*
Megaptera novaeangliae, **79**, 558, 569, 587
Mesoplodon bahamondi, see *Mesoplodon traversii*
Mesoplodon bidens, **149**, 561, 573
Mesoplodon bowdoini, 144, **162**, 560, 573
Mesoplodon carlhubbsi, **144**, 162, 560, 573
Mesoplodon densirostris, **125**, 147, 560, 573
Mesoplodon europaeus, **152**, 560, 573
Mesoplodon gervaisi, see *Mesoplodon europaeus*
Mesoplodon ginkgodens, **136**, 138, 560, 573
Mesoplodon grayi, **132**, 163, 168, 559, 573
Mesoplodon hectori, **140**, 163, 560, 573
Mesoplodon hotaula, 130, 136, **138**, 560, 573
Mesoplodon layardii, **158**, 163, 168, 560, 573
Mesoplodon mirus, 150, **155**, 560, 573
Mesoplodon pacificus, see *Indopacetus pacificus*
Mesoplodon perrini, 130, **142**, 561
Mesoplodon peruvianus, **146**, 560, 573
Mesoplodon stejnegeri, **126**, 164, 560, 573
Mesoplodon traversii, **130**, 167, 561, 573
Mirounga angustirostris, **478**, 590, 597
Mirounga leonina, **472**, 590, 597
Monachus monachus, **451**, 590, 597
Monachus schauinslandi, see *Neomonachus schauinslandi*
Monachus tropicalis, *Neomonachus tropicalis*
Monodon monoceros, **172**, 562, 576
Neobalaena marginata, see *Caperea marginata*
Neomeris asiaeorientalis, see *Neophocaena phocaenoides*
Neomonachus schauinslandi, **446**, 590, 597
Neomonachus tropicalis, **553**, 590, 597
Neophoca cinerea, **431**, 589, 594
Neophoca hookeri, see *Phocarctos hookeri*
Neophocaena phocaenoides, **339**, 342, 561, 575
Neophocaena asiaeorientalis, 339, 341, **342**, 561, 575
Neophocaena asiaeorientalis asiaeorientalis, **342**
Neophocaena asiaeorientalis sunameri, **342**
Neophocaena sunameri, see *Neophocaena asiaeorientalis*

Odobenus rosmarus, **440**, 588, 592
Odobenus rosmarus rosmarus, **440**
Odobenus rosmarus divergens, **440**
Odobenus rosmarus laptevi see *Odobenus rosmarus*
Ommatophoca rossi, see *Ommatophoca rossii*
Ommatophoca rossii, **455**, 590, 598
Orcaella brevirostris, **183**, 533, 566, 577
Orcaella fluminalis, see *Orcaella brevirostris*
Orcaella heinsohni, **180**, 183, 234, 566, 577
Orcinus glacialis, see *Orcinus orca*
Orcinus nanus, see *Orcinus orca*
Orcinus orca, **186**, 563, 581
Orcinus rectipinna, see *Orcinus orca*
Otaria byronia, see *Otaria flavescens*
Otaria flavescens, **426**, 589, 593

Pagophilus groenlandicus, **500**, 591, 597
Pagophilus groenlandicus groenlandicus, **500**
Pagophilus groenlandicus oceanicus, **500**
Peponocephala electra, **207**, 563, 579
Phoca caspica, see *Pusa caspica*
Phoca fasciata, see *Histriophoca fasciata*
Phoca groenlandica, see *Pagophilus groenlandicus*
Phoca largha, **509**, 591, 596
Phoca sibirica, see *Pusa sibirica*
Phoca vitulina, **504**, 509, 591, 596
Phoca vitulina concolor, see *Phoca vitulina*,
Phoca vitulina mellonae, **504**
Phoca vitulina richardii, **504**
Phoca vitulina stejnegeri, see *Phoca vitulina*

Phoca vitulina vitulina, **504**
Phocarctos hookeri, **435**, 589, 594
Phocoena communis, see *Phocoena phocoena*
Phocoena dioptrica, **330**, 561, 575
Phocoena obtusata, see *Phocoena dioptrica*
Phocoena phocoena, **326**, 561, 576
Phocoena phocoena phocoena, **326**
Phocoena phocoena relicta, **326**
Phocoena phocoena vomerina, **326**
Phocoena relicta, see *Phocoena phocoena*
Phocoena sinus, **336**, 561, 576
Phocoena spinipinnis, **333**, 561, 576
Phocoena vomerina, see *Phocoena phocoena*
Phocoenoides dalli, **322**, 561, 575
Phocoenoides dalli dalli, **323**
Phocoenoides dalli truei, **323**
Phocoenoides truei, see *Phocoenoides dalli*
Physeter catodon, see *Physeter macrocephalus*
Physeter macrocephalus, **88**, 559, 571
Platanista gangetica, **346**, 562, 573
Platanista gangetica gangetica, **346**
Platanista gangetica minor, **346**
Platanista indi, see *Platanista gangetica*
Platanista minor, see *Platanista gangetica*
Pontoporia blainvillei, **355**, 562, 574
Pseudorca crassidens, **200**, 563, 581
Pusa caspica, **521**, 590, 596
Pusa hispida, **513**, 591, 596
Pusa hispida botnica, **513**
Pusa hispida hispida, **513**
Pusa hispida ladogensis, **513**
Pusa hispida ochotensis, **513**
Pusa hispida saimensis, **513**
Pusa sibirica, **518**, 590, 596

Sagmatias amblodon, see *Lagenorhynchus australis*
Sagmatias australis, see *Lagenorhynchus australis*
Sibbaldus musculus, see *Balaenoptera musculus*
Sotalia brasiliensis, see *Sotalia guianensis*
Sotalia fluviatilis, **214**, 217, 565, 578
Sotalia guianensis, 214, **217**, 565, 578
Sotalia pallidus, see *Sotalia fluviatilis*
Sotalia tucuxi, see *Sotalia fluviatilis*
Sousa borneensis, see *Sousa chinensis*
Sousa chinensis, **227**, 234, 565, 578
Sousa lentiginosa, see *Sousa chinensis*
Sousa plumbea, 224, 227, **230**, 234, 565, 578
Sousa sahulensis, 227, 230, **234**, 565, 578
Sousa teuszii, **224**, 234, 565, 578
Stenella attenuata, **246**, 566, 582
Stenella attenuata attenuata, **246**, 566, 582
Stenella attenuata graffmani, **246**, 566, 582
Stenella clymene, **261**, 565, 583
Stenella coeruleoalba, **264**, 565, 583
Stenella dubia, see *Stenella attenuata*
Stenella euphrosyne, see *Stenella coeruleoalba*
Stenella frontalis, **251**, 565, 583
Stenella graffmani, see *Stenella attenuata*
Stenella longirostris, **255**, 566, 583
Stenella longirostris centroamericana, **256**, 566, 583
Stenella longirostris longirostris, **256**, 566, 583
Stenella longirostris orientalis, **256**, 566, 583
Stenella longirostris roseiventris, **256**, 566, 583
Stenella plagiodon, see *Stenella frontalis*
Stenella styx, see *Stenella coeruleoalba*
Steno bredanensis, **220**, 564, 577
Steno rostratus, see *Steno bredanensis*
Stenodelphis blainvillei, see *Pontoporia blainvillei*

Tasmacetus shepherdi, **122**, 559, 571
Trichechus inunguis, **529**, 599
Trichechus manatus, **524**, 599, 600
Trichechus manatus latirostris, **524**
Trichechus manatus manatus, **524**
Trichechus senegalensis, **527**, 599, 600
Tursiops aduncus, 237, 242, **243**, 565, 580
Tursiops catalania, see *Tursiops aduncus*
Tursiops gephyreus, see *Tursiops truncatus*
Tursiops gillii, see *Tursiops truncatus*
Tursiops nuuanu, see *Tursiops truncatus*
Tursiops truncatus, **237**, 243, 244, 245, 276, 564, 579

Ursus maritimus, **242**
Ursus maritimus marinus, **242**
Ursus maritimus maritimus, **242**

Zalophus californianus, **413**, 421, 551, 589, 594
Zalophus californianus californianus, see*Zalophus californianus*
Zalophus californianus japonicus, see *Zalophus japonicus*
Zalophus californianus wollebaeki, see *Zalophus wollebaeki*
Zalophus japonicus, **551**, 589, 594
Zalophus wollebaeki, **421**, 589, 594
Ziphius cavirostris, **109**, 560, 572

Photo Credit Suplementary Information Robin Baird—Photos taken under NMFS permit No. 731-1774. • Isabel Beasley—Credit: I. Beasley, courtesy of INPEX. • Vicky Beaver—Credit: V. Beaver/Ocean Alliance. • Brenda Becker—Photos taken under NMFS permit No. 848-1695. • Laurent Bouveret—Credit: L. Bouveret/OMMAG. • Peter Boveng—Photos taken under NMFS permit No. 15126. • Jon Brack—Photos taken under NMFS permit No. 10137. • Gavin Brady—Photos taken under NMFS permit No. 782-1676. • Gavin Brady—Photos taken under NMFS permit No. 782-1765. • Tom Brereton—Credit: T. Brereton/Marinelife. • Amelia Brower—Photos taken under NMFS permit No. 14245. • Michael Cameron—Photos taken under NMFS permit No. 782-1765. • Michael Cameron—Photos taken under NMFS permit No. 358-1787. • Gustavo Cardenas—Credit: courtesy J. Urban, PRIMMA/UABCS. • Mark Carwardine—All photos copyright Mark Carwardine. • Sal Cerchio—Underwater photos from video taken by SCWCS videographer. • Ted Cheeseman—Credit: T. Cheeseman/Cheeseman's Ecology Safaris. • Cynthia Christman—Photos taken under NMFS permit No. 775-1600-09. • Tom Collopy—Wildnorthphoto.com • Marta Cremer—Credit: Projecto Toninhas/UNIVILLE. • Shawn Dahle—Photos taken under NMFS permit No. 782-1676. • Shawn Dahle—Photos taken under NMFS permit No. 782-1765. • Shawn Dahle—Photos taken under NMFS permit No. 358-1787. • Wayne Davis—Credit: W. Davis, OceanAerials.com. • Aaron Dietrich—Photos taken under NMFS permit No. 848-1695. • Rob Digiovanni—Photos taken under NMFS permit No. 775-1600-09. • David Donnelly—Credit for *Tursiops* photo: D. Donnelly, Dolphin Research Institute. • Photos taken and used courtesy AAD (Shepherd's beaked whale). • Peter Duley—Photos taken under NMFS permit No. 775-1600-09. • John Durban—Credit: J. Durban/Bahamas Marine Mammal Research Organization. • Michael Ellis—Footloose Forays • Simon Elwen—Credit: S. Elwen, Namibian Dolphin Project, University of Pretoria. • Lloyd Edwards—Credit: L. Edwards, Raggy Charters. • Pail Ensor—Photos taken and used courtesy IWC (southern bottlenose whale). • Holly Fearnbach—Credit: H. Fearnbach/Bahamas Marine Mammal Research Organization. • John Gibbens—Sealimages.com • Peter Gill—Credit: P. Gill/Blue Whale Study. • Jan Haelters—Credit: J. Haelters/RBINS. • David Holley—Credit: D. Holley, Department of Parks and Wildlife, Australia. • Charles Howell—Credit: C. Howell, WFO Images. • Troels Jacobsen—Credit: Polar Images • M. Jarman—Credit: M. Jarman, Office of Environment and Heritage, NSW, Australia. • John Jansen—Photos taken under NMFS permit No. 782-1676. • John Jansen—Photos taken under NMFS permit No. 782-1765. • John Jansen—Photos taken under NMFS permit No. 15126. • Chrsitin Kahn—Credit: NOAA/NEFSC/C. Kahn. Photos taken under MMPA permit No. 775-1600. • Hertha Kashevarof—Photos taken under NMFS permit No. 932-1905/MA-009526. • Lucy Keith—Credit: Senegal River Basin Authority. • Amy Knowlton—Photos taken under NMFS permit. • William Koski—Photos taken under NMFS permit No. 782-1791. • Josh London—Photos taken under NMFS permit No. 358-1787. • Dan Luers—Photos taken under NMFS permit No. 848-1695. • Kara Mahoney Robinson—Photos taken under NMFS permit No. 655-1652-01. • Koa Matsuoka—Photos taken under NMFS permit No. 10137. • Roger McDonell—Credit: R. McDonnell, Stubbs Island Whale Watching. • Marine Meunier—Credit: M. Meunier, OMMAG. • Cedric Millon—Credit: C. Millon/Guadeloupe Evasion Decouverte, Deshaies. • Stephen Moore—Credit: S. Moore, Office of Environment and Heritage, NSW, Australia. • Danny Moussa—Credit: D. Moussa/AAMP. • L. Mouysset—Credit: L. Mouysset, Globice-Reunion. • Jim Nahmens—Credit: J. Nahmens, Nature's Spirit Photography. • Misty Niemeyer—Photos taken under NMFS permit. • Ilana Nimz—Photos taken under NMFS permit No. 10137. • Michael S. Nolan—All photos copyright Michael S. Nolan. • Paula Olson—Photos taken and used courtesy IWC (blue, Antarctic minke, humpback, southern bottlenose, and Gray's beaked, and long-finned pilot whales; southern right whale dolphin; and spectacled porpoise. Photos taken and used courtesy NOAA/NMFS/SEFSC (false killer whale; Atlantic spotted and striped dolphins). Photos taken and used courtesy AAD (Shepherd's beaked whale). • Orionna—Orionna@Dreamstime.com • Steven Prorak—Stevenprorak@dreamstime.com. • Dymytro Pylypenko—Dymytropylypenko@dreamstime.com. • Brenda Rone—Photos taken under NMFS permit No. 775-1600-09. • M. Rosso—Credit: M. Rosso, CIMA Research Foundation, Italy. • Oliver Shipley—Credit: O. Shipley and O. O'Shea, Cape Eleuthera Institute, Bahamas. • Olga Shpak—Credit: White Whale Program/IPEE RAS. • Jenny Smith—Credit: J. Smith, Murdoch University. • Mari Smultea—Photos taken under NMFS permit No. 14451. • Mark Sullivan—Photos taken under NMFS permit No. 10137. • Adam U—Photos taken under NMFS permit No. 15240. • M. G. Velasco—Credit: M. G. Velasco, Universidad de La Laguna. • Kim Westerskov—Natural Images. • David Withrow—Photos taken under NMFS permit No. 782-1765. • David Withrow—Photos taken under NMFS permit No. 358-1787. • Xvaldes—Xvaldes@dreamstime.com.

Front Cover Photos Main photo: Common minke whale (W. Osborn); Left to right: Amazon River dolphin, (M. Carwardine); Polar bear (M. Nolan); Killer whale (J. Durban); South American sea lion (M. Jørgensen)

Back Cover Left to right: Southern right whale (S. Dawson); Bearded seal (T. Jacobsen); Dugong (L. Dinraths); Cuvier's beaked whale (G. Ocio)